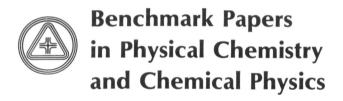

Benchmark Papers
in Physical Chemistry
and Chemical Physics

Series Editors: Joyce J. Kaufman
Walter S. Koski
The Johns Hopkins University

**Benchmark Papers
in Physical Chemistry and
Chemical Physics /3**

A BENCHMARK® Book Series

ION-MOLECULE REACTIONS, PART I
Kinetics and Dynamics

Edited by

J. L. FRANKLIN

Rice University

Dowden, Hutchinson
& Ross, Inc.

STROUDSBURG, PENNSYLVANIA

81 80 79 1 2 3 4 5
Manufactured in the United States of America.

LIBRARY OF CONGRESS CATALOGING IN PUBLICATION DATA
Franklin, Joseph Louis, 1906-
Ion-molecule reactions.
    (Benchmark papers in physical chemistry and chemical physics; 3)
    Includes indexes.
    CONTENTS: pt. 1. Kinetics and dynamics.—pt. 2. Elevated
pressures and long reaction times.
    1.   Chemical reaction, Conditions and laws of.
2. Ions. 3. Molecules. I. Title.
QD501.F7454   1978   541'.39   78-16358
ISBN 0-87933-331-6 (pt. 1); 0-87933-337-5 (pt. 2)

Distributed world wide by Academic Press,
a subsidiary of Harcourt Brace Jovanovich,
Publishers.

SERIES EDITORS' FOREWORD

The Benchmark Series in Physical Chemistry and Chemical Physics is an effort to cull through, organize, and make accessible to the reader the highlights of the evolution of the field by presenting carefully selected significant publications that have led to the development of the various disciplines along with a critical evaluation of the significance of the various papers. It thus enables the seasoned investigator who wishes to step into a new field of research effective means for doing so. To the student and the professor the series presents in compact form the major works that underlie current research activities. Such collected works also give the reader an idea of the history of the field. It is hoped, therefore, that such a series will have value from both the reasearch and the pedagogical points of view.

Each volume in the series is edited by a recognized authority in the area covered by the book. An introduction to the subject, comments on the papers, and a subject index are prepared by the volume editor.

This volume, *Ion-Molecule Reactions, Part I: Kinetics and Dynamics*, has been edited by Joe L. Franklin, Emeritus Professor of Chemistry and formerly Robert A. Welch Professor of Chemistry at Rice University, who has made many pioneering contributions to the field. Consisting of forty-one papers and editor's comments, the volume highlights the development of the growing field of ion-molecule reactions from the initial discoveries to the current investigations.

JOYCE. J. KAUFMAN
WALTER S. KOSKI

PREFACE

The mass spectrometer has proved to be one of the most flexible and broadly useful of all scientific instruments. Over the years, emphasis on various uses for the instrument has changed, but it is interesting that even in its earlier days, a considerable number of its long-term applications were in evidence. In the first quarter of this century, these instruments were used for chemical analysis, ionization and appearance potential measurements, atomic mass determination, and studies of ion-molecule reactions. Of all these, perhaps the most broadly useful has proved to be ion-molecule reactions. It is to this aspect that the present volumes are directed.

After several very early studies of ion-molecule reactions, the subject lay dormant for some 30 years until its revival in the 1950s, when it expanded almost explosively. As a result of the renewed interest and obvious opportunities for a wide variety of applications, a number of new and unique kinds of instruments came into being and permitted expanding both the theoretical studies and practical applications of ion-molecule reactions. Thus the ion-cyclotron resonance, flowing afterglow, tandem, molecular beam, and chemical ionization instruments were developed for specific purposes and have made possible more probing investigations of the nature of chemical reactions and broader applications of ion-molecule reactions to a variety of problems.

The purpose of the Benchmark series is to identify and assemble in one convenient place the most important papers in the field and those most representative of its development. Through a review of the literature for such papers, it quickly became evident that ion-molecule reactions could not be presented satisfactorily in a single volume; it was therefore decided to expand the presentation to two volumes.

Broadly speaking, the field of ion-molecule reactions has been divided into investigations of (1) the nature of the collisions and reactions of ions with molecules and (2) the effects of elevated pressures and/or long reaction times on ion-molecule chemistry. The two volumes are divided along these lines and, I think, give a reasonably complete and balanced summary of the field. Of course, a sharp division along these lines was not possible, and the reader will detect several papers that do not lend themselves to such compartmentalization. Such departures were made deliberately in order to make an orderly and coherent development of a topic even though the broad division was not adhered to. To those who find this offensive, I apologize.

Each volume is divided into several sections on more limited topics. In each of these, I have gathered papers of importance to that topic and presented them in approximately chronological order. Naturally, some of the papers could be assigned to more than one category. Where this occurred, I have included the paper in the section where I thought it would contribute most effectively.

Probably other editors would have chosen other papers for inclusion in these volumes. I can say only that the papers I have selected represent my personal evaluation, with the concurrence of Professors Koski and Kaufman, the series editors. Since I have been involved with ion-molecule reactions for the past 25 years, I have many friends who are active in the field. I hope that my selections will not offend any of them and, if so, I wish to express my sincere regrets. The great volume of the literature and the limitation of the volumes' size have made it impossible to include many papers that amply deserve to be included.

Finally, I wish to express my gratitude to Professors Koski and Kaufman for their support and counsel and to my long-suffering secretary, Sushila Kapadia, for her patience and her hard and skillful work, without which these books never would have been finished, and to my wife who gave me invaluable support and help in the preparation of the manuscript.

J. L. FRANKLIN

CONTENTS

Contents

EFFECTS OF ENERGY IN VARIOUS MODES

Contents

CONTENTS BY AUTHOR

Contents by Author

ION-MOLECULE REACTIONS

INTRODUCTION

Since the mid-fifties, the reaction of ions with neutral molecules has grown into one of the major topics of current research in chemistry and physics. There are several reasons. Because of their electric charge, ions are much more easily manipulated and detected than are neutral species. Consequently, far fewer ions than neutrals are needed for detection and quantitative measurement. Further, as will be discussed in several papers, the cross section for collision of ions with neutrals is usually about an order of magnitude greater than that for two neutrals of similar mass. Thus collision reactions of ions and neutrals are readily studied at very small concentrations of the reactants—so small, in fact, that the reactions can be made clearly second-order—and so the strictly elementary reaction can be studied. Further, ion-molecule reactions lend themselves readily to study over much wider ranges of pressure (concentration) and temperature than do most reactions in condensed phases. As a result, it is possible to unequivocally detect successions of many reactions originating with a certain primary ionic species. These properties make it possible to determine reaction rates and cross sections and to examine the effects of various molecular parameters and reaction conditions on the rate constant. Further, the great reactivity of ions with neutrals makes it possible to perform single-beam, crossed-beam, and merging-beam studies, which give profound insights into the dynamics of reactions.

Thus hindsight makes it quite clear why ion-molecule reactions have proved to be of great interest in contemporary molecular physics and chemistry. A few ion-molecule reactions have been known for many years, however. Thus the identity of a particle of $m/e = 3$ was established as H_3 by J. J. Thomson in 1913 (see Paper 1), but its formation by the reaction

$$H_2^+ + H_2 \rightarrow H_3^+ + H$$

was not established until 1925 by Hogness and Lunn (Paper 3) and by Smyth.[1]

1

After these pioneer studies, very little attention was paid to ion-molecule reactions until the 1950s, when workers in several laboratories almost simultaneously became interested in the subject. Probably the first of these was Tal'roze in Russia, but he was followed closely by workers in several U.S. laboratories. Most of these studies were performed by laboratories having a considerable emphasis on the chemistry of hydrocarbons, so it was natural that methane would be a compound of interest. In all these laboratories, the reaction

$$CH_4^+ + CH_4 \rightarrow CH_5^+ + CH_3$$

was observed, and its rate constant was found to be in the neighborhood of 10^{-9} cm³/molecule s. The existence of a strongly bonded compound CH_5^+ was completely unanticipated and led to an intensified interest in ion-molecule reactions.

The initial studies in the 1950s were performed, at least in the United States, by mass spectrometrists in commercially available instruments designed primarily for quantitative analysis of complex mixtures (especially of hydrocarbons) but employed in some instances for the determination of the heats of formation of ions and free radicals from measured appearance potentials. While they were useful for detecting a number of reactions and measuring approximate values of rate constants, these instruments were quickly seen to be far from satisfactory for any more probing investigation of ion-molecule reactions. Thus the reacting ion falling through the electric field in the source was accelerated from its point of formation in the electron beam to the exit slit. Consequently, the rate constant so determined was an average over the path, and hence over the kinetic energy of the ion. Since the effect of collision energy on reaction rate was not known, results obtained by this method were open to serious question. Further, many of the reactant ions were formed with internal (vibrational and perhaps electronic) energy. The amount of this internal energy and its effect on reaction rate were completely unknown and thus contributed further uncertainty as to just what was being measured. Early studies depended mainly on matching the appearance potential of a given secondary ion with that of various primaries to establish the ionic precursor and hence the reaction. This was usually satisfactory if there was only one precursor. If there was more than one, however, those of higher appearance potential usually went undetected. The commercial instruments were designed to operate at source pressures near 10^{-6} torr and at whatever temperature the source attained at operating conditions. Pressures approaching 10^{-2} torr were difficult to reach because the open sources resulted in excessive scattering in the analyzer and very rapid destruction of the filament. A source capable of operating at higher pressures (up to about 1 torr) was obviously much needed.

Thus, while the early work employing open, single-source instruments was provocative, it was primarily important in showing the possibilities of the field. Consequently, much subsequent emphasis was placed on the design and construction of instruments that would be better suited to the solution of specific problems in ion-molecule reactions. These have included the pulsed ion source, high-pressure and controlled temperature sources, drift tubes fitted for mass analysis, flowing afterglow, ion-cyclotron resonance, and tandem mass spectrometers with intervening collision chamber, crossed-beam, and merging-beam instruments. Each of these instruments has proved useful for solving specific problems, but none is suited for solving all problems.

With the great array of new and sophisticated instruments and techniques, the study of ion-molecule reactions has proceeded at an accelerated pace for nearly 20 years with no sign as yet that it is slowing down. Important contributions have been made to the understanding of chemical reactions, especially of their rates and mechanisms; a very important analytical method, chemical ionization, has been developed; great advances have been made in the understanding of the solvation of ions and its effect on reactions in solution; and a large body of thermochemical values of ions and neutral species has been accumulated. Thus ion-molecule reactions have made major contributions to the chemistry and physics of reactions.

Early on, the study of ion-molecule reactions divided into two broad classes: one devoted primarily to the detailed nature of the collision process and the effect of various conditions on it, and the other devoted primarily to the effect of pressure, time, and temperature on the reaction. The first class may be considered as more nearly the physics and the other as the chemistry of the collision process. In view of the great abundance of interesting and important papers on ion-molecule reactions, it became apparent that a single volume could not do justice to the subject. Consequently, it was decided to present it in two volumes whose contents would be divided along these lines.

This volume is devoted to the physics of the collision process. In addition to some early, historically important papers, it includes papers concerned with the dynamics of reactive collisions; the effects of kinetic, vibrational, and electronic energies on rates and

3

cross sections; and some consideration of the theory of ion-molecule reaction rates. It is an attempt to present the most important papers illustrating the development of various aspects of the physics of ion-molecule reactions.

The interest generated by a field of research is probably well measured by the number of times it is reviewed. From 1959 on, reviews of all or parts of the subject have appeared in books or journals, and conferences or symposiums have been held to discuss it at least annually. Several of the most important of these are listed below [2-32] as sources of more extensive information than can be presented here.

As of this writing, the study and application of ion-molecule reactions continues to thrive. The number of laboratories devoted to one or another aspect of the subject continues to increase, although perhaps at a somewhat slower pace, and new applications continue to appear. Thus studies of the photolysis and of the spectra of secondary ions are being made. Quite recently, there has been speculation and very strong inference that many of the interstellar molecules detected by radio astronomy are the result of ion-molecule reactions. There is, of course, abundant evidence that ion-molecule reactions play a major part in the chemistry of the upper atmosphere as well as in flames, electric discharges, and radiation phenomena. The use of chemical ionization as a means of identifying very small amounts of high-molecular-weight compounds, especially in samples of biological and medical interest, appears to be of very great importance and will almost certainly continue to be important for some time.

The prophecy of future results and directions of research in active fields of science is a very uncertain undertaking. In a field such as this, too many able and curious scientists, often using powerful new instruments, are engaged in research in all aspects of the subject for interesting and important discoveries and uses not to be made. I am confident of this. But what these will be is beyond my ability to predict, and I gladly resign the post of prophet to whoever is willing to assume it. But I am certain that ion-molecule reactions will continue to be of great interest for a considerable time to come.

REFERENCES

1. Smyth, H. D., *Phys. Rev.* **25**: 452 (1925).
2. Field, F. H., and J. L. Franklin, *Electron Impact Phenomena*, Academic Press, New York (1957).
3. Rosenstock, H. M., U.S. At. Energy Comm. Rep. JLI-650-3-TID-4500 (1959).

4. Franklin, J. L., F. H. Field, and F. W. Lampe, *Adv. Mass Spectrom.* **1**: 308 (1959).
5. Durup, J., *Les Reactions Entre Ions Positits et Molecules en Phase Gazeuse,* Gauthier-Villares, Paris, (1960).
6. Franklin, J. L., F. H. Field, and F. W. Lampe, *Prog. React. Kinet.* **1**: 169 (1961).
7. Tal'roze, V., *Pure Appl. Chem.* **5**: 455 (1962).
8. Pahl, M., *Ergeb. Exakt. Naturwiss.* **34**: 182 (1962).
9. Melton, C. E., in F. W. McLafferty (ed.), *Mass Spectrometry of Organic Ions,* Academic Press, New York. p. 65 (1963).
10. Hasted, J. B., *Physics of Atomic Collisions,* Butterworth's, London (1964).
11. Henchman, M., *Annu. Rep. Chem. Soc.* (London) **62**: 39 (1965).
12. Tal'roze, V. I., and G. V. Karachevtsev, *Adv. Mass Spectrom.* **3**: 211 (1966).
13. Giese, C. F., *Adv. Chem. Phys.* **10**: 247 (1966).
14. Ausloos, P. J., (ed.), *Adv. Chem. Ser.* **58** (1966).
15. Ferguson, E. E., *Rev. Geophys.* **2**: 305 (1967).
16. Ferguson, E. E., *Adv. Electron. Phys.* **24**: 1 (1968).
17. Friedman, L., *Annu. Rev. Phys. Chem.* **19**: 273 (1968).
18. Futrell, J. H., and T. O. Tiernan, in P. Ausloos (ed.), *Fundamental Processes in Radiation Chemisty,* Wiley-Interscience, New York, 1968.
19. Ferguson, E. E., F. C. Fehsenfeld, and A. L. Schmeltekopf, *Adv. At. Mol. Phys.* **5**: 1 (1969).
20. Henglein, A., in C. Schlier (ed.), *Proc. Int. Sch. Phys.* "Enrico Fermi," Course XLIV, Academic Press, New York, p. 139 (1970).
21. McDaniel, E. W., V. Čermǎk, A. Dalgano, E. E. Ferguson, and L. Friedman, *Ion-Molecule Reactions,* Wiley-Interscience, New York (1970).
22. Ausloos, P., *Prog. React. Kinet.* **5**: 113 (1970).
23. Melton, C. E., *Principles of Mass Spectrometry and Negative Ions,* Marcel Dekker, New York (1970).
24. Ferguson, E. E., *Acc. Chem. Res.* **3**: 402 (1970).
25. Parker, J. E., and R. S. Lehrle, *Int. J. Mass Spectrom. Ion Phys.* **7**: 421 (1971).
26. Friedman, L., and B. G. Reuben, *Adv. Chem. Phys.* **19**: 33 (1971).
27. Franklin, J. L. (ed.), *Ion-Molecule Reactions,* Vols. 1 and 2, Plenum Press, New York (1972).
28. Fehsenfeld, F. C., A. L. Schmeltekopf, D. B. Dunkin, and E.E. Ferguson, ESSA Tech. Rep. ERL 135-A1 3, 1969; Ferguson, E., *At. Data Nucl. Data Tables* **12**: 159 (1973).
29. Ausloos, P. (ed.), *Interactions Between Ions and Molecules,* Plenum Press, New York (1975). (Lectures presented at NATO Advanced Study Institute on Ion-Molecule Interactions, Biarritz, France, June/July 1974.)
30. Beauchamp, J. L., *Annu. Rev. Phys. Chem.* **22**: 527–556 (1971).
31. Franklin, J. L., and P. W. Harland, *Annu. Rev. Phys. Chem.* **25**: 485–526 (1974).
32. Koyano, Inosuke, in Bamford, C. H., and C. F. H. Tipper (eds.), *Comprehensive Chemical Kinetics,* Volume 18, Elsevier, Amsterdam, pp. 293–428 (1976).

DISCOVERY OF ION-MOLECULE REACTIONS

Editor's Comments
on Papers 1, 2, and 3

1   **THOMSON**
     On the Nature of X₃, the Substance Giving the "3" Line

2   **DEMPSTER**
     The Ionization and Dissociation of Hydrogen Molecules and the Formation of H₃

3   **HOGNESS AND LUNN**
     The Ionization of Hydrogen by Electron Impact as Interpreted by Positive Ray Analysis

In the late nineteenth and early twentieth centuries, physicists were profoundly interested in the nature of electricity and its effect on gases. Thus electric discharges and their chemical results were studied with the aid of the tools available. The discovery of canal rays by Goldstein[1] in 1886 yielded a device that permitted experimenters for the first time to separate ions formed in a discharge according to their mass-to-charge ratio. This enabled experimenters to begin to gain some insight into the ions present in electric discharges and to infer their mode of formation. One intriguing observation in these early studies was a particle of $m/e = 3$ at conditions and from materials that could not produce such a particle by direct ionization or fragmentation. The first paper in this section presents Thomson's reasoning, which led to the conclusion that the particle was H_3. The succeeding papers by Dempster and by Hogness and Lunn present results obtained with mass spectrometers and thus with instruments much better suited to the study of ions than Thomson's cathode ray device. Thus the third paper, by Hogness and Lunn, shows definitely that H_3^+ is formed by the reaction

$$H_2^+ + H_2 \rightarrow H_3^+ + H$$

A paper by Smyth[2] at about the same time also reached this conclusion.

Other papers published between 1916 and 1931 presented observations of ions of unexpected masses and considered their mode of formation. Perhaps the one giving the most definite concclusions and probably the most nearly like current studies of ion-molecule reactions is that by Hogness and Harkness,[3] in which both the positive and negative ions found in iodine were shown to result from ion-molecule reactions; thus

$$I^+ + I_2 \rightarrow I_2^+ + I$$
$$I_2^+ + I_2 \rightarrow I_3^+ + I$$

$$I^- + I_2 \rightarrow I_2^- + I$$
$$I_2^- + I_2 \rightarrow I_3^- + I$$

Smyth reviewed the state of mass spectrometry, including ion-molecule reactions, in a paper published in 1931.[4] After that, very little was done on this subject until the paper by Tal'roze and Lyubimova (Paper 4, Part II) initiated the current enthusiasm for it.

REFERENCES

1. Goldstein, E., *Sitzungsber. Berliner Akad. Wiss.* **39**: 691 (1886) *Wiedermann's Ann.*, **LXIV**: 38 (1898).
2. Smyth, H. D., *Phys. Rev.* **25**: 452 (1925).
3. Hogness, T. R., and R.W. Harkness, *Phys. Rev.* **32**: 784 (1928).
4. Smyth, H. D., *Rev. Mod. Phys.* **3**: 347 (1931).

ON THE NATURE OF X$_3$, THE SUBSTANCE GIVING THE "3" LINE

J. J. Thomson

[*Editor's Note:* In the original, material precedes this excerpt.]

The only known substances which could give the line with the value of m/e three times that of the hydrogen atom are: (1) an atom of carbon charged with four units of electricity, and (2) a molecule containing three atoms of hydrogen. The first of these alternatives must be abandoned for the following reasons: (1) we have seen that a line corresponding to a multiple charge on an atom is accompanied, unless the pressure is exceedingly low, by certain peculiarities in the line corresponding to the atom with one charge ; for example, if there were a line corresponding to a carbon atom with two charges, the line corresponding to the carbon atom with one charge would be prolonged until its extremity was only one-half the normal distance from the vertical axis ; if there were a line corresponding to the carbon atom with three charges, the ordinary carbon line would be prolonged until its distance from the vertical was only one-third of the normal distance, while a carbon atom with four charges would prolong the ordinary carbon line to within one-quarter of the distance from the axis. Again the greater the charge the less the intensity, so that a line due to a quadruply charged carbon atom would be accompanied by a stronger line due to a triply charged atom, a still stronger one due to a doubly charged atom, while the normal carbon line would be the strongest of all.

Now in the case of the 3 line we do not find any of these characteristics ; the carbon line is not prolonged to within one-quarter of the normal distance, and so far from the line being accompanied by a stronger line due to a doubly charged carbon atom, in many of the cases where the 3 line is strongest the line due to the doubly charged atom is not strong enough to be detected ; indeed in some of these cases the 3 line is stronger than the normal carbon line.

Again since the gas giving the 3 line can be stored in the vessel A for days after the bombardment has ceased, if this line were due to carbon with four charges it must be because some carbon compound is formed by the bombardment, which when introduced into the discharge tube gives, when the discharge passes through it, carbon atoms with four charges. Now experiments have been made with a great variety of carbon carbons introduced directly into the discharge tube, CH_4, CO_2,

11

CO, C_2H_4, C_2H_2, $COCl_2$, CCl_4, and many others, and none of these have given this line: we must therefore abandon this solution of the problem.

I find too that whenever large amounts of X_3 are produced spectroscopic examination shows that considerable quantities of hydrogen are liberated by the bombardment; in fact the brightness in the spectroscope of the hydrogen lines in the bombardment vessel may be taken as giving a rough indication of the brightness to be expected of the X_3 line in the positive ray photograph.

Let us next consider the connexion between the production of X_3 and the nature of the substance bombarded by the cathode rays. We get more definite conditions for the bombarded body if we use soluble salts instead of pieces of metal or minerals. The latter may have absorbed X_3 and contain stores of this gas in the absorbed state which are liberated when the solid is bombarded by cathode rays. If we could subject the solid before the bombardment to some process by which we could free it from absorbed gas we should expect that if the source of the X_3 were gas absorbed by the solid, the bombardment of a substance which had been treated in this way would not give rise to any X_3.

The most effective way of liberating the absorbed gas would seem to be to dissolve the solid in a suitable solvent and then evaporate the solution to dryness. Those salts which are soluble in water or alcohol can readily be treated in this way. I have therefore made experiments on a large number of soluble salts bombarding them before and after they have been dissolved and evaporated to dryness, and testing by the positive ray photographs the yield of X_3 in each case. Sal-ammoniac made by allowing streams of ammonia and hydrochloric acid gas was found to give X_3 when bombarded; in this case the possibility that X_3 was absorbed in the salt would seem to be excluded.

12

I find that salts may be divided into two classes with respect to the way in which their evolution of X_3 is affected by solution and evaporation. One class of salts which includes KI. Li_2CO_3. KCl give a very much smaller output of X_3 after this treatment than they did before; the other class which includes KOH, LiCl, LiOH, $CaCl_2$ give much the same output after solution as they did before even though they are dissolved and evaporated over and over again. The salts of the first class do not contain hydrogen, while those of the second either contain hydrogen or are deliquescent and thus can absorb water from the atmosphere on their way to the bombardment chamber after evaporation. The fact that some salts continue to give supplies of X_3 after repeated solution and evaporation shows I think that X_3 can be manufactured from substances of definite chemical composition by bombardment with cathode rays, and the fact that such salts contain hydrogen either as part of their constitution or in water of crystallization suggests that X_3 consists of hydrogen and is represented by the formula H_3. The other alternative is that it is an element produced by the disintegration of some or other of the elements in the salt, but this view would not explain why its production is so closely associated with the presence of hydrogen.

One of the most convenient ways of preparing X_3 is to bombard potash, KOH, by cathode rays. I bombarded a few grammes of potash for several months pumping off after each day's running the gases liberated by the bombardment, these consisted of H_2. O_2 and X_3 and at the end of the time I could not detect any falling off in the rate of production of X_3. By bombarding the potash, supplies of X_3 mixed with hydrogen and oxygen were obtained and a series of experiments made with them with the object of discovering some of the properties of this gas. The method used to test for the presence of X_3 after the mixed gases of which it was a constituent had been subjected to any treatment was to introduce a small quantity of the mixed gases into the discharge tube, take a positive ray photograph and estimate the brightness of the X_3 line.   Before such experiment the discharge tube was well washed out with oxygen and a test photograph taken to make certain that no X_3 was in the tube before the introduction of the gas which was to be tested.   One property of this gas, that of combining with mercury vapour when an electric discharge passes through a mixture of these gases, has already been mentioned.   Another property was discovered accidentally : the mixed gases obtained by bombarding the potash were drawn off day by day and stored up for further tests.   It was soon noticed that some of the samples kept much better than others and it seemed possible that this difference might be due to differences in the brightness of the light to which the samples had been exposed.   To test this a piece of magnesium wire was burnt in front of a sample which was known to contain a considerable quantity of X_3, with the result that the X_3 almost disappeared.   The gas exposed to the light was a mixture of hydrogen, oxygen and X_3, if the oxygen is taken out of the mixture by absorbing it with charcoal cooled with liquid air, exposure to light produces no effect on the X_3 ; the conclusion we draw is that under the influence of the light the X_3 combines with oxygen.   If the mixture is kept in the dark or if the oxygen is taken out of it the X_3 lasts for a long time, certainly for several weeks.   Again if a strong spark is sent through the mixture containing oxygen so that a vigorous explosion takes place the X_3 disappears, presumably combining with the oxygen.   If the oxygen is removed the mixture of hydrogen and X_3 will stand a good deal of sparking without any considerable diminution in the amount of X_3.

The fact that sparking with oxygen destroys the X_3, makes the removal of the hydrogen, which is by far the largest constituent of the mixture, a matter of considerable difficulty. The most effective way I know of increasing the proportion of X$_3$ is first to remove the oxygen, then to put the mixture of H$_2$ and X$_3$ into a vessel to which a palladium tube is attached ; when the palladium is heated to redness the hydrogen diffuses through it much more rapidly than the X$_3$, though some of this gas can get through the palladium. The result is that the gas left behind in the vessel contains a much greater proportion of X$_3$ than it did before. The preponderance of H$_2$ in the original mixture is, however, so great that even by this means I have not been able to prepare any sample in which the hydrogen was not greatly in excess.

Another interesting property is that the X$_3$ almost disappears when placed in a quartz tube with some copper oxide and the whole heated to a red heat.

In the absence of oxygen and copper oxide X$_3$ may be heated to a high temperature without destruction. Summing up the results we see that when hydrogen is present in the substance bombarded a continuous supply of X$_3$ can be obtained, while from substances which do not contain hydrogen the supply is soon exhausted. Again, under certain conditions such as exposure to bright light, vigorous sparking in the presence of oxygen, contact with glowing copper oxide, X$_3$ combines with oxygen. These results seem to me to point to the conclusion that X$_3$ is tri-atomic hydrogen H$_3$. If this is so, its properties are very interesting. Unlike O$_3$ its existence cannot be reconciled with the ordinary views about valency. If, however, we regard an atom of hydrogen as consisting of a positive nucleus and one negative corpuscle, it will exert forces analogous to those excited by a magnet and I can see no reason why a group of three of these arranged so that their axes form a closed ring should not form a stable arrangement.

15

The stability of X_3 is much greater than that of ozone O_3, the latter does not persist for nearly as long as X_3, it breaks up at a moderate temperature which would have no effect on X_3, and it disappears under a kind of sparking which would leave X_3 undecomposed. In fact X_3 seems more stable than any known allotropic form of an element.

Many attempts have been made to obtain spectroscopic evidence of X_3 by putting mixtures of this gas and hydrogen in a quartz tube and photographing the spectrum obtained when a discharge was sent through the tube, the electrodes were pieces of tin-foil placed outside the tube. No lines which could be ascribed to X_3 were detected, the first and second spectra of hydrogen were bright, and in spite of efforts to get rid of mercury vapour the mercury lines were visible.

Bombardment by cathode rays is not the only method of obtaining X_3. I heated by an electric current in a good vacuum a fine tantalum wire, such as are used for metallic filament lamps, until it fused and found that a considerable amount of X_3 was given off. Some time ago I found that when the discharge from a Wehnelt cathode was sent through an exhausted tube X_3 was liberated. I have found subsequently that it is not necessary to send the discharge through the tube, the heating of the cathode is sufficient to liberate this gas. Again if a considerable quantity of potash is placed in a vacuum and left for some time an appreciable quantity of X_3 is liberated.

[*Editor's Note:* Material has been omitted at this point.]

2

THE IONIZATION AND DISSOCIATION OF HYDROGEN MOLECULES AND THE FORMATION OF H$_3$

A. J. Dempster

Exhibition Scholar of the University of Toronto

BY the analysis of positive rays, J. J. Thomson has shown that in a discharge-tube containing hydrogen there are present charged atoms, charged molecules, and sometimes a constituent with a mass three times that of the atom of hydrogen. The pressure used was about ·003 mm. of mercury, and consequently the potential necessary was of the order of 20,000 volts. In the following experiments a different method of getting the positive rays was used. Electrons from a Wehnelt cathode C are accelerated in the field CA. They ionize the gas, and the positive particles produced are given a velocity which carries them past the edge of C (2 mm. wide) and through the narrow tube T. These positive particles are then deflected by magnetic and electric fields, and fall on a screen which has a parabolic slit S. Each constituent of the rays is drawn out into a parabolic curve, and, by increasing the magnetic field, the various parabolas may successively be brought to fall on the slit. When this occurs the charged particles pass through S and give up their charges to the Faraday chamber F. The

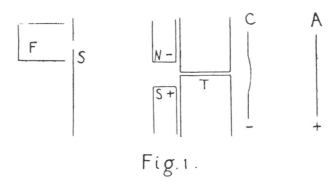

Fig. 1.

advantages obtained by using a Wehnelt cathode are, that any desired potential may be used, in particular low potentials, and that the pressure of the gas may be made as low as desired and may be varied without changing the potential. The change that occurs in the constituents of the rays as the pressure is decreased is described in this paper.

With 800-volt rays and hydrogen at a pressure of about ·01 mm. the curve in fig. 2 was obtained. The abscissæ represent the strength of the deflecting magnetic field, the zero being to the left of the origin, and the ordinates give the charge obtained in the Faraday chamber as the field was increased so as to bring the three lightest constituents of the rays over the parabolic slit. There are present hydrogen atoms, hydrogen molecules, and very much of the constituent with atomic weight 3. The curve obtained with a pressure of ·0017 mm. is given in fig. 3 ; there is a noticeable decrease in the relative amount of both H_1 and H_3. Fig. 4 gives the curve when a charcoal bulb with liquid air was used. Here the pressure was less than ·0005 mm., and H_1 and H_3 have practically disappeared. That this change was due to decreasing pressure, and not to the removal of some constituent of the gas by the charcoal, was shown by the fact that when hydrogen was admitted, while the charcoal and liquid air were on, H_1 and H_3 regained their original relative intensities.

Since in the high vacuum the free path of the molecules is very great, the positives which are still formed by the dense stream of corpuscles coming from the Wehnelt cathode make very few collisions with the hydrogen molecules.

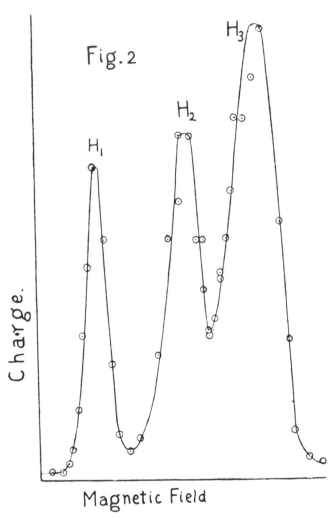

Hence these positive molecules are analysed in the condition in which they are just after they have been ionized. We must conclude, then, that electrons ionize only by detaching a single electron from the molecule, and are not able to dissociate the molecule into atoms. When the pressure is greater, some of the positive molecules collide with the molecules of the gas before the cathode, and this collision

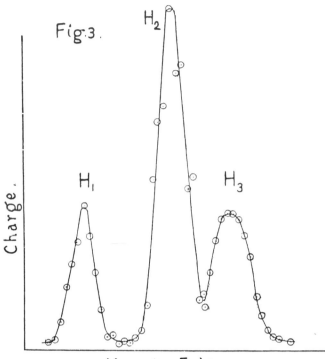

Fig. 3.

results in a dissociation of the gas into atoms. A positive atom thus formed may attach itself to a neutral molecule and give rise to H_3.

The experiments thus confirm for electrons of this speed the conclusion reached by Millikan from his experiments with oil-drops (Phil. Mag. xxi. p. 753 (1911)) that gaseous ionization produced by β-rays, or by X-rays of all hardnesses, consists in the detachment from a neutral molecule of a single elementary charge. They also verify the theory advanced by J. J. Thomson to account for the results obtained in his experiments on positive rays (Phil. Mag. xxiv. p. 234 (1912)), that the electrons and positive rays produce different types of ionization.   The results show also that H_3 cannot be regarded as a stable gas, but that it is a temporary complex formed only when the hydrogen is in a dissociated state.   In his paper in Phil. Mag. xxiv. p. 241 (1912), J. J. Thomson finds that H_3 " occurs under certain conditions

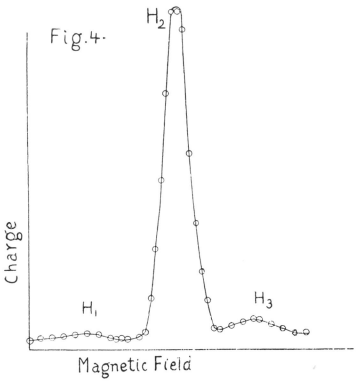

of pressure and current," but later ('Rays of Positive Electricity') regards it as a stable gas which, among other properties, combines with oxygen and mercury when under the influence of the electric discharge.   It is possible that the disappearance of H_3 in the latter case is due to a higher vacuum, for I find that the pressure is much reduced when the discharge is passed through mercury-vapour and hydrogen.

Summary.

Electrons of 800 volts speed ionize hydrogen by detaching a single elementary charge from the molecule.   They are not able to dissociate the gas.

The positive molecules so formed are able to dissociate the gas. When this occurs the complex H_3 is formed. H_3 cannot be regarded as a stable gas, since it is not present when there is no dissociation of the hydrogen molecules.

3

Reprinted from *Phys. Rev.* 26:44–55 (1925)

THE IONIZATION OF HYDROGEN BY ELECTRON IMPACT AS INTERPRETED BY POSITIVE RAY ANALYSIS*

By T. R. Hogness and E. G. Lunn

Abstract

Ions produced in hydrogen by electron impact.—Using an apparatus previously described in which positive ions formed by impact of electrons of definite energy (V_1+V_2) are accelerated and then deflected magnetically around a semi-circle into a Faraday cylinder, the relative numbers of ions of types H^+, H_2^+ and H_3^+ were measured *as a function of pressure* from $<0.1\times10^{-4}$ mm to .006 mm, and also *as a function of impact energy* (V_1+V_2) to 60 volts. At low pressures only H_2^+ ions are formed; as the pressure is increased the percentage of H_3^+ increases in proportion to the pressure. In the apparatus used, the percentage of H^+ increased with pressure but did not exceed 4 percent, while the percentage of H_3^+ ions reached 60. These results confirm the conclusions of Dempster and Smyth that the primary process in the ionization of hydrogen is the ionization of the molecule without dissociation. The previously measured ionization potential at 16 volts (confirmed in this work) is that for the formation of H_2^+. H_3^+ also appears at this potential, but as a result of a secondary process. It appears that the H_2^+ ion is readily dissociated by collision and that the H^+ ion formed may unite with the H_2 molecule collided with or with some other molecule to form H_3^+. The interpretation of ionization potentials reported by other observers is discussed in the light of these results.

Ions produced in helium containing hydrogen, by electron impact.—The percentage of H^+ ions found was greater even than the percentage of H_2^+, while no H_3^+ ions were observed. Evidently the primary ions H_2^+ are readily dissociated by impact with He atoms. Evidence was found for the ions HeH^+ and also for an ion with $m/e=6$, perhaps HeH_2^+.

$\mathbf{I}$N a recently published preliminary report[1] the authors have described the apparatus employed in this investigation and have given the conclusions that could be drawn from the scanty results then available. By changing the experimental procedure and widening the range of experimental conditions employed they have since made a more detailed study of the problem. The results of this study have made untenable the conclusion formerly drawn that in the ionization of hydrogen by electron impact there are two independent primary processes

$$H_2 = H_2^+ + \epsilon \tag{1}$$

$$H_2 = H^+ + H + \epsilon \tag{2}$$

---

* When this paper was first submitted the authors were informed by the editor that an article by H. D. Smyth, "Primary and Secondary Products of Ionization in Hydrogen," was already in press (Phys. Rev. **25**, 452, April 1925). Through the kindness of the editor we have had the advantage of reading proof of that article and have accordingly revised the discussion in this paper to give recognition to Smyth's work.

[1] Hogness and Lunn, Proc. Nat. Acad. Sci., **10**, 398 (1924).

and have given evidence that Eq. (1) represents the only primary process, the formation of H^+ and H_3^+ being secondary. This is the conclusion drawn by Dempster[2] in a much overlooked and neglected paper from his investigation of the ionization of hydrogen by high-voltage electrons, and also by Smyth.[3]

DESCRIPTION OF THE APPARATUS AND METHOD

The apparatus (Fig. 1) is essentially an ionization potential tube so arranged that the products of ionization can be analyzed by Dempster's[4] positive-ray method. The preliminary report of this work gives a detailed description of the apparatus which need not be repeated here. Consideration of the equation $e/m = 2V_4/H^2r^2$ by means of which

TABLE I

Electrometer readings for H^+, H_2^+ and H_3^+ peaks, at various pressures.

Pressure (10^{-4} mm)	Readings			Percent		
	H^+	H_2^+	H_3^+	H^+	H_2^+	H_3^+
<0.1	0	360	0	0	100	0
	0	241	0	0	100	0
2.0	5	520	1.3	0.9	96.7	2.4
	3	434	0.8	0.7	97.5	1.8
5.4	17	850	35	1.9	94.2	3.9
	17	630	33	2.5	92.6	4.9
	18	900	49	1.9	93.0	5.1
11	8	500	94	1.3	83.1	15.6
	5	320	45	1.3	86.5	12.2
12	7	520	103	1.1	82.6	16.3
	10	512	115	1.6	80.1	18.3
15	15	530	110	2.3	81.0	16.7
22	1(?)	220	69	0.3(?)	75.9	23.8
	33	1000	265	2.5	77.1	20.4
28	28	560	260	3.3	66.1	30.6
	27	560	240	3.3	67.7	29.0
39	.61	720	540	4.6	54.5	40.9
	27	520	360	3.0	57.3	39.7
56	40	470	690	3.3	39.2	57.5
	36	460	680	3.1	39.1	57.8
He and H_2	153	70	0	68.6	31.4	0

[2] Dempster, Phil. Mag. **31**, 438 (1916).

[3] H. D. Smyth, Phys. Rev. **25**, 452 (April 1925). See also Proc. Roy. Soc. **102A**, 283 (1922); **104A**, 121 (1923); **105A**, 116 (1924); Nature, **111**, 810 (1923); **114**, 124 (1924); Phys. Rev. **23**, 297 (1924); J. Franklin Inst. **198**, 795 (1924).

[4] Dempster, Phys. Rev. **11**, 316 (1918).

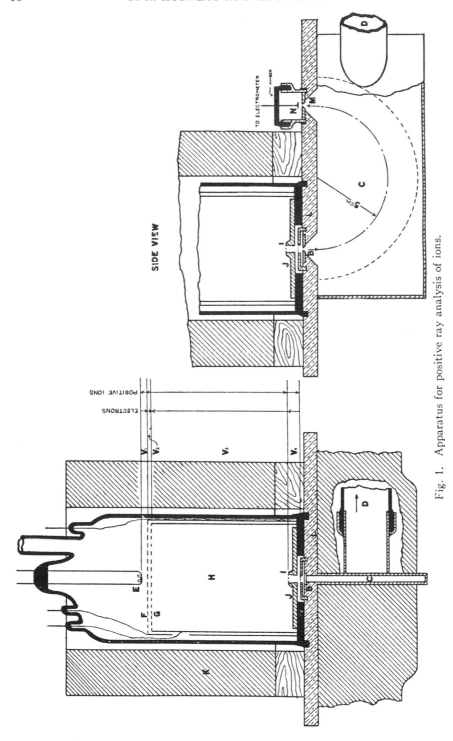

Fig. 1. Apparatus for positive ray analysis of ions.

the specific charges are determined, shows that two experimental procedures are open, (1) that of keeping the magnetic field H constant and focussing the ion beam by varying the accelerating potential V_4, and (2) that of varying the magnetic field while V_4 is kept constant. The former procedure was used in searching for new ions and for measuring their specific charges, the latter in measuring the ionic intensities since its use permitted constancy of electrical conditions in the tube and gave consistent and reproducible results.

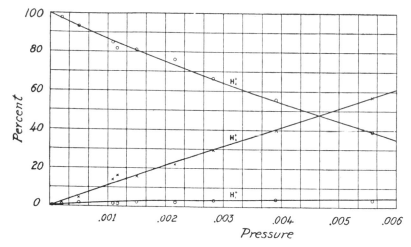

Fig. 2. Percentage of each of ionic species (H⁺, H₂⁺, H₃⁺) as a function of pressure.

Of the present experiments those that seem to throw most light on the problem of interpreting the processes of ionization are the ones on the change of the relative intensities of the ions H⁺, H₂⁺ and H₃⁺ with change in pressure. The data from these experiments are recorded in Table I. The electrometer deflections given were obtained with the constant deflection method. The right-hand side of Table I gives the percentage of each ionic species present as measured by the electrometer deflection ratio. The averages of the values for each pressure given in the table are plotted in Fig. 2, in which the pressures are recorded as abscissas and the percentages as ordinates.

It is evident from the table and from Fig. 2 that as the pressure is lowered the percentages of H⁺ and H₃⁺ decrease in a regular manner and approach zero at zero pressure. This clearly indicates that the formation of H₂⁺ is the primary process and that H⁺ and H₃⁺ are formed from H₂⁺ as the result of secondary collisions with gas molecules. H₂⁺ is apparently metastable and is disrupted when it collides with a neutral gas molecule in one of two ways

$$H_2^+ + H_2 = H_3^+ + H \tag{3}$$

$$H_2^+ = H^+ + H \tag{4}$$

If Eq. (4) represents the sole collision reaction then the formation of H_3^+ would be the result of the tertiary process

$$H^+ + H_2 = H_3^+ \tag{5}$$

Now if H_3^+ were formed only by the reaction (5), in order to account for the fact that the intensity of H^+ was very small at all pressures, it would be necessary to conclude that the proton has a much smaller mean free path than would be predicted from kinetic theory considerations. Hence it seems reasonable to suppose that when H_2^+ ions collide with neutral molecules the reaction taking place in the majority of cases is (3), and that reaction (4) takes place less often.

When the field V_3 for drawing the positive ions from the ionization chamber was increased to such an extent that only those electrons in the upper portion of that chamber had ionizing energy, then the intensity of $(H^+ + H_3^+)$ increased with respect to that of H_2^+ as might have been expected since the H_2^+ ions formed had greater chance of collision having had a longer average path to traverse. Moreover the ratio of the intensity of H^+ to that of H_3^+ increased considerably. This can be explained by assuming that under these conditions H_2^+ breaks up more often in accordance with reaction (4). This would happen if when the H_2^+ ion acquires sufficient energy before colliding, the H^+ formed by disruption were not trapped by the H_2 molecule to form H_3^+ but passed on into the resolving chamber.

There remains the possibility, however, that H_2^+ is stable and must acquire a definite velocity before a collision would result in disruption into H^+ and H. To examine this the change in the percentage of $(H_3^+ + H^+)$ with small variations of V_3 was studied. It was found that as V_3 was gradually lowered from 4.5 volts to 0.1 volt, $(V_1 + V_2)$ being 48 volts, the percentage of $(H_3^+ + H^+)$ increased very slightly instead of decreasing as might be expected if energy were required to cause H_2^+ to break up. Although inconclusive, this result seems to indicate that H_2^+ is energetically unstable. This conclusion is in accord with the following, deduced by Sommerfeld on theoretical grounds. "The H_2^+ ion is *unstable energetically*; it can dissociate into H and H^+, giving up energy. At the same time it follows from this for the ionization of the H_2 molecule that if this happens in the sense of scheme [Eq. (1)] it requires a greater ionization potential than if it proceeds according to the scheme [Eq. (2)]. This conclusion is independent of any assumptions about the model of the neutral H_2, and also remains preserved if we pass from the H_2^+ molecule considered so far to a far more general molecule."[5]

In the experiments plotted in Fig. 2, 48 volts were applied to the impact electrons. The effects of slower speed electrons were investigated at one arbitrarily chosen pressure by obtaining the percentages of the several ions as a function of $(V_1 + V_2)$ (Fig. 3). The increase in the percentage of H_3^+ as $(V_1 + V_2)$ was decreased may be explained as due to the characteristics of the discharge tube. The now greater retarding potential V_3 decreased the velocity of those electrons that had penetrated into the lower part of the ionization chamber (*II* of Fig. 1) to such an extent that they could not ionize the gas. The H_2^+ ions were then formed at a greater average distance from the gauze *I*, and having had a longer

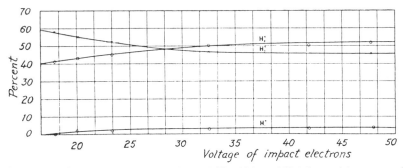

Fig. 3. Variation in relative numbers of H^+, H_2^+ and H_3^+ ions with energy of impact electrons.

average path to traverse had greater chance to collide. The curves of Fig. 3 could not be extended below 18 volts with accuracy because of the small intensity of each of the ionic species. Little significance, moreover, can be attached to the H^+ percentage curve at low voltages because of the inaccuracy of measurement of the small H^+ ion current in this region. From these observations of Fig. 3, we conclude that the processes described above are also true for impact electrons of smaller velocities.

Some experiments were made on relative ionic intensities in mixtures of helium with a relatively small amount of hydrogen. It was found that the percentage of H^+ was much greater than in pure hydrogen at any pressure employed (see the last line of Table I). Under these conditions the reactions designated by Eqs. (3) and (5) are much less probable and the H_2^+ on collision with the He atom disrupted to form H^+.

[5] Sommerfeld, "Atomic Structure and Spectral Lines," 1st Eng. Ed. Appendix 14, page 605.

With mixtures of helium and hydrogen in the tube two particularly interesting intensity peaks of $m/e = 5$ and a less definite one at about $m/e = 6$ were observed repeatedly. A typical run showing these is plotted in Fig. 4. Although the small amount of these ions did not permit of a study of their origin, there is little doubt that the $m/e = 5$ ion is the ion of helium hydride, HeH^+, while the other may be HeH_2^+.

The Ionization Potential

To determine the ionization potential for the formation of H_2^+, V_4 was set to give the peak of the H_2^+ intensity curve and with the magnetic

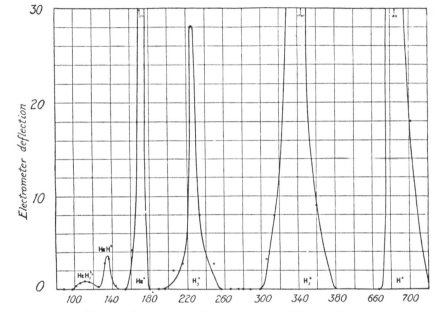

Fig. 4. Peaks obtained with mixture of He and H_2.

field, the filament current, V_1 and V_3 all held constant, the electrometer current was noted as V_2 was decreased in steps of 0.3 volt. When the rate of deflection of the electrometer was plotted against $V_1 + V_2$, curves like that of Fig. 5 (labelled H_2^+) were obtained. The point at which each curve cut the zero ordinate was taken as the approximate (uncorrected) value of the ionization potential. Then by setting V_1 and V_2 immediately above and below this, and observing the electrometer deflection over a long period of time, a more accurate value was obtained.

To obtain the necessary correction to be applied because of the initial velocity of the electrons, the potential drop along the filament, and contact differences of potential, and to take into account the sensitivity of the apparatus, helium was introduced into the tube and the ionization potential curve for He+ obtained in a manner analogous to that described for H_2^+. (See Fig. 5, He+.) 24.5 volts was taken as its true ionization potential. In order that the corrections thus found should have any significance it was necessary to choose the hydrogen and helium pressures such that (1) the "saturation" intensities of He+ and H_2^+ be the same, and (2) the ionization potential curves for helium and for hydrogen be of approximately the same shape. This choice of pressures was made as follows. It was found that 50-60 volts was above the "saturation" voltage. The intensity of the He+ line in the calibration run when $V_1 + V_2$ was about 60 volts was therefore noted, the helium then pumped out, and hydrogen introduced at such a pressure as gave a H_2^+ line of this same intensity. The smaller figure of Fig. 5 shows the appearance of the ionization potential curves for the He+ and H_2^+ when the respective pressures were so chosen. The two curves are evidently almost superposable.

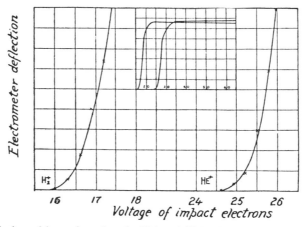

Fig. 5. Variation of intensity of peaks H_2^+ and He+ with energy of impact electrons.

The corrected values for the ionization potential of hydrogen found in successive runs are: 16.3, 16.5, 15.7, 15.8, 16.1, 15.9, 15.8, 15.8. The average value is 16.0.

To show that H_3^+ and H_2^+ appear at the same ionization potential, the pressure was so regulated that the H_3^+ and H_2^+ lines were of equal intensity and the potentials for the formation of the two ions compared. Under these conditions the two potentials were identical, but at lower pressures where H_2^+ predominated, H_3^+ and H^+ were not detected at as low voltages as was H_2^+.

The several lines of investigation presented above unite then in giving evidence that, in the ionization of hydrogen by electron impact, the primary process of ionization is the detachment of an electron from a molecule $(H_2 = H_2^+ + \epsilon)$ and that the H_2^+ thus formed is probably energetically unstable and can break up on collision to give H^+ (and H). The evidence that the formation of H^+ is the result of a secondary process may be briefly summarized. (1) At very low pressures no H^+ was found; (2) H^+ was never found with impact electrons having a velocity lower than the minimum necessary for the production of H_2^+; and (3) with mixtures of helium and hydrogen large amounts of H^+ were found.

The following then are the conditions for producing in a discharge tube each of the different ions of hydrogen in predominating amounts: For H^+, large potential difference and the addition of some inert gas such as helium; for H_2^+, low pressures; for H_3^+, higher pressure and small potential difference.

DISCUSSION

It is of interest to note that the above results and the main conclusions derived therefrom are exactly in accord with the following conclusion of Dempster[2] who used 800 volt electrons in his experiments: ". . . . electrons ionize only by detaching a single electron from the molecule, and are not able to dissociate the molecule into atoms." Graphical analysis of his intensity ratios shows also that at zero (extrapolated) pressure no H^+ or H_3^+ would be formed. The present conclusions also agree in the main with those of Smyth,[3] differing therefrom only in detail as to the secondary process. He concludes that the formation of H_3^+ is a tertiary process while we believe it to be largely a secondary one (Eq. 3). Smyth's Fig. 6 shows in a striking manner the increase of H_3^+/H^+ with decrease of V_3. This we believe gives support to our view. In discussing the stability of H_2^+, Smyth maintains that if that ion were stable with respect to H^+ and H, the additional energy necessary for its disruption would have to be supplied by the impacting electron; but this energy could also be supplied by the kinetic energy of the H_2^+ acquired from the electrical field. The possibility may be pointed out here that, contrary to the conclusions drawn from theoretical considerations, H_2^+ may be stable with respect to H^+ and H, and yet be unstable with respect to H_3^+, i.e. the reaction (3) may take place with evolution of energy while that of equation (5) may not. There does not seem to be as yet sufficient experimental evidence to test this possibility.

The recently reported spectroscopic investigations of Richardson and Tanaka[6] on low voltage arcs in hydrogen lead them also to conclude that the primary process of ionization is the formation of H_2^+.

[1] Richardson and Tanaka, Proc. Roy. Soc. **106A**, 663 (1924).

It may be well to discuss the bearing of the results of the positive ray studies on critical potential measurements. Table II gives the measurements of several observers on the critical potentials which they

TABLE II

Ionization potentials of hydrogen.

Observers	Values in volts				
Davis and Goucher[7]	11		15.8		
Bishop[5]	11		15.7		
Found[9]			15.8		
Compton and Olmstead[10]	10.8		15.9		
Krüger[11]	11.5		16.4		29.7
Boucher[12]		13.6	15.6		
Foote and Mohler[13]			16.0		
Mohler, Foote and Kurth[14]			16.0		
Olmstead[15]	11.5	13.6	16.0		
Horton and Davies[16]		13.5	15.9	22.8	29.4
Mackay[17]			15.8		
Olson and Glockler[18]			16.7		
Olmstead and Compton[19] (2800°C)		13.5			
Smyth[3]			16.0		
Mean			16.0		

ascribe to ionization of hydrogen. All are in agreement concerning the existence of an ionization potential at about 16 volts, but all except Smyth have interpreted this potential as that at which ionization plus dissociation (Eq. 2) takes place. This interpretation is obviously no longer tenable. The 13.5 volt point observed has been ascribed to ionization of the hydrogen atom, the atomic hydrogen presumably being formed by thermal dissociation or by collision of an excited hydrogen molecule with an unexcited one (see below). The potential at about 11 volts which has been interpreted by some observers as that for the process $H_2 = H_2^+ + \epsilon$, is probably due to excitation, or, as suggested by Horton and Davies,[20] to the ionization of mercury vapor. A new interpretation must

[7] Davis and Goucher, Phys. Rev. **10**, 101 (1917).

[8] Bishop, Phys. Rev. **10**, 244 (1917).

[9] Found, Phys. Rev. **16**, 41 (1920).

[10] Compton and Olmstead, Phys. Rev. **17**, 45 (1921).

[11] Krüger, Ann. der Phys. **64**, 288 (1921).

[12] Boucher, Phys. Rev. **19**, 189 (1922).

[13] Foote and Mohler, Origin of Spectra, p. 68.

[14] Mohler, Foote and Kurth, Phys. Rev. **19**, 414 (1922).

[15] Olmstead, Phys. Rev. **20**, 613 (1922).

[16] Horton and Davies, Phil. Mag. **46**, 872 (1923).

[17] Mackay, Phil. Mag. **46**, 828 (1923); Phys. Rev. **24**, 319 (1924).

[18] Olson and Glockler, Proc. Nat. Acad. Sci. **9**, 122 (1923).

[19] Olmstead and Compton, Phys. Rev. **22**, 559 (1923).

[20] Horton and Davies, loc. cit.[16] In their helium studies they were able to detect mercury vapor spectroscopically in spite of precautions taken to prevent its entrance.

also be sought for the 22.8 volts potential of Horton and Davies, and for the 29.4 and 29.7 points of these observers and of Krüger respectively. The first was ascribed to molecular ionization, the second to molecular dissociation with ionization of both atoms, $H_2 = 2 H^+ + 2\epsilon$. If such a process as this last took place the lower curve of Fig. 3 would show an increase in the percentage of H^+ at about 30 volts. Olson and Glockler[18] found nine critical potentials in the interval between 14.86 and 16.68 volts, five of which, in addition to the ionization potential, are each 3.16 volts greater than a resonance potential or ionization potential of atomic hydrogen. They interpret these five critical potentials as measuring the energy necessary to dissociate the molecule and resonate one of its atoms, the 3.16 volts being then the heat of dissociation. The three remaining lines which have no apparent relation to the Lyman series they ascribe to molecular excitation not accompanied by dissociation. Hughes[21] in considering Smyth's earlier work in the light of his own experiments, concluded that excitation by electron impact is often accompanied by dissociation. If this view is accepted the interpretation of Olson and Glockler's work, with the exception of the ionization potential, remains unchanged. Von Keussler,[22] however, concluded from a study of spectral data, including his own, that dissociation is an effect secondary to molecular excitation and occurs if the excited molecule is disturbed, as by collision, before it has time to radiate. Such dissociation by collision results only if the excited molecule has energy in excess of that necessary to dissociate the molecule and produce the excited atom. In view of the present experiments and those of Smyth, it appears that von Keussler's theory is the more logical one. There can be very little difference between an ionized molecule and one in the higher stages of excitation. If dissociation does not accompany ionization it would hardly be expected to accompany excitation. If von Keussler's view is adopted, the interpretation of Olson and Glockler's work follows immediately. The 3.16 volts represents, then, the energy of the excited molecule in excess of the energy of the excited atom. If the excited molecule is energetically unstable with respect to its dissociation products and radiates energy or produces kinetic energy on dissociation, the heat of dissociation must then be less than 3.16 volts. While this excess energy of the excited molecule may not be the same for all stages of excitation, any differences that exist may be small enough to fall within the limits of error of the measurements of Olson and Glockler.

[21] Hughes, Phil. Mag. **48,** 56 (1924).
[22] Von Keussler, Zeits. f. Physik, **14,** 19 (1923).

It may be of interest to note that the results of the experiments on mixtures of helium and hydrogen described above point toward an explanation of Merton and Nicholson's[23] experiments on the extension of the Balmer series in which they observed more lines of this series in mixtures of helium with relatively small amounts of hydrogen than could be detected in pure hydrogen.

Taking the mean of the experimental data recorded in Table II gives 16.0 volts as the ionization potential of hydrogen for the formation of H_2^+. As the H_2^+ ion is probably unstable with respect to H and H^+, it must give off energy when it dissociates. If ionization could take place according to the scheme in Eq. (2) the ionization potential for such a process would then be less than 16.0 volts, assuming this to be the correct value, and the heat of dissociation would then be less than $16.0 - 13.5 = 2.5$ volts, equivalent to 57,500 calories per mol. Langmuir,[24] Isnardi,[25] and Wohl,[26] by indirect methods obtained 84,000, 95,000 and 90,000 calories per mol respectively for this value. In view of the probable uncertainty of the values obtained by both these lines of investigation, we are not yet ready to assume that this discrepancy is a real one.

The several theoretical models for H_2^+ give values for the ionization potential of H_2 higher than the observed 16.0 volts. The Bohr model, considered untenable for other reasons, leads to the value 17.85 volts (2.20 Rh − 0.88 Rh), the Pauli[27] model to 23.7 volts, and the various models of Niessen[28] to values 23.5 to 28.8 volts.

Department of Chemistry,
University of California,
February 9, 1925.

[23] Merton and Nicholson, Proc. Roy. Soc. **96A**, 112 (1919).

[24] Langmuir, J. Am. Chem. Soc. **34**, 860 (1912); Langmuir and Mackay, ibid. **36**, 1708 (1914); Langmuir, ibid. **37**, 417 (1915).

[25] Isnardi, Zeits. Elektrochem. **21**, 404 (1915).

[26] Wohl, Zeits. Elektrochem. **30**, 49 (1924).

[27] Pauli, Ann. der Physik **68**, 117 (1922).

[28] Niessen, Physica **2**, 345 (1922); Ann. der Physik **70**, 129 (1923).

BEGINNINGS OF ION-MOLECULE REACTIONS AS A MAJOR FIELD OF CURRENT STUDY

Editor's Comments
on Papers 4 Through 9

After the very early studies of a few ion-molecule reactions, culminating in the late 1920s, the field was almost abandoned for about 25 years. Mann, Hustrulid, and Tate[1] observed H_3O^+ in an electron impact study of water and concluded that it was formed by the reaction

$$H_2O^+ + H_2O \rightarrow H_3O^+ + OH$$

Mitchell, Perkins, and Coleman[2] detected HCO_2^+ in very small intensity in mixtures of CO_2 with hydrogen and water. And interestingly, in 1936 Eyring, Hirschfelder, and Taylor (Paper 12, Part III), using quasi-equilibrium theory, derived an equation for the rate constant for the reaction

$$H_2^+ + H_2 \rightarrow H_3^+ + H$$

which yielded a numerical value in good agreement with experimental values obtained 20 or more years later.

The literature reveals little more on ion-molecule reactions until the 1950s, when the subject suddenly became extremely popular, with a deluge of papers that has continued to increase up to the present.

Part II includes several of the most important of these early studies, starting with the landmark Paper 4 by Tal'roze and Lyubimova. All of these employed conventional single-source, electron impact mass spectrometers. In most instances, only a single gas was present in the source. A considerable number of reactions are reported—far more than the total number ever previously reported—and several of these were surprising indeed. The various reactions were determined primarily by matching the appearance potentials of primary and secondary ions. It was apparent from the low pressures and short times needed for the reactions to be observed that their rate constants were very large, approximately 10^{-9} cm^3/molecule s. It seemed probable that little or no activation energy could be involved and hence that the reaction must be exothermic to be detected. This, of course, does not mean that all ion-molecule reactions are exothermic or that none has an activation energy. However, the recognition that a reaction under these conditions had to be exothermic to be detected proved very useful in identifying which reactions could and could not occur.

All these early investigators were interested in the rates of the reaction and made serious efforts to measure rate constants. All except Paper 7 employed continous ion extraction, so the reaction time for primary ions of a given mass was taken as the time required for the ion to fall from the point of formation in the electron beam to the exit slit under the influence of the repeller field. Since the pressure and path length were small, only single collisions could be expected. Hence time, calculated in this way, was valid and about as accurate as the field strength and flight path were known. Since retention time was constant, rates had to be determined by varying the source pressure. This, of course, is a valid method, but chemists were rather uncomfortable with it since rates are usually determined by measuring changes in intensity with time. Indeed, Field et al. (Paper 5) attempted to vary the retention time by varying the field strength but found that the rate constants determined in this way decreased markedly with increasing field strength, i.e., with increasing average collision energy. Since the collision energy varied over the ion path, no satisfactory quantitative interpretation was possible. Stevenson and Schissler (Paper 6) also studied several ion-molecule reactions and measured the reaction cross section as a function of field strength; i.e., of average collision energy. They found $Q \alpha E^{-1/2}$ where Q is the cross section and E is field strength. The results of Field et al. (Paper 5) gave $Q \alpha E^{-1}$ approximately. Although the reaction studied in the two laboratories were not the same, the results were sufficiently different to arouse considerable interest in the dependence of cross section and rate constant of ion-molecule reactions on collision energy. The papers by Boelrijk and Hamill (Paper 8) and by Ryan and Futrell (Paper 9) represent early, single-source studies directed toward this end.

Continuous ion extraction had the obvious defect that the rate constant was an average over the path length. To get around this, Tal'roze and Frankevich (Paper 7) and later Ryan and Futrell (Paper 9) employed a pulsed ion source, in which ions were formed by a pulse of electrons in the absence of a repeller field. After a variable time following the electron pulse, the ions were swept out of the ion chamber by a strong repeller field. Thus the primary ions were of constant average energy, and reaction could be determined as a function of delay time. Ryan and Futrell thus were able to compare rate constants determined by the pulse mode with those determined with continuous withdrawal. All the reactions studied involved either proton or hydrogen atom transfer. It will be observed that some of the rates are enhanced, some reduced, and some unaffected by the presence of a field during the course of reaction.

Several other experimenters have employed the pulsed mode for studies of ion-molecule reactions, including those who have employed time of flight instruments, which must be pulsed to be effective.

REFERENCES

1. Mann, M. M., A. Hustrulid, and J. T. Tate, *Phys. Rev.* **58**: 340 (1940).
2. Mitchell, J. J., R. H. Perkins, and F. F. Coleman, *J. Chem. Phys.* **16**: 835 (1948).

4

SECONDARY PROCESSES IN THE ION SOURCE OF A MASS SPECTROMETER

V. L. Tal'roze and A. K. Lyubimova

This article was translated expressly for this Benchmark volume by Carole O. Mancini and Semyon Kukes from Dokl. Akad. Nauk SSSR **86**:909–912 (1952).

We investigated processes in the ion source of a mass spectrometer at conditions in which saturated (methane, ethane, propane, butane) and unsaturated (ethylene, propylene, isobutylene) hydrocarbons and water were subjected to ionization.

Upon increasing the source pressure of the substance being investigated, the intensity of ions having a mass one unit greater than the parent begins to increase. Thus in water, propylene, and isobutylene, the ions of mass 19 (H_3O^+), mass 43 ($C_3H_7^+$), and mass 57 ($C_4H_9^+$), respectively, were observed and exhibited such a pressure dependence.

In the case of ethylene, such ions were not detected. Similarly, such "increased" ions having the general formula C_nH_{2n+3} were not found with the saturated hydrocarbons, with the interesting exception of methane, the lowest member of the homologous series. Increasing the methane pressure in the source results in the formation of the ion CH_5^+.

Experiments were conducted with an ordinary gas ion source[1] in the form of a nichrome box 20 x 10 x 5 mm with 7 x 1 mm slits at opposite ends for the admission of electrons from an incandescent filament and their outlet to an electron trap. Gas was admitted to the chamber through a 2-mm orifice in the wall. The ions were impelled out of the chamber through a 10 x 1 mm slit, by means of a weak electric field perpendicular to the electron beam. The electron emission current in various experiments was 0.1 to 1.0 MA. Pressure was measured either by an ionization gauge attached to the chamber around the source * or by a mercury manometer attached to the gas inlet tube. In the latter case, the source pressure was obtained by applying independently determined values for the pressure drop between the capillary and the ionization chamber.

Figure 1 shows a linear dependence on pressure of the ratio, η, of the intensity of the "increased" (M + 1) ion to that of the corresponding parent ion (M) for water (3), propylene (1), and methane (2). The electron energy, U_e, employed was 70 eV. The curves for water, propylene, and methane represent, respectively, the ratios of the intensities of ions of mass 19 to 18, 43 to 42, and 17 to 16. The intersection with the ordinate axis gives in each case the ratio of the intensities of the parent ion having the natural abundance of C^{13} to that of the parent haveing only C^{12}. Thus for propylene and methane, η at the ordinate is 3.3 percent and 1.1 percent, respectively.

*The accuracy of measurement did not justify correction for the difference in pressure inside and outside the ionization chamber.

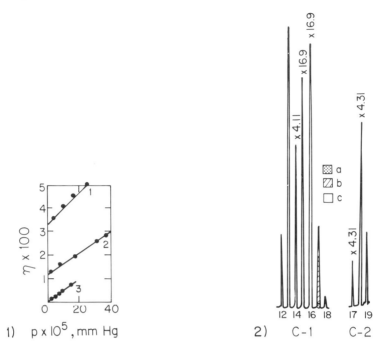

1) $p \times 10^5$, mm Hg

2)   C-1   C-2

Since water was always present, it was necessary to correct the ion of mass 17 for the contribution from OH⁺. This could be done with great accuracy since the intensity of mass 18 (H_2O^+) was known and the relative intensities of the OH⁺ and H₂O⁺ ions were carefully measured in advance. Figure 2 shows recorder tracings of the mass spectra of methane (C-1) and of water (C-2) obtained in the absence of methane when employing 60-eV electrons. The methane spectrum was obtained at a source pressure of approximately 10^{-4} mm Hg. The relative intensities of masses 17 and 18 are shown in C-2, along with the intensity of the ion of mass 19, which is attributed to H_3O^+. The various shadings in the ion peak at mass 17 in C-1 indicate the contributions of OH⁺ (a), $C^{13}H_4^+$ (b), and $C^{12}H_5^+$ (c).

The existence of H_3O^+ ions in solution is generally known and is confirmed by its formation in the ion source.[2] The existence of this ion and the formation of the ions $C_nH^+_{2n+1}$ from the olefins and the observed proportionality of η and p are in accordance with the following reactions:

$$(H - O - H^+) + H_2O \rightarrow \left[\begin{array}{c} H \diagdown \diagup H \\ O \\ | \\ H \end{array} \right]^+ + OH \qquad (1)$$

$$(H - \overset{\overset{\displaystyle H}{|}}{\underset{\underset{\displaystyle H}{|}}{C}} - \overset{\overset{\displaystyle H}{|}}{C} - \overset{\overset{\displaystyle H}{|}}{C} - H)^+ + C_3H_6 - (H - \overset{\overset{\displaystyle N}{|}}{\underset{\underset{\displaystyle H}{|}}{C}} - \overset{\overset{\displaystyle H}{|}}{\underset{\underset{\displaystyle H}{|}}{C}} - \overset{\overset{\displaystyle H}{|}}{C} - H)^+ + C_3H_5 \qquad (2)$$

40

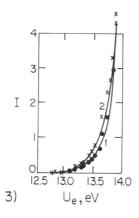

3)

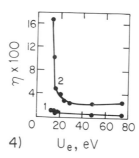

4)

In H_3O^+, the proton attaches to the $2p$ electrons of one of the lone pairs of oxygen to form the new bond.

In a somewhat different class is the formation of the ion CH_5^+, which we would call the methanium ion. Since the participation of 1s electrons is excluded, we apparently have to deal here with a structure similar to that of the ion H_3^+, in which three protons are bonded by two electrons.

With the purpose of ascertaining the nature of the secondary process leading to the formation of CH_5^+, the ionization efficiency curves of CH_4^+ and CH_5^+ were determined. The lower portions of these curves are given in figure 3 (1 - CH_4^+ current; 2 - CH_5^+ * current on an expanded scale; with the ion currents in arbitrary units) and show the coincidence of the two appearance potentials to within the accuracy of the experiment ($\pm 0.05v$). From this, we conclude that CH_5^+ is formed by the reaction

$$CH_4^+ + CH_4 \rightarrow CH_5^+ + CH_3 \qquad (3)$$

From this point of view, the relative intensity, $\eta = f(U_e)$, which decreased abruptly with an increase of U_e was unexpected. This abrupt change may be attributed to the presence in the source of a weak magnetic field coinciding in direction with that of the electron beam. Figure 4 shows the dependence of η on U_e for methane, both in the absence and in the presence of the magnetic field. Similar behavior is observed with isobutylene but is absent with propylene.

This behavior can be accounted for by the following considerations. Let CH_5^+ be formed by reaction (3). A stationary state condition then gives

$$\frac{d(CH_5^+)}{dt} = k_3 (CH_4^+) (CH_4) - k_2 (CH_5^+) = 0 \qquad (4)$$

where the parentheses indicate concentrations, k_3 is the rate constant for reaction (3), and k_2 is the "rate constant" for the flight of CH_5^+ out of the source and is dependent on the space charge and the field strength in the source. Rearranging gives

$$\eta \equiv \frac{(CH_5^+)}{(CH_4^+)} = \frac{k_3 (CH_4)}{k_2} \qquad (5)$$

*Corrected for $C^{13}H_4^+$.

41

The increase in η is caused by the magnetic field results from a decrease in k_2 caused by the lengthening and distorting of the trajectory of the scattered ions. The sudden increase in η near the ionization potential is perhaps attributable to the greatly reduced number of positive ions and hence to a reduced compensation of space charge by the positive ions present. The influence of the magnetic field would greatly accentuate this effect at electron energies near the ionization potential. In the case of propylene, the failure of either the electron energy U_e or of the magnetic field to influence η is perhaps attributable to the exoergicity of the reaction, as was true of reaction (1). The large initial velocity of the secondary ion in this case may be so great that neither the small magnetic field (H = 250 oersteds) nor the change in space charge was sufficient to affect the ion's trajectory. The absence of a peak for $C_2H_5^+$ in the spectrum of C_2H_4 when pressure is increased may be explained by the fact that the reaction

$$C_2H_4^+ + C_2H_4 \rightarrow C_2H_5^+ + C_2H_3$$

is endoergic by some 0.9 eV. [3-6]

If our inference concerning the cause of the effect of magnetic field upon η is confirmed for other reactions, a method will be provided for obtaining a direct assessment of the heat of the secondary reaction under investigation. This could be done by observing the effect on η (that is, on k_2) of magnetic fields of various intensities. We also investigated the formation of H_3O^+ when a mixture of CH_4 and H_2O was ionized. In this experiment, a linear relation between $(H_3O^+)/(H_2O^+)$ and pressure was observed, which shows that the reaction

$$CH_4 + H_2O^+ \rightarrow H_3O^+ + CH_3$$

occured.

The authors express their appreciation to V. N. Kondrat'ev for his valuable advice and participation in discussions of the results.

BIBLIOGRAPHY

1. Cf. Detection and Identification of Radioactive Tracers, a Collection of Articles Translated from English.
2. Mann, M. M., A. Hustrulid, and J. T. Tate, *Phys. Rev.* **58**: 340 (1940).
3. Honig, R. E., *J. Chem. Phys.* **16**: 105 (1948).
4. Eltenton, G. C., *ibid.* **15**: 454 (1947).
5. Korolov, V. V., and A. V. Frost, "Free Energies of Formation of Organic Compounds," May 1949.
6. Szwarc, M., *Chem. Rev.* **47**: 75 (1950).

5

Reprinted from Am. Chem. Soc. J. 79:2419–2429 (1957)

Reactions of Gaseous Ions. I. Methane and Ethylene

By F. H. Field, J. L. Franklin and F. W. Lampe

[Contribution from the Research and Development Division, Humble Oil & Refining Company]

At elevated pressures in a mass spectrometer ion source reactions occur between certain ions and the neutral species present. We have studied the various secondary ions formed in methane and ethylene at elevated pressures and have determined the reactions by which they are formed and the rates of these reactions. The rates are all extremely fast. The reaction rates have been treated by classical collision theory and it has been shown that to a fair approximation the cross-sections and reaction rate constants can be predicted from a simple balance of rotational and polarization forces.

Introduction

Secondary processes in mass spectrometers were first observed by early workers in the field[1] but were treated, for the most part, as nuisances due to experimental difficulties that had to be overcome in the development of analytical mass spectrometry. In recent years, however, a number of studies of secondary processes have been reported[2] which are of considerable interest because of the information afforded concerning the gaseous reactions of ions with molecules. Tal'roze and Lyubimova[3] reported the formation of $CH_5{}^+$ in methane. Stevenson and Schissler[4] published specific reaction rates for the formation of $CD_5{}^+$ in deuterated methane, $D_3{}^+$ in deuterium, and for the reactions of A^+ with H_2, D_2 and HD. In a recent note Schissler and Stevenson[5] have reported many more ion–molecule reactions included in which are reactions forming $C_2H_5{}^+$ and $C_2D_5{}^+$ in methane and deuterated methane, respectively, and $C_2H_5{}^+$, $C_3H_3{}^+$ and $C_3H_5{}^+$ in ethylene. These reactions were reported to exhibit cross-sections that decreased to zero at finite values of ion-energy and small negative temperature coefficients.

This paper comprises a detailed study of the ion–molecule reactions taking place in methane and ethylene when these compounds are subjected to ionization by electron impact in a mass spectrometer.

Theoretical

Consider the reaction of a primary ion with a neutral molecule to consist of the formation of a transition-state ion which then decomposes unimolecularly to various product ions and neutral fragments. That is

$$P^+ + M \xrightarrow{k_1} PM^+ \tag{R1}$$

$$PM^+ \xrightarrow{k_{S_j^+}} S^+{}_j + F_j \tag{R2}$$

(1) H. D. Smyth, Rev. Mod. Phys., **3**, 347 (1931).
(2) For a complete review see F. H. Field and J. L. Franklin, "Electron Impact Phenomena and the Properties of Gaseous Ions," Academic Press, New York, N. Y., in press.
(3) V. L. Tal'roze and A. K. Lyubimova, Doklady Akad. Nauk. S.S.S.R., **86**, 909 (1952).
(4) D. P. Stevenson and D. O. Schissler, J. Chem. Phys., **23**, 1353 (1955).
(5) D. O. Schissler and D. P. Stevenson, ibid., **24**, 926 (1956).

where there will be a set of the above reactions for each primary ion that reacts with neutral molecules. If the time of decomposition of the transition-state ion is short compared with ionic residence times in the ionization chamber, the number of secondary ions of the jth type that are formed will be

$$n_{S_j^+} = \frac{k_{S_j^+}}{\sum_j k_{S_j^+}} n_{PM^+} \tag{1}$$

where the n's are the number of ions of the various kinds formed per unit time. Since the primary ions are formed in the electron beam at a constant rate, in the following for the sake of simplicity we shall in general refer to the n's as the number of ions formed, the rate aspect of the process to be kept in mind at all times. The number of transition-state ions formed is equal to the product of the number of primary ions formed, the number of collisions made by one primary ion with neutral molecules during its ionization chamber residence time, and the collision efficiency or

$$n_{PM^+} = fQ[M]n^0{}_{P^+} \tag{2}$$

where

$n^0{}_{P^+}$ = number of primary ions formed
f = collision efficiency for the formation of PM^+
Q = total no. of collisions made by a single primary ion with neutral molecules at unit concn.
$[M]$ = no. of molecules per unit volume

Combining (1) and (2) and introducing τ, the time in which the primary ion makes collisions with neutrals (the primary ion residence time), gives

$$n_{S_j^+} = \frac{k_{S_j^+}}{\sum_j k_{S_j^+}} n^0{}_{P^+} f \frac{Q}{\tau_{P^+}} [M] \tau_{P^+} \tag{3}$$

Q/τ_{P^+} is the time-average collision rate, and the product of this quantity and the collision efficiency is the rate constant for the formation of the transition-state ion. Recognizing this and rearranging (3) gives

$$\frac{n_{S_j^+}}{n^0{}_{P^+}} = \frac{k_{S_j^+}}{\sum_j k_{S_j^+}} k_1 [M] \tau_{P^+} \tag{4}$$

The number of primary ions formed will be proportional to the number of primary ions collected plus the number of secondary ions (derived from the primary ions) collected. Thus, if we assume equal collection efficiencies for all ions we can write (remembering that the n's are really numbers of ions per unit time)

$$\frac{I_{S_j^+}}{I_{P^+} + \sum_j I_{S_j^+}} = \frac{k_{S_j^+}}{\sum_j k_{S_j^+}} k_1 [M] \tau_{P^+} \qquad (5)$$

where the I's are the observed ion-currents. In addition, it is easily shown that

$$\frac{I_{S_j^+}}{I_{S_k^+}} = \frac{k_{S_j^+}}{k_{S_k^+}} \qquad (5a)$$

To determine specific reaction rates from experimental measurements it is necessary to calculate the average residence time of the primary ions in the ionization chamber. In these calculations an idealized model of the ionization chamber is used, in that we ignore the effects of: (1) the magnetic field in the ionization chamber, (2) potential penetrations through the various ionization chamber orifices, (3) space charge effects, and (4) surface charge effects. The electron beam is assumed to be of infinitesimal thickness and the potential gradient in the ionization chamber (due to the ion-repeller) is taken as uniform.

Consider the primary ion to be formed in the electron beam at a distance d_0 from the ion-exit slit and at the moment of its formation to be moving with average thermal velocity in a direction making an angle ϕ with the perpendicular directed from the electron beam to the slit. It easily can be shown from the laws of motion that the residence time, τ, of an ion reaching the ion-exit slit is given by

$$\tau = \frac{1}{A} \sqrt{2Ad_0 + B^2 \cos^2 \phi} - \frac{B}{A} \cos \phi \qquad (6)$$

where
$A = eV/m$
$B = \sqrt{8kT/\pi m}$
e = electronic charge
V = voltage gradient in the ionization chamber
m = mass of the ion

Since all values of ϕ are equally probable, the average ion residence time is obtained by integration as

$$\bar{\tau} = \frac{1}{4\pi} \int_0^\tau \tau(2\pi \sin \phi) \, d\phi \qquad (7a)$$

or

$$\bar{\tau} = \frac{1}{2A} \sqrt{B^2 + 2Ad_0} + \frac{d_0}{2B} \ln \frac{\sqrt{B^2 + 2Ad_0} + B}{\sqrt{B^2 + 2Ad_0} - B} \qquad (7b)$$

This equation does not apply accurately at ionization chamber voltage gradients less than about 1 volt/cm. since at lower fields some of the ions will be lost by striking the ion repeller.

The reaction rate constant k_1 can be evaluated from the observed ion currents by the use of equations 5, 5a and 7b.

By means of the following analysis the observed ion currents can be used to determine the cross section for the reaction $P^+ + M \rightarrow PM^+$. By combining equations 1 and 2 and converting to currents we obtain

$$\frac{n_{S_j^+}}{n_{p^+}} = \frac{I_{S_j^+}}{I_{P^+} + \sum I_{S^+}} = \frac{k_{S_j^+}}{\sum k_{S^+}} fQ[M] \qquad (8a)$$

Q, the total number of collisions made by a single primary ion with neutral molecules at unit concentration can be written

$$Q = \sigma Q' \qquad (8b)$$

where

Q' = total no. of collisions made by a primary ion during its ionization chamber lifetime with neutral molecules at unit concn. taking the collision cross-section to be unity
σ = collision cross-section for a primary ion and neutral molecule

Substituting (8b) in (8a) gives

$$\frac{I_{S_j^+}}{I_{p^+} + \sum_j I_{S_j^+}} = \frac{k_{S_j^+}}{\sum \Sigma_{S_j^+}} k f\sigma Q'[M] \qquad (8c)$$

For a reaction with no activation energy, $f\sigma$ constitutes the reaction cross-section. Equation 8c permits the evaluation of this quantity from experimental measurements if Q' is known.

To obtain an expression for Q', we again consider the primary ion to be formed in the electron beam at a distance d_0 from the slit and at the moment of its formation to be moving with average thermal velocity in a direction making an angle ϕ with the perpendicular directed from the electron beam to the side of the chamber containing the slit. We assume that the number of ions of each type emerging from the slit is proportional to the number of that type in the ionization chamber. If this is true the position of the ion, at the instant of its formation, along an axis parallel to the electron beam is arbitrary and we take this coördinate to be zero. The velocity of any ion at time t after its formation can then be shown to be

$$v = \sqrt{B^2 + A^2 t^2 + 2ABt \cos \phi} \qquad (9)$$

where the symbols are as defined previously. The average ion velocity is obtained by integration over all directions as before and is

$$\bar{v}_1 = \left(B + \frac{A^2 t^2}{3B} \right) \qquad (10)$$

for the region in which $B > At$.

$$\bar{v}_2 = \left(At + \frac{B^2}{3At} \right) \qquad (11)$$

for the region in which $At > B$.

Similarly, the average relative velocity of two species α and β moving at different velocities is

$$\bar{\xi}_1 = \left(v_\alpha + \frac{v_\beta^2}{3v_\alpha} \right) \text{ for } v_\alpha > v_\beta \qquad (12)$$

$$\bar{\xi}_2 = \left(v_\beta + \frac{v_\alpha^2}{3v_\beta} \right) \text{ for } v_\beta > v_\alpha \qquad (13)$$

where ξ denotes relative velocity.

Substituting (10) and (11) for the ion velocity, v_β, and B for the average thermal velocity, v_α, of the neutral molecules into (12) and (13) results in

$$\bar{\xi}_1 = \frac{1}{3} \left(4B + \frac{2A^2 t^2}{3B} + \frac{A^4 t^4}{9B^3} \right) \text{ for } At < B \qquad (14)$$

$$\bar{\xi}_2 = \frac{1}{3} \left(3At + \frac{B^2}{At} + \frac{3A^3 t^3}{3A^2 t^2 + B^2} \right) \text{ for } At > B \qquad (15)$$

Therefore, for unit cross-section, the total number of collisions made by one primary ion with neutral molecules at unit concentration during the time $\bar{\tau}$ is

$$Q' = \int_0^{B/A} \xi_i \, dt + \int_{B/A}^\tau \xi_i \, dt \qquad (16)$$

where B/A is the time at which the ion and molecule velocities are equal. Integration of (16) gives

$$Q' = \frac{A\bar{r}^2}{2} + \frac{B^2}{A}\left[\frac{247}{382} + \frac{1}{3}\ln\frac{1A\bar{r}}{\sqrt{B^2 + 3A^2\bar{r}^2}}\right] \qquad (17)$$

where $\bar{r}$ is given by (7b).

In the above derivation it was assumed the neutral molecules were all moving with average thermal velocity. Actually there exists a distribution of velocities. However, it has been shown[6] that consideration of this distribution leads to a small difference in relative velocities that is well within our experimental error.

Experimental

Three types of measurements were made: (1) the variation of ion intensities with pressure (pressure studies), (2) the variation of ion intensities with ion repeller voltage (repeller studies), and (3) the appearance potentials of various ions. The first two types of measurements were made with a slightly modified Consolidated Electrodynamics Corp. (CEC) Model 21-620 cycloidal focussing mass spectrometer. The appearance potential measurements were made with a Westinghouse Type LV mass spectrometer.

The pertinent dimensions and operating characteristics of the CEC instrument are as follows. The distance between the ion repeller electrode and the ion exit slit is 1.0 mm., and the electron beam (thickness = 0.17 mm.) passes midway between the two. The length of the repeller electrode in the direction of the electron beam is 5.63 mm. The instrument as received from the manufacturer was modified by by-passing the pressure sensitive filament protection circuit and by altering the ion repeller circuit to permit the application of constant, predetermined repeller voltage. It is not provided with an adjustment for the electron accelerating voltage, and consequently all measurements were made at the manufacturer's pre-set value of about 75 volts. Similarly, no control over the ionization chamber temperature is provided (although the temperature can be measured), and temperature variations of 20-25° occurred from one experiment to another and even in the course of a single experiment. The actual value of the temperature may be taken as 150 ± 10°. The electron current was maintained at 2 μamp.

The methane and ethylene were of Phillips reagent grade, and they were condensed in liquid nitrogen and subjected to a bulb-to-bulb distillation. A middle fraction constituting about half the material condensed was collected for the measurements.

Pressure Studies.—The pressure studies involved the determination of the mass spectra of methane and ethylene at different values of the reservoir pressure. In the CEC mass spectrometer secondary ions begin to be formed in detectable amounts at a reservoir pressure of about 1 mm. In most of the experiments the reservoir pressure was varied between 1 and 30 mm. in steps of 2-5 mm. The reservoir pressure was determined by a small mercury manometer read with a cathetometer, and values obtained in this way were reproducible to 0.1 mm. In all pressure studies the repeller voltage was maintained at 1.0 volt, an arbitrarily selected value.

The quantitative interpretation of the results requires a knowledge of the pressure in the ionization chamber. To determine this quantity as a function of the more easily measured reservoir pressure, the ion repeller of the mass spectrometer was biased negatively with respect to the ionization chamber and the collected ion current measured with an RCA millimicroammeter. The repeller currents (I_i) for various reservoir pressures (p_r) were determined for argon and neon, and plots of I_i/p_r against p_r were constructed. For reservoir pressures up to about 4 mm. I_i/p_r remained constant with pressure and then increased. From this behavior and from a rough comparison of the atomic mean free paths with the known diameter of the gas leak into the

ionization chamber (1 μ), we conclude that for relatively low molecular weight gases at reservoir pressures up to about 4 mm. the gas flow through the leak is molecular in character, but above this pressure mass flow occurs. In the region where molecular flow occurs, the ionization chamber atom concentration (N_i) can be expressed as $N_i = \gamma p_r$, and substituting in the relation $I_i/I_e = N_i Q_i l$ we get $\gamma = I_i/I_e Q_i l p_r$ (I_e = electron current, Q_i = total cross-section for ionization, and l = length of the repeller). From this equation, values of Q_i obtained from the data tabulated by Massey and Burhop,[7] and our experimental currents we find $\gamma_{Ne} = 7.7 \times 10^{12}$ and $\gamma_A = 7.3 \times 10^{12}$. These are equal to within experimental error, as is to be expected theoretically.

To establish the total cross-section for ionization for methane and ethylene it is assumed that in the molecular flow range of pressures the average of the values of γ for argon and neon can be applied to methane and ethylene. Then the Q_i values for these substances can be calculated from the observed currents in this pressure range. Knowing the Q_i values the ionization chamber concentration corresponding to any reservoir pressure can be calculated from the observed currents.

The Q_i values obtained are $Q_i(CH_4) = 4.8_3 \times 10^{-16}$ cm.^{2} and $Q_i(C_2H_4) = 6.9_2 \times 10^{-16}$ cm.^{2}. Recently Otvos and Stevenson[8] have determined the relative total ionization cross-sections for a number of compounds, and from their data one calculates that $Q_i(C_2H_4)/Q_i(CH_4) = 1.55$. Our value for this ratio is 1.43, in satisfactory agreement. However, from Otvos and Stevenson, $Q_i(CH_4)/Q_i(Ne) = 4.28$, whereas our value is 7.75. Partially as a consequence of this discrepancy, we determined the total ionization cross-section of acetylene, obtaining the value $5.6_9 \times 10^{-16}$ cm.^{2}. This agrees reasonably well with the value of 4.98×10^{-16} cm.^{2} quoted by Massey and Burhop. Unfortunately, Otvos and Stevenson do not give a value for the total ionization cross-section for acetylene by low energy electrons. However, their relative value for ethylene can be converted to an absolute value of 4.13×10^{-16} cm.^{2} using the Massey and Burhop value for $Q_i(Ne)$, and this ethylene value is appreciably lower than both our acetylene value and that given by Massey and Burhop. This seems quite suspicious, and consequently we have used our values of the methane and ethylene cross-sections in the determination of the ionization chamber concentrations.

An example of the variation of the intensities of primary and secondary ions as a function of the ionization chamber molecular concentrations, N_i, is given in Fig. 1. We think that the observed deviations from first- and second-order behavior are to be attributed to scattering of the ion beam in the analyzer and, if this be true, the intensity variation of $CH_4{}^+$, for example, should be represented by the relations

$$I_{CH_4{}^+} = a N_i e^{-\sigma_s L N_a} \qquad (18a)$$

$$\log (I_{CH_4{}^+}/N_i) = -\sigma_s L N_a/2.303 + \log a \qquad (18b)$$

where σ_s is the scattering cross-section, L is the path length in the analyzer (taken to be 8.0 cm. in our instrument), and N_i and N_a the molecular concentrations in the ionization chamber and the analyzer, respectively. Since the experimental plot of $\log (I_{CH_4{}^+}/N_i)$ against N_a is approximately a straight line (a small amount of curvature exists), we conclude that indeed the form of the $CH_4{}^+$ curve in Fig. 1 is for practical purposes completely accounted for by scattering in the analyzer, and we infer that all observed deviations from expected first- or second-order behavior can be attributed to this source. The values of N_a used in the plot of eq. 18b were calculated from pressures indicated by an ionization gage located on the envelope of the analyzer. The gage was not calibrated. From the mean slope of the plot we calculate that for the scattering of $CH_4{}^+$ in CH_4 in our instrument $\sigma_s = 4.0 \times 10^{-15}$ cm.^{2}.

To enable us to assess the extent to which scattering affects the values of the current ratios $I_s/(I_p + I_s)$ needed in the kinetic calculations, we have derived[9] an approximate expression for the scattering of an ion beam in the analyzer of a mass spectrometer. The scattering is assumed to re-

(6) L. B. Loeb, "The Kinetic Theory of Gases," McGraw–Hill Book Co., New York, N. Y., Ind. Ed. 1934, p. 96

(7) H. S. W. Massey and E. H. S. Burhop, "Electronic and Ionic Impact Phenomena," Oxford Univ. Press, 1952, p. 38.

(8) J. W. Otvos and D. P. Stevenson, THIS JOURNAL, **78**, 546 (1956).

(9) Details of the derivation are available on request.

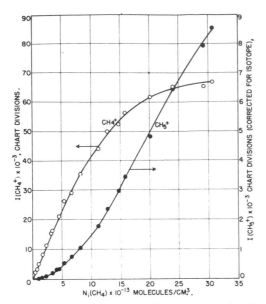

Fig. 1.—Primary and secondary ion currents against ion chamber concentration.

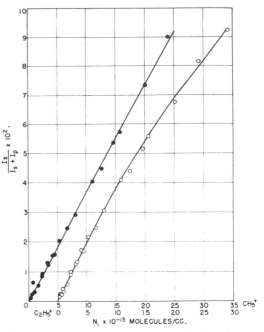

Fig. 2.—Typical curves of current ratios against ion chamber concentration.

sult solely from ion–molecule polarization interaction. The expression is

$$\sigma_s = \left(\frac{3\pi^3 \alpha e L}{16 S_{eff} \; V_{esu}} \right)^{1/2} = 41.8 \left(\frac{\alpha e L}{S_{eff} \; V_{pract.}} \right)^{1/2} \quad (19)$$

where

α = polarizability of the scattering gas
e = electronic charge
L = ion path length in analyzer in cm.
V = ion accelerating voltage
S_{eff} = av. distance from center of collector slit to edges of slit in cm.
$S_{eff} \cong \lambda/4$ where λ = length of collector slit in cm.

For the scattering of $CH_4{}^+$ in CH_4 in our mass spectrometer we calculate from equation 19 that $\sigma_s = 2.0 \times 10^{-15}$ cm.^{2}, to be compared with the experimental value of 4.0×10^{-15} cm.^{2}. The agreement is sufficiently satisfactory to lead us to believe that the theoretical dependence of σ_s on α and V is substantially correct, and thus we write $\sigma_s = b(\alpha/V)^{1/2}$, and the semi-empirical expression $\sigma_s = 1.45 \times 10^{-2} (\alpha/V)^{1/2}$ results.

For the formation of the $CH_5{}^+$ ion from methane we have the relation

$$\frac{I_{CH_5}}{I_{CH_4{}^+} + I_{CH_5{}^+}} \approx \frac{I_{CH_5{}^+}}{I_{CH_4{}^+}} \; \alpha N_i e^{-N_a L(\sigma_{CH_5{}^+} + \sigma_{CH_4{}^+})} \quad (20)$$

and using scattering cross-section values calculated from our semi-empirical expression we calculate that a graph of $I_{CH_5{}^+}/(I_{CH_4{}^+} + I_{CH_5{}^+})$ versus N_i should not depart from linearity until $N_a = 8.2 \times 10^{13}$ molecules/cc. Since the N_a value corresponding to the highest N_i value used in our experiments is 3.4×10^{13}, noticeable deviations of the current ratio from linearity should not be observed. As Fig. 2 shows, this prediction is correct. Similar calculations lead to the prediction that the graph of $I_{C_2H_5{}^+}/(I_{CH_4{}^+} + I_{C_2H_5{}^+})$ vs. N_i should depart from linearity by 10% at $N_i = 4 \times 10^{13}$ molecules/cc., which is in approximate agreement with the behavior of the experimental graph given in Fig. 2. From these and similar results we feel that differential scattering effects are not important at ion source concentrations below about $N_i = 5 \times 10^{13}$ molecules/cc. The rate constants reported in this paper are calculated only from current ratios at concentrations below this value. Two other factors which will affect the accuracy of $I_s/(I_s + I_p)$ are (1) mass discrimination in the

analyzer and (2) primary and secondary ions are formed in different regions of the ionization chamber. From discrimination information reported by Robinson[10] we estimate that the largest discrimination effect occurring in our experiments is 12%. The second factor might give rise to a difference in collection efficiency for the two types of ions, but we do not know the magnitude of the resulting error, if any, in the current ratio value.

Repeller Studies.—The repeller studies involved the determination of primary and secondary ion mass spectra at different values of the ion repeller voltage (V_R) at a known, essentially constant value of the reservoir pressure (about 5 mm.). Three types of variation of ion intensities with increasing repeller voltage are to be observed (Fig. 3). That for primary ions is illustrated by the plot of $I_{CH_4{}^+}$, while the two types observed for secondary ions are illustrated by the plots of $I_{CH_5{}^+}$ and $I_{C_2H_5{}^+}$. We are unable to explain these differences in behavior.

None the less, the forms of the intensity variations with repeller voltage can be used in interpreting the observed secondary spectrum of methane. By comparing Figs. 3 and 4 we conclude that the mass 29 and 17 ions are formed by one type of process, namely, a bimolecular gas phase ionic reaction, but that the mass 26, 27, 28 and 16 ions are formed by another type of process, namely, ionization by electron impact in the electron beam. Since the intensity of the mass 28 ion shows a first-order pressure dependence, it seems likely that it results from a small amount of impurity, probably N_2. However, the mass 26 and 27 intensities are second order in methane pressure and must be formed from methane by some kind of secondary reaction. The appearance potential of the mass 26 ion is 11.5 e.v., which, within the limits of experimental error, is equal to the ionization potential of acetylene. This suggests that the methane undergoes a bimolecular reaction (probably on the filament) to produce acetylene, which diffuses into the ionization chamber and undergoes ionization in the electron beam. The mass 27 ion probably is formed in the same way, although the observed appearance potential (12.1 e.v.) is of no help in elucidating the nature of the ion or the reaction by which it is formed.

(10) C. F. Robinson and L. G. Hall, Rev. Sci. Instruments, 27, 504 (1956).

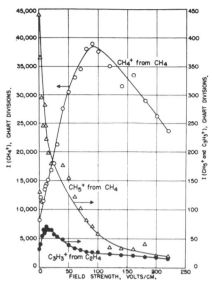

Fig. 3.—Primary and secondary ion currents against ionization chamber field strength.

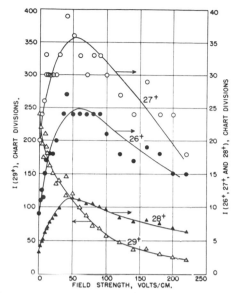

Fig. 4.—Plots of various ion currents from CH₄ against field strength.

Results obtained from repeller studies are subject to an uncertainty stemming from the possible variation in ion collection efficiency with repeller voltage. In addition, results at quite low values of V_R (less than 1 volt) are particularly suspect because the conditions in the ion source at these low voltages are probably ill-defined. Potential penetrations through the ion exit slit and the electron beam slits, the potential depression in the electron beam, surface charges, etc., could cause the actual potential in the ionization chamber to be several tenths of a volt different from that calculated on the basis of only the applied repeller voltage.

An observation which may be related to problems of ion collection efficiency is that at values of $V_R > 2.5$ volts the observed intensity of the mass 29 ion from ethylene becomes smaller than the intensity of the $C^{12}C^{13}H_4{}^+$ ion as calculated from the observed intensity of the mass 28 ion. On the surface this means that the secondary ion contribution to the mass 29 intensity vanishes above $V_R = 2.5$ volts and is probably erroneous even below this value. We can offer no explanation for this behavior.

Appearance Potential Measurements.—Measurements of appearance potential were made in a Westinghouse Type LV mass spectrometer using the vanishing current technique. For most of the measurements O_2 introduced along with the material under investigation was used to calibrate the electron energy scale, but in a few measurements CO_2 was used. The methane or ethylene pressure in the gas reservoir was kept at about 10 mm., which gave about maximum secondary ion currents. Experimental difficulties prevented the determination of the ionization chamber pressure to which this reservoir pressure corresponds. The electron current was maintained at 3.0 μamp. and the ion repeller at 1.9 volts.

The ionization efficiency curves sometimes show breaks (discontinuities) which are attributable to the high pressures. These arise from two sources: (1) contributions to the ion current at a given mass from ions one or two mass units different as a result of scattering and (2) pyrolysis of the sample yielding products which upon ionization contribute to the ion current at the mass being investigated. The difficulty stemming from the first process can be circumvented by obtaining the appearance potentials by repeatedly scanning across the mass spectral region of interest at small increments of the ionizing voltage. The ion intensities free from scattered contributions can then be obtained by reading the peak heights from the means of the adjacent valleys. The occurrence of pyrolysis can be inferred from the form of the ionization efficiency curve, from the value

of the lowest appearance potential, and from some knowledge of the chemical properties of the substance under investigation.

From the observed agreement of replicate determinations we estimate the uncertainty in the appearance potential values to be about 0.3 volt except for the values for the mass 50 and 51 ions from ethylene, which are probably uncertain to about 1 volt.

Results[11]

Typical high pressure mass spectra (reservoir pressure = 10 mm.) of methane and ethylene are given in Table I. Table II contains the appearance potentials of the more important ions and the

TABLE I

HIGH PRESSURE MASS SPECTRA OF METHANE AND ETHYLENE ($P_r = 10$ MM., $V_R = 1$ V.)

CH₄		C₂H₄	
m/e	Intensity, chart div.	m/e	Intensity, chart div.
12	530	24	670
13	1500	25	2230
14	3580	26	17600
15	33700	27	22500
16	39400	28	37200
17	1780	29	1250
26	58	37	8
27	167	38	20
28	163	39	190
29	990	40	17
30	30	41	830
		42	24
		50	26
		51	33
		52	14
		53	102
		54	6
		55	72

(11) When discrepancies appear between this paper and a previous note (THIS JOURNAL, **78**, 5697 (1956)), the present results supersede the earlier ones.

TABLE II
APPEARANCE POTENTIALS AND REACTIONS

m/e	A_1 (e.v.)	Probable process	A_2 (e.v.)	Probable process
		Ions from methane		
17	10.9	$CH_4 + N_2 \rightarrow NH_3^+ \ldots$ $NH_3 \rightarrow NH_3^+$ $(I = 10.5^a)$?	12.8	$CH_4 \rightarrow CH_4^+$ $CH_4^+ + CH_4 \rightarrow CH_5^+ + CH_3^{b}$
26	11.5	$2CH_4 \rightarrow C_2H_2 + 2H_2$ $C_2H_2 \rightarrow C_2H_2^+$ $(I = 11.4)$		
27	12.1	?		
28	11.4	?		
29	12.6	?	13.9	$CH_4 \rightarrow CH_3^+ + H$ $(A = 14.4)$ $CH_3^+ + CH_4 \rightarrow C_2H_5^+ + H_2.^c$ $\Delta H = -19^d$
		Ions from ethylene		
24	11.7	$C_2H_4 \rightarrow C_2 + 2H_2$ $C_2 \rightarrow C_2^+$ $(I = 11.5)$		
26	11.4	$C_2H_4 \rightarrow C_2H_2 + H_2$ $C_2H_2 \rightarrow C_2H_2^+$ $(I = 11.4)$	12.9	$C_2H_4 \rightarrow C_2H_2^+ + H_2$ $(A = 13.5)$
28	10.3	$C_2H_4 \rightarrow C_2H_4^+$ $(I = 10.6)$		
29	..	$C_2H_3^+ + C_2H_4 \rightarrow C_2H_5^+ + C_2H_2,^c \Delta H = -16$		
39	11.8	$C_2H_4 \rightarrow C_2H_2 + H_2$ $C_2H_2 \rightarrow C_2H_2^+$ $(I = 11.4)$ $C_2H_2^+ + C_2H_4 \rightarrow C_3H_3^+ + CH_3, \Delta H = -15$	13.8	$C_2H_4 \rightarrow C_2H_2^+ + H_2$ $(A = 13.5)$ $C_2H_2^+ + C_2H_4 \rightarrow C_3H_3^+ + CH_3,^c$ $\Delta H = -15$
41	10.1	$C_2H_4 \rightarrow C_2H_4^+$ $(I = 10.6)$ $C_2H_4^+ + C_2H_4 \rightarrow C_3H_5^+ + CH_3,^c \Delta H = -17$		
50	24.5	$C_2H_4 \rightarrow C_2^+ + H_2 + H(?)$ $(A = 26.5)$ $C_2^+ + C_2H_4 \rightarrow C_4H_2^+ + H_2,$ $(\Delta H = -162)$		
51	19.5	$C_2H_4 \rightarrow C_2H^+ + H_2 + H$ $(A = 19.3)$ $C_2H^+ + C_2H_4 \rightarrow C_4H_3^+ + H_2, \Delta H = -92$		
53	12.7	$C_2H_4 \rightarrow C_2H_2^+ + H_2$ $(A = 13.5)$ $C_2H_2^+ + C_2H_4 \rightarrow C_4H_5^+ + H, \Delta H = -13(?)$		
55	10.8	$C_2H_4 \rightarrow C_2H_4^+$ $(I = 10.6)$ $C_2H_4^+ + C_2H_4 \rightarrow C_4H_7^+ + H, \Delta H = -15$		

[a] Ionization potentials (I) and appearance potentials (A) in e.v. Values taken from tabulation given in reference 2. [b] Reaction previously observed by Tal'roze and Lyubimova, reference 3. [c] Reaction previously observed by Schissler and Stevenson, reference 5. [d] Heats of reaction in kcal./mole. Values are calculated from ionic heats of formation tabulated in reference 2.

reactions by which they are formed. A_1 and A_2 refer to the lower and higher critical potentials found in a given ionization efficiency curve. The identification of the reactant ions has been made by considering the energetics of possible reactions, the appearance potentials of the various ions, and the dependence of the ion abundance upon pressure and repeller voltage. All the reactions involving the gas phase reactions of ions show second-order pressure dependence, and their repeller voltage dependence corresponds to that of a secondary ion. The accepted values of the primary ion ionization and appearance potentials with which the secondary ion appearance potentials are to be compared are given parenthetically in Table II. It is assumed that only exothermic secondary reactions will be observed, and this assumption can be justified by detailed considerations. The heats for the secondary reactions thought to occur are given in Table II.

Additional Comments on Table II: $m/e = 24$ and 26 from C_2H_4.—The appearance potentials of these ions were determined to learn the technique of making appearance potential measurements at high pressures. The ethylene mass 28 ion was

used to calibrate the ionizing voltage scale. The two high pressure critical potentials observed for the mass 26 ion agree closely with the accepted values for the ionization potential of C_2H_2 and the appearance potential for the electron impact process $C_2H_4 \rightarrow C_2H_2^+ + H_2$, respectively; and this agreement indicates strongly that the lower high pressure value refers to acetylene molecules formed by pyrolysis from ethylene. Similarly, since the appearance potential for the formation of mass 24 ion from ethylene by fragmentation under electron impact is 26.5 volts,[2] the high pressure appearance potential for this ion refers to C_2 formed by pyrolysis and really constitutes a measure of the ionization potential of the C_2 molecule. As such it confirms the value of $I(C_2) = 11.5 \pm 1.0$ volts found by Chupka and Inghram.[12] By contrast, we found no evidence that products of mass 25 or 27 are produced in the pyrolysis of ethylene.

$m/e = 39$ from C_2H_4.—Two appearance potentials of $C_3H_3^+$ were found, the lower one corresponding to, although slightly greater than, $I(C_2H_2)$, and the upper falling midway between $A(C_2H_2^+)$ and

(12) W. A. Chupka and M. G. Inghram, *J. Chem. Phys.*, **21**, 371 (1953).

48

$A(C_2H_3^+)$. Because of this uncertainty, the $C_3H_3^+$ abundance was studied with a small, fixed concentration of ethylene and various concentrations of acetylene. The $C_3H_3^+$ abundance was linear with the acetylene concentration, and we conclude that this ion is formed from $C_2H_2^+$.

$m/e = 29$ from C_2H_4.—This peak in part is attributable to $C^{12}C^{13}H_4^+$. Its dependence upon repeller voltage was different from that of any other ion we have observed in that the total abundance of $m/e = 29$ decreased to that of $C^{12}C^{13}H_4^+$ (and even considerably lower) when the repeller voltage gradient increased above about 25 volt/cm. In view of this peculiar behavior no appearance potential was determined, but the only reasonable reaction is $C_2H_3^+ + C_2H_4 \rightarrow C_2H_5^+ + C_2H_2$, $\Delta H = -16$ kcal./mole.

Experimental Reaction Rate Constants.—From equation 5 it is seen that the slope of the linear portion of the plot of $I_S/(I_P + I_S)$ vs. ionization chamber molecular concentration is equal to $(k_{S_i^+}/\sum_j k_{S_i^+})k_1 \tau_{P^+}$, and using values of τ_{P^+} calculated from equation 7b, $(k_{S_i^+}/\sum_j k_{S_i^+}) k_1$ is readily obtained.

When only one product ion is formed from a given transition state ion $(k_{S_i}/\sum_j k_{S_i^+})k_1$ reduces to k_1, the bimolecular rate constant. When several product ions are formed k_1 is obtained by taking $\sum_j (k_{S_i^+}/\sum_j k_{S_i^+})k_1$. Table III contains the rate constants for all the reactions of methane and ethylene subjected

TABLE III

RATE CONSTANTS FROM PRESSURE STUDIES (V_R = 1.0 VOLT)

Reaction	$(k_{S_i^+}/\Sigma k_{S_i^+})k_1$ (j) $\times 10^{10}$ (cc./molecule sec.)	$k_1 \times 10^{10}$ (cc./molecule sec.)
	8.9		
$CH_4^+ + CH_4 \rightarrow CH_5^+ + CH_3$	9.4	8.93	8.93
	8.5		
	10.2		
$CH_4^+ + CH_4 \rightarrow C_2H_5^+ + H_3$	9.9	9.67	9.67
	8.9		
	3.6		
$C_2H_3^+ + C_2H_4 \rightarrow C_2H_5^+ + C_2H_2$	3.8	3.82	3.82
	3.9		
	4.0		
	2.6		
$C_3H_2^+ + C_2H_4 \rightarrow C_3H_3^+ + CH_3$	2.4	2.50	
	1.6		
$C_3H_2^+ + C_2H_4 \rightarrow C_4H_5^+ + H$	1.6	1.60	
$C_3H_2^+ + C_2H_4 \rightarrow (C_4H_6^+)$			4.10
$C_3H_4^+ + C_2H_4 \rightarrow C_3H_5^+ + CH_3$	4.7	4.80	
	4.9		
$C_3H_4^+ + C_2H_4 \rightarrow C_4H_7^+ + H$	0.54	0.495	
	0.45		
$C_3H_4^+ + C_2H_4 \rightarrow (C_4H_8^+)$			6.30
$C_2^+ + C_2H_4 \rightarrow C_4H_2^+ + H_2$	10.5	10.2	10.2
	9.8		
$C_2H^+ + C_2H_4 \rightarrow C_4H_3^+ + H_2$	4.7	5.8	5.8
	6.9		

to pressure studies. In these measurements the repeller voltage was kept at 1.0 volts.

The percentage average deviations from average for replicate determinations of the rate constants for the eight reactions subjected to pressure studies ranged from 0 to 19%, with most of the deviations of the order 3–5%. We guess that an upper limit to the absolute uncertainty of the values is perhaps 15%.

When the pressure is held constant and the repeller voltage varied it is convenient to calculate the rate constant for each value of the voltage. Figure 5 is a typical curve showing the dependence

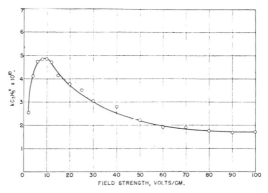

Fig. 5.—Rate constant for $C_2H_4^+ + C_2H_4 \longrightarrow C_3H_5^+ + CH_3$ against ion chamber field strength.

of the rate constant upon the ionization chamber electric field strength, and Table IV contains the rate constants at several field strengths, for all of the reactions subjected to repeller studies. The reservoir pressure was kept at 5 mm. The values tabulated are the averages of replicate measurements. As a matter of interest, measurements on the reaction $CH_4^+ + CH_4 \rightarrow CH_5^+ + CH_3$ were made at field strengths up to 220 volts/cm. The form of the curve up to a field strength of 100 volts/cm. was like that of Fig. 5, and above 100 volts/cm. the value of the rate constant remained essentially constant with a value of about 2.2×10^{-10} cc./molecule sec.

The agreement between replicate determinations of the rate constants from repeller voltage studies (making comparisons at 10 volts/cm.) is generally satisfactory, the percentage deviations from average ranging from 1 to 15% with most of the deviations of the order of 5–10%. The agreement between the rate constant values determined from repeller studies (Table IV, 10 volt/cm. values) and pressure studies (Table III) is acceptable, the percentage differences between average values ranging from 0 to 15% for all values but one (42%) with most of the values falling between 0 and 10%.

It is interesting to note from Tables III and IV that the values of k_1 of all of the reactions differ from each other by at most about a factor of 2.5. These rate constants correspond to extremely fast reactions, and such fast reactions cannot have any appreciable energy of activation. This is in accordance with observations of Stevenson and Schissler[4] and of Schissler and Stevenson,[5] who measured

Table IV

Reaction Rate Constants at Various Field Strengths ($P_R \approx 5$ Mm.)

Reaction	100	60	40	20	10	8	6	2
$CH_4^+ + CH_4 \rightarrow CH_5^+ + CH_3$	2.2	4.8	7.0	8.4	8.5	8.5	8.0	5.0
$CH_3^+ + CH_4 \rightarrow C_2H_5^+ + H_2$	3.4	5.4	7.4	8.8	8.4	7.8	7.0	4.8
$C_2H_3^+ + C_2H_4 \rightarrow C_2H_5^+ + C_2H_2{}^a$				0.9	3.0	3.2	3.4	2.2
$C_2H_2^+ + C_2H_4 \rightarrow C_3H_3^+ + CH_3$	1.6	1.5	1.8	1.9	2.4	2.3	2.2	1.2
$C_2H_2^+ + C_2H_4 \rightarrow C_4H_5^+ + H$	0.44	0.46	0.72	1.2	1.6	1.6	1.6	0.70
$C_2H_2^+ + C_2H_4 \rightarrow (C_4H_6^+)$	2.0	2.0	2.5	3.1	4.0	3.9	3.8	1.9
$C_2H_4^+ + C_2H_4 \rightarrow C_3H_5^+ + CH_3$	1.5	1.8	2.5	3.3	4.6	4.6	4.4	2.4
$C_2H_4^+ + C_2H_4 \rightarrow C_4H_7^+ + H$	0.13	0.14	0.20	0.34	0.50	0.54	0.49	0.24
$C_2H_4^+ + C_2H_4 \rightarrow (C_4H_8^+)$	1.6	1.9	2.7	3.6	5.1	5.1	4.9	2.6
$C_2^+ + C_2H_4 \rightarrow C_4H_2^+ + H_2$	5.8	7.1	7.2	8.8	10.1	10.2	9.3	3.6
$C_2H^+ + C_2H_4 \rightarrow C_4H_3^+ + H_2$	1.6	2.5	2.4	2.9	3.8	3.7	3.0	1.5

Column heading above values: $10^{10}(ks^+j/\Sigma jks^+j)k_1$, cc./molecule sec. at indicated V/d

^{a} Values very uncertain.

Table V

Cross-sections, σ, in Cm.$^2 \times 10^{16}$ at Various Field Strengths

Reaction		100	60	40	20	10	8	6	4	2
$CH_4^+ + CH_4 \rightarrow CH_5^+ + CH_3$	$f\sigma$ expt.	5.6	15	27	45	61	66	68	71	55
	σ calcd.	10	13	15	20	26	27	29	32	36
$CH_3^+ + CH_4 \rightarrow C_2H_5^+ + H_2$	$f\sigma$ expt.	8.4	17	28	46	58	59	58	59	51
	σ calcd.	10	13	15	20	26	27	29	32	36
$C_2H_2^+ + C_2H_4 \rightarrow (C_4H_6^+) \rightarrow$ products	$f\sigma$ expt.	6.4	8.2	12	21	36	38	41	40	27
	σ calcd.	13	16	19	25	32	34	37	40	45
$C_2H_4^+ + C_2H_4 \rightarrow (C_4H_8^+) \rightarrow$ products	$f\sigma$ expt.	5.4	9.2	14	26	48	52	55	54	38
	σ calcd.	13	16	19	25	32	34	37	40	45
$C_2^+ + C_2H_4 \rightarrow C_4H_2^+ + H_2$	$f\sigma$ expt.	18	28	35	58	89	97	98	82	49
	σ calcd.	13	16	19	25	32	34	37	40	45
$C_2H^+ + C_2H_4 \rightarrow C_4H_3^+ + H_2$	$f\sigma$ expt.	5.1	10	12	19	34	36	32	29	21
	σ calcd.	13	16	19	25	32	34	37	40	45

the rates of several such reactions at various temperatures. Unfortunately, we were unable to vary the temperature in our reaction zone. Stevenson and Schissler[4] have reported a value of $1.3_8 \times 10^{-9}$ cc./molecule sec. for the rate constant for the reaction $CD_4^+ + CD_4 \rightarrow CD_5^+ + CD_3$. The ionization chamber field strength is not stated. The agreement with our value for the reaction of CH_4 is good.

Experimental Reaction Cross-sections.—The reaction rates can be expressed in terms of reaction cross-sections, $f\sigma$. Experimental values of $f\sigma$ have been calculated from the current ratios using equation 8c taking Q' values from equation 17, and these values (the averages of replicate determinations) are listed in Table V. Theoretical values of the cross-sections calculated from equation 29 described below are also listed in Table V. A graph of experimental and theoretical cross-sections against ionization chamber field strength is given in Fig. 6.

Schissler and Stevenson[5] report the following cross-sections

Reaction	Cross-sections ($\times 10^{16}$) in cm.^{2} at ion energy	
	0.10 e.v.	1.00 e.v.
$CH_3^+ + CH_4 \rightarrow C_2H_5^+ + H_2$	165 ± 5	39 ± 1
$C_2H_4^+ + C_2H_4 \rightarrow C_3H_5^+ + CH_3$	112 ± 3	21 ± 1
$C_2H_2^+ + C_2H_4 \rightarrow C_3H_3^+ + CH_3$	24 ± 3	6 ± 1

The agreement with our experimental results (comparing Schissler and Stevenson's 0.10 and 1.00 e.v. values with our values at 4 and 40 volts/cm.) is not bad, particularly when it is remembered that the reactions to be compared are slightly different since

we consider the total reactions for $C_2H_4^+$ and $C_2H_2^+$. Schissler and Stevenson[5] report that for the reactions tabulated the cross-sections decrease to zero at finite values of the average speed of the ion. We cannot deduce exactly how they calculate their average speed, and so we cannot make a definite comparison of our results with theirs. However, within the voltage range that we investigated, the cross-sections for all reactions but one remained finite. The exception is the formation of $C_2H_5^+$ from ethylene. Its behavior was very strange, and although we do not understand it, we are inclined to attribute it to unknown instrumental effects.

Our reaction cross-sections are quite large and relatively invariant from one reaction to another. By comparison, the gas kinetic cross sections of CH_4 and C_2H_4 at $423°K$. are[13] 47.8×10^{-16} cm.^{2} and 66.5×10^{-16} cm.^{2}, respectively.

Theoretical Reaction Cross-sections.—It is of interest to attempt to explain these large cross-sections theoretically. Following the general approach introduced by Eyring, Hirschfelder and Taylor,[14] we assume that the cross-section is determined by the distance, r, at which the centrifugal force tending to separate ion and molecule is exactly counterbalanced by the attractive force due

(13) Landolt-Börnstein Zahlenwerte und Functionen, 6 Auflage, "Atom und Molecularphysik," 1 Teil, Springer-Verlag, Berlin, 1950, p. 370.

(14) H. Eyring, J. O. Hirschfelder and H. S. Taylor, *J. Chem. Phys.*, **4**, 479 (1936). See also S. Glasstone, J. K. Laidler and H. Eyring, "The Theory of Rate Processes," McGraw-Hill Book Co., New York, N. Y., 1941, pp. 220 ff.

to polarization of the molecule by the ion. In our calculation it is assumed that all of the initial energy of the system (KE of ion and molecule and potential energy of the polarization interaction) is converted into rotational energy of the ion–molecule complex; we neglect any translational energy of the center of gravity of the complex. Thus, the theory is approximate, and cross-sections obtained from it constitute lower limits to the true values.

The rotation is treated classically. The centrifugal force away from center of mass must be the same for both ion and molecule, so

$$\frac{m_1 v_1^2}{r_1} = \frac{m_2 v_2^2}{r_2} \qquad (21)$$

The center of mass of the system is defined by $m_1 r_1 = m_2 r_2$ with $r_1 + r_2 = r$, the distance between ion and molecule. From the definition of the center of mass, r, and equation 21

$$m_1 v_1 = m_2 v_2 \qquad (22)$$

that is, the linear momentum of the molecule and ion must be equal when rotating. The energy must be conserved so

$$\frac{\alpha e^2}{2 r^4} + T_1^0 + T_2^0 = T_1^f + T_2^f \qquad (23)$$

where T_1^0 and T_2^0 are the initial average kinetic energy of the ion and molecule, respectively, T_1^f and T_2^f are the kinetic energies of the ionic and neutral components, respectively, of the transition state, and α is the polarizability of the molecule. From (22) and (23)

$$T_1^f = \frac{m_2}{m_1 + m_2}\left[\frac{\alpha e^2}{2 r^4} + T_1^0 + T_2^0\right] \qquad (24)$$

The average initial kinetic energies are

$$\overline{T_1^0} = \left[\frac{eVd_0}{3} + \frac{3kT}{2}\right] \qquad (25)$$

$$\overline{T_2^0} = 3/2 kT \qquad (26)$$

In (25) and (26) k is the Boltzmann constant, T is the temperature, and the other terms are as defined previously. Combining equations 24, 25 and 26

$$T_1^f = \frac{m_2}{m_1 + m_2}\left[\frac{\alpha e^2}{2 r^4} + 3kT + \frac{eVd_0}{3}\right] \qquad (27)$$

The attractive force due to polarization is given by $2\alpha e^2/r^5$, and equating polarization force to centrifugal force

$$\frac{2\alpha e^2}{r^5} = \frac{2 T_1^f}{r_1} \qquad (28)$$

Substituting for T_1^f and r_1 and taking $\sigma = \pi r^2$

$$\sigma = \frac{e\pi\sqrt{\alpha}}{\sqrt{6kT + \frac{2eVd_0}{3}}} \qquad (29)$$

Theoretical cross-section values calculated from equation 29 are given in Table V and a plot is given in Fig. 6. The polarizability values used for CH_4 and C_2H_4 are 2.55×10^{-24} cm.³ and 4.06×10^{-24} cm.³, respectively, as calculated from molar polarization values given in Landolt–Börnstein.[15]

The theoretical and experimental cross-sections disagree at worst by about a factor of two, which is satisfactory for calculations of this kind. We

(15) Ref. 13, 3 Teil, p. 515.

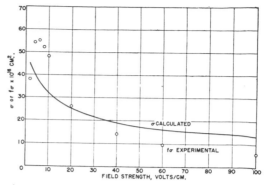

Fig. 6.—Cross-sections for $C_2H_4^+ + C_2H_4^+ \longrightarrow [C_4H_8^+]$.

conclude, then, that the theoretical treatment is for the most part valid and that the dominant factor in determining the cross-sections for these reactions is the ion–molecule polarization interaction. However, in addition to the disagreement in absolute values, theory and experiment do not agree well with respect to (1) the detailed functional dependence of the cross-sections upon ionization chamber field strength and (2) the fact that theory predicts that the cross-sections for the reactions of all ions with a given molecule will be the same, while in actuality some variation from one reaction to another does occur. While we are inclined to the belief that these disagreements result largely from inadequacies in the theory, some of the difficulty may result from experimental error and/or a variable experimental efficiency factor, f. In particular, we do not think that the maximum observed in the experimental $f\sigma$ vs. field strength relation at low field strengths is significant, but rather that it results from a lack of knowledge of the true conditions existing in the ionization chamber at low repeller voltages.

Stevenson and Schissler[4] and Schissler and Stevenson[5] find a small negative temperature coefficient (for the cross-section ?) for some reactions, while for others the rate constant is independent of the temperature. It may be seen from equation 29 that the cross-sections theoretically should depend inversely on the temperature, and at low ionization chamber field strengths the cross-sections are approximately proportional to $T^{-1/2}$. From equations 5, 7b, 8c, 17 and 29 it may be seen that the dependence of the rate constant, k_1, on temperature will be very complicated and cannot be predicted. However, the theory cannot account for the occurrence of temperature coefficients in some reactions but not in others.

Rate Constants for Thermal Speed Ions.— We are not aware of the existence of any values based on experiment of bimolecular rate constants of gaseous thermal speed ions, and it is of considerable interest to attempt to get such values. We do not think that the rate constants we have obtained at zero repeller voltage correspond to thermal speed ions because of the previously discussed lack of definition of the potentials in the ionization chamber at very low voltages. Consequently, we prefer to obtain the rates by extrapolation from our more reliable higher field strength measurements.

For this purpose we require an extrapolation function which more adequately represents the field strength variation of the experimental $f\sigma$ values than does equation (29). As an empirical (but reasonable) procedure which is justified to the extent that it gives satisfactory results, we assume that σ varies according to equation 29 and the efficiency factor, f, varies inversely with the velocity p, of the ion in the activated complex, i.e., $f = \theta/p$ where θ is a proportionality constant. The kinetic energy of the ion in the complex is given by equation 27, from which p is easily derived. Writing $r^4 = \sigma^2/\pi^2$ in the resulting expression for p and taking σ from equation 29, we obtain

$$f = \frac{1}{2}\left(\frac{m_1(m_1 + m_2)}{m_2}\right)^{1/4} \frac{\theta}{(3kT + eVd_0/3)^{1/2}} \quad (30)$$

where the indices 1 and 2 refer to the ion and the molecule, respectively. Then $f\sigma$, the experimental cross-section, is

$$f\sigma = \frac{1}{2}\left(\frac{m_1(m_1 + m_2)}{m_2}\right)^{1/4} \frac{\theta e\pi\alpha^{1/2}}{(3kT + eVd_0/3)} \quad (31)$$

The proportionality constant θ can be evaluated from one experimental value of $f\sigma$ or preferably from the slope of a plot of $f\sigma$ vs. $1/(3kT + eVd_0/3)$. The resulting function represents nicely the field strength variation of $f\sigma$ except for the dubious points at low field strengths, as is illustrated by the typical plot given in Fig. 7.

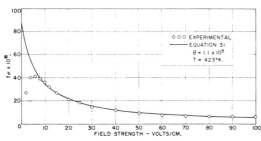

Fig. 7.—Typical plot of $f\sigma$ vs. field strength for $C_2H_2^+ + C_2H_4 \longrightarrow [C_4H_6^+]$.

The relationship between rate constant and cross-section is

$$k = f\sigma\bar{\xi} \quad (32)$$

where $\bar{\xi}$ is the relative velocity of the ion and molecule. Taking $f\sigma$ from equation 31 for $V = 0$ and $\bar{\xi}$ from equation 12 evaluated for thermal velocities, we obtain for the rate constant for thermal speed ions

$$k_{V=0} = \frac{\theta e}{6}\left(\frac{8\pi\alpha}{kT}\right)^{1/2}\left(\frac{m_\alpha + m_\beta}{m_\beta}\right)^{1/2} \quad (33)$$

The index α always refers to the lighter particle.

Eyring, Hirschfelder and Taylor[14] have developed an expression for the rate constant of thermal speed ions, namely

$$k = 2\pi\kappa e\alpha^{1/2}\left(\frac{m_\alpha + m_\beta}{m_\alpha m_\beta}\right) \quad (34)$$

where κ is the transmission coefficient. In Table VI we list thermal cross-sections and rate constants

calculated by our extrapolation method and rate constants calculated from the Eyring, Hirschfelder and Taylor[14] expression taking $\kappa = 1$.

TABLE VI
CROSS-SECTIONS AND RATE CONSTANTS FOR THERMAL REACTIONS

Reaction	Extrl. experimental $f\sigma_{298}°K$. (cm.^{2}) $\times 10^{16}$	$k_{298}°K$. (cc./mole sec.) $\times 10^9$	Theoretical (EH&T) k (cc./mole sec.) $\times 10^9$
$CH_4^+ + CH_4 \rightarrow CH_5^+ + CH_3$	221	2.8	1.3
$CH_4^+ + CH_4 \rightarrow C_2H_5^+ + H_2$	251	3.2	1.3
$C_2H_2^+ + C_2H_4 \rightarrow (C_4H_6^+)$	123	1.2	1.3
$C_2H_4^+ + C_2H_4 \rightarrow (C_4H_8^+)$	155	1.5	1.3
$C_3H^+ + C_3H_4 \rightarrow C_4H_5^+ + H_2$	131	1.2	1.3
$C_2^+ + C_2H_4 \rightarrow C_4H_3^+ + H_2$	284	3.5	1.3

TABLE VII
COMPARISON OF PRIMARY AND SECONDARY MASS SPECTRA
(BASED ON LARGEST OF THE PEAKS COMPARED)

Ion	1-Butene	Primary cis-Butene-2	Isobutene	Cyclobutane	Secondary from $C_2H_4^+ + C_2H_4$
$C_3H_4^+$	6.5	7	11	6.5	1.5
$C_3H_5^+$	100	100	100	100	100
$C_4H_6^+$	2.5	4	2.5	3	0.2
$C_4H_7^+$	18	22	16	21	8.8

Ion	1,2-Butadiene	1,3-Butadiene	1-Butyne	2-Butyne	From $C_2H_2^+ + C_2H_4$
$C_3H_3^+$	100	100	100	57	100
$C_4H_4^+$	28	11	10	23	7
$C_4H_5^+$	100	59	57	100	54

With thermal rate constants and cross-sections of this magnitude it is self-evident that the occurrence of ionic reactions must be seriously considered in any process in which the formation of ions is a possibility, e.g., in radiation chemistry processes. The agreement between our rate constants and those from Eyring, Hirschfelder and Taylor[14] is satisfactory, but it should be noted that our treatment predicts a weak ($T^{-1/2}$) dependence of the rate constant on temperature, which is not found in the EH&T expression. We do not know which of these predictions is correct.

Comparison of Secondary and Primary Mass Spectra.—In those cases where a known intermediate such as $C_4H_8^+$ or $C_4H_6^+$ is formed it is interesting to compare the primary mass spectrum from various C_4H_8 and C_4H_6 compounds with our observed secondary spectrum. Thus the intensities of those peaks in the secondary mass spectrum of ethylene which might have resulted from the reaction of $C_2H_4^+$ and C_2H_4 are compared to those of the same peaks in the primary spectra of various C_4H_8 compounds. Similarly the secondary spectra that might have resulted from the reaction of $C_2H_2^+$ and ethylene are compared to those of several C_4H_6 compounds. Such comparisons are made in Table VII taking the primary spectra from the API tabulation. There is an approximate correspondence which indicates that the intermediate ions (activated complex) must be at least qualitatively similar to the respective parent ions in the primary spectra. In fact, it is tempting to decide that the $C_4H_6^+$ intermediate is more similar to the 1,3-butadiene or 1-butyne ions than to the other two primaries shown.

Acknowledgment.—We wish to express our appreciation to Mr. Burl L. Clark for his invaluable help in making the measurements and calculations.

BAYTOWN, TEXAS

Reactions of Gaseous Molecule Ions with Gaseous Molecules. IV. Experimental Method and Results

D. P. STEVENSON AND D. O. SCHISSLER

Shell Development Company, Emeryville, California

INTRODUCTION

IN two recent short communications[1,2] the authors outlined a method of deducing the rates of bimolecular reactions between gaseous ions and molecules from measurements of the relative intensities of secondary to primary ions in the mass spectra of pure compounds and/or mixtures. From the mass spectrometric point of view, the existence of such reactions and the concomitant appearance of secondary ions in mass spectra is a definite source of annoyance. However, as several authors[3–5] have noted, the very large specific rates that many of these reactions have, $\sim 10^{-9}$ cm^3/molecule sec, indicate such reactions may play a significant role in reactions induced by ionizing radiation in gases. Furthermore, the ability to measure the rates of such very fast, elementary reactions has interesting implications for the development of theories of absolute reaction rates.

It is the purpose of this paper to describe the methods of measurement and provide a more complete account of the experiments that have been briefly described in the preliminary notes referred to above.[1,2] The present paper is concerned only with the phenomenological aspects of the study. The accompanying paper[6] presents a theoretical interpretation of the experiments.

EXPERIMENTAL METHOD AND INSTRUMENT CALIBRATION

The experimental work was carried out with a Westinghouse, Type LV, $\pi/2$-sector analyzer mass spectrom-

eter tube. The electronic circuits for supplying regulated potentials to the various tube elements, regulated current to the magnets, and regulated cathode emission were designed and constructed by Messrs. W. H. Thurston and R. L. DeVault and are all of conventional design.

The mass spectrometer is operated with the ion source and ion collector at ground potential and with the analyzer section at a negative potential. This eliminates from the observed mass spectra the diffuse background and peaks of nonintegral mass-to-charge ratio that result from the dissociation of ions during and after their acceleration to the mass analysis kinetic energies.[7]

The basic mass spectral measurements were made in the following manner: With the gas inlet reservoir charged with the appropriate (measured) quantities of the gas or gases being studied, the mass spectrum of the sample flowing from the reservoir through the ion source was scanned through the "mass range" of interest by continuously decreasing the analyzing magnetic field. Successive scans of the mass spectrum were made for various settings of the ion repeller potential, usually in a sequence such as, ion repeller potential, $V_r = 0.50$, 1.00, 2.00, 4.00, 8.00, 16.00, 16.00, 8.00, 4.00, 2.00, 1.00, and 0.50 volts. The ion analyzing potential was -1000 volts for all ions of mass-to-charge ratio $m/q \geq 12$ and -2000 volts for $m/q \leq 12$.

The ratio of intensity of the secondary ion, S$^+$, to that of the primary ion, P$^+$, defined by the reaction,

$$P^+ + R \rightarrow S^+ + Q \cdots, \tag{1}$$

as computed from measured intensities on the "records" was corrected to zero time for the decay in pressure of the reactant, R, in the reservoir. The "leak" connecting the reservoir to the mass spectrometer ion source is effusive over the range of reservoir pressures employed and its speed is such that the hydrogen pres-

[1] D. P. Stevenson and D. O. Schissler, J. Chem. Phys. **23**, 1353 (1955).

[2] D. O. Schissler and D. P. Stevenson, J. Chem. Phys. **24**, 926 (1956).

[3] Meisels, Hamill, and Williams, J. Chem. Phys. **25**, 790 (1956).

[4] Symposium on Radiation Chemistry of Organic Molecules, Div. Phys. and Inorg. Chem., Am. Chem. Soc. Meeting, Miami, April 7, 1957. See collected papers, J. Phys. Chem., November, 1957.

[5] Field, Franklin, and Lampe, J. Am. Chem. Soc. **79**, 2419 (1957).

[6] G. Gioumousis and D. P. Stevenson, J. Chem. Phys. **28**, 294 (1958).

[7] R. E. Fox and J. A. Hipple, Rev. Sci. Instr. **19**, 462 (1948).

sure in the 2.5-liter reservoir decays 1.07% per min. Checks of the leak speed made with hydrogen, neon, argon, and krypton showed the speed to decrease as [molecular weight]$^{-\frac{1}{2}}$ as is required for an effusive leak. During the recording of the successive mass spectra the recorder paper ran continuously at a speed of 60 mm/min providing a direct time scale. The results of a typical experiment, in the reaction of the argon ion (Ar$^+$) with hydrogen molecules, are shown in Table I.

In experiments with low molecular weight gases (as $D_2^+ + D_2$), the time that elapsed between the recording of successive ions (such as $m/q=6$ and 5) was sufficiently great (the order of one minute between $m/q=6$ and $m/q=4$) that all observed intensities were corrected to zero time before relative intensities of secondary to primary ions were computed.

In order to be able to compute a reaction cross section for a reaction such as (1) from observations of the relative yield of secondary ions it is necessary to know the length, l, of the path of the primary ions through the reactant gas, R, and the concentration of the reactant gas, n_r. The apparent reaction cross section is given by

$$Q = (i_s/i_p)[l n_r]^{-1} \qquad (2)$$

where i_x is the mass spectral current (corrected) of X^+. In writing this equation as a definition of the reaction cross section, we assume that either the efficiency of mass spectral measurement of S$^+$ is equal to that of P$^+$ or a suitably computed correction has been made for variation in ion collection efficiency. It is further assumed that a negligibly small fraction of the primary ions are "lost" by reaction. Since in all experiments $i_s/i_p < 0.01$, the latter condition was certainly fulfilled. With respect to possible variations in the efficiency of collecting and measuring secondary and primary ions, it was necessary to assume there are no differences other than those originating in the mass dependent discrimination characteristic of the slit systems of the ion beam collimating and accelerating system.[8] For mass spectra recorded by scanning the magnetic field, mass dependent discrimination is only a second-order effect for ions of $m/q > 20$ in the mass spectrometer employed in this investigation. For light ions, $m/q \leq 12$, there is a large effect. By means of measurements on H$_2^+$, HD$^+$, and D$_2^+$ at various analyzing potentials it has been found that the ion collection efficiency is accurately represented by $[1 - A(MV_a)^{-\frac{1}{2}}]$ where A is an empirical constant, M the mass-to-charge ratio of the ion and V_a the analyzing potential. This form, with the empirically evaluated, A, was used to correct to "constant" efficiency mass spectral measurements of the intensities of ions, D$_3^+$, HD$_2^+$, D$_2^+$, H$_2$D$^+$, HD$^+$, etc. The form of the efficiency factor for "light" ions is that suggested by an approximate analysis of discrimination between colinear accelerating slits in a weak magnet field.

[8] N. Coggeshall, J. Chem. Phys. **12**, 19 (1944).

TABLE I. Typical experiment on reaction in argon-hydrogen system.

Time (seconds)	V_r (volts)	$(ArH^+/Ar^+)_{obs}$ $\times 10^4$	$(ArH^+/Ar^+)_{corr}$ $\times 10^4$
0	0.50	64.4	64.4
94	1.00	42.5	42.2
178	2.00	29.2	30.1
272	4.00	21.5	22.6
366	8.00	14.7	15.7
464	16.00	11.2	12.2
594	16.00	10.8	12.0
704	8.00	14.2	16.1
820	4.00	18.6	21.5
925	2.00	25.0	29.5
1048	1.00	34.6	41.8
1149	0.50	53.0	65.0

$(ArH^+/Ar^+)_{corr} = (ArH^+/Ar^+)_{obs} \exp(0.0107t)$; t in minutes; reservoir, $p_r^0(H_2) = 330$ microns; $p_r^0(Ar) = 99.2$ microns.
Ion source temperature = 384°K, 75-volt ionizing electrons.

The ion path length, l, is approximately known from the geometry of the ion source. This distance is to be associated with that from the point of origin of the ions to the exit slit from the ion source. The point of origin of ions is in the electron beam that may be assumed to be a plane containing the cathode. The nominal distance from the plane containing the cathode to the ion source exit slit is 0.27 cm. Approximate calculation of the potential distribution in the ion source indicates that the electrons traversing the ion source at such a distance from the exit slit should acquire 71% of the ion repeller potential in addition to the energy acquired from the electron acceleration electrode. Measurements of the variation of the apparent appearance potentials of the argon ion, Ar$^+$, and the neon ion, Ne$^+$, with ion repeller potential indicated the electrons to acquire 69±1% of this energy, in excellent agreement with the geometrical estimate. Thus we conclude that the basic ion path length may be taken to be $l = 0.27$ cm.

Actually the ionizing electron beam is not confined even approximately to a plane. The thickness of the electron beam can be estimated from the variation in the measured resolving power of the mass spectrometer with the ion repeller potential. For ion repeller fields of the magnitude of those employed in the present work, the reciprocal resolving power (ion beam breadth at ion collector slit), $\Delta M/M$, increases $2.3 \pm 0.3 \times 10^{-4}$ per volt of ion repeller potential when the ion analyzing potential is -1000 volts. This corresponds to an electron beam thickness of 0.09 cm or 33% of the mean distance (0.27 cm = l above) of the electron beam from the ion source exit slit. Hence, actually, $l = 0.27$ cm ±17%. However, as is shown in the accompanying paper,[6] the neglect of such a spread in point of origin of the reacting ions introduces only a small second-order error, provided the distribution of points of origin of the ions is symmetric with respect to the mean position of origin.

The relation between the gas concentration, n_r, in the ion source and the externally measured pressure of the gas in the reservoir was determined by measurement of the total ion current produced in helium, neon, and argon in the ion source per unit ionizing electron current and the use of the absolute cross sections of these gases for ionization by single electron impact reported by Smith.[9] Because of the rather low (ca 150 gauss) electron collimating magnetic field employed with this mass spectrometer, calculations were made to ascertain that the actual length of the electron path through the ionization chamber does not differ significantly from the linear dimension ($s = 1.77$ cm) of the ion chamber in the direction of the electron beam. For electrons of kinetic energy, eV^-, moving parallel to a magnetic field, B_a, in the presence of a transverse electric field, E_r, the length of the electron path, $l(e^-)$, is

$$l(e^-) = \int_0^{l_0} \left[v_0{}^2 + 4\frac{\alpha^2}{\beta^2} \sin^2\frac{\beta l}{2} \right]^{\frac{1}{2}} dt \quad (3)$$

where

$$l_0 = \frac{s}{v_0}, \quad v_0 = \left(\frac{e}{m}\frac{V^-}{150}\right)^{\frac{1}{2}}, \quad \alpha = \frac{e}{m}\frac{E_r}{300}, \quad \beta = \frac{e}{m}\frac{B_a}{c},$$

e/m is the ratio of charge to mass of the electron, and c is the speed of light. The integral for $l(e^-)$ is

$$l(e^-) = E[\text{am}(\beta_s/2v_0), \text{mod}(2i\alpha/v_0\beta)]; \quad [i = (-1)^{\frac{1}{2}}] \quad (4)$$

or for $(2\alpha/v_0\beta)^2 < 1$, the elliptic integral becomes

$$l(e^-) = s \left[1 + 2.84\frac{(E_r)^2}{V^-(B_a)^2} - 9.58\frac{(E_r)^2}{[V^-(B_a)^2]^{\frac{1}{2}}} \sin\frac{0.527 B_a}{(V^-)^{\frac{1}{2}}} + \cdots \right]. \quad (5)$$

Calculations by means of the rapidly convergent series, (5), show that for $B_a > 100$ gauss and transverse electric fields, $E_r < 65$ volts/cm, $l(e^-)$ is less than 0.6% greater than s.

In measuring the ion production in the ion source, the ion repeller electrode was connected to ground through the input resistance of a micromicroammeter and the electrode carrying the exit slit was raised ca 10 volts positive with respect to ground by means of a "B" battery across a 50 $K\Omega$ potentiometer. This gave an ion collecting field (E_r) of 26 volts/cm. This value of the ion collecting field was chosen after it was established that the total ion current was collected for all fields equal to or greater than $\sim 50\%$ of this magnitude. Too great a field gradient must, of course, be avoided because of possible unpredictable effects on the electron trajectories. Measurements of ion source positive ion currents were made with helium, neon, and argon for various inlet system pressures between 50 and 500 μ Hg and ionizing electron (75-volt) currents between 5 and 20 μa with the ion source at 380°K. From the observed ratios of positive ion to electron currents, $l(e^-) = 1.77$ cm, and the ionization cross sections quoted in the foregoing,[9] there were found for the instrument constant, C, defined by the equation,

$$n_r[\text{molecules/cm}^3] = C(380/T)^{\frac{1}{2}} p_r \text{ (microns Hg)}, \quad (6)$$

the values $2.4_0 \times 10^9$ (ex helium), $2.6_8 \times 10^9$ (ex neon), and 2.52×10^9 (ex argon). The best mean value (giving double weight to the more accurate determination ex argon) is,

$$C = 2.53 \pm 0.11 \times 10^9 \text{(molecules)}/(\text{cm}^3 \text{ micron}).$$

The factor $(380/T)^{\frac{1}{2}}$ in Eq. (6) arises from the fact that for gas effusing at a constant rate through a volume, the density (concentration) varies inversely with the square root of the temperature of the volume.

The rather good agreement between the values of the calibration constant, C, determined on three gases spanning a range of a factor of ten in molecular weight shows that there is a satisfactorily high ratio of pumping speed to gas conductance of the "circuit" connecting the ion source to the pumps.[10]

Since much of the discussion above involves the geometry of the Type LV mass spectrometer ion source, two sectional views of this source are shown in Fig. 1.[11]

TREATMENT OF THE DATA

As was indicated above, the primary data consist of measurements of the pressure(s) of the gas(es) in the reservoir and the ratio of intensity of a secondary ion to that of a primary ion in the mass spectrum of the gas. The measured pressures along with knowledge of the temperature of the ion source (measured by means of a thermocouple attached to the walls of the ion source)[12] and the time that has elapsed since the gas started to flow through the ion source, permits the calculation of the concentration of gas, n_r, in the ion source by means of Eq. (6). If it is assumed that the intrinsic efficiency

[9] P. T. Smith, Phys. Rev. 36, 1293 (1930). For ionization by 75-volt electrons, Smith gives for helium, $0.31_8 \times 10^{-16}$, neon, $0.55_1 \times 10^{-16}$, and argon, 3.54×10^{-16} cm²/atom.

[10] Barnard, Modern Mass Spectrometry (Institute of Physics, London, 1953), p. 100.
[11] The authors are indebted to Dr. J. A. Hipple for the detailed information on the construction of this ion source as shown in Fig. 1.
[12] The gas flows into the ion source of the Type LV mass spectrometer tube through a slit from a hollow metal block. The canal-shaped slit provides sufficient restriction to the gas flow that the gas is in thermal equilibrium with the block that forms the top and two sides of the ion chamber. The thermocouple referred to in text is welded directly to this block, and thus provides a reliable measure of the temperature of the gas in the ion source. See Fig. 1 and J. Chem. Phys. 17, 101 (1949).

of collection of secondary ions (S$^+$) is the same as that for primary ions (P$^+$) and that the relative intensity, i_s/i_p has been corrected for mass-dependent discrimination, there can be calculated the apparent phenomenological reaction cross section, Q, for the formation of the secondary ion,

$$Q = (i_s/i_p)[n_r l]^{-1}; \qquad l = 0.27 \text{ cm.} \qquad (7)$$

As was first noted by Washburn, Berry, and Hall,[13] and may be seen in the data shown in Table I above, Q for these ion source reactions is a decreasing function of the ion repeller field strength. This cross section may be expected to vary with the energy of the ionizing electrons in some range or ranges of the latter, and, if the reaction has a significant activation energy, Q could be temperature dependent.

As may be seen in Eq. (7) the exact magnitude of the reaction cross section, Q, depends on the exact definition of the reactants, i.e., the reaction giving the secondary ion, S$^+$, might be either

$$P^+ + R \rightarrow S^+ + \cdots \qquad (8a)$$

or

$$P + R^+ \rightarrow S^+ + \cdots. \qquad (8b)$$

Hence before any attempt is made to interpret the apparent magnitude of a reaction cross section, it is necessary to ascertain the reactants. It may frequently happen that it is inconvenient or impossible to satisfactorily record the mass spectra in such a manner that the reactant ion intensity is observed essentially simultaneously with that of the secondary ion. In such circumstances an alternate ion may be used as a reference primary ion to compute an apparent reaction cross section, Q_{app}, that will be related to the true reaction cross section by a multiplicative constant that is equal to the ratio of the cross section for formation (by electrons) of the reference ion to that of the reactant. That is, if the apparent reaction cross section has been computed for Eq. (8a), while the actual reaction is (8b), then

$$Q = \frac{i_s \cdot l^{-1}}{n_p i_r} = \frac{q_p}{q_r} \cdot \frac{i_s \cdot l^{-1}}{i_p n_r} = \frac{q_p}{q_r} Q_{app}, \qquad (9)$$

where q_p and q_r are the cross sections for the formation of P^+ and R^+, respectively, by electron impact.

Even though it may be at times an ambiguous quantity, we shall tabulate in the section of this paper giving experimental results the quantity, Q_{app}, that is calculated directly from the primary mass spectral data. In those cases that this quantity is not the actual reaction cross section, this will be indicated in the legend to the table and the multiplier, q_p/q_r, will be given. The units of Q (or Q_{app}) are 10^{-16} cm^2 per molecule.

Three methods are available for the identification of the reactants leading to a particular secondary ion.

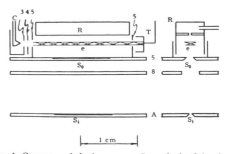

FIG. 1. Geometry of the ion source. C = cathode; 3,4 = electron drawing out and focusing electrodes; 5 = ion chamber reference electrode; R = ion repeller electrode and gas inlet; T = electron trap; S_0 = exit slit from ion chamber; 8 = ion focus electrode; S_1 = ion accelerating slit; cross-hatched region, e, represents the approximate volume of the ionizing electron beam; $S_0 = 0.45_6$ mm; $S_1 = 0.20_3$ mm.

One of these methods is experimental while the other two depend on theoretical inference. The experimental method for the determination of the reactant primary ion [i.e., Eq. (8a) or (8b)] involves the comparison of the dependence of the intensity of the secondary ion on the energy of the ionizing electrons with that of various possible reactant primary ions, in the range of low ionizing electron energies near the appearance potentials of the primary ions. The intensity ratio, i_s/i_p, will be found to be independent of ionizing electron energy when P^+ is the reactant primary ion, and this ratio will either increase or decrease at low electron energies if P^+ is not the reactant ion. Whether the ratio increases or decreases will be determined by the relative order of the appearance potentials of the true and reference primary ion. If the appearance potential of the actual reactant primary ion is less than that of the reference primary, the ratio $i_s/i_{p_{ref}}$ will increase sharply as the electron energy decreases toward the appearance potential of P^+_{ref} and conversely. This method of identification will of course fail if two or more possible reactant ions have nearly the same appearance potential and similarly shaped ionization efficiency curves.

The two methods of identifying the reactant ion that may be described as theoretical methods depend on the theory of the rates of these ion-molecule reactions. As has been previously noted[1] for many ion-molecule reactions, and for all the reactions considered in this paper, the empirical reaction cross section Q (Eq. 7) is found to be, within the experimental error, of the form

$$Q = Q'[E_r]^{-\frac{1}{2}}, \qquad (10)$$

where Q' is a temperature independent constant with the units, cm^2/molecule (volts/cm)$^{\frac{1}{2}}$. As is shown in the accompanying paper,[6,14] for a reaction whose cross

---

[13] Washburn, Berry, and Hall, "Mass spectrometry in physics research," Natl. Bur. Standards, Circ. 522, p. 141.

[14] See also reference (5) where a different method from that employed in reference (6) is used to obtain formally similar results to those of reference (6).

section is largely determined by relatively long-range interactions of the ion-induced dipole, r^{-5}, force law type, measured under the conditions that obtain in a mass spectrometer source like that of the Type LV, the specific (bimolecular) rate, k_0, is given by

$$k_0 = Q \cdot (eE_r l / 600 M_p^+)^{\frac{1}{2}} \quad (11)$$

where M_p^+ is the mass of the reactant ion and the other quantities have been defined already. Substituting the appropriate numbers for our experiments

$$k_0 = 0.36_0 \times 10^6 Q'[M_p^+]^{-\frac{1}{2}} \text{ cm}^3/\text{molecule sec} \quad (12)$$

for $M_p^+ =$ the mass of the reactant ion in A.W.U. Furthermore, the theoretical value of the specific rate, k_t, is

$$k_t = 2\pi e(\alpha/\mu)^{\frac{1}{2}} = 14.7_5 \times 10^{-10} (P_E/\mu_a)^{\frac{1}{2}} \text{ cm}^3/\text{molecule sec}, \quad (13)$$

where P_E is the molecular refraction (Lorentz-Lorenz) of the reactant molecule and μ_a is the reduced mass of the activated complex in A.W.U.

If there is marked disparity between the molecular weights of possible reactants, such as is the case in the nitrogen-hydrogen, carbon monoxide-hydrogen, and oxygen-hydrogen systems considered later, the k_0 calculated from the observed Q' will be strongly dependent on the reactant ion assumed; the alternate choice of H_2^+ or N_2^+ would cause a difference of almost a factor of four in the k_0. Comparison of the alternate k_0's with the theoretical k_t that is relatively insensitive to the choice of reactant may then permit a conclusion with respect to the more probable reactant pair.

In order for the specific rate to be of the form of Eq. (13), it is necessary that the reaction be *not* endoergic. Frequently there is available from appearance potential and thermal data, sufficient information to permit calculation of the energy change involved in various possible reactions that could give a particular secondary ion. Those that turn out endoergic may then be eliminated from further consideration. This provides a second theoretical means of selection of the reactant ion giving rise to a particular secondary ion. Because of the necessity of energetic knowledge concerning reactants and products, this method is most useful when the reactants are hydrocarbons and their ions and the secondary product is a conventional hydrocarbon ion. No use of this method of identification of reactants is made in this paper.

SOURCES OF ERROR AND ACCURACY

It is the purpose of this section to consider the two aspects of the question of the accuracy of the reaction cross sections and/or specific reaction rates of formation of secondary ions that we deduce from the mass spectral data. The two separable aspects are, (1) the relative accuracy of determination of the cross sections

for different reactions, and (2) the absolute accuracy of an individual cross section.

The relative accuracy of the measurement of the cross sections for the formation of different secondary ions is, in the absence of large background interference with the secondary ion, essentially the same as the relative accuracy of measurement of mass spectral sensitivities (specific ion currents). For the present mass spectrometer this relative accuracy is $\pm 1\%$ when the accuracy of the measurement of ion currents is not limited by signal-to-noise ratio, i.e., dynamic range of the ion detector-recorder system. The effective dynamic range of the detector-recorder as used in the present investigation was *ca* 1.5×10^4, hence for virtually all our measurements the accuracy was determined by the signal-to-noise ratio associated with the secondary ion since measurements were restricted to conditions such that $i_s/i_p < 10^{-2}$. Since very few measurements were made under conditions such that the signal-to-noise ratio was less than 10, we believe the relative accuracy of all cross sections is better than $\pm 10\%$, and in general that the relative error is less than $\pm 5\%$.

The general nature of repeatability of measurement that is a direct measure of the relative accuracy of the reaction cross sections is indicated in Table II where there are shown the results of experiments made at the same ion repeller field strength, $E_r = 10.3$ volts/cm, and various initial reservoir pressures of hydrogen and argon, and various ionizing electron currents. The ionizing electron energy was 75 volts in all experiments. The uncertainty of the individual values of Q for ArH$^+$, and the apparent relative abundance of ^{36}Ar, indicated in this table are those estimated from the accuracy of measurement of the "weak" signal determined by the signal-to-noise ratio of the detector. The uncertainties indicated for the average $Q(\text{ArH}^+)$, ± 4.4, and for ^{36}Ar$^+/^{40}$Ar$^+$, ± 1.0, are the standard deviations. These may be compared with the standard deviations to be expected for the estimated "accuracies" of the individual measurements, ± 3.5 and ± 1.0, respectively.

The problem of assaying the absolute accuracy involves a number of questions, some of which are imponderable. The three factors that enter in addition to the accuracy of measurement of the relative intensity of secondary to primary ion current are: (1) the precision of the calibration constant, C, of Eq. (6), that relates the externally measured pressure to the concentration of gas in the ion source, (2) the accuracy to which the ion path length, l, in the ion source is known, and (3) the accuracy of the ion repeller field strength.

The agreement between the values of C determined by measurements on three different gases suggests this calibration constant to be accurate to $\pm 5\%$. This leaves begging the question of the absolute accuracy of the constant since our calibration constant can be no more accurate than Smith's[9] determination

TABLE II. Effects of ionizing electron current and gas pressure on $Q(\mathrm{ArH^+})$.

	Ion repeller field, $E_r = 10.3$ volts/cm				
	Reservoir pressure microns		Electron current		
Exp.	H_2	Ar	μa	$Q(\mathrm{ArH^+}) \times 10^{16}$	$^{36}\mathrm{Ar^+}/^{40}\mathrm{Ar^+} \times 10^4$
41–1a, 41–5a	214.0	130.0	5.0	82.4±7	31.5±2
41–1b, 41–5b	214.0	130.0	10.0	85.7±7	32.6±2
41–4a	214.0	64.3	10.0	78.7±7	30.5±2
41–4b	214.0	64.3	20.0	87.5±7	31.4±2
41–5	103.0	130.0	10.0	76.0±7	32.8±2
			Average	82.1±4.4	31.8±1

of the absolute ionization cross sections of the three rare gases. Smith made no estimate of the probable error of his measurements. In view of the care with which Smith's measurements were made, it seems reasonable to assume his measurements were no less accurate than our ability to repeat his relative cross sections for the three gases. Thus we estimate our constant, C, to be accurate to ±10% on an absolute basis.

The mean distance from the electron beam to the exit slit of the ion source, i.e., the average distance traversed by the ions is probably accurate to ±3%. This conclusion is based on the agreement between the probable mean position of the electron beam calculated from the geometry of the ion source and that measured by means of the mean energy the ionizing electrons acquire from the ion repeller field. The ion beam breadth measurements indicate the electron beam to have a thickness that is a quite sensible fraction of the mean distance of the beam from the ion source exit slit. As the calculations in the accompanying paper[6] show, relatively little error is introduced by the neglect of the finite thickness of the electron beam, i.e., the spread in ion path lengths, provided the half-thickness of the beam is less than $1/5$ of the mean distance, l, and the distribution of electrons in the beam is symmetrical with respect to the mean. The first of the conditions with respect to the electron beam is satisfied, its half-breadth is the order of $1/6$ the mean distance, l. In the absence of any real knowledge of the distribution of electrons in the section of the electron beam, we can only guess that the resultant uncertainties in the absolute value of l are no greater than ±10%.

The ion repeller voltage was measured by means of a Westinghouse Type PX5 precision (±1%) voltmeter. Thus, this measurement is certainly not limiting in the accuracy of our knowledge of the ion repeller field strength E_r. The imponderable source of error in the determination of the field strength E_r lies in the indeterminant contact potentials associated with the surfaces of the electrodes forming the walls of the ion source. The surface effects should be small since all the surfaces are of gold and these surfaces remain quite clean as shown by the very slight mass spectral background that results when the ion source temperature is rapidly raised from 380° to 580°K. Hipple[15] has estimated in one case that the surface potentials could be as great as +0.2 volt for an ion source of this type. If this estimate is accepted as a typical limit, the error in the ion repeller field strength could be as great as 40% at the lowest value (1.3 volts/cm) we have employed. Since the effectiveness of this source of error decreases rapidly with the magnitude of the applied potential, and we find little or no systematic divergence of the reciprocal square root dependence of the apparent reaction cross section at all repeller fields, we believe the constant potential error to be considerably less than this estimate of Hipple[15] would suggest. The calculations in the accompanying paper[6] indicate that the second source of divergence of the true ion repeller field from the apparent value indicated by the voltmeter reading, the space charge of the electron beam, is also negligible. Thus we believe ±10% to be a conservative estimate of the absolute accuracy of ion repeller field strength we report.

Since the various sources of error discussed in the foregoing are independent we may combine them in the usual fashion and conclude that a reasonable estimate of the over-all, absolute accuracy of the reaction cross sections is ±25%. The general quality of the agreement found between theoretically calculated and observed specific rates of many of the ion-molecule reactions suggests this estimate of accuracy to be not unrealistic.

EXPERIMENTAL RESULTS

Reaction of Argon and Hydrogen

The reaction of the argon ion, $\mathrm{Ar^+}$, with hydrogen molecules was studied in greater detail than any other reaction. This emphasis arose from the particular convenience of this reaction. Argon has a relatively large ionization cross section,[9] and is virtually monoisotopic with the predominant isotope that of greatest mass, $^{40}\mathrm{Ar} = 99.6\%$ natural abundance. The ions of the less

[15] J. A. Hipple, Phys. Rev. 71, 594 (1947).

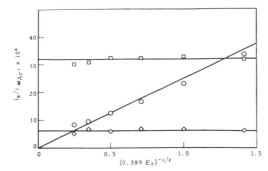

FIG. 2. Effect of ion repeller potential on the relative intensities of ions. Ionizing electrons: 10 μa, 75 volts. $p_r(Ar) = 130$ microns Hg, $p_r(H_2) = 214$ microns Hg. □, $^{36}Ar^+/^{40}Ar^+$; ◇, $^{38}Ar^+/^{40}Ar^+$; ○, $^{40}ArH^+/^{40}Ar^+$.

abundant isotopes, $^{38}Ar = 0.07\%$, and $^{36}Ar = 0.32\%$, do not interfere or provide a background against which the reaction product must be measured (as is the case with neon, see below), while at the same time providing intensities on the recorded mass spectra that serve as reference checks on the measurement of primary ion currents under varying conditions of mass spectral measurement. These statements are illustrated in Fig. 2 where there are shown the results of a typical experiment on the argon-hydrogen reaction. In Fig. 2 there are plotted three intensity ratios, $^{36}Ar^+/^{40}Ar^+$, $^{38}Ar^+/^{40}Ar^+$, and $^{40}ArH^+/^{40}Ar^+$, against the reciprocal square root of the ion repeller field strength for the conditions specified in the legend to the figure. The essential invariance of the apparent relative abundances of the argon isotopes with variation of the ion repeller field over a factor of thirty, clearly demonstrates that the decrease of $^{40}ArH^+/^{40}Ar^+$ with increasing field strength is a *real effect*, not an instrumental artifact. Furthermore, the essential agreement between the average apparent relative abundances of the argon isotopes with the best values carefully determined by Nier,[16] shows that mass dependent discrimination is indeed negligible over ranges of molecular weight of the order of $\pm 10\%$.

That the reaction giving the argonium ion, ArH^+, is one between argon ion, Ar^+, and hydrogen molecules, H_2, was established by measurements of the relative yield of ArH^+ at low ionizing electron energy with a constant ion repeller field strength. As is shown in Fig. 3, the ratio ArH^+/Ar^+, remains essentially constant as the ionizing electron energy is decreased to the lowest value compatible with measurement of ArH^+ (i.e., permitted by the signal-to-noise ratio of the ion detector). On the other hand, the ratio ArH^+/H_2^+, decreases rapidly when the ionizing electron energy becomes less than 30 volts. For electron energies greater than *ca* 30 volts, the ionization efficiency curves for

both Ar^+ and H_2^+ are essentially flat and parallel. For electron energies of less than 30 volts the argon ionization efficiency curve decreases more rapidly toward the appearance potential than does that of hydrogen. The constancy of ArH^+/Ar^+ demonstrates that the ionization efficiency curve of ArH^+ parallels that of Ar^+, and that the argon ion, Ar^+, is the reactant.

The effect of gas temperature on the reaction forming argonium ions was studied by measurements with the ion source at various temperatures across the range 333 to 614°K. As may be seen in Table III, there is no trend and the variation in the reaction cross section across this temperature range is no greater than the standard deviation.

The reactions of the argon ion, Ar^+, with the isotopic species of hydrogen, HD and D_2 were studied in the same manner, but less extensively, as was the reaction with H_2. Only two temperatures, 380° and 575°K were investigated and only 75-volt ionizing electrons were employed.

There are summarized in Table III the results of the measurements on the three species of hydrogen. In this Table III there are included the "least square" values, Q', for the reaction cross section of Ar^+ with each hydrogen species and the standard deviation of the Q for this "best value".

Reaction of Krypton and Hydrogen

The krypton-hydrogen reacting system was studied at three temperatures, 380°, 492°, and 615°K, while the krypton-deuterium system was studied at only the single temperature, 380°K, and at low ionizing electron

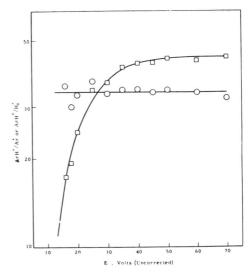

FIG. 3. Variation with ionizing electron energy (E^-) of the ion reaction functions, ArH^+/Ar^+, ○, and ArH^+/H_2^+, □. Ionizing electron current = 15 μa, Ion repeller potential (E_r) = 5 14 v/cm.

16 A. O. Nier, Phys. Rev. **77**, 789 (1950). For Airco spectroscopically pure argon Nier found $^{36}Ar = 0.33_3\%$ and $^{38}Ar = 0.06_5\%$.

TABLE III. Summary of measurements of reaction cross section for Ar^+ and the hydrogen molecules, H_2, HD, and D_2.
(75-volt ionizing electrons)

No. of Exps.	"Hydrogen"	$T°K$	$Q(ArH^+) \times 10^{16}$ cm² per molecule					
			$E_r = 1.28$	2.57	5.14	10.2$_8$	20.5$_6$	41.1$_2$ v/cm
2	H_2	333	251.	181.	130.	97.$_5$		
3	H_2	377	248.	179.	133.	97.$_0$		
2	H_2	380	294.	195.	139.	101.$_6$	74.$_4$	59.$_0$
2	H_2	425	240.	177.	132.	96.$_7$		
2	H_2	489	243.	181.	131.	99.$_8$		
2	H_2	537	235.	176.	131.	98.$_5$		
2	H_2	575	266.	188.	134.	97.$_6$	72.$_6$	55.$_8$
2	H_2	614	253.	185.	140.	108.$_1$		
	Average Q		254.	182.	134.	99.$_5$	73.$_2$	57.$_5$

Least square[a] $Q' = 259 \pm 13 \times 10^{-16}$

No. of Exps.	"Hydrogen"	$T°K$	$E_r = 1.28$	2.57	5.14	10.2$_8$	20.5$_6$	41.1$_2$ v/cm
4	HD	380 $\{$ Q / ArH^+/ArD^+	228. / 0.88$_3$	153. / 0.87$_9$	103.$_2$ / 0.86$_8$	74.4 / 0.86$_8$	50.$_5$ / 0.86	29.$_8$ / 0.7$_0$
2	HD	575 $\{$ Q / ArH^+/ArD^+	224. / 0.85$_9$	156. / 0.86$_1$	107.$_2$ / 0.89$_9$	82.$_3$ / 0.82	53.$_7$ / 0.8$_3$	46.$_6$ / 0.8$_4$
	Average Q		226.	155.	105.$_2$	77.$_5$	51.$_7$	35.$_2$
	Average ArH^+/ArD^+		0.87$_6$	0.87$_3$	0.87$_7$	0.85$_2$	0.85	0.7$_3$

Least square $Q' = 251 \pm 7 \times 10^{-16}$
Best $ArH^+/ArD^+ = 0.85_8 \pm 0.025$

No. of Exps.	"Hydrogen"	$T°K$	$E_r = 1.28$	2.57	5.14	10.2$_8$	20.5$_6$	41.1$_2$ v/cm
3	D_2	380	225.	144.	101.$_8$	72.$_6$	52.$_1$	33.$_3$
2	D_2	574	202.	140.	103.$_8$	75.$_9$	54.$_5$	35.$_8$
	Average		212.	142.	102.$_2$	74.$_1$	53.$_0$	34.$_1$

Least square $Q' = 237 \pm 6 \times 10^{-16}$

[a] In this and subsequent tables, least squares Q' is the value of Q' that minimizes the sum,

$$\sum_{i=1}^{n} (Q_i - Q'[E_r^{-\frac{1}{2}}]_i)^2, \quad \text{i.e.} \quad \sum_{i=1}^{n} Q_i [E_r^{-\frac{1}{2}}]_i \bigg/ \sum_{i=1}^{n} [E_r^{-\frac{1}{2}}]_i^2.$$

energy. That the reaction forming the kryptonium ion, KrH^+, is

$$Kr^+ + H_2 \rightarrow KrH^+ + H \qquad (14)$$

was established by measurements at low ionizing electron energies. As may be seen in Fig. 4, the ratio $i(KrH^+)/i(Kr^+)$ remains more or less constant with decreasing electron energy to the lowest values of the latter that could be used while still giving sufficient KrH^+ intensity for reasonable measurement. On the other hand, for ionizing electrons with less than nominally 20 volts energy the ratio KrH^+/H_2^+ increases markedly. The greater ionization potential of H_2 than of Kr causes the ionization efficiency curve of H_2^+ to decrease more rapidly than does that of Kr^+ for electrons with less than 20 volts energy.

The computation of $i(^{84}KrH^+)/[n(^{84}Kr)i(H_2^+)]$ from the actually observed $i(^{84}Kr^+)n(H_2)$ was carried out in the following manner. For 75-volt ionizing electrons the relative total ionization cross sections of krypton and argon are in the ratio of 1.48 to 1.00.[17] From the data of Smith,[9] and Tate and Smith[18] the

relative total ionization cross sections of argon and hydrogen are in the ratio of 1.00 to 0.279, also for 75-volt ionizing electrons. Since the fraction of the krypton ion current in the form of singly charged ions (for 75-volt ionizing electrons) is 0.835,[19,20] the effective ratio of the cross section for formation of Kr^+ and H_2^+ by 75-volt electrons is $q(Kr^+)/q(H_2) = (1.48/0.270) \times 0.835 = 4.58$, where allowance has been made for the small, ca 2%, fraction of the total hydrogen ion current that is in the form of H^+. The values of $q(Kr^+)$ and $q(H_2^+)$ at low ionizing electron energies relative to the values for 75-volt electrons were specially measured, and used in conjunction with the relative value, 4.58, for 75-volt electrons to compute the $KrH^+/n_{Kr} \cdot H_2^+$ shown in the second part of Table IV.

In the first part of Table IV there are given the observed cross sections for the formation of $^{84}KrH^+$ as a function of the ion repeller field for 75-volt ionizing electrons at the three temperatures studied. These Q's are for Kr^+ the reactant as the low energy ionizing electron data indicate.

Measurements on the formation of the kryptonium

[17] J. W. Otvos and D. P. Stevenson, J. Am. Chem. Soc. 78, 546 (1956).
[18] J. T. Tate and P. T. Smith, Phys. Rev. 39, 270 (1932).
[19] D. P. Stevenson, J. Chem. Phys. 18, 1347 (1950), and further unpublished measurements by the present authors.
[20] J. T. Tate and P. T. Smith, Phys. Rev. 46, 773 (1934).

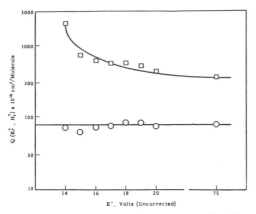

FIG. 4. Variation with ionizing electron energy (E^-) of the apparent reaction cross section, Q, for Kr^+ the reactant, $\bigcirc$, and for H_2^+ the reactant, $\square$. Ionizing electron current = 15 μa, $E_r = 2.57$ v/cm.

ion at low ionizing electron energies demonstrated that the space-charge potential of the electron beam can become of sufficiently significant proportions to distort the results of such measurements. The measurements reported in Table IV on the apparent cross sections for formation of KrH^+ from Kr^+ at low electron energies were made with an ion repeller field of 2.57 volts/cm and an ionizing electron beam of 15 μa. Similar sets of measurements were made with the ion repeller field equal to 1.28 volts/cm and (a) 15 μa and (b), 20 μa of ionizing electrons. There are shown in Table V the apparent reaction cross sections for the formation of $^{84}KrH^+$, $^{86}KrH^+$, and $^{86}KrD^+$ under the conditions of very low ion repeller field, high ionizing electron current, and low ionizing electron energy. The inherent repeatability of measurements of this type is demonstrated in the columns 5 and 7 labeled $S(^{86}Kr)$ where there are given the specific intensities (sensitivities) of the $^{86}Kr^+$ ion in arbitrary units (positive ion current per unit pressure in the inlet system reservoir).

From the results of measurements at greater ion repeller fields and low ionizing electron energies, there would be expected $Q(KrH^+)$ to equal 102×10^{-16} cm²/molecule for $E_r = 1.28$ volts/cm. The much greater values listed in Table V are believed to be due to space-charge neutralization of the ion repeller field with resultant increase in reaction cross section due to its inverse dependence on this field strength.

Reaction of Neon and Hydrogen

Relatively few experiments were made on the neon-hydrogen system. As already indicated, the low ionization cross section of neon, combined with an unfavorable distribution of natural isotopic abundances ($^{20}Ne = 90.9\%$, $^{21}Ne = 0.26\%$, and $^{22}Ne = 8.82\%$)[21] make

[21] A. O. Nier, Phys. Rev. **79**, 450 (1950).

it difficult to make even semiquantitative measurements of the cross section for the formation of the neonium ion, NeH^+. There are shown in Fig. 5 the results of measurements of the ratio of the intensity of the ion of $m/q = 21$ to that of $m/q = 20$ in the neon mass spectrum as a function of ion repeller field for (I), no hydrogen present, and (II), hydrogen concentration equal to 0.75×10^{12} molecules/cm³ in the ion source. The best line (II) through the $(E_r)^{-1} = 0$ intercept of (I) gives, $Q' = 34 \pm 4 \times 10^{-16}$ for the reaction,

$$Ne^+ + H_2 = NeH^+ + H. \qquad (15)$$

Since the ionization cross section of neon and hydrogen for 75-volt electrons are in the ratio, 0.62 to 1.00,[9,18] the reaction cross section would be reduced to, $Q' = 21 \pm 3 \times 10^{-16}$ if the reaction were that of H_2^+ plus Ne. In view of the facts that the ionization potential of Ne is *ca* 6 volts greater than that of H_2, and that in the cases of argon and krypton it is the rare gas ion that reacts with H_2, it seems reasonable to assume that the neon ion is the reactant.

Reaction of Nitrogen and Hydrogen

The study of ionic reactions in nitrogen-hydrogen mixtures was limited to establishing the existence of a reaction leading to the formation of the nitrogenium ion, N_2H^+, with no effort being expended in establishing directly the reactant ion. Deuterium was used in order

TABLE IV. Summary of measurements of reaction cross section for Kr^+ and hydrogen molecules. (75-volt electrons)

$T°K$	$E_r = 1.28$	2.57	5.14	10.2₈	20.5₆ volts/cm
		$Q \times 10^{16}$ cm²/molecule			
380		69.₄	57.₆	34.₇	25.₃
380		84.₁	56.₅	40.₆	27.₁
385	99.₅	77.₁	56.₅	36.₄	27.₂
492	100.₅	72.₄	52.₄	34.₁	20.₈
616	122.₄	90.₀	66.₄	47.₈	15.₄
Average	107.₆	78.₈	57.₉	38.₇	23.₁

Least squares $Q' = 123.9 \pm 2.6 \times 10^{-16}$

	For $E_r = 2.57$		
		Apparent Q for	
V^- [a]	Kr^+ [b]	H_2^+ [c]	$Q(KrH^+)/Q(KrD^+)$ [d]
75.0	$78.₈ \times 10^{-16}$	$361. \times 10^{-16}$	...
20.0	73.0	433.	1.6₉
19.0	81.₂	514.	1.6₁
18.0	83.₀	562.	1.5₃
17.0	75.₄	560.	1.6₇
16.0	71.₂	610.	1.6₉
15.0	61.₈	729.	1.5₆
14.0	70.₆	2080.	1.6₂

[a] V^- = apparent ionizing electron energy.
[b] Q for Kr^+ the reactant.
[c] Q for H_2^+ the reactant.
[d] Measurements on $^{86}KrH^+$ and $^{86}KrD^+$, see Table V.

TABLE V. Space charge effects on apparent reaction cross sections.[a]
Ion repeller field, 1.28 v/cm

Electron energy nominal volts	$Q(^{84}KrH^+)$ (15 μa)	$Q(^{84}KrH^+)$ (20 μa)	$Q(^{86}KrH^+)$ (20 μa)	$S(^{86}Kr^+)$ 20 μa	$Q(^{86}KrD^+)$ 20 μa	$S(^{86}Kr^+)$ 20 μa
20.0	168.	399.	400.	63.0	238.	62.6
19.0	168.	442.	435.	56.9	284.	54.8
18.0	175.	514.	506.	50.0	334.	49.5
17.0	196.	666.	660.	41.8	397.	41.6
16.0	234.	878.	890.	33.2	532.	33.3
15.0	302.	1350.	1340.	22.6	868.	23.1
14.0	470.	2250.	2290.	11.8	1425.	12.4

[a] The apparent reaction cross sections are for Kr+ the reactant ion and the units are 10^{-16} cm²/molecule.

to avoid the necessity for correcting for background due to naturally occurring $^{14}N-^{15}N$ $(m/q=29)$. The natural abundance of ^{15}N $(ca\ 0.38\%)$ is sufficiently small that the ratio of $^{15}N_2/^{14}N_2$, $\sim 0.1\times10^{-4}$ is effectively below the signal-to-noise ratio of the ion detector. Two experiments were carried out with the ion source at 380°K using 75-volt ionizing electrons. In the first of these the nitrogen-deuterium concentration ratio was unity, in the second the ratio was $\sim 1:2$. The results are summarized in Table VI.

As given in Table VI Q' for N_2^+ is $(253\pm6)\times10^{-16}$. From the relative ionization cross sections of nitrogen and hydrogen given by Tate and Smith,[18]

$$q(N_2)/q(H_2) = 2.7$$

(neglecting the contributions of N_2^{++}, N^+, etc., to the total nitrogen ion current), this would correspond to $Q'=680\times10^{-16}$ if D_2^+ not N_2^+ were the reactant ion. The corresponding specific reaction rates would be,

$$k_{obs}(N_2^+) = 17.2\times10^{-10} \text{ cm}^3/\text{molecule sec,}$$
$$k_{obs}(D_2^+) = 173.\times10^{-10} \text{ cm}^3/\text{molecule sec.} \quad (16)$$

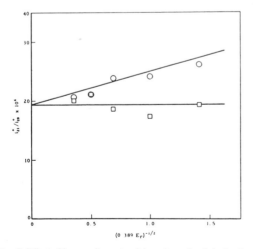

FIG. 5. Effect of ion repeller potential on the ratio of the ion intensities at $m/q=21$ and 20 for I, □ neon alone and II, ○ neon+H₂.

The theoretical rates [Eq. (13)] would be

$$k_t(N_2^+) = 11.2\times10^{-10} \text{ cm}^3/\text{molecule sec,}$$
$$k_t(D_2^+) = 16.5\times10^{-10} \text{ cm}^3/\text{molecule sec,} \quad (17)$$

where the molar refractions of nitrogen $(P_e=4.39)$ and deuterium $(P_e=2.04)$ are taken from the compilation in Landolt-Börnstein Tables.[22] Comparison of the theoretical rate constants with the possible k_{obs} strongly favors N_2^+ as the reactant.

Reaction of Carbon Monoxide and Hydrogen

A single experiment on the system carbon monoxide-hydrogen was carried out at 380°K with 75-volt electrons. Again deuterium was used to minimize isotopic interference. After correction for the apparent natural abundance of $^{12}C^{18}O$ determined in an experiment on pure carbon monoxide, there were obtained the apparent reaction cross sections (for CO+ the reactant) shown in Table VI. Since the ionization cross sections of nitrogen and carbon monoxide are very nearly equal,[21] as are their polarizabilities,[22] it is apparent that the argument that led to the conclusion that N_2^+ was the reactant ion will also lead to the conclusion that CO+ is the reactant ion giving COD+ (or DCO+).

TABLE VI. Reactions of nitrogen and carbon monoxide with deuterium.
380°K, 75-volt ionizing electrons

Gases	$E_r=1.28$	2.57	5.14	10.2₈	20.5₆
		$Q\times10^{-16}$ cm²/molecule sec.			
$N_2+D_2(1/1)$	229.	...	...	77.5	...
$N_2+D_2(1/2)$	230.	150.	104.	75.0	53.6

Least squares $Q'(N_2^+) = 253\pm6\times10^{-16}$

| CO+D₂(1/2) | 215. | 151. | 106. | 70.5 | 45.9 |

Least squares $Q'(CO^+) = 240\pm4\times10^{-16}$

[22] Landolt-Börnstein, *Zahlenwerte und Funktionen* (Springer-Verlag, Berlin, 1951), sixth edition, Vol. 1, Part 3.

TABLE VII. Reactions of O_2 and hydrogen at 380°K.

Electron energy (nominal volts)	E_r volts/cm	$Q(O_2^+)$ ×10¹⁶ cm²/molecule	$Q(H_2^+)$
		Hydrogen (H_2)	
75	1.29	98.2	267.
75	10.28	28.3	77.
20	1.29	194.	312.
18	1.29	167.	295.
16	1.29	127.	345.
		Hydrogen (D_2)	
75	1.29	63.6	173.
75	2.57	47.8	130.
75	5.14	32.5	88.4
75	10.28	22.1	60.1
75	20.56	10.7	29.1

$Q'(O_2H^+) = 297 \pm 12 \times 10^{-16}$ for H_2^+
$Q'(O_2D^+) = 198 \pm 7 \times 10^{-16}$ for D_2^+

Reaction of Oxygen and Hydrogen

The results of experiments on the systems oxygen-hydrogen and oxygen-deuterium are summarized in Table VII. Here we tabulate in two columns apparent reaction cross sections for O_2^+, the reactant, and for H_2^+, the reactant. Data of Tate and Smith[18] were combined with our mass spectrometric data to fix the ratio of the oxygen to hydrogen ionization cross sections. We note that the apparent cross section for O_2H^+ formation, calculated for O_2^+ as the reactant, decreases markedly across the electron energy range, 20–16 volts. The apparent increase over the value for 75-volt electrons is undoubtedly an electron beam space-charge effect of the type described above under krypton. Across this range of electron energies, 20–16 volts, the apparent cross section calculated for H_2^+ the reactant, shows a small increase of the sort that would be expected from our previous observations on the krypton system. It may be noted that the experiments summarized in Table VII were carried out with a 10-μ ionizing electron beam, thus a smaller electron beam space-charge effect would be expected than was encountered in the krypton case where higher beam currents were employed. The reaction apparently involves H_2^+ ions rather than O_2^+ ions.

Reaction of Hydrogen Ions with Hydrogen

A number of experiments on the reaction of deuterium ions (D_2^+) with deuterium (D_2) were carried out at various temperatures in the range 380–615°K, and a few preliminary experiments were made on the hydrogen deuteride ion (HD^+) with hydrogen deuteride reaction at 380°K. No experiments involving the common hydrogen (H_2^+, H_2) reaction were performed. As may be seen in Table VIII, the Q', for the $D_2^+ + D_2 \rightarrow D_3^+ + D$ reaction, as measured at the three

temperatures, 380, 494, and 615°K, agree well within the standard deviations.

In correcting the measured $i(D_3^+)/i(D_2^+)$ ratios for mass dependent discrimination, the following procedure was employed: the apparent sensitivities of H_2, HD, and D_2 were measured at various mass analyzing voltages in the range −750 to −2000 volts. When plotted against $(MV)^{-\frac{1}{2}}$ these sensitivities fell on an excellent straight line. By extrapolation of this line it was found that for −2000 volts analyzing voltage the efficiencies of detection of ions of $m/q=6$ and 5 are 1.10 and 1.06, respectively, relative to that of $m/q=4$ equal to unity. On this same scale the relative efficiencies for ions of $m/q=3$ and 2 are 0.914 and 0.768, respectively. From these results the observed $i(D_3^+)/i(D_2^+)$ were reduced by the factor $(1.10)^{-1}$ in computing the "Q's" shown in Table VIII.

From measurements at relatively low ion source pressures and high ion repeller fields it was concluded that the sample of HD, prepared by the reaction of D_2O (Stewart Oxygen Company) with lithium aluminum hydride[23] in ether solution, contained 0.62% D_2, corresponding to a $i(D_2^+)/i(HD^+)$ ratio of 67.5×10^{-4}. These data, along with the relative efficiencies of measurements of $m/q=5$, 4, and 3 given above, were used to "correct" the observed ratios

$$i(m/q=5)/i(m/q=3),$$

and $i(m/q=4)/i(m/q=3)$.

There are shown in Table IX the apparent reaction cross sections for D_2H^+ ($m/q=5$) and DH_2^+ ($m/q=4$) formation found in two experiments. Although neither of the apparent reaction cross sections, $Q(D_2H^+)$ nor $Q(DH_2^+)$, is linear in $(E_r)^{-\frac{1}{2}}$, the sum of these Q's is linear in $(E_r)^{-\frac{1}{2}}$, with a total $Q'=337\pm30\times10^{-16}$. The total Q' to be expected from the values found for D_3^+ formation ($Q'=79.3\pm7\times10^{-16}$, Table VIII), would be 92×10^{-16}. This disagreement, combined with the marked discrepancy between the cross sections for formation of $m/q=5$ and 4, suggests that an additional reaction takes place, namely,

$$HD^+ + HD \rightarrow D_2^+ + H_2 \qquad (18)$$

and presumably also

$$HD^+ + HD \rightarrow D_2 + H_2^+. \qquad (19)$$

Alternately it may be that D_2 formation takes place at the cathode and that there is insufficient differential pumping between the cathode and ion chamber. This explanation for the anomalously high $m/q=4$ sensitivity does not account for the observed strong dependence on the ion repeller field.

Further study of the HD reaction system is clearly indicated to be needed.

---

[23] Wender, Friedel, and Orchin, J. Am. Chem. Soc. **71**, 1140 (1949)

TABLE VIII. Reaction of D_2^+ with D_2.
(75-volt ionizing electrons)

$T°K$	No. of exps.	$Q\times10^{16}$ cm²/molecule					$Q'\times10^{16}$
		$E_r=1.29$	2.57	5.14	10.2$_8$	20.5$_6$ v/cm	
380	4	69.$_4$	49.$_3$	34.$_3$	23.$_0$	15.$_5$	78.$_0\pm4$
494	2	66.$_5$	49.$_9$	38.$_4$	26.$_6$	14.$_8$	78.$_5\pm7$
615	2	66.$_8$	53.$_7$	41.$_1$	27.$_6$	15.$_2$	81.$_4\pm9$

$$Q'=79._3\pm7\times10^{-16}$$

Reactions of Hydrogen Halides and Rare Gases

The systems HCl-Argon and HBr-Krypton were examined for evidence of self reactions of the hydrogen halide ion with molecules and for possible hydrogen halide-rare gas (ion) reactions. Evidence for the reactions,

$$HCl^+ + HCl \rightarrow H_2Cl^+ + Cl \tag{20}$$

$$HBr^+ + HBr \rightarrow H_2Br^+ + Br, \tag{21}$$

was found, but there were no indication of reactions yielding argonium ion (in reaction with HCl), or kryptonium ion (in reaction with HBr). Studies involving hydrogen halides are rendered difficult by the fact that these substances elute adsorbed substances from the glassware, inducing uncomfortably high "background." The "background" ions are distinguished from the products of ion-molecule reactions by means of the different relation between ion intensity and ion repeller field that obtain for primary and secondary ions.

From the slopes of the lines,

$$i(m/q=39)/i(H^{37}Cl^+)n(HCl)$$

and $i(m/q=83)/i(H^{81}Br^+)n(HBr)$ vs $(E_r)^{-\frac{1}{2}}$ we deduce for reactions (20) and (21) the Q', $73\pm20\times10^{-16}$ and $55\pm6\times10^{-11}$, respectively. There is a strong possibility that the Q' are "low" because the ion source concentration of hydrogen halide may not have achieved the equilibrium value corresponding to the inlet system pressure, even though a quite extended conditioning period was employed prior to making the runs with these substances.

For the ion repeller field equal to 2.57 volts/cm it was found that

$$\frac{ArH^+}{Ar^+n(HCl)l} \quad \text{and} \quad \frac{ArH^+}{n(Ar)HCl^+l}$$

are $<3\times10^{-16}$ cm²/molecule

and

$$\frac{KrH^+}{Kr^+n(HBr)l} \quad \text{and} \quad \frac{KrH^+}{n(Kr)HBr^+l}$$

are $<5\times10^{-16}$ cm²/molecule.

Thus, for the possible reactions,

$$R^+ + HX \rightarrow RH^+ + X, \tag{22}$$

$$R + HX^+ \rightarrow RH^+ + X, \tag{23}$$

where $R=$ argon or krypton and $X=$ chlorine or bromine, respectively, the specific rates of reactions of thermal ions, k_0, are

$$k_0 < 0.1\times10^{-10} \text{ cm²/molecule sec}$$

for HCl-argon or HBr-krypton systems.

DISCUSSION

For the theory of ion-molecule reactions of the type described in this paper, the important points are the dependence of the reaction cross section on (1) the magnitude of the ion repeller field, E_r, across the mass spectrometer ion source, (2) the temperature of the gas in the ion source, and (3) the mass of the reactants. For these points the most significant experiments described above are those involving reactions in the systems, argon and the isotopes of hydrogen, krypton and the isotopes of hydrogen, and deuterium. For these systems it is found that the reaction cross sections can be represented by the form, $Q=Q'(E_r)^{-\frac{1}{2}}$, over ranges

TABLE IX. Reaction of HD$^+$ and HD.
380°K, 75-volt ionizing electrons

	$E_r=1.29$	2.57	5.14	10.2$_8$	20.5$_6$ v/cm
$Q(D_2H^+)\times10^{16}$ cm²/molecule	90.$_8$	38.$_0$	24.$_3$	13.$_7$	8.$_6$
	78.$_1$	39.$_5$	23.$_3$	15.$_2$	8.$_6$
$Q(DH_2^+)=10^{16}$ cm²/molecule	231.	165.	125.	91.$_0$	47.$_4$
	201.	173.	130.	99.$_5$	56.$_0$

of E_r spanning factors of 15 to 30, with standard deviation in Q that is less than our best estimate of the reliability of measurements of the relative values of mass spectral intensity ratios of the magnitude of those involved in the determination of the reaction cross sections (Q's). There is no systematic variation of the reaction cross section with temperature of the gas in the range 380–600°K, and the reaction cross section decreases with the reduced mass of the isotopic reactant pairs.

As is shown in the accompanying paper, these are the behaviors to be expected for the reactions of charged with uncharged simple particles when the reactions involve no activation energies. Details of the comparison of the theory and experiment are to be found in the companion paper.

Reprinted from *Russ. J. Phys. Chem.* **34**:1275–1279 (1960)

PULSE METHOD OF DETERMINING THE RATE CONSTANTS OF ION-MOLECULE REACTIONS *

V. L. Tal'roze and E. L. Frankevich

There has been much recent study of reactions between ions and molecules, first demonstrated in the ionisation chamber of the mass spectrometer, since these processes are important in reactions induced by ionising radiation [1-3]. Early studies of processes such as $CH_4 + CH_4^+ = CH_5^+ + CH_3$ showed them to be frequently faster than normal gas reactions [4]. Mass spectrometry showed that processes involving proton or hydrogen atom transfer had no activation energy, unless endothermic [4-7]. The collision cross-sections involved in many such reactions have recently been determined [6-9] with the mass spectrometer (Fig. 1). Electrons emitted by the cathode *1* are accelerated and ionise gas in the ionisation chamber *2*; at the pressure of the gas 10^{-4}–10^{-5} mm Hg, the ions have a 0.1–10% probability of colliding with molecules to form secondary ions. Ions are withdrawn from the chamber by the repulsive electrical field set up between it and the electrode *3*. The collision cross-section σ for the ion–molecule reaction is determined by means of the expression

$$\sigma = (n\,l)^{-1} I_2/I_1,$$

where I_2 and I_1 are the measured currents of secondary and primary reacting ions, n is the concentration of reacting molecules in the chamber, and l is the length of the primary ions' path in the chamber, equal to the total number of collisions undergone there by a primary ion with molecules of unit concentration, calculated for unit collision cross-section [10].

The concentration of molecules in the chamber is usually calculated from the pressure, measured at the gas inlet; gas enters through a calibrated diaphragm. The least accurate calculation is that of l; Stevenson and Schissler [8] took it as the perpendicular distance from the electron beam to the exit slit of the chamber. Field, Franklin, and Lampe [7,10] refined the calculation by considering the curvature of the paths of the primary ions. To the error in the assumed distance from the ion source to the exit slit is added a small systematic error in determining the dependence of σ on the field intensity in the chamber, since the distance from the electron beam [electron source ? (Ed. of Translation)] depends on this intensity.

There is always an electrical field in the reaction zone conveying kinetic energy to the ions, 10–10^3 times greater

---

* Reported at the 8th Mendeleev Conference on General and Applied Chemistry [14,15].

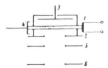

Fig. 1.   Ionisation chamber of the mass spectrometer: 1) cathode; 2) chamber; 3) repulsion electrode; 4) electron collector; 5) ion-focusing electrode; 6) accelerating electrode.

than the heat energy corresponding to the temperature of the gas and the walls (400°–500°K). However, values of σ are required for ion energies characteristic of reactions induced by ionising radiation or occurring in the glow discharge, at a few hundred degrees *; the ways in which σ depends on E, the energy imparted to the ions by the withdrawal field, or on the ions' velocity, v, are found at high energies, and extrapolation gives σ values at lower energies. In view of the various possible relations between σ and v, the extended extrapolation is a possible source of error. Not every elementary consideration [11], in which the ion–molecule reaction is considered as an interaction of the ion with a dipole induced by it, leads to a relation such as $\sigma \propto 1/v$, indicated by Stevenson and Schissler's data [8] for reactions of A^+, N_2^+, and CO^+ with H_2, among others. The surprising relation $\sigma \propto v^{-2}$ was clearly established by Field, Franklin, and Lampe [7] for several ion–molecule reactions of hydrocarbons, including that between $C_2H_2^+$ and C_2H_4. The σ–v relation might well change as v does; the ion–dipole interaction examined by Giomousis and Stevenson [11] gave the relation $\sigma \propto 1/v$, which probably changes as v increases and σ decreases to a critical value, designated [11] πr_c^2. σ may be not only a function of the translational velocity of the ion, but also of the vibrational, rotational, and electronic temperatures of both ion and molecule. In high-pressure reactions, to calculate the total speed of which the constituent velocity constants must be known, the energy distribution between the degrees of freedom corresponds more closely to equilibrium than in the usual determinations of σ, in which an excited ion sometimes bypasses a molecule which is in all respects at high temperature.

This article concerns the pulse method, a new mass-spectroscopic determination of the rate constants of ion-molecule reactions by direct kinetic measurements in the absence of an electric field. Short electronic pulses ionise the gas in the chamber periodically. There is no field withdrawing ions from the chamber during ionisation, and the primary ions formed in the path of the electron beam have a thermal velocity ** fixed by the wall temperature [12]. At an interval t_z after ionisation, primary and secondary ions are withdrawn from the chamber by short electrical pulses and analysed.

_____
* Excluding hot ions, acquiring excess energy on ionic dissociation.

** This is true only for ions which do not receive initial kinetic energy on ionisation.

Assuming no decay during ion formation by the electronic pulse, the rate of change of secondary-ion concentration n_2^* is given by

$$dn_2^*/dt = k n_0 n_1^*$$ (1)

where n_1^* and n_0 are the mean concentrations of primary ions and molecules in the chamber, and k is the rate constant of the reaction. Eqn. (1) applies after the pulse until the primary ions reach the walls. After an interval t_z, the concentration of secondary ions becomes

$$n_2^* = k n_1^* n_0 t_z.$$ (2)

Mean ion currents out of the chamber, not concentrations, are measured. The probabilities of withdrawal for ions of different mass vary, but only very little in reactions involving proton or hydrogen atom transfer since the masses of primary and secondary ions differ by only one unit. They also vary little on account of differences in the initial velocities of the ions; we showed that secondary ions possess little kinetic energy even when formed in highly exothermic reactions [2]. Hence, assuming equal withdrawal probabilities,

$$I_2^T = k_T I_1 n_0 t_z,$$ (3)

where I_1 and I_2^T are mean currents of primary ions and secondary ions formed by thermal ion–molecule collisions, and k_T is the rate constant for the reactions of ions with thermal speeds. Eqn. (3) is generally approximate, since I_1 represents the concentration only of primary ions which have not reacted, but at pressures of about 10^{-4} mm Hg so few primary ions react that the error is negligible. During the time t_z between ionisation and withdrawal, no electric field is present and secondary ions are formed by collisions of the molecules with thermal ions, but during the withdrawal pulse accelerated ions react with molecules. The mean current of secondary ions at withdrawal is given by

$$I_2^E = k_E I_1 n_0 t_w,$$ (4)

where k_E is the rate constant for ions reacting in the electric field. In the duration t_l of the ionising pulse secondary ions also form, giving a current

$$I_2^i = \tfrac{1}{2} k_i I_1 n_0 t_l.$$ (5)

The space charge caused by passage of electrons through the chamber may cause k_i to differ from k_T. The currents I_2^T, I_2^E, and I_2^i are measured as a combined current:

$$I_2 = I_1 n_0 (\tfrac{1}{2} k_i t_l + k_T t_z + k_E t_w).$$ (6)

The plot of I_2/I_1 against t_z is linear, with slope $k_T n_0$, so that k_T may be determined experimentally.

We have assumed ion currents to be proportional to the mean ion concentrations in the chamber, but Eqns. (3)–(6) will not be true unless the primary- and secondary-ion concentrations are distributed identically in the chamber. This is because the primary ions only move away from the region of their formation, after the ionising pulse, so that their concentration there is soon very small. But secondary ions, formed at the edge of this region, may accumulate in it; their relative concentration would be greater than the mean, so that preferential withdrawal from this region would raise the apparent rate constant. However, it is shown below that this effect is negligible.

Eqn. (6) shows that the duration of the ionising and withdrawal pulses does not affect the slope of the I_2/I_1 vs. t_z graph. Ionisation must not be too prolonged, however, since ions very distant from their origin after the time t_l will not react during the interval t_z. 1 μsec was chosen for t_l,

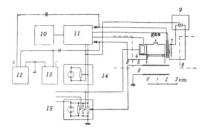

Fig. 2. Ionisation source: 1) focusing electrode; 2) guard electrode; 3) withdrawal grid; 4) ionisation chamber; 5) ion-focusing electrode; 6) accelerating electrode; 7) screened cathode; 8)magnet; 9) cathode supply; 10) supply circuit; 11) compensation circuit; 12) and 13) 26-I generators; 14) stablised ionising voltage 0-50 V; 15) stablised accelerating voltage 1800 V. [1-3 were missing in the Russian text and could not be identified with certainty. (Ed. of Translation)].

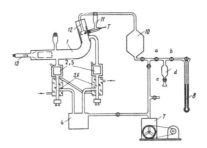

Fig. 3. Vacuum system of mass spectrometer: 1) analyser tube, 2) and 5) liquid-nitrogen traps; 3) and 6) oil diffusion pumps; 4) fore-vacuum space; 7) fore-vacuum pump; 8) inlet manometer; 9) small inlet reservoir; 10) large inlet reservoir; 11) ionisation manometer tube; 12) inlet diaphragm; 13) secondary-electron multiplier; T — thermocouple.

TABLE 1.

$p \times 10^4$, mm Hg	T, °K	$n_0 \times 10^{-13}$, cm^{-3}	$a \times 10^{-4}$, sec^{-1}	$k_T \times 10^{10}$, cm^3 mole^{-1} sec^{-1}
4.0	357	10.8	1.24	11.5
5.0	367	13.1	1.5	11.5
5.1	386	12.8	1.47	11.5
6.05	390	14.9	1.79	12.0
7.4	361	19.8	2.3	11.6
8.5	356	23	2.57	11.2
			av. $k_T = 11.6 \times 10^{10}$ cm^3 mole^{-1} sec^{-1}		

Note. p and T are the chamber pressure and temperature, and n_0 is the calculated concentration of molecules therein; α is the slope of the I_2/I_1 vs. t_z graph.

TABLE 2.

t_w, μsec	$p \times 10^4$, mm Hg	T, °K	$n_0 \times 10^{-13}$, cm^{-3}	$a \times 10^{-4}$, sec^{-1}	$k_T \times 10^{10}$, cm^3 mole^{-1} sec^{-1}
3	8.8	349	24.4	2.58	10.6
4	8.45	349	23.4	2.58	11.0
5	8.2	350	22.6	2.58	11.4
7	7.9	350	21.8	2.58	11.8

* For legend see Table 1 (Ed. of Translation).

in which time ions of mass ~16 atomic units and mean velocity corresponding to the wall temperature, 400°K, travel about 1 mm, or an eighth of their path in the chamber. The time taken for ions of average thermal velocity to travel to the walls fixed the maximum value of t_z at 8 μsec. The duration of withdrawal pulses was determined by the time needed to withdraw the most distant ions travelling away from the exit slit, calculated at 2.5 μsec (t_w) for ions of mass 16 atomic units in a potential gradient of ~10 V cm^{-1}.

EXPERIMENTAL

Apparatus. The mass spectrometer was described earlier[13] when used to determine the appearance potentials of primary and secondary ions, and only the ion source was modified in the pulse method (Fig. 2). The source and chamber electrodes are of nichrome foil 0.3 mm thick and the apertures in electrodes 4 and 3 contain copper grids of transparency ~70%, with openings 120 μ in diameter. Gas is let into the chamber through apertures of diameter 20 μ in a copper diaphragm. The tungsten cathode 7, heated by a direct current, emits electrons which pass through the slits of the focussing electrode 1 and the guard electrode 2 into the chamber 4. The electron beam is focussed by a longitudinal magnetic field of ~200 oersted, set up by the magnet 8. The duration of the ionising pulse is t_i. While there is no ionisation, a bias is put on 2 which keeps all but the fastest electrons out of the chamber. A positive charge pulse from the 26-I generator then passes onto 2 to admit the electron beam; it also sets off the second 26-I generator, which puts a negative charge on the withdrawal grid 3 during the interval t_z. 10^4 pulses sec^{-1} are passed. A permanent positive bias, relative to the chamber, is put on 3 to compensate the small potential difference between 3 and the accelerating electrode 6, which withdraws ions from the chamber. To increase the registration sensitivity, an FEU-30 "Ekran" louver-type secondary-electron multiplier was used, giving an amplification factor of 10^4 with a potential difference of 200 V between the dynodes. An ionisation manometer directly connected to the ionisation chamber measured the pressure therein; the temperature of the walls, determining the velocity of the ions, was measured with a nichrome–Constantan thermocouple T (Fig. 3).

Procedure. The apparatus was adjusted onto a background line at 18 atomic units during continuous ionisation and ion withdrawal. The electron beam was then arrested by means of a negative bias on the guard electrode (Fig. 2, 2) and the ionising pulse turned on. A permanent field intensity for ion withdrawal was then chosen; if the ion current was not fully arrested, a permanent positive bias was put on the withdrawal grid 3 to compensate the field in the ionisation region. Withdrawal pulses were then passed at an amplitude U_w such that the ion current was a maximum *. The dependence of I_1 on t_z was a very sensitive index of electric-field compensation in the ionisation chamber. For a potential gradient of 1 V cm^{-1}, primary ions of mass 16 atomic units travel, at $\sim 4 \times 10^5$ cm sec^{-1}, from their place of origin to the walls in 1.5 μsec. With a suitable compensating current, however, the primary-ion current was almost constant for t_z values up to 3 μsec, and then decreased slowly almost to zero when $t_z = 11$ μsec. This indicates that the speeds acquired by ions in the residual field do not exceed thermal velocities by more than 30% **.

TABLE 3.

$p \times 10^4$, mm	T, °K	$n_0 \times 10^{13}$, cm^{-3}	$\alpha \times 10^{-4}$, sec^{-1}	$k_T \times 10$, cm^3 mole^{-1} sec^{-1}
9.1	411	21.3	1.82	8.55
9.1	411	21.3	1.83	8.6
9.3	408	21.9	1.86	8.5
9.4	410	22.1	1.86	8.4

$$k_T = 8.5 \times 10^{10} \text{ cm}^3 \text{ mole}^{-1} \text{ sec}^{-1}$$

* For legend see Table 1 (Ed. of Translation).

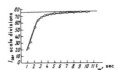

Fig. 4.   Variation of ion current with duration of withdrawal pulse; t_z = 1 sec; $p = 9 \times 10^{-4}$ mm Hg; $T = 375°$K.

Withdrawal pulses were prolonged enough to remove all the ions from the chamber. A typical ion current–t_w graph is shown (Fig. 4); virtually all the ions are withdrawn when $t_w = 5$ μsec, which agrees with calculation. The following values were used: $t_w = 4.5$ μsec, $U_w = 45$–75 V, $t_1 = 1$ μsec, $t_z = 1$–7 μsec, and energy of ionising electrons = 40 eV.

The rate constant of the reaction $CH_4 + CH_4^+ = CH_5^+ + CH_3$ was measured *. Fig. 5 is a typical graph of t_z against the ratio of the ion currents of CH_5^+ and CH_4^+. Results were obtained at different pressures and t_w values (Tables 1 and 2 respectively). Increase in t_w corresponds to extension of the region from which ions are withdrawn. Different distributions of primary- and secondary-ion concentrations would make the observed rate constant closely dependent on t_w; such difference is shown to be slight (Table 2).

The rate constant of the reaction $H_2O + H_2O^+ = H_3O^+ + OH$ was also measured (Table 3). A linear graph of I_2/I_1 vs. t_z is shown in Fig. 6.

The accuracy of the determination of rate constants depends on that of the measurements of pressure and t_z. The maximum error of the ionisation manometer readings was ±15%, and that of t_z values was ±10%, whence that of the rate constant was ±25%. Normal methods of mass spectrometry give values of σ at relatively high energies. Since the dependence of σ on thermal velocity is not well known, extrapolation of values for σ to thermal velocities and the subsequent calculation of corresponding rate constants are ill-founded. In order to calculate the overall rate of a reaction made up of component ion–molecule processes, the rate constants of these must be known. Our method gives the rate constants of ion–molecule reactions directly, at the temperature of the chamber, and these may be used directly for this calculation.

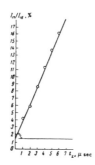

Fig. 5.   Variation of I_2/I_1 with t_z for the reaction $CH_4 + CH_4^+ = CH_5^+ + CH_3$. $T = 356°$K; $t_w = 4.5$ μ sec; slope = $nv\sigma$.

Fig. 6.   Variation of I_2/I_1 with t_z for the reaction $H_2O + H_2O^+ = H_3O^+ + OH$. $T = 408°$K; $t_w = 4.5$ μ sec; $p = 9.3 \times 10^{-4}$ mmHg.

* Two opposing factors cause the maximum to occur as U_w increases; the region from which ions are withdrawn grows, and the defocussing effect of the withdrawal field becomes evident.
** In addition to the field from the slit, there may also be a field due to contact potentials and the surface charge of the chamber.

* Ya. A. Andreev (student of the Leningrad Polytechnic Institute) took part in these experiments.

The values obtained by Field, Franklin, and Lampe by extrapolation of high velocity σ values to thermal velocities are [10] $K_{CH_4^+} = 2.8 \times 10^{-9}$ cm^3 mole^{-1} sec^{-1} and [9] $K_{D_3O^+} = 2.2 \times 10^{-9}$ cm^3 mole^{-1} sec^{-1}, at 423° ± 10°K. These are ~2.5 times larger than our values, which may indicate that at low ion speeds the relation $\sigma \propto v^{-2}$, used for extrapolation is not true. Such a conclusion is, however, premature, since the different technique and principles of our method may simply lead to a more accurate absolute value of k. Measurement of the dependence of k on the temperature of the ionisation chamber should solve this problem. The combination of our method with the use of quasi-monochromatic electron beams [13] and measurement near the ionisation potential would indicate the dependence of k on the excitation energy of primary ions.

SUMMARY

1. A pulse method has been developed for measuring the rate constants of reactions between ions and molecules at thermal velocities.

2. The advantages of this method over the usual determination of collision cross-section have been examined.

3. The rate constant for the reaction $CH_4 + CH_4^+ = CH_5^+ + CH_3$ has been found to be 11.6×10^{-10} cm^3 mole^{-1} sec^{-1} at 370° ± 20°K, and for $H_2O + H_2O^+ = H_3O^+ + OH$ to be 8.5×10^{-10} cm^3 mole^{-1} sec^{-1}, at 410°K. The energy of the ionising electrons was 50 eV.

1. G.G.Meisels, W.H.Hamill, and R.R.Williams, Jnr., J.Chem. Phys., 25, 790 (1956).
2. V.L.Tal'roze and E.L.Frankevich, "Trudy I Vsesoyuznogo Soveshchaniya po Radiatsionnoi Khimi, 1957" (Proceedings of the First All-Union Conference on Radiation Chemistry, 1957), Izd. Akad. Nauk SSSR, Moscow, 1958, pp.13-18.
3. D.P.Stevenson, J.Phys.Chem., 61, 1453 (1957).
4. V.L.Tal'roze and A.K.Lyubimova, Dokl.Akad. Nauk SSSR, 86, 909 (1952).
5. V.L.Tal'roze and E.L.Frankevich, Dokl.Akad.Nauk SSSR, 111, 376 (1956).
6. D.O.Schissler and D.P.Stevenson, J.Chem.Phys., 24, 926 (1956); 23, 1353 (1955).
7. F.H.Field, J.L.Franklin, and F.W.Lampe, J.Amer.Chem. Soc., 78, 5697 (1956).
8. D.P.Stevenson and D.O.Schissler, J.Chem.Phys., 29, 282 (1958).
9. F.W.Lampe, F.H.Field, and J.L.Franklin, J.Amer.Chem. Soc., 79, 6132 (1957).
10. F.H.Field, J.L.Franklin, and F.W.Lampe, J.Amer.Chem. Soc., 79, 2419 (1957).
11. G.Giomousis and D.P.Stevenson, J.Chem.Phys., 29, 294 1958.
12. V.L.Tal'roze and E.L.Frankevich, Zhur.Tekh.Fiz., 29, 497 (1956).
13. Idem, Pribory i Tekh.Eksper., 2, 48 (1957).
14. Idem, " Referaty Dokladov i Soobshchenii No.12, VIII Mendeleevskii S'ezd" (Report of Proceedings No.12, Eighth Mendeleev Congress), 1958, p.59.
15. Idem, Izv.Akad.Nauk SSSR, Otd.Khim.Nauk, 7, 1351 (1959).

Copyright © *1962 by the American Chemical Society*

Reprinted from *Am. Chem. Soc. J.* **84**:730–734 (1962)

Effects of Relative Velocity upon Gaseous Ion–Molecule Reactions; Charge Transfer to the Neopentane Molecule[1]

By N. Boelrijk and W. H. Hamill

[Contribution from the Department of Chemistry of the University of Notre Dame, Notre Dame, Indiana]

Introduction

In order to understand radiation chemistry in terms of ionic processes it is necessary to consider, among others, the phenomenon of charge transfer. A recently reported example[2] is

$$C_2H_2^+ + C_6H_6 \longrightarrow C_2H_2 + C_6H_6^+$$

which accounts for the previously observed retarding effect of benzene on the alpha-induced polymerization of acetylene.[3] The same study[2] also indicates that charge transfer may not occur when an efficient ion–molecule reaction also is allowed.

The present study was undertaken with neopentane as the common electron donor in various binary mixtures in a conventional mass spectrometer. This choice of donor molecule was dictated by the impossibility of detecting an increment in the ion current of the molecular (unfragmented) ion when it also was produced, even in fairly low abundance, by electron impact. The ion $C_5H_{12}^+$ from neopentane has a relative abundance of only *ca.* 0.01%, and even small contributions from bimolecular processes can be detected efficiently. Incidental to the study of neopentane in various mixtures for evidence of charge transfer, it was observed that the yield of $C_5H_{11}^+$ ions (0.04%) also was enhanced by bimolecular processes. This effect also has been observed by Field and Lampe.[4]

Experimental

Measurements were made in a CEC 21–103A mass spectrometer with a model 31–402 ion source. The repeller potential was manually adjusted using dry cells and step-potentiometers adjustable by 0.1 v. increments over the range 0–24 v. The ionizing voltage was continuously adjustable over the range 5–73 v. and in steps of 0.1 v. over the range 5–28 v. Electron current was 10 μ-amp. The distance from the center of the electron beam, which is 0.018 cm. thick, to the exit slit is 0.130 cm. and to the repeller plates is 0.120 cm. The electron energy therefore is increased by one-half the applied repeller voltages.

The leak consists of two fine holes in a thin gold foil, and observation shows gas flow to be effusive within the range of pressures used. Pumping of gas from the source also is effusive and the ratio of pressures between the reservoir and the source is constant and independent of the gas used. By calibration,[5] the concentration of gas in the ion source is 1.58×10^{10} mol./cc. per micron of reservoir pressure.

(1) Presented at the 136th National Meeting of the American Chemical Society, Atlantic City, September, 1959. This work was performed under the auspices of the Radiation Laboratory, University of Notre Dame, supported in part by the U. S. Atomic Energy Commission under contract AT(11-1)-38.

(2) P. S. Rudolph and C. E. Melton, *J. Chem. Phys.*, **32**, 586 (1960).

(3) S. C. Lind and P. S. Rudolph, *ibid.*, **26**, 1768 (1957).

(4) F. H. Field and F. W. Lampe, *J. Am. Chem. Soc.*, **80**, 5587 (1958).

(5) D. P. Stevenson and D. O. Schissler, *J. Chem. Phys.*, **29**, 282 (1958).

Results

The dependence upon ion repeller field of the cross section Q of the reaction

$$Ar^+ + H_2 \longrightarrow ArH^+ + H$$

was measured on the Notre Dame instrument to allow comparison with published results. We find, in terms of the equation $Q = 2\,\sigma_L/E^{1/2} + \sigma_k$ that $\sigma_L = 67 \times 10^{-16}$ cm.^{2} ev.$^{1/2}$ and $\sigma_k = 11 \times 10^{-16}$ cm.^{2} From our graph of the data of Stevenson and Schissler,[5] who used a Westinghouse Type LV, $\pi/2$-sector analyzer mass spectrometer, we obtain $\sigma_L = 71 \times 10^{-16}$ cm.^{2} ev.$^{1/2}$ and $\sigma_k = 13 \times 10^{-16}$ cm.^{2} The theoretical value of σ_L is 70×10^{-16} ev.$^{1/2}$ for a reaction probability of unity.

Pressure Dependence.—A necessary condition for ion-molecule reactions is a second order pressure dependence for the secondary ion, whereas a primary ion exhibits first order pressure dependence. We have found that at fairly low repeller field strength this is not the case. Thus, for CD_4^+ and CD_5^+ ions in CD_4, and at 4 v. cm.$^{-1}$, the order is 0.88 and 1.66, respectively. At 8 and 40 v. cm.$^{-1}$, however, the expected integral order dependences actually were observed. Measurements of cross sections on instruments of the type used in this work often are not dependable at less than 10 v. cm.$^{-1}$ repeller field strength.

Neopentane.—An examination of many substances for evidence of charge transfer in appropriate mixtures resulted in choosing neopentane because of its particularly small 72-ion abundance. Incidental to this study we found, as did Field and Lampe,[4] evidence for secondary 71-ion formation from neopentane. Such parent-mass-minus-one secondary ions are of common occurrence, both by unimolecular decomposition and by "hydride ion" transfer.

In the present work it was observed that the 71-ion current increased as the 1.66 power of the pressure of neopentane, both at 12 and at 40 v. cm.$^{-1}$ This suggests a combination of primary and secondary processes. The i_{71}-ion current may be referred arbitrarily to the primary i_{65}-ion current, as a matter of convenience, since they are of the same magnitude. For later reference we note that $i_{65}/i_{57} = 0.0354\%$. It is to be expected then that $i_{71} = a'\,p_1 + b'\,p_1^2$ and $i_{65} = a''\,p_1$ where p_1 is the pressure of neopentane. It was found that $i_{71}/i_{65} = 0.63 + 1.03 \times 10^{-2}\,p_1$ at a repeller field 12 v. cm.$^{-1}$ and $i_{71}/i_{65} = 0.58 + 3.91 \times 10^{-3}p_1$ at 40 v. cm.$^{-1}$ expressing p in microns reservoir pressure.

In neopentane mixtures it may be expected that $i_{71} = a'p_1 + b'\,p_1{}^2 + c'\,p_1\,p_2$, where p_2 is the pressure of additive. The ratio i_{71}/i_{65} should be linear in p_2. Ethane, propane and carbon disulfide give such results. Keeping $p_1 = 105$ microns and E_e, the ion energy at the exit slit, at 1.8 ev., the results for added ethane are described by the equation

$$i_{71}/i_{65} = 1.51 + 3.8 \times 10^{-3}\,p_2$$

and for added propane by

$$i_{71}/i_{65} = 1.55 + 3.4 \times 10^{-3}\,p_2$$

In mixtures with carbon disulfide it was not possible to refer i_{71} to i_{65} because of interference. The results for these mixtures at $E_e = 1.56$ ev. and 418 microns constant total pressure may be expressed by

$$i_{71}/p_1 = 0.42 + 2.1 \times 10^{-3}\,p_1$$

Since the parameters of this equation are not derivable from those for pure neopentane by a common factor, it appears that an ion from carbon disulfide (shown to be $CS_2{}^+$) also induces the formation of $C_5H_{11}{}^+$.

Table I

Effects of Additives Upon 71- and 72-Ion Abundances Relative to the Primary 65-Ion at $E_e = 0.4$ e.v.

Additive (M)	$\Delta\,\dfrac{i_{71}}{i_{65}}$	$\Delta\,\dfrac{i_{72}}{i_{65}}$	$\Delta\,\dfrac{i_{72}{}^b}{i_{65}i_M{}^+}$	I (M)
IIe	0.0	0.0		24.58
D$_2$	.0	.0		15.44
N$_2$	.5	.0		15.56
NH$_3$	.0	.0		10.23
CO$_2$	− .3	.0		13.79
CD$_4$	2.7	.0		13.26
C$_2$H$_2$	0.0	.0		11.41
C$_2$H$_4$	1.7^a	.10	0.16	10.51
C$_2$H$_6$	2.7	.21	.23	11.65
C$_2$D$_6$	2.8	.28	.27	11.65
C$_3$H$_6$	3.7	.0		9.7
c-C$_3$H$_6$	4.8	.0		10.09
C$_3$H$_8$	10.6^a	.18		11.21
i-C$_4$H$_8$	7.1	.0		9.35
1-C$_4$H$_8$	4.5	.0		9.72
n-C$_4$H$_{10}$	4.8	.1		10.8
n-C$_4$D$_{10}$	4.8	.1		10.8
CH$_3$Cl	2.0	.0		11.22
CH$_3$Br	3.4	.26	.26	10.54
CH$_3$I	1.1	.0		9.54
C$_2$H$_5$Cl	2.5	.03	.064	10.97
C$_2$H$_5$Br	4.5	.45	.65	10.29
C$_2$H$_5$I	5.9	.0		9.33
CH$_3$OH	1.9	.0		10.85
Pyridine	1.9	.0		9.8
CS$_2$	−0.2	2.40	1.3	10.08

^{a} Measured at $E_e = 0.25$ ev. ^{b} $i_M{}^+$ is the current of the molecular ion of additive except for C$_2$H$_6$, C$_2$D$_6$ when it refers to C$_2$H$_4{}^+$, C$_2$D$_4{}^+$.

The 72-ion current for neopentane, as received, was not altered by gas chromatographic purification of the material. Neither does this current arise from a secondary reaction, since 22 measurements of i_{72}/i_{65} (corrected for the isotopic component from i_{71}) at various p_1 and repeller fields showed no significant variation. The results of tests with many additives are reported in Table I. No ion with ionization potential I greater than 11

ev. enhanced i_{72}. Propane is an apparent exception, but for this and all other hydrocarbons tested there is a fairly good correlation between $\Delta\,i_{72}/i_{65}$ and i_{28}. The largest effect observed for any additive was produced by carbon disulfide.

Secondary Ion Current–Repeller Dependence.— Carbon disulfide increases i_{71} rather less than i_{72}, but the dependence of i_{71} upon E_e is unlike that for any other additive; the results for carbon disulfide and for ethane-d_6 are compared in Table II.

Table II

Effect of Repeller Field

Added subs.	E_e	0.42	1.04	1.56	2.60	5.20
CS$_2$	$\Delta i_{72}/i_{65}$	2.40	0.99	0.67	0.38	0.18
CS$_2$	$\Delta i_{71}/i_{65}$	−0.20	.50	.40	.26	.18
C$_2$D$_6$	$\Delta i_{72}/i_{65}$	0.282	.138	.085	.064	.029
C$_2$D$_6$	$\Delta i_{71}/i_{65}$	3.10	1.31	.95	.59	.35

The dependence of the 71-ion current upon E_e, the terminal ion energy, relative to the 65-ion current, is reported in Table III for pure neopentane and for several mixtures. The parameters M and C refer to the empirical equations $i_{71}/i_{65} = C + ME_e{}^{-1/2}$, etc., as indicated.

Table III

Dependence of Secondary-to-primary Ion Currents Upon Repeller Field Strength

Additive	Reservoir p, microns	Range E_e, e.v.	C	M
None	418	1.3– 5.2	−0.97	7.36^c
None	217	1.9– 5.2	− .52	4.18^c
None	217	1.6– 8.3	.09	3.41^{a,c}
None	217	5.2–12.5	.75	4.4^{a,e}
C$_2$H$_6$	209 + 209	1.3– 5.2	− .85	5.65^c
C$_2$H$_6$	105 + 313	1.6– 6.2	− .35	4.14^c
C$_2$H$_6$	105 + 313	5.0–12.5	.61	4.44^e
C$_2$D$_6$	209 + 209	1.0– 5.2	− .47	1.79^d
C$_2$D$_6$	209 + 209	1.0– 5.2	− .36	1.57^d
C$_2$D$_6$	209 + 209	1.0– 5.2	.0	0.14^{b,f}
C$_2$D$_6$	209 + 209	1.0– 5.2	.0	0.15^{b,f}
n-C$_3$H$_6$	209 + 209	1.0– 5.2	− .78	3.13^d
C$_3$H$_8$	209 + 209	1.6– 4.3	−1.28	7.77^d
CH$_3$Cl	209 + 209	1.0– 5.2	−0.25	1.37^d
CS$_2$	209 + 209	1.0– 5.2	−0.02	1.06^{b,f}

^{a} The accelerating voltage was 1500 v. for this series and 600 v. for all others. This may account for the discrepancy between the intercepts for this and the preceding series in the plot vs. $E_e{}^{-1/2}$. ^{b} Corrected for isotopic contributions from C$_5$H$_{11}{}^+$. ^{c} Parameters refer to $i_{71}/i_{65} = C + ME_e{}^{-1/2}$. ^{d} Parameters refer to $\Delta i_{71}/i_{65} = C + ME_e{}^{-1/2}$. ^{e} Parameters refer to $i_{71}/i_{65} = C + ME_e{}^{-1}$. ^{f} Parameters refer to $\Delta i_{72}/i_{65} = C + ME_e{}^{-1}$.

Appearance Potentials.—Appearance potentials were measured relative to xenon, at 0.7 to 0.8 repeller voltage. The results appear in Table IV.

Table IV

Appearance Potentials of Neopentane, Alone and in Mixtures

Additive:	None	CS$_2$	C$_2$H$_6$	C$_3$H$_4$
C$_5$H$_{11}{}^+$	12.5		11.8	10.7
C$_5$H$_{12}{}^+$	12.4	10.1	12.1	10.6

The appreciable primary 71-ion current made it impossible to establish the primary ion from neopentane responsible for the second order contribution to i_n/i_{46}. Field and La pe[4] have reported on the basis of appearance potential measurements, that the 41-ion was responsible. They emphasized that such "hydride ion" reactions are probably rather general, and the present results support this view for neopentane.

The appearance potentials of CS_2^+, CS^+, S^+ and C^+ from CS_2 are 10.10, 14.8, 14.1 and 25.1, respectively.[6] The charge transfer process in mixtures with carbon disulfide is considered to be

$$CS_2^+ + C_5H_{12} \longrightarrow CS_2 + C_5H_{12}^+$$

The only appearance potentials below 13.5 ev. for ethane and ethylene are: $C_2H_6^+$, 11.65; $C_2H_5^+$, 12.84; $C_2H_4^+$, 12.09 from ethane and[7,8] $C_2H_4^+$, 10.51 from ethylene.[9]

The indicated reactions in mixtures with ethylene are

$$C_2H_4^+ + C_5H_{12} \longrightarrow C_5H_{12}^+ + C_2H_4 \quad \text{(A)}$$
$$C_2H_4^+ + C_5H_{12} \longrightarrow C_5H_8 + C_5H_{11}^+ \quad \text{(B)}$$

A second break in the ionization efficiency curve of $C_5H_{11}^+$ for the neopentane–ethylene mixture was observed at 13.0 ev. This may be attributed to a reaction of $C_3H_5^+$ from neopentane itself. The reported appearance potential for the primary 41-ion[10] is 13.13 ev.

In mixtures with ethane it is difficult to assign the primary ions with assurance but the best agreement is found for the processes

$$C_2H_6^+ + C_5H_{12} \longrightarrow C_5H_{11}^+ + C_2H_6 + H \quad \text{(C)}$$
$$C_2H_5^+ + C_5H_{12} \longrightarrow C_5H_{11}^+ + C_2H_6 + H_2 \quad \text{(D)}$$

The reactions A and B probably also take place in these mixtures.

Discussion

Ion-induced dipole forces lead to energy-dependent collision cross sections[11,12]

$$\sigma(E) = \pi b_0^2 = 2\pi(e^2\alpha/\mu g^2)^{1/2} \quad (1)$$

where α, μ and g are the electric polarizability of the molecule, the reduced mass and the relative velocity of the collision pair. As an adequate approximation the molecule may be considered to be at rest and the energy E of the ion of mass m_1 taken as $E = \frac{1}{2}m_1 g^2$. The current of secondary ions i_s formed from i_p primary ions over a path d_0 at a concentration of N molecules/cc. is expressible in terms of the phenomenological cross section Q

$$Q = i_s/i_p N d_0 \quad (2)$$

Since we also have

$$Q = E_e^{-1} \int_0^{E_e} \sigma(E)dE \quad (3)$$

it follows that

$$Q = 2\pi e(2m_1\alpha/\mu E_e)^{1/2} = 2\sigma_1 E_e^{-1/2} \quad (4)$$

(6) H. D. Smyth and J. P. Blewett, *Phys. Rev.*, **46**, 276 (1934).
(7) D. P. Stevenson and J A. Hipple, *J. Am. Chem. Soc.*, **64**, 1588 (1942).
(8) K. Watanabe, *J. Chem. Phys.*, **26**, 542 (1957).
(9) W. C. Price and W. T. Tutte, *Proc. Roy. Soc.* (London), **A174**, 207 (1940).
(10) F. W. Lampe and F. H. Field, *J. Am. Chem. Soc.*, **81**, 3238 (1959).
(11) P. Langevin, *Ann. chim. phys.*, **5**, 245 (1905).
(12) G. Gioumousis and D. P. Stevenson, *J. Chem. Phys.*, **29**, 294 (1958).

where E_e is the ion energy at the exit slit and σ_L represents the collected constants or the value of $\sigma(E)$ at one ev. This equation has been used to describe the results of several ion-molecule reactions.[5,12] There are, however, many more reactions which depart considerably from the predicted linear dependence of Q upon $E_e^{-1/2}$, in particular reactions of hydrocarbons for which[13] Q varies as E_e^{-1}.

Re-examination of Previous Work.—In the derivation of equation 4 by Gioumousis and Stevenson[12] the most serious restriction imposed is that the gas kinetic collision cross section σ_K shall be small compared to πb_0^2. "Thus the analysis is most likely to be valid for some such reaction as that between a noble gas ion and hydrogen, or hydrogen with hydrogen."[12] On the other hand, if the preceding description is to be consistent with short range repulsive forces, it is to be expected that as the relative energy of the collision partners increases the Q of equation 4 becomes eclipsed by a nearly energy-independent cross section, σ_K, whose value should approximate the gas kinetic collision cross section.

Some of the relevant parameters are listed herewith in Table V for a few reactions.

TABLE V

COMPARISON OF σ_L AND σ_K

Ion	Molecule	$\alpha \times 10^{24}$ cc.	$\frac{m_1 + m_2}{m_2}$	$\sigma_L \times 10^{16}$ ev.$^{1/2}$ cm.^{2}	$\sigma_K \times 10^{16}$ cm.^{2}	$(\sigma_L/\sigma_K)^2$
$H_2^+(D_2^+)$	$H_2(D_2)$	0.78	2	21	21	1.0
Ar^+	H_2	.79	21	69	24	8.2
Kr^+	H_2	.79	43	97	26	14.0
H_2^+	Kr	2.48	1.0	33	26	1.6
CH_4^+	CH_4	2.6	2	38	33	1.4
$C_2H_4^+$	C_2H_4	4.26	2	49	40	1.5
O_2^+	D_2	0.79	9	45	24	6.6
D_2^+	O_2	1.60	1.1	22	24	0.8

Values of σ_K are based on van der Waals radii. The last column lists values of the particular ion energy E_t at which $\sigma_L/E_t^{1/2} = \sigma_K$. It is

$$E_t = (\sigma_L/\sigma_K)^2 = 2\pi^2 e^2 \alpha(m_1 + m_2)/m_2\sigma_K^2 \quad (5)$$

It appears that equation 4 is inadequate to describe $Q(E)$ for many reactions, even at relatively low ion energy, and also for reactants as small as H_2^+ and H_2. Let us represent the *reaction cross section* by

$$\sigma(E) = [\sigma_1/E^{1/2} - \sigma_0] + \sigma_k \quad (6)$$

The term in square brackets represents the effective area for glancing collisions; σ_0 and σ_k correspond to head-on collisions. In the language of collision theory, each sigma involves the product of an area and an efficiency factor. Tentatively we may identify σ_1 with $P_L\sigma_L$ and σ_0 with $P_L\sigma_K$ where σ_K is a gas kinetic collision cross section. Similarly, σ_k becomes $P_K\sigma_K$; the reaction probability factors P_L and P_K need not be equal, so that σ_0 and σ_k may be quite different and either may well be zero. For lack of information σ_0 and σ_k are taken to be independent of energy. The equation is valid for $0 < E < E_t$ for which the term in

(13) F. H. Field, J. L. Franklin and F. W. Lampe, *J. Am. Chem. Soc.*, **79**, 2419 (1957).

brackets equals or exceeds zero. For $E > E_t$ we assume

$$\sigma(E) = \sigma_k \qquad (7)$$

Integration over the low energy range gives

$$Q = E_e^{-1} \int_0^{E_e} \sigma(E)dE = 2\sigma_1 E_e^{-1/2} + \sigma_{k0}; \; E_e < E_t \quad (8)$$

where $\sigma_{k0} = \sigma_k - \sigma_0$. At higher energy

$$Q = E_e^{-1} \int_0^{E_t} \sigma(E)dE + E_e^{-1} \int_0^{E_e} \sigma(E)dE =$$
$$E_e^{-1}[2\sigma_1 E_t^{1/2} - \sigma_0 E_t] + \sigma_k = E_e^{-1}\Sigma + \sigma_k \quad (9)$$

If the preceding considerations are valid, the dependence of Q upon E_e for the reaction

$$D_2^+ + D_2 \longrightarrow D_3^+ + D$$

should obey equations 8 and 9, rather than 4. The results of Stevenson and Schissler,[5] for this and other reactions, appear in Fig. 1, all expressed in terms of Q vs. $E_e^{-1/2}$. The cross section for ArH$^+$ obeys equation 8 rather than 4, while that for D$_3^+$ and DO$_2^+$ indicate departure from equation 8 at fairly low E_e. Since we attribute greater precision to these measurements than did the observers, further consideration is desirable.

There is a necessary relationship between the parameters of equations 8 and 9 for the particular functional dependence of Q upon E which we have postulated. Considering that insufficient data are available to establish empirically the correctness of the postulated functions by graphing, we may take the concordance of the observed and calculated slopes as a limited test of the adequacy of equations 8 and 9 to describe the facts. The results of this comparison appear in Table VI. Values of E_t, σ_1 and Σ are all in fair agreement with theory. The present data do not permit a critical test of the alternative descriptions. Other work[14-16] provides further evidence of the effects implied in equations 6–10.

Table VI

	$E_t{}^a$, e.v.	$\sigma_k{}^a$	$\sigma_{k0}{}^a$	$\sigma_1,{}^a$ obsd.	$\sigma_1,{}^b$ calcd.	$\Sigma,$ obsd.	$\Sigma,{}^c$ calcd.
D$_2^+$ + D$_2 \rightarrow$							
D$_3^+$ + D	1.7	6	11	17	21	50	42
D$_2^+$ + O$_2 \rightarrow$							
DO$_2^+$ + D	1.5	0	9	16	22	62	44

[a] From Fig. 2. [b] By eq. 4, identifying σ_L and σ_1. [c] Assuming $\sigma_0 = 0$ and $\sigma_1 E_t^{1/2} = \sigma_L^2/\sigma_K$.

When $\sigma(E) > 0$ at low values of E_e and $\sigma(E) = 0$ above a critical limit, then for measurements in the interval above this limit the expression QE_e = const. must apply. Thus, it already has been found that $Q \propto E_e^{-1}$ for the formation of the persistent collision complex.[17] In such instances one expects the complex to be rather unstable and the rate of decomposition to increase rapidly with increasing ion energy. With fixed collection time in a mass spectrometer, this would be expressed by a limiting, critical energy E_c beyond which $\sigma(E)$ rapidly approaches zero. If, for simplicity,

(14) L. P. Theard and W. H. Hamill, J. Am. Chem. Soc., in press.
(15) R. F. Pottie, A. J. Lorquet and W. H. Hamill, ibid., **84**, 529 (1962).
(16) Don Kubose, Andrée Lorquet and Thomas Moran, this Laboratory.
(17) R. F. Pottie and W. H. Hamill, J. Phys. Chem., **63**, 877 (1959).

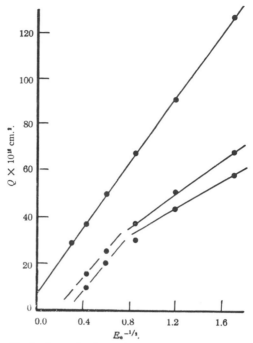

Fig. 1.—Ion–molecule reaction cross section Q as function of ion energy from the work of Stevenson and Schissler.[5] From top to bottom, curves refer to ArH$^+$, D$_3^+$ and DO$_2^+$.

this limit is treated as a discontinuity, then equation 9 also should apply to this case. The data for charge transfer to the 72-ion in neopentane are of this type and all the facts strongly suggest that C$_5$H$_{12}^+$ is very unstable. We conclude that it is susceptible to unimolecular decomposition following charge transfer when a critical ion–molecule relative velocity has been exceeded. We obtain from the preceding equations for ion-induced dipoles, at fairly low energy

$$Q = E_e^{-1} \int_0^{E_e} \sigma(E)dE + E_e^{-1} \int_{E_e}^{E_e} \sigma(E)dE$$
$$= (2\sigma_1 E_e^{-1/2} - \sigma_0 E_e)/E_e; \; E_e < E_t \quad (10)$$

It is evident that the relation

$$QE_e = \text{constant} \qquad (11)$$

will apply regardless of the force law.

Charge Exchange.—The qualitative dependence of $\Delta i_{72}/i_{65}$ upon the ionization potential of the added gas is evident. Concerning fragment ions we can, of course, eliminate at once all for which the difference of ionization potentials is endothermic (e.g., C$_2$H$_3^+$, C$_2$H$_5^+$ and C$_3$H$_7^+$). Comparison of results for C$_2$H$_4$ and C$_2$H$_6$ and their ion abundances strongly indicates C$_2$H$_4^+$ as the effective primary ion. This is supported by the results with added n-C$_4$H$_{10}$ and n-C$_4$D$_{10}$ but not by those with i-C$_4$H$_8$ and 1-C$_5$H$_8$. The disagreement may arise from differences in structure of the C$_2$H$_4^+$-ions.

The possibility that the increased 72-ion current in mixtures is due to reaction rather than simple charge exchange was tested with C$_2$D$_6$ and with C$_4$D$_{10}$, for which the sum of all ion currents from $m/e = 73$ to 78 was $0.14 \, i_{72}$ and $0.52 \, i_{72}$ respectively. For comparison, in mixtures with C_2H_6 and C_4H_{10} the values were $0.23 \, i_{72}$ and $0.45 \, i_{72}$. Also, such a hypothesis would not explain the dependence of Q upon E_e^{-1}.

TABLE VII

ENERGY DEPENDENCE FOR REACTION CROSS SECTIONS

Primary ion	Secondary ion	σ_l $\times 10^{16}$ cm.^{2} e.v.$^{1/2}$	σ_0 $\times 10^{16}$ cm.^{2}	QE_e $\times 10^{16}$ cm.^{2} e.v.
CS_2^+	$C_5H_{12}^+ + C_5H_{11}^+$	5.3	3.1	
$C_3H_6^{+a}$	$C_5H_{11}^+$	30	15	
$C_2H_4^{+b}$	$C_5H_{11}^+$	7	4	
$C_2D_4^{+c}$	$C_5H_{12}^+$			1.5
$C_2H_4^+ + C_3H_6^{+d}$	C_5H_{12}			1.5
$C_2H_4^{+e}$	$C_5H_{11}^+$			1.0
CS_2^+	$C_5H_{12}^+$			5.3

^{a} For pure neopentane or added propylene. ^{b} Combined average for $C_2H_4^+$ and $C_2D_4^+$ from runs with C_2H_6 or with C_2D_6. ^{c} From C_2D_6. ^{d} Combined average for $C_2H_4^+$ and $C_3H_6^+$ from C_3H_8, assuming equal efficiencies. ^{e} From C_2H_6.

The combined current, $\Delta \, i_{72} + \Delta \, i_{71}$, should be a measure of the primary process F, provided decomposition does not proceed beyond $C_5H_{11}^+$. In fact equation 8 describes the results very well on this basis, and it would be expected that the primary process F obey such an energy dependence.

$$CS_2^+ + C_5H_{12} \longrightarrow CS_2 + C_5H_{12}^+ \qquad (F)$$

$$C_5H_{12}^+ \longrightarrow C_5H_{11}^+ + H \qquad (G)$$

The results appear in Table VII. On the other hand, the net $\Delta \, i_{72}$ obeys equation 11. Considering all the evidence, we conclude that $\Delta \, i_{72}$ also obeys the particular law, 10. If one accepts this interpretation, the critical energy for the decomposition reaction G can be obtained from the measured values of the parameters in equations 8 and 10 if we assume that $\sigma_k = 0$. We have $QE_e = 2\sigma_l E_c^{1/2} - \sigma_0 E_c = 5.3 \times 10^{-16}$ cm.^{2} e.v. and $2\sigma_l = 10.6 \times 10^{-16}$ cm.^{2} e.v.$^{1/2}$; $- \sigma_{k0} = \sigma_0 = 3.1 \times 10^{-16}$ cm.^{2} One obtains $E_c = 0.39$ e.v. for $C_5H_{12}^+$. Finally, using the same parameters we find $E_t = (5.3/1.3)^2 = 2.9$ e.v. for $C_5H_{11}^+$ and $C_5H_{12}^+$ combined which places an approximate upper limit on the range of validity of the measurements suitable for testing the preceding equation.

The preceding data in Table VI also support the crude description by collision theory mentioned above, $viz.$ $\sigma_l/\sigma_0 = \sigma_L/\sigma_K$, assuming that reaction P-factors cancel. For the first three entries in Table VII we find ratios of 1.7, 2.0 and 1.8. The corresponding calculated ratios of σ_L/σ_K are 1.5, 1.1 and 1.0.

Neopentane.—It appears that no single ionic species is responsible for the enhanced 71-ion formation, either in mixtures or in neopentane alone. Comparison of the results in Table I with the mass spectral patterns suggests that CH_3^+, $C_2H_5^+$ and $C_2H_4^+$ are effective. The parameters σ_l and σ_0 for the more efficient reactions appear in Table VII. It is significant that every additive which enhances i_{72} also enhances i_{71}, while the converse appears not to hold. This suggests that, in addition to the mechanism previously proposed by Lampe and Field,[10] there is also a spontaneous decomposition following charge transfer. The effect is most pronounced for carbon disulfide for which we postulate, as suggested by the data of Table II, the consecutive reactions F and G.

Acknowledgment.—We are grateful to Dr. D. P. Stevenson for critical comment.

9

Reprinted from J. Chem. Phys. **42**:824–829 (1965)

Effect of Translational Energy on Ion–Molecule Reaction Rates. I

Keith R. Ryan* and Jean H. Futrell

Aerospace Research Laboratories, Office of Aerospace Research, Wright–Patterson Air Force Base, Ohio

INTRODUCTION

FOR many of the ion–molecule reactions reported in the literature, values are given for the specific reaction rate as well as the phenomenological cross section. In most cases the reactions have been studied in a conventional ion source of the mass spectrometer. Consequently, the observed reaction involves a beam of ions whose translational energy changes from thermal at the point of origin to some value E_F at the exit slit. The value of E_F depends on the magnitude of the repeller voltage used to extract the ions from the source.

For simple reactions systems with few internal degrees of freedom such as

$$CO^+ + H_2 \rightarrow COH^+ + H, \qquad (1)$$

it has been demonstrated experimentally[1] that the phenomenological cross section can be expressed as

$$Q \propto (E_r)^{-\frac{1}{2}},$$

where E_r is the field strength. Using the Langevin form of the classical collision cross section, Gioumousis and Stevenson[2] derived the following expression for the phenomenological cross section,

$$Q = 2\pi \exp[(\alpha_r/\mu)(2M_p/eE_r l)]^{\frac{1}{2}}, \qquad (2)$$

where α_r is the polarizability of the neutral particle, M_p is the mass of the ion, μ is the reduced mass of the pair, and l the source path length. The specific reaction rate k can be related to Q by

$$k = (eE_r l/2M_p)^{\frac{1}{2}} Q. \qquad (3)$$

Thus, in systems for which the relationship expressed in (2) holds k will indeed be a constant independent of E_r and hence the speed of the ion.

For reactions between ions and molecules which possess many internal degrees of freedom it appears

that the dependence of Q on E_r is not that given by (2).[3] Specific reaction rates for systems of this kind have been found to depend on the speed of the ions. The experiments of Giese and Maier[4] using a preselected ion beam of controlled energy indicate that even in simple systems the manner in which k varies with ion velocity depends strongly on the process involved. Unfortunately, instrumental considerations limited the experiments to ions with energies of 4 eV or higher.

Tal'roze and Frankevich[5] have described a technique which allows the determination of the specific reaction rate for an ion–molecule reaction under conditions where both the reacting partners have thermal energies only. A beam of electrons flows into the ion source for about a microsecond, during which time no field exits to extract ions formed by the electron beam. The ionizing beam is switched off and at some known time later an ion-extracting pulse removes ions generated by the electron beam as well as those formed from ion-molecule reactions. A knowledge of the time delay between the ionizing pulse and the ion-extracting pulse as well the ratio of secondary to primary ion beam readily allows the specific reaction rate to be calculated.

In the experiments of Tal'roze and Frankevich the following systems were studied:

$$CH_4^+ + CH_4 \rightarrow CH_5^+ + CH_3, \qquad (4)$$

and

$$H_2O^+ + H_2O \rightarrow H_3O^+ + OH; \qquad (5)$$

these workers concluded that within the experimental error the specific reaction rates obtained for these two systems were the same as those obtained by the earlier techniques at an ion-extraction field of 10 V/cm. This finding is in contradiction to the estimated thermal rate constants of Field et al.[1] who predicted that the specific reaction rate at zero repeller for Reaction (4) is about three times the value obtained at 10 V/cm.

---

* Visiting Research Associate. Permanent address: Department of Chemistry, University of Canterbury, Christchurch 1, New Zealand.

[1] D. P. Stevenson and D. O. Schissler, J. Chem. Phys. **29**, 282 (1958).
[2] G. Gioumousis and D. P. Stevenson, J. Chem. Phys. **29**, 294 (1958).
[3] F. H. Field, J. L. Franklin, and F. W. Lampe, J. Am. Chem. Soc. **79**, 2419 (1957).
[4] C. F. Giese and W. B. Maier, J. Chem. Phys. **39**, 739 (1963).
[5] V. L. Tal'roze and E. L. Frankevich, Zh. Fiz. Khim. **34**, 2709 (1960).

It is the purpose of this paper to present specific reaction rates obtained by the technique of Tal'roze and Frankevich for several systems and to compare these with reaction rates at several values of ion-extraction voltage.

EXPERIMENTAL

The instrument used in this study was a Consolidated Electrodynamics 21 103C to which some modifications were made. Because of the reduced ion intensity when operating in the pulsed ionizing current mode, an electron muliplier was used to increase the signal sensitivity. The multiplier had a minimum gain of 2×10^4, and a maximum gain of 2×10^6 over the gain of the electrometer supplied with the instrument. As required, the electron multiplier could be bypassed and the electrometer used as the detector.

Two precautions were taken to minimize the effects of stray fields in the ionization region. The electron trap which normally operates about 250 V positive to the ionization region was operated at 1.5 V positive. This potential was maintained by a dry cell. In addition the ion exit slit as well as the aperture through which the electrons enter the source were covered with a fine nickel mesh of 100 lines/in., with a transmission of some 70%.

The dimensions of the ion source are such that an ion of 30 amu generated in the electron beam and moving with thermal energy only will reach the walls in about 3 μsec. An ion of mass 2 amu with thermal energy will travel the same distance in approximately 0.7 μsec. Considerations of this kind determine the maximum allowed repeller delay time for a given ion and hence the type of pulse circuitry required. Ionizing and repeller pulses are derived from a free-running multivibrator timer. The output of this timer is differentiated and fed into two separate one-shot multivibrators, one of which has a variable-width square-wave output. Differentiation and comparison of the trailing edge of the two signals serves as the time base for the repeller pulse. The ionizing pulse and the repeller pulse are each amplified and clipped in order to restore a square wave appearance to the pulse. Each pulse has a rise time of 0.05 μsec.

A filament assembly, Serial No. 25597, of the type formerly used in the C.E.C. M407 instrument has been used in this study. This assembly includes a control grid adjacent to the filament. The electron beam is cut off by holding this grid 20 V negative to the filament. The ionizing pulse, which is about 1 μsec in duration and has a repetition frequency of about 10 kc/sec, raises the potential of this grid to two volts positive to the filament. In order to extract the ions, a repeller pulse of 15 V and 1.5 μsec duration can be switched on any time from 0 to 10 μsec after the ionizing pulse is switched off. If desired both pulses may be operated simultaneously. For the results reported here the

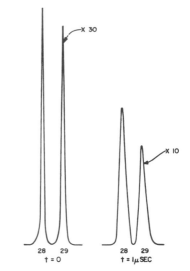

Fig. 1. Effect of repeller delay time on peak shape. Peak shapes for the ions CO^+ and COH^+ at 0- and 1.0-μsec repeller delay times. To obtain the ratio of 29:28 one must divide the 29 peak height by the factor shown on the diagram. This corrects for the different detector sensitivities used for recording the ion intensities.

maximum repeller delay used was 1.0 μsec, while the increments in delay time used were 0.1 μsec.

In addition to the specific reaction rates determined by the pulse method, the systems were also studied by the more conventional ion–molecule techniques described, for example, by Field et al.[3] When operating in this mode we have used two values of electron current. Results were obtained for currents of 10 and 0.1 μA, respectively. At the lowest currents the emission regulator was bypassed and the current control was obtained by means of a constant-voltage transformer. The current stability under these conditions was within 2%.

RESULTS AND DISCUSSION

Before any measurements of specific reaction rates for thermal ion reactions could be made, a study of the source characteristics in the pulse mode was necessary. Of principal importance were the primary ion-current dependence on repeller delay, the effect of ion-accelerating voltage on primary current and the effect of repeller delay on ion peak shape.

The effect of repeller delay time on peak shape is shown in Fig. 1. The ion peak shapes shown are for the two ions involved in Reaction (1), namely, CO^+ and COH^+. Clearly, a delayed repeller pulse has a severe effect upon the peak shape. We have taken the position that, although the peak shapes have changed, both ion profiles have changed in the same way, and we have assumed that the ratio of peak heights gives the required ratio of ion intensities.

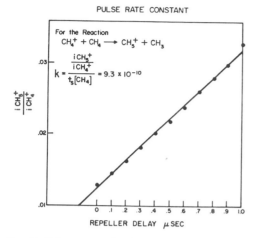

PULSE RATE CONSTANT

For the Reaction
$$CH_4^+ + CH_4 \longrightarrow CH_5^+ + CH_3$$

$$k = \frac{\dfrac{i\,CH_5^+}{i\,CH_4^+}}{t_s [CH_4]} = 9.3 \times 10^{-10}$$

FIG. 2. Pulse rate results for the reaction $CH_4^+ + CH_4 \rightarrow CH_5^+ + CH_3$. Results shown are for one particular experiment. Further the observed rate is uncorrected for discrimination effects.

The amount of decay during the period of 1 μsec is in all cases much higher than would be expected from the source lifetimes of thermal ions. This might seem to cast doubt on the feasibility of applying the pulse technique to the ion source used in the present study. It must be remembered, however, that in order to calculate the specific reaction rate the quantity required is the ratio of secondary to primary ion current. Following Tal'roze and Frankevich we may write

$$I_s = k I_p n_0 t_d, \qquad (6)$$

where I_s is the number of secondary ions formed in delay time t_d, k is the specific reaction rate, I_p the primary ion current and n_0 the concentration of neutral reactants. Equation (6) assumes that $I_s \ll I_p$, an assumption which is certainly true for all experiments described here. In general secondary ions will be formed throughout the duration of the ionizing pulse, during the pulsed ion extraction time and during the controlled field-free reaction time. Secondary ion formation during the first two time intervals will not be dependent upon the repeller delay time. Consequently, in a plot of I_s/I_p against delay time, the formation of secondary ions during the first two time intervals will be detected as a positive intercept on the I_s/I_p axis.

One is, of course, unable to measure I_p or I_s, but only the mass-resolved components of these ionic species. Decay curves show that the primary-ion intensity which reaches the collector drops off sharply with time. However, the important condition is that the ratio of secondary ions to primary ions in the volume of the source from which the analyzer selects ions represents the state of the reaction at the time of

sampling. As long as this condition is satisfied, poor decay curves are not an indication of the failure of the technique but rather are related to the focusing properties of the source.

It would appear from the foregoing discussion that a necessary condition for the success of the pulse technique is that a plot of I_s/I_p collected ion currents has a linear dependence of repeller delay. In Fig. 2 we show our results for the reaction

$$CH_4 + CH_4^+ \rightarrow CH_5^+ + CH_3. \qquad (7)$$

We feel that the quality of the results is sufficient justification for assuming that the technique can be successfully applied in our source. The value obtained for the specific reaction rate was 9.6×10^{-10} cm^3 molecule^{-1} sec^{-1} and was reproducible to within 5%.

It is important to establish the effect of ion-accelerating voltage on the measured specific reaction rate. One might expect that field penetration from the accelerating voltage could effect the specific reaction rate at least two ways. First, stray fields could considerably shorten the source residence time thereby lowering the observed reaction rate. Further, if the specific reaction rate is dependent upon the ion velocity, one would expect the observed rate to depend on accelerating voltage in the event of considerable field penetration. Figure 3 shows the results obtained for the specific reaction rate as a function of ion-accelerating voltage. We conclude from this that there is very little field penetration into the source.

For the reaction systems studied it was found that if the precursor and product ion had only a small percentage mass difference, good linearity could be obtained for plots of the kind shown in Fig. 2. Results

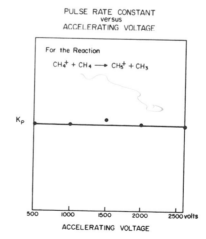

PULSE RATE CONSTANT
versus
ACCELERATING VOLTAGE

For the Reaction
$$CH_4^+ + CH_4 \longrightarrow CH_5^+ + CH_3$$

K_p

ACCELERATING VOLTAGE

FIG. 3. The effect of ion-accelerating voltage on the observed thermal specific reaction rate for the reaction $CH_4^+ + CH_4 \rightarrow CH_5^+ + CH_3$.

for the reaction

$$NH_3^+ + NH_3 \rightarrow NH_4^+ + NH_2 \qquad (8)$$

are given in Fig. 4. However, for the reaction

$$H_2^+ + H_2 \rightarrow H_3^+ + H, \qquad (9)$$

linearity for the plot of $I(H_3^+)/I(H_2^+)$ against delay time could only be obtained for repeller delays up to 0.7 μsec. In Fig. 5 we show our results for this system. It is necessary to comment here that the results obtained at 0.9 and 1.0 μsec continue the sharp upward trend indicated by the 0.8-μsec results.

Another reaction for which the product ion has a

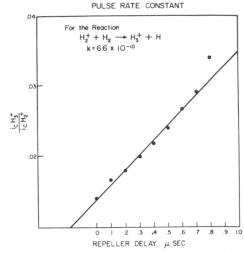

PULSE RATE CONSTANT

FIG. 5. Experimental results for the reaction $H_2^+ + H_2 \rightarrow H_3^+ + H$.

A series of measurements was made for the specific reaction rate as a function of the ion repeller field strength with field strengths ranging from 10 V/cm down to 2 V/cm. The results obtained in this way showed similar characteristics for all the systems examined. The specific reaction rate appeared to pass through a broad maximum at low repeller field strengths. Repeated runs showed that the values obtained at low field strengths were not reproducible and that the position of the maximum varied from one experiment to another. The results of one set of experimental observations for the reaction

$$C_2H_3^+ + C_2H_4 \rightarrow C_2H_5^+ + C_2H_2 \qquad (11$$

are shown in Fig. 7. Field *et al.* reported the field

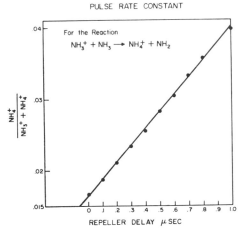

PULSE RATE CONSTANT

FIG. 4. Experimental results for the reaction $NH_3^+ + NH_3 \rightarrow NH_4^+ + NH_2$.

mass 50% greater than the precursor ion is

$$D_2^+ + D_2 \rightarrow D_3^+ + D, \qquad (10)$$

and one might expect a similar departure from linearity as that shown for Reaction (9) in Fig. 5. The results for Reaction (10) shown in Fig. 6 indicate that the linear relation holds for at least 1 μsec. From this we conclude that the departure from linearity in the case of Reaction (9) is a direct result of the appreciable thermal velocity of the H_2^+ ion, rather than the large difference in mass of the precursor and product ion.

For all systems studied by the pulse mode, a parallel set of measurements were made using the techniques described by Field *et al.*[3] The object of the additional measurements was to allow a comparison of the thermal rate constant with values obtained when the ionic reactant possessed considerable velocity. Initially these experiments were run with an ionizing current of 10 μA.

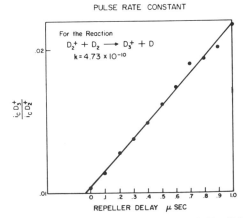

PULSE RATE CONSTANT

FIG. 6. Experimental results for the reaction $D_2^+ + D_2 \rightarrow D_3^+ +$

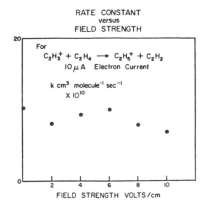

FIG. 7. The observed specific reaction rates for the reaction $C_2H_3^+ + C_2H_4 \rightarrow C_2H_5^+ + C_2H_2$ as a function of repeller field strength. Results were obtained with an ionizing current of 10 μA.

strength dependence of the specific reaction rate for this reaction and noted that, in common with several other systems studied, the value obtained appeared to pass through a maximum at low field strength. These maxima were attributed to experimental effects arising from an imprecise knowledge of the conditions in the ion source.

It seemed to us that space charge effects might account for some of these nonreproducible results at low repeller field. In a second series of investigations we reduced the ionizing current by a factor of 100 and repeated the measurements using 0.1-μA ionizing current. The results were now found to be reproducible and the previously observed maxima did not occur. In Fig. 8 we show these low-current results for Reaction (11).

In Table I we tabulate our results for the specific reaction rates of several systems for thermal ions as well as the 10-V/cm results. Also shown is the ratio

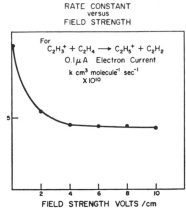

FIG. 8. The observed specific reaction rates for the reaction $C_2H_3^+ + C_2H_4 \rightarrow C_2H_5^+ + C_2H_2$ as a function of repeller field strength. Results were obtained with an ionizing current of 0.1 μA.

of the thermal value to the 10-V/cm value. In compiling Table I we have taken into account the chemical discrimination factor of the electron multiplier. To do this we have made use of the fact that the 10-V/cm result is insensitive to the magnitude of the ionizing current within the range of 0.1 to 10.0 μA. Therefore, we have determined the specific reaction rate using the electron multiplier as well as the electrometer at 10-V/cm field strength. We then multiply our thermal result by the ratio of the value obtained on the electrometer to that obtained on the multiplier. This has a strong effect on the observed pulse value in some cases. For example, in methane the value was changed from 9.6×10^{-10} to 17.0×10^{-10} cm³ mole⁻¹ sec⁻¹ by this correction.

For the reaction

$$H_2^+ + Ne \rightarrow NeH^+ + H, \qquad (12)$$

the experimental ratio measured was NeH^+/Ne^+ in

TABLE I. Comparison of pulse rates and 10-V/cm rates.

Reaction	Pulse rate K_p cm³ mol⁻¹ sec⁻¹ $\times 10^{10}$	10-V/cm rate K_{10} cm³ mol⁻¹ sec⁻¹ $\times 10^{10}$	K_p/K_{10}
$H_2 + H_2^+ \rightarrow H_3^+ + H$	6.6	20.0	0.33
$D_2 + D_2^+ \rightarrow D_3^+ + D$	4.73	14.0	0.33
$CO^+ + H_2 \rightarrow COH^+ + H$	15.3	14.1	1.1
$CH_4^+ + CH_4 \rightarrow CH_5^+ + CH_3$	16.9	10.7	1.57
$NH_3^+ + NH_3 \rightarrow NH_4^+ + NH_2$	32.6	13.0	2.1
$H_2O^+ + H_2O \rightarrow H_3O^+ + OH$	21.9	18.2	1.20
$C_2H_3^+ + C_2H_4 \rightarrow C_2H_5^+ + C_2H_2$	15.3	6.5	2.35
$CH_3CHOH^+ + CH_3CH_2OH$ $\rightarrow CH_3CH_2OH_2^+ + CH_3CHO$	31.4	30.3	1.04
$Ne + H_2^+ \rightarrow NeH^+ + H$	1.20	1.25	1.04

order to minimize discrimination introduced during mass analysis. To calculate the specific reaction rate the ratio of Ne^+/H_2^+ in the ion source was required. This was obtained by measuring the relative collection efficiency of the instrument for H_2^+ and Ne^+ and then determining the ratio Ne^+/H_2^+ in the mixture.

In a study of the reaction

$$H_2^+ + He \rightarrow HeH^+ + H \qquad (13)$$

in the pulse mode it was found that the specific reaction rate was approximately $\frac{1}{20}$th of the rate for the reaction

$$H_2^+ + H_2 \rightarrow H_3^+ + H. \qquad (9)$$

This is in fair agreement with the results of Von Koch and Friedman[6] who found a rapid decline in the specific reaction rate in the hydrogen–helium system as the repeller field approached zero.

[6] H. von Koch and L. Friedman, J. Chem. Phys. **38**, 1115 (1963).

From Table I it can be seen that the thermal rate constants for Reactions (9) and (10) are much lower than their respective 10-V/cm values. These systems are unique amongst those studied in that the product ion was 50% heavier than the ion used to monitor the progress of the reaction. It is possible that the observed lower value of the thermal rate constant is an instrumental effect caused by mass discrimination introduced when the ion source is operated under the conditions described above. For the other reactions studied, one compares the ratio of ion currents of ions which do not differ in mass by more than one part in 14, under which conditions one might expect discrimination effects to be considerably reduced.

Apart from Reactions (9) and (10) the systems appear to divide themselves into two classes, those for which $K_p \cong K_{10}$ and those for which K_p is significantly greater than K_{10}. For this second class we suggest that the transition-state ion–molecule complex at low field strength differs from the structure which predominates at 10 V/cm. We should point out that Tal'roze and Frankevich concluded that for Reaction (7) the specfic reaction rate at thermal energies is much the same as at 10 V/cm. The same conclusion could be drawn from the results of Hand and von Weyssenhoff,[7] who obtained a specific reaction rate of 11×10^{-10} cm^3 molecule^{-1} sec^{-1} for Reaction (7) using a pulse technique. Our results indicate that the thermal rate constant for Reaction (7) is significantly higher than the 10-V/cm value when corrections are made for mass-discrimination effects of the electron multiplier. In our case the correction changed the value of k for Reaction (7) from 9.6×10^{-10} to 16.9×10^{-10} cm^3 molecule^{-1} sec^{-1}. It is interesting to note that Field et al.[3] in their early publication predicted thermal reaction rates which were significantly higher than the value obtained at 10 V/cm.

ACKNOWLEDGMENT

The authors are pleased to acknowledge the assistance of Dean Miller, who designed and maintained the pulse circuitry used in this study. He also assisted in obtaining many of the experimental results.

[7] C. W. Hand and H. von Weyssenhoff, Can. J. Chem. **42**, 195 (1964).

THEORY

Editor's Comments
on Papers 10 Through 15

It is surprising that the first theoretical paper that is relevant to the rates of ion-molecule reactions was published by Paul Langevin [1] in 1905, considerably before ion-molecule reactions had been identified. A translation by Earl McDaniel of Langevin's paper is given here (Paper 10). This is a classical impact parameter treatment assuming structureless point particles approaching each other at low velocities. The paper by Gioumousis and Stevenson (Paper 11) applies Langevin's method to ion-molecule reactions at conditions existing in the ion source of a mass spectrometer. Both papers assume the neutral to have no permanent dipole. Thus the important potential energy term is that between the ion and the charge-induced dipole of the neutral. The fragment by Eyring, Hirschfelder, and Taylor (Paper 12) approaches the interaction of an ion and a neutral from the viewpoint of quasi-equilibrium theory. The approaching ion and neutral are assumed to surmount a pseudo-potential barrier resulting from the balance of rotational and potential energies. The absolute rate theory formalism leads to the rate constant

$$k = 2 \Pi e \left(\frac{\alpha}{\mu} \right)^{1/2}$$

where α is the polarizability of the neutral, μ is reduced mass, and e is the unit of electric charge. This is just the result obtained by Gioumousis and Stevenson.

Since both treatments are valid only for molecules without a permanent dipole, it was natural that the theories be extended to include the effect of a permanent dipole. Moran and Hamill (Paper 13) were the first to do this. Their treatment assumes the optimum orientation of the dipole toward the ion and results in the expression

$$k = 2 \Pi e \left[\left(\frac{\alpha}{\mu} \right)^{1/2} + \frac{\mu_D}{\mu v} \right]$$

where μ_D is the dipole moment and v is velocity.

It was obvious from Moran and Hamill's treatment that the assumption that the dipole of the neutral lined up with the field of the charge represented an extreme or limiting case and led to results that were not always satisfactory. Ten years later, Su and Bowers (Paper 14) directed their attention toward this problem with results that have proved to be a considerable improvement over the fixed dipole treatment.

All these treatments, however, result in rates of collision, not rates of reaction. The assumption made by Gioumousis and Stevenson and others that reaction will occur for any collision of an ion with a neutral for which the collision radius is less than a certain critical value is not always correct. Results calculated in this way often agree with experiment, but jast as often they do not. The treatment applies only to collisions involving low energies and to reactions having no energy of activation, and yielding only one set of products.

In view of these deficiencies, Light and his colleagues have developed the phase space theory [2,3] and have applied it to several ion-molecule reactions (Paper 15). Light's theory assumes that a tight complex is formed and that the fraction of the reaction occurring through a certain exit channel is determined by the ratio of the phase space for that channel to the sum of the phase spaces for all channels. The method has given good results for several triatomic systems [2,4] in addition to those given in Paper 15, but it is extremely difficult to apply to more complex systems.

The phase space theory has been extended to include quantum mechanical considerations and the quantum mechanical theory has been applied to ion-molecul reactions. [5-9]

REFERENCES

1. Langevin, Paul, *Ann. Chim. Phys.* **5**: 245 (1905).
2. Light, J. C., *J. Chem. Phys.* **40**: 3221 (1964).
3. Pechukas, P., and J. C. Light, *J. Chem. Phys.* **42**: 3281 (1965).
4. Wolf, F. A., *J. Chem. Phys.* **44**: 1619 (1966).
5. Lin, J., and J. C. Light, *J. Chem. Phys.* **45**: 2545 (1966).
6. Nikitin, E., *Teor. i Eksp. Khim.* **1**: 135 (1965).
7. Pechukas, P., J. C. Light, and C. Rankin, *J. Chem. Phys.* **44**: 794 (1966).
8. Truhlar, D. G., and A. Kupperman, *J. Phys. Chem.* **73**: 1722 (1969).
9. Truhlar, D. G., *J. Chem. Phys.* **51**: 4617 (1969); **56**: 1481 (1972).

10

LANGEVIN'S CALCULATION

OF THE DIFFUSION

AND MOBILITY COEFFICIENTS*

E. W. McDaniel

This appendix consists of Paul Langevin's calculation of the coefficients of mutual diffusion and mobility. It is taken from his classic paper "Une Formule fondamentale de théorie cinétique," which appeared in *Annales de chimie et de physique*, Series 8, **5**, 245–288 (1905). Many changes in notation and terminology have been made in order to put the material in a form more familiar to the modern reader. A similar treatment of the problem of mutual diffusion is to be found in the kinetic theory text by Present.[1]

1. THE MOMENTUM TRANSFER EQUATION

Consider a mixture of gases containing molecules of two species, of masses m_1 and m_2, in numbers equal, respectively, to N_1 and N_2 per unit volume. The number densities N_1 and N_2 are assumed to vary from point to point in the gas. Suppose the velocities of the molecules of species 1 with components (ξ_1, η_1, ζ_1) are distributed according to the Maxwell distribution about their mean value, which represents the velocity of mass

---

* Langevin's paper, which is the subject of this appendix, was translated by E. W. McDaniel and published in Air Force Office of Scientific Research Document No. TN-60-865 (1960) [Georgia Institute of Technology, Atlanta, Ga.].

motion of the first gas. Likewise, the velocities of the molecules m_2, with components (ξ_2, η_2, ζ_2), are distributed according to the same law about their mean value, the velocity of mass motion of the second gas, which is different, in general, from that of the first. The difference in these velocities of mass motion, that is, the mass velocity of one gas relative to the other, is a measure of the intensity of the diffusive flow which results from the gradients in the number densities.

Let (u_1, v_1, w_1) be the velocity of mass motion of the first gas and (u_2, v_2, w_2) that of the second. The distribution law gives the number of molecules per unit volume of the first species whose velocities lie between (ξ_1, η_1, ζ_1) and $(\xi_1 + d\xi_1, \eta_1 + d\eta_1,$ and $\zeta_1 + d\zeta_1)$ as

$$dN_1 = c_1 \exp\{-hm_1[(\xi_1 - u_1)^2 + (\eta_1 - v_1)^2 + (\zeta_1 - w_1)^2]\}\, d\xi_1\, d\eta_1\, d\zeta_1$$
$$= f_1\, d\tau_1 \tag{1}$$

where

$$d\tau_1 = d\xi_1\, d\eta_1\, d\zeta_1$$
$$f_1 = c_1 \exp -hm_1[(\xi_1 - u_1)^2 + (\eta_1 - v_1)^2 + (\zeta_1 - w_1)^2] \tag{2}$$
$$c_1 = N_1\left(\frac{hm_1}{\pi}\right)^{3/2} \tag{3}$$

and

$$h = \frac{1}{2kT} \tag{4}$$

We may easily verify that

$$\bar{\xi}_1 = \frac{\int \xi_1\, dN_1}{\int dN_1} = u_1 \quad \text{and} \quad \tfrac{1}{2}m_1\overline{(\xi_1^2 + \eta_1^2 + \zeta_1^2)} = \frac{3kT}{2}$$

We have a similar expression for the molecules of the second species:

$$dN_2 = c_2 \exp\{-hm_2[(\xi_2 - u_2)^2 + (\eta_2 - v_2)^2 + (\zeta_2 - w_2)^2]\}\, d\xi_2\, d\eta_2\, d\zeta_2$$
$$= f_2\, d\tau_2$$

The partial pressure of the first gas on an element of surface equals the momentum transferred during unit time across unit area of this surface by the molecules of this gas, the surface element, of course, being supposed to move with the velocity of mass motion of the gas. For a surface element perpendicular to the X axis at a point at which the number density of molecules is N_1 and the partial mass density is $\rho_1 = N_1 m_1$, the components of the partial pressure are

$$p_{1xx} = \rho_1\overline{(\xi_1 - u_1)^2} = m_1 \int f_1(\xi_1 - u_1)^2\, d\tau_1$$
$$p_{1xy} = \rho_1\overline{(\xi_1 - u_1)(\eta_1 - v_1)} = m_1 \int f_1(\xi_1 - u_1)(\eta_1 - v_1)\, d\tau_1 \tag{5}$$
$$p_{1xz} = \rho_1\overline{(\xi_1 - u_1)(\zeta_1 - w_1)} = m_1 \int f_1(\xi_1 - u_1)(\zeta_1 - w_1)\, d\tau_1$$

The relations necessary for the equilibrium of an element of volume of the gas are

$$p_{1xy} = p_{1yx}; \qquad p_{1xz} = p_{1zx}; \qquad p_{1yz} = p_{1zy}$$

The mean pressure of the gas is defined by $p_1 = \frac{1}{3}(p_{1xx} + p_{1yy} + p_{1zz})$. It is easy to show that if the distribution law is Maxwellian the tangential pressure vanishes and

$$p_{1xx} = p_{1yy} = p_{1zz} = p_1 = N_1 kT \tag{6}$$

Consider now a fixed element of volume ($dx\,dy\,dz$) with its center (x, y, z) at a point at which the partial density is ρ_1. The conservation of matter is expressed by the equation

$$\frac{\partial \rho_1}{\partial t} + \frac{\partial(\rho_1 u_1)}{\partial x} + \frac{\partial(\rho_1 v_1)}{\partial y} + \frac{\partial(\rho_1 w_1)}{\partial z} = 0 \tag{7}$$

Let us seek the time rate of change of momentum of the molecules contained in this volume element. Considering first only the x component of the momentum, we see that there will be an increase due to the flow of molecules into the volume element through one of the faces $dy\,dz$. The increase is $dy\,dz \int \xi_1 f_1 m_1 \xi_1\, d\tau_1$. If we express this quantity in the algebraically identical form $m_1 dy\,dz[\int f_1(\xi_1 - u_1)^2\, d\tau_1 + 2u_1 \int f_1(\xi_1 - u_1)\, d\tau_1 + N_1 u_1^2]$, the use of (5) shows its magnitude to be $dy\,dz(p_{1xx} + \rho_1 u_1^2)$. The middle term has vanished, since the average value of $(\xi_1 - u_1)$ is zero by definition.

The excess of the x component of momentum entering the volume element through one $dy\,dz$ face over that leaving through the opposite face, per unit volume and unit time, is thus

$$-\frac{\partial p_{1xx}}{\partial x} - \frac{\partial(\rho_1 u_1^2)}{\partial x}$$

On taking the analogous quantities for the other dimensions, we have for the momentum increase provided by the flow of molecules per unit volume and unit time

$$-\frac{\partial p_{1xx}}{\partial x} - \frac{\partial p_{1xy}}{\partial y} - \frac{\partial p_{1xz}}{\partial z} - \frac{\partial(\rho_1 u_1^2)}{\partial x} - \frac{\partial(\rho_1 u_1 v_1)}{\partial y} - \frac{\partial(\rho_1 u_1 w_1)}{\partial z}$$

Any externally applied force will make another contribution to the momentum change. If X_1, Y_1, Z_1 denote the components of the applied force per unit mass, this contribution to the net increase of the x component of the momentum, per unit volume and unit time, is evidently $\rho_1 X_1$.

If a single gas is present, there will be only the two above mentioned factors in the increase of the momentum; but, if two gases are mixed, collisions between molecules of the two species produce an exchange of momentum in the x direction which is proportional to the difference $(u_2 - u_1)$ of the x components of the velocities of mass motion of the two gases, that is, to their relative velocities of mass motion. Without making any hypothesis concerning its form at the present time, let us designate by $\mathscr{B}(m_1\xi_1)$ the x component of momentum transferred per unit volume and unit time to the molecules of the first species during their collisions against molecules of the second species. Collisions between molecules of the same species evidently do not change the total momentum. This quantity $\mathscr{B}$ plays the essential role in the theory of diffusion, and it is on its exact calculation that we now turn our effort.

The x component of momentum per unit volume is $m_1 \int f_1 \xi_1 \, d\tau_1 = \rho_1 u_1$. On equating its time derivative to the total increase due to the various causes, we obtain

$$\frac{\partial(\rho_1 u_1)}{\partial t} = -\frac{\partial p_{1xx}}{\partial x} - \frac{\partial p_{1xy}}{\partial y} - \frac{\partial p_{1xz}}{\partial z}$$

$$- \frac{\partial(\rho_1 u_1{}^2)}{\partial x} - \frac{\partial(\rho_1 u_1 v_1)}{\partial y} - \frac{\partial(\rho_1 u_1 w_1)}{\partial z} + \rho_1 X_1 + \mathscr{B}(m_1\xi_1)$$

Using the conservation equation (7) and introducing the mobile time derivative

$$\frac{Du_1}{Dt} = \frac{\partial u_1}{\partial t} + u_1 \frac{\partial u_1}{\partial x} + v_1 \frac{\partial u_1}{\partial y} + w_1 \frac{\partial u_1}{\partial z}$$

to express the time rate of change of u_1 in a moving element following the mass motion of the gas, we find the time rate of change of the x component of momentum of the moving element to be

$$\rho_1 \frac{Du_1}{Dt} + \frac{\partial(p_{1xx})}{\partial x} + \frac{\partial(p_{1xy})}{\partial y} + \frac{\partial(p_{1xz})}{\partial z} = \rho_1 X_1 + \mathscr{B}(m_1\xi_1) \qquad (8)$$

This is the momentum transfer equation that determines the motion of the first gas. If this mass motion is sufficiently slow, (u_1, v_1, w_1) will be very small compared with the average values of (ξ_1, η_1, ζ_1), the thermal agitation velocities of the molecules. Under these conditions, deviations from the Maxwell distribution are extremely small, and we may say that the distribution is Maxwellian and thus isotropic about the velocity of mass motion. Equation 8 then becomes

$$\rho_1 \frac{Du_1}{Dt} + \frac{\partial p_1}{\partial x} = \rho_1 X_1 + \mathscr{B}(m_1\xi_1) \qquad (9)$$

Likewise, for the second gas we have

$$\rho_2 \frac{Du_2}{Dt} + \frac{\partial p_2}{\partial x} = \rho_2 X_2 - \mathscr{B}(m_1\xi_1) \qquad (10)$$

2. CALCULATION OF THE MOMENTUM TRANSFER
   PRODUCED BY COLLISIONS

Now let us calculate the quantity $\mathscr{B}(m_1\xi_1)$, which represents the exchange of momentum in collisions between molecules of the different species. Without altering the equations of motion of the molecules, and consequently without changing the results, we may give the axes a uniform translational motion (u_1, v_1, w_1) to annul the velocity of mass motion of

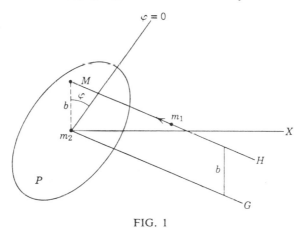

FIG. 1

the first gas; (u_2, v_2, w_2) then represents the relative velocity of mass motion, and we thus simplify the form of the equations.

Boltzmann[2] showed that $\mathscr{B}(m_1\xi_1)$ may be written in the form

$$\mathscr{B}(m_1\xi_1) = m_1 \iiint_0^\infty \int_0^{2\pi} f_1 f_2(\xi_1' - \xi_1) v_0 b \, d\tau_1 \, d\tau_2 \, db \, d\varphi \qquad (11)$$

where $(\xi_1', \eta_1', \zeta_1')$ is the velocity assumed after the collision by a molecule (ξ_1, η_1, ζ_1) which encounters a molecule (ξ_2, η_2, ζ_2). The relative velocity of the molecules before the collision is v_0, its magnitude being

$$v_0 = \sqrt{(\xi_2 - \xi_1)^2 + (\eta_2 - \eta_1)^2 + (\zeta_2 - \zeta_1)^2}$$

The dynamics of the collision are determined by the relative velocity and the impact parameter b as follows: referring to Fig. 1, the line m_2X passes through molecule m_2 and is parallel to the X axis. The molecule m_1 moves with respect to m_2 along the line MH, which is parallel to m_2G. The plane P is drawn through m_2 and is perpendicular to m_2G. In the absence of forces m_1 would pass within a distance b of m_2 and would encounter the plane P at point M, the azimuth angle having the value φ. The $\varphi = 0$ axis is defined by the intersection of planes P and m_2GX. When a central force is introduced, the relative trajectory of m_1 is curved but lies entirely in the plane Gm_2M. Its shape is determined by the relative velocity and the law of force between the molecules.

The relative trajectory of m_1 is composed of two branches that are symmetrical with respect to the line joining the apses, points A and E in

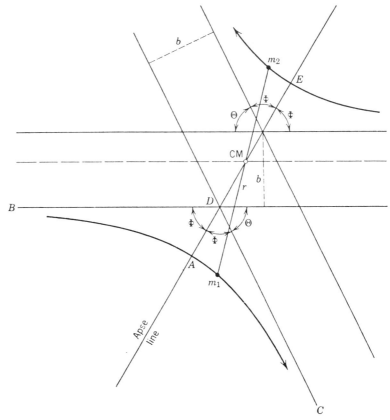

FIG. 2. $\Theta = \pi - 2\Phi =$ scattering angle in CM system; $b =$ impact parameter; CM = center of mass; and $r =$ distance between m_1 and m_2.

Fig. 2. The apses are the points of closest approach to the center of mass for m_1 and m_2, respectively. After the collision the relative velocity DC makes the same angle Φ with respect to AE as the initial relative velocity BD. The change in the velocity component ξ_1 is given as a function of Φ and the initial velocities by

$$\xi_1' - \xi_1 = \frac{m_2}{m_1 + m_2}\left[2(\xi_2 - \xi_1)\cos^2\Phi + \sqrt{v_0^2 - (\xi_2 - \xi_1)^2}\sin 2\Phi \cos\varphi\right]$$

If we introduce the reduced mass

$$M_r = \frac{m_1 m_2}{m_1 + m_2} \tag{12}$$

and let V represent the mutual potential energy of m_1 and m_2, we may calculate Φ by the equation (see Chapter 3, Section 3-4)

$$\Phi = \int_0^{\rho_0} \frac{d\rho}{\sqrt{1 - \rho^2 - (2V/M_r v_0^2)}} \tag{13}$$

Here $\rho = b/r$, where r is the distance of separation of the molecules, and ρ_0 is the smallest positive root of the expression under the radical, that is, the outermost zero of the denominator. It is assumed here that the force law is a continuous function of r and that m_1 and m_2 do not actually collide.

If, on the other hand, we suppose m_1 and m_2 to collide elastically but assume that they interact only at the instant of impact, we find that the relative trajectory consists simply of two straight line segments and that Φ is given by

$$\Phi = \sin^{-1} \frac{b}{D_{12}} \tag{14}$$

where D_{12} is the sum of the radii of the molecules m_1 and m_2, both of which are considered to be spherical.

If, finally, it is assumed that the molecules attract one another and then collide elastically, the relative trajectory remains symmetrical, but it has a discontinuity in the first derivative at A at the instant of collision; Φ is now given by

$$\Phi = \int_0^{\rho^*} \frac{d\rho}{\sqrt{1 - \rho^2 - (2V/M_r v_0^2)}} \tag{15}$$

ρ^* being equal to b/D_{12} and corresponding to the value of ρ at the instant when the impact occurs.

If $(\xi_1' - \xi_1)$ in (11) is replaced by its value above, the result is

$$\mathscr{B}(m_1 \xi_1) = 4\pi M_r \iiint_0^\infty f_1 f_2 v_0 (\xi_2 - \xi_1) \cos^2 \Phi b \, d\tau_1 \, d\tau_2 \, db$$

We have here seven consecutive integrations to perform, since $d\tau_1$ and $d\tau_2$ each correspond to a product of three differentials.

To simplify further the expressions for f_1 and f_2, we may choose the X axis, until now undetermined, to be parallel to the relative velocity of the two gases. The relative velocity is then written as $(u_2{}^*, 0, 0)$, and we have for f_1 and f_2

$$f_1 = c_1 \exp -hm_1(\xi_1{}^2 + \eta_1{}^2 + \zeta_1{}^2)$$

$$f_2 = c_2 \exp -hm_2[(\xi_2 - u_2{}^*)^2 + \eta_2{}^2 + \zeta_2{}^2]$$

The angle Φ, for a given law of force, depends uniquely on b and v_0, which determine the relative trajectory. Define a cross section* $q(v_0)$ by the equation

$$q(v_0) = \int_0^\infty \cos^2 \Phi b \, db \tag{16}$$

Then

$$\mathscr{B}(m_1\xi_1) = 4\pi M_r \iint f_1 f_2 v_0 q(v_0)(\xi_2 - \xi_1) \, d\tau_1 \, d\tau_2 \tag{17}$$

In the special case of the inverse-fifth-power force law treated by Maxwell $v_0 q(v_0)$ is a constant and the relative velocity v_0 disappears from the integral. Nothing remains but

$$\iint f_1 f_2(\xi_2 - \xi_1) \, d\tau_1 \, d\tau_2 = N_1 N_2 u_2{}^* = N_1 N_2(u_2 - u_1)$$

The problem is then reduced to the calculation of the constant $v_0 q(v_0)$, which does not present any particular difficulty.

In the general case the presence of v_0 in the integral necessitates the following artifice: hold v_0 constant and associate with each velocity (ξ_2, η_2, ζ_2) only those values of (ξ_1, η_1, ζ_1) that correspond to the values of v_0 between v_0 and $v_0 + dv_0$. This domain of (ξ_1, η_1, ζ_1) depends on two parameters, and we can without difficulty perform the five integrations that correspond to variations of these two parameters and of (ξ_2, η_2, ζ_2) and reserve until last the sixth integration with respect to v_0. It is important to choose a convenient order for performing the integrations.

The discussion is simplified if we represent each velocity by a point with coordinates (ξ_1, η_1, ζ_1) or (ξ_2, η_2, ζ_2) with respect to an origin 0 (Fig. 3). Let v_2 be the point (ξ_2, η_2, ζ_2). The points (ξ_1, η_1, ζ_1) which we can associate with it are contained between two spheres with center v_2 and radii v_0 and $v_0 + dv_0$. Let r_1 and r_2 be the distances Ov_1 and Ov_2,

---

* This cross section equals the diffusion cross section $q_D = 2\pi \int_0^\infty (1 - \cos \Theta) b \, db$ divided by 4π (see Section 9-2).

which are the magnitudes of the velocities (ξ_1, η_1, ζ_1) and (ξ_2, η_2, ζ_2), respectively.

The angles α (between Ov_2 and $O\zeta$), β (between the planes v_1Ov_2 and ζOv_2), γ (between v_2v_1 and Ov_2), and the azimuth δ of v_2 with respect to

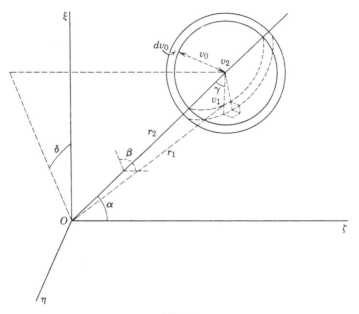

FIG. 3

$O\zeta$ comprise with r_2 the five parameters that we associate with v_0. We then write

$$d\tau_1 = v_0^2 \sin \gamma \, d\gamma \, d\beta \, dv_0$$

$$\xi_1^2 + \eta_1^2 + \zeta_1^2 = r_1^2 = r_2^2 + v_0^2 - 2r_2v_0 \cos \gamma$$

and

$$\xi_2 - \xi_1 = v_0 \cos \gamma \frac{\xi_2}{r_2} + v_0 \sin \gamma \cos \beta \left(1 - \frac{\xi_2^2}{r_2^2}\right)^{1/2}$$

Then

$$\frac{\mathscr{B}(m_1\xi_1)}{4\pi M_r} = \iint (\xi_2 - \xi_1)v_0 q(v_0) f_1 f_2 \, d\tau_1 \, d\tau_2$$

$$= \int f_2 \, d\tau_2 \int v_0^3 q(v_0)(\xi_2 - \xi_1)c_1$$

$$\times \exp \left[-hm_1(r_2^2 + v_0^2 - 2r_2v_0 \cos \gamma)\right] \sin \gamma \, d\gamma \, d\beta \, dv_0$$

Replacing $(\xi_2 - \xi_1)$ by its value given above and noting that the term in $\cos \beta$ disappears, we have

$$\frac{\mathscr{B}(m_1\xi_1)}{4\pi M_r} = 2\pi c_1 \int f_2 \frac{\xi_2}{r_2} \, d\tau_2 \int v_0{}^4 q(v_0) e^{-hm_1(r_2^2 + v_0^2)} \, dv_0$$

$$\times \int e^{2hm_1 r_2 v_0 \cos \gamma} \sin \gamma \cos \gamma \, d\gamma$$

$$= \frac{\pi c_1}{2h^2 m_1{}^2} \int v_0^2 q(v_0) \, dv_0 \int \frac{\xi_2}{r_2^3} [(2hm_1 r_2 v_0 - 1)e^{-hm_1(r_2 - v_0)^2}$$

$$+ (2hm_1 r_2 v_0 + 1)e^{-hm_1(r_2+v_0)^2}] f_2 \, d\tau_2$$

Now write

$$\xi_2 = r_2 \cos \alpha$$

and

$$d\tau_2 = 2\pi r_2{}^2 \cos \alpha \, dr_2 \, d\alpha$$

where we have integrated over δ between 0 and 2π.
Also write

$$f_2 = c_2 \exp -hm_2[r_2^2 - 2r_2 u_2{}^* \cos \alpha + u_2^{*2}]$$

Then

$$\frac{\mathscr{B}(m_1\xi_1)}{4\pi M_r} = \frac{\pi^2 c_1 c_2}{h^2 m_1^2} \int_0^\infty v_0{}^2 q(v_0) \, dv_0$$

$$\times \int_0^\infty [(2hm_1 r_2 v_0 + 1) \, e^{-hm_1(r_2+v_0)^2} + (2hm_1 r_2 v_0 - 1)e^{-hm_1(r_2-v_0)^2}]$$

$$\times e^{-hm_2(r_2^2 + u_2^{*2})} \, dr_2 \int_0^\pi e^{2hm_2 r_2 u_2{}^* \cos \alpha} \cos \alpha \sin \alpha \, d\alpha$$

or

$$\frac{\mathscr{B}(m_1\xi_1)}{4\pi M_r} = \frac{\pi^2 c_1 c_2}{4h^4 m_1{}^2 m_2{}^2 u_2{}^2} \int_0^\infty v_0^2 q(v_0) \, dv_0$$

$$\times \int_0^\infty [(2hm_1 r_2 v_0 + 1)e^{-hm_1(r_2+v_0)^2} + (2hm_1 r_2 v_0 - 1)e^{-hm_1(r_2-v_0)^2}]$$

$$\times [(2hm_2 r_2 u_2{}^* + 1)e^{-hm_2(r_2+u_2{}^*)^2} + (2hm_2 r_2 u_2{}^* - 1)$$

$$\times e^{-hm_2(r_2-u_2{}^*)^2}] \frac{dr_2}{r_2{}^2}$$

The integration with respect to r_2 is effected by developing the product in the parentheses, utilizing the formula

$$\int_{-\infty}^\infty e^{-(ax^2 + 2bx + c)} \, dx = \frac{e^{(b^2/a)-c}}{\sqrt{a}} \int_{-\infty}^\infty e^{-x^2} \, dx = \sqrt{\pi} \, \frac{e^{(b^2/a)-c}}{\sqrt{a}}$$

and noting that

$$\int_{-\infty}^{\infty} e^{-(ax^2+2bx+c)} \frac{dx}{x^2} = -2b \int_{-\infty}^{\infty} e^{-(ax^2+2bx+c)} \frac{dx}{x} - 2a \int_{-\infty}^{\infty} e^{-(ax^2+2bx+c)} dx$$

To finish the calculation, it remains after this fivefold integration to replace c_1 and c_2 by their values given in (3). Then

$$\mathscr{B}(m_1\xi_1) = 2N_1N_2 \left(\frac{\pi m_1 m_2}{h(m_1+m_2)} \right)^{1/2} \frac{1}{u_2^{*2}}$$

$$\times \int_0^{\infty} \left\{ \left(2h \frac{m_1 m_2}{m_1+m_2} v_0 u_2^* + 1 \right) \exp \left[\frac{-hm_1 m_2}{m_1+m_2} (v_0 + u_2^*)^2 \right] \right.$$

$$\left. + \left(\frac{2hm_1 m_2}{m_1+m_2} v_0 u_2^* - 1 \right) \exp \left[- \frac{hm_1 m_2}{m_1+m_2} (v_0 - u_2^*)^2 \right] \right\} v_0^2 q(v_0) \, dv_0$$

Putting

$$z = v_0 \left(\frac{hm_1 m_2}{m_1+m_2} \right)^{1/2} = \left(\frac{M_r v_0^2}{2kT} \right)^{1/2}$$

and

$$\varepsilon = u_2^* \left(\frac{hm_1 m_2}{m_1+m_2} \right)^{1/2} = \left(\frac{M_r u_2^{*2}}{2kT} \right)^{1/2} \tag{18}$$

we have

$$\mathscr{B}(m_1\xi_1) = 2N_1N_2 \sqrt{2\pi M_r kT} \, u_2^* \int_0^{\infty} q(v_0) e^{-(z^2+\varepsilon^2)}$$

$$\times [(2\varepsilon z - 1)e^{2\varepsilon z} + (2\varepsilon z + 1)e^{-2\varepsilon z}] \frac{z^2 \, dz}{\varepsilon^3}$$

It is easy to show that εz is always a very small quantity except for very large values of the relative velocity v_0. These values do not interest us because the number of molecules with such velocities is small due to the fact that the exponential term in the Maxwell distribution decreases so rapidly above the mean square velocity. We have

$$\varepsilon z = v_0 u_2^* \frac{M_r}{2kT}$$

The mean square value of v_0 is such that

$$\tfrac{1}{2} M_r \overline{v_0^2} = \frac{3kT}{2} \quad \text{and} \quad \varepsilon z = u_2^* \frac{3v_0}{2\overline{v_0^2}}$$

It would be necessary that v_0 attain the enormous value $v_0 = 2\overline{v_0^2}/3u_2^*$ in order that εz cease to be extremely small. Therefore we may replace the function of εz by its expansion in a series cut off at its first term and write finally

$$\mathscr{B}(m_1\xi_1) = \tfrac{3}{3}^2 N_1 N_2 \sqrt{2\pi M_r kT} \, u_2{}^* \int q(v_0) e^{-z^2} z^5 \, dz \qquad (19)$$

(Here we take account of the fact that ε^2 is very small in comparison with z^2.)

Equation 19 properly gives for the exchange of momentum between the molecules of the two gases an expression proportional to the relative velocity of mass motion, $u_2{}^*$, as it should when $u_2{}^*$ is small in relation to the mean molecular agitation velocity. It remains to perform a single integration which ordinarily requires a graphical calculation similar to that Maxwell performed in the case of the inverse-fifth-power repulsive force.

Before passing to the applications of this general formula, let us first verify that it leads exactly to the result obtained by Maxwell for the case of the inverse-fifth-power force. To recover Maxwell's result, assume a repulsive force inversely proportional to the $(n + 1)$th power of the separation distance:

$$f(r) = \frac{c}{r^{n+1}}, \quad \text{and} \quad V = \frac{c}{nr^n}$$

(Here $n = 4$). Then if we put

$$\alpha = \left(\frac{M_r v_0{}^2}{c/b^n}\right)^{1/n}$$

(13) gives the angle Φ as a function of α alone:

$$\Phi = \int_0^{\rho_0} \frac{d\rho}{\sqrt{1 - \rho^2 - (2/n)(\rho/\alpha)^n}} = \Phi(\alpha)$$

and we have

$$q(v_0) = \int_0^\infty \cos^2 \Phi b \, db = \left(\frac{c}{M_r v_0{}^2}\right)^{2/n} \int_0^\infty \cos^2 \Phi \alpha \, d\alpha$$

The integral remaining is a numerical constant, calculated graphically by Maxwell for the case $n = 4$ and called A_1 when multiplied by 4π:

$$q(v_0) = \left(\frac{c}{M_r v_0{}^2}\right)^{2/n} \frac{A_1}{4\pi}$$

We proceed here, in the same manner as Maxwell, to calculate this constant for arbitrary n. Maxwell[3] put

$$A_3 = A_1 \left[\frac{c}{m_1 m_2 (m_1 + m_2)}\right]^{2/n}$$

from which

$$q(v_0) = \left(\frac{m_1 + m_2}{v_0}\right)^{4/n} \frac{A_3}{4\pi} = \frac{A_3}{4\pi}\left[\frac{m_1 m_2(m_1 + m_2)}{2kT}\right]^{2/n} z^{-4/n}$$

Substituting in the general formula (19), we obtain

$$\mathscr{B}(m_1\xi_1) = \frac{8A_3}{3\pi} N_1 N_2 \sqrt{2\pi M_r kT}\left[\frac{m_1 m_2(m_1 + m_2)}{2kT}\right]^{2/n} u_2{}^* \int_0^\infty e^{-z^2} z^{(5-4/n)}\, dz$$

Note that this expression contains T to the power $(\frac{1}{2} - 2/n)$ and that if the gases are maintained under constant pressure as the temperature varies N_1 and N_2 vary inversely with the absolute temperature. The value of $\mathscr{B}(m_1\xi_1)$ is thus proportional to $T^{-(3/2+2/n)}$. We use this result later.

If we put $n = 4$ to recover Maxwell's result, the equation for $\mathscr{B}(m_1\xi_1)$ becomes

$$\mathscr{B}(m_1\xi_1) = \frac{8A_3}{3\sqrt{\pi}} \rho_1 \rho_2 u_2{}^* \int_0^\infty e^{-z^2} z^4\, dz$$

Now

$$\int_0^\infty e^{-z^2} z^4\, dz = \frac{3\sqrt{\pi}}{8}$$

so

$$\mathscr{B}(m_1\xi_1) = A_3 \rho_1 \rho_2 u_2{}^* = A_3 \rho_1 \rho_2 (u_2 - u_1)$$

which is exactly the result obtained by Maxwell.

3. THE MUTUAL DIFFUSION COEFFICIENT FOR THE ELASTIC SPHERE MODEL

Let us compare the predictions of the mean free path and momentum transfer methods for the mutual diffusion of two gases whose molecules are assumed to be elastic spheres which interact only at the instant of impact. Using the mean-free-path method, Boltzmann[4] obtained for the diffusion of molecules m_1 in gas m_2 the expression

$$(\mathscr{D}_{12})_0 = \frac{2}{3\pi D_{12}{}^2 N_2}\left(\frac{2kT}{\pi(m_1 + m_2)}\right)^{1/2} \tag{20}$$

where the concentration of m_1 is assumed to be negligible in comparison with that of m_2.

We now apply the momentum transfer method to this problem. We assume that no external forces are applied and that the mass motion is slow enough to allow neglect of the acceleration term in the equation of motion (9), which then becomes

$$\frac{\partial p_1}{\partial x} = \mathscr{B}(m_1\xi_1)$$

To calculate the cross section $q(v_0)$ for the elastic sphere model, put

$$\Phi = \sin^{-1} \frac{b}{D_{12}}$$

Then

$$q(v_0) = \int_0^\infty \cos^2 \Phi b \, db = \int_0^x \left(1 - \frac{b^2}{D_{12}^2}\right) b \, db = \frac{D_{12}^2}{4}$$

and

$$\mathscr{B}(m_1\xi_1) = {}^{32}_3 N_1 N_2 \sqrt{2\pi M_r kT} \, (u_2 - u_1) \frac{D_{12}^2}{4} \int_0^\infty e^{-z^2} z^5 \, dz$$

or

$$\mathscr{B}(m_1\xi_1) = \tfrac{8}{3} N_1 N_2 \, D_{12}^2 \sqrt{2\pi M_r kT} \, (u_2 - u_1) = A N_1 N_2 (u_2 - u_1)$$

Then we have for the diffusion equation:

$$u_1 - u_2 = -\frac{1}{A N_1 N_2} \frac{\partial p_1}{\partial x}$$

But from (6) $p_1 = N_1 kT$, and

$$u_1 - u_2 = -\frac{kT}{A N_2} \frac{1}{p_1} \frac{\partial p_1}{\partial x}$$

When compared to the equation defining the diffusion coefficient $\mathscr{D}_{12}$,

$$u_1 - u_2 = -\frac{\mathscr{D}_{12}}{p_1} \frac{\partial p_1}{\partial x}$$

this gives

$$\mathscr{D}_{12} = \frac{kT}{A N_2} = \frac{3}{16 D_{12}^2 N_2} \left(\frac{2kT}{\pi M_r}\right)^{1/2} \tag{21}$$

Except for the numerical coefficient, which is of little importance, (21) differs from (20) principally in the substitution of M_r for $(m_1 + m_2)$. The difference is most pronounced when the masses m_1 and m_2 are very different, for the diffusion coefficient furnished by the momentum transfer method is much larger. We have, in fact,

$$\frac{\mathscr{D}_{12}}{(\mathscr{D}_{12})_0} = \frac{9\pi}{32}\left(x + \frac{1}{x}\right) \quad \text{where} \quad x^2 = \frac{m_1}{m_2}$$

The minimum of this ratio corresponds to $x = 1$, that is, to equality between the masses m_1 and m_2. The minimum value is $\mathscr{D}_{12}/(\mathscr{D}_{12})_0 = 9\pi/16 = 1.767$. Thus the coefficient furnished by the mean-free-path method is much too small, and the difference is increased when x departs from 1 in either direction, increasing indefinitely with the difference between the masses m_1 and m_2.

A formula similar to (21) can be deduced from the results obtained by Maxwell[5] in one of his early papers on the kinetic theory, where for the first time the dynamical conditions of the collision were introduced to complete the purely statistical arguments of the method of free paths. The formula to which these results lead can be written, in the notation used here,

$$\mathscr{B}(m_1\xi_1) = 2N_1N_2D_{12}^2\sqrt{2\pi M_r kT}\,(u_2 - u_1)$$

From this we get the diffusion coefficient

$$\mathscr{D}_{12} = \frac{1}{4D_{12}^2N_2}\left(\frac{2kT}{\pi M_r}\right)^{1/2}$$

which differs from the exact value only in the ratio $\frac{4}{3}$. This numerical difference is due to Maxwell's assumption that the velocity is the same for all molecules of the same species. It is evident that the argument used by Maxwell is rigorous, since it makes use of the dynamical conditions of the collision and leads to the correct result if we take account of the distribution of velocities, as we have done for an arbitrary force law.

4. THE INFLUENCE OF TEMPERATURE

Equation 21 indicates proportionality to $T^{3/2}$ for the variation of the diffusion coefficient with temperature at constant pressure. This is the same kind of variation as in a law of force inversely proportional to a very high power of the distance. We have, in fact, seen that in a force inversely proportional to the $(n + 1)$th power of the distance the quantity $\mathscr{B}(m_1\xi_1)$ varies for a constant total pressure of the gas mixture as $T^{-(3/2+2/n)}$, that is, that $\mathscr{D}_{12}$, being inversely proportional to $\mathscr{B}$, varies as $T^{(3/2+2/n)}$, which gives $T^{3/2}$ for very large n. For the fifth-power law $n = 4$, and we find, with Maxwell, proportionality to T^2.

The method of integration which has permitted this solution of the problem of the mutual diffusion of two gases does not appear to be applicable to the calculation of the viscosity or the thermal conductivity of a gas. The difference is that in diffusion the departures from the Maxwellian velocity distribution are not essentially important. On the contrary, these differences play a significant role in the other phenomena.

5. CALCULATION OF THE MOBILITY

We have demonstrated a formula generalizing the results of the dynamical method introduced by Maxwell in the kinetic theory of gases and applied it to the particularly simple case in which the molecules repel each other according to the inverse-fifth power of the distance. We have shown that it is possible, whatever the interaction, to calculate the exchange of momentum in a mixture of two gases due to molecular collisions by the formula

$$\mathscr{B}(m_1\xi_1) = \tfrac{3}{3}^2 N_1 N_2 \sqrt{2\pi M_r kT}\,(u_2 - u_1)\int_0^\infty q(v_0)\,e^{-z^2}z^5\,dz \qquad (19)$$

Let us now apply this formula to the calculation of the mobility of an ion of finite size moving through a gas whose molecules are attracted toward the ion because of polarization forces. In those cases in which the number of ions is extremely small in comparison with the number of neutral molecules, there is no reason to consider mutual collisions between the ions in calculating their mobility.

If K represents the dielectric constant of the gas m_1, at pressure p_1, containing N_1 molecules per unit volume, the attractive force on a molecule by an ion of charge e at distance r is approximately

$$f = \frac{K-1}{2\pi N_1}\frac{e^2}{r^5}$$

and corresponds to a potential energy

$$V = -\frac{K-1}{8\pi N_1}\frac{e^2}{r^4}$$

If we ascribe to the ion of mass m_2 a finite size, so that the sum of its radius and that of a molecule is D_{12}, we must consider the curvature in the trajectory due to the attraction as well as the deflection that is produced at the instant of impact.

It is necessary to calculate $\mathscr{B}(m_1\xi_1)$ in order to obtain the mobility of the ions. If we neglect the effects of acceleration and diffusion, (10) becomes

$$\rho_2 X_2 = \mathscr{B}(m_1\xi_1) = \tfrac{3}{3}^2 N_1 N_2 \sqrt{2\pi M_r kT}\,(u_2 - u_1)\int_0^\infty q(v_0)e^{-z^2}z^5\,dz$$

where $\rho_2 X_2$ is the external force acting on the N_2 ions contained in unit volume of the gas. If E is the electric field intensity and e is the ionic charge,

$$\rho_2 X_2 = N_2 eE$$

Then, $(u_2 - u_1)$ being the relative velocity of the ions with respect to the gas, the mobility $\mathscr{K}$ of the ions is given by

$$\mathscr{K} = \frac{u_2 - u_1}{E} = \frac{e}{A}$$

where

$$A = \tfrac{3}{3}^2 N_1 \sqrt{2\pi M_r kT} \int_0^\infty q(v_0) e^{-z^2} z^5 \, dz$$

Now

$$q(v_0) = \int_0^\infty \cos^2 \Phi b \, db$$

where the angle Φ is given by (13):

$$\Phi = \int_0^{\rho^*} \frac{d\rho}{\sqrt{1 - \rho^2 + (2V/M_r v_0^2)}}$$

The smallest positive root ρ_0 of the quantity under the radical (i.e., the outermost zero of the denominator) is equal to ρ^* if an elastic impact does not take place, whereas ρ^* assumes the value b/D_{12} in the contrary case, since r, the distance between centers, cannot be less than D_{12}.

Now

$$\frac{2V}{M_r v_0^2} = \frac{(K-1)}{8\pi N_1 kT} \frac{e^2 \rho_1^4}{z^2 b^4} = \frac{K-1}{8\pi p_1} \frac{e^2 \rho_1^4}{z^2 b^4}$$

and, putting

$$\mu^2 = \frac{K-1}{8\pi p_1} \frac{e^2}{D_{12}^4} \quad \text{and} \quad b^2 = \frac{2\mu D_{12}^2}{z} \beta^2 = \frac{2}{z}\left[\frac{(K-1)e^2}{8\pi p_1}\right]^{1/2} \beta^2 \quad (22)$$

we obtain

$$\Phi = \int_0^{\rho^*} \frac{d\rho}{\sqrt{1 - \rho^2 + (\rho^4/4\beta^4)}}$$

There are two cases to distinguish, according to whether or not an elastic impact takes place, that is according to the value of ρ_0, the root of the radical, with respect to

$$\frac{b}{D_{12}} = \beta\left(\frac{2\mu}{z}\right)^{1/2} \quad (23)$$

Changing the variable ρ by putting

$$\rho = y\sqrt{2\beta^2}$$

we get

$$\Phi = \sqrt{2\beta^2} \int_0^{y^*} \frac{dy}{\sqrt{1 - 2\beta^2 y^2 + y^4}}$$

y^* being equal to the smallest root of the radical if it exists and is less than $\sqrt{\mu/z}$; y^* being equal to $\sqrt{\mu/z}$ in the contrary case, which corresponds to an elastic impact.

We have for the calculation of the integral Φ two different methods, according to whether the quantity under the radical has real or imaginary roots, that is, whether $\beta > 1$ or $\beta < 1$. In the first case the calculation of Φ gives elliptic functions and can be carried out by using tables that give values of

$$F_\psi(\varphi) = \int_0^\varphi \frac{d\varphi}{\sqrt{1 - \sin^2 \psi \sin^2 \varphi}}$$

for all values of φ and ψ, in particular the values of the complete function

$$F_\psi' = \int_0^{\pi/2} \frac{d\varphi}{\sqrt{1 - \sin^2 \psi \sin^2 \varphi}}$$

(1) If an elastic impact does not take place, we shall have, on putting $2\beta^2 = \sin \psi + (1/\sin \psi)$ and $y = \sqrt{\sin \psi \sin \varphi}$,

$$\Phi = \sqrt{1 + \sin^2 \psi} \int_0^{\pi/2} \frac{d\varphi}{\sqrt{1 - \sin^2 \psi \sin^2 \varphi}} = \sqrt{1 + \sin^2 \psi} \, F_\psi'$$

(2) If an elastic impact does take place, the limiting value φ^* of φ is given by

$$\sqrt{\sin \psi} \sin \varphi^* = y^* = \left(\frac{\mu}{z}\right)^{1/2} \quad \text{or} \quad \varphi^* = \sin^{-1}\left(\frac{\mu}{z \sin \psi}\right)^{1/2}$$

Then

$$\Phi = \sqrt{1 + \sin^2 \psi} \int_0^{\sin^{-1}\sqrt{\mu/z\sin\psi}} \frac{d\varphi}{\sqrt{1 - \sin^2 \psi \sin^2 \varphi}}$$

$$= \sqrt{1 + \sin^2 \psi} \, F_\psi\left[\sin^{-1}\left(\frac{\mu}{z \sin \psi}\right)^{1/2}\right]$$

In the second case, if the roots y_0 are imaginary, an elastic impact will always take place, for otherwise the ion and molecule draw indefinitely closer together if they are considered reduced in size to points located at their centers. We always have

$$\Phi = \sqrt{2\beta^2} \int_0^{\sqrt{\mu/z}} \frac{dy}{\sqrt{1 - 2\beta^2 y^2 + y^4}}$$

The values for the integral have been calculated for various values of β between 0 and 1 and $\sqrt{\mu/z}$ between the same limits. When $\sqrt{\mu/z}$ is greater than 1, we may utilize the result of the same calculation, since it is easy to verify that

$$\int_0^{\sqrt{\mu/z}} \frac{dy}{\sqrt{1 - 2\beta^2 y^2 + y^4}} = 2 \int_0^1 \frac{dy}{\sqrt{1 - 2\beta^2 y^2 + y^4}} - \int_0^{\sqrt{z/\mu}} \frac{dy}{\sqrt{1 - 2\beta^2 y^2 + y^4}}$$

$\sqrt{z/\mu}$ being less than unity when $\sqrt{\mu/z}$ is larger than unity.

104

For real roots the occurrence of an elastic impact depends on the value of $\sqrt{\mu/z}$ in relation to the smallest root of the equation

$$1 - 2\beta^2 y^2 + y^4 = 0$$

which is

$$y_0 = \sqrt{\sin \psi}$$

This quantity is always less than one. Therefore an elastic impact will occur for all values of ψ if $\mu/z > 1$. On the contrary, if $\mu/z < 1$ and we put $\mu/z = \sin \varepsilon$, elastic impacts will not take place if $\psi < \varepsilon$ but will occur if $\psi > \varepsilon$.

We then obtain the following table of calculations:

$$\frac{\mu}{z} = \sin \varepsilon < 1 \left\{ \begin{array}{l} \beta < 1, \quad 2\Phi = \sqrt{2\beta^2} \displaystyle\int_0^{\sqrt{\mu/z}} \frac{dy}{\sqrt{1 - 2\beta^2 y^2 + y^4}} \\[2ex] \beta = \sqrt{\tfrac{1}{2}}\left(\sin \psi + \dfrac{1}{\sin \psi}\right) > 1 \\[2ex] \text{where} \left\{ \begin{array}{l} \psi < \varepsilon, \quad \Phi = \sqrt{1 + \sin^2 \psi}\, F_\psi{}' \\[1.5ex] \psi > \varepsilon, \quad \Phi = \sqrt{1 + \sin^2 \psi}\, F_\psi\left[\sin^{-1}\left(\dfrac{\sin \varepsilon}{\sin \psi}\right)^{1/2}\right] \end{array} \right. \end{array} \right.$$

$$\frac{\mu}{z} > 1 \left\{ \begin{array}{l} \beta < 1, \quad 2\Phi = \sqrt{2\beta^2}\left(2\displaystyle\int_0^1 \frac{dy}{\sqrt{1 - 2\beta^2 y^2 + y^4}} + \displaystyle\int_0^{\sqrt{z/\mu}} \right. \\[3ex] \hspace{6em} \left. \times \frac{dy}{\sqrt{1 - 2\beta^2 y^2 + y^4}}\right) \\[2ex] \beta = \sqrt{\tfrac{1}{2}[\sin \psi + (1/\sin \psi)]} > 1, \quad \Phi = \sqrt{1 + \sin^2 \psi}\, F_\psi{}' \end{array} \right.$$

For an ion and molecule of given dimensions D_{12} is determined, and μ follows from (22):

$$\mu^2 = \frac{K - 1}{8\pi p_1} \frac{e^2}{D_{12}^4}$$

Under these conditions, a value of z from (18) corresponds to each value of the relative velocity v_0:

$$z = \left(\frac{M_r v_0{}^2}{2kT}\right)^{1/2}$$

Similarly, to each value of v_0 there corresponds a value of μ/z.

Then, varying β from 0 to ∞, we have values of Φ for the various

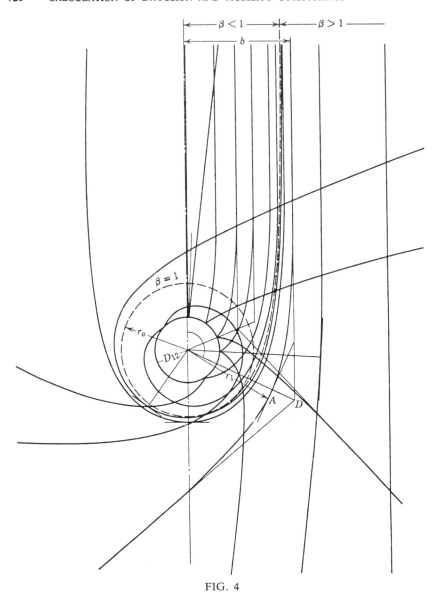

FIG. 4

trajectories that correspond to the same value of the relative velocity. Figures 4, 5, and 6 give the form of these trajectories for very different values of μ/z.

Take first of all a small value for the relative velocity, such as $\sqrt{\mu/z} = 2$, which corresponds to a reduced relative velocity $\mu/z = 4$. The attractions then play an important role, and the relative trajectories are strongly

106

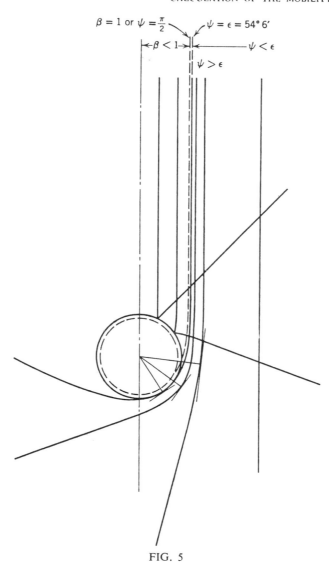

FIG. 5

curved inward, as in Fig. 4. This figure shows curves for various values of β, each of which corresponds, according to (23), to a reduced initial distance

$$\frac{b}{D_{12}} = \beta \left(\frac{2\mu}{z}\right)^{1/2}$$

Now assume that the ion is stationary. The relative trajectory gives the motion of the molecule when it is considered to be reduced to point

107

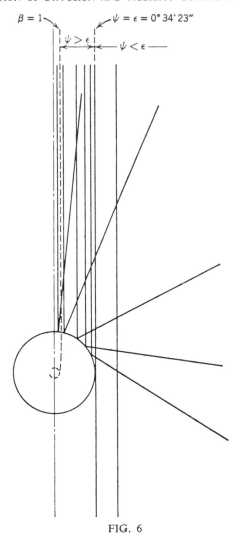

FIG. 6

size and the ion is assumed to have for its radius the sum D_{12} of the real radii. The circle of radius D_{12} is drawn with a full line.

For a sufficiently large β an elastic impact does not take place—the trajectory is simply curved inward and consists of two symmetrical portions separated by the apse A whose position is determined by the value Φ given in the preceding table and by the corresponding value r_1 of the distance to the center of attraction:

$$\frac{r_1}{D_{12}} = \left(\frac{\mu}{z \sin \psi}\right)^{1/2} .$$

108

For a value of $\mu/z > 1$ an elastic impact does not take place for any value of β greater than unity.

For $\beta = 1$ or $\psi = \pi/2$ the molecule revolves about the ion following a dotted circle of radius r_0, where

$$\frac{r_0}{D_{12}} = \left(\frac{\mu}{z}\right)^{1/2}$$

The trajectories corresponding to $\beta > 1$ lie entirely outside this circle.

For $\beta < 1$ the molecule penetrates the interior of the circle and, if an elastic impact does not occur, draws indefinitely nearer the center of attraction. An elastic impact produces a deflection that gives to the trajectory a second branch symmetrical with the first about the radius passing through the point of impact.

As the relative velocity increases, μ/z decreases, the trajectories are less curved, and elastic impacts become more important than the attraction from the point of view of exchange of momentum. Figure 5 corresponds to $\mu/z = 0.9$. The dotted circle penetrates inside the circle of radius D_{12}, so that a molecule may undergo an elastic impact even when, in the absence of this impact, it does not approach indefinitely close to the center of attraction. As indicated in the table, this requires the decomposition of the variation of β into three regions: the region $\beta < 1$, for which an elastic impact takes place in all cases, then a second region ($\psi > \varepsilon$), in which an elastic impact takes place before the molecule attains the apse, and finally a third ($\psi < \varepsilon$), in which an elastic impact does not occur.

Finally, if the velocity becomes very large, as in Fig. 6, in which $\mu/z = 0.01$ and $\sqrt{\mu/z} = 0.1$, elastic impacts play the essential role and the curvature of the trajectories due to the attractive force is no longer significant.

To obtain the total momentum exchange between the two gases, it is necessary to calculate the quantity $q(v_0)$ for each velocity. From (16) we have

$$q(v_0) = \int \cos^2 \Phi b \, db = \frac{\mu D_{12}^2}{2} \int \cos^2 \Phi \, d\beta^2$$

For each value of μ/z we calculate the values of Φ for all values of β and construct a curve with β^2 as the abscissa and $\cos^2 \Phi$ as the ordinate. The area under this curve is $y = \int \cos^2 \Phi \, d\beta^2$, the area being measured by a graphical procedure.

In drawing a similar curve for different values of μ/z, we have y as a function of μ/z or of z if μ is given by

$$q(v_0) = \mu D_{12}^2 \frac{Y}{z}$$

109

Then

$$A = \tfrac{3}{3}\tfrac{2}{}N_1\sqrt{2\pi M_r kT}\mu D_{12}^2 \int_0^\infty ye^{-z^2}z^4\,dz \tag{24}$$

and putting

$$Y = f(\mu) = \int_0^\infty ye^{-z^2}z^4\,dz \tag{25}$$

we finally get the mobility

$$\mathcal{K} = \frac{e}{A} = \frac{3}{16Y\sqrt{(K-1)\rho_1}}\left(\frac{m_1+m_2}{m_2}\right)^{1/2} \tag{26}$$

To each value of μ, that is, to each size of the ion since

$$\mu = \left[\frac{(K-1)}{8\pi p_1}\frac{e^2}{D_{12}^4}\right]^{1/2}$$

there corresponds a value of Y and, consequently, of $\mathcal{K}$. The graph in Fig. 7 represents the results of all these calculations, that is, it gives $3/16\,Y$ on the ordinate and

$$\frac{1}{\mu} = \left[\frac{8\pi p_1}{(K-1)e^2}\right]^{1/2}D_{12}^2 \tag{27}$$

on the abscissa.

In order to verify the results of this calculation, note that, for small values of μ, μ/z is very small throughout the range of interesting values of z; that is, impacts play the essential role, for the polarization attraction

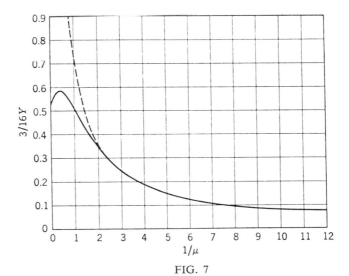

FIG. 7

becomes very weak. In this case we have, as for pure elastic impacts,

$$\int \cos^2 \Phi b \, db = \frac{D_{12}^2}{4}$$

Then

$$Y = \int \cos^2 \Phi \, d\beta^2 = \frac{z}{4\mu}$$

so

$$Y = \int_0^\infty y e^{-z^2} z^4 \, dz = \frac{1}{4\mu} \int_0^\infty e^{-z^2} z^5 \, dz = \frac{1}{4\mu}$$

Finally

$$\frac{3}{16Y} = \frac{3\mu}{4} = \frac{3}{4(1/\mu)} \tag{28}$$

On the contrary, when μ is very large, Y then takes a limiting value Y_p which corresponds to the ordinate of the origin of the full curve in Fig. 7. Then $3/16Y_p = 0.505$. This extreme case corresponds to a negligible influence of elastic impacts, that is, to motion in the gas of a charged particle of extremely small dimensions whose drift is impeded principally by its attraction for the molecules. We have for the corresponding mobility

$$\mathcal{K}_p = \frac{0.505}{\sqrt{(K-1)\rho_1}} \left(\frac{m_1 + m_2}{m_2}\right)^{\frac{1}{2}} \tag{29}$$

This expression for the limit of the mobility of the ions when the polarization attraction plays the essential role does not contain the charge e of the ion. This is because the force on the molecules which tends to retard the motion makes $\mathcal{B}(m_1\xi_1)$ proportional to the charge, whereas the motive force in an electric field is itself proportional to the charge.

Let us now seek to deduce from our results the probable size of the ions. Denote by x the unknown ratio of the diameter of an ion to that of a molecule D_1. We have

$$D_{12} \fallingdotseq D_1 \frac{x+1}{2} \quad \text{and} \quad \mu = \frac{e}{D_1^2}\left[\frac{2(K-1)}{\pi P}\right]^{\frac{1}{2}} \frac{1}{(x+1)^2}$$

and μ, and consequently $3/16Y$, can be calculated as a function of x. Since

$$\frac{m_1 + m_2}{m_2} = 1 + \frac{1}{x^3}$$

$$\mathcal{K} = \frac{3}{16Y\sqrt{(K-1)\rho}} \left(\frac{m_1 + m_2}{m_2}\right)^{\frac{1}{2}} = \frac{3}{16Y\sqrt{(K-1)\rho}} \left(1 + \frac{1}{x^3}\right)^{\frac{1}{2}}$$

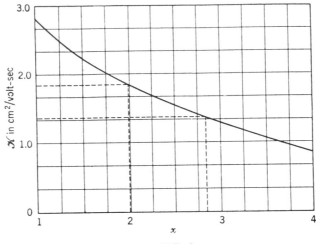

FIG. 8

Thus we have $\mathscr{K}$ as a function of x. We can then draw a curve showing x as the abscissa and the theoretical value of $\mathscr{K}$ as the ordinate (Fig. 8).

REFERENCES

1. R. D. Present, *Kinetic Theory of Gases*, McGraw-Hill, New York, 1958, Chapters 8 and 11.
2. L. Boltzmann, *Vorlesungen über Gastheorie*, Vol. 1, J. A. Barth, Leipzig, 1896, p. 119.
3. L. Boltzmann, *op. cit.*, p. 197.
4. L. Boltzmann, *op. cit.*, p. 96.
5. J. C. Maxwell, *Phil. Mag.* **19,** 19 (1860); **20,** ?1 (1860).

11

Reprinted from *J. Chem. Phys.* **29**:294–299 (1958)

Reactions of Gaseous Molecule Ions with Gaseous Molecules. V. Theory

George Gioumousis and D. P. Stevenson

Shell Development Company, Emeryville, California

(Received November 29, 1957)

Ion-molecule reactions of the sort observed as secondary reactions in mass spectrometers have been treated by the methods of the modern kinetic theory; that is, the rate of reaction is expressed in terms of the velocity distribution functions of the reactants and the cross section for the reaction. The cross section, which is calculated by means of the properties of the classical collision orbits, is found to have an inverse square root dependence on energy. The ion distribution function, which is far from Maxwellian, is found by means of an explicit solution of the Boltzmann equation. A simple relation is given which relates the mass spectrometric data to the specific rate of the same reaction under thermal conditions. For the simpler molecules, this rate may be calculated completely *a priori*, with excellent agreement with experiment.

IN a recent communication,[1] it was reported that for a number of reactions of the form

$$X^+ + YH \rightarrow XH^+ + Y,$$

which are observed in mass spectrometers when the ion source pressure is greater than normally used in analytical work, the phenomenological reaction cross sections (defined in references 1 and 2) are inversely proportional to the square root of the ion repeller voltage and of the reduced mass μ of the reactants and are independent of the temperature. In part, these findings are in accord with the theory of such reactions developed by Eyring *et al.*,[2] that is, the Eyring theory predicts independence of temperature and inverse square root dependence on reduced mass for the reaction rate.

However, the Eyring theory, based as it is on equilibrium statistical mechanics, requires that the reactants possess a Maxwellian distribution in velocity space, while here the ions have been accelerated by the ion repeller field and so are far from Maxwellian.

Field, Franklin, and Lampe[3] recently presented a theory which takes account of the acceleration of the ions, but only approximately, by the methods of the old kinetic theory. The concept of a reaction cross section is introduced, and estimated by a classical analogy with the method of the preceding work. It is recognized that the cross section is a function of energy, but the average is taken as the value at the average energy. Such a procedure, which is characteristic of the old kinetic theory, is adequate for a sharply peaked distribution, but less so for the ionic distribution, which, as will be shown, has a peak whose position is space-dependent. Moreover, a heuristically derived efficiency factor is introduced which as will also be, shown, is unnecessary.

It is the purpose of this communication to present a theory based on the rigorous kinetic theory of gases in terms of the notion of reaction cross section. Such a theory is independent of any assumption of near-equilibrium, but rather is valid for any distribution, and in particular the rather peculiar one of interest. An accompanying paper is devoted to a description of the experimental details.[4]

In general, the rate of reaction p at a point may be expressed in terms of the velocity distributions f_1, f_2 of the two reactants and of the microscopic cross section σ (defined in the fashion usual in nuclear physics),

[1] D. P. Stevenson and D. O. Schissler, J. Chem. Phys. **23**, 1353 (1955).

[2] Eyring, Hirschfelder, and Taylor, J. Chem. Phys. **4**, 479 (1936).

[3] Field, Franklin, and Lampe, J. Am. Chem. Soc. **79**, 2419 (1957).

[4] D. P. Stevenson and D. O. Schissler, J. Chem. Phys. **29**, 282 (1958).

with result

$$p = \iint [f_1(\mathbf{v}_1)][f_2(\mathbf{v}_2)][g\sigma(g)] d\mathbf{v}_1 d\mathbf{v}_2. \quad (1)$$

Here g is the relative velocity $|\mathbf{v}_1 - \mathbf{v}_2|$.

Of the three factors in the integral above, the ionic distribution function has the least familiar form. Consider the situation in the ion source. The major constituents are neutral molecules at an extremely low pressure. A uniform electric field exists in a direction which we may take as the z axis. At right angles to this is an electron beam localized over a short distance in z. Within the electron beam molecules are ionized, then accelerated by the electric field. Thus the distribution of ions in velocity space is a function of their positions in the ion chamber. It is possible to solve the Boltzmann equation for this function subject to the assumption of zero electron beam thickness.

The neutral molecules, on the other hand, are in thermal equilibrium essentially undisturbed by the small number of electronic collisions, so that their distribution is Maxwellian, and thus presents no difficulty.

The cross sections are found by means of an unusual classical argument, based on the properties of orbits in central force fields. The treatment appears rigorous for a certain class of molecules and leads to the result that the cross section is independent of the specific chemical nature of the reactants, but rather is a function only of such classical parameters as mass, charge, and electric polarizability.

THE DISTRIBUTION FUNCTION

The variation of a molecular distribution function with time and position is given by the Boltzmann equation[5]

$$(\partial f/\partial t) + \mathbf{v} \cdot (\partial f/\partial \mathbf{r}) + (\mathbf{F}/m) \cdot (\partial f/\partial \mathbf{v}) = Q \quad (2)$$

where Q is the collisional term. In this problem the gas is so dilute that the only contribution of any consequence to Q is the primary ionization process. The electron beam in the ion chamber is thin and flat and roughly monoenergetic. Let the coordinate system be chosen in such a way that the plane $z = 0$ is within the beam, and let it be assumed that the beam is strictly planar. Furthermore, it is an excellent approximation to take the molecular velocity as unchanged by ionization, since the electron mass is so small that momentum transfer is negligible. Then the number of ions formed, per unit volume, per unit velocity range is

$$B\delta(z)(m/2\pi kT)^{3/2} \exp -mv^2/2kT,$$

where m and v refer to the ion, $\delta(z)$ is the Dirac delta function, and B is the number of ionizations per unit area on the plane.

The ion repeller field is at right angles to the electron beam so that only the z component of the field is not zero. Furthermore, since only the steady state solution is desired, the time derivative term vanishes. The distribution function f then factors into three parts, of which those for v_x and v_y are simply one-dimensional Maxwellian distributions. The z-component equation for $f(v_z, z)$,

$$v_z(\partial f/\partial z) + (eE/m)(\partial f/\partial v_z)$$
$$= B\delta(z)(m/2kT)^{\frac{1}{2}} \exp -mv_z^2/2kT \quad (3)$$

where E is the field and e the charge of the ion, remains to be solved.

The substitution

$$u = (1/kT)(\tfrac{1}{2}mv_z^2 - eEz)$$

leads to the equation

$$z(\partial f/\partial z) = +G\delta(z) \qquad \text{for } v_z \geq 0, \quad (4)$$
$$= -G\delta(z) \qquad \text{for } v_z < 0, \quad (4')$$

where

$$G = \frac{B}{\pi^{\frac{1}{2}}} \frac{m}{2kT} \frac{e^{-u}}{u} \qquad \text{for } u > 0$$
$$= 0 \qquad \text{for } u < 0.$$

In terms of z and v_z the solution is

$$f(v_z, z) = \nu(v_z, z) \frac{B}{\pi^{\frac{1}{2}}} \frac{m}{2kT} \frac{\exp[-(1/kT)(\frac{1}{2}mv_z^2 - eEz)]}{(1/kT)^{\frac{1}{2}}(\frac{1}{2}mv_z^2 - eEz)} \quad (5)$$

where ν has a value zero, one or two, as follows

$$\nu = 1 \qquad \text{for } z < 0,$$
$$= 2 \qquad \begin{cases} \text{for } z > 0, \\ v_z > 0, \\ \tfrac{1}{2}mv_z^2 - eEz > 0, \end{cases}$$
$$= 0 \qquad \text{for } v_z < 0,$$
$$= 0 \qquad \text{for } \tfrac{1}{2}mv_z^2 - eE_z < 0.$$

It is instructive to calculate the flux of ions given by this distribution. For positive z, a simple integration gives the value B independently of the value of z, which means simply that all ions formed eventually reach this region. For negative z the flux is zero. The flux in one definite direction is not zero, but rather has a maximum of $B/2$ when $z = 0$ and drops very rapidly from that value. As an illustration, the flux is $B/20$ when the dimensionless quantity $-Eez/kT$ is 1.16, and $B/200$ when it is 1.82. Thus, the number of ions remains appreciable only for a short distance back of the plane of formation.

[5] Hirschfelder, Curtiss, and Bird, *Molecular Theory of Gases and Liquids* (John Wiley and Sons, Inc., New York, 1949), p. 502.

THE CROSS SECTION

The long-range potential between ions and neutral spherical molecules is of the inverse fourth power sort, that is,

$$\phi(r) = -e^2\alpha/2r^4, \qquad (6)$$

where e is the charge of the ion and α the electric polarizability of the molecule. The orbits of such a potential have been given by Langevin[6] in connection with a calculation of the mobilities of ions in gases. Typical cases are shown in Fig. 1.

An orbit is determined by two parameters, the relative initial velocity g and the impact parameter b, which is the distance between the initial part of the orbit and a parallel line which goes through the center. For these orbits there is a critical value of b, dependent on g,

$$b_0 = (4e^2\alpha/\mu g^2)^{\frac{1}{4}}, \qquad (7)$$

such that orbits for which $b < b_0$ pass through the origin while orbits for which $b \geq b_0$ come no closer than

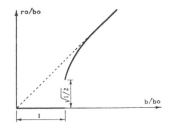

FIG. 2. Dependence of distance of closest approach r_0 upon the impact parameter b. The energy enters only through the critical impact parameter b_0, which is a scale factor.

$b_0/\sqrt{2}$. The dependence of the distance of closest approach r_0 on b is given in Fig. 2.

Suppose now that there exists a critical radius r_c such that reaction is impossible if the ion-molecule distance be greater than this, and practically certain if less. Then, so long as r_c be between zero and $b_0/\sqrt{2}$, all collisions for which $b < b_0$ must lead to reaction, and so the reaction cross section[7] is

$$\sigma(g) = \pi b_0^2 = (\pi/g)(4e^2\alpha/\mu)^{\frac{1}{2}}. \qquad (8)$$

Since b_0 varies inversely as the square root of the relative velocity, there must be some collisions at the higher velocities for which it is not true that $r_c < b_0/\sqrt{2}$, but the above expression will still be valid if these are sufficiently few. Thus the polarizability must be large and the mass of the molecule small so that b_0 may be large. Similarly the molecule should be "small," i.e., of limited extent of electron cloud, so that r_c may be small. In addition, energy may be lost to the internal degrees of freedom, so that these are best few and with widely spaced energy levels. Thus the analysis is most likely to be valid for some such reaction as that between a noble gas ion and hydrogen, or hydrogen with hydrogen.[8]

REACTION RATE

The rate of production, per unit volume, of secondary ions is given by Eq. (1), but what is desired is the rate of production integrated over the whole volume. If A be the cross-sectional area of the ion chamber at right angles to the electric field, then this rate is given by

$$P = A \int_{-\infty}^{l} \left(\iint f_1(\mathbf{v}_1) f_2(\mathbf{v}_2) g\sigma(g) d\mathbf{v}_1 d\mathbf{v}_2 \right) dz, \qquad (9)$$

where the lower limit may be taken as minus infinity to excellent approximation, and the upper limit l is the distance from the electron beam to the exit slit. From Eq. (8) it may be seen that the factor, $g\sigma(g)$ in Eq. (9), involving the cross section is independent of g,

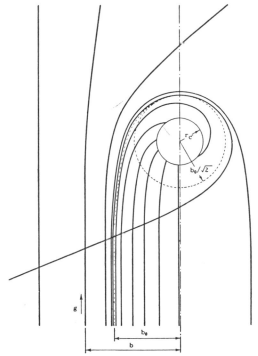

FIG. 1. A typical family of trajectories as a function of the impact parameter b. The dotted trajectory is the critical one for $b = b_0$ and approaches the circle $r = b_0/\sqrt{2}$. Only trajectories with $b < b_0$ will enter the reaction sphere if it has radius r_c less than this, i.e., $r_c < b_0/\sqrt{2}$. Thus the cross section for entrance into the reaction sphere depends on b_0 but not, within limits, r_c. On the assumption that all molecules which enter the reaction sphere do react, the same may be said of the reaction cross section.
These curves are a modified form of those given by Langevin.[6]

[6] P. Langevin, Ann. chim. phys. 5, 245 (1905).

[7] Field et al. (see reference 3) use the distance of closest approach in place of the impact parameter in calculating the cross section, with result that their value is incorrect by a factor of two.
[8] It should be recognized that the inverse square root dependence on energy of the cross section is not universal, but rather depends on the particular form of the potential chosen. For an inverse nth power potential, $\phi(r) = -\alpha r^{-n}$ (with $n > 2$), the cross section varies as $E^{-2/n}$.

so the integrations over $\mathbf{v}_1$ and $\mathbf{v}_2$ are independent. Since most of the factors in the integrand are Maxwellian distributions, the integration of these factors may be performed by inspection, leaving

$$P = n_2 A \left[2\pi e(\alpha/\mu)^{\frac{1}{2}} \right] \left[(B/\pi^{\frac{1}{2}}) \cdot (m_1/2kT) \right]$$

$$\left(\int_{-\infty}^{l} \int_{-\infty}^{\infty} \nu(v_{1z}, z) \frac{\exp[-(1/kT)(\frac{1}{2}m_1 v_{1z}^2 - eEz)]}{[(1/kT)(\frac{1}{2}m_1 v_{1z}^2 - eEz)]^{\frac{1}{2}}} \right) dv_{1z} dz,$$

(10)

where n_2 is the number density of the molecules. A change of variable

$$u = (1/kT)(\tfrac{1}{2}m_1 v_{1z}^2 - eEz)$$

(11)

and integration with respect to z yields

$$P = n_2 A B (2\pi^{\frac{1}{2}}) e(\alpha/\mu)^{\frac{1}{2}} (2m_1 kT)^{\frac{1}{2}} (l/eEl) F[(eEl/kT)^{\frac{1}{2}}],$$

(12)

where if

$$a = (eEl/kT)^{\frac{1}{2}},$$

the function $F(a)$ is given by

$$F(a) = \int_0^{\infty} (u + a^2)^{\frac{1}{2}} (e^{-u}/u)^{\frac{1}{2}} du.$$

(13)

Experimental results are given in terms of a quantity Q, the phenomenological cross section, defined by[4]

$$i(S^+) = i(P^+) n_2 l Q,$$

(14)

where $i(P^+)$ and $i(S^+)$ are the ion currents due to the primary and secondary ions, respectively. Now $i(S^+)$ is proportional to P and $i(P^+)$ to AB, so that

$$Q = (2\pi)(2m_1 \alpha e^2/\mu)^{\frac{1}{2}} \cdot (kT/eEl)^{\frac{1}{2}} \cdot F(eEl/kT)^{\frac{1}{2}}.$$

(15)

The function $F(a)$ is unity when a is zero, but for large a is asymptotically proportional to a, that is,

$$F(a) \sim a\pi^{\frac{1}{2}}$$

(16)

Since the potential energy due to the electric field eEl would always be much greater than the thermal energy kT, a would be much greater than unity, and we have

$$Q = 2\pi (2m_1 \alpha e^2/\mu)^{\frac{1}{2}} \cdot 1/(eEl)^{\frac{1}{2}}.$$

(17)

This cross section, Q, is independent of temperature and varies inversely as the square root of the reduced mass and of the ion repeller voltage.

It is evident from the above that the microscopic cross-section σ and the phenomenological cross-section Q are distinct entities dependent on different variables, and the relation between them is not of a simple form. This distinction is not clear in the previously mentioned paper of Field et al.[3]

It is very important, especially in problems involving radiation chemistry, to be in a position to compare the above cross section with the rate constant for the thermal reaction, that is, for the reaction when the ions as well as the molecules satisfy a Maxwellian distribution. Since $g\sigma(g)$ is a constant, if both f_1 and f_2 are Maxwellian the rate of production of secondary ion (3) is

$$dn_3/dt = n_1 n_2 g\sigma(g) = 2\pi n_1 n_2 (e^2 \alpha/\mu)^{\frac{1}{2}},$$

(18)

and the rate constant is[9]

$$k = 2\pi (e^2 \alpha/\mu)^{\frac{1}{2}}.$$

(19)

In terms of Q, k is then given by

$$k = (eEl/2m_1)^{\frac{1}{2}} Q.$$

(20)

This equation, it should be noted, has a wider range of validity than either Eq. (17) or Eq. (19). In other words, Eq. (20) would hold approximately so long as $g\sigma(g)$ were nearly constant, even if the value were very different from the theoretical value derived from Eq. (8). Thus, in the cases where the inverse square root law holds for the dependence of cross section on ion repeller voltage, Eq. (20) gives a rigorous interpretation of the mass spectrometric data in terms of ordinarily observable reactions. This is particularly important since many of these reactions are of the sort postulated as intermediate steps in the reactions of radiation chemistry.

One point which remains to be considered is to what extent our conclusions in the foregoing are in error because of the nonzero thickness of the electron beam. It is evident that such a distributed ion source leads to a distribution in the values of z_0. In order to set an upper limit to the error, let it be assumed that the distribution is uniform for $z_0 \pm 20\%$ and zero beyond. Then the cross section may be calculated by a simple integration, with result that it agrees with the value assuming a plane source at z_0 to within 0.5%. Since the actual distribution is likely to be peaked about z_0, and has been estimated[4] to be of lesser extent than assumed above, it would appear that the approximation of zero beam thickness is a good one.

COMPARISON WITH EXPERIMENT

The comparison with the experimental work of Stevenson and Schissler[1] is given in Table I. The first column lists the polarizabilities[10] of the neutral reactants. Those for the various isotopic hydrogens are theoretical and the remaining ones experimental. The second column contains the reduced mass of the reactants in atomic mass units; the third column the ratio of the maximum to the minimum components of the polarizability tensor, and thus gives an estimate of how gross is the error in our assumption of an iso-

[9] Except for an obvious misprint, Eq. (8c) of reference 2 is the same as Eq. (19). It appears surprising that a classical and quantum mechanical argument should agree exactly, but the Eyring treatment uses a classical approximation at a late stage, so both are really classical.

[10] Hirschfelder et al., reference 5, p. 950.

TABLE I. Comparison with experimental rate constants.

Reaction	$\alpha \cdot 10^{24}$ cm³	μ (atomic mass units)	$\dfrac{\alpha_{max}}{\alpha_{min}}$	$k \cdot 10^9$ m³/molecule sec Exptl.	$k \cdot 10^9$ m³/molecule sec Theoret.
$Ar^+ + H_2 \rightarrow ArH^+ + H$	0.7894	1.919	1.40	1.68	1.50
$Ar^+ + HD \rightarrow \begin{cases} AH^+ + D \\ AD^+ + H \end{cases}$	0.7829	2.810	1.40	1.43	1.23
$Ar^+ + D_2 \rightarrow ArD^+ + D$	0.7749	3.661	1.40	1.35	1.09
$Kr^+ + H_2 \rightarrow KrH^+ + H$	0.7894	1.969	1.40	0.48₇	1.47
$Kr^+ + D_2 \rightarrow KrD^+ + D$	0.7749	3.845	1.40	0.30₄	1.05
$Ne^+ + H_2 \rightarrow NeH^+ + H$	0.7894	1.832	1.40	0.27₄	1.53
$N_2^+ + D_2 \rightarrow N_2D^+ + D$	0.7749	3.523	1.40	1.72	1.10
$CO^+ + D_2 \rightarrow COD^+ + D$	0.7749	3.523	1.40	1.63	1.11
$O_2 + H_2^+ \rightarrow O_2H^+ + H$	1.60	1.897	1.94	7.56	2.16
$O_2 + D_2^+ \rightarrow O_2D^+ + D$	1.60	3.579	1.94	3.56	1.52
$D_2 + D_2^+ \rightarrow D_3^+ + D$	0.7749	2.015	1.40	1.43	1.45
$HCl^+ + HCl \rightarrow H_2Cl^+ + Cl$	2.63	17.994	1.31	0.43₈	0.89
$HBr^+ + HBr \rightarrow H_2Br^+ + Br$	3.61	41.001	1.27	0.22₁	0.67

tropic tensor. Finally, there are the rate constants: the first calculated from the experimental Q's via Eq. (20), and the second the theoretical value of Eq. (19).

As has been shown by Stevenson and Schissler,[1] all of these reactions satisfy the inverse square root law for the dependence of cross section on ion repeller voltage and are temperature independent. Thus the microscopic cross section must have nearly the functional form derived earlier, that is, one varying inversely with the relative velocity, though the constant factors may not be in agreement. For those reactions where the experimentally derived and calculated rate constant are in substantial agreement the inference is that the ion-induced dipole model for the determining stage of the reaction is essentially correct. Of the remaining reactions there are some for which the model is evidently incorrect. Thus, oxygen has a polarizability tensor which is far from isotropic, while hydrogen chloride, hydrogen bromide, and methanol have electric dipole moments.[11] There is another group of molecules which has been investigated, that of hydrocarbon ions and molecules[3,4] for which the phenomenological cross section is a linear function of the reciprocal voltage and for which there is a small temperature dependence. Certainly the functional form of the microscopic cross section in these reactions is not of the sort considered here.

A delicate test of whether the errors are due to some

[11] In general, one would expect the reaction cross sections for molecules with permanent electric dipole moments, to be greater than those given above, because the net effect would be an attractive force. However, the potential depends on the orientation of the molecule, so that some suitable average potential must be found. If a Boltzmann factor is used, the resulting potential has an inverse fourth power dependence on distance, and so could be included within the model we have assumed. However, the potential as so derived assumes thermal equilibrium, while here, as noted already, we are far from thermal equilibrium. The matter is speculative, and must be considered further.

factor consistently over or underestimated, say a non-unit transmission coefficient or perhaps a polarizability that must be taken as the maximum component of the tensor rather than the mean, lies in the examination of a series of isotopes. Here the ratios of the rate constants should depend only on the variation of the reduced mass and a slight variation of the polarizability. On referring to Table II one sees that the calculated ratios agree with experiment in two cases where the ions are the same, but not in one where the ions are different. This indicates that our view of an ion as a structureless point charge is overly simplified.

CONCLUSIONS

The result of greatest validity in the foregoing work is Eq. (20), the relation between the specific rate of the thermal reaction and the phenomenological cross section, as discussed in the previous section.

Essentially, the mechanism herein proposed is that the course of the reaction depends only on the long range forces between the reactants, as was postulated by Eyring, Hirschfelder, and Taylor.[2] The limitations of such a view are evident; in the reaction of HD with

TABLE II. Ratios of rate constants for isotope effect.

Reaction	k/k' Exptl.	k/k' Theoret.
$Ar^+ + H_2 \rightarrow ArH^+ + H$	1.24	1.38
$Ar^+ + HD \rightarrow \begin{cases} ArD^+ + H \\ ArH^+ + D \end{cases}$	1.06	1.13
$Ar^+ + D_2 \rightarrow ArD^+ + D$	1.00	1.00
$Kr^+ + H_2 \rightarrow KrH^+ + H$	1.60	1.40
$Kr^+ + D_2 \rightarrow KrD^+ + D$	1.00	1.00
$Kr^+ + H_2 \rightarrow KrH^+ + H$	1.78	0.96
$Ne^+ + H_2 \rightarrow NeH^+ + H$	1.00	1.00

HD$^+$, for example, it can only predict the sum of HD$_2^+$ and H$_2$D$^+$ but not the ratio. Our conclusions then are:

(i) that formation of the activated complex depends only on the long range forces,

(ii) that our method of calculation of the probability of formation is valid if certain conditions are met, to wit: no dipole moment, $r_c \leq b_0/\sqrt{2}$, etc., and

(iii) that while in some cases the activated complex always goes on to react, so that our expression for the rate constant is very close to the experimental, this need not always be so.

ACKNOWLEDGMENT

One of us (GG) wishes to acknowledge several discussions with Professor Farrington Daniels of the University of Wisconsin which served as precursors to the point of view developed herein.

12

Reprinted from pp. 481–482 of *J. Chem. Phys.* 4:479–491 (1936)

THE THEORETICAL TREATMENT OF CHEMICAL REACTIONS PRODUCED BY IONIZATION PROCESSES. PART I. THE ORTHO-PARA HYDROGEN CONVERSION BY ALPHA-PARTICLES

H. Eyring, J. O. Hirschfelder, and H. S. Taylor
Frick Chemical Laboratories, Princeton, N.J.

[*Editor's Note:* In the original, material precedes this excerpt.]

The Formation of H_3^+ by Secondary Collision

The separation, d, between the centers of gravity of H_2 and of H_2^+ at the activated state is so large, because of the nature of their interaction energy, that each of the molecules may be assumed to rotate as freely as they do when they are far apart.

It is known that the interaction of a hydrogen atom with a hydrogen molecule requires about 7 kcal. of activation energy.[9] The activation energy of $H_2 + H_2^+$ is probably small compared to this value since, at large distances, the molecule is attracted by ion-polarization forces to the ion and this attraction may persist to small distances. If it does so, an apparent activation will still arise from the rotational energy of the system as a whole. For a system rotating with a quantized angular momentum, $jh/2\pi$ the energy is increased, by the centrifugal forces, to an amount $E_{rot} = j(j+1)h^2/8\pi^2 m_H d^2$.[10] Since the ion polarization energy is $E_{pol} = -\alpha e^2/2d^4$, these two potentials superposed give an activation energy and an activated state for every value of j different from $j = 0$. The quantity α is the polarizability of the hydrogen molecule. Following Eyring,[11] we obtain

---

[9] A. Farkas, Zeits. f. physik. Chemie **B10**, 419 (1930); Geib and Harteck, ibid., Bodenstein Festband, 849 (1931).

[10] In this paper m with an appropriate subscript indicates the mass of the atom or molecule in grams, while M is the mass in atomic weight units.

[11] Eyring, J. Chem. Phys. **3**, 107 (1935); Eyring, Gershinowitz and Sun, J. Phys. Chem. **3**, 786 (1935).

for the rate of reaction

$$k = \frac{\kappa(2\pi m_{H_4}kT)^{\frac{3}{2}}h^{-3}[1+\sum_{J=1}^{\infty}(2J+1)\exp -(E_{rot}{}^* - E_{pol}{}^*)(kT)^{-1}]kTh^{-1}}{(2\pi m_{H_2}kT)^{\frac{3}{2}}h^{-3}(2\pi m_{H_2{}^+}kT)^{\frac{3}{2}}h^{-3}}.$$

In this expression, the rotational and vibration partition functions for H_2 and $H_2{}^+$ have already been cancelled in the numerator and denominator. The starred energy symbols refer to the respective energies at the activated state, for which state the expression, $j(j+1)h^2/8\pi^2 m_H d^2 - \alpha e^2/2d^4$, has a maximum value. Hence, the distance, d^*, has the value,

$$d^* = (j(j+1)h^2/8\pi^2 m_H \alpha e^2)^{-\frac{1}{2}}, \tag{5}$$

and at this distance the activation energy becomes

$$E_{rot}{}^* - E_{pol}{}^* = j^2(j+1)^2 h^4/128\pi^4 m_H{}^2\alpha e^2. \tag{6}$$

Setting $a^2 = h^4/128\pi^4 m_H{}^2\alpha e^2 kT$ we can write the quantity A inside the square brackets in the velocity equation as

$$A = 1 + \sum_{j=1}^{\infty}(2j+1)\exp -j^2(j+1)^2 a^2. \tag{7a}$$

Since the numerical value of a^2 is small, $2.74\times10^{-3}T^{-1}$, the summation, A, may be replaced by the integral

$$A = \int_0^{\infty}(2j+1)\exp -j^2(j+1)^2 a^2 dj = \int_0^{\infty}(1/a)\exp(-y^2)dy = \sqrt{\pi}/2a = (\pi^{\frac{1}{2}}/2)(2\alpha kT)^{\frac{1}{2}}8\pi^2 m_H e h^{-2}. \tag{7b}$$

Hence, our expression for the velocity of reaction becomes:

$$k = \kappa\left(\frac{m_{H_4}}{m_{H_2}m_{H_2{}^+}}\right)^{\frac{3}{2}}\frac{kT}{h}\cdot\frac{h^3}{(2\pi kT)^{\frac{3}{2}}}\cdot\frac{2\pi^2 m_{H_2}e}{h^2}(2\pi kT)^{\frac{1}{2}}$$

or $k = \kappa 2\pi m_H{}^{\frac{1}{2}}e\alpha^{\frac{1}{2}} = 2.069\times10^{-9}\kappa$ cc per molecule per sec. or 1.25×10^{15} cc per mole per sec. It is evident that the velocity of this reaction is very high even when the possible rotation of the system is taken into account.

[*Editor's Note:* Material has been omitted at this point.]

13

Reprinted from *J. Chem. Phys.* **39**:1413–1422 (1963)

Cross Sections of Ion–Permanent-Dipole Reactions by Mass Spectrometry*

THOMAS F. MORAN AND WILLIAM H. HAMILL

Department of Chemistry and Radiation Laboratory, University of Notre Dame, Notre Dame, Indiana

(Received 31 May 1963)

Proton transfer is the most probable type of reaction observed in a number of ion–molecule interactions involving permanent dipoles. Integrated cross sections, which are unusually large, are described in terms of an ion–dipole pair oriented near the position of minimum energy at low relative velocity. As the relative velocity is increased, this alignment becomes less likely. The energy-dependent cross section is eclipsed at high relative velocity by the "hard" cross section which does not vary with energy. Proton affinities are estimated for several alkyl cyanides. Some of the reported reactions appear to involve excited states of the primary ions.

INTRODUCTION

THE integrated cross section for collision of an ion with a polarizable molecule in the ion source of a mass spectrometer has been described by Gioumousis and Stevenson[1] in terms of point particles. A modification of this treatment allows for the physical size of the colliding species.[2] Experimental results from this laboratory[3–5] have been adequately explained in terms of this description, both for large and for small molecules.

The present study examines the long-range forces between an ion and a molecule possessing a permanent dipole making use of the previous results concerning the hard-core cross section of the colliding species.

The potential V between an ion and polarizable molecule having a dipole moment $\mathbf{\mu}$ which is oriented in the position of minimum energy is

$$-V = e\mu/r^2 + e^2 a/2r^4, \qquad (1)$$

where e is the electronic charge, a is the polarizability, and r is the distance between ion and molecule. The Lagrangian equations of motion for this system were solved by standard techniques[6] and the approximation that the relative velocity of the colliding pair is given by the velocity of the ion. The collision cross section as a function of ion energy E is

$$\sigma(E) = \sigma_D E^{-1} + \sigma_L E^{-\frac{1}{2}}, \qquad (2)$$

where, in terms of reduced mass μ,

$$\sigma_D = \pi e\mu m_1/\mu, \qquad (3)$$

$$\sigma_L = \pi e(2am_1/\mu)^{\frac{1}{2}}, \qquad (4)$$

the integrated cross section Q is

$$Q = E_e^{-1} \int_{E_i}^{E_e} \sigma(E)\, dE, \qquad (5)$$

where E_e is the energy of the primary ions at the ion exit slit and E_i is the initial energy of the ions. Designating the transmission coefficients for chemical reactions to be P_D for interactions corresponding to σ_D and P_K for knockon encounters in the cross-sectional area σ_K, and taking

$$\sigma(E) = P_D(\sigma_D E^{-1} + \sigma_L E^{-\frac{1}{2}} - \sigma_K) + P_K \sigma_K, \qquad (6)$$

we obtain, from Eqs. (5) and (6), when $E_e \gg E_i$ and $E_t < E_e$ (see below):

$$Q = E_e^{-1} P_D \sigma_D \ln E_e/E_i + 2P_D \sigma_L E_e^{-\frac{1}{2}} + \sigma_K(P_K - P_D), \qquad (7)$$

with the assumption that the dipole axis and the molecule are colinear. The latter assumption is required by the very large measured reaction cross sections.

As the ion energy increases, the energy-dependent cross section of Eq. (6) becomes smaller than, and is eclipsed by, the hard molecular cross section σ_K arising from short-range repulsive forces at a characteristic transitional energy E_t, viz.,

$$\sigma_K = \sigma_D E_t^{-1} + \sigma_L E_t^{-\frac{1}{2}}. \qquad (8)$$

At ion energy above E_t, the integral of Eq. (5) between limits E_i to E_t and E_t to E_e becomes

$$Q = E_e^{-1}[P_D \sigma_D(\ln E_t/E_i - 1) + P_D \sigma_L E_t^{\frac{1}{2}}] + P_K \sigma_K. \qquad (9)$$

* This article is based upon a thesis submitted by T. F. Moran in partial fulfillment of the requirements for the Ph.D. degree at the University of Notre Dame. This work was supported in part by The Radiation Laboratory of the University of Notre Dame operated under contract with the U. S. Atomic Energy Commission.

[1] G. Gioumousis and D. P. Stevenson, J. Chem. Phys. **29**, 294 (1958).
[2] N. Boelrijk and W. H. Hamill, J. Am. Chem. Soc. **84**, 730 (1962).
[3] R. F. Pottie, A. J. Lorquet, and W. H. Hamill, J. Am. Chem. Soc. **84**, 529 (1962).
[4] L. P. Theard and W. H. Hamill, J. Am. Chem. Soc. **84**, 1134 (1962).
[5] D. A. Kubose and W. H. Hamill, J. Am. Chem. Soc. **85**, 125 (1963).
[6] For example see Goldstein, *Classical Mechanics* (Addison-Wesley Press Publishing Company, Inc., Reading, Massachusetts, 1959).

At the minimum of potential energy the dipole is aligned with the r vector, although libration through a small angle leads to essentially the same cross section. If all orientations of the dipole are taken to have equal probability at a distance where the cross section is determined by the ion–induced dipole forces, the cross section will be of the form $\sigma(E) = \sigma_L E^{-\frac{1}{2}}$. However, the measured Q is an integrated cross section and there will be collisions at all ion energies to E_e. An expression for Q must take into account the possibility that at very low ion energies the molecule can assume the position of minimum energy and Eq. (7) applies.

As the ion–molecule relative velocity increases the phase angle will lag, until, at some energy E_ϕ, the collision cross section no longer obeys Eq. (4) and becomes simply $\sigma(E) = \sigma_L E^{-\frac{1}{2}}$: The transition is taken to be sharp, for simplicity. The integral of Eq. (5) from E_i to E_ϕ and E_ϕ to E_e, for ion exit-energy settings above E_ϕ, gives

$$Q = E_e^{-1}(P_D \sigma_D \ln E_\phi / E_i) + 2P_D \sigma_L E_e^{-\frac{1}{2}} + \sigma_K (P_K - P_D).$$

$$(10)$$

The transition at E_ϕ will be succeeded by another at E_t when $\sigma_K = \sigma_L E_t^{-\frac{1}{2}}$. That is for $E_e > E_t > E_\phi$ the integral of Eq. (5) from E_i to E_ϕ, E_ϕ to E_t and E_t to E_e becomes

$$Q = E_e^{-1}(P_D \sigma_D \ln E_\phi / E_i + P_D \sigma_L E_t^{\frac{1}{2}}) + P_K \sigma_K. \quad (11)$$

EXPERIMENTAL

The measurements were made with a C.E.C. 21–103A mass spectrometer with a 31–402 ion source. The ionizing voltage and repeller voltage circuits have been modified and described.[2] A vibrating reed and recorder in tandem gave a dynamic range greater than 10^6. Methods of determining pressure dependence and appearance potentials and the basis for choosing the following instrument settings are described elsewhere.[7] The settings were: 70 eV electron energy, 370 V accelerating voltage, 10.5 μA, and a total pressure in the ion source corresponding to 1.2×10^{13} molecules/cc. Focus settings were optimized for the particular ion being examined.

The chemicals were the best grade available and were further purified using standard techniques. Other chemicals were prepared as follows: cyanogen by heating mercuric cyanide; hydrocyanic acid from potassium cyanide and phosphoric acid; methyl fluoride from methyl iodide over potassium fluoride at 420° in an all-glass container.

[7] Detailed descriptions of the instrument modifications and of the techniques used for the various measurements are given in "Notes on Techniques for Studying Ion-Molecule Reactions" by T. F. Moran and D. A. Kubose, which is available at no cost from W. H. Hamill.

TABLE I. Ion-molecule reactions forming protonated molecules.

Primary ion	Molecule	Secondary ion	$\dfrac{Q_{thresh}}{Q_{70\ eV}}$	σ_D calc (Å² eV)	$P_D \sigma_D$ obs (Å² eV)	E_ϕ (eV)	E_i (eV)	High-energy region $\dfrac{\text{calc slope}}{\text{obs slope}}$ for Q vs E_e^{-1}	σ_K graph (Å²)	σ_K from $\sigma_L/E_i^{\frac{1}{2}}$ (Å²)	van der Waals σ_K (Å²)	Molar refraction (Å²)	$P_K \sigma_K$ (Å²)
$CH_3OCH_3^+$	CH_3OCH_3	$CH_3OCH_3H^+$	0.75	26	27	1.5	3.0	0.95	21	28	47	41	0
CH_3OH^+	CH_3OH	CH_3OHH^+	0.40	31	30	2.3	2.9	0.91	15	23	43	27	6
CH_3CN^+	CH_3CN	CH_3CNH^+	0.86	74	53	...	2.5	0.97	55	60ᵃ	63	37	0
HCN^+	HCN	$HCNH^+$	1	56	32	1.4	3.3	1.10	<35, >20	20	55	24	<35
$C_2D_2^+$	CH_3CN	CH_3CND^+	~1	62	31	1.4	3.4	1.00	22	22	49	31	0

ᵃ $\sigma_K = \sigma_D/E + \sigma_L/E^{\frac{1}{2}}$

TABLE II. Reaction of methane ion and alkyl cyanide to give protonated cyanide: $CD_4^+ + RCN \rightarrow RCND^+ + CD_3$.

Molecule	σ_{Dcalc} (Å² eV)	σ_{Lcalc} (Å² eV½)	E_t (eV)	E_ϕ (eV)	P_D	Slope of region II (Å² eV)	High-energy region calc slope / obs slope for Q vs E_e^{-1}	σ_K graph (Å²)	σ_K from $\sigma_{L/E}$ (Å²)	van der Waals σ_K (Å²)	Molar refraction σ_K (Å²)	$P_K\sigma_K$ (Å³)
$CH_3(CH_2)_3CN$	48	55	1.2	0.1	1	68	1.06	60	50	61ᵃ	39	0
$CH_3(CH_2)_2CN$	50	50	1.2	0.1	1	64	0.97	45	45	56	36	0
CH_3CH_2CN	52	46	1.9	0.2	1	96	0.91	25	33	46	33	6
CH_3CN	56	40	2.7	0.5	1	165	0.91	20	24	48	26	15
HCN	48	32	3.3	0.8	1	155	0.92	15	19	43	24	10

ᵃ Value taken is the mean between C_3H_7CN and $C_6H_{11}CN$.

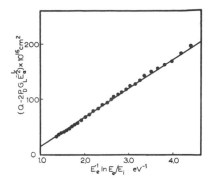

FIG. 1. $Q - 2P_D\sigma_L E_e^{-\frac{1}{2}}$ vs $E_e^{-1} \ln E_e/E_i$ for the reaction $CH_3CN^+ + CH_3CN \rightarrow CH_3CNH^+ + CH_2CN$ at low ion velocity for $E\phi > E_t$ [see Eq. (7)].

RESULTS AND DISCUSSION

Parent–daughter relationships in ion–molecule reactions are commonly assigned on the basis of appearance potential measurements. Since several primary ions may contribute[8] when cross-section measurements are made with 70-eV electrons the method can identify only the reacting ion of lowest appearance potential. The extent to which a molecular ion contributes to form a given secondary ion can be determined by comparing cross sections just above threshold and at 70 eV electron energy, except in the case of excited ion–molecule reactions as reported by Henglein[9]: In the present work it has been found that the most probable ionic reaction product in our ion–dipole systems is a protonated molecule and that both molecular and fragment primary ions may undergo this type of reaction. Thus, Q measured at 70 eV must be cor-

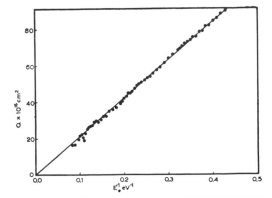

FIG. 2. Q vs E_e^{-1} for the reaction $CH_3CN^+ + CH_3CN \rightarrow CH_3CNH^+ + CH_2CN$ at high ion velocity for $E\phi > E_t$ [see Eq. (8)].

[8] H. VonKoch and E. Lindholm, Arkiv Fysik 19, 123 (1961).
[9] A. Henglein, Z. Naturforsch. 17a, 37 (1962).

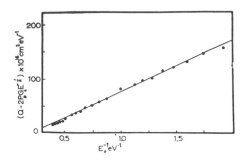

FIG. 3. $Q-2P_D\sigma_L E_e^{-\frac{1}{2}}$ vs E_e^{-1} for the reaction $CD_4^+ + CH_3CH_2CN \rightarrow CH_3CH_2CND^+ + CD_3$ at intermediate ion velocity for $E_\phi < E_t$ [see Eq. (10)].

Fig. 2 and the slope is predictable using parameters established in the lower-energy region.

When E_ϕ lies below E_t, Eqs. (10) and (11) can be applied as in Figs. 3 and 4. When energies lower than E_ϕ are in an accessible range, Eq. (7) is applicable and the data are handled in the manner of Fig. 1, where the slope is $P_D\sigma_D$. The intercepts of the high- and low-energy regions are $P_K\sigma_K$ and $\sigma_K(P_K - P_D)$. σ_K is determined from these intercepts knowing P_D from the low-energy region and it is also given by $\sigma_K = \sigma_L E_t^{-\frac{1}{2}}$. E_ϕ is the highest energy at which $Q - 2\sigma_L E_e^{-\frac{1}{2}}$ vs $E_e^{-1} \ln E_e / E_i$ deviates from linearity and the lowest energy at which $Q - 2\sigma_L E_e^{-\frac{1}{2}}$ vs E_e^{-1} deviates. E_ϕ can also be determined from the slope of $Q - 2\sigma_L E_e^{-\frac{1}{2}}$ vs E_e^{-1} in the intermediate range (referred

rected by the factor $Q_{\text{threshold}}/Q_{70\text{ eV}}$. The fragmentation pattern is one of the factors that determines the importance of contributions from ions other than the molecular ion. Since E_t and E_ϕ are affected by the mass of the primary ion, the energy dependence is' quantitative only when all important contributing primary ions differ by no more than a few mass units.

When E_ϕ exceeds E_t Eq. (7) can be tested in the low-energy range as in Fig. 1 by plotting $Q - 2P_D\sigma_D E_e^{-\frac{1}{2}}$ vs $E_e^{-1} \ln E_e / E_i$, where E_i is determined by the temperature of the ion source, 250°. The slope is $P_D\sigma_D$ and the intercept $\sigma_K(P_K - P_D)$ according to Eq. (7). Theoretical values of σ_L and σ_D were used and trial values of P_D in $Q - 2P_D\sigma_L E_e^{-\frac{1}{2}}$ chosen until consistency with the required slope was achieved. The intercepts of the high- and low-energy regions according to Eqs. (7) and (9) are $P_K\sigma_K$ and $\sigma_K(P_K - P_D)$. σ_K can be determined from these intercepts knowing P_D from the low-energy region and it is also given by Eq. (8). Above E_t, Q follows a linear E_e^{-1} dependence as in

TABLE III. Reactions of CD_3I ion and alkyl cyanide to give protonated cyanide.

Molecule	Secondary ion	Appearance potential (eV)	Q 1.5 eV ion energy (Å)
HCN	HCND+	10.4±0.5	3
CH₃CN	CH₃CND+	10.4±0.5	8
CH₃CH₂CN	CH₃CH₂CND+	10.2±0.3	23
CH₃(CH₂)₂CN	CH₃(CH₂)₂CND+	10.1±0.3	14
CH₃(CH₂)₃CN	CH₃(CH₂)₃CND+	10.0±0.3	9

to as the slope of Region II in Table II), which is $P_D\sigma_D \ln E_\phi / E_i$, knowing the value of $P_D\sigma_D$ from the low-energy region. The various methods of determining the parameters agree. Above E_t, Q follows an E_e^{-1} dependence and the slope can be predicted using the parameters from the lower-energy region.

Synthetic cases in which the angle between the dipole and the r vector change slowly at a given energy, for example 5° for every 0.1 eV ion energy up to E_t, demonstrate that quantitative data on the energy dependence of η in the potential $V = -e\mu r^2 \cos\eta$ cannot be achieved experimentally in a narrow region approximating the interval between E_ϕ and E_t. The experimental data indicate that η goes from a small angle toward 90°.

Values of the parameters obtained in the manner described for Eqs. (7), (9), (10), and (11), and illustrated by Figs. 1–4, appear in Tables I and II. Molecular cross sections from van der Waals radii and molar refraction are included for comparison.

In the series of reactions

$$CD_4^+ + RCN \rightarrow RCND^+ + CD_3$$

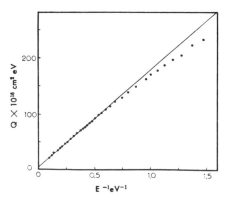

FIG. 4. Q vs E_e^{-1} for the reaction $CD_4^+ + CH_3CH_2CN \rightarrow CH_3CH_2CND^+ + CD_3$ at high ion velocity for $E_\phi < E_t$ [see Eq. (11)].

TABLE IV. Self-reactions in $CH_3(CH_2)_3CN$, $CH_3(CH_2)_2CN$, CH_3CH_2CN to give (molecule+H)$^+$.

System	Primary ions m/e	Relative intensity of primary ion compared to largest in the spectrum	(Sum of primary ions) / (Total analyzed current)	$Q_{threshold}$ / Q_{70eV}
CH_3CH_2CN	55	10		0.80
	54	62	0.63	
	28	100		
$CH_3(CH_2)_2CN$	41	100	0.62	0.85
	29	62		
$CH_3(CH_2)_3CN$	55	21		
	54	52	0.61	0.87
	43	97		
	41	100		

only two regions are accessible: the high-energy region, which has a lower energy limit of E_t; and the intermediate region. The measured values of $P_D\sigma_D$ and $P_D\sigma_L$ equal the theoretical σ_D and σ_L. The value of Q at 0.8 eV ion energy in the reaction with HCN requires that P_D be unity and this same value has been assigned for the other reactions involving CD_4^+ with RCN. E_ϕ, which is determined from the slope of the intermediate region, decreases with increasing moment of inertia of the molecule. As the size of the molecule increases, E_t decreases as expected.

The reactions

$$RCN^+ + CD_4 \rightarrow RCND^+ + CD_3$$

do not occur at threshold for propionitrile, butyronitrile, and valeronitrile. This can be demonstrated because the ionization potentials of the RCN are lower than that of CD_4. With hydrogen cyanide and acetonitrile, abstraction of a D atom from CD_4 by RCN^+ [10] cannot be ruled out, although the dependence of Q on E_e does not indicate a major contribution of this sort.

There is a reduced P_D for the last three reactions in Table I involving cyanides and this is interpreted to be a geometric effect due to ion–dipole forces. The lowest ionization potentials of acetonitrile, hydrogen cyanide and acetylene have been postulated to result in removal of an electron from the triple bond.[11] If at contact distances the ion and the molecule occupy a position close to that of the potential minimum given by Eq. (1), assuming that the negative end of the dipole is the CN group, any process leading to $RCND^+$ must occur over distances of several bond lengths. This effect also arises with CD_3Br^+, to be discussed later.

Appearance potentials of secondary ions and measured values of Q for the reaction

$$CD_3I^+ + RCN \rightarrow RCND^+ + CD_2I$$

appear in Table III. The ionization potential of CD_3I refers to removal of a nonbonding electron from the I atom[12] leading to doublet levels at 9.5 eV and 10.2 eV. Within experimental error, the appearance potentials of the secondary ions of the above reaction corre-

TABLE V. Self-reactions in $CH_3(CH_2)_3CN$, $CH_3(CH_2)_2CN$, $CH_3(CH_2CN)$ to give (molecule+H)$^+$.

System	E_t (eV)	σ_K from $\dfrac{\sigma_L}{E_t^{\frac{1}{2}}}$ (Å^2)	σ_K van der Waals (Å^2)	σ_K molar refraction (Å^2)	E_ϕ (eV)	σ_D (Å^2 eV)	σ_L (Å^2 eV$^{\frac{1}{2}}$)	$P_K\sigma_K$ (Å^2)
CH_3CH_2CN	1.7	39	55–60	38–43	0.5	59–76	52–60	0
$CH_3(CH_2)_2CN$	2.5	37	63–72	41	1.7	53–61	58–61	6
$CH_3(CH_2)_3CN$	2.5	43	69–93	43–48	1.8	57–64	65–68	6

[10] T. W. Martin and C. E. Melton, J. Chem. Phys. 32, 700 (1960).
[11] J. A. Cutler, J. Chem. Phys. 16, 136 (1948).
[12] W. C. Price, J. Chem. Phys. 4, 539 (1936).

spond to the 10.2-eV state. Exclusion of the $^2P_{\frac{1}{2}}$ state of the ion from reaction may arise from a quantum-mechanical restriction on angular momentum or from an energy requirement for reaction. Since the reaction

$$CD_3Br^+ + RCN \rightarrow RCND^+ + CD_2Br$$

occurs, and the appearance potential of the secondary ion corresponds to the $^2P_{\frac{1}{2}}$ state of the CD_3Br ion, the restriction due to the angular momentum can be ruled out. Thus, there is evidence that excitation of the primary ion fulfills an energy requirement for proton transfer to the RCN molecule. Since the total CD_3I^+ current is taken for i_p in the computation of Q, although only part of the CD_3I ions react, the correct value of Q is larger than the apparent value.

The reaction

$$C_6D_{12}^+ + RCN \rightarrow RCND^+ + C_6D_{11}$$

(excluding HCN) occurs, and the value of Q at threshold indicates an ion–dipole interaction. The appearance potential of the secondary ion equals the ionization potential of C_6D_{12}. This reaction is only slightly more favorable, energetically, than the reaction

$$CD_3I^+ \text{ (ground state)} + RCN \rightarrow RCND^+ + CD_2I,$$

which does occur. Using schemes as

	ΔH (kcal/mole)
$C_6D_{12}^+ + CH_3CN \rightarrow CH_3CND^+ + C_6D_{11},$	≤ 0
$C_6D_{12} \rightarrow C_6D_{12}^+ + e,$	228
$e + D^+ \rightarrow D,$	-313
$D + C_6D_{11} \rightarrow C_6D_{12},$	-99
$D^+ + CH_3CN \rightarrow CH_3CND^+$	$\Delta H \leq -184$ kcal/mole

it appears that the proton affinity for RCN (acetonitrile to valeronitrile) is between -184 and -191 kcal/mole. In these calculations the C–D bond dissociation energy for C_6D_{12} and CD_3I is taken as 99 kcal/mole. The reaction between $C_6D_{12}^+$ and HCN does not give $HCND^+$ or any other observable products although the reaction between HCN^+ and C_6D_{12} forms $HCND^+$. Thus, the proton affinity of HCN can only be given a rough limit of -170 kcal/mole from these reactions.

The reaction

$$C_6D_6^+ \text{ (ground state)} + CH_3(CH_2)_2CND$$
$$\rightarrow CH_3(CH_2)_2CND^+ + C_6D_5$$

TABLE VI. Self-reactions in $CH_3(CH_2)_3CN$, $CH_3(CH_2)_2CN$, CH_3CH_2CN to give (molecule $+H$)$^+$.

System	E_a (eV)	Q_{obs} (Å^2)	Q_{pred} (Å^2)
CH_3CH_2CN	1	196	201
	2	102	103
	4	50	52
	8	25	26
$CH_3(CH_2)CN$	1	274	275
	2	160	157
	4	84	81
	8	46	44
$CH_3(CH_2)_3CN$	1	270	280
	2	162	167
	4	89	86
	8	42	46

does not take place. The product $CH_3(CH_2)_2CND^+$ is formed only at an appearance potential of 12.6 ± 0.5 eV. The ionization efficiency curve for this product is very different from the usual secondary ions and suggests an excited state reaction. The appearance potential of this product $CH_3(CH_2)_2CND^+$ approximates a measured threshold of 11.7 eV for an excited state of $C_6H_6^+$.[13] These results confirm the estimates of proton affinity.

The reaction

$$C_2D_4^+ \text{ (from ethylene)} + HCN \rightarrow HCND^+ + C_2D_3$$

was not observed. However, HCN reacts with an ion from ethane to form $HCND^+$ which has an appearance potential identical to the $C_2D_4^+$ fragment ion. Assuming that $C_2D_4^+$ is the primary ion, P_D is found to be 0.25. This could indicate that 25% of the $C_2D_4^+$ ions from ethane have a different structure or energy distribution from the $C_2D_4^+$ ions from ethylene and are able to react. Since the appearance potentials of $C_2D_6^+$ and $C_2D_4^+$ are only 0.5 eV, apart an excited $C_2D_6^+$ ion cannot be eliminated as a primary precursor.

The compounds propionitrile, butyronitrile, and valeronitrile undergo self ion–molecule reactions with a large cross section in which the major reaction product is the (molecule$+H$)-ion. Fragment ions in the spectrum contribute to the (molecule$+H$)-ion and since these ions cover a large mass range, E measured represents an averaged value. If one ion were to predominate, then the behavior of Q vs E_e could be identified with a particular ion but this is not completely the case with these reactions. Appearance potential measurements identify the process at threshold. Many ions of major abundance in the primary spectrum have appearance

---

[13] R. E. Fox and W. N. Hickam, J. Chem. Phys. **22**, 2059 (1954).

potentials close to the (molecule+H)-ion and these ions are grouped under the heading "Primary ions" in Table IV. The Q due to these ions is determined by measuring Q at the threshold and 70 eV electron energy. Other major ions have higher appearance potentials and Q at threshold is measured before their onset.

The kinetic energy of these reacting primary ions has been determined because mass spectrometers discriminate against them.[14] This discrimination would reduce the analyzed primary ion current and give an abnormally high Q. Measurements of kinetic energy were made by applying a negative voltage to the repeller plates and measuring the peak height at a particular m/e as a function of the repeller voltage, using a molecular ion for reference. Other instrumental settings were identical to those outlined in the experimental section. The measured kinetic energy of every primary ion in Table IV does not exceed kT at source temperature.

Only the high-ion-energy region was examined graphically in these self-ion–molecule reactions. Q in this energy range is given by Eq. (11) with E_t determined from the lowest energy at which Q vs E_e^{-1} deviates from linearity. From the slope of the high-energy region E_ϕ is estimated using averaged theoretical values of σ_L and σ_D. σ_K is given by $\sigma_K = \sigma_L E_t^{-1}$. Knowing E_t and E_ϕ, Eqs. (7), (10) and (11) are applied in their respective regions to calculate Q at various values of E_e. Table V gives the parameters for these reactions and Table VI gives the predicted values of Q.

The reactions

$$CD_3Br^+ + RCN \rightarrow RCND^+ + CD_2Br$$

occur, where RCN ranges from HCN to $CH_3(CH_2)_3CN$, as expected from the estimates of proton affinity for RCN. Appearance potentials of the secondary species all equal the lower ionization potential of CD_3Br, viz., for the $^2P_{\frac{1}{2}}$ state. This sets limits to the proton affinity of HCN between -169 to -191 kcal/mole.

The heavy CD_3Br^+ would be expected to favor orientation of RCN molecules to the position of minimum energy and the dependence of Q on ion exit energy E_e confirms this expectation. That is, E_ϕ occurs at an energy greater than E_t. The two energy regions are described by Eqs. (7) and (9). The parameters appear in Table VII; transmission coefficients are seen to change markedly with R.

With methyl fluoride, appearance potentials indicate the reaction

$$CH_3F^+ + HCN \rightarrow HCNH^+ + CH_2F$$

For this reaction $E_\phi > E_t$ and the transmission coeffi-

---

[14] N. D. Coggeshall, J. Chem. Phys. **36**, 1640 (1962).

TABLE VII. Reactions of methyl bromide with alkyl cyanides to give protonated cyanides: $CD_3Br^+ + RCN \rightarrow RCND^+ + CD_2Br$.

Molecule	$Q_{threshold}$ / $Q_{70\,eV}$	D_{calc} (Å^2 eV)	L_{calc} (Å^2 eV$^{\frac{1}{2}}$)	$P_D\sigma_D$ (Å^2 eV)	P_D	$E_{t\,obs}$ (eV)	obs slope for Q vs E_t	calc slope σ_K from Eq. (6) (Å^2)	van der Waals σ_K (Å^2)	Molar refraction σ_K (Å^2)	σ_K graphical (Å^2)	$P_K\sigma_K$ (Å^2)
								High-energy region				
$CH_3(CH_2)_3CN$	0.95	86	78	82	0.95	2.9	1.00	75	64	48	70	22
$CH_3(CH_2)_2CN$	1.0	94	76	100	1.1	2.8	0.96	78	59	47	79	6
$CH_3(CH_2)CN$	0.94	106	71	83	0.78	2.9	0.98	77	51	42	80	5
CH_3CN	0.95	127	65	63	0.50	3.1	0.95	78	55	37	60	6
HCN	0.91	129	59	9	0.07	3.3	0.96	69	48	31	63	0

Table VIII. Reaction of cyclohexane with alkyl cyanides: $C_6D_{12}^+ + RCN \rightarrow RCND^+ + C_6D_{11}$.

Molecule	$\dfrac{Q_{threshold}}{Q_{70\ eV}}$	$\sigma_{D\,calc}$ $(Å^2\ eV)$	$\sigma_{L\,calc}$ $(Å^2\ eV^{1/2})$	$P_D\sigma_D$ $Å^2\ eV$	P_D	σ_K from Eq.(8) $(Å^2)$	E_t (eV)	σ_K graph $(Å^2)$	van der Waals σ_K $(Å^2)$	Molar refraction σ_K $(Å^2)$	$P_K\sigma_K$ $(Å^2)$
CH_3CN	0.45	124	66	83	0.67	70	3.6	83	68	48	17
CH_3CH_2CN	0.45	104	70	94	0.90	76	3.0	50	65	53	10
$CH_3(CH_2)_2CN$	0.41	93	75	89	0.96	70	3.3	70	75	57	7
$CH_3(CH_2)_3CN$	0.42	84	77	...	~1	...	~3.0		83	61	0

cient P_D approximates unity, in contrast to the reaction involving CD_3Br^+.

Ionization of CD_3Br presumably involves removal of one of the p electrons perpendicular to the C–Br bond.[12] Since proton transfer to an alkane is not an efficient process, the proton is most likely attached to the CN group in alkyl cyanides. As the RCN approaches the vacant p orbital in Br, short-range repulsive forces of the CD_3 group deflect the R group leaving the CN preferentially oriented toward the CD_3. The transfer of D to CN would occur more easily with greater repulsion, that is, with larger R. HCN may be an extreme example with N experiencing greater repulsion from CD_3 than the lone H atom in which case transfer of D would have to occur over a distance greater than the C–H bond distance, thereby leading to a reduced transmission coefficient. In CH_3F the bonding electron has been removed and p orbitals in F contribute to the repulsive forces. At contact distances HCN may be perpendicular to the C–F bond with the C–N group close to CD_3 than in the corresponding CD_3Br case.

The parameters for the reactions

$$C_6D_{12}^+ + RCN \rightarrow RCND^+ + C_6D_{11}$$

(excluding HCN) are given in Table VIII. The ratio of cross section at threshold to that at 70 eV is approximate because other ions from the fragmentation of $C_6D_{12}^+$ have appearance potentials fairly close to that of the molecular ion. Since there is a large mass disparity between the molecular ion and other ions in the spectrum which must contribute to give the protonated ion as evidenced by $Q_{threshold}/Q_{70\ eV}$, the parameters are not as accurate as in other examples but still show an energy trend in the cross section.

Reactions of RCN molecules with ions heavier than CD_4^+ but lighter than $C_6D_{12}^+$ have E_ϕ approximating $\frac{1}{2}E_t$ as shown in Table V. CD_3Br^+ is able to align RCN and thus $E_\phi \geq E_t$. If the relative velocity of the colliding pair determined by the mass of the ion and moment of inertia of the molecule are important factors in determining the orientation of species at distances where the cross section is determined, reactions of $C_6D_{12}^+$ should be an intermediate type with E_ϕ close to E_t. Table VIII presents the results of the $C_6D_{12}^+$ reactions where $E_\phi \geq E_t$. Only with valeronitrile, which has the largest moment of inertia of the RCN, is the ion not able to align the molecule before the hard core dominates the cross section.

Reactions of the type $RCN^+ + D_2 \rightarrow RCND^+ + D$ occur for HCN[5] and CH_3CN only and obey the description appropriate to ion-induced dipole reactions.[5] Values of $\sigma_{L,\ calc}$ and $P_L\sigma_{L,\ obs}$ are 42 and 48×10^{-16} $cm^2\ eV^{1/2}$ for HCN, and 51 and 9×10^{-16} $cm^2\ eV^{1/2}$ for CH_3CN, respectively. It had been noted earlier[10] that higher RCN did not react with H_2, although it appears that the reactions are thermochemically allowed.

Using the previously determined proton affinities of the alkyl cyanides:

$$D^+ + RCN \rightarrow RCND^+,$$

$$\Delta H_{rx} = -187 = \Delta H_f(RCND)^+$$

$$- \Delta H_f(RCN) - \Delta H_f(D)^+,$$

$$\Delta H_f(RCND)^+ = -187 + \Delta H_f(RCN)$$

$$+ \Delta H_f(D) + I.P.(D).$$

$$RCN^+ + D_2 \rightarrow RCND^+ + D,$$

$$\Delta H_{rx} = \Delta H_f(RCND)^+$$

$$+ \Delta H_f(D) - \Delta H_f(RCN)^+$$

$$\Delta H_{rx} = -187 + \Delta H_f(RCN) + \Delta H_f(D)$$

$$+ I.P.(D) + \Delta H_f(D) - \Delta H_f(RCN)^+$$

$$= -187 + 2\Delta H_f(D) + I.P.(D)$$

$$- I.P.(RCN)$$

$$= -187 + 104 + 313 - I.P.(RCN),$$

$$\Delta H_{rx} = 226 - I.P.(RCN),$$

$$\Delta H_{rx} = 9.8 \text{ eV} - I.P.(RCN).$$

The I.P. of the RCN > 11.0 eV and the reactions are allowed.

Transmission coefficients for reaction of various ions with RCN molecules as a function of exothermicity appear in Fig. 5. In this graph proton affinities were assumed to be 187 kcal/mole for RCN molecules from CH_3CN to C_4H_9CN and 172 kcal/mole for HCN. C–D bond dissociation energies were taken to be 98 kcal/mole except for C_2D_4 and C_2D_2, where values of 96 and 104 kcal/mole were chosen. No reactions were observed below onsets given by the respective curves. Transmission coefficients are reduced in some cases although the internal energy of the ion is sufficient for reaction. Steric effects of an ion–dipole pair oriented at contact distances may be responsible for this reduction in reaction probability, however the internal energy of the reacting ions appears to be the most important factor in determining the probability of reaction in simple proton transfer reactions. The appearance potentials of $RCND^+$ resulting from reaction with $CD_3I^+(^2P_{\frac{1}{2}})$ did not change with increasing translational energy.

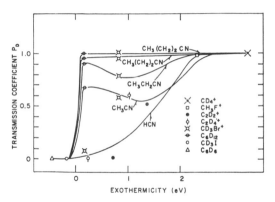

Fig. 5. Transmission coefficient P_D as a function of exothermicity for reaction $R'D^+ + RCN \rightarrow RCND^+ + R'$.

Although this is not an extremely sensitive method to test energy conversion, since Q is a decreasing function of energy, translational energy appears unable to overcome the energy barrier to reaction. Reaction of $C_2D_4^+/C_2D_4$ with HCN to form $HCND^+$ or any other intense ion–molecule product, could not be induced by increasing the translational energy of the ion.

In order to test possible effects of chemistry peculiar to the RCN ion–molecule reactions which might influence the dependence of Q on E_e, reactions involving C_2N_2 were examined. Using C_2D_4 as an ion source, the reaction

$$C_2D_3^+ + C_2N_2 \rightarrow C_2N_2D^+ + C_2D_2$$

occurs and at low ion-exit energies Q has the dependence appropriate to the ion-induced dipole model. This is expected because C_2N_2 has no permanent dipole moment. The parameters obtained from this treatment[5] are:

$$\sigma_L = 48 \text{ Å}^2, \qquad P_L\sigma_L = 26 \text{ Å}^2,$$

$$E_t = 2.5 \text{ eV}, \qquad \text{and } \sigma_K = 25 \text{ Å}^2.$$

With C_2D_6 as ion source, only the reaction

$$C_2D_6^+ \text{ (and other ions)} + C_2N_2 \rightarrow C_2N_2D^+ + \text{fragments}$$

can be demonstrated by appearance potentials, but Q becomes very much smaller as the electron energy approaches threshold indicating contributions from other ions. The parameters derived from the repeller-voltage dependence are larger than predicted when only $C_2D_6^+$ is taken for the primary ion, although Q obeys the ion-induced dipole force law. The reaction

$$(CD_4 + C_2N_2)^+ \rightarrow C_2N_2D^+ + CD_3$$

also follows a Q vs $E_e^{-\frac{1}{2}}$ dependence at low energies.

Butyronitrile and valeronitrile have low-intensity parent-minus-one mass peaks at 70-eV electron energy, thus allowing an estimate of the importance of hydride ion transfer.

The reactions

$$CD_3^+ + CH_3(CH_2)_n CN \rightarrow CH_2(CH_2)_n CN^+ + CD_3 H,$$

for $n=2$ or 3, are allowed energetically but do not occur with large transmission coefficients. In fact, P_D must be less than 0.1 if reaction occurs at all. Lindholm[15] has also found a comparatively rare occurrence of hydride ion transfer among reactions of various primary ions with CH_3OH. Henglein[16] found that the exchange reaction

$$OH^+ + H_2O \rightarrow H_2O + OH^+$$

did not occur, possibly because of competing charge exchange.

---

[15] P. Wilmenius and E. Lindholm, Arkiv Fysik **21**, 97 (1962).
[16] A. Henglein and G. A. Muccini, Z. Naturforsch. **17a**, 452 (1962).

14

Reprinted from J. Chem. Phys. 58:3027–3037 (1973)

Theory of ion–polar molecule collisions. Comparison with experimental charge transfer reactions of rare gas ions to geometric isomers of difluorobenzene and dichloroethylene*

Timothy Su and Michael T. Bowers

Department of Chemistry, University of California, Santa Barbara, California, 93106

(Received 24 August 1972)

A classical theory for the ion–permanent dipole interaction is developed that takes into consideration the thermal rotational energy of the polar molecule. The theory is formulated in terms of an r-dependent average orientation angle $\bar{\theta}(r)$ between the dipole and the line of centers of collision. The technique allows quantitative determination of the capture cross section as a function of ion–polar molecule relative velocity. In addition, capture limit rate constants are readily calculated both at thermal energies and as a function of relative energy. Charge transfer rate constants from various rare gas ions to difluorobenzene and dichloroethylene isomers have been measured at thermal energies using ion cyclotron resonance spectroscopy. Rate constants are considerably larger than predicted by the capture theories for both polar and nonpolar isomers indicating long range electron jump is a prevalent mechanism for charge transfer in these systems. The average-dipole-orientation theory developed in this paper adequately accounts for the effects of the permanent dipole moment on the rate constant.

INTRODUCTION

It has been pointed out that some ion–molecule reactions in which the neutral molecule has a dipole moment[1,2,3] have reaction cross sections greater than those predicted by the Langevin theory of the collision of a point charge with a polarizable molecule.[4,5] In the Langevin treatment, the long range potential between the ion–molecule pair is

$$V_L(r) = -\tfrac{1}{2}(\alpha q^2/r^4),\qquad(1)$$

where q is the charge of the ion, α is the polarizability, and r is the ion–molecule separation distance. An expression for the capture reaction rate constant k_{cap} is derived to be

$$k_{cap} = 2\pi q(\alpha/\mu)^{1/2},\qquad(2)$$

where μ is the reduced mass. Theard and Hamill,[1] and Moran and Hamill,[6] treated the reaction between an ion and a polar molecule by introducing the ion–permanent dipole potential

$$V_D = (-q\mu_D/r^2)\cos\theta\qquad(3)$$

in addition to the charge–induced dipole potential, where μ_D is the dipole moment of the molecule and θ is the angle the dipole makes with r. These authors used the simplifying assumption that the dipole "locks in" on the ion with $\theta=0°$. This results in an additional term in the rate constant:

$$k = \pi[(4q^2\alpha/\mu)^{1/2} + (2q\mu_D/\mu v)],\qquad(4)$$

where v is the relative velocity at infinite separation. At thermally averaged velocities, the rate constant becomes[2]

$$\bar{k} = (2\pi q/\mu^{1/2})[\alpha^{1/2} + \mu_D(2/\pi kT)^{1/2}],\qquad(5)$$

where in this instance $\bar{k}$ is the thermally averaged rate constant and k is Boltzmann's constant. Experimental rate constants for ion–dipole reactions are much less than those predicted by Eqs. (4) or (5). Gupta et al.[2] suggested that "locking in" of the dipole does not necessarily occur.

A recent communication from this laboratory[7] has reported the thermal energy rate constants from various rare gas ions to the three geometric isomers of difluoroethylene. This set of isomers was chosen because each member has essentially the same polarizability[8] and ionization potential[9] but considerably different dipole moments[10]: trans = 0.0 D, cis = 2.42 D, and 1,1 = 1.38 D. Two major phenomena were observed: (1) All of the trans rate constants are substantially larger than predicted by the charge–induced dipole theory. This was interpreted to imply a long range electron jump mechanism contributed to the rate constant.[11] (2) The rate constants increased with increasing dipole moment but the increase was substantially less than expected for a locked dipole. If a parameter c with values between 0 and 1 is introduced to Eq. (5) to compensate for the effectiveness of the charge "locking in" the dipole, Eq. (5) becomes

$$k = (2\pi q/\mu^{1/2})[\alpha^{1/2} + c\mu_D(2/\pi kT)^{1/2}].\qquad(6)$$

By comparison of the experimental and theoretical cis/trans and 1,1/trans ratios an average value of c was obtained to be 0.05 ± 0.02 which corresponds to an average librational angle of the dipole of about 87° with respect to the line of centers of collision.

Dugan et al.[12–15] solved numerically the equation of motion for the collision of an ion with a rotating polar molecule. The results of their calculations indicate that "locking in" of the dipole is not likely to occur.

In this work the charge transfer rate constants from various rare gas ions to two sets of geometric isomers are reported, *trans-*, 1,1-, and *cis-*dichloroethylene and *para-*, *meta-*, and *ortho-*difluorobenzene. For each set of isomers, as for those of difluoroethylene, the members have approximately the same polarizability and ionization potential but different dipole moments. A classical theoretical model considering the thermal molecular rotational energy is developed to estimate the capture cross sections and rate constants for ion–dipole reactions and to compare with experimental results.

THEORY

Before going into the details of the theoretical treatment it is useful to state explicitly the approximations of the theory and to point out the implications these approximations have with regard to the details of the ion–polar molecule collision. The first and most important assumption concerns the angle θ, the angle the dipole of the polar molecule makes with the line of centers of the collision. In a given microscopic event, the rotor can have any initial orientation angle θ at collision time $t=0$, (i.e., when the distance r between the ion and polar molecule is large). As t increases, r decreases and the particles begin to be influenced by the potentials of Eqs. (1) and (3). At sufficiently small r, the rotational motion of the dipole is hindered by the field of the ion causing a net orientation of the dipole and oscillatory momentum transfer between the rotor and the system as a whole. In order to relate these microscopic collisions with macroscopic experimental data it is necessary to average the collisions over the various allowed initial rotor orientations and energies and over an initial distribution of impact parameters. A momentum transfer collision could then be defined by an ion trajectory undergoing a significant deflection and a capture collision by an ion trajectory that penetrated to within a certain preset value of r. Such an approach to ion–polar molecule collisions has been taken by Dugan and co-workers.[12-15] This approach is very useful in the investigation of details of trajectories of *nonreactive* ion–polar molecule phenomena, including energy transfer studies. It does not provide a clear cut criteria for defining capture collisions, however, and thus is quantitatively ambiguous in the area of determining upper limits on ion–molecule reaction rate phenomena similar to the role the Langevin theory plays for ion–nonpolar molecule collisions. It is in precisely this area the model to be proposed here hopes to play an important role. In the model proposed here the angle θ is treated in the "average sense". The average orientation angle $\bar{\theta}$ is determined as a function of r. At large r (e.g., 50 Å) $\bar{\theta} = 90°$ and the rotating polar molecule is unaffected by the ion. The value of r is then incrementally decreased to ≈ 5 Å and a value of $\bar{\theta}$ calculated at each incremental value of r. It is this average $\bar{\theta}(r)$ that is then used in the potential function. This technique has the decided advantage of computa-tional simplicity and it generates capture collision criteria in a manner analogous to the model used for ion–nonpolar molecule collisions. There is no need to arbitrarily define a critical capture approach distance nor to average over initial orientation or impact parameters.

A second assumption of the present theory is that on the average there is no net transfer of angular momentum between the rotating molecule and the system as a whole. Dugan and co-workers have shown that for nonreactive ion–polar molecule collisions some angular momentum is transferred in an oscillatory way between the rotating molecule and the system as a whole. The angular momentum transfer occurs because the ion induces changes in the angular velocity $\dot{\theta}$ of the dipole as either the negative or positive end of the dipole approaches the ion. When the negative end of the dipole approaches a positive ion, it is accelerated and angular momentum is transferred from the system to the rotor. As the negative end rotates away from the ion, the positive end of the dipole begins to rotate toward the ion causing a deceleration of the rotor and a transfer of angular momentum from the rotor to the system as a whole. The rotor velocity (and hence angular momentum) thus oscillates in the plane of the collision. *The net effect, however, is that on the average very little angular momentum is transferred from rotor to system or vice versa* (at values of r greater than 5–7 Å). The effect of the ion on the rotor angular momentum in this average sense is taken into account in the detailed calculation of $\bar{\theta}$ as a function of r to be described later in this paper. The possibility of explicitly including the coupling of the system and rotor angular momentum into the present model is also under active investigation.

A third assumption of the theory is concerned with the rotational degrees of freedom of the rotor. In the average-dipole-orientation model presented here, it is assumed the rotor has two rotational degrees of freedom in the plane of collision and that the third degree of freedom is aligned along the dipole and does not contribute. This assumption is rigorously valid for diatomic molecules but is not exactly correct for more complicated systems. We believe the deviations from reality imposed by this assumption are small, however. An initial estimate of the average contribution of out of plane rotations indicates they will change the final rate constant by less than 5%. Explicit incorporation of more complicated rotors is being further investigated, however, and will be reported in the future.

Finally, it should be emphasized the theory presented here is intended to be primarily useful in calculating phenomenological capture rate constants and cross sections. These capture phenomena occur at impact parameters in the range of 8–15 Å. The model is not intended to describe phenomena that occur at impact parameters less than 5 Å nor to apply to microscopic phenomena. These areas are better investigated by a

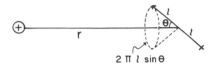

FIG. 1. Coordinate system for the interaction between an ion and a linear polar molecule.

more detailed trajectory model such as that proposed by Dugan. The details of the theory follow.

Consider a polar molecule having a permanent dipole moment μ_D, with separation $2l$ between the poles. A point positive ion with charge q is at a distance r from the center of mass of the polar molecule making an angle θ with the direction of the dipole as shown in Fig. 1. For $r \gg l$, the potential V_d of this system is given by Eq. (3). Combining both the charge–induced dipole potential [Eq. (1)] and the charge–permanent dipole potential, the total average potential V between the pair is

$$V(r) = -(q^2\alpha/2r^4) - (q\mu_D/r^2)\cos\bar\theta(r), \quad (7)$$

where $\bar\theta(r)$ is the average angle of θ at a distance r. The method of calculating $\bar\theta(r)$ will be discussed shortly.

The phenomenon of the "capture collision" can best be understood by considering the effective radial potential $V_{\rm eff}$. According to our model, the average effective radial potential of the ion–dipole pair is[16]

$$V_{\rm eff}(r) = (L^2/2\mu r^2) - (q^2\alpha/2r^4) - (q\mu_D/r^2)\cos\bar\theta(r), \quad (8)$$

where L is the translational angular momentum. No coupling of angular momentum between the system and rotor is included. At the critical distance of approach r_k, $V_{\rm eff}$ is a maximum. Thus

$$\left(\frac{\partial V_{\rm eff}(r)}{\partial r}\right)_{r_k} = 0 = -\frac{L^2}{\mu r_k^3} + \frac{2q^2\alpha}{r_k^5}$$

$$+ \frac{2q\mu_D}{r_k^3}\cos\bar\theta(r=r_k) - \frac{q\mu_D}{r_k^2}\left(\frac{\partial\cos\bar\theta}{\partial r}\right)_{r_k}$$

or

$$-\frac{L^2}{2\mu r_k^2} + \frac{q^2\alpha}{r_k^4} + \frac{q\mu_D}{r_k^2}\cos\bar\theta(r=r_k)$$

$$+ \frac{q\mu_D}{2r_k}\sin\bar\theta(r=r_k)\left(\frac{\partial\theta}{\partial r}\right)_{r_k} = 0. \quad (9)$$

The critical energy E_k for surmounting the barrier is obtained by evaluating $V_{\rm eff}(r)$ at r_k:

$$V_{\rm eff}(r=r_k) = E_k = (L^2/2\mu r_k^2) - (q^2\alpha/2r_k^4)$$

$$- (q\mu_D/r_k^2)\cos\bar\theta(r=r_k). \quad (10)$$

In center of mass coordinates the angular momentum about the center of mass is

$$L = \mu bv, \quad (11)$$

where b is the impact parameter. The analogous kinetic energy is

$$T = \tfrac{1}{2}\mu v^2 = E_k. \quad (12)$$

It should be noted that Eqs. (9)–(12) assume that the barrier in the potential function is an explicit function of the relative velocity of the ion and polar molecule only. The rotational energy of the polar molecule appears only implicitly in that it will determine the value of $\bar\theta$ at a given separation r. This fact is a direct result of assuming there is no net coupling of the system and rotor angular momentum as discussed earlier. Combining Eqs. (9)–(12) and rearranging

$$b_k^2 = r_k^2 + (q^2\alpha/r_k^2\mu v^2) + (2q\mu_D/\mu v^2)\cos\bar\theta(r=r_k), \quad (13)$$

where b_k is the critical impact parameter. In terms of the angle averaged microscopic capture cross section $\langle\sigma\rangle$,

$$\langle\sigma\rangle = \pi b_k^2 = \pi r_k^2 + (\pi q^2\alpha/r_k^2\mu v^2)$$

$$+ (2\pi q\mu_D/\mu v^2)\cos\bar\theta(r=r_k) \quad (14)$$

and

$$v^2 = (q^2\alpha/\mu r_k^4) + (q\mu_D/r_k\mu)\sin\bar\theta(r=r_k)(\partial\bar\theta/\partial r)_{r_k}. \quad (15)$$

Suppose $\bar\theta(r)$ is known. Then there are three variables in Eqs. (14) and (15): v, $\langle\sigma\rangle$, and r_k. If r_k is specified, v and $\langle\sigma\rangle$ can be determined. By varying r_k, the relation between $\langle\sigma\rangle$ and v can be evaluated. Notice that when $\mu_D = 0$, and eliminating r_k, Eqs. (14) and (15) reduce to

$$\sigma(v) = (2\pi q/v)(\alpha/\mu)^{1/2}, \quad (16)$$

which is the same as the microscopic cross section derived by Gioumousis and Stevenson[5] from the charge–induced dipole interaction of ions and nonpolar molecules.

The problem now is to evaluate $\bar\theta(r)$ as a function of r. The average value of θ at a certain ion–molecule separation r is

$$\bar\theta = \int\theta P(\theta)\,d\theta / \int P(\theta)\,d\theta, \quad (17)$$

where $P(\theta)$ is the probability that the dipole is at an orientation θ. To find $P(\theta)$, there are two primary factors to be considered: (1) an energy factor. In this case $P(\theta)$ is inversely proportional to the angular velocity in the plane of collision, $\dot\theta$. (2) An orientation (spatial) factor: The number of ways of arranging the dipole such that it makes an angle θ with r is proportional to the circumference of the circle generated by rotating the negative pole around r (see Fig. 1), which is $2\pi l \sin\theta$. Therefore,

$$P(\theta)\alpha(\sin\theta/\dot\theta). \quad (18)$$

Since the rotational kinetic energy of the polar molecule in the plane of collision, $KE_{\rm rot}$, is proportional to $\dot\theta^2$ $(KE_{\rm rot} = \tfrac{1}{2}I\dot\theta^2)$, then

$$P(\theta)\alpha[\sin\theta/(KE_{\rm rot})^{1/2}]. \quad (19)$$

Applying conservation of energy,

$$E_{rot} = KE_{rot} + V_D, \qquad (20)$$

where E_{rot} is the total energy of the rotating system in plane of collision and V_D is the ion–permanent dipole potential as defined in Eq. (3). It is assumed that the total energy of the rotating system is the thermal rotational energy of the molecule at infinite separation from the ion ($V_D = 0$), and the charge–induced dipole potential does not affect the orientation of the polar molecule. Thus

$$P(\theta) \alpha [\sin\theta / (E_{rot} - V_D)^{1/2}]. \qquad (21)$$

Substituting Eqs. (3) and (21) into Eq. (17), we get

$$\bar{\theta} = \int \frac{\theta \sin\theta d\theta}{[E_{rot} + (q\mu_D/r^2) \cos\theta]^{1/2}} \bigg/ \int \frac{\sin\theta d\theta}{[E_{rot} + (q\mu_D/r^2) \cos\theta]^{1/2}}. \qquad (22)$$

In evaluating Eq. (22) there are two cases to be considered: (1) When $E_{rot} = E_1 < q\mu_D/r^2$, the motion is an oscillating one. The maximum value K of θ is given by

$$E_1 = -(q\mu_D/r^2) \cos K. \qquad (23)$$

(2) For $E_{rot} = E_2 > q\mu_D/r^2$, the motion is nonoscillatory. The dipolar molecule has enough energy to swing around in a complete circle. In this case, θ oscillates between a maximum and a minimum value. The motion is still a periodic one, the dipole making one complete revolution each time θ increases or decreases by 2π.

For case (1), substituting Eq. (23) into (22),

$$\bar{\theta}_1 = \int_0^K \frac{\theta \sin\theta d\theta}{(\cos\theta - \cos K)^{1/2}} \bigg/ \int_0^K \frac{\sin\theta d\theta}{(\cos\theta - \cos K)^{1/2}}. \qquad (24)$$

The denominator of Eq. (24) can easily be integrated to be $2(1 - \cos K)^{1/2}$. The numerator has to be evaluated in terms of elliptic functions.[17] Letting $\sin\phi = (1/a) \times \sin(\theta/2)$, where $a = \sin(K/2)$, the numerator of Eq. (24) becomes

$$\int_0^{\pi/2} \frac{4\sqrt{2} a^2 \cos^2\phi d\phi}{(1 - a^2 \sin^2\phi)^{1/2}}, \qquad (25)$$

which can be integrated by infinite series. Let

$$A = \int_0^{\pi/2} \frac{a^2 \cos^2\phi d\phi}{(1 - a^2 \sin^2\phi)^{1/2}}$$

$$= \int_0^{\pi/2} a^2 \cos^2\phi \left(1 + \tfrac{1}{2} a^2 \sin^2\phi + \frac{1 \times 3}{2 \times 4} a^4 \sin^4\phi + \cdots \right.$$

$$\left. + \frac{1 \times 3 \cdots (2n-1)}{2 \times 4 \cdots (2n)} a^{2n} \sin^{2n}\phi + \cdots \right) d\phi$$

$$= a^2 \left\{ \tfrac{1}{4}\pi + \tfrac{1}{2}\pi \left[\frac{1}{2^2} \frac{1}{4} a^2 + \left(\frac{1 \times 3}{2 \times 4}\right)^2 \frac{1}{6} a^4 + \cdots \right. \right.$$

$$\left. \left. + \left(\frac{1 \times 3 \cdots (2n-1)}{2 \times 4 \cdots (2n)}\right)^2 (2n+2)^{-1} a^{2n} + \cdots \right] \right\}. \qquad (26)$$

It can be proven that A is a convergent series. If two hundred terms are taken, the remainder is less than 0.1%. Therefore,

$$\bar{\theta}_1 = 2\sqrt{2} A / (1 - \cos K)^{1/2} \qquad (27)$$

and

$$\frac{d\bar{\theta}_1}{dr} = \frac{-\sqrt{2} A \sin K (dK/dr)}{(1 - \cos K)^{3/2}} + \frac{2\sqrt{2} (dA/dr)}{(1 - \cos K)^{1/2}}; \qquad (28)$$

dA/dr is also a convergent infinite series.

In case (2)

$$\bar{\theta}_2 = \int_0^\pi \frac{\theta \sin\theta d\theta}{[E_2 + (q\mu_D/r^2) \cos\theta]^{1/2}} \bigg/ \int_0^\pi \frac{\sin\theta d\theta}{[E_2 + (q\mu_D/r^2) \cos\theta]^{1/2}}. \qquad (29)$$

If B and C are the integrals in the numerator and denominator, respectively, of Eq. (29), then

$$\bar{\theta}_2 = B/C; \qquad (30)$$

C can be integrated directly:

$$C = (2r^2/q\mu_D) \{ [E_2 + (q\mu_D/r_2)]^{1/2} - [E_2 - (q\mu_D/r^2)]^{1/2} \}. \qquad (31)$$

Using the method of integration by parts and letting $\phi = \theta/2$, B is converted to

$$B = H + P \int_0^{\pi/2} (1 - \gamma^2 \sin^2\phi)^{1/2} d\phi, \qquad (32)$$

where

$$H = -(2E_1^{1/2} r^2 / q\mu_D) \pi [1 - (q\mu_D / E_2 r^2)]^{1/2},$$

$$P = (4E_2^{1/2} r^2 / q\mu_D) [1 + (q\mu_D / E_2 r^2)]^{1/2},$$

$$\gamma^2 = 2q\mu_D / (q\mu_D + E_2 r^2).$$

The integral in Eq. (32) is a standard elliptic integral which can be integrated in terms of an infinite series similar to case (1). Taking the derivative of Eq. (30),

$$\frac{d\bar{\theta}^2}{dr} = C^{-1}\frac{dB}{dr} - \frac{B}{C^2}\frac{dC}{dr}. \tag{33}$$

The probability of finding a molecule with internal energy between ϵ_r and $\epsilon_r + d\epsilon_r$ is[18]

$$P(\epsilon_r)d\epsilon_r = (1/kT)\exp(-\epsilon_r/kT)d\epsilon_r, \tag{34}$$

where to a good first approximation ϵ_r is given by

$$\epsilon_r = J(J+1)h^2/8\pi^2 I; \tag{35}$$

J is the rotational quantum number and I the moment of inertia. The above form of ϵ_r is only exactly valid for diatomic molecules and spherical top molecules. It is approximately valid for symmetric tops and for asymmetric tops where one axis is along the dipole. The effects of these "three-dimensional" rotors on the final values of $\bar{\theta}(r)$ and ultimately $\langle\sigma\rangle$ and k will be investigated in a future paper.

If E_1 is the average energy in the plane of collision of all those molecules such that $E_{rot} \leq q\mu_D/r^2$, and E_2 the average energy of those with $E_{rot} \geq q\mu_D/r^2$, then

$$\bar{\theta}(r) = \bar{\theta}_1(r)F_1 + \bar{\theta}_2(r)F_2 \tag{36}$$

and

$$\frac{d\bar{\theta}}{dr} = \bar{\theta}_1\frac{dF_1}{dr} + F_1\frac{d\bar{\theta}_1}{dr} + \bar{\theta}_2\frac{dF_2}{dr} + F_2\frac{d\bar{\theta}_2}{dr}. \tag{37}$$

F_1 and F_2 are the fractions of molecules that have E_{rot} less than and greater than $q\mu_D/r^2$, respectively. E_1, E_2, F_1, and F_2 are calculated from Eq. (34).

The microscopic rate constant $k(v)$ is given by

$$k(v) = v\langle\sigma(v)\rangle. \tag{38}$$

For thermal velocities, the macroscopic rate constant is

$$\bar{k}_{therm} = \int_0^\infty v\langle\sigma(v)\rangle P(v)\,dv, \tag{39}$$

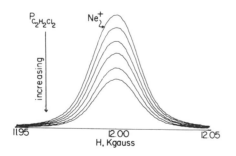

FIG. 2. ICR lineshape data for the reaction Ne+ with *cis*-C₂H₂Cl₂. Neon pressure 5×10^{-5} torr, C₂H₂Cl₂ pressure (max) 7×10^{-6} torr, $\Delta H = 27\pm0.6$ G.

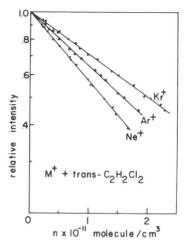

FIG. 3. Logarithm of relative intensities of rare gas ions as a function of *trans*-C₂H₂Cl₂ concentration n. Reaction time is 1.85×10^{-3} sec.

where $P(v)$ is the thermal velocity distribution function, and is given by

$$P(v) = 4\pi(\mu/2\pi kT)^{3/2}\exp(-\mu v^2/2kT)v^2. \tag{40}$$

Theoretical thermal rate constants are obtained by integrating Eq. (39) numerically. The parametric values of v and $\langle\sigma(v)\rangle$ are found graphically from Eqs. (14) and (15) using the above determined values of $\bar{\theta}$ and $\partial\theta/\partial r$. Further details on the method of calculating these quantities are given in the Appendix.

EXPERIMENTAL

A homemade ion cyclotron resonance (ICR) spectrometer of conventional design is used for studying the charge transfer reactions. The detailed description of the instrument is given elsewhere.[19,20] All experiments were carried out at about 300°K. The total ion currents were maintained less than 5×10^{-12} A.

The detecting rf level was maintained at the absolute minimum to obtain near thermal reactant ion energies. Pressures were measured on a Granville–Phillips ion gauge, calibrated by an MKS Baratron capacitance manometer. Reaction times were determined by the trapping ejection technique.[21] The measured drift times were very close to calculated drift times in all cases. Primary ions were formed by impact of 30 eV electrons. The gases were introduced into the ICR cell through two separate Granville–Phillips variable leaks.

Experimental rate constants were obtained as follows. The rare gas was admitted into the ICR cell at a pressure between 1×10^{-5} and 1×10^{-4} torr and kept constant at that pressure for the entire period of the experiment. After tuning up on the rare gas ion peak, the target gas was introduced into the cell at various

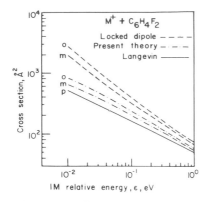

FIG. 4. Theoretical angle averaged cross sections for difluorobenzene isomers as a function of ion–molecule (IM) relative translational energy ϵ at 300°K.

pressures. The magnetic field was swept through the rare gas ion peak for each pressure of the target gas. Figure 2 shows the signal of Ne^+ at various pressures of cis-$C_2H_2Cl_2$ as an example. Notice that the peak shape and the position of the peak was essentially the same at the various target gas pressures. The half-width at half-height in Fig. 2 is 13.7 ± 0.3 G. The peak height of rare gas ion at zero target gas pressure remained the same at the beginning and the end of experiment. The relative peak heights are taken to be the relative intensities of the rare gas ion at the various target gas pressures. The rate constant is calculated from the slope of a plot of the logarithm of the relative rare gas ion intensity vs concentration (in molecules/cm³) of the target gas. The concentration of the target molecule is calculated from the ideal gas law. Figure 3 shows such plots for Ne^+, Ar^+, and Kr^+ with $trans$-$C_2H_2Cl_2$ as the target gas. Each straight line is obtained from at least three separate experiments. Figures 2 and 3 are representative of all data. Each experiment has been repeated several times over a period of several months. The deviation of the rate constants obtained is generally less than 5%. The absolute accuracy of the rate constants is of the order of $\pm15\%$.

The rare gases were purchased from Matheson and were used without further purification. The difluorobenzene isomers were obtained from Aldrich Chemical Co. and the dichloroethylene isomers from Chemical Sample Co. The isomers of both sets were thoroughly outgassed before use.

RESULTS

The theoretical predictions of microscopic cross sections as a function of ion–molecule relative kinetic energy for difluorobenzene (DFB) and dichloroethylene (DCE) geometric isomers are given in Figs. 4 and 5, respectively, together with the "locked dipole" cross sections predicted from Eq. (4) ($\sigma = k/v$). Polarizabili-

ties are estimated by the LeFevre method[22] to be 8.8 and 7.81 Å³ for DFB and DCE, respectively. Dipole moments are as follows:

DFB: $para = 0.00$ D; $meta = 1.58$ D [10]; $ortho = 2.40$ D [23]

DCE: $trans = 0.00$ D; $1,1 = 1.34$ D [10]; $cis = 1.90$ D.[10]

Figures 6 and 7 show the average polar angle $\bar{\theta}$ at various ion–dipole separations r.

Tables I and II list the experimental thermal energy charge transfer rate constants from rare gas ions to DFB and DCE isomers, respectively, along with the present theoretical predicted rate constants calculated from Eqs. (39) and (40), and those predicted from Eq. (5) for the "locked dipole". The last column is the ratio of experimental and theoretical rate constant for nonpolar isomers ($trans$-DCE and p-DFB).

DISCUSSION

Reaction Mechanism

By comparing the experimental rate constants of nonpolar isomers with charge–induced dipole theory (last column of Tables I and II), several phenomena are apparent: (1) All nonpolar rate constants are substantially larger than predicted by charge–induced dipole theory. (2) Except for Xe^+, the rate constants exhibit a $\mu^{1/2}$ dependence. (3) There is no apparent dependence on energy defect.[24] The large rate constant and lack of dependence on energy defect have been previously interpreted in the difluoroethylene communication.[7] Masson, Birkenshaw and Henchman report that $\approx50\%$ of the reactive charge exchange collisions of Ar^+ on CH_4 proceed *without momentum transfer* between 1.5 and 4.2 eV Ar^+ ion energy. Recent papers by Bowers and Elleman[26] and Gauglhofer and Kevan[27] suggest that thermal energy charge transfer

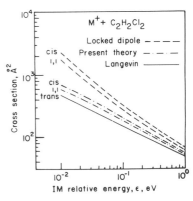

FIG. 5. Theoretical angle averaged cross sections for dichloroethylene isomers as function of IM relative translational energy ϵ at 300°K.

rate constants reflect the Franck–Condon manifold for ionization of the target molecule (with some distortion due to the ion–molecule collision). That is, reactive, thermal energy charge transfer collisions prefer to proceed via a sudden, nonadiabatic electron jump. Since the electron is a single, very light particle, transfers can take place at long range beyond the maximum impact parameter for a capture collision. The experimental evidence presented here, and elsewhere,[7,26,27] is in agreement with this statement. For the sake of argument, let us separate the reactive charge transfer collisions into two categories: (1) those outside the maximum impact parameter for capture collisions and (2) those within the capture limit. The first class of reactions presumably proceeds via nonadiabatic electron jump and occur only with favorable Franck–Condon factors. The second class of reactions may or may not proceed via an electron jump mechanism at large r. It seems reasonable, however, that the matrix elements for the electron transfer would be functions of r only. Thus, at thermal velocities, for systems with favorable Franck–Condon factors the mechanism should be independent of impact parameter and both classes of reaction proceed via the same mechanism. For systems with less favorable Franck–Condon factors than those reported here, long range charge transfer is less likely and ion–molecule collisions at small impact parameters are necessary to physically distort the neutral target molecule.

The $Xe^{+\cdot}$ rate is interesting in that it is somewhat lower than expected on a purely $\mu^{1/2}$ basis in both difluorobenzene and dichloroethylene. In each case the $He^{+\cdot}$, $Ne^{+\cdot}$, $Ar^{+\cdot}$, and Kr^{+} data follow $\mu^{1/2}$ closely. These latter four ions lead to dissociation via the reactions

$$(C_2H_2Cl_2^{+\cdot})^* \rightarrow C_2H_2Cl^{+} + \cdot Cl, \quad (41)$$

$$(C_6H_4F_2^{+\cdot})^* \rightarrow C_6H_3F^{+\cdot} + HF, \quad (42)$$

while in both cases $Xe^{+\cdot}$ leads only to the nondissoci-

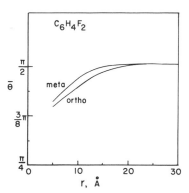

FIG. 6. Average dipole orientation angle $\bar{\theta}$ as a function of IM separation r, for *meta*- and *ortho*-difluorobenzene.

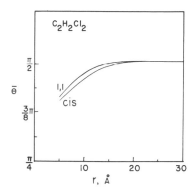

FIG. 7. Average dipole orientation angle $\bar{\theta}$ as a function of IM separation r, for 1, 1- and *cis*-dichloroethylene.

ated parent ion. The above reactions have been observed in the ICR and confirmed using resonant ejection double resonance. One possible explanation is that the longest range electron jump reactions are forbidden for $Xe^{+\cdot}$ because insufficient distortion has occurred in the neutral molecule to make the Franck–Condon factor for a transition to a bound state of the molecular ion sufficiently large. At somewhat smaller impact parameters the necessary distortion occurs and the reaction proceeds. The Franck–Condon factors to the dissociative states are large in all cases as the photoelectron spectra indicate.[28]

The raw experimental data as well as the theoretical predictions are presented in Tables I and II. Clearly the data for the polar isomers are approximately a factor of two larger than predicted by the current average-dipole-orientation theory. This fact is not surprising since the rate constants for the nonpolar isomers are also about a factor of two larger than the pure polarization theory predicts. It is clear that a large percentage of the collisions occur outside the capture limit as previously discussed. The current theory is a capture collision theory and is thus not suitable for direct comparison with the raw charge transfer data. It should be noted, however, that even the raw data exhibit about the amount of increase with dipole the average-dipole-orientation theory predicts and a much smaller increase than predicted by the locked dipole limit.

In order to better compare theory with experiment, consider the following model. Again let us break the total charge transfer rate into two components:

$$k_T = k_{cap} + k_{LR}, \quad (43)$$

where k_{cap} is a capture component and k_{LR} is a long range component occurring outside the capture limit; k_{cap} would be the quantity to ideally compare with theory. From (43), $k_{cap} = k_T - k_{LR}$; k_{LR} can be estimated from the nonpolar isomer data by subtracting the the-

TABLE I. Experimental and theoretical rate constants for rare gas ion–difluorobenzene charge transfer reactions at 300°K (all rate constants $\times 10^9$ cm³ molecule⁻¹·sec⁻¹).

| | Experiment | | | Theory | | | | | |
| | | | | Induced[a] dipole | Present theory | | Locked dipole[b] | | para/Induced dipole |
	para	meta	ortho		meta	ortho	meta	ortho	
He⁺	7.15	8.24	9.05	3.56	4.57	5.56	11.09	14.98	2.01
Ne⁺	3.41	3.76	4.21	1.70	2.18	2.64	5.28	7.15	2.00
Ar⁺	2.58	3.08	3.26	1.28	1.65	2.01	4.00	5.41	2.01
Kr⁺	2.02	2.30	2.47	1.00	1.29	1.57	3.12	4.22	2.02
Xe⁺	1.36	1.44	1.63	0.90	1.15	1.40	2.79	3.78	1.51

[a] Calculated from Eq. (2).
[b] Calculated from Eq. (5).

oretical capture rate constant from the experimental number. If it is assumed that the resultant long range component is approximately the same for all of the isomers, the numbers listed under experiment in Tables III and IV are obtained. What effectively has been done is to "normalize" the data to the nonpolar capture limit to see how accurately the theory predicts the effect of the dipole. The agreement in Tables III and IV is excellent. Considering the approximations made in obtaining the "experimental" data, the degree of agreement is no doubt fortuitous. It is encouraging, however, that the theory predicts dipole effects of approximately the correct magnitude.

A more valid area of comparison is in the observed and calculated values of the average orientation angle $\langle \bar{\theta} \rangle$. This can be done in the following way. The param-

eter c in Eq. (6) can be determined by the relation

$$\frac{k_{\text{dipolar}}}{k_{\text{nonpolar}}} = \frac{(2\pi q/\mu^{1/2})[\alpha^{1/2} + c\mu_D (2/\pi kT)^{1/2}]}{2\pi q(\alpha/\mu)^{1/2}}, \quad (44)$$

where k_{dipolar} stands for the 1,1, cis, ortho, or meta species and k_{nonpolar} stands for the trans or para species. Table V compares the experimental and theoretical c values obtained from the rate constants in Tables III and IV and the corresponding overall average librational angles ($c = \cos\langle \bar{\theta} \rangle$). (In the previous study of difluoroethylene isomers,[7] the parameter c was calculated in a similar manner but the long range contributions to the rate constant were not subtracted out. The total rate constants were used resulting in a value of $\langle \bar{\theta} \rangle \cong 87°$. If the larger than capture contributions were subtracted, c for the difluoroethylenes would be

TABLE II. Experimental and theoretical rate constants for rare gas ion–dichloroethylene charge transfer reactions at 300°K (all rate constants $\times 10^9$ cm³ molecule⁻¹·sec⁻¹).

| | Experiment | | | Theory | | | | | |
| | | | | Induced[a] dipole | Present theory | | Locked dipole[b] | | trans/Induced dipole |
	trans	1,1	cis		1,1	cis	1,1	cis	
He⁺	6.65	7.75	8.20	3.35	4.13	4.79	9.76	12.44	1.98
Ne⁺	3.00	3.48	3.72	1.61	1.99	2.31	4.69	5.98	1.86
Ar⁺	2.40	2.75	2.93	1.23	1.52	1.76	3.59	4.63	1.95
Kr⁺	1.85	2.15	2.27	0.98	1.21	1.40	2.86	3.63	1.89
Xe⁺	1.20	1.32	1.68	0.88	1.08	1.26	2.56	3.26	1.36

[a] Calculated from Eq. (2).
[b] Calculated from Eq. (5).

~0.125 corresponding to $\langle\bar{\theta}\rangle = 82.8°$.) Experimental values of $\langle\bar{\theta}\rangle$ are nearly $\pi/2$ and decrease slightly with increasing dipole moment. Theoretical predictions agree with the experimental results although theory predicts a slightly greater dependence on dipole moment. As may be seen from Figs. 6 and 7, $\bar{\theta}$ is always close to $\pi/2$ for all r in the range of interest. It becomes slightly larger than $\pi/2$ at larger r where the potential energy is small and most molecules can rotate in a complete circle. Since at $\theta = 180°$ the potential energy is maximum and the kinetic energy is minimum, the negative end of the dipole spends more time on the half sphere where $\theta > \pi/2$ causing $\bar{\theta}$ to be slightly larger than $\pi/2$. As r approaches infinity, $\bar{\theta}$ will eventually approach $\pi/2$. Qualitatively, the average thermal rotational energy is comparable to the potential energy of the ion-dipole pair even at rather small separations (5–10 Å). The polar molecule is always rotating and only slightly hindered by the ion–dipole potential energy. As the

TABLE III. Capture collision rate constants of rare gas-difluorobenzene charge transfer reactions at 300°K (all rate constants$\times 10^9$ cm^3 molecule^{-1}·sec^{-1}).

	Experiment[a]			Theory		
	para	meta	ortho	para	meta	ortho
He$^+$	3.56	4.65	5.46	3.56	4.57	5.56
Ne$^+$	1.70	2.05	2.50	1.70	2.18	2.64
Ar$^+$	1.28	1.78	1.96	1.28	1.65	2.01
Kr$^+$	1.00	1.30	1.47	1.00	1.29	1.57

[a] See text for discussion of how these data were obtained.

dipole moment increases, the hindering becomes more effective and hence a smaller value of $\langle\bar{\theta}\rangle$ is obtained.

SUMMARY

The present theory assumes that the rotational energy is high enough so that it is within the classical limit. It does not apply to very low temperatures. The theory does not consider the precise motion of each individual polar molecule but treats the rotational motion of the molecules in a sense of overall average. Further, it assumes there is no net angular momentum transfer between the system and the rotor and that the rotor can be approximated by a two-dimensional model. It turns out that the calculated rate constant is independent of the moment of inertia of the molecule. From the experimental results of this work and the previous study of difluoroethylene,[7] it seems that there is no large experimental dependence on the moment of inertia. The theory also assumes that the ion-molecule separation is large compared to the molecular size. The rate is thus predicted to be independent of the size of the molecule. This only applies to low ion

TABLE IV. Capture collision rate constants of rare gas-dichloroethylene charge transfer reactions at 300°K (all rate constants$\times 10^9$ cm^3 molecule^{-1}·sec^{-1}).

	Experiment[a]			Theory		
	para	meta	ortho	para	meta	ortho
He$^+$	3.35	4.45	4.90	3.35	4.13	4.79
Ne$^+$	1.61	2.09	2.33	1.61	1.99	2.31
Ar$^+$	1.23	1.58	1.76	1.23	1.52	1.76
Kr$^+$	0.98	1.28	1.40	0.98	1.21	1.40

[a] See text for discussion of how these data were obtained.

energies where the capture cross section is relatively large. For high ion energies, the molecular size should be taken into account. However, at high energies, the capture concept may not be important and other theoretical models may be appropriate.

Finally, the theory is compared to dipolar effects observed in charge transfer collisions. The fact that the observed rate constants are considerably larger than the capture limit makes a detailed comparison with theory difficult. It is apparent, however, that even in these systems the theory predicts the dipole effect with some accuracy. When the raw data are "normalized" to the capture limit, excellent agreement is obtained. Other types of reaction are being investigated where such corrections are unnecessary. Proton transfer reactions from H_3^+ and CH_5^+ to the DCE, DFE, and DFB isomers give excellent agreement with theory when the raw data are used.[29] In these cases the proton transfer reactions proceed at essentially the capture limit. Since the current theory is a "capture limit" theory, such reactions should provide a useful test. Finally, the energy dependence of the rate constant (cross section) provides an important test of the theory. Preliminary results indicate the average-dipole-orientation theory correctly represents the energy dependence of several proton transfer reactions at energies below a few eV.[30,31]

ACKNOWLEDGMENTS

The support of the National Science Foundation under Grant GP-15628 is gratefully acknowledged. We

TABLE V. Orientation of the dipole.

	Experiment		Theory	
	C	$\langle\bar{\theta}\rangle$	C	$\langle\bar{\theta}\rangle$
1,1-$C_2Cl_2H_2$	0.154±0.006	81.2±0.3°	0.124	82.9°
cis-$C_2Cl_2H_2$	0.167±0.005	80.5±0.3°	0.160	80.8°
m-$C_6H_4F_2$	0.13±0.01	82.5±0.5°	0.135	82.2°
o-$C_6H_4F_2$	0.15±0.01	81.5±0.5°	0.177	79.8°

also gratefully acknowledge many useful discussions with Dr. John Dugan with regard to many aspects of the theoretical development.

APPENDIX

The relation between capture cross section and ion–molecule relative velocity is given by Eqs. (A1) and (A2):

$$\langle\sigma\rangle = \pi r_k^2 + (\pi q^2\alpha/r_k^2\mu v^2) + (2\pi q\mu_D/\mu v^2)\cos\bar\theta\,(r=r_k), \tag{A1}$$

$$v^2 = (q^2\alpha/\mu r_k^4) + (q\mu_D/r_k\mu)\sin\bar\theta\,(r=r_k)\,(d\bar\theta/dr)r_k, \tag{A2}$$

where

$$\bar\theta(r) = \bar\theta_1(r)F_1 + \bar\theta_2(r)F_2, \tag{A3}$$

$$\frac{d\bar\theta}{dr} = \bar\theta_1\frac{dF_1}{dr} + F_1\frac{d\bar\theta_1}{dr} + \bar\theta_2\frac{dF_2}{dr} + F_2\frac{d\bar\theta_2}{dr}. \tag{A4}$$

The following equations are used to calculate $\bar\theta_1(r)$ and $d\bar\theta_1/dr$:

$$\bar\theta_1 = 2\sqrt{2}A/(1-\cos K)^{1/2} \tag{A5}$$

and

$$\frac{d\bar\theta_1}{dr} = \frac{-\sqrt{2}A\,\sin K(dK/dr)}{(1-\cos K)^{3/2}} + \frac{2\sqrt{2}(dA/dr)}{(1-\cos K)^{1/2}}, \tag{A6}$$

where

$$A = \tfrac{1}{2}\pi\,\sin^2\tfrac{1}{2}K\left[\tfrac{1}{2} + \sum_{n=1}^{\infty}\left(\frac{1\times3\cdots(2n-1)}{2\times4\cdots(2n)}\right)^2 (2n+2)^{-1}\sin^{2n}\tfrac{1}{2}K\right], \tag{A7}$$

$$\frac{dA}{dr} = \tfrac{1}{2}\pi\,\sin\tfrac{1}{2}K\,\cos\tfrac{1}{2}K\,\frac{dK}{dr}\left[\tfrac{1}{2} + \sin\tfrac{1}{2}K\sum_{n=1}^{\infty}\left(\frac{1\times3\cdots(2n-1)}{2\times4\cdots(2n)}\right)^2\frac{n}{2n+2}\sin^{2n-1}\tfrac{1}{2}K\right.$$
$$\left. + \sum_{n=1}^{\infty}\left(\frac{1\times3\cdots(2n-1)}{2\times4\cdots(2n)}\right)^2 (2n+2)^{-1}\sin^{2n}\tfrac{1}{2}K\right], \tag{A8}$$

$$K = \cos^{-1}(-E_1 r^2/q\mu_D), \tag{A9}$$

$$\frac{dK}{dr} = \frac{(2rE_1/q\mu_D) + (r^2/q\mu_D)(dE_1/dr)}{\sin K}, \tag{A10}$$

$$E_1 = 0.841\{kT - [zE_m/(1-z)]\}, \tag{A11}$$

$$\frac{dE_1}{dr} = -0.841\left[\frac{E_m}{(1-z)^2}\frac{dz}{dr} + \frac{z}{1-z}\frac{dE_m}{dr}\right], \tag{A12}$$

$$E_m = q\mu_D/0.841r^2, \tag{A13}$$

$$dE_m/dr = -2q\mu_D/0.841r^3, \tag{A14}$$

$$z = \exp(-E_m/kT), \tag{A15}$$

and

$$dz/dr = -(z/kT)(dE_m/dr). \tag{A16}$$

The following equations are used to calculate $\bar\theta_2(r)$ and $d\bar\theta_2/dr$:

$$\bar\theta_2 = B/C \tag{A17}$$

and

$$\frac{d\bar\theta_2}{dr} = C^{-1}\frac{dB}{dr} - \frac{B}{C^2}\frac{dC}{dr}, \tag{A18}$$

where

$$B = H + P\left(\tfrac{1}{2}\pi - \tfrac{1}{2}\pi\sum_{n=1}^{\infty}\frac{1\times3\cdots(2n-1)}{2^{2n}(n!)^2(2n-1)}\gamma^{2n}\right), \tag{A19}$$

$$\frac{dB}{dr} = \frac{dH}{dr} + \frac{dP}{dr}\left(\tfrac{1}{2}\pi - \tfrac{1}{2}\pi\sum_{n=1}^{\infty}\frac{1\times3\cdots(2n-1)}{2^{2n}(n!)^2(2n-1)}\gamma^{2n}\right) + P\left(\pi\sum_{n=1}^{\infty}\frac{n[1\times3\cdots(2n-1)]^2}{2^{2n}(n!)^2(2n-1)}\gamma^{2n-1}\frac{d\gamma}{dr}\right), \tag{A20}$$

$$H = -(2E_2^{1/2}r^2/q\mu_D)\pi(1-q\mu_D/E_2 r^2)^{1/2}, \tag{A21}$$

$$P = (4E_2^{1/2}r^2/q\mu_D)(1 + q\mu_D/E_2r^2)^{1/2}, \tag{A22}$$

$$\gamma^2 = 2p\mu_D/(q\mu_D + E_2r^2), \tag{A23}$$

$$\frac{dH}{dr} = -\frac{\pi}{q\mu_D}\left(\frac{r}{E_2}\frac{dE_2}{dr} + 4\right)(E_2r^2 - q\mu_D)^{1/2} \tag{A24}$$

$$-\pi\left(\frac{2}{r} + E_2^{-1}\frac{dE_2}{dr}\right)\bigg/\left(E_2 - \frac{q\mu_D}{r^2}\right)^{1/2}, \tag{A25}$$

$$\frac{dP}{dr} = \frac{2}{q\mu_D}\left(\frac{r}{E_2}\frac{dE_2}{dr} + 4\right)(E_2r^2 + q\mu_D)^{1/2} \tag{A26}$$

$$-2\left(\frac{2}{r} + E_2^{-1}\frac{dE_2}{dr}\right)\bigg/\left(E_2 + \frac{q\mu_D}{r^2}\right)^{1/2}, \tag{A27}$$

$$\frac{d\gamma}{dr} = \frac{-2q\mu_D[2E_2r + r^2(dE_2/dr)]}{(q\mu_D + E_2r^2)^2}, \tag{A28}$$

$$C = (2r^2/q\mu_D)\{[E_2 + (q\mu_D/r^2)]^{1/2} - [E_2 - (q\mu_D/r^2)]^{1/2}\} \tag{A29}$$

$$\frac{dC}{dr} = (q\mu_D)^{-1}\left(\frac{r}{E_2}\frac{dE_2}{dr} + 4\right)[(E_2r^2 + q\mu_D)^{1/2} - (E_2r^2 - q\mu_D)^{1/2}] - \left(E_2^{-1}\frac{dE_2}{dr} + \frac{2}{r}\right)\left[\left(E_2 + \frac{q\mu_D}{r^2}\right)^{-1/2} + \left(E_2 - \frac{q\mu_D}{r^2}\right)^{-1/2}\right], \tag{A30}$$

$$E_2 = 0.841(E_m + kT), \tag{A31}$$

and

$$dE_2/dr = 0.841(dE_m/dr). \tag{A32}$$

F_1, F_2, dF_1/dr, and dF_2/dr are as follows:

$$F_1 = 1 - z, \qquad dF_1/dr = -dz/dr \tag{A33}$$

and

$$F_2 = z, \qquad dF_2/dr = dz/dr. \tag{A34}$$

*Supported by Grant GP-15628 of the National Science Foundation.

[1]L. P. Theard and W. H. Hamill, J. Am. Chem. Soc. **84**, 1134 (1962).

[2]S. K. Gupta, E. G. Jones, A. G. Harrison, and J. J. Myher, Can. J. Chem. **45**, 3107 (1967).

[3]K. R. Ryan, J. Chem. Phys. **52**, 6009 (1970).

[4]M. P. Langevin, Ann. Chim. Phys. **5**, 245 (1905).

[5]G. Gioumousis and D. P. Stevenson, J. Chem. Phys. **29**, 294 (1958).

[6]T. F. Moran and W. H. Hamill, J. Chem. Phys. **39**, 1413 (1963).

[7]M. T. Bowers and J. B. Laudenslager, J. Chem. Phys. **56**, 4711 (1972).

[8]J. A. Bevan and L. Kevan, J. Chem. Phys. **73**, 3860 (1969).

[9]R. Bralsford, P. V. Harris, and W. C. Price, Proc. R. Soc. A **258**, 459 (1960).

[10]R. D. Nelson, Jr., D. R. Tide, Jr., and A. A. Maryott, Natl. Stand. Ref. Data Ser. **10**, (1967).

[11]D. Rapp and W. E. Francis, J. Chem. Phys. **37**, 2631 (1962).

[12]J. V. Dugan and J. L. Magee, NASA Report TN-D-3229, February, 1966.

[13]J. V. Dugan, J. H. Price, and J. L. Magee, Chem. Phys. Lett. **3**, 323 (1969).

[14]J. V. Dugan and J. L. Magee, J. Chem. Phys. **47**, 3103 (1967).

[15]J. V. Dugan, 18th Ann. Conf. Mass Spec. Allied Topics, San Francisco, p. 395 (1970).

[16]E. W. McDaniel, *Collision Phenomena in Ionized Gases* (Wiley, New York, 1964).

[17]W. E. Byerly, *Integral Calculus* (Stechert, New York, 1964), 2nd ed., pp. 215–282.

[18]T. L. Hill, *Statistical Thermodynamics* (Addison Wesley, London, 1962), pp. 153–159.

[19]M. T. Bowers and P. R. Kemper, J. Am. Chem. Soc. **93**, 5352 (1971).

[20]M. T. Bowers, D. H. Aue, and D. D. Elleman, J. Am. Chem. Soc. **94**, 4225 (1972).

[21]T. B. McMahon and J. L. Beauchamp. Rev. Sci. Instrum. **42**, 1632 (1971).

[22]R. J. W. LeFevre, Adv. Phys. Org. Chem. **3**, 1 (1965).

[23]A. L. McClellan, *Tables of Experimental Dipole Moment* (Freeman, San Francisco, 1963).

[24]H. S. W. Massey and E. H. S. Burhop, *Electronic and Ionic Impact Phenomena* (Clarendon, Oxford, England, 1965).

[25]A. J. Masson, K. Birkenshaw, and M. J. Henchman, J. Chem. Phys. **50**, 4112 (1969).

[26]M. T. Bowers and D. D. Elleman, Chem. Phys. Letters (to be published).

[27]J. Gauglhofer and L. Kevan, Chem. Phys. Letters (to be published).

[28]D. W. Turner, C. Baker, A. D. Baker, and C. R. Brundle, *Molecular Photoelectron Spectroscopy* (Wiley-Interscience, New York, 1970).

[29]T. Su and M. T. Bowers, J. Amer. Chem. Soc. (to be published).

[30]M. T. Bowers, T. Su, and V. G. Anicich, J. Chem. Phys. (to be published).

[31]M. T. Bowers and T. Su, Adv. Elec. Electron. Phys. (to be published).

15

Reprinted from *J. Chem Phys.* **43**:3209–3219 (1965)

Phase-Space Theory of Chemical Kinetics. II. Ion–Molecule Reactions*

J. C. LIGHT AND J. LIN

The revised phase-space (or statistical) theory of chemical kinetics has been used to compute cross sections for simple ion–molecule reactions. Reaction rates and isotope effects were investigated for bimolecular exchange reactions of the types

$$X^+ + HD \rightarrow \begin{cases} X^+ + HD \\ XH^+ + D, \\ XD^+ + H \end{cases} \quad (X \text{ is He, Ne, Ar, Kr})$$

$$X + HD^+ \rightarrow \begin{cases} X + HD^+ \\ XD^+ + H \\ XH^+ + D. \end{cases}$$

Dissociative charge-transfer reactions of rare-gas ions with CO were also investigated. Reasonable agreement with experiment, with no adjustable parameters used in the calculations, indicates that the phase-space theory has considerable predictive value in *ab initio* calculations of reactive cross sections for low-energy ion–molecule reactions.

INTRODUCTION

THE phase-space theory of chemical kinetics recently proposed by one of the authors[1] is based on the following hypothesis:

The probability of formation of any given product in a "strong coupling" collision if proportional to the ratio of the phase space available to that product divided by the total phase space available with conservation of energy and angular momentum.

Since the possible justifications of the validity of the hypothesis are given in (I), they are not repeated here. The purposes of this paper are to present the revised[2] formulas for computing the phase space available, to investigate the possibilities of computing dissociation cross sections (where a strict interpretation of "phase space available" would lead to ridiculous results), and to examine the theory by a rather thorough comparison of theory and experiment.

In the first section we present the explicit formulas for the classical phase space available to a particular product in a particular vibrational quantum state. These differ from the formulas of (I) where a "box normalization" was used since it was found that a "time normalization' must be used to obtain cross sections which satisfy microscopic reversibility. That the present theory is the correct and essentially unique statistical theory satisfying microscopic reversibility

was shown in (II). The modifications of the formula are relatively minor, and the essential definitions "strong coupling collisions" remain unchanged from (I

In the second section the dissociative charge-transf reactions of rare-gas ions with CO are discussed. Th somewhat arbitrary definitions of dissociation are i vestigated, and it is found that the model proposed (I) for this process gives reasonable agreement wi experiment. In the third section we consider the ap plication of the theory to the computation of cro sections and isotope ratios for the reactions of rare-ga atoms (and ions) with HD+ (and HD). The conserv tion of angular momentum poses rather severe restri tions on these reactions, and these effects are di cussed. The last section is a brief summary of the ap plicability of the phase space hypothesis to chemic kinetics.

I. PHASE-SPACE THEORY

As in (I), the transformation of the phase-spa hypothesis into equations proceeds in two distin steps: the reduction of the complete phase space to t hypersurface on which energy and angular momentu are conserved; and finding the limits of integratic over this hypersurface which are determined by t definition of strong coupling collisions. The limits r main unchanged from (I), but the phase-space el ment given in Eq. (I.8)[3] (after cancellation of d leads to cross sections which do not satisfy microscop reversibility. As was shown in (II), the elements d and dr must be transformed to the canonical pair d and dt, where r and p_r are the radial coordinate a conjugate momentum of the diatomic molecule and t

* This research was supported by grants from the National Science Foundation.

[1] J. C. Light, J. Chem. Phys. **40**, 3221 (1964), hereafter referred to as I.
[2] P. Pechukas and J. C. Light, J. Chem. Phys. **42**, 3281 (1965), hereafter referred to as II.
[3] For example, (I. 8) refers to Eq. (8) of Ref. 1.

third body, and E and t are the total energy and time. If this is done, the element dt may be canceled in numerator and denominator of the expression (I.1) for the probability of reaction, and the resulting cross sections satisfy microscopic reversibility.

In the notation of (I), if $\Gamma_\beta(E, \mathbf{J}, i_\beta)$ is the phase space available to the product β in vibrational state i_β for total energy E and angular momentum $\mathbf{J}$ of the complex, then the probability of formation of β, i_β is given by (I.1):

$$P(\beta, i_\beta \mid E, \mathbf{J}) = \Gamma_\beta'(E, \mathbf{J}, i_\beta) / \sum_{K=\alpha,\beta\cdots} \sum_{i_K} \Gamma_K(E, \mathbf{J}, i_K), \tag{1}$$

where $\Gamma_\beta'(E, \mathbf{J}, i_\beta)$ excludes those regions of phase space in which the product β, i_β may dissociate. To calculate the available phase space Γ, the entire (three-particle) phase space is reduced as follows. The phase space element of three particles is

$$d\tau = d^2\mathbf{R}d^3\mathbf{P}[drdp_r d J_{\mathrm{orb}} d J_{z\,\mathrm{orb}} d\alpha_{\mathrm{orb}} d\beta_{\mathrm{orb}}]$$
$$\times [dr'dp_r'][d J_{\mathrm{rot}} d J_{z\,\mathrm{rot}} d\alpha_{\mathrm{rot}} d\beta_{\mathrm{rot}}], \tag{2}$$

where R, P are the center-of-mass (barycentric) coordinates; r and p_r are the radical coordinate and momentum of the third body with respect to the center of mass of the diatomic; r' and p' refer to the vibrational coordinates of the diatomic molecule and are treated quantum mechanically; J_{orb}, α_{orb}, $J_{z\,\mathrm{orb}}$, and β_{orb} are the orbital angular momentum, its z component, and the two angle variables conjugate to them. The variables subscripted "rot" are those for the rotational angular momentum of the diatomic.

The phase-space element on the hypersurface of constant energy and angular momentum is found by transforming $drdp_r d J_{\mathrm{orb}} d J_{z\,\mathrm{orb}} d\beta_{\mathrm{orb}}$ to $dEdtdJdJ_zd\beta = d\tau'$. The Jacobian of this transformation is just J/J_{orb}, so the volume element of the hypersurface is

$$d\Gamma = \frac{Jd\alpha_{\mathrm{orb}}d\alpha_{\mathrm{rot}}d\beta_{\mathrm{rot}}d J_{\mathrm{rot}}d J_{z\,\mathrm{rot}}}{[J^2 + J_{\mathrm{rot}}^2 - 2J J_{z\,\mathrm{rot}}]^{\frac{1}{2}}},$$
$$d\tau = d\Gamma d\tau', \tag{3}$$

where we have chosen the z axis in the direction of $\mathbf{J}$. Since only the ratios of phase space volumes enter into Eq. (1), and the integrals over α_{orb}, α_{rot}, and β_{rot} are independent of channel and vibrational state, the ratio in (1) will be given correctly if we take

$$\Gamma_\beta(E, \mathbf{J}, i_\beta) = \iint \frac{d J_{\mathrm{rot}}d J_{z\,\mathrm{rot}}}{[1 + (J_{\mathrm{rot}}/J)^2 - 2J_{z\,\mathrm{rot}}/J]^{\frac{1}{2}}}. \tag{4}$$

Note that this differs from (I.9) by the absence of the factor $(1/v) = (\mu/\epsilon)^{\frac{1}{2}}[1 - J_{\mathrm{rot}}^2/2I\epsilon]^{-\frac{1}{2}}$ in the integral. This is due to the fact that we have transformed the canonical pair $drdp_r$ to dE_Tdt. This formulation, as shown in (II), does satisfy microscopic reversibility.

The limits of the integral in Eq. (4) must now be determined from our definition of a "strong coupling" collision and a definition of what constitutes a stable product. It is in these definitions that all the physics of the situation enter since, for each channel and vibrational state, the actual value of the integral in (4) is determined completely by them. The definitions must be used symmetrically (for each channel), that is, if a system $\{\beta\}$, described by β, i_β, E, $\mathbf{J}$, J_{rot}^β, v^β could not enter into a strong coupling collision, then a strong coupling complex formed from state $\{\alpha\}$, cannot dissociate into $\{\beta\}$.

For reactions without activation energy at low translational energy, the definition of a strong coupling collision used in (I) seems adequate. That is, the system must have enough translational energy to surmount the rotational barrier determined by the sum of the long-range attractive potential and the repulsive orbital angular momentum potential. Since this is the criterion used by Gioumousis and Stevenson,[4] and it was described fully in (I), the details are not presented here. Suffice it to say that if the long-range attractive forces are given by $V(r) = -a/r^n (n=4$ for ion–molecule reactions), then the maximum impact parameter leading to a strong coupling collision is

$$b_m = (na/2E_{\mathrm{transl}})^{\frac{1}{2}}[\tfrac{1}{2}(n-2)(a/E_{\mathrm{transl}})]^{(2-n)/2n}, \tag{5}$$

where E_{transl} is the initial translational energy. Since we chose the z axis in the direction of the total angular momentum vector, we have a minimum value of $J_{z\,\mathrm{rot}}$ below which the complex cannot separate into products given by

$$J_{z\,\mathrm{rot}} \geq [J^2 + J_{\mathrm{rot}}^2 - F^2(E, J, J_{\mathrm{rot}})]/2J, \tag{6}$$

where

$$F^2(E, J, J_{\mathrm{rot}}) = 2\mu[\tfrac{1}{2}(na)]^{2/n}$$
$$\times \{[n/(n-2)][E - E_{\mathrm{vib}} + Q - (J_{\mathrm{rot}}^2/2I)]\}^{1-(2/n)} \tag{7}$$

and μ is the reduced mass of the diatomic molecule and third body, I is the moment of inertia of the diatomic molecule, and Q is the exothermicity of the channel. As in (I), the lower limit of the integration over $J_{z\,\mathrm{rot}}$ is given by the larger of $(-J_{\mathrm{rot}})$ or Eq. (6), and the limits of the integration over J_{rot} are given by the roots of the equation

$$(J - J_{\mathrm{rot}})^2 - F^2(E, J, J_{\mathrm{rot}}) = 0. \tag{8}$$

If J_1 and J_2 are the roots of Eq. (8), $(J_2 > J_1)$ the phase-space integral becomes:

$$\Gamma = A \int_{\phi'}^{X_2} dX \int_{\phi}^{X} dX_z [1 + (X/Z)^2 - 2X_z/Z]^{-\frac{1}{2}}, \tag{9}$$

[4] G. Gioumousis and D. P. Stevenson, J. Chem. Phys. 29, 294 (1958).

where

$$A = 2I(E - E_{vib} + Q),$$

$$\phi' = |X_1| H(X_1),$$

$$\phi = \begin{cases} -X & X < |X_1| \\ (2Z)^{-1}[Z^2 + X^2 - C(1-X^2)^\gamma], & X > |X_1|, \end{cases}$$

$$\gamma = 1 - 2/n,$$

$$C = [\mu n/(n-2)I][(n-2)aI/A]^{2/n},$$

$$H(\chi) = \begin{cases} 0 & \chi < 0 \\ 1 & \chi > 0, \end{cases}$$

$$\text{sgn}(\chi) = \begin{cases} -1 & \chi < 0 \\ 1 & \chi > 0. \end{cases}$$

$$X_1 = J_1 A^{-\frac{1}{4}},$$

$$X_2 = J_2 A^{-\frac{1}{4}},$$

$$Z = J A^{-\frac{1}{4}},$$

All the computations of the following sections start with the solutions of Eq. (8) and the evaluation of integrals like Eq. (9). These computations were performed on the University of Chicago IBM 7094 computer. (A brief outline of the typical program used is given in the appendix.)

The integral can be done analytically except for one portion. The result is expressible compactly using the Heaviside and sgn functions:

$$\Gamma/A = I_1 + I_2 H(-X_1) + I_3 \,\text{sgn}(Z - X_2)$$
$$+ I_4 H(X_1)\,\text{sgn}(Z - X_1) - Z^2 H(X_2 - Z) H(Z - X_1), \quad (10)$$

where

$$I_1 = -C^{\frac{1}{4}} \int_{|X_1|}^{X_2} dX(1-X^2)^\gamma,$$

$$I_2 = Z|X_1| + \tfrac{1}{2}X_1^2,$$

$$I_3 = -ZX_2 + \tfrac{1}{2}X_2^2,$$

$$I_4 = Z|X_1| - \tfrac{1}{2}X_1^2.$$

This set of equations is the replacement for Eqs. (I.20.2) and (I.21) of the reference (I). It should be noted that Γ of Eq. (10) above may include regions of phase space in which the product diatomic molecule has more energy in rotation and vibration than the dissociation energy. We now turn to the problem of how to treat a system in which the total energy is large enough to dissociate the diatomic molecule.

Strictly speaking, the statistical theory of chemical reactions breaks down if dissociation is allowed since the discrete vibrational levels join the continuum and an integration over the phase space available to this continuum will diverge. This will happen even if the vibration is treated classically since the integration over r' will diverge. If this is done, the phase-space theory will predict complete dissociation whenever it is energetically possible. Practically speaking, however, the phase-space theory can be modified to allow both

dissociation and exchange reactions to occur with reasonable probabilities.

To do this we make a subsidiary assumption: Dissociation occurs, not by direct excitation to the vibrational continuum, but by excitation of the rotation of a discrete vibrational level above a critical value. That reasonable results can be obtained by this method is shown in the next section. The choice of the critical value of the rotational angular momentum for dissociation to occur is, again, somewhat arbitrary as is the definition of a strong coupling collision itself. It is necessary, however, to handle the dissociation process in a reasonable manner for ion–molecule reactions since most reactions are observed at energies where this is possible, if not probable.

Mathematically, the assumption above requires the definition of J_{rot}', a value of the rotational angular momentum (which may depend on the vibrational state) above which the product dissociates. We must, therefore, calculate the phase-space volumes $\Gamma_\beta'(E, \mathbf{J}, i_\beta)$ corresponding to stable products by limiting the integral over J_{rot} to values $J_{rot} < J_{rot}'$.

Setting

$$X' = J_{rot}'/A^{\frac{1}{4}}$$

and

$$Y = \begin{cases} X' & \text{if } X' < X_2, \\ X_2 & \text{if } X' \geq X_2, \end{cases}$$

we may write

$$\Gamma_\beta'(E, \mathbf{J}, i_\beta) = \Gamma(A, X_1, Z, Y) \quad X' > |X_1|, \quad (11)$$

where $\Gamma(A, X_1, Z, Y)$ is given by Eq. (10) with Y substituted for X_2 (where necessary). Since X' may be less than $|X_1|$ (which X_2 cannot be), we also have three new cases:

$$X' < X_1 \qquad \Gamma' = 0, \qquad\qquad\qquad (a)$$

$$X' < |X_1|; \quad X' < Z \quad \Gamma' = A(X')^2, \qquad (b)$$

$$X' < |X_1|; \quad X' > Z \quad \Gamma' = A(2ZX' - Z^2). \quad (c) \quad (12)$$

These formulas may also be used to determine the distribution of rotational energy, since Γ' corresponds to the phase space available with rotational energy less than J_{rot}'.

To convert the probabilities calculated from Eqs. (1)–(12) into cross sections, we make the approximation that the total angular momentum of the complex, J, is equal to the initial orbital angular momentum, i.e.,

$$J = \mu_i v_i b, \qquad (13)$$

where b is the impact parameter, v_i the initial velocity, and μ_i the reduced mass of the incoming channel. This assumption is very good for the cases considered here since the reduced mass of the initial diatomic molecules is not large compared with the reduced mass of the system in the initial channel. This leads

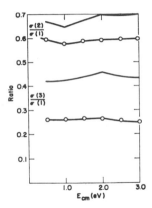

FIG. 1. Mass effect in model system. Channel (1), $M^4+(M^1M^2)^+$; (2), $M^1+(M^4M^1)^+$; (3), $M^1+(M^2M^4)^+$. Ratio of cross sections vs translational energy. ——, Present work; $\bigcirc$, Ref. 1.

to the cross section

$$\sigma_{\alpha,i_\alpha \to \beta,i\beta}(E_{\text{transl}})$$

$$= 2\pi \int_0^{b_m(\alpha,E_{\text{transl}})} b\,db\,P(\beta,i_\beta \mid E_{\text{transl}} + E_{i_u}, \mu, v, b), \quad (14)$$

where $b_m(\alpha, E_{\text{transl}})$ is the maximum impact parameter leading to a strong coupling collision defined in Eq. (5), and E_{i_α} is the initial vibrational energy.

The total cross section for the reaction $\alpha \to \beta$ is then the sum over initial and final vibrational states

$$\sigma_{\alpha \to \beta}(E_{\text{transl}}) = \sum_{i_\alpha} W_{i_\alpha} \sum_{i_\beta} \sigma_{\alpha,i_\alpha \to \beta,i\beta}(E_{\text{transl}}), \quad (15)$$

where W_{i_α} is the probability that the initial vibrational state was i_α. For all the reactions considered in this paper, Eqs. (13)–(15) were used to convert probabilities to cross sections.

Since the phase-space element was revised, the model computations described in (I) were repeated to determine the effects on the cross sections of the heat of reaction, the force constants (polarizabilities), and the masses of the atoms. Surprisingly enough, the only noticeable changes from the previous calculations were in the mass effects. Although the channel with the highest reduced mass is still favored, it is by a lesser amount. The results are shown in Fig. 1 for the system

$$M^4+(M^1M^2)^+ \to \begin{cases} M^4+(M^1M^2)^+ & \text{(a)} \\ M^2+(M^1M^4)^+ & \text{(b)} \\ M^1+(M^2M^4)^+, & \text{(c)} \end{cases}$$

with the diatomic molecules having identical vibrational levels (eight levels spaced 0.25 eV apart), dissociation energies $(= 2 \text{ eV})$ and the atoms having identical polarizabilities $(= 1 \text{ Å}^3)$.

II. DISSOCIATIVE CHARGE-TRANSFER REACTIONS

Since the only major ambiguity in the phase-space theory for ion–molecule reactions is the treatment of dissociative processes, this point was tested by comparison with experiment before more extensive calculations were made. The reactions chosen (largely because of the availability of excellent experimental data) were the dissociative charge-transfer reactions of CO and the rare-gas ions recently studied by Giese and Maier.[5]

The pertinent data used in the calculations are listed in Table I. Only the ground electronic states of CO, CO^+, and the rare-gas ions were considered. The energy levels of the diatomic molecules were computed from ω_e and D_0 assuming a Morse potential function. The CO was assumed to be in the ground vibrational state initially. The reactions considered are listed below together with the relevant heats of reaction to the ground vibrational states:

$$He^+ + CO \to \begin{cases} He^+ + CO \\ He + CO^+ (+10.46 \text{ eV}) \\ \quad \to He + C^+ + O (+2.10 \text{ eV}), \quad (A) \end{cases}$$

$$Ne^+ + CO \to \begin{cases} Ne^+ + CO \\ Ne + CO^+ (+7.45 \text{ eV}) \\ \quad \to Ne + C^+ + O (-0.91 \text{ eV}), \quad (B) \end{cases}$$

$$Ar^+ + CO \to \begin{cases} Ar^+ + CO \\ Ar + CO^+ (+1.64 \text{ eV}) \\ \quad \to Ar + C^+ + O (-6.72 \text{ eV}). \quad (C) \end{cases}$$

The phase space corresponding to stable products (CO^+ or vibrationally excited CO) was determined with two choices of J_{rot}'. The first choice, used in (I), was

$$J_{\text{rot}}' = [2I(D_0 - E_{\text{vib}})]^{\frac{1}{2}}. \quad (16)$$

This corresponds to assuming dissociation occurs for every molecule for which the sum of the rotational and vibrational energy exceeds the dissociation energy D_0. This choice allows no molecules to be metastable to dissociation by rotational excitation.

A more restrictive choice, allowing metastable molecules, is to assume that dissociation by rotation occurs only when the rotational barrier of the diatomic can be overcome. The potential energy curve of a diatomic ion at large internuclear separation can be expected to consist of a polarization term and angular

[5] C. F. Giese and W. B. Maier II, J. Chem. Phys. **39**, 197 (1963).

momentum term. We take

$$V'(r) = -(\alpha_N e^2/2r^4) + (J_{rot}^2/2\mu_D r^2), \quad (17)$$

where μ_D is the reduced mass of the diatomic ion molecule, α_N is the polarizability of the neutral atom of the diatomic ion, and e is the electronic charge. The maximum of this potential V_m is at the distance r_m where

$$r_m = [2\alpha_N e^2 \mu_D / J_{rot}^2]^{\frac{1}{2}},$$

$$V_m = J_{rot}^4/8\mu_D^2 \alpha_N e^2. \quad (18)$$

We assume dissociation for those molecules for which the sum of the vibrational and rotational energy exceeds the sum of the dissociation energy and V_m, i.e., dis-

TABLE I. Atomic and molecular constants.

	α (Å^3)	I_p (eV)[d]	D_0^0 (eV)	ω_e (cm^{-1})[g]	r_e (Å)[g]
CO	1.95[a]	14.01	11.11[e]	2170.2	1.128
CO$^+$			8.36[f]	2214.2	1.115
He	0.205[b]	24.46			
Ne	0.390[b]	21.47			
Ar	1.63[b]	15.68			
Kr	2.46[b]	13.93			
C		11.26			
O	0.579[c]				
H$_2$[g]			4.476	4395	0.74

[a] H. H. Landolt and R. Börnstein, *Zahlenwerte und Functionen* (Springer-Verlag, Berlin, 1951), Vol. 1.
[b] H. Margenau, Rev. Mod. Phys. **11**, 1 (1939).
[c] Estimated using screening constants. See J. O. Hirschfelder, C. F. Curtiss, and R. B. Bird, *Molecular Theory of Gases and Liquids* (John Wiley & Sons, Inc., New York, 1954), p. 951 ff.
[d] *Handbook of Chemistry and Physics*, edited by C. Hodgman (Chemical Rubber Publishing Company (These values are for the ground electronic states, i.e., $^2P_{\frac{3}{2}}$ states of He, Ne, Ar, and Kr).
[e] L. M. Branscomb and S. J. Smith, Phys. Rev. **98**, 1127 (1955); C. F. Giese and W. B. Maier II, J. Chem. Phys. **39**, 197 (1963).
[f] $D_e^0(CO^+) = D_e^0(CO) + I(C) - I(CO)$.
[g] G. Herzberg, Ref. 7.

sociation starts when J_{rot} satisfies the equation

$$D_0 - E_{vib} + J_{rot}^4/8\alpha_N\mu_D^2 e^2 - J_{rot}^2/2\mu_D r_e^2 = 0, \quad (19)$$

where we have set $E_{rot} = J_{rot}^2/2I$. This equation gives

$$J_{rot}' = (2\alpha_N e^2 \mu_D/r_e^2)^{\frac{1}{2}}\{1 - [1 - 2r_e^4(D_0 - E_{vib})/\alpha_N e^2]^{\frac{1}{2}}\}^{\frac{1}{2}}. \quad (20)$$

Physically, (20) favors dissociation from high vibrational levels, reducing to (16) for $(D_0 - E_{vib})r_e^4/\alpha_N e^2 \ll 1$. Equation (16) gives a criterion for dissociation in which all vibrational levels are treated equivalently. The expected result, that (20) gives less dissociation than (16) is found.

The cross sections for dissociative charge transfer for Reactions (A), (B), and (C) are presented in

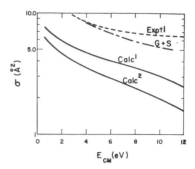

FIG. 2. Dissociative charge transfer, He$^+$+CO. "Exptl" from Ref. 5. G and S is total cross section, Ref. 4. "Calc.[1]," Eq. (16) "Calc.[2]," Eq. (20).

Figs. 2–4 with the experimental results of Giese and Maier.[5] Inspection shows that the use of Eq. (16) gives cross sections about a factor of 2 higher than Eq. (20) for all energies. In general, Eq. (16) seems to agree somewhat better with experiment. One may speculate that the use of the more liberal condition for rotational dissociation compensates to some extent for the fact that we do not allow direct vibrational transitions to the continuum.

Comparison of the three curves as a whole reveals the following pattern: as the energy above threshold increases above a few electron volts, the calculated results fall off more rapidly than the experimental results. This may be attributed to two factors. First, the continued use of the criterion for strong coupling

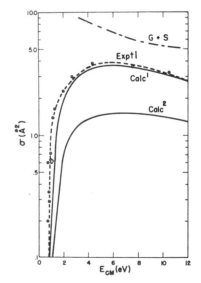

FIG. 3. Dissociative charge transfer Ne$^+$+CO. Curves labeled as in Fig. 2.

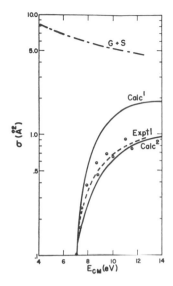

FIG. 4. Dissociative charge transfer, Ar$^+$+CO. Curves labeled as in Fig. 2.

collisions of Gioumousis and Stevenson at higher translational energies underestimates the *total* cross section. (Above about 4 to 6 eV for these systems, the critical impact parameter defining the limit of a strong coupling collision, $b_m{}^2 = [2\alpha e^2/E_{transl}]^{\frac{1}{2}}$ is about equal to the size of the CO molecule.)

In addition, the restriction to rotational dissociation means that the reduced mass of the initial system probably plays a more important role in the calculations than in fact. The reaction with He$^+$, for which the calculated results are too low by factors of 2 or 3, is the one with the lowest reduced mass. The requirement of conservation of angular momentum with low total angular momentum inhibits the production of CO$^+$ in high rotational states, thereby reducing the calculated dissociation cross sections. When the initial reduced mass is increased, in going from Reaction (A) to (C) (the reduced mass increases from 3.5 to 16.5 amu), the average total angular momentum increases and the calculated total dissociation cross section increases from about 0.6 to about 1.8 times the experimental result.

The over-all errors are, except for He$^+$, rather small in absolute magnitude, corresponding to less than 20% of the total cross section. We may tentatively conclude that the mechanism assumed for dissociation [using Eq. (20)] while oversimplifying the physical process does give reasonable results, particularly in the threshold region. Equation (20) is therefore used to define the dissociation region throughout the rest of the paper.

It should be emphasized that the computations reported in this section were purely *ab initio*. No adjustable parameters are used, the data being from spectroscopic sources only. Considering this fact, the agreement with experiment must be considered quite satisfactory.

III. ION-MOLECULE EXCHANGE REACTIONS— ISOTOPE EFFECTS

The phase-space theory was originally developed for bimolecular exchange reactions without activation energy. Surprisingly enough, the experimental information on such reactions is more complete for reactions between neutral species than for ion–molecule reactions, and computations on these systems will be published shortly.[6] There is, however, one set of ion–molecule reactions which has been studied rather extensively in the energy range of interest: the reactions between rare-gas ions (or atoms) and hydrogen (ions or molecules). It is somewhat disappointing that the experimental data from different sources are not in particularly good agreement, but enough information is available to show whether or not the phase-space theory gives reasonable results.

We concentrate on two aspects of the kinetics; the energy dependence of the over-all cross sections and the isotope ratios of the products when HD (or HD$^+$) is one of the reactants. The reactions studied are of three types

$$X + H_2{}^+ \to \begin{cases} X + H_2{}^+ \\ XH^+ + H, \end{cases} \tag{D}$$

$$X + HD^+ \to \begin{cases} X + HD^+ \\ H + XD^+ \\ D + XH^+, \end{cases} \tag{E}$$

$$X^+ + HD \to \begin{cases} X^+ + HD \\ H + XD^+ \\ D + XH^+; \end{cases} \tag{F}$$

(X is He, Ne, Ar, Kr).

The spectroscopic constants for the ions XH$^+$ are, unfortunately, unknown for X as Ar, or Kr. For the helium and neon hydride ions various wavefunctions have been computed and the molecular parameters needed were taken from these computations. For the argon and krypton hydrides, rough estimates were made and computations were performed for several values of the dissociation energy. The additional molecular parameters used are listed in Table II. Again, the molecular ions were assumed to be bound by a Morse potential, and the energy-level spacings were calculated in this way from the values of D_e and ω_e. Only the ground electronic states of the ions were considered. The vibrational spacing of the isotopic species (HeD$^+$, for instance) were computed in the normal way from $\omega_e(XH^+)$.[7]

[6] P. Pechukas, J. C. Light, and C. Rankin, J. Chem. Phys. (to be published).
[7] G. Herzberg, *Spectra of Diatomic Molecules* (D. Van Nostrand Company, Inc., Princeton, New Jersey, 1950).

TABLE II. Molecular ion constants.

	D_0[a] (eV)	r_e (Å)	ω_e (cm^{-1})
HeH$^+$	1.727[b]	0.76[b]	3286[b]
NeH$^+$	2.023[c,d]	0.972[d]	3053[d]
ArH$^+$	3.03[c,e]	1.14[e]	2214[g]
KrH$^+$	4.09[e,f]	1.27[e]	1963[g]

[a] All dissociation energies are from the ground vibrational state.

[b] B. Anex, J. Chem. Phys. 38, 1651 (1963). [Computations by S. Peyerimhoff (to be published) are in substantial agreement with these.]

[c] These three values of D_0 were varied by ±0.2 eV to check on how critical the value of D_0 is to the computation.

[d] S. Peyerimhoff, J. Chem. Phys. (to be published).

[e] T. F. Moran and L. Friedman, J. Chem. Phys. 40, 860 (1964).

[f] Given as an upper limit (Ref. e). R. E. Stanton, J. Chem. Phys. 39, 2368 (1963) gives 4.07 eV as a lower limit for the binding energy of KrH$^+$.

[g] Estimated from the vibrational frequency of the corresponding isoelectronic hydrogen halide using $\omega_e(XH^+) = \omega_e(HY)/1.35$, where when X is Ar, Y is Cl; when X is Kr, Y is Br. This relation holds very well for NeH$^+$ and HF as well as HeH$^+$ and H$_2$. Although the over-all reactive cross section may vary somewhat if these values are in error, the isotopic ratios will not.

For the reactions (D) and (E), the initial vibrational energy distribution [the W_{i_a} of Eq. (15)] must be used since the cross sections are quite sensitive to the initial vibrational state. Since the ionization is be fairly fast electrons (~70 eV), the vibrational distribution should be given by the Franck–Condon factors.[8] The values used were calculated for H$_2^+$ and HD$^+$ by Wacks.[9]

He+H$_2^+$

The total cross section [Eq. (15)] for this reaction was computed for barycentric energies between 1 and 9 eV. The reaction has been studied experimentally by both Giese and Maier[10] and by von Koch and Friedman.[11] Our results for the cross section as a function of energy are given in Fig. 5 along with the experimental curve of Giese et al. The cross sections should be directly comparable since the experiments were carried out with a nearly monochromatic beam of ions. The differences may be due to (a) failure of the theory; (b) different values of $W_{H_2^+}$ in the experiment from those used in the calculation; or (c) experimental errors. It is unlikely, in this case, that statistically significant errors have been made in either the density of energy levels or the exothermicity. Also, an error in the amount of dissociation cannot be responsible for the descrepancy at low energies, since the theory predicts almost no dissociation. It seems possible that

[8] The recent results of Rapp et al. [D. D. Briglia and D. Rapp, Bull. Am. Phys. Soc. 10, 181 (1965); Phys. Rev. Letters 14, 245 (1965)] suggest, however, that anomalies may exist for the H$_2$–H$_2^+$ ionization.

[9] M. E. Wacks, J. Res., Natl. Bur. Stds. 68A, 631 (1964).

[10] C. F. Giese and W. B. Maier II, J. Chem. Phys. 39, 739 (1963).

[11] M. von Koch and L. Friedman, J. Chem. Phys. 38, 1115 (1963).

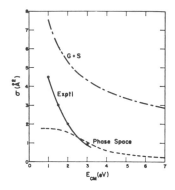

FIG. 5. He+H$_2^+$→HeH$^+$+H. "Exptl" from Ref. 10.

all three factors above may enter. It should be noted, however, that the phase-space theory gives results which differ by no more than a factor of $\frac{1}{2}$ from the experimental results of Giese and Maier.

In Fig. 6 we plot the values of $v\sigma$ and of $\bar{v}Q$ calculated as well as $\bar{v}Q$ from the experiments of von Koch and Friedman. Q is the phenomenological cross section obtained from measurements made in their mass spectrometers. It is related to the microscopic cross section by the equation[12]

$$Q = \frac{m_i}{\mathcal{E}_{el}} \int_0^{v_{max}} v_i \sigma(v_i)\,dv_i = E_{max}^{-1} \int_0^{E_{max}} \sigma(E)\,dE,$$

where $\mathcal{E}$ is the repeller field strength; l, the ion path length (maximum); e, the charge on electron; v_{max}, the maximum ion velocity, $(2E_{max}/m_i)^{\frac{1}{2}}$; and m_i is the mass of ion. The second equality follows since E_{max} (laboratory) $= e\mathcal{E}l = \frac{1}{2}(m_i v_i^2)$ and $dE = m_i v_i dv_i$. The average velocity is $\bar{v} = (E_{max}/m_i)^{\frac{1}{2}}$. The experimental curve is taken directly from von Koch and Friedman's article[11] and may not, therefore, be exact.

It can easily be seen that the phase-space theory results are in excellent agreement with the experi-

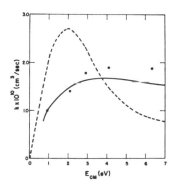

FIG. 6. He+H$_2^+$→HeH$^+$+H. - - - -, $v(v)$; ——, $\bar{v}\langle(v)\rangle$; experimental points from Ref. 11.

[12] J. C. Light, J. Chem. Phys. 41, 586 (1964).

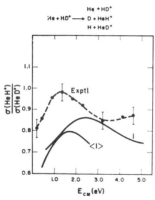

FIG. 7. Isotope effect vs translational energy. He+HD$^+$ "Exptl." from Ref. 13; Curve 1, σ(HeH$^+$)/σ(HeD$^+$); Curve $\langle 1 \rangle =$ $\langle \sigma$(HeH$^+$) $\rangle / \langle \sigma$(HeD$^+$) $\rangle$. Reaction:

$$\text{He}+\text{HD}^+ \rightarrow \begin{array}{l} \text{He}+\text{HD}^+ \\ \text{D}+\text{HeH}^+ \\ \text{H}+\text{HeD}^+. \end{array}$$

mental results of von Koch and Friedman. The interpretation of the results is very simple. At low translational energies there is considerably more phase space available to the nonreactive channel, even with the initial H$_2^+$ in excited vibrational states, since there are more vibrational levels available in the initial channel than in the reactive channel. As the initial kinetic energy is increased, the number of states available to both products and reactants increases at about the same rate, thereby reducing the relative amount of phase space available to reactants as compared to products. This increases the probability of reactive scattering. When the energy is increased above 2 eV, the dissociation process begins to come in with appreciable probability, reducing the probability of reactive scattering to HeH$^+$. The ratio of reactive to nonreactive scattering is also decreased somewhat for this reaction since the reduced mass of the product system is less than that of the reactants (see Fig. 1). This effect is probably minor when compared with the effect of the endothermicity of the reaction. It should, perhaps, be noted that the total "strong coupling" cross sections are about two to five times as large as the cross sections for reaction, depending on which set of experimental data is used. The probability of reaction is greater from excited vibrational states of H$_2^+$, but some reaction occur whenever it is energetically possible. The over-all cross section for reactions is quite sensitive to the initial internal energy distribution of H$_2^+$.

He+HD$^+$, Ne+HD$^+$

The isotope effects as a function of translational energy for these systems were measured by both Klein and Friedman[13] and by Giese and Maier.[14] Again their

13 F. S. Klein and L. Friedman, J. Chem. Phys. 41, 1789 (1964).
14 C. F. Giese and W. B. Maier II (private communication).

results are in disagreement, particularly as to the shape of the curve of $\sigma(\text{XH}^+)/\sigma(\text{XD}^+)$ vs E at low energies. The phase-space theory was used to calculate these quantities using the data in Table II, with the vibrational frequencies adjusted to the different reduced masses of the diatomic species. The initial vibrational populations used were those given by Wacks.

The results compared with the experimental data of Klein and Friedman are shown in Figs. 7 and 8. The averaged ratio of the cross sections necessary to compare with Klein and Friedman's results is given by

$$\langle R \rangle = \int_0^{E_{max}} \sigma_{\text{XH}^+}(E)\,dE \Big/ \int_0^{E_{max}} \sigma_{\text{XD}^+}(E)\,dE$$

$$= Q(\text{XH}^+)/Q(\text{XD}^+).$$

The agreement with the experimental results is not spectacular. However, the fact that the isotope ratios, predicted and found, are below unity for HeH$^+$/HeD$^+$, is in itself quite significant. The absolute-reaction-rate theory predicts ratios on the order of 1.3[13] for this reaction. In order to reduce the isotope ratio, Klein and Friedman invoked two additional processes, the statistical theory of unimolecular decomposition of the complex, and the possible dissociation of the product molecular ion into He and H$^+$ (or D$^+$).

The phase-space theory is in as good agreement with the results of Klein and Friedman as are the computations they made, and their physical arguments on the decomposition of the complex and dissociation of the products are supported, to some extent, by the phase-space calculations. We find, that at low ion velocities, the conservation of angular momentum does not pose severe requirements, and the effect of the greater density and lower energy of XD$^+$ states is predominant. As the velocity increases, the larger reduced mass (see Fig. 1) of the final state favors XH$^+$ formation since angular momentum is more easily conserved, therefore increasing the isotope ratio. Finally, when the dissociative processes become important, the XH$^+$ tends to dissociate more easily than XD$^+$ since it needs less angular momentum to do so.

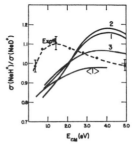

FIG. 8. Isotope effect vs translational energy. Ne+HD$^+$ "Exptl." from Ref. 13; Curve 1, $D_0 = 2.02$ eV; 2, $D_0 = 1.82$ eV; 3, $D_0 = 2.22$ eV. $\langle 1 \rangle = \langle \sigma$(NeH$^+$) $\rangle / \langle \sigma$(NeD$^+$) $\rangle$, $D_0 = 2.02$ eV.

The effect of the requirement of conservation of angular momentum in isotope effects is clearly illustrated in Fig. 9. The derivative of the cross sections for formation of HeH+ and HeD+ in Reaction (E), $d\sigma/db$, is plotted against the impact parameter b. At low impact parameter (and J), the effect of the higher density of internal states of XD+ predominates. As the initial angular momentum increases, however, the smaller reduced mass of the XD++H system means it cannot carry off, in orbital angular momentum, the total initial angular momentum. Therefore the rotational angular momentum of the HeD+ must be oriented along the initial angular momentum vector if the conservation requirements are to be satisfied. This, of course, reduces the magnitude of the phase-space integrals of Eq. (9). Finally, at the largest impact parameters for strong coupling, the angular momentum requirements cannot be satisfied in the HeD++H channel at all, and the only reaction product is HeH++D (and nonreactive scattering).

The phase-space theory apparently mixes somewhat to thoroughly all the available states, thus weighting XD+ (with its higher density of rotational and vibrational states) somewhat too heavily. In the Monte Carlo studies of Blais and Bunker,[15] for instance, it was found that if the initial orbital angular momentum was small, the final rotational angular momentum was, in some cases, small also, indicating that large volumes of the rotational phase space available to the products were not used. The over-all errors in using the phase-space theory for isotope effects are not, however, too large. The cross sections to the XH+ are, if the experiments are correct, about 10% too large, while the XD+ cross sections are about 10% too small at worst. The general shifts of the isotope effects in going from He to Ne are reproduced well by the phase-space theory. Finally, it might again be noted that the phase-space

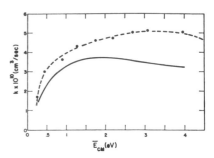

FIG. 10. Ne+HD+, $D_0=2.02$ eV. ——, $[\langle\sigma(NeH^+)\rangle+\langle\sigma(NeD^+)\rangle]\bar{v}_{HD^+}$; - - - -, exptl. (Ref. 16) $\times v_{HD^+}/v_{H_2^+}$.

theory, with no adjustable parameters, gives results as reasonable as those calculated by Klein and Friedman (where an energy parameter, as well as the configuration of the complex, was adjusted).

The Ne+HD+ reaction is qualitatively similar to the He+HD+ reaction. Three computations were done with binding energies of NeH+ of 2.02, 1.82, and 2.22 eV labeled 1, 2, and 3, respectively. The differences are surprisingly small, the trend being toward larger isotope ratios for smaller binding energies (more endothermic reaction), except at very low velocities where the differences in zero-point energy predominate. The shift of the peak of the ratios to higher energy is probably due to the increase in the equilibrium internuclear distance of NeH+ and NeD+. Since we allow dissociation of the product only by rotation, the increased moment of inertia of NeH+ and NeD+ (over HeH+ and HeD+) reduces the amount of dissociation at a given energy. Thus the effect of dissociation on the isotope ratios does not become apparent until higher translational energies.

The absolute values of the cross section for the Ne+HD+ reaction, Fig. 10 (for $D_0=2.02$ eV), are in quite good agreement with the experimental results of Moran and Friedman[16] on Ne+H2+. Again, the reduction of $\bar{v}Q$ at low energies may be attributed to a low probability of reaction due to small volumes of phase space for the products rather than to an activation energy,[16] and the phase-space theory fits the data within 25%. The ion velocity was adjusted to take into account the higher reduced mass of the Ne+HD+ system studied here, and the sum of the "rate" constants (NeH++NeD+) is plotted in Fig. 9 for comparison with Moran and Friedman's results.

Ar++HD, Kr++HD

These reactions, studied by Klein and Friedman,[18] provide, because of the very large mass ratios, information on the isotope effects at very low barycentric energies. The energies are low enough so that no com-

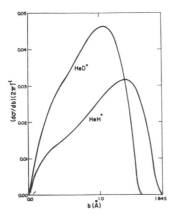

FIG. 9. He+HD+. $d\sigma/db(=bP)$ vs b (impact parameter) $b_{max}=$ 1.845 Å.

[15] N. Blais and D. Bunker, J. Chem. Phys. 39, 315 (1963).

[16] T. F. Moran and L. Friedman, J. Chem. Phys. 39, 2491 (1963).

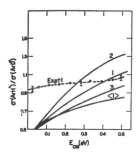

FIG. 11. Isotope effect, $Ar^+ + HD$. Curve 1, $D_0 = 3.03$ eV; 2, $D_0 = 2.83$ eV; 3, $D_0 = 3.23$ eV. $\langle 1 \rangle = \langle \sigma(ArH^+) \rangle / \langle \sigma(ArD^+) \rangle$; $---$, exptl. (Ref. 13).

plete dissociation can take place. Due to the uncertainties in the dissociation energies of the products, calculations were done for three values of D_0 (ArH^+), 3.03, 2.83, and 3.23 eV, the first value being that estimated by Moran and Friedman.[17] For KrH^+, the values of 4.09 and 4.29 eV were used.

Only the ground states of the ions were considered, since the proportion of the higher-energy states is unknown. If reaction does occur predominantly from the $^2P_{\frac{1}{2}}$ states, the reactions would be more exothermic and the cross-section curves would shift in the manner indicated by the higher values of D_0. This is significant only in the Kr^+ case in which the extraordinary behavior of $\sigma(KrH^+)/\sigma(KrD^+)$ for $D_0 = 4.09$ eV is due to the fact that only the first two vibrational levels of the products can be occupied. The dip in the ratio at $E_{cm} = 0.17$ eV is due to the fact that the second vibrational level of KrD^+ begins to be populated here. By $E_{cm} = 0.22$ eV, the second level of KrH^+ is also available, and the ratio increases again.

The results are plotted in Figs. 11 and 12. For comparison with experiment, the ratios of the averaged cross sections were computed for $D_0(ArH^+) = 3.03$ eV and $D_0(KrH^+) = 4.09$ eV. The $Ar^+ + HD$ reaction is one for which the statistical theory apparently fails rather badly (the ratio being off by a factor of 0.6). One cause for this may be that the attractive force is not adequately represented by a $-c/r^4$ potential. The ionization potentials of Ar and HD are nearly equal, and this may lead to stronger or longer-range potentials than r^{-4}.[18] If this is the case, then we have excluded from consideration some higher-angular-momentum complexes which would dissociate, as shown in Fig. 9, preferentially to XH^+. It has been reported that the total cross section for the $Ar^+ + H_2$ reaction exceeds the Gioumousis and Stevenson limit by as much as a factor of 2,[10] which is rather suggestive of stronger or longer-range forces.

IV. DISCUSSION

The purpose of this paper was to evaluate, as well as possible, the validity of the phase-space theory of chemical kinetics as applied to diverse ion–molecule reactions. For such a test, it is necessary to make *ab initio* computations, using the best available data from fields other than kinetics to evaluate the necessary constants. The weakest point of the theory is the treatment of dissociative processes where restrictions of some sort must be placed on the phase space in order to avoid divergences. The basic restriction used here—dissociation by rotation only from bound vibrational levels—gives only semiquantitative results, erring by as much as a factor of 2 in either direction. No other *a priori* theory, however, is any better.

The results for ion–molecule exchange reactions are, on the whole, very good, with deviations of more than 25% being rare. In addition, the phase-space theory accounts rather well for the velocity behavior of He (and Ne) + H_2^+ reactions, including the deviations from the Gioumousis and Stevenson cross sections at both low and high energies. The results on isotope ratios are in reasonable agreement with experiment, with some systematic error in favor of the heavier product (XD^+). Again, however, the velocity behavior of the isotopic ratios is accounted for quite well by the phase-space theory.

We conclude, therefore, that the phase-space theory of chemical kinetics provides reasonable *a priori* estimates of the cross sections of three-body ion–molecule reactions. In addition, phase space considerations allow physically reasonable interpretations of the rate constants without invoking activation energies, linear and bent complexes, etc. We end, however, on a note of caution. For several recently reported reactions ($He^+ + H_2$, $K^+ + N_2O$, $O^+ + N_2$), some reactive channels appear to be closed where, from energy and spin considerations only, they should be open.[19] No simple theory can (as

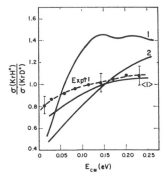

FIG. 12. Isotope effect, $Kr^+ + HD$. Curve 1, $D_0 = 4.09$ eV; Curve 2, $D_0 = 4.29$ eV. $\langle 1 \rangle = \langle \sigma(KH^+) \rangle / \langle \sigma(KD^+) \rangle$. $---$, exptl. (Ref. 13).

[17] T. F. Moran and L. Friedman, J. Chem. Phys. **40**, 860 (1964).
[18] C. F. Giese, *Advances in Mass Spectrometry*, edited by J. D. Waldren (Pergamon Press, Inc., New York, 1965).
[19] L. Friedman and T. F. Moran, J. Chem. Phys. **42**, 2624 (1965).

yet, at least) determine for what reactions this will happen. In these cases, obviously, the phase-space theory will fail, and if they occur often, the utility of the phase space theory as a predictive tool will be considerably reduced. As yet, however, such cases appear to be rather rare, and it seems that the phase space theory can be applied with reasonable expectation of success to ion–molecule reactions.

ACKNOWLEDGMENTS

The authors gratefully acknowledge the general support of the Institute for the Study of Metals by the Advanced Research Projects Agency.

APPENDIX

Table III gives an outline of our program.

TABLE III. Outline of program used in phase-space calculations.

Main program	Subroutines	
	FORTRAN	FAP
Input: Binding energies ω_e, heats of reaction, masses, velocity range, step sizes, long-range force constants, equilibrium internuclear distances of diatomics		
	Calculate Morse energy levels, reduced masses; dissociate energies from vibration levels.	
1. Set velocity Exit if $V > V_{max}$		
	Calculate velocity dependent constants for each vibrational state, maximum impact parameter for strong coupling	
Set cross sections, etc. to zero 2. Set impact parameter Exit to (3) if $b > b_{max}$		
	Compute total phase-space volumes for each channel and vibrational level.	Compute roots of Eq. (8)
Compute probabilities from phase-space volumes. Add bP to cross sections. Cycle to (2).	Compute phase-space volumes for stable products. Add to get denominator of Eq. (1).	
		Evaluate integral from series expansion
3. Write cross sections, etc., Cycle to (1). Exit.		

Editor's Comments
on Papers 16 and 17

16 GURNEE and MAGEE
Interchange of Charge Between Gaseous Molecules in Resonant and Near-Resonant Processes

17 BOHME, HASTED, and ONG
Calculation of Interchange Reaction Rates by a 'Nearest Resonance' Method

Charge exchange is usually considered a collision process in which the only change consists of the transfer of one or more electrons between an ion and a neutral atom or molecule. Several kinds of charge exchange are recognized, but the ones of interest to us are (1) symmetrical resonance reactions $A^+ + A \rightarrow A + A^+$ and (2) asymmetric reactions $A^+ + B \rightarrow A + B^+$. Of course, analogous reactions of negative ions also occur and can be classified under (1) or (2). There is much literature on the theory of various aspects of charge exchange reactions, but unfortunately, it is satisfactory only for certain symmetric resonance reactions that are of only slight interest to the subject of ion-molecule reactions. Some progress, mainly of a semi-empirical nature, has been made on the theory of asymmetric reactions, but this has been most successful when A and B are atomic species.

For a qualitative understanding of charge exchange, Massey[1] has formulated the adiabatic hypothesis. Suppose that in process (2) above there is a difference ΔE in the ionization energy of A and B. Assuming that normally the velocities of the orbital electrons will be much larger than the velocity of approach of the ion and neutral, there will be ample time for the electrons to adjust to the perturbation so that very little reaction will occur; i.e., the cross section for reaction will be small. By analogy to the case of applying a disturbing force to a vibrating oscillator, it would be expected that there would be very little effect unless the time during which the two were sufficiently close to affect each other was comparable to the period of oscillation. Thus for maximum cross sections, $\tau \nu \approx 1$ where ν is frequency and τ is the time of the collision. τ can be taken as l/v where l is the distance over which the electrical perturbation is effective and v is the velocity of approach. In quantum mechanical terms, $v = \Delta E/h$ so the collision for maximum cross section will be $l \Delta E/hv \approx 1$. This gives the approximate conditions for maximum charge exchange but does not give any information about the cross section for reaction. It also suffers from the fact that l is not well defined.

A valid theory of charge exchange will necessarily rest on quantum mechanics, but unfortunately, such treatments become extremely complicated for reactions involving polyatomic systems. To surmount this difficulty, some semitheoretical treatments of charge exchange have been developed and have proved fairly successful. Two such treatments are given in this section. In Paper 16, the relative motion of the colliding particles is treated classically and all other degrees of freedom are treated by quatum mechanics. Both symmetric and asymmetric processes are considered. In Paper 17, the adiabatic criterion is extended to permit the calculations of the cross section at various energies. More extensive theoretical treatments of charge exchange may be found in the following publications. [2-5]

REFERENCES

1. Massey, H. S. W., *Rep. Prog. Phys.* **12**: 248 (1949).
2. Massey, H. S. W., and E. H. S. Burhop, *Eletronic and Ionic Impact Phenomena*, Oxford University Press, Oxford (1952).
3. Hasted, J. B., *Physics of Atomic Collisions*, Butterworth's London (1964).
4. Hasted, J. B., in D.R. Bates (ed.), *Charge Transfer and Collisional Detachment in Atomic and Molecular Processes*, Academic Press, New York (1962).
5. McDaniel, E. W., *Collisional Phenomena in Ionized Gases*, John Wiley & Sons, New York (1964).

16

*Reprinted from J. Chem. Phys. **26**:1237–1248 (1957)*

Interchange of Charge between Gaseous Molecules in Resonant and Near-Resonant Processes*

E. F. GURNEE† AND J. L. MAGEE

Department of Chemistry, University of Notre Dame, Notre Dame, Indiana‡

(Received September 20, 1955)

Interchange of charge between monatomic and diatomic ions and molecules in gaseous systems is considered theoretically by an "impact parameter" method. Formulas for charge transfer cross sections for resonant and near-resonant cases are developed, and comparisons are made between experimental and theoretical results. The validity of the impact parameter method is considered by comparison with the more rigorous method of scattered waves.

I. INTRODUCTION

THE theory of charge interchange between gaseous ions and molecules has been previously discussed.[1-4] The most rigorous theory applies only to charge exchange between monatomic ions and atoms, and general application of the theory to cases of interest in physical and chemical problems has not been achieved. An approximate theoretical treatment which gives the correct dependence of this process on the various parameters and can be used in applications to polyatomic systems is needed. An impact parameter method seems to be indicated for such a treatment, and in this paper we have used such a method to develop formulas for charge transfer cross sections.

Holstein[5] has applied the impact parameter method to charge transfer in rare gas systems. Similar treatments, in which the relative motion of the system is treated classically, have been used by Kallmann and London,[6] Rice,[7] and Zener[8,9] for transfer of excitation between atoms. This latter process is formally equivalent to the charge transfer process.

The validity of the impact treatment has been considered in detail by Mott and Frame.[10] This approximation will always be valid when the de Broglie wavelength of the incident particle is sufficiently small; the classical trajectory of the particle will then be a good description of the wave-packet motion.

To put the matter in a quantitative form, classical mechanics can be used for the relative motion in a

* Condensed from a dissertation presented by E. F. Gurnee in partial fulfillment for the degree of Doctor of Philosophy at the University of Notre Dame.

† Now at The Dow Chemical Company, Midland, Michigan.

‡ A contribution from the Radiation Project of the University of Notre Dame, supported in part by the U. S. Atomic Energy Commission under contract AT(11–1)–38.

[1] H. Massey and R. Smith, Proc. Roy. Soc. (London) **A142**, 142 (1933).

[2] E. C. G. Stueckelberg, Helv. Phys. Acta **5**, 369 (1932).

[3] N. Mott and H. Massey, *The Theory of Atomic Collisions* (Oxford University Press, New York, 1949).

[4] H. S. W. Massey and E. H. S. Burhop, *Electronic and Ionic Impact Phenomena* (Oxford University Press, New York, 1952).

[5] T. Holstein, J. Phys. Chem. **56**, 832 (1952).

[6] H. Kallmann and F. London, Z. physik. Chem. **B2**, 207 (1929).

[7] O. K. Rice, Phys. Rev. **38**, 1943 (1931).

[8] C. Zener, Proc. Roy. Soc. (London) **A137**, 696 (1933).

[9] C. Zener, Phys. Rev. **38**, 277 (1931).

[10] N. F. Mott and J. W. Frame, Proc. Cambridge Phil. Soc. **27**, 511 (1931).

collision problem if the trajectories of the particles are well-defined both before and after the collision. Thus, the wavelengths associated with the *momentum* and the *momentum transfer* must be small compared to the impact parameter, R_0. (See Fig. 1.) This is expressed mathematically by the two relations

$$R_0 M v \gg \hbar$$

$$R_0 \Delta p \gg \hbar$$

where M is the mass of the particle, v is the velocity, and Δp is the momentum transfer in the collision.

The second restriction could also be expressed in a slightly different form by noting that Δp will have an order of magnitude given by V/v where V is the effective potential energy of interaction. The collisions treated in this paper satisfy these restrictions.

In the considerations of this paper we also assume that the particle velocity is sufficiently smaller than the classical electron velocities of the bound electron so that "adiabatic" conditions are almost maintained, i.e., the Born-Oppenheimer approximation can be used.

It is of interest to compare the impact parameter method with the more rigorous method of scattered waves, and this has been done in Appendix II. There it is shown that approximations made by Massey and Smith[1] in the numerical evaluation of charge transfer cross sections by this more rigorous method for monatomic systems, make their results identical with those obtained by the impact parameter treatment.

II. THE RESONANT CASE

The transfer of charge in a resonant process can be represented by the equation

$$A + A^+ \rightarrow A^+ + A. \tag{1}$$

In the initial state the system has a wave function Ψ_i and in the final state a wave function Ψ_f; these correspond to the left- and right-hand sides of Eq. (1), respectively, with the molecule and the ion at infinite separation. Zeroth-order wave functions will be used throughout these calculations; first-order energies of interaction are used. The electron transfer is, therefore, treated as a problem in first-order time-dependent perturbation theory. The relative motion of the system is not included in these wave functions.

The total wave function for the system may be written as

$$\Psi = (c_i \Psi_i + c_f \Psi_f) \exp(-iEt/\hbar) \tag{2}$$

where c_i and c_f are complex numbers which are functions of the time t; E is the total energy of the system exclusive of the relative motion of the molecule and ion. The functions Ψ_i and Ψ_f are written as the product of internal and electronic wave functions. The total Hamiltonian is

$$H = H_{int}^+ + H_{int}^0 + H_{el}^+ + H_{el}^0 + H' \tag{3}$$

where the superscripts 0 and + refer to the molecule and ion, respectively, int refers to internal energies, el to electronic energies, and H' is that part of the Hamiltonian which represents interaction between the molecule and the ion. Substitution of this Hamiltonian and the wave function (2) into the time-dependent Schrödinger equation yields

$$c_i H' \Psi_i + c_f H' \Psi_f = i\hbar (\dot{c}_i \Psi_i + \dot{c}_f \Psi_f) \tag{4}$$

where the dot refers to differentiation with respect to time.

A set of coupled equations is now formed by multiplication of this last equation in turn by Ψ_i^* and Ψ_f^* and integration over all space coordinates:

$$c_i H_{ii}' + c_f H_{if}' = i\hbar(\dot{c}_i + \dot{c}_f F) \tag{5}$$

$$c_i H_{fi}' + c_f H_{ff}' = i\hbar(\dot{c}_i F + \dot{c}_f) \tag{6}$$

where

$$H_{ij}' = \int_\tau \Psi_i^* H' \Psi_f d\tau$$

$$F = \int_\tau \Psi_i^* \Psi_f d\tau = \int_\tau \Psi_f^* \Psi_i d\tau.$$

The model used to calculate the charge transfer cross section is shown in Fig. 1. In the initial state the molecule is at the point a, located at a distance R_0 from the x axis, and the ion is at the point b, which is located at $x = -\infty$. The point b moves along the x axis with the constant velocity v. In the final state the ion is at the point a and the molecule is at point b, where b is now located at $x = +\infty$. The deviation from a straight line path can be shown to be unimportant.

If $P(R_0)$ represents the probability that the charge has been transferred in the above described collision, then the total charge transfer cross section can be expressed as

$$\sigma = 2\pi \int_0^\infty P(R_0) R_0 dR_0. \tag{7}$$

The value of $P(R_0)$ is given by $c_f^* c_f$ evaluated at $x = +\infty$; it is convenient to denote this quantity by $c_f^* c_f(\infty)$. It is now necessary to solve Eqs. (5) and (6) for the coefficient c_f.

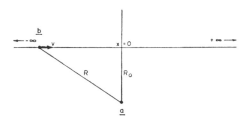

FIG. 1. The model chosen to represent the collision. The atom is located at the point a, a fixed distance R_0 from the x axis. The ion, located at b, moves along the x axis with the constant velocity v from $x = -\infty$ to $x = +\infty$.

Since these coefficients are complex numbers, we can let

$$c_i = \rho e^{i\omega} \qquad (8)$$

where the amplitude ρ and the phase ω are functions of time. It can be shown (Appendix I) that the coefficient c_f is always out of phase with c_i by $\pi/2$. Thus

$$c_f = i(1-\rho^2)^{\frac{1}{2}} e^{i\omega}. \qquad (9)$$

Substitution of these equations into either Eq. (5) or (6) and equating the real parts and the imaginary parts, leads to the equations

$$\dot{\omega} = \frac{F\dot{\rho}}{(1-\rho^2)^{\frac{1}{2}}} - \frac{H_{ii}'}{\hbar} \qquad (10)$$

$$\frac{H_{if}'}{\hbar} + F\dot{\omega} = \frac{\dot{\rho}}{(1-\rho^2)^{\frac{1}{2}}}. \qquad (11)$$

These equations are readily integrated, with the use of $vdt = dx$ to give

$$P(R_0) = c_f^* c_f(\infty)$$

$$= \sin^2 \int_{-\infty}^{\infty} \frac{1}{\hbar v} \frac{H_{fi}' - FH_{ii}'}{1 - F^2} dx = \sin^2 Q \quad (12)$$

where Q is defined as the definite integral. The total cross section can now be found by graphical integration of Eq. (7).

III. EXAMPLES OF SINGLE CHARGE TRANSFER IN RESONANT SYSTEMS

As a simple example, the above results will be applied to the reaction

$$H + H^+ \rightarrow H^+ + H. \qquad (13)$$

The wave functions may be written as

$$\Psi_i = \psi^0_{a\,el} = \frac{1}{\sqrt{\pi}} e^{-r_a}$$

$$\Psi_f = \psi^0_{b\,el} = \frac{1}{\sqrt{\pi}} e^{-r_b}$$

and the interaction Hamiltonian H', which operates on Ψ_i, is

$$H' = \frac{1}{R} - \frac{1}{r_b}.$$

The matrix element of this interaction Hamiltonian is well known.[11] It is conveniently written in terms of

$$S(R) = (\psi^0_{b\,el}, \psi^0_{a\,el}) = e^{-R}(\tfrac{1}{3}R^2 + R + 1)$$

$$J(R) = \left(\psi^0_{b\,el}, \frac{1}{r_b}\psi^0_{a\,el}\right) = e^{-R}(R+1)$$

[11] H. Hellmann, *Quantenchemie* (Franz Deuticke, Leipzig, 1937).

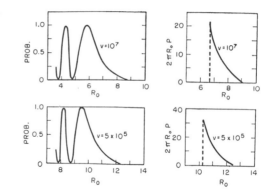

FIG. 2. The probability of charge transfer as a function of R_0 for the reaction $H^+ + H \rightarrow H + H^+$ at two different velocities. This probability oscillates rapidly between 0 and 1 for values of R_0 smaller than those shown. The graphical method of integration to obtain the contribution to the cross section from the exponential tail is also shown.

and is found to be

$$H_{fi}' = \frac{S(R)}{R} - J(R) = e^{-R}(-\tfrac{2}{3}R + 1/R).$$

The evaluation of the probability for charge transfer thus involves the calculation of integrals of the type

$$\int_{-\infty}^{+\infty} R^n e^{-R} dx \quad n = 0, 1, \cdots.$$

For values of R large compared to x, this integral may be approximated with the aid of the expansion

$$R = (R_0^2 + x^2)^{\frac{1}{2}} \approx R_0 + x^2/2R_0 + \cdots$$

as

$$R_0^n e^{-R_0} (2\pi R_0)^{\frac{1}{2}} \left(1 + \frac{n}{2R_0}\right).$$

With this approximation we have

$$P(R_0) = \sin^2 \left\{ \frac{(2\pi R_0)^{\frac{1}{2}}}{\hbar v} e^{-R_0} \left[-\frac{2}{3}R_0 - \frac{1}{3} + \frac{1}{R_0} - \frac{1}{2R_0^2} \right] \right\}. \quad (14)$$

The probability for charge transfer as a function of R_0 is plotted in Fig. 2 for velocities of 10^7 and 5×10^5 cm/sec. For values of R_0 smaller than those shown in the figure, the probability function oscillates rapidly between zero and one. The probability is assumed to be $\frac{1}{2}$ through the entire oscillatory region, and the exponential tail is integrated graphically to complete the integration of Eq. (7).

The calculated values of the cross section for Eq. (13) are listed in Table I.

It is desirable to use this same type of calculation for more complicated systems. The use of a one-electron approximation for the wave function makes such an

TABLE I. Cross sections for resonant charge transfer in system H⁺+H.

v (cm/sec)	(A²)
10^7	27.4
5×10^6	57.6

extension possible. A simple analytical form which seems indicated is the nodeless wave function[12]:

$$\psi = N r^{n-1} e^{-\alpha r}. \tag{15}$$

Such a wave function satisfies the wave equation for large values of r if α is taken as $(2I)^{\frac{1}{2}}$, where I is the ionization potential in atomic units. At such large distances the shielding by the other electrons is complete so that the wave function is certainly hydrogen-like. For smaller distances the curvature given by Eq. (15) is too small if one takes $\alpha = (2I)^{\frac{1}{2}}$ since the shielding is progressively less complete. Consequently one would expect "effective" values of α to be somewhat larger than $(2I)^{\frac{1}{2}}$ for moderate distances of interaction.

If the one electron approximation is used, with the wave function (15), then the value of α is the only undetermined parameter in the cross section calculation. There is actually no great sensitivity in this choice, however, and so the cross sections are quite well determined for a given case. We have taken values of α so that the one electron wave equation is approximately satisfied for the most important range of r rather than treating α as an adjustable parameter.

The cross section calculated in a manner analogous to the proton case is shown as a function of velocity for He⁺, Ne⁺, and A⁺ in Figs. 3, 4, and 5 as solid lines. Experimental values[13-17] are indicated as circles.

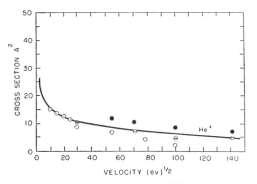

FIG. 3. The calculated cross section (solid line) for the reaction He⁺+He→He+He⁺ as a function of velocity (ev)$^{\frac{1}{2}}$. Experimental values are indicated by circles. ⊗ Dempster[13]; ○ Smith[14]; ⊖ Berry[15]; ⊙ Wolf[16]; ● Keene.[17]

[12] J. C. Slater, Phys. Rev. 36, 57 (1930).
[13] A. J. Dempster, Phil. Mag. 3, 115 (1937).
[14] R. A. Smith, Proc. Cambridge Phil. Soc. 30, 514 (1934).
[15] H. W. Berry, Phys. Rev. 74, 848 (1948).
[16] F. Wolf, Ann. Physik 29, 33 (1937).
[17] J. P. Keene, Phil. Mag. 40, 369 (1949).

Table II gives the values of α that have been used in these calculations.

IV. DOUBLE CHARGE TRANSFER IN RESONANT SYSTEMS

The charge transfer cross sections for reactions of the type

$$A + A^{++} \rightarrow A^{++} + A \tag{16}$$

can be calculated using the same nodeless wave functions as before. In this case, we arrive at

$$c_f{}^* c_f(\infty) = \sin^2 \int_{-\infty}^{+\infty} \frac{1}{\hbar v} \cdot 4 S(R) \left[\frac{S(R)}{R} - J(R) \right] dx \tag{17}$$

where the S and J integrals have been previously defined. This may be written in terms of an average value

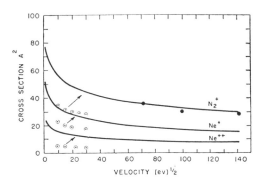

FIG. 4. The calculated cross sections (solid lines) as a function of velocity, (ev)$^{\frac{1}{2}}$, for the reactions:

$$Ne^+ + Ne \rightarrow Ne + Ne^+$$
$$Ne^{++} + Ne \rightarrow Ne + Ne^{++}$$
$$N_2{}^+ + N_2 \rightarrow N_2 + N_2{}^+.$$

Experimental values are given by circles. ● Berry[15]; ⊙ Wolf.[16,18]

of $S(R)$ as

$$c_f{}^* c_f(\infty) = \sin^2 4 \langle S(R) \rangle_{Av} \int_{-\infty}^{+\infty} \frac{1}{\hbar v} \left[\frac{S(R)}{R} - J(R) \right] dx. \tag{18}$$

We now assume

$$\langle S(R) \rangle_{Av} = \tfrac{1}{2} S(R_0)$$

and we thus arrive at

$$c_f{}^* c_f(\infty) = \sin^2 2 S(R_0) \int_{-\infty}^{+\infty} \frac{1}{\hbar v} \left[\frac{S(R)}{R} - J(R) \right] dx. \tag{19}$$

The integral involved in this last expression is the same as that in the single charge transfer case, and the entire expression is thus readily evaluated. The calculated cross sections for double charge transfer for Ne⁺⁺ and A⁺⁺ are shown in Figs. 4 and 5 along with experimental values.[18]

[18] F. Wolf, Ann. Physik 34, 341 (1938).

In addition to the double-charge transfer indicated by Eq. (16), it is also possible for this reaction to proceed in two steps:

$$A^{++}+A \rightarrow A^{+}+A^{+} \rightarrow A+A^{++}.$$

In principle, this is also a resonant process since the initial and final electronic energies are identical; the individual steps, however, are nonresonant. It therefore seems that this two-stage process would have a much smaller cross section than Eq. (16) for atoms. In the case of molecules, in which intersecting potential energy curves are common, the two stage process might very well become important; but here molecular dissociation often predominates over charge transfer.

V. DIATOMIC MOLECULES IN EXACT RESONANCE

In the resonant charge transfer between diatomic molecules, there is the additional factor of the internal

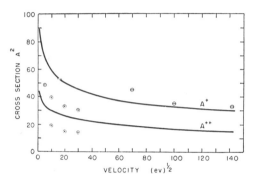

FIG. 5. The calculated cross sections (solid lines) as a function of velocity, $(ev)^{\frac{1}{2}}$, for the reactions

$$A^{+}+A \rightarrow A+A^{+}$$
$$A^{++}+A \rightarrow A+A^{++}.$$

Experimental values are indicated by circles. ⊖ Berry[15]; ⊙ Wolf.[16,18]

motion of the nuclei to be considered. Consider the reaction

$$H_2+H_2^{+} \rightarrow H_2^{+}+H_2 \qquad (20)$$

for collisions involving values of R_0 appreciably greater than the internuclear separations. The initial and final wave functions may be written as:

$$\Psi_i = \Psi^0_{a\ int}\Psi^0_{a\ el}\Psi^+_{b\ int}\Psi^+_{b\ el}$$
$$\Psi_f = \Psi^+_{a\ int}\Psi^+_{a\ el}\Psi^0_{b\ int}\Psi^0_{b\ el}.$$

The centers of mass of the molecules are located at the points a and b of Fig. 1. The electronic wave functions used are the usual Heitler-London functions, and the internal motions are described by the harmonic oscillator and rigid rotator functions. With these wave functions we have

$$H_{ji}' = (\Psi^0_{a\ int}, \Psi^+_{a\ int})(\Psi^0_{b\ int}, \Psi^+_{b\ int})H'_{f\,i\,el}. \qquad (21)$$

TABLE II. Values of the screening constant α and exponent n.

Atom	n	α
H	1	1.0
He	1	1.4
N	2	1.2
Ne	2	1.4
A	3	1.3
Kr	4	1.2

The two integrals multiplying the electronic matrix element in the last equation are identical, due to the symmetry of the problem. They are essentially overlap integrals between the initial rotational and vibrational levels with those for the final state. The rotational wave function for the free rotator is just the spherical harmonic function, Y_l^m. For the distant collisions which determine the charge transfer cross section, the rotational motion is still free, and the orthogonal properties of the spherical harmonic functions require that a molecule with magnetic and azimuthal quantum numbers m and l form an ion with the same quantum numbers according to this simplified model. Similarly, the ion that is neutralized maintains its rotational state. The only factor in $(\Psi^0_{a\ int}, \Psi^+_{a\ int})$ different from unity will be the overlap of the vibrational wave functions.

Since the molecule and the ion are assumed to be in the lowest vibrational states initially, they will still be in the ground states after charge transfer in a resonant process. In general, the molecule and the ion do not have the same internuclear separation and vibrational frequencies; hence the overlap of their vibrational wave functions will not be unity, and

$$H'_{fi} = p^2 H'_{f\,i\,el} \qquad (22)$$

where p is the vibrational overlap integral. The calculated values for p for vibrational levels of interest are listed in Table III.

The quantity $H'_{f\,i\,el}$ is a matrix element which involves orbitals on four centers. Several three-center integrals appear, and we have used reasonable approximations to these integrals in terms of $S(R)$ and $J(R)$. These are summarized in Table IV. The matrix element becomes

$$H'_{f\,i\,el} = \frac{[1+S(\rho)]^2}{[1+S^2(\rho)]}\left\{\frac{S(R)}{R}-J(R)\right\} \qquad (23)$$

where ρ represents the average of the internuclear distances for the molecule and the ion.

TABLE III. Values of overlap between vibrational wave functions.

System	Overlap (p)
$H_2-H_2^{+}$	0.312
$HD-HD^{+}$	0.264
$D_2-D_2^{+}$	0.195
$N_2-N_2^{+}$	0.940
$H_2-(H_2^{+})'$	−0.540

TABLE IV. Approximation to three-center integrals. The centers (c,d) (at a separation ρ) and the centers (f,g) (at a separation ρ) are located a distance R apart, where $R \gg \rho$.

$$\int_{\tau_1} \frac{1}{r_{c1}} f(1)g(1)d\tau_1 \approx \frac{1}{R} S(\rho)$$

$$\int_{\tau_1} \frac{1}{r_{c1}} d(1)g(1)d\tau_1 \approx J(R)S(\rho)$$

If

$$L_{cd,fg} = \int_{\tau_1,\tau_2} c(1)d(1)\frac{1}{r_{12}} f(2)g(2)d\tau_1 d\tau_2$$

then

$$L_{fg,cc} \approx \frac{1}{R} S(\rho)$$

$$L_{fg,cd} \approx \frac{1}{R} S^2(\rho)$$

$$L_{fd,cc} \approx \frac{1}{R} S(R)S(\rho)$$

$$L_{fd,cd} \approx \frac{1}{R} S(R)S^2(\rho)$$

This expression reduces correctly to the He case as $\rho \to 0$.

Expression (23) must be averaged over all possible relative orientations of the molecule and ion. Consider the center of mass of the ion and the molecule fixed at the points a and b, respectively, of Fig. 6. If the equilibrium separations of the nuclei are $2\rho_1$ and $2\rho_2$ for the ion and molecule, then the separation of the two interacting nuclei r varies as the nuclei take on all possible positions on spheres a and b. Thus, $J(R)$ is actually some average of $J(r)$ and similarly with $S(R)$. Since both of these integrals vary as re^{-r}, we have, after averaging re^{-r} over both spheres,

$$\langle re^{-r} \rangle_{\text{Av}} = \frac{1}{4} \cdot \frac{e^{\rho_1} - e^{-\rho_1}}{\rho_1} \cdot \frac{e^{\rho_2} - e^{-\rho_2}}{\rho_2} \cdot Re^{-R}.$$

Thus, the value of H_{fi}' to be used in the cross section calculation is given by expression (23) multiplied by the factor q:

$$q = \frac{1}{\rho_1 \rho_2} \cdot \sinh\rho_1 \cdot \sinh\rho_2.$$

The charge transfer probability is

$$P(R_0) = \sin^2\left\{ q p^2 \frac{[1+S(\rho)]^2}{1+S^2(\rho)} \right.$$
$$\left. \times \int_{-\infty}^{+\infty} \frac{1}{\hbar v} \left[\frac{S(R)}{R} - J(R) \right] dx \right\}. \quad (24)$$

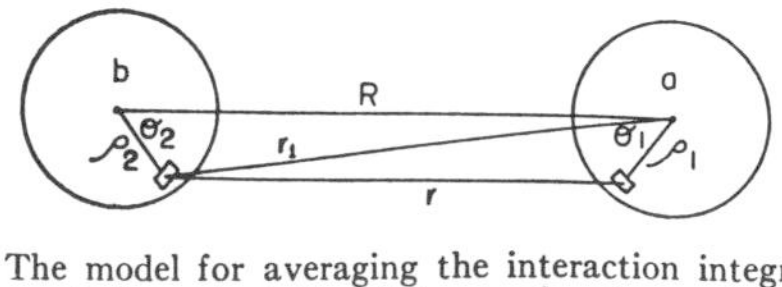

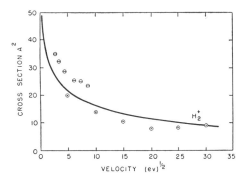

FIG. 6. The model for averaging the interaction integrals over various relative orientations between the atom and the ion.

The charge transfer cross sections for H_2^+, D_2^+, and HD^+ in their parent gases can readily be calculated from this formula. They differ only in the value of p. The calculated cross section for H_2^+ is shown in Fig. 7 along with experimental values.[16,19] Figure 8 shows a comparison between the calculated values for the cross section for H_2^+, D_2^+, and HD^+.

In addition, Eq. (24) can be used for the reaction

$$N_2 + N_2^+ \rightarrow N_2^+ + N_2. \quad (25)$$

Here we use an electronic wave function for the molecule that corresponds to the Heitler-London function for H_2 except that now the atomic functions involved are Slater's nodeless functions for atomic nitrogen with the appropriate value for α. The calculated results for this reaction are shown in Fig. 4.

An interesting comparison is made in Table V. In this table we list the values of the cross sections for the

FIG. 7. The calculated cross section (solid line) as a function of velocity, $(ev)^{\frac{1}{2}}$, for the reaction

$$H_2^+ + H_2 \rightarrow H_2 + H_2^+.$$

Experimental values are indicated by circles. ⊙ Wolf[16]; ⊖ Simons.[19]

ions H^+, H_2^+, N^+, N_2^+ in their parent gases. It is seen that the values of N_2^+ are greater than those for N^+, while H_2^+ gives cross sections lower than those for H^+. This shift in order is explained by the fact that the harmonic oscillator overlap for the N_2^+ case is approximately unity and the molecule has a larger cross section than the atom due to the factor

$$q \frac{[1+S(\rho)]^2}{[1+S^2(\rho)]}.$$

In the case of H_2^+, this multiplying factor is reduced to less than unity by the fact that the harmonic oscillator overlap is quite small.

Table V also shows that for a given velocity, the cross section for the H^+ case is about equal to that for N^+ and considerably greater than that for He^+. This effect can be qualitatively understood from the relatively high

[19] J. Simons et al., J. Chem. Phys. 11, 312 (1943).

effective charge of the helium nucleus which shrinks the associated electron cloud as compared with hydrogen atoms. Thus, the charge distribution about the hydrogen atom is more similar to the distribution around the nitrogen atom. This is essentially the same as saying that the ionization potential of H and N are about the same, while that for He is considerably greater.

VI. MOBILITY

The mobility of a positive ion in a gas of the parent molecule is considerably lower than that expected from the mass of the ion.[20] This lower mobility is qualitatively explained as an effect of charge transfer. Since for thermal energies the cross section for charge transfer is about two or three times the kinetic theory cross section for the atoms, it is not surprising that this effect is of considerable magnitude.

A rather complete quantum mechanical treatment of the mobility of He+ in He has previously been given.[5,21] Here we will show that the Langevin formula for mobility can be used in the resonant case provided the collision cross section involved is somewhat modified.[22]

According to the Langevin formula, the mobility of an ion in its parent gas, assuming large interaction cross

TABLE V. Comparison of charge transfer cross sections (A²) for ions in the parent gases.

v (cm/sec)	H+	H₂+	N+	N₂+	He+
10^7	27.4	16.4	32.9	41.8	13.9
5×10^5	57.6	42.2	63.0	74.4	29.1

sections, is given by

$$\mu = \frac{3}{8} \frac{e}{\sigma_L n} \left(\frac{N\pi}{MkT} \right)^{\frac{1}{2}} \tag{26}$$

where

μ is the mobility,
e is the electronic charge,
σ_L is the cross section,
n is the number of molecules/cm³,
N is Avagodro's number,
M is the molecular weight,
k is Boltzmann's constant, and
T is the absolute temperature.

In the derivation of this formula, σ_L is actually the kinetic theory cross section. However, in cases where charge transfer is possible it seems reasonable to let σ_L also include the thermal cross section for charge transfer σ. Since one-half of the collisions involved in the kinetic theory cross section σ_V result in charge transfer, they

[20] A. M. Tyndall, *Mobility of Positive Ions in Gases* (Cambridge University Press, New York, 1938).
[21] H. Massey and C. Mohr, Proc. Roy. Soc. (London) **A144**, 188 (1934).
[22] J. A. Hornbeck and G. H. Wannier, Phys. Rev. **82**, 458 (1951).

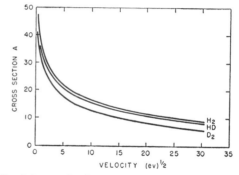

FIG. 8. A comparison between the calculated values of the cross section for the reactions

$$H_2^+ + H_2 \rightarrow H_2 + H_2^+$$
$$HD^+ + HD \rightarrow HD + HD^+$$
$$D_2^+ + D_2 \rightarrow D_2 + D_2^+.$$

have already been included in σ. Thus, let

$$\sigma_L = \sigma + \tfrac{1}{2}\sigma_V. \tag{27}$$

The values of mobilities of singly charged ions in their respective gases, as calculated by this method are shown in Table VI for 273°K.

The first column of observed values is at 0°C while the second is at 18°C. The agreement between calculated and observed values is quite satisfactory except in the case of helium. The accepted value of the mobility of He+ in He is about 11, which is essentially the same as that calculated by Massey and Mohr. The higher value 21.4 is attributed to the ion He₂+.[23]

While this method of treating the charge transfer cross section as a physical reality is certainly open to objections, the fact that the results obtained are of the proper order of magnitude is interesting.

VII. THE NEAR-RESONANT CASE

The transfer of charge in a near-resonant system may be represented by the equation

$$A + B^+ \rightarrow A^+ + B. \tag{28}$$

Proceeding in the same manner as in Sec. II, we arrive at the following equation, similar to Eq. (2), for the total wave function of the system, exclusive of the

TABLE VI. Comparison of observed and calculated mobilities for ions in their parent gases.

	σ	v	μ (calc)	μ (observed) (reference 22)	(reference 20)
He+	37.2	14.9	17.5	10.8	21.4
Ne+	66.0	21.3	4.56	4.4	6.23
A+	110.5	42.2	1.89	1.63	1.93
N₂+	105.5	45.0	2.32		2.67
H₂+	53.2	23.3	17.1		14.7

[23] R. Meyerott, Phys. Rev. **66**, 242 (1944).

relative motion:

$$\Psi = c_i \Psi_i \exp(-E_i t/\hbar) + c_f \Psi_f \exp(-iE_f t/\hbar). \quad (29)$$

In this case we no longer have the simplification that is possible at exact resonance when $E_i = E_f$. If we define $E_f - E_i = \epsilon$, and then follow the same procedure used to construct the coupled equations for the resonant case, we arrive at

$$c_i H_{ii}' + c_f H_{fi}' \exp(-i\epsilon x/\hbar v)$$
$$= i\hbar v[\dot{c}_i + \dot{c}_f F \exp(-i\epsilon x/\hbar v)] \quad (30)$$

$$c_i H_{fi}' \exp(+i\epsilon x/\hbar v) + c_f H_{ii}'$$
$$= i\hbar v[\dot{c}_i F \exp(+i\epsilon x/\hbar v) + \dot{c}_f] \quad (31)$$

where the dot refers to differentiation with respect to x. Since it is assumed that the energy associated with the relative motion of the particles is large compared with the energy difference ϵ of the two states, it is permissible to introduce an average velocity v. Furthermore, it has been assumed that $H_{ii}' = H_{ff}'$ and $H_{fi}' = H_{if}'$. These matrix elements are equal for the case of exact resonance using zeroth-order wave functions, but for the case of near-resonance using zeroth-order wave functions, the matrix elements are not strictly equal.

Even with these assumptions it is not possible to solve Eqs. (30) and (31) in the simple manner that was used for Eqs. (5) and (6), because it is not apparent that the $\pi/2$ phase relationship between the coefficients is generally valid.

It will be useful to write the explicit form for the solutions of these equations for the resonant case, i.e., $\epsilon = 0$. At any point x_1 the resonant solutions are

$$c_i(x_1) = -\exp\left(i \int_{-\infty}^{x_1} B dx\right) \cdot \cos \int_{-\infty}^{x_1} A dx \quad (32)$$

$$c_f(x_1) = i \exp\left(i \int_{-\infty}^{x_1} B dx\right) \cdot \sin \int_{-\infty}^{x_1} A dx \quad (33)$$

where

$$A = \frac{1}{\hbar v} \cdot \frac{H_{fi}' - F H_{ii}'}{1 - F^2}$$

$$B = \frac{1}{\hbar v} \cdot \frac{F H_{fi}' - H_{ii}'}{1 - F^2}.$$

Since we are only interested in the asymptotic values of the coefficients, i.e., $x_1 = \infty$, we will postulate the following asymptotic forms for the solutions of Eqs. (30) and (31) for the limit $x_1 \to \infty$:

$$c_i(x_1) = -\exp\left(i \int_{-\infty}^{x_1} B dx\right) \cos \int_{-\infty}^{x_1} A e^{-i\gamma x} dx \quad (34)$$

$$c_f(x_1) = i \exp\left(i \int_{-\infty}^{x_1} B dx\right) \sin \int_{-\infty}^{x_1} A e^{+i\gamma x} dx \quad (35)$$

where

$$\gamma = \frac{\epsilon}{\hbar v}.$$

These expressions are in agreement with the differential equations and they satisfy the normalization condition. They also reduce to the resonant solution when $\epsilon = 0$. Due to the symmetry of the quantity A it is readily seen that these postulated solutions are out of phase by $\pi/2$ at the point $x_1 = \infty$. In the case of exact resonance this phase relationship was found to hold for all values of x.

From Eq. (35) the probability for charge transfer in a near-resonant system is given by

$$c_f^* c_f(\infty) = \sin^2 \int_{-\infty}^{\infty} \frac{1}{\hbar v} \frac{H_{fi}' - F H_{ii}'}{1 - F^2} \cos\left(\frac{\epsilon}{\hbar v} x\right) dx. \quad (36)$$

Equations (30) and (31) can be solved under the assumption that the probability for charge transfer is extremely small. In this case, we can consider $c_i = 1$, $\dot{c}_i = 0$, and $c_f = 0$. Integrating Eq. (31) directly, gives at any point x_1

$$c_f(x_1) = -i \int_{-\infty}^{x_1} \frac{H_{fi}'}{\hbar v} \exp(i\epsilon x/\hbar v) \quad (37)$$

and hence

$$c_f^* c_f(\infty) = \left[\int_{-\infty}^{\infty} \frac{H_{fi}'}{\hbar v} \cos\left(\frac{\epsilon x}{\hbar v}\right) dx\right]^2. \quad (38)$$

This expression is identical with Eq. (36) since $F H_{ii}' \ll H_{fi}'$, $F^2 \ll 1$ and $\sin\theta = \theta$ for $\theta \ll 1$. The postulated solutions to the coupled equations thus give the correct value for the probability of charge transfer in the asymptotic limit of small probability.

VIII. EXAMPLE OF NEAR-RESONANT CHARGE TRANSFER

As an example of charge transfer in a near-resonant process consider the discharging of argon ions in molecular hydrogen:

$$H_2 + A^+ \to H_2^+ + A. \quad (39)$$

The integrals involved in Eq. (36) are the same as the integrals for the resonant case multiplied by the factor

$$\exp(-\rho_0 \beta^2/2\alpha^2) \quad (40)$$

where

$$\beta = \frac{\epsilon}{\hbar v}$$

$$\rho_0 = \alpha R_0.$$

The value of this exponential quantity has been plotted as a function of ϵ (in ev) in Fig. 9, for several different velocities using $\rho_0 = 10$ and $\alpha = 1$. This figure shows that for a given energy difference ϵ, a decrease in velocity tends to give a decrease in the probability for charge transfer. At the same time, however, the usual increase in probability found in the formula for the resonant case, is operating; the net effect is a maximum in the probability of charge transfer at some particular velocity. It should also be noted that ϵ appears as the square and hence the charge transfer probability depends only on the magnitude of ϵ and not on its sign. From the figure it is also apparent that for large values of ϵ this method of charge transfer would only be possible at very high velocities.

For reaction (39), the ionization potential of H_2 is 15.42 ev and that of A is 15.76 ev, and consequently the value of ϵ is 0.32 ev. From Fig. 9 it can be estimated that at this value of ϵ the exponential factor would be small except at very high velocities. Thus, reaction (39) would be expected to occur by this mechanism only at very high velocities when the charge transfer probability would be small anyway.

However, consider the reaction

$$H_2 + A^+ \rightarrow (H_2^+)' + A. \qquad (41)$$

Here the H_2^+ formed is in the first vibrational level instead of the ground state. Since ϵ was defined as the difference between the total energy of the initial and final states (excluding the energy of the relative motion of the two particles) we now have, for reaction (41)

$$\epsilon = 0.32 \text{ ev} - h\nu \qquad (42)$$

where h is Planck's constant and ν is the vibrational frequency of H_2^+. Substitution of the appropriate values for h and ν leads to

$$\epsilon = 0.04 \text{ ev}.$$

This latter reaction will thus have a much larger cross section than reaction (39) due to the larger exponential term as well as the more favorable vibrational overlap (Table III). We will now calculate the cross section for reaction (41) using a model similar to the resonance case.

The hydrogen nuclei c and d are centered on the point a and the argon nucleus is located at b (see Fig. 1);

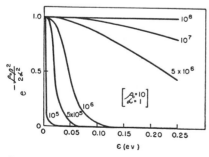

Fig. 9. The factor $\exp[-(\rho_0\beta^2)/(2\alpha^2)]$ as a function of the resonance defect ϵ for $\rho_0 = 10$ and $\alpha = 1$, and for relative velocities of 10^5 to 10^8 cm/sec.

we now have

$$\Psi_i = \Psi_{i\,\text{vib}} \frac{1}{\{2[1+S^2(\rho)]\}^{\frac{1}{2}}} [c(1)d(2) + c(2)d(1)] \qquad (43)$$

$$\Psi_f = \Psi_{f\,\text{vib}} \frac{1}{\{2[1+S(\rho)]\}^{\frac{1}{2}}} a(1)[c(2) + d(2)] \qquad (44)$$

where $c(1)$, $d(1)$, $c(2)$, and $d(2)$ are hydrogen $1s$ functions and $a(1)$ is the Slater nodeless wave function for argon. The perturbation term operating on Ψ_i is

$$H' = \frac{2}{R} - \frac{1}{r_{a1}} - \frac{1}{r_{a2}}. \qquad (45)$$

Using the same approximations for the three center integrals as was used in the resonance case, we have

$$H_{fi}' = (\Psi_{f\,\text{vib}}, \Psi_{i\,\text{vib}}) \left[\frac{1+S(\rho)}{1+S^2(\rho)} \right]^{\frac{1}{2}} \left[\frac{S(R)}{R} - J(R) \right] \qquad (46)$$

where

$$J(R) = \int_{\tau_1} \frac{1}{r_{a1}} a(1)c(1)d\tau_1.$$

By including the factor to allow for all possible orientations of the diatomic molecule we have

$$H_{fi}' = pq \left[\frac{1+S(\rho)}{1+S^2(\rho)} \right]^{\frac{1}{2}} \left[\frac{S(R)}{R} - J(R) \right], \qquad (47)$$

where p is the harmonic oscillator overlap between the ground state of H_2 and the first vibrational level of H_2^+. The orientation factor is given by

$$q = \frac{2}{\rho} \sinh\frac{\rho}{2}.$$

The S and J integrals for wave functions of this type are considerably complicated by the fact that the exponential is not the same in both wave functions. We will assume that the interaction between a $1s$ function and nodeless $3s$ function is approximately the same as the interaction between two $2s$ nodeless functions. The cross section for reaction (41) is now calculated using Eq. (36) to calculate the charge transfer probability at different values of R_0, and following the same techniques that were used in the exact resonance case. These results are plotted in Fig. 10 along with experimental values.[24] Due to the small energy difference between the two states, the cross sections are of the same order of magnitude as resonance cross sections and the maximum is located at a rather low velocity.

IX. DISCUSSION

The treatment presented above was the simplest which the authors thought could reasonably be ex-

[24] F. Wolf, Ann. Physik 27, 543 (1937).

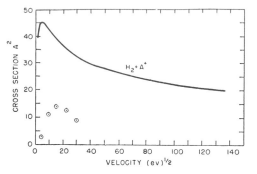

FIG. 10. The calculated cross section (solid line) as a function of velocity, $(ev)^{\frac{1}{2}}$, for the reaction

$$H_2 + A^+ \rightarrow (H_2^+)' + A.$$

The experimental values for the discharging of A^+ in H_2 as given by Wolf[24] are indicated by ⊙.

pected to explain charge transfer phenomena semi-quantitatively. Although the model is admittedly crude, the agreement which has been obtained with the limited amount of experimental data available is encouraging. From this comparison it appears that the data of Wolf is too low, or at least it is unlikely that the data of Wolf and Berry are consistent. Thus, there is no indication of a systematic deviation of calculated cross sections as a function of velocity. Such a deviation might suggest an inadequacy in the one electron model with a constant screening factor. The somewhat arbitrary choice of this factor α is a weakness of the theory. This is particularly true in the case of near-resonance where the quantity α appears in an exponential factor.

The transfer of charge in near-resonant systems with noncrossing potential energy curves will have appreciable probability only when the energy difference between the two states is extremely small. It is, therefore, essential that the systems have vibrational degrees of freedom so that a sufficient number of states are available to insure a small resonance defect ϵ. Such a situation arises in the discharging of ions in organic vapors. In these relatively complicated molecules the vibrational states have a considerable density and hence the transfer of charge by this "near-resonant" process is quite probable.

For relatively close collisions in the near-resonance case it is also possible that the exchange interaction will be sufficiently large to give a splitting of the energy states that will result in a crossing. This mechanism has been previously discussed by Magee[25] for the discharging of A^+ in H_2; it probably predominates for close collisions while the above discussed method is applicable for distant collisions.

It should also be noted that the reverse reaction, the discharge of H_2^+ in A, should have a smaller cross section since the potential energy of the reactants is now

at a lower level than the potential energy of the products. The resonance defect ϵ for the higher vibrational states of the products would even be greater than that for the ground state. This conclusion is observed experimentally,[24] although this reverse reaction still has an appreciable cross section.

APPENDIX I. PHASE RELATION BETWEEN COEFFICIENTS c_i AND c_f

If Eqs. (5) and (6) are added we have

$$i\hbar(\dot{c}_i + \dot{c}_f)(1+F) = (c_i + c_f)(H_{ii}' + H_{if}'). \tag{I-1}$$

Introducing the variable x in place of t, this equation becomes

$$\frac{d(c_i + c_f)}{dx} = -\frac{i}{\hbar v}\frac{(H_{ii}' + H_{if}')}{(1+F)}(c_i + c_f). \tag{I-2}$$

This may be integrated directly to give the value of $(c_i + c_f)$ at any point x as

$$(c_i + c_f)_x = e^{iK_x} \tag{I-3}$$

where we have made use of the initial conditions $c_i = 1$, $c_f = 0$ at $x = -\infty$, and where

$$K_x = \int_{-\infty}^{x} \frac{1}{\hbar v}\frac{(H_{ii}' + H_{if}')dx}{(1+F)}. \tag{I-4}$$

Since K_x is a real quantity, for any value of x we have

$$|c_i + c_f| = 1. \tag{I-5}$$

Let $c_i = \rho e^{i\omega}$ and $c_f = \rho' e^{i\omega'}$. Then, from Eq. (I-5)

$$\rho^2 + \rho'^2 + 2\rho\rho' \cos(\omega - \omega') = 1. \tag{I-6}$$

Furthermore, from the fact that Eq. (2) must be normalized we have

$$\rho^2 + \rho'^2 + 2\rho\rho'F \cos(\omega - \omega') = 1. \tag{I-7}$$

Equations (I-6) and (I-7) lead directly to

$$2\rho\rho'[1 - F] \cos(\omega - \omega') = 0. \tag{I-8}$$

But certainly F cannot equal unity for all values of x. Hence,

$$(\omega - \omega') = \pm \pi/2. \tag{I-9}$$

This establishes the phase relationship between the coefficients. It should also be noted that

$$\rho' = (1 - \rho^2)^{\frac{1}{2}}. \tag{I-10}$$

This last expression is true for all values of x regardless of the value of the overlap integral F, since $\cos(\omega - \omega') = 0$.

APPENDIX II. COMPARISON OF THE IMPACT PARAMETER METHOD WITH THE METHOD OF SCATTERED WAVES

The impact parameter method of calculation treats the relative motion of the colliding molecules by

[25] J. L. Magee, J. Phys. Chem. 56, 555 (1952).

classical mechanics and introduces quantum mechanics for all other degrees of freedom. It is interesting to make a comparison of this treatment with a purely quantum-mechanical method of scattered waves.[1]

Consider the reaction of a proton with a hydrogen atom. The hydrogen atom is considered fixed at the point a of Fig. 1, and a steady stream of protons, represented by a plane wave, is scattered into two waves: one represents elastic (billard ball) scattering, and the other represents inelastic scattering due to charge transfer. Let ψ_1 and ϕ_1 be the electronic and nuclear wave functions, respectively, for the system with the electron at a, and let ψ_2 and ϕ_2 be the corresponding functions for the system after electron transfer.

The nuclear functions ϕ_1 and ϕ_2 are subject to the following boundary conditions at large values of R:

$$\phi_1 \sim e^{ikx} + f_1(\theta,\varphi)\frac{e^{ikR}}{R} \qquad (II\text{-}1)$$

$$\phi_2 \sim f_2(\theta,\varphi)\frac{e^{ikR}}{R} \qquad (II\text{-}2)$$

where $k = (Mv/\hbar)$, M is the mass of the proton; $f_1(\theta,\varphi)$ and $f_2(\theta,\varphi)$ are coefficients for elastic and inelastic scattering, respectively. Since the system has axial symmetry, these coefficients are actually independent of φ and may be considered as functions of θ only.

The total wave function for the system maybe written as

$$\Psi = \psi_1\phi_1 + \psi_2\phi_2 \qquad (II\text{-}3)$$

and the total Hamiltonian as

$$H = -\frac{\hbar^2}{2M}\nabla^2 + H_{el} + H' \qquad (II\text{-}4)$$

where ∇^2 is the Laplacian operator for the nuclear motion, H_{el} is the Hamiltonian for the electronic energy of the atom, H' is the interaction Hamiltonian for the interaction between the atom and the proton. With this Hamiltonian and wave function, the following two equations can be constructed (m = mass of electron):

$$\left\{\nabla^2 + k^2 - \frac{2M}{m}\frac{(H_{12}' + H_{11}')}{(1+F)}\right\}(\phi_1+\phi_2) = 0 \quad (II\text{-}5)$$

$$\left\{\nabla^2 + k^2 + \frac{2M}{m}\frac{(H_{12}' - H_{11}')}{(1-F)}\right\}(\phi_1-\phi_2) = 0. \quad (II\text{-}6)$$

When these equations are solved subject to the asymptotic solutions given by Eqs. (II-1) and (II-2) we arrive at

$$f_2(\theta) = \frac{1}{4ik}\sum_{l=0}^{\infty}(2l+1)(e^{i2\eta_l} - e^{i2\delta_l})P_l(\mu) \qquad (II\text{-}7)$$

where l is the angular momentum quantum number defined by $(MvR_0)^2 = l(l+1)\hbar^2$, $P_l(\mu) = P_l(\cos\theta)$ is the Legendre polynomial, and the phase shifts η_l and δ_l are given by

$$\eta_l = -\frac{\pi M}{m}\int_0^{\infty}\frac{H_{12}' + H_{11}'}{(1+F)}[J_{(l+\frac{1}{2})}(kR)]^2 R dR \quad (II\text{-}8)$$

$$\delta_l = \frac{\pi M}{m}\int_0^{\infty}\frac{H_{12}' - H_{11}'}{(1-F)}[J_{(l+\frac{1}{2})}(kR)]^2 R dR. \quad (II\text{-}9)$$

The total cross section for charge transfer σ is readily found by integrating over all values of θ. Thus

$$\sigma = 2\pi\int_0^{\pi}|f_2(\theta)|^2\sin\theta d\theta \qquad (II\text{-}10)$$

or

$$\sigma = \frac{\pi}{k^2}\sum_{l=0}^{\infty}(2l+1)\sin^2(\delta_l - \eta_l). \qquad (II\text{-}11)$$

The impact parameter method for the calculation of the charge transfer cross section gave the expression

$$\sigma = 2\pi\int_0^{\infty}\sin^2 Q \cdot R_0 dR_0. \qquad (II\text{-}12)$$

From Eqs. (II-11) and (II-12) we see that the contribution to the total cross section from some particular value of l is

$$\sigma_l = \frac{\pi}{k^2}(2l+1)\sin^2(\delta_l - \eta_l) \qquad (II\text{-}13)$$

while the comparable contribution from a particular value of R_0 is

$$R_0 = 2\pi R_0 \Delta R_0 \sin^2 Q. \qquad (II\text{-}14)$$

It is readily shown, from the definition of l and k that

$$2\pi R_0 \Delta R_0 = \frac{\pi}{k^2}(2l+1). \qquad (II\text{-}15)$$

Thus, to show the equivalence of these two methods, it is necessary to show that

$$Q = \delta_l - \eta_l. \qquad (II\text{-}16)$$

Now Q can be determined by integrating over R from R_0 to ∞ a matrix element multiplied by

$$\frac{1}{k}\cdot\frac{1}{(R^2 - R_0^2)^{\frac{1}{2}}} \qquad (II\text{-}17)$$

while the difference in the phase shifts is determined from the same matrix element multiplied by

$$\pi[J_{(l+\frac{1}{2})}(kR)]^2 \qquad (II\text{-}18)$$

and integrated from $R=0$ to $R=\infty$.

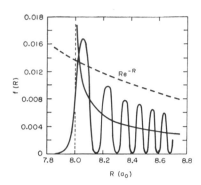

FIG. 11. A comparison between the probability factor obtained from the "semiclassical" treatment (solid decreasing curve) and that obtained by the method of scattered waves (decaying oscillatory function). The function Re^{-R} is shown for comparison (dashed decreasing curve).

These last two functions are now plotted [designated as $f(R)$] in Fig. 11 for $k=100$ and $R_0=8$ in atomic units. For the reaction between a proton and a hydrogen atom this value of k corresponds to a velocity of

1.2×10^7 cm/sec. Figure 2 shows that for this value of R_0 and velocity, the probability for charge transfer will be small and we are justified in using the above formulas [(II-8) and (II-9)] for the phase shifts. The curve for the Bessel function was sketched in after locating the position of the zeros, and the position and magnitude of the maxima, as well as two points in the region of exponential buildup.[26] The dashed sloping line in the figure is the function Re^{-R}; it is plotted to show the general behavior of the matrix element.

For values of R greater than those shown in the figure, the comparison may be made analytically by using the asymptotic expression for the Bessel function.

It is interesting to note that in the evaluation of the phase shifts in an actual calculation, Massey and Smith[1] replaced the Bessel function by its average value. Thus, to evaluate the cross sections in any relatively simple calculation it appears necessary to introduce classically the relative motion of the particles, either explicitly in the beginning (impact method) or implicitly at the end of the calculation (scattered waves).

[26] E. Jahnke and F. Emde, *Funktionentafein* (Dover Publications, New York, 1943).

17

Calculation of interchange reaction rates by a 'nearest resonance' method

D. K. BOHME†, J. B. HASTED and P. P. ONG‡

University College London

1. Introduction

It is well known that thermal energy gas reaction rates for exothermic processes of the type

$$X + YZ \rightarrow XY + Z \tag{1}$$

where both X and YZ may be neutral, but one may be ionized, may differ from each other by orders of magnitude for not easily discernable reasons. In this paper we discuss these processes, ion–atom interchange as well as charge transfer and atom–atom interchange, in terms of the Massey near-adiabatic criterion (Massey 1949). The magnitude of the cross section for a reaction is then determined essentially from a comparison of the time $\tau_t = h/|\Delta E|$ of the electronic transition involving an energy change equal to the energy defect ΔE of the reaction and the time $\tau_c = a/v$ of collisions. Here a is a collision length known as the adiabatic parameter and v is the velocity of impact. Only for $\tau_t \ll \tau_c$ is adiabatic adjustment of the collidant wave functions possible and the probability of transition between initial and final states small. But although a qualitative behaviour of a particular cross section may be obtained from this criterion, it gives no information on absolute cross sections.

The adiabatic criterion was found to hold in the form of a maximum rule (Hasted 1951) for atomic charge-transfer processes, with $a = 7$ Å. The maximum cross section occurs when $\tau_t = \tau_c$. This rule is consistent (Lee and Hasted 1965) with semi-classical impact parameter calculations (Rapp and Francis 1962) which, although relatively crude in that semi-empirical atomic orbitals are used, give surprisingly good agreement with experiment, even in the near-thermal energy region.

Although the near-adiabatic resonance criterion has proved to be a qualitatively useful guide for atomic collisions, there has been some doubt as to its applicability to gaseous reactions of the type of equation (1). Many such reactions involving electronic transitions, as well as many ion–molecule dissociative charge-transfer reactions, have been found to occur rapidly at thermal energies. These results are contrary to what is expected on the basis of Massey's adiabatic hypothesis, on the assumption that the collision products are formed without internal excitation.

† Now at ESSA, Boulder, Colorado, U.S.A.
‡ Now at University of Singapore.

It is the proposal of the authors that, owing to added rotational and vibrational degrees of freedom, low energy collisions involving molecules, in which the rotational and vibrational levels are merely broadened and not made to overlap, are dominated by 'accidental resonance' conditions in which $\tau_t \simeq \tau_c$. Since many thermal energy collisions take place with inward spiralling orbits except at the largest impact parameters, τ_c is larger than is the case in non-spiralling collisions; therefore for accidental resonance ΔE must be very small, probably only a few wave numbers. The electronic transition in this case involves an energy change equal to the energy defect corresponding to the closest exothermic resonance with a given vibrational–rotational energy level of the reactant molecule, and not equal to the ground-state energy defect of the reaction. As a result, the adiabatic region is often subthermal. In exothermic processes it is often possible to find an accidental near-resonance when the products are formed in vibrationally and rotationally excited states, but it is more likely when the nearest product rotational quantum number J is smallest, since the separation of levels increases with J.

Impact parameter calculations of charge transfer (Rapp and Francis 1962) show that as the impact velocity decreases from the value for which $\tau_t = \tau_c$ the cross section falls monotonically. A simple approximate expression for the cross section σ has been found to fit experimental atomic charge-transfer data (Hasted 1959) in this impact velocity region:

$$\sigma = A \exp\left(-\frac{k'a|\Delta E|}{hv}\right) \tag{2}$$

with A, k' constants. Since this expression is only justifiable empirically and in any case does not hold in the far adiabatic region, it has not been widely used, except for two-electron capture processes (Kozlov and Bondor 1966). It is nevertheless particularly suitable for rough approximation to processes dominated by accidental resonance.

The maximum value of the cross section, σ_m, may for ion–molecule reactions be calculated directly from the inward-spiralling orbit equation (Gioumousis and Stevenson 1958)

$$\sigma_m = \frac{\pi}{v}\left(\frac{4\alpha e^2}{\mu}\right)^{1/2} \tag{3}$$

for collisional reduced mass μ, electronic charge e and target polarizability α (Rothe and Bernstein 1959, Parkinson 1960). This is substituted in equation (2):

$$\sigma_m(v_m) = A \exp\left(-\frac{k'a|\Delta E|}{hv_m}\right). \tag{4}$$

For neutral reactions σ_m is calculated using the impact parameter method developed for charge transfer (Rapp and Francis 1962), as discussed in § 2.5 below.

The velocity v_m at which the maximum cross section is reached may be calculated from the equality $h/|\Delta E| = a/v_m$, with adiabatic parameter a determined empirically, but $a \geqslant 7$ Å on account of the spiralling orbits. For $v \geqslant v_m$, $\sigma = \sigma_m$ is taken as a sufficiently good approximation. For $v \leqslant v_m$, A is eliminated from equations (2) and (4) to give

$$\sigma = \sigma_m(v_m) \exp k'\left(1 - \frac{a|\Delta E|}{hv}\right). \tag{5}$$

In the present treatment of molecular processes ΔE is taken to be the smallest energy defect, corresponding to the closest exothermic resonance with a given vibrational–rotational energy sublevel of the reactant molecule. Equation (5), therefore, approximates the absolute cross section for the transition between specific energy sublevels of the reactant and product molecules.

2. Calculation of the near-resonance cross section

The data available in the literature on atomic and molecular constants used in these calculations are given in tables 1(*a*) and 1(*b*). Ionization potentials, dissociation energies and vibrational and rotational constants have been taken from Herzberg (1950) unless otherwise indicated.

All calculations were carried out on the University College IBM 360/65 computer.

Table 1(*a*). Molecular constants

Molecule	I (ev)	Dissociation energy (ev)	B_e	Rotational and vibrational constants			
				ω_e	$\omega_e x_e$	$\omega_e y_e$	α_e
$N_2(x\,^1\Sigma_g^+)$	15·576	9·756	1·9987	2358·07	14·188	−0·0124†	0·0187
$O_2(x\,^3\Sigma_g)$	12·063	5·080	1·4457	1580·36	12·073	+0·0546	0·0158
$NO(x\,^2\Pi_{3/2})$	9·266¶	6·487	1·7046	1904·03	13·97	−0·0012	0·0178
$NO(x\,^2\Pi_{1/2})$				1903·68			
$CO(x\,^1\Sigma^+)$	14·013‡	11·108	1·9314	2170·21	13·461	+0·0308	0·0175
$N_2^+(x\,^2\Sigma_g^+)$			1·9322	2207·19	16·136	−0·0400	0·0202
$O_2^+(x\,^2\Pi_g)$			1·6722	1876·4	16·53	0·0	0·0198
$NO^+(x\,^1\Sigma^+)$			2·002	2378·70₃	16·647§	0·0	—
$CO^+(x\,^2\Sigma^+)$			1·9772	2214·24	15·164	−0·0007	0·0190

† Lofthus 1960.

‡ Callomon 1967, private communication.

§ Miescher 1967, private communication; W. R. Simmons, *University of Colorado* AFCRL 65–670.

¶ Dressler and Miescher 1966.

Table 1(*b*). Atomic energy levels

Atomic state	Ionization potential I (ev)	Atomic state	Term level (cm⁻¹)
$N(^4S^\circ_{3/2})$	14·545	$O(^3P_2)$	0·0
$O(^3P_2)$	13·615	$O(^3P_1)$	158·5
$Ar(^2P_{3/2})$	15·756	$O(^3P_0)$	226·5

Atomic state	Term level (cm⁻¹)		
$N(^4S^\circ_{3/2})$	0·0	$O(^1D)$	15867·7
$N(^2D_{5/2})$	19223	$O(^1S)$	33792·4
$N(^2D_{3/2})$	19231	$O^+(^4S^\circ_{3/2})$	0·0
$N(^2P^\circ_{3/2,1/2})$	28840	$Ar(^1S_0)$	0·0
$N^+(^3P_0)$	0·0	$Ar^+(^2P_{3/2})$	0·0
$N^+(^3P_1)$	49·1	$Ar^+(^2P_{1/2})$	1432
$N^+(^3P_2)$	131·3		

2.1. *Energy levels*

The vibrational quantum states of the reactant and product molecule or ion are calculated from the term values of the anharmonic oscillator, which are given by

$$G(v) = \omega_e(v+\tfrac{1}{2}) - \omega_e x_e(v+\tfrac{1}{2})^2 + \omega_e y_e(v+\tfrac{1}{2})^3. \qquad (6)$$

Here v is the vibrational quantum number and ω_e, $\omega_e x_e$ and $\omega_e y_e$ are the vibrational constants of the electronic level.

The rotational quantum states of the vibrational level are calculated from the term values of the rigid rotator, which are given by

$$F_v(J) = B_e J(J+1) \qquad (7)$$

169

and the vibrating rotator, which are given by

$$F_v(J) = B_v J(J+1).$$ (8)

Here J is the rotational quantum number and B_v is the rotational constant in the vibrational level v, given by

$$B_v = B_e - \alpha_e(v + \tfrac{1}{2})$$ (9)

where B_e is the rotational constant corresponding to the equilibrium separation r_e and α_e is a third rotational constant.

2.2. *Population distribution*

For a neutral or ionic molecular gas in thermal equilibrium the number of molecules N_v in the vibrational level v is given by

$$N_v = \frac{N}{Q_v} \exp\left\{ -\frac{G(v)hc}{kT} \right\}.$$ (10)

Here h, c, k and T are respectively Planck's constant, the velocity of light, the Boltzmann constant and the absolute temperature. N is the total number of molecules in all vibrational levels, and Q_v is a normalization factor corresponding to the state sum (or partition function) which is given by

$$Q_v = \sum_{v=0}^{\infty} \exp\left\{ -\frac{G(v)hc}{kT} \right\}.$$ (11)

The number of molecules N_{vJ} in the rotational level J of the vibrational level v at the temperature T is given by

$$N_{vJ} = \frac{N}{Q_v Q_r} (2J+1) \exp\left[-\frac{\{G(v)+F_v(J)\}hc}{kT} \right]$$ (12)

where each state has a $(2J+1)$-fold degeneracy and Q_r is the rotational state sum. This simplifies to

$$N_{vJ} = \frac{N_v}{Q_r} (2J+1) \exp\left\{ -\frac{F_v(J)hc}{kT} \right\}$$ (13)

which implies that the distribution over the rotational levels in each vibrational level is the same, but the absolute population of all the rotational levels in the particular vibrational level is considerably smaller than for the lowest vibrational level, corresponding to the factor $\exp\{-G(v)hc/kT\}$.

2.3. *Determination of the contributing near-resonances*

For the general case of an exothermic process of the type

$$X + YZ(v, J) \rightarrow XY(v', J') + Z$$ (14)

the energy defect is given by the equation

$$\frac{\Delta E}{hc}(v, J \rightarrow v', J') = \frac{\Delta E}{hc}(0, 0 \rightarrow 0, 0) - \{G'(v') - G'(0)\}$$

$$+ \{G(v) - G(0)\} - \{F'(J') - F(J)\}$$ (15)

where $\Delta E(0, 0 \rightarrow 0, 0)$ represents the energy defect at infinite separation for the case when both the reactant and product molecules are in the ground state vibrationally and rotationally.

Since $\Delta E(0, 0 \to 0, 0)$ is actually a function of the separation of the colliding particles, the correct value of ΔE should correspond to the separation at which the transition occurs or is most probable. However, for reasons discussed at the end of this paper the value of $\Delta E(0, 0 \to 0, 0)$ at infinite separation has been employed in the initial calculations.

In the actual calculation for an exothermic process with a given energy defect $\Delta E(0, 0 \to 0, 0)$ a computer programme is set up to choose the appropriate vibrational–rotational levels in the product molecule which will give the six smallest energy defects $\Delta E(v, J \to v', J')$ corresponding to the six closest exothermic resonances with a given vibrational–rotational energy level of the reactant molecule. These six energy defects $n = 1$–6 all contribute to the cross section at a given impact velocity, but nearly always the nearest one or two resonances dominate. Calculations taking into account only the nearest four resonances give almost identical results.

This procedure is repeated for each vibrational–rotational energy level of the reactant molecule weighted with the distribution function

$$f(v, J) = \frac{1}{Q_v Q_r} (2J+1) \exp\left[-\frac{\{G(V) + F_v(J)\}hc}{kT} \right] \tag{16}$$

calculated between the limits $J = 0$ and $J = 30$, $v = 0$ and $v = 20$.

2.4. The cross section for equilibrium and non-equilibrium conditions

The contribution to the total cross section of a vibrational–rotational level of the reactant molecule is given by

$$\sigma_{v,J}(v) = \sigma_m(v_m) f(v, J) \exp\left[k' \left\{ 1 - \frac{a|\Delta E|(v, J \to v', J')}{hv} \right\} \right], \qquad v \leqslant v_m \tag{17}$$

$$= \sigma_m(v_m) f(v, J) \qquad v \geqslant v_m.$$

The total cross section for the reaction is then

$$\sigma(v) = \sum_{v,J} \left\{ \sum_{n=1}^{n=6} \sigma_{v,J} f(v, J) \right\} \Big/ \sum_{v,J} f(v, J). \tag{18}$$

The variation of the total cross section for reaction with temperature under conditions of translational, rotational and vibrational temperature equilibrium can now be calculated, since the velocity distribution function is dependent only on temperature.

Furthermore, the vibrational and rotational parts of the exponential in the expression for the distribution function can be separated, thus allowing the treatment of non-equilibrium conditions. In special cases the reactants are not in complete thermodynamic equilibrium; the kinetic, rotational and vibrational gas temperatures (respectively T, T_r, T_v) may all be different. The distribution function (equation (16)) will then take the form

$$f(v, J) = \frac{1}{Q_v Q_r} (2J+1) \exp\left\{ -\frac{G(v)hc}{kT_v} \right\} \exp\left\{ -\frac{F_v(J)hc}{kT_r} \right\} \tag{19}$$

2.5. Calculation of σ_m

The proposals of Rapp and Francis (1962) for symmetrical resonance charge transfer at thermal energies have been shown (Hasted 1968) to be successful in approximate terms. The approach adopted in the present calculations is that the chemical transition is between two different states of the collision pseudo-molecule; usually it involves the transition of one electron, as is the case in the charge-transfer collision; therefore the same proposals should be applicable.

Rapp and Francis (1962) chose at each impact velocity the largest of two cross sections, the spiralling-orbit cross section σ_0 (equation (3)) and the rectilinear-orbit cross section σ_{RF} of equation (20) below. At the temperatures with which we are concerned here, σ_0 is usually the larger, but not by very much. For ion–molecule reactions, we choose $\sigma_m = \sigma_0$, taking no account of the rectilinear-orbit cross section. It is found that this gives best agreement with the experimental temperature functions. For neutral reactions, it is probable that orbiting does not contribute nearly so strongly, and we therefore choose $\sigma_m = \sigma_{RF}$:

$$\sigma_{RF}{}^{1/2} = 10^{-8} \exp\{1 \cdot 208(3 \cdot 466 - \lg \mathscr{I})\} - 5 \cdot 627 \times 10^{-8} \mathscr{I}^{-1/2} \lg v \qquad (20)$$

where $\mathscr{I}$ is the average ionization potential (ev) of the collision pair and v is the relative velocity given by equation (25) with mean velocity

$$\bar{v} = 1 \cdot 45 \times 10^4 \left(\frac{T}{\mu}\right)^{1/2} \text{cm s}^{-1} \qquad (21)$$

where μ is the reduced mass (a.m.u.) and T the kinetic temperature. By analogy with the average ionization potential $\frac{1}{2}\{I(X) + I(YZ)\}$, for charge-transfer collisions, we write for ion–atom interchange

$$\mathscr{I} = \frac{1}{2}\left\{I(X) + \frac{I(Y) + I(Z)}{2}\right\} \qquad (22)$$

and for atom–atom interchange

$$\mathscr{I} = \frac{1}{3}\{I(X) + I(Y) + I(Z) + 2E_a(X) + E_a(Y) + E_a(Z) + E_a(YZ)\} \qquad (23)$$

where E_a are electron affinities in electron volts (negative numbers). Equation (22) is actually superfluous to the present calculations, since we take $\sigma_m = \sigma_0$ for ion–atom interchange.

3. Results

The computer calculations have been carried out for a number of different values of the adiabatic parameter a and the constant k'. The most suitable values for the parameters a and k' have been determined empirically. It is found (Bohme *et al.* 1967 a) that for ion–atom interchange and charge transfer $a = 100$ å and $k' = 2$ and for atom–atom interchange $a = 200$ å and $k' = 2$ are the most suitable values.

The constant k' is a measure of the 'sharpness' of the maximum in the individual cross section function. For high-energy atomic-charge-transfer data $k' \simeq 1$ is appropriate (Hasted 1959), but $k' = \frac{1}{4}$ is sometimes incorrectly assumed. The present calculations fit the data best with $k' = 2$. Low-energy charge-transfer maxima are often 'sharper' (Edmonds and Hasted 1964) than those at high energies, but no detailed theoretical reason has been advanced; possibly curve-crossing is involved.

3.1. *Room-temperature reaction rate constants*

In thermal equilibrium the rate constant k is related to the collision cross section

$$k = \frac{\int f(v)\sigma(v)v \, dv}{\int f(v) \, dv} \qquad (24)$$

$$f(v) = \frac{4}{\pi^{1/2}} \frac{v^2}{\bar{\bar{v}}^3} \exp\left(-\frac{v^2}{\bar{\bar{v}}^2}\right) \qquad (25)$$

where the most probable velocity

$$\bar{\bar{v}} = \left(\frac{2kT}{\mu}\right)^{1/2}.$$

Thermal energy rate constants have been computed using the rigid rotator model for the processes tabulated in table 2, and the results are compared graphically in figure 1 with experimental data (Bohme *et al.* 1967 b, Clyne and Thrush 1961, Fehsenfeld *et al.* 1965, 1966 a, b, Ferguson *et al.* 1965 a, b, Goldan *et al.* 1966, Phillips and Schiff 1962, Sparks 1966, unpublished, Warneck 1967). For certain processes which terminate in more than

Table 2

(1) $Ar^+(^2P^0_{3/2,1/2}) + CO(x\ ^1\Sigma^+) \rightarrow CO^+(x\ ^2\Sigma^+) + Ar(^1S_0)$

(2) $Ar^+(^2P^0_{3/2,1/2}) + NO(x\ ^2\Pi_r) \rightarrow NO^+(x\ ^1\Sigma^+,\ a\ ^3\Sigma^+) + Ar(^1S_0)$

(3) $Ar^+(^2P^0_{3/2,1/2}) + N_2(x\ ^1\Sigma_g^+) \rightarrow N_2^+(x\ ^2\Sigma_g^+) + Ar(^1S_0)$

(4) $Ar^+(^2P^0_{3/2,1/2}) + O_2(x\ ^3\Sigma_g^-) \rightarrow O_2^+(x\ ^2\Pi_g) + Ar(^1S_0)$

(5) $N^+(^3P_{0,1,2}) + CO(x\ ^1\Sigma^+) \rightarrow CO^+(x\ ^2\Sigma^+) + N(^4S^0_{3/2})$

(6) $N^+(^3P_{0,1,2}) + O_2(x\ ^3\Sigma_g^-) \rightarrow O_2^+(x\ ^2\Pi_g) + N(^4S^0_{3/2},\ ^2D^0_{5/2,3/2})$

(7) $N^+(^3P_{0,1,2}) + NO(x\ ^2\Pi_r) \rightarrow NO^+(x\ ^1\Sigma^+,\ a\ ^3\Sigma^+) + N(^4S^0_{3/2},\ ^2D^0_{5/2,3/2},\ ^2P^0_{3/2,1/2})$

(8) $O^+(^4S^0_{3/2}) + O_2(x\ ^3\Sigma_g^-) \rightarrow O_2^+(x\ ^2\Pi_g) + O(^3P_{2,1,0})$

(9) $O^+(^4S^0_{3/2}) + NO(x\ ^2\Pi_r) \rightarrow NO^+(x\ ^1\Sigma^+) + O(^3P_{2,1,0},\ ^1D_2,\ ^1S_0)$

(10) $N^+(^3P_{0,1,2}) + O_2(x\ ^3\Sigma_g^-) \rightarrow NO^+(x\ ^1\Sigma^+,\ a\ ^3\Sigma^+) + O(^3P_{2,1,0},\ ^1D_2,\ ^1S_0)$

(11) $O^+(^4S^0_{3/2}) + N_2(x\ ^1\Sigma_g^+) \rightarrow NO^+(x\ ^1\Sigma^+) + N(^4S^0_{3/2})$

(12) $N_2^+(x\ ^2\Sigma_g^+) + O(^3P_{2,1,0}) \rightarrow NO^+(x\ ^1\Sigma^+) + N(^4S^0_{3/2},\ ^2D^0_{5/2,3/2})$

(13) $O_2^+(x\ ^2\Pi_g) + N(^4S^0_{3/2}) \rightarrow NO^+(x\ ^1\Sigma^+) + O(^3P_{2,1,0},\ ^1D_2,\ ^1S^0)$

(14) $N_2^+(x\ ^2\Sigma_g^+) + O(^3P_{2,1,0}) \rightarrow O^+(^4S^0_{3/2}) + N_2(x\ ^1\Sigma_g^+)$

(15) $N_2^+(x\ ^2\Sigma_g^+) + N(^4S^0_{3/2}) \rightarrow N_2(x\ ^1\Sigma_g^+) + N^+(^3P_{0,1,2})$

(16) $N(^4S^0_{3/2}) + NO(x\ ^2\Pi_r) \rightarrow N_2(x\ ^1\Sigma_g^+) + O(^3P_{2,1,0},\ ^1D_2)$

(17) $N(^4S^0_{3/2}) + O_2(x\ ^3\Sigma_g^-) \rightarrow NO(^2\Pi_r) + O(^3P_{2,1,0})$

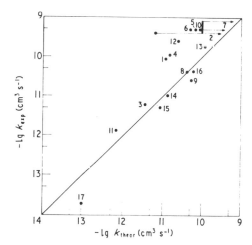

Figure 1. Comparison of experimental and calculated rate constants for processes (1)–(17) (key in table 2). Where experimental limits are published, the points are represented as vertical lines. Horizontal arrows attached to points 2, 7 and 10 indicate that an enhancement of the calculated rate is to be expected, because no account has been taken of excited product NO^+ $a\ ^3\Sigma^+$, whose spectroscopic data are not known with sufficient precision for calculations to be made. For process (17), $T_{exp} = 500\ °K$, $T_{calc} = 300\ °K$.

one electronic level the computed rate is a summation of the rates for each, except where the statistical weight factor is zero, as is the case for certain excited products of processes (9), (13) and (16). The collidant fine-structure sub-levels are equally weighted (Ar^+ $J = \frac{3}{2}, \frac{1}{2}$), but this is open to criticism, since the participation of the higher energy sub-levels is sometimes precluded, for incompletely understood reasons (Scott and Hasted 1964, Hussain and Kerwin 1965). All product fine-structure sub-levels are included as separate channels.

In atomic charge transfer it is usual (Rapp and Francis 1962) to multiply the calculated cross section by a statistical weight factor, equal to the ratio of number of configurations of the quasi-molecule of collision able to pass directly to the product states to the total number of configurations which can be formed from the collidant states (Herzberg 1966). We find that such a procedure does not weaken the importance of the adiabatic criterion in interpreting the experimental data, but it does lower nearly all the calculated rate constants until they are smaller than the experimental data. In these calculations no statistical weight factor has been included, in the belief that this is indirect evidence that transitions between different configurations of the quasi-molecule can occur in trinuclear systems.

It is arguable that processes (8), (10), (11), (13), (15) and (17) are able to proceed in two ways, namely with interchange of either of the two atoms of the homologous molecular species; however, the calculated rates have not been multiplied by a factor of 2 to conform with this argument.

3.2. *The reaction rate constant as a function of vibrational temperature*

As has been indicated in § 2.4, it is possible to calculate the dependence of thermal energy reaction rates on vibrational excitation only. In this case the translational and rotational temperatures of the reactant molecules are held constant at $T = 300\ ^\circ\text{K}$.

Recent laboratory experiments have shown that the $T = 300\ ^\circ\text{K}$ rate constant for the ion–atom interchange reaction (11),

$$\text{O}^+(^4\text{S}^o_{3/2}) + \text{N}_2(\text{x}\,^1\Sigma_g^+) \rightarrow \text{NO}^+(\text{x}\,^1\Sigma^+) + \text{N}(^4\text{S}^o_{3/2}) \tag{27}$$

is greatly increased when the N_2 is vibrationally excited (Schmeltekopf *et al.* 1967). Figure 2 compares the calculated reaction rate constant as a function of the N_2 vibrational temperature with the experimental data. There is reasonable agreement. It is significant that the

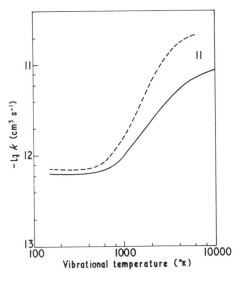

Figure 2. Comparison of experimental and calculated rate constants for process (11), with varying vibrational temperature. Broken curve, experimental data (Schmeltekopf *et al.* 1967); full curve, calculated values.

experimental evidence indicates that the vibrational distribution of N_2 created by the microwave discharge is reasonably close to a Boltzmann distribution so that an effective temperature can be implied. It is difficult to assess the importance of rotational excitation in these experiments, although the relatively high rotational relaxation rate ensures rapid equilibration to room temperature.

Although the rigid rotator calculations give fair agreement with the experimental results (figures 1, 2, 7 and 8), this is not so true in the case of vibrating rotator calculations. The rotational quantum levels of the vibrational level v, when calculated from the term values of the vibrating rotator, are closer together than those calculated from the term values of the rigid rotator. The overall effect is, for example, that the total summed rate constant for reaction (11) will be larger at all temperatures considered (figure 8). The rate constant for this reaction is dependent on the value of α_e (equation (9)) especially at lower vibrational temperatures. An accurate spectroscopic value of α_e for NO^+ was not available in the literature. Hence computer calculations have been made for reaction (11) in which α_e was varied over a large range (0·0–0·0500) with special emphasis being given to the region of values in the vicinity of the expected value of $\alpha_e(\sim 0·0200)$. The corresponding values for the rate constant at 300 °K varied over two orders of magnitude non-monotonically. It is a striking feature of these calculations that the rigid rotator model seems to be the most appropriate in this and other cases.

A classical consideration of the collision complex in which the primary ion is spiralling around the neutral particle in nearly circular orbits under the influence of an inverse fourth-power polarization field yields a value for the period of one complete revolution $\tau_0 \sim 10^{-14}$ s. The period for one vibration of a typical molecule is $\tau_v \geqslant 10^{-14}$ s. The spiralling ion therefore makes one or more complete revolutions during the period of one vibration and 'sees' effectively a rigid rotator during that time. The orbiting ion may damp the vibration of the molecule through the coupling of its angular momentum with the internal energy of the molecule.

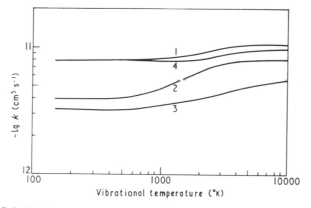

Figure 3. Calculated rate constants for processes (1)–(4), with varying vibrational temperature.

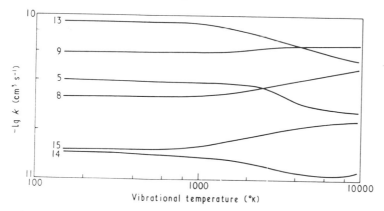

Figure 4. Calculated rate constants for processes (5), (8), (9) and (13)–(15) with varying vibrational temperature.

The dependence of the rate constant on the vibrational temperature of the primary molecule or ion is shown in figures 3 and 4 for other processes. The rotational and translational temperatures are constant at $T = 300$ °K. A comparison of these rigid rotator calculations with experiment is, as yet, not possible.

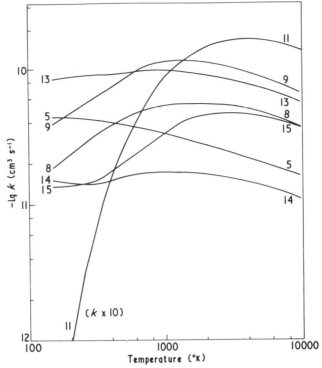

Figure 5. Calculated rate constants for processes (5), (8), (9), (11) and (13)–(15) with full temperature variation.

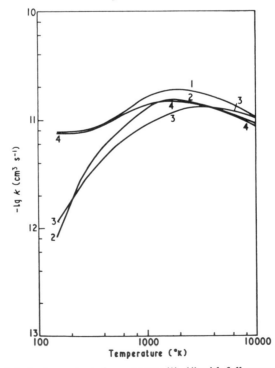

Figure 6. Calculated rate constants for processes (1)–(4) with full temperature variation.

176

3.3. *Rate constants for varying kinetic temperature and for full temperature variation*

Calculations have been made with full temperature variation (kinetic collisional, vibrational and rotational). The results are shown in figures 5 and 6. There is at present very little experimental data with which they can be compared, but flowing afterglow data will shortly become available, involving temperature variation of the entire experiment.

Calculations have also been made with variation only of the kinetic temperature (rotational and vibrational temperature held at 300 °K). These are comparable not only with mass spectrometer source measurements, but with drift tube data (Bohme *et al.* 1967 b,

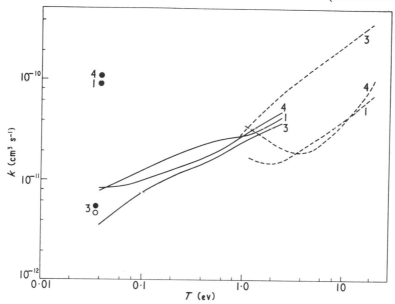

Figure 7. Comparison of experimental and calculated rate constants for processes (1), (3) and (4) with kinetic temperature variation. Broken curves, unpublished experimental data of the authors. Full curves, calculated values. Full circles, experimental afterglow data (Fehsenfeld *et al.* 1966 a). Open circle, Knewstubb (1968, private communication).

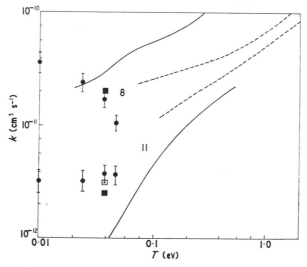

Figure 8. Comparison of experimental and calculated rate constants for processes (8) and (11), with kinetic temperature variation. Full lines, calculated values. Broken curves and full circles, Nakshbandi and Hasted (1967). Open square, Fehsenfeld *et al.* 1965; full square, Copsey *et al.* 1966.

Ong and Hasted 1968) in which the kinetic temperature is varied electrically whilst the gas temperature is held constant. Results are shown in figures 7 and 8.

4. Discussion

The empirical choice of k' and a, as well as the decision to put $\sigma_m = \sigma_0$ and not choose for σ_m the largest of the two values σ_0 and σ_{RF}, and the decision not to include statistical weight factors (Rapp and Francis 1962), have all been dictated by lengthy comparisons of calculations with experimental data. In the preliminary calculations (Bohme *et al.* 1967 a) only 300 °K data were taken into account, but in the present work it has been found that data with kinetic temperature variation are a much more sensitive test. Our original values of k' and a still stand, but it has been found that choosing $\sigma_m = \sigma_0$ is necessary if the rate of increase of k with temperature is to be kept down to within reach of the experimental values. In particular the experimental falling of k in processes (1), (4) and (8) is not reproduced in these calculations, and, since it is faster than the orbiting equation (3), it could only be explained by some feature not envisaged in the present calculations, possibly pseudo-crossing of potential energy surfaces.

It might be argued that endothermic resonances in which the energy difference is made up from the kinetic energy of impact will contribute to the summations. We have made full calculations involving this assumption and report that the correspondence with experiment is adversely affected. In particular processes (11) (both T and T_v dependence), (17) (full temperature dependence), as well as the calculated cross sections in figure 1, are raised by amounts which cause us to reject the endothermic resonance contribution hypothesis. Conversion of collisional orbiting momentum to internal energy would appear to be difficult, as is the case at energies just above threshold in phase-space calculations (Wolf 1966) of ion–atom interchange.

Nevertheless, there is reason to be encouraged by the present, rather crude, methods of calculation. It will be possible to combine them with similar non-spiralling-orbit collision calculations, computing classically the dependence of collision time upon impact parameters, and inferring a suitable distribution of adiabatic parameters.

A significant prediction of the present approach is that the product molecule will be dominantly in the vibrational–rotational state determined by the energy difference ΔE of the exothermic collision. Many exothermic atom–atom interchange reactions of type (1) have been shown by experiment to yield vibrational excitation in the product molecule. The vibrationally excited molecules have been studied by absorption or emission spectroscopy, chemical reactivity, and crossed molecular beams. The production of vibrationally excited molecules in the ground electronic states by collisional processes has been discussed in detail by Dalgarno with specific reference to aeronomy (Dalgarno 1963). There is sufficient evidence to indicate that atom–atom interchange reactions usually result in the formation of strongly vibrating molecules. However, few spectrosco picstudies were performed in which the mean collision time was much longer than the radiative lifetime of the product molecule. As a result, collisional deactivation was sufficient to perturb an absolute measurement of vibrational excitation in the products. Considerably more accurate information on the excitation energy of the products can be obtained from the measurement of the angular distribution of the products. Datz and Schmidt (1967) have made a crossed molecular beam study of the reaction

$$D + Br_2 \rightarrow DBr^v + Br. \tag{28}$$

From their experimental results they inferred that the bulk of the DBr observed is excited to the maximum attainable vibrational–rotational state (in this case the ninth vibrational level). Similar results were obtained for alkali–halogen systems (Minturn *et al.* 1966).

In crossed-beam ion–molecule reaction experiments it has not yet been possible to infer the excitation of the products quantitatively, although one experiment (Doverspike *et al.* 1966) has come close to it, and the stimulus to solve this problem experimentally is great.

A consideration of a simple valence-bond resonance description of the activated complex in atom–atom interchange reactions, coupled with experimental and theoretical evidence concerning the efficiency of transfer of vibrational energy at a collision, has led Polanyi

(1959) to predict that almost the entire exothermicity of reaction will be contained in vibration of the bond being formed, and to discuss ten particular cases of exothermic atom–atom interchange reactions, in each of which some experimental evidence exists for the presence of more than equilibrium vibration in the bond formed.

On the other hand Smith (1959), on the basis of a purely classical treatment, predicts that the maximum fraction of the exothermicity of an atom–atom interchange reaction available for high vibrational modes is limited by a kinematic factor $\sin^2\beta$ where β is the angle of rotation required to take a coordinate system suitable for describing the reactants into one suitable for describing the products. The kinematic factor depends only on the masses m_X, m_Y and m_Z of the atoms involved and can be calculated from the relation

$$\tan^2\beta = \frac{m_Y}{m_X} + \frac{m_Y}{m_Z} + \frac{m_Y^2}{m_X m_Z}. \tag{29}$$

This model is successful in predicting the highest vibrational level produced in most of the reactions considered by Smith. From the expression for $\tan^2\beta$, it is evident that transfer into high vibrational states will be most efficient when the atom being exchanged is much heavier than at least one of its partners; and, conversely, that transfer will be least efficient when the central atom is much lighter than both its neighbours.

A criticism that is frequently made of the usefulness of the adiabatic criterion is that in the usual formulation, adopted in § 1 of this paper, the energy defects ΔE_{eff} which are effective during the collision process are assumed equal to those maintaining at infinite nuclear separation, ΔE_∞. This is obviously not correct, and may lead to particularly serious errors for accidental resonance processes such as are considered here. An attempt was made by Hasted and Lee (1962) to take account of polarization and coulombic interaction, and we might proceed along similar lines. But thermal ion–atom interchange collisions differ in two respects from the medium energy charge-transfer collisions considered earlier.

(i) The inward spiralling orbits seriously modify the nuclear separations r over which the transitions take place. However, the longest time is spent at the critical separation r_c; a crude approximation might be to consider the interaction at this single separation.

(ii) The centrifugal potential $P^2/2\mu r^2$ (where P is the angular momentum of impact and μ the reduced mass) is no longer unchanged during collision, since in interchange reactions there is a change in μ.

Considering only centrifugal and polarization interactions, the defect $\Delta E(0, 0 \to 0, 0)$ which is written in equation (15), may be crudely modified as follows:

$$\Delta E(0, 0 \to 0, 0)_{r_c} = \frac{\mu_i \Delta \mu}{\mu_f} v^2 + \frac{\Delta \alpha e^2}{2 r_c^4} + \Delta E(0, 0 \to 0, 0)_\infty \tag{30}$$

where

$$\Delta \mu = \mu_f - \mu_i \tag{31}$$

$$\Delta \alpha = \alpha_i - \alpha_f \tag{32}$$

and α is the polarizability of the neutral molecule; the subscripts i and f indicate respectively initial and final. To obtain this relation the collisional angular momentum is taken as

$$P = \mu v b_c \tag{33}$$

where b_c is the critical impact parameter, such that $r_c = b_c/\sqrt{2}$. The assumption is made that $v_f = \mu_i v_i/\mu_f$, i.e. that there is no conversion of internal to kinetic energy.

Consideration of the archetypal process (11) (table 2) at $T = 300\,^\circ\text{K}$ shows that the first right-hand side term in equation (30) ($-0 \cdot 003$ ev) approximately balances the second right-hand side term ($+0 \cdot 004$ ev). The balance of centrifugal and polarization interactions results in the use of $\Delta E(0, 0 \to 0, 0)$ being unexpectedly successful. However, future calculations will no doubt include these two interactions. The rotational quantum number of the pseudo-molecule is of order $J = 80$; this is of such magnitude that at $T = 300\,^\circ\text{K}$

the rotational energy of the collidant molecule contributes only marginally, even if it were to couple efficiently. At high temperatures this is no longer the case and it will be necessary to treat the problem in greater detail.

It will be seen from the comparison of our calculations with experiment for processes (1), (3), (8) and (11) that in these instances there is a significant rise in rate constant with decreasing temperature. The present 'nearest resonance' calculations are too crude to reproduce this effect; presumably potential energy surface crossings become dominant.

References

BOHME, D. K., HASTED, J. B., and ONG, P. P., 1967 a, *Chem. Phys. Lett.*, **1**, 259–62.

BOHME, D. K., ONG, P. P., HASTED, J. B., and MEGILL, L. R., 1967 b, *Planet. Space Sci.*, **15**, 1777–80.

CLYNE, M. A. A., and THRUSH, B. A., 1961, *Proc. R. Soc. A*, **261**, 259–73.

COPSEY, M. J., SMITH D., and SAYERS, J., 1966, *Planet. Space Sci.*, **14**, 1047–56.

DALGARNO, A., 1963, *Planet. Space Sci.*, **10**, 19–28.

DATZ, S., and SCHMIDT, T. W., 1967, *Proc. 5th Int. Conf. Physics of Electronic and Atomic Collisions, Leningrad* (Leningrad: Nauka), pp. 247–9.

DOVERSPIKE, L. D., CHAMPION, R. L., and BAILEY, T. L., 1966, *J. Chem. Phys.*, **45**, 4385–91.

DRESSLER, K., and MIESCHER, E., 1966, *Astrophys. J.*, **141**, 1266–71.

EDMONDS, P. H., and HASTED, J. B., 1964, *Proc. Phys. Soc.*, **84**, 99–109.

FEHSENFELD, F. C., FERGUSON, E. E., and SCHMELTEKOPF, A. L., 1966 a, *J. Chem. Phys.*, **45**, 404–5.

FEHSENFELD, F. C., SCHMELTEKOPF, A. L., and FERGUSON, E. E., 1965, *Planet. Space Sci.*, **13**, 219–23.

—— 1966 b, *J. Chem. Phys.*, **44**, 4537–8.

FERGUSON, E. E., FEHSENFELD, F. C., GOLDAN, P. D., and SCHMELTEKOPF, A. L., 1965 a, *J. Geophys. Res.*, **70**, 4323–9.

FERGUSON, E. E., FEHSENFELD, F. C., GOLDAN, P. D., SCHMELTEKOPF, A. L., and SCHIFF, H. I., 1965 b, *Planet. Space Sci.*, **13**, 823–7.

GIOUMOUSIS, G., and STEVENSON, D. P., 1958, *J. Chem. Phys.*, **29**, 294–9.

GOLDAN, P. D., SCHMELTEKOPF, A. L., FEHSENFELD, F. C., SCHIFF, H. E., and FERGUSON, E. E., 1966, *J. Chem. Phys.*, **44**, 4095–103.

HASTED, J. B., 1951, *Proc. R. Soc. A*, **205**, 421–37.

—— 1959, *J. Appl. Phys.*, **30**, 25–7.

—— 1968, *Advances in Atomic and Molecular Physics*, **4**, 237–66 (New York: Academic Press).

HASTED, J. B., and LEE, A. R., 1962, *Proc. Phys. Soc.*, **79**, 702–9.

HERZBERG, G., 1950, *Molecular Spectra and Molecular Structure: I. Spectra of Diatomic Molecules*, 2nd edn (New York: Van Nostrand).

—— 1966, *Molecular Spectra and Molecular Structure: III. Electronic Spectra and Electronic Structure of Polyatomic Molecules*, 1st edn (New York: Van Nostrand).

HUSSAIN, M., and KERWIN, L., 1965, *Abstr. 4th Int. Conf. Physics of Electronic and Atomic Collisions, Quebec* (New York: Science Bookcrafters).

KOZLOV, V. F., and BONDAR, S. A., 1966, *Zh. Eksp. Teor. Fiz.*, **50**, 197–202.

LEE, A. R., and HASTED, J. B., 1965, *Proc. Phys. Soc.*, **85**, 673–7.

LOFTHUS, A., 1960, *University of Oslo Spectrosc. Rep.*, No. 2.

MASSEY, H. S. W., 1949, *Rep. Prog. Phys.*, **12**, 248–69.

MINTURN, R. E., DATZ, S., and BECKER, R. L., 1966, *J. Chem. Phys.*, **44**, 1149–59.

NAKSHBANDI, M. M., and HASTED, J. B., 1967, *Planet. Space Sci.*, **15**, 1781–6.

ONG, P. P., and HASTED, J. B., 1968, to be published.

PARKINSON, D., 1960, *Proc. Phys. Soc.*, **75**, 169–73.

PHILLIPS, L. F., and SCHIFF, H. I., 1962, *J. Chem. Phys.*, **36**, 1509–17.

POLANYI, J. C., 1959, *J. Chem. Phys.*, **31**, 1338–51.

RAPP, D., and FRANCIS, W. E., 1962, *J. Chem. Phys.*, **37**, 2631–45.

ROTHE, E. W., and BERNSTEIN, R. B., 1959, *J. Chem. Phys.*, **31**, 1619–27.

SCHMELTEKOPF, A. L., FEHSENFELD, F. C., GILMAN, G. I., and FERGUSON, E. E., 1967, *Planet. Space Sci.*, **15**, 401–6.

SCOTT, J. T., and HASTED, J. B., 1964, *Proc. A.S.T.M. Mass-Spectrometry Symposium* (Paris: SERMA).

SMITH, F. T., 1959, *J. Chem. Phys.*, **31**, 1352–9.

WARNECK, P., 1967, *J. Chem. Phys.*, **46**, 502–12.

WOLF, F. A., 1966, *J. Chem. Phys.*, **44**, 1619–28.

DYNAMICS OF ION-MOLECULE REACTIONS

Editor's Comments
on Papers 18, 19, and 20

18 **POTTIE and HAMILL**
Persistent Ion-Molecule Collision Complexes of Alkyl Halides

19 **MATUS et al.**
Kinematic Investigations of Ion-Neutral Collision Mechanisms at ~ 1 eV

20 **ALLEN and LAMPE**
Ion-Molecule Reactions in Monosilane-Benzene Mixtures. Long-Lived Collision Complexes

In the various theoretical treatments presented in Part III, it is assumed that a complex is formed in which the ion and neutral rotate around a common center, normally for one or more complete revolutions. The phase space theory assumes that the complex persists for sufficient time for the internal energy to reach an equilibrium distribution among the various modes. For example, a complex of a small ion and a small molecule will persist for a time greater than 10^{-13} to 10^{-12} s. Further, the products leaving the complex will be distributed uniformly in all directions around the center of mass.

The earliest report of such long-lived complexes was that of Pottie and Hamill (Paper 18), who observed the attachment of the parent ions of several alkyl halides to the corresponding neutral molecules. Since these complexes persisted long enough to reach the collector, their lifetimes must have been on the order of 10 or more microseconds. In his study of ion-molecule reactions in ethylene at elevated pressures (up to about 0.2 Torr), Field[1] observed several reactions that were kinetically third-order with rate constants around 10^{-27} cc^2/molecule^2 s. Since the rate of collision of the primary ion with the neutral and that of the stabilizing collision of the complex with the neutral will be in the order of 10^{-10} to 10^{-9} cc/molecule s, the lifetime of the complexes must have been 10^{-7} to 10^{-9} s. Subsequent studies have detected many other third-order reactions having rate constants of 10^{-30} to 10^{-25} cc^2/molecules^2, which thus involved complexes with lifetimes of 10^{-12} to 10^{-5} s.

Consequently, the existence of such complexes as intermediates in cerain reactions cannot be doubted.

One would expect the formation of a long-lived complex to be favored at low collision energies and to become less favorable as the collision energy is increased. It is somewhat surprising that complexes in some systems are formed even at relatively large energies.

In Paper 19, a complex pulsing technique is employed with a single-source mass spectrometer to study reactions of $CD_4^+ + CD_4$ and $HD^+ + HD$. The reactions were shown to proceed through a complex even at laboratory collision energies of 1.7 and 0.36 eV, respectively.

In Paper 20, two tandem in-line mass spectrometers with an intervening collision chamber were used to study reactions of ions from benzene with SiH_4 and of ions from SiH_4 with benzene. In several instances, the collision complex persists for a sufficient time to be collected—a matter of some 8 μs. Since these reactions are exothermic and the energy of the impacting ions was varied from 1 to 7 or 8 eV, the energy of the complex was quite large. The persistence of these complexes for such long times is very surprising.

REFERENCE

1. Field, F. H., *J. Amer. Chem. Soc.* **83**: 1523 (1961).

18

Reprinted from *J. Phys. Chem.* **63**:877–879 (1959)

PERSISTENT ION-MOLECULE COLLISION COMPLEXES OF ALKYL HALIDES[1]

By Roswell F. Pottie and William H. Hamill[2]

In the course of a study of ion–molecule reactions of ethyl iodide in the ionization chamber of a mass spectrometer we found evidence for an ionic species at, or close to, $m/e = 312$. This mass corresponds to $C_4H_{10}I_2^+$ which suggests its formation in the bimolecular process

$$C_2H_5I^+ + C_2H_5I \longrightarrow C_4H_{10}I_2^+ \qquad (1)$$

This reaction, if verified, would be of particular interest as the first example (to our knowledge) of a persistent collision complex, or "sticky collision."[3]

Although previously unobserved in mass spectrometry, such collision complexes are not unanticipated. In conventional gaseous systems the dimeric ion should occur more frequently because it is susceptible to collisional stabilization by removal of excess energy. Similarly, it can be expected to grow by accretion of additional molecules and this concept is the basis of Lind's "cluster theory."[4]

According to to a recent simple treatment[5] of ion-molecule "sticky collisions" the lifetime τ of the complex is given by

$$\tau = \nu^{-1}(1 - E_b/E)^{1-\alpha} \qquad (2)$$

where E is the total internal energy, E_b the energy required to dissociate the complex and α the number of effective degrees of vibrational freedom. To be detected in a mass spectrometer the complex must have a lifetime approximating 10^{-6} sec.

Results

The general experimental procedure for this work follows Stevenson and Schissler.[6] Details have been described elsewhere.[7]

In order to confirm the reaction 1, as indicated by preliminary experiments, it was next established that peak height tentatively assigned to $C_4H_{10}I_2^+$ varied as the square of the inlet pressure of ethyl iodide. Further, its ion intensity varied inversely with the repeller field strength at values in excess of 10 v. cm.$^{-1}$ both for 70 v. and for 10.5 v.

electrons. The lower value corresponds to the position of the maximum in Fig. 1. A number of ion–molecule reactions are known to exhibit such an inverse dependence of ion abundance upon repeller field strength,[8] although an inverse square root dependence is considered normal.

The resolving power at $m/e = 312$ is unfavorable and special care was necessary to identify the ion. This was accomplished, in part, by using the metastable suppressor electrode to improve the resolution. The mass spectrum in this region was calibrated at mass-to-charge ratios of 296, 310 and 338 using propylene diiodide, tetramethylene diiodide and hexamethylene diiodide, respectively. In this manner it was definitely established that an ion–molecule reaction product of $m/e = 312$ had been formed from ethyl iodide. Furthermore, it follows that reaction (1) is occurring.

As a final, confirmatory test we compared the ionization efficiency curve for the parent molecule–ion $C_2H_5I^+$ with that of the daughter. The vanishing current method gave an appearance potential at $m/e = 312$ which agreed, within experimental error, with that at m/e 156. Rather unexpectedly, the curves show a striking difference. The ionization efficiency curve for the primary ion (see Fig. 1) is normal. In contrast, the curve for $C_4H_{10}I_2^+$ rises sharply from the onset of ionization to a well-defined maximum about 1.5 v. higher. Then, following a small subsequent decrease, the curve behaves normally over an interval of several volts and reaches a plateau only 6 v. above its appearance potential, decreasing slightly at 70 v. The similarity and difference of the two curves in Fig. 1 have been emphasized by choosing an appropriate scale factor to equalize ion abundances at higher ionizing voltages.

A similar reaction has been found yielding $C_4H_{10}Br_2^+$ from ethyl bromide, $C_6H_{14}I_2^+$ from propyl iodide and $C_3H_8I_2^+$ from a mixture of methyl and ethyl iodides. No reaction was found for methyl iodide, methyl bromide or propyl chloride. The latter is an unfavorable case for test because of a low abundance of $C_3H_7Cl^+$. The calculated cross sections σ for reactions in one-component systems at 70 v. ionizing voltage and 4 v. cm.$^{-1}$ repeller field strength appear in Table I.

The ion abundance curves for $C_2H_5Br^+$ and $C_4H_{10}Br_2^+$ strongly resemble the corresponding curves in Fig. 1. For $C_6H_{14}I_2^+$ there is slight evidence of structure in the ion abundance curve and the abundance ratio of primary to secondary ion is substantially constant over a 10 v. range above

(1) Contribution from the Radiation Project operated by the University of Notre Dame and supported in part under Atomic Energy Commission Contract AT-(11-1)-38.

(2) To whom correspondence and requests for reprints should be sent.

(3) Dr. N. A. I. M. Boelrijk, working in these laboratories, has examined many systems for evidence of this phenomenon, without success. J. L. Franklin, F. H. Field and F W. Lampe, in a preprint of a paper presented at the Joint Conference on Mass Spectrometry, London, September, 1958, have remarked upon the absence of peaks attributable to such species in mass spectra containing evidence of other ion–molecular reactions.

(4) S. C. Lind, "The Chemical Effects of Alpha Particles and Electrons," 2nd ed., Chemical Catalog Co., New York, N. Y., 1929.

(5) M. Burton and J. L. Magee, This Journal, **56**, 842 (1952).

(6) D. P. Stevenson and D. O. Schissler, *J. Chem. Phys.*, **29**, 282 (1958).

(7) R. F. Pottie, R. Barker and W. H. Hamill, *Rad. Res.*, in press.

(8) G. Gioumousis and D. P. Stevenson, *J. Chem. Phys.*, **29**, 294 (1958).

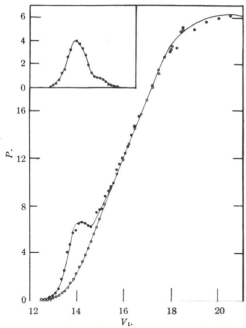

Fig. 1.—Peak height P vs. ionizing voltage V_1 for $C_4H_{10}I_2{}^+$ (●) from $C_2H_5I^+$ (○) at 4 v. cm.$^{-1}$ ion repeller field strength.

TABLE I

CROSS-SECTIONS FOR PERSISTENT COLLISION COMPLEXES

	$\sigma \times 10^{16}$ cm.^{2} at 4 v. cm.$^{-1}$
$C_2H_5I^+ + C_2H_5I \rightarrow C_4H_{10}I_2{}^+$	1.5
$C_2H_5Br^+ + C_2H_5Br \rightarrow C_4H_{10}Br_2{}^+$	3.2
$C_3H_7I^+ + C_3H_7I \rightarrow C_6H_{14}I_2{}^+$	12

their common appearance potential. The ion abundance curve for $C_3H_8I_2{}^+$ resembles Fig. 1 in having a maximum 1.5 v. above the appearance potential. In addition, there is a second maximum about 3.5 v. above the appearance potential, followed by the plateau some 5 v. higher which persists to 70 v.

For the three reactions in Table I the difference between the ionization potential of the primary ion and the appearance potential of the secondary ion does not exceed 0.2 v. For $C_3H_8I_2{}^+$ the primary ion cannot be distinguished because the ionization potentials of methyl iodide and of ethyl iodide differ by less than 0.2 v.

Although no difference could be distinguished between appearance potentials of primary and secondary ions, the maxima in the ion abundance curves suggested contributions to the observed reactions from resonance excitations of parent molecules. Reactions of the type

$$A^* + A \longrightarrow A_2{}^+ + e^-$$

are well known for atoms and have been observed by mass spectrometry.[9] For comparison with the alkyl halides we measured the $Ar_2{}^+$ intensity from argon as a function of repeller field strength.

(9) J. A. Hornbeck and J. P. Molnar, *Phys. Rev.*, **84**, 621 (1951).

At 20 v. ionizing voltage, corresponding to the maximum in the ionization efficiency curve, the observed ion intensity dependence for $m/e = 80$ was characteristic of primary ions, as expected. In contrast, for ethyl iodide at 70 v. ionizing voltage, where the contribution from the low energy component should be negligible, the peak height for the ion at $m/e = 312$ decreased in a manner characteristic of secondary ions with increasing repeller field strength. When the ionizing voltage was then decreased to a value corresponding to the maximum in Fig. 1, the peak height again decreased in a similar manner with increasing repeller field strength. We conclude that the energy-rich precursor for the ion at $m/e = 312$ is an ion, and not an excited neutral molecule, over the entire range of ionizing voltage.

Ions at double the parent mass were not found for methyl bromide or methyl iodide at 70 v. ionizing voltage but there is still the possibility of a resonance process at some lower electron energy. The appropriate mass regions were therefore scanned repeatedly, over small voltage intervals, from 2 v. below to 10 v. above the appearance potentials of the parent molecule ions. No collision complex ion was detected in either instance.

Discussion

Stevenson and Schissler[6] have pointed out that parent ion–daughter ion relationship is better established by constancy of their abundance ratios at low ionizing voltages than by simple agreement of their appearance potentials. The ions $C_4H_{10}I_2{}^+$ and $C_4H_{10}Br_2{}^+$ are exceptional in this respect over the lower range of voltage while agreeing excellently over the medium and higher range. The curves in Fig. 1 strongly suggest two processes and, if so, the facts require that $C_2H_5I^+$ is involved in each. To account for the results it is postulated that there are two isomeric ionic species of the empirical formula $C_2H_5I^+$ and of unequal reaction cross-sections. If one species behaves normally in giving a constant parent ion–daughter ion ratio, the contribution of the abnormal variety can be isolated by subtracting the parent curve from the daughter curve. This difference curve is plotted in the panel of Fig. 1 and suggests a resonance process.

Before attempting to explain this effect it is first necessary to consider the possibility that it is merely an artifact. Appearance potentials were measured at repeller field strengths of 4 v. cm.$^{-1}$ in order to maximize secondary ion current, but at the risk of instability. The ionization efficiency curves for primary ions were all normal, as for the one illustrated, while those for collision complex ions exhibit the features mentioned. Also, ionization efficiency curves for many secondary ions which are not collision complexes have been determined with the same instrument and at the same repeller field strength but with no evidence of the anomalies reported here, not excluding such ions resulting from alkyl halides.

There is no necessary inconsistency between the structure reported for the secondary ion abundance curves and the apparent lack of structure in the corresponding primary ion curves. The electron energy resolution of the instrument is so low that negative evidence is simply inconclusive. One is inclined to attempt to correlate the maxima in the secondary ion abundance curves with the double series of electronic states observed for the alkyl halides.[10] These series lead to ionization potentials separated by 0.64 and 0.59 v. for methyl and ethyl iodides and by 0.33 and 0.33 v. for methyl and ethyl bromides. A serious deficiency of this proposal is that it cannot account for the break in the curve for $C_3H_8I_2{}^+$ at an electron energy *ca.* 3.5 v. above the ionization potential. It would also be necessary to assume that formation of one of these states was sharply resonant, although the probability of ion formation by electron impact is normally linear in the excess energy.

Whatever may be the detailed explanation of these effects, it is rather likely that the sensitivity of the collision complex to small energy differences will amplify the effect, particularly so if only a small fraction of complexes survive long enough to be collected. Referring to equation 2 let

$$E = E_b + E_f + E_v + 3kT$$

E_f is the kinetic energy of the primary ion arising from the repeller field. Its maximum value was 0.5 e.v. and since σ varies as $E_f{}^{-1/2}$ we take $E_f = 0.1$ e.v. as representative. E_v is the vibrational energy of the primary ion resulting from vertical ionization. The ionization potential of ethyl iodide by electron impact[11] is 9.47 e.v. and the spectroscopic value[10] is 9.34 e.v., giving $E_v = 0.12$ e.v. At the temperature of the ionization chamber (250°), $3kT$ is 0.14 e.v. The value of E_b is very uncertain but 1 e.v. is a plausible value. Letting

(10) W. C. Price, *J. Chem. Phys.*, **4**, 539, 547 (1936).

(11) J. D. Morrison and A. J. C. Nicholson, *ibid.*, **20**, 1021 (1952).

$\tau = 10^{-6}$ sec. and $\nu = 10^{13}$ sec.$^{-1}$ leads to a required minimum 13 degrees of vibrational freedom. Considering the high internal energy, heavy atoms and relatively weak C–C bonds, this is an acceptable value. For these values of the parameters, decreasing the internal energy of the collision complex by only 0.03 e.v. would more than double the value of τ. If the very small cross-section for $C_4H_{10}I_2{}^+$ can be interpreted as inefficient collection due to extensive decomposition of the collision complex, then doubling the value of τ will have a large effect upon σ.

Considering the sensitivity of the value of τ to the internal energy of the complex we may consider that τ changes discontinuously to zero at some critical value, E_c, of the translational energy of the primary ion. There is a corresponding critical limit, l_c, of the length of the ion track (measured from the electron beam toward the exit slit), along which a viable complex can form. This distance is shorter, the greater the repeller field strength, F. The yield of complexes is proportional to

$$\int_0^{l_c} \sigma(E)\,dl = \int_0^{E_c} \sigma(E)F^{-1}dE$$

The cross-section for collision, $\sigma(E)$, is presumed to depend upon $E^{-1/2}$ but it is clear that regardless of the functional dependence, the yield of secondary ions varies inversely with field strength, provided that E_c is less than the maximum energy attainable by the primary ion.[12]

Acknowledgment.—It is a pleasure to acknowledge helpful conversations with Professor John L. Magee.

(12) N. A. I. M. Boelrijk and W. H. Hamill, to be published in further detail.

19

Reprinted from *Faraday Soc. Discuss.* **44**:146–156 (1967)

Kinematic Investigations of Ion-Neutral Collision Mechanisms at ~1 eV

By L. Matus,* I. Opauszky,† D. Hyatt,‡ A. J. Masson,§ K. Birkinshaw§ and M. J. Henchman§

Dept. of Physical Chemistry, The University, Leeds, 2.

Received 16th June 1967

A new pulsing technique is described for the study of ion-neutral collision processes at energies of ~1 eV. Short, high voltage pulses applied to the repeller plate of a mass-spectrometer ion source can produce reactant ion beams of constant velocity. Velocity analysis of both reactant and product ions is achieved through pulsed de-focusing of the ion beam, using a time-of-flight technique.

The collision mechanisms of the reactions

$$CD_4^+ + CD_4 \rightarrow CD_5^+ + CD_3$$

$$HD^+ + HD \begin{cases} \rightarrow H_2D^+ + D \\ \rightarrow HD_2^+ + H \end{cases}$$

are shown to involve complex formation at lab. energies of 1·7 and 0·36 eV respectively. The following non-reactive collision processes have been studied:

$$He^+ + H_2 \rightarrow He^+ + H_2$$

$$C_2H_4^+ + C_2H_6 \rightarrow C_2H_4^+ + C_2H_6$$

The latter forms a complex which dissociates to give back the original reactants whereas the former does not proceed via a complex. This is explained by the inability of the former system to convert relative translational energy into internal energy of the complex. For the latter process, this conversion is very much easier.

Investigations of ion-neutral reactions have been fruitful in pursuing several of the principal objectives in the study of chemical reaction under " single collision " conditions. Examples of this success are the first observation of spectator stripping,[1] the direct investigation of reaction thresholds [2] and of the energy dependence of reaction cross sections.[3] Yet results have not been forthcoming in the energy region below about 1-2 eV, in which the corresponding neutral-neutral investigations have produced such dramatic results.[4] Paradoxically the accessible energy range for neutral-neutral studies has been, until recently, the inaccessible range for the corresponding ion-neutral ones. The recent cross-beam experiments of Herman *et al.*[5] at energies below 1 eV constitute a major advance in the field of ion-neutral reactions.

This paper presents some preliminary and qualitative results using a different technique,[6] which cannot produce the kind of detailed data available from crossed-beam experiments but which nevertheless promises to be useful, particularly for the study of those ion-neutral collision processes at low energy where the velocity of

*on leave of absence (1965-6) from the Central Research Institute for Physics, Hungarian Academy of Sciences, Budapest, Hungary.

†on leave of absence (1966-7) from the Central Research Institute for Physics, Hungarian Academy of Sciences, Budapest, Hungary.

‡present address: Hahn Meitner Institut fur Kernforschung Berlin, Berlin-Wannsee, West Germany.

§present address: Dept. of Chemistry, Brandeis University, Waltham, Massachusetts 02154, U.S.A.

the product ion is low.   Detection efficiences will have to be improved considerably before such systems can be investigated by crossed-beam techniques.

This multiple-pulse technique consists of another variation on the basic mass-spectrometer ion source technique.   The difficulty in the traditional use of this technique is that the reactive events occur throughout an energy range and this has precluded any kinematic investigation of collision mechanism.   The multiple-pulse technique avoids this limitation by pulsing the electron beam (20 nsec) to produce a cluster of ions in the electron beam region of the ionization chamber.   As soon as the electron beam gate is shut, a short, high voltage, positive pulse is delivered to the repeller, during which time the ions are strongly accelerated. Subsequently they travel to the ion-exit slit at a constant velocity in a field-free region. The availability of pulse generators capable of producing 50 V/20 nsec pulses means that the time for ion formation and acceleration can be held to less than 10 % of the total residence time in the chamber.

The production of constant velocity ion beams is one necessary preliminary for kinematic studies ; the second requirement is a means for velocity analysis.   This is effected by further pulsing techniques, in which the first electrode in the ion accelerating system is a split electrode which may be operated as a gate by pulsing one half of it. Velocity analysis is thus achieved by a time-of-flight technique in which the residence time of the ion in the chamber is measured.   Applied to primary ions, it provides a means of testing the impulse technique for producing the constant velocity ion beams. Applied to a secondary ion, it provides a means of investigating the kinematics of the collision which produced the secondary.

This approach has several limitations which either restrict its use or complicate the analysis of the data : (i) there is no mass analysis of the primary ions.   (ii) Before acceleration the primary ions will in general have the same Maxwellian velocity distribution as the molecules from which they are formed.   Thus, after acceleration, the constant velocity primary ion beam will have a distribution of velocities.   (iii) Primary ions will be formed throughout the whole width of the electron beam region. Thus, primary ions with the same velocity will exhibit a range of residence times. (iv) The collision region, i.e., the primary ion path length, is comparatively long and thus secondary ions formed with the same velocity will exhibit a wide distribution of residence times in the ionization chamber.   (v) The ion-exit slit has a finite width and it will not accept secondary ions scattered through large angles, particularly those formed early in the primary ion track.

On the other hand there are features in its favour : (vi) velocity analysis is performed immediately the ions leave the collision region.   (vii) Any secondary ions, which are either formed with low velocity or scattered through large angles, may be driven from the chamber by superimposing a low positive d.c. voltage on the repeller pulse.   Thus their identity and relative intensity may be monitored.

These multiple-pulse techniques can be applied, at ion energies of ~ 1 eV, to any collision process where the velocity of the ion is changed.   Processes which are accessible to the technique thus include both ion-neutral reactions, charge transfer (both symmetric and asymmetric) and non-reactive scattering (both elastic and inelastic).   This paper reports some preliminary investigations of some of these problems.

EXPERIMENTAL

The mass spectrometer is a 90°-sector instrument (6 in. radius of curvature) fitted with specially developed ion sources.   The source region is a cylindrical steel can (11 in. diam. and 4 in. long) which is differentially pumped.   The gas pressure in the ionization chamber is measured directly by a McLeod gauge.   Fuller details may be found elsewhere.[7]

The pulse circuitry is shown in fig. 1.   Pulses are provided from the following types of pulse generator : Datapulse model 102 (max. voltage 60 V, min. pulse width 50 nsec), Datapulse model 108 (50 V, 20 nsec), Hewlett Packard model 214 A (100 V, 50 nsec).   The pulse generators are held at ground potential and pulses are supplied to the ion source, held at ~ 2 kV, through capacitance couplings.   50 Ω coaxial leads, properly terminated, are used outside the vacuum envelope but unshielded leads within it.   The pulse programme is

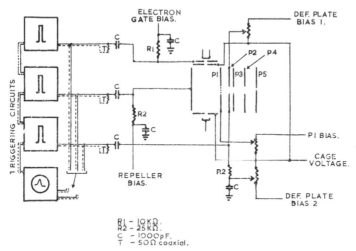

R1 – 10KΩ.
R2 – 25kΩ.
C – 1000pF.
T – 50Ω coaxial.

Fig. 1.—Schematic diagram of the pulse circuitry, illustrated for ion source B (not drawn to scale). For the experiments reported in this paper, P1 and the cage were held at the same potential.

monitored with a Tektronix oscilloscope (model 547 with 1A1 plug-in) and use of the unshielded leads causes no significant pick-up or distortion of the pulse shapes.

Two different ion sources have been employed and relevant sections of these are shown in fig. 2.   Source A has a short ion path length 0·22 cm (centre of the electron beam to ion-exit slit) and has been used for most of the exp riments reported here.   P1 is the electrode

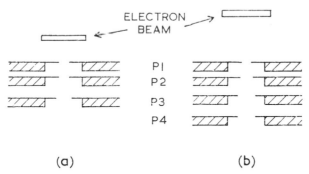

(a)   (b)

Fig. 2.—Scale drawings of cross sections of the two ion-sources, A and B, viewed parallel to the electron beam.   Ion path lengths 0·22 and 0·40 cm, respectively.   In each case the final ion accelerating electrode is omitted.

containing the ion-exit slit and P2 is the split electrode used in the measurement of ion residence times.   P3 provides partial acceleration and P4, held at ground, the final acceleration of the ions.   The ion-optics of this source are such that the two halves of P2 must be biased negative to P1 for the signal to be maximized.   An essential requirement of the technique is that penetrating fields must be excluded from the ionization chamber ; to mınimize this, fine tungsten meshes (0·001 in. diam. wire, 0·005 in. spacing) are spread over the ion-exit slit and el ctron-exit slit in the ionization chamber.

The distribution of primary ion residence times is measured using the pulse programme shown in fig. 3a.   The electron beam is pulsed (20 nsec), a suitable pulse is then fed immediately to the repeller and the potentials on the various electrodes in the ion lens are adjusted to maximize the intensity of the primary ion signal.   A 50 V/5 μsec pulse (positive or negative) is next applied to one half of P2 such that t is the time between the electron beam pulse and the leading edge of the deflection pulse. $I(t)$, the measured ion intensity for a delay time t, is then determined as a function of t by movement of the deflection pulse along the time axis.   A repetition rate of 20 kHz is used.   This and the high transmission efficiency of the source give large enough ion intensities for Faraday-cup detection to be employed.

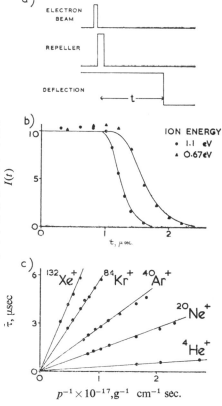

FIG. 3.—(a) Pulse programme used to obtain deflection curves. (b) Comparison between theoretical deflection curves and experimental data for $^{20}Ne^+$ primary ions at average energies of 0·67 and 1·1 eV. $I(t)$, the normalized intensity of those ions with residence times greater than t, is plotted against t, the delay time in μsec (data obtained with source B). (c) Plots, for different primary ions, of the measured average residence time $\bar{\tau}$ in μsec against p^{-1}, where p is the repeller impulse in g cm sec^{-1}. The straight lines indicate the theoretically predicted dependence (data obtained with source B).

The deflection pulse deflects all ions with residence times less than t and thus $I(t)$ is the measured intensity of all ions with residence times greater than t.   Deflection curves of $I(t)$ against t have the shape shown in fig. 3b and the primary ions exhibit a range of distribution times due to the Maxwellian velocity distribution of the ions prior to acceleration and the finite thickness of the electron beam.

The measurement can also be performed in another way, using a gating technique. Here one half of P2 is permanently biased with respect to the other to deflect all ions.   A 50 nsec pulse is then applied a time t after the electron beam pulse to offset the bias, open the gate and transmit those ions with a residence time t.   Gating curves give the distribution of residence times directly and are the differential form of the deflection curves.   In practice, they are unreliable quantitatively, giving residence times which are too long, since the gate, when open, transmits a bunch of ions in the vicinity of P2 and not those just emerging through the slit.

The deflection curves yield an average value for the residence time for a primary ion for a particular repeller impulse and this may be compared directly with a theoretical value computed from the mass of the ion, the magnitude of the repeller impulse and the relevant ion source dimensions. Source A has consistently given residence times which are lower than those predicted theoretically and conclusive evidence now exists that this is due to field penetration by the ion accelerating field, despite the presence of the protective mesh over the ion-exit slit.

A new source B has recently been commissioned (ion path length 0·40 cm) with specially designed features to overcome this field penetration. The incorporation of an additional split electrode now allows P2, the defocusing electrode, to be held at the same potential as P1. Results from this source are encouraging and fig. 3c shows the excellent agreement between theoretical and experimental average residence times for a wide range of ion masses and repeller impulses. Moreover distributions, generated theoretically, agree extremely well with the experimental curves. Examples of this fit are shown for neon in fig. 3b for two different energies; for source B, the theory considers only the Maxwellian correction since the electron beam is narrow compared to the ion path length.

The technique for producing constant velocity ion beams now seems well established from the experiments using source B. The available energy range is limited at present to ~0·2-3 eV but its use is restricted by the width of the energy distribution which is ~10 % at 3 eV, increasing to 50 % at 0·2 eV.

KINEMATIC STUDIES : REACTIVE COLLISIONS

In principle, the deflection curves of both primary and secondary ions may be transformed to give a velocity analysis of the reactive collision. This in turn yields information on the collision mechanism, i.e., the angular distribution of products in the centre-of-mass framework.

ELEMENTARY THEORY

The simplest model ignores the initial Maxwellian velocity distribution of the target molecules, the finite thickness of the electron beam, the relative translational energies of the products in the c.m. system and non-reactive scattering of the primary ion beam. Primary ions with a singular residence time $\bar{\tau}_p$ produce secondary ions with a range of residence times bounded by $\bar{\tau}_p < t < 2\bar{\tau}_s - \bar{\tau}_p$, where $\bar{\tau}_s$ is the average residence time of the secondary ions. By holding the amount of reaction below 10 %, attenuation of the primary ion beam may be ignored. Hypothetical deflection and gating plots, appropriately normalized, which are predicted by this model, are shown in fig. 4 for both primary and secondary ions. The total primary and secondary ion intensities, I_p and I_s are related by

$$I_s = I_p N k \bar{\tau}_p$$

where N is the target molecule concentration in the ionization chamber and k is the rate constant for the reaction.

Three extreme types of collision mechanism can be suggested for the ion-neutral reaction. An intermediate-complex mechanism requires the distribution of the incident ion momentum among all the nuclei in the collision complex. Alternatively, reactions involving the transfer of a neutral fragment Y, according to

$$X^+ + YZ \rightarrow XY^+ + Z,$$

may proceed by a pick-up mechanism while those involving the transfer of a charged fragment Y$^+$, according to

$$XY^+ + Z \rightarrow X + YZ^+,$$

may proceed by a stripping mechanism. The important feature of stripping and pick-up mechanisms is the absence of momentum transfer from X to Z. The relationships between $\bar{\tau}_p$ and $\bar{\tau}_s$ for these three models are easily shown to be :

COMPLEX :　　　　　　　　$\bar{\tau}_s/\bar{\tau}_p = 1 + m_m/2m_p,$　　　　　　　　　　(1)

PICK-UP :　　　　　　　　$\bar{\tau}_s/\bar{\tau}_p = \frac{1}{2}(1 + m_s/m_p),$　　　　　　　　(2)

STRIPPING :　　　　　　　$\bar{\tau}_s/\bar{\tau}_p = (2m_s - m_m)/2(m_s - m_m),$　　　　(3)

where m_p, m_s and m_m are the masses of the primary ion, secondary ion and target molecule respectively.

In practice, the deflection curve of a primary ion (fig. 3b) differs significantly from the predictions of this simple model (fig. 4), principally due to the initial Maxwellian velocity

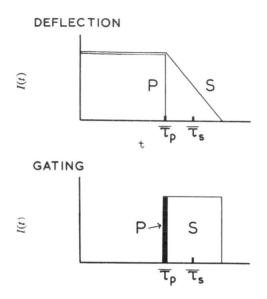

FIG. 4.—Hypothetical deflection and gating curves for primary (P) and secondary (S) ions according to the predictions of the simple model. $I(t)$ is plotted against t, the delay time, where $I(t)$ is the normalized intensity of all ions with residence times greater than t for a deflection curve and of all ions with residence time t for a gating curve. $\bar{\tau}_p$ and $\bar{\tau}_s$ are the average residence times of primary and secondary ions respectively.

distribution of the ions.   Nevertheless, for certain types of ion-neutral reaction, the predictions of expressions (1), (2) and (3) are sufficiently different that insight into the collision mechanism may be gained by application of the simple model.

COLLISION MECHANISM OF $CD_4^+ + CD_4 \rightarrow CD_5^+ + CD_3$

Deflection curves are shown in fig. 5 for primary and secondary ions in the reactions

$$CD_4^+ + CD_4 \rightarrow CD_5^+ + CD_3 \tag{4}$$

$$CD_3^+ + CD_4 \rightarrow C_2D_5^+ + D_2 \tag{5}$$

for the following primary ion energies (lab.): $CD_3^+ = 1.7$ eV, $CD_4^+ = 1.9$ eV.   Average residence times, $\bar{\tau}_p$ and $\bar{\tau}_s$ are estimated from the half-heights of the deflection curves.* Reaction (5) serves as a control experiment since the predicted values of $\bar{\tau}_s/\bar{\tau}_p$ from expressions (1), (2) and (3) are 1·55 (complex), 1·44 (pick-up) and 1·71 (stripping) and are thus similar. The experimental value of 1·73 is compatible with these predictions.

Simple discrimination experiments suggest that reaction (4) occurs by a stripping mechanism at ~10 eV (lab.) and above but that a transition to a complex mechanism occurs at lower energies.[1]   The energy dependence of the isotope effect, found for this reaction with $CH_2D_2^+$ and CD_4 suggests that the transition occurs around 4 eV (lab.).[8]   The deflection

*This is an approximation due to the asymmetry of both primary and secondary deflection curves. The error is < 5 % provided $p/(mkT)^{\frac{1}{2}} \gtrsim \sim 6$, where p and m are primary ion impulse and mass and T is the source temperature.

curves yield a value of $\bar{\tau}_s/\bar{\tau}_p = 1\cdot53$ for reaction (4), to be compared with theoretical predictions of $1\cdot50$ (complex), $1\cdot05$ (D atom pick-up) and $6\cdot00$ (deuteron stripping). Complex formation thus seems to be the predominant collision mechanism.

In order to be able to exclude a stripping mechanism, the technique must be shown to produce proper deflection curves for a reaction where the secondary ion velocity is much less than that of the primary. A suitable system for this test is the reaction

$$H_2^+ + Ne \rightarrow NeH^+ + H \tag{6}$$

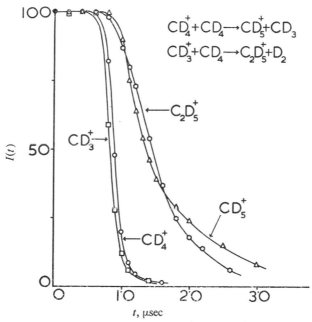

$$CD_4^+ + CD_4 \rightarrow CD_5^+ + CD_3$$
$$CD_3^+ + CD_4 \rightarrow C_2D_5^+ + D_2$$

FIG. 5.—Normalized deflection curves for primary CD_3^+ and CD_4^+ ions and secondary CD_5^+ and $C_2D_5^+$ ions; plots of $I(t)$, the relative intensity of ions with residence times greater than t, against t, the delay time in μsec (data obtained with source B).

for which $\bar{\tau}_s/\bar{\tau}_p \geqslant 6$, for both complex and stripping mechanisms. Experimentally this is found and confirms that the technique would have revealed a stripping mechanism for reaction (4), had this been the case.

Reaction (4) thus proceeds by complex formation at low energies yet isotopic experiments reveal that the identity of the collision partners is not lost since hydrogen ion transfer is twice as frequent as hydrogen atom transfer at 0·3 eV (lab.).[8, 9] Paradoxically the complex must be bound strongly enough to permit momentum distribution and energy equilibration[10] but sufficiently loosely to prevent the electron transfer which would result in loss of identity.

COLLISION MECHANISM OF $H_2^+ + H_2 \rightarrow H_3^+ + H$

Similar experiments on both

$$H_2^+ + H_2 \rightarrow H_3^+ + H \tag{7}$$
$$D_2^+ + D_2 \rightarrow D_3^+ + D \tag{8}$$

reveal that this reaction too proceeds through a complex mechanism at primary ion energies of 0·2-0·5 eV (lab.). In this case, however, isotopic labelling can provide confirmatory evidence from a study of

$$HD^+ + HD \begin{cases} \nearrow H_2D^+ + D & \tag{9} \\ \searrow HD_2^+ + H & \tag{10} \end{cases}$$

for which the simple model predicts identical deflection curves for both product ions.   Data for reactions (9) and (10), shown in fig. 6 for a primary ion energy of 0·36 eV (lab.), show that this is approximately so, the $\bar{\tau}_s/\bar{\tau}_p$ ratios being slightly above and below that predicted for complex formation (see table 1).   Here again the mechanism is energy dependent since

TABLE 1.—COMPARISON BETWEEN EXPERIMENTAL VALUES OF $\bar{\tau}_s/\bar{\tau}_p$ AND
THEORETICAL VALUES PREDICTED BY THE SIMPLE MODEL

| | predictions of simple model | | | |
reaction	complex	stripping	pick-up	expt.
$HD^+ + HD \rightarrow H_2D^+ + D$	1·50	2·50	1·17	1·43
$HD^+ + HD \rightarrow HD_2^+ + H$	1·50	1·75	1·33	1·58

Durup [11] has found that reaction (8) proceeds by deuteron stripping above 4 eV (lab.) and Doverspike et al.[12] find evidence for D^+ stripping and D atom pick-up down to 2 eV (lab.). Once more the chemical identity of the collision partners appears not to be lost at 0·3 eV (lab.).[13]

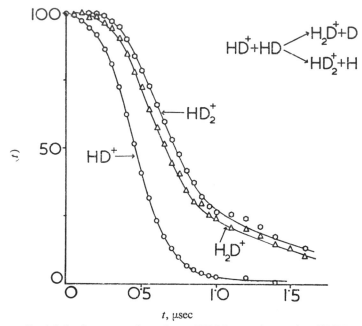

FIG. 6.—Normalized deflection curves for primary HD^+ ions and secondary H_2D^+ and HD_2^+ ions; plots of $I(t)$, the relative intensity of ions with residence times greater than t, against t, the delay time in μsec (data obtained with source A).

More detailed information on the collision kinematics, evidenced by the differences in the deflection curves of the product ions in fig. 5 and 6, require comparison between theoretically generated curves and experimental curves, obtained without field penetration.

KINEMATIC STUDIES: NON-REACTIVE COLLISIONS
The reaction

$$He^+ + H_2 \rightarrow HeH^+ + H + 8·6 \text{ eV} \qquad (11)$$

is an example of a highly exothermic reaction which is not observed [3, 14, 15] and the other reaction path

$$He^+ + H_2 \rightarrow He + H^+ + H \qquad (12)$$

occurs with a very low cross-section, $\sigma < 6 \times 10^{-17}$ [14], $< 10^{-18}$ [15], $< 3 \times 10^{-17}$ cm² (this study), less than 1 % of the orbiting cross-section calculated from the ion-induced dipole interaction.[16] Friedman [17] has rationalized this lack of reaction [12] in terms of the non-intersection of the He^+-H_2 and $He-H_2^+$ potential energy surfaces in the range of separation possible.* Light,[19] on the other hand, has applied his " phase-space " theory to reaction (11) obtaining a cross section of $\sim 10^{-16}$ cm^2.

Further insight comes from velocity analysis of the He$^+$ after its collision wi h the H$_2$ to examine if a complex was formed—a complex which then dissociated to give the original reactants. If this were so, the deflection curve of the helium would become increasingly extended as the hydrogen pressure in the ionization chamber is increased since the curve would now be a composite of primary and secondary deflection curves. Fig. 7 shows the results for such a study together with a theoretical deflection curve, predicting a lower limit for the curve which should have been formed had complex formation been occurring. This theoretical curve should be broader still since the theory neglects the velocity distribution of the hydrogen molecules and the weighting of the collisions towards the beginning of the ion path—an important effect at high pressures.[20]

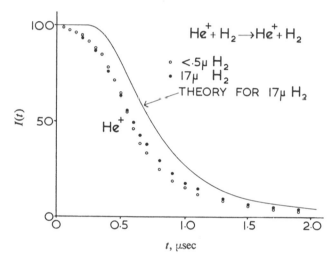

FIG. 7.—Dependence of $I(t)$, the normalized relative intensity of He$^+$ ions with residence times greater than t, against t, the delay time in μsec for < 0.5 and 17 microns of hydrogen in the ionization chamber. The curve is a theoretically predicted lower limit for the He deflection curve expected for complex formation at 17 μ H$_2$. Primary He$^+$ ion energy $= 0.29$ eV (lab.) (data obtained with source A).

Clearly no complex is being formed ; thus the strong-coupling requirement of Light's theory does not obtain and it is not surprising that its predicted cross-section is at variance with the experimental results. The slight change in shape of the deflection curve is almost certainly due to elastic scattering.

Fig. 8 suggests a reason why a complex should not be formed for this reaction. For an impact parameter $b < b_0$, the critical impact parameter, the helium ion is elastically scattered from the centrifugal barrier ; for $b = b_0$, orbiting will occur ; but for $b < b_0$, the ion will pass over the centrifugal barrier. Then, two courses become possible. Either it is elastically scattered from the repulsive wall or it is held by the hydrogen as a complex, in which case enough translational energy must be transformed into internal energy of the complex for the system to be held within the well by the centrifugal barrier. It seems plausible that such a conversion could well be difficult since it would involve vibrational excitation of the hydrogen molecule.

*application of the Koopman theorem to the He—H$_2$ surfaces calculated by F. A. Mies and M. Krauss.[18]

This model receives support from a similar study on another unreactive system,[21]

$$C_2H_4^+ + C_2H_6 \rightarrow C_2H_4^+ + C_2H_6 \tag{13}$$

Here the change in shape of the deflection curve for the $C_2H_4^+$ ion (fig. 9) studied as a function of the ethane pressure in the ionization chamber is compatible with that predicted for complex formation.   In this case, conversion of translational energy into internal energy of the

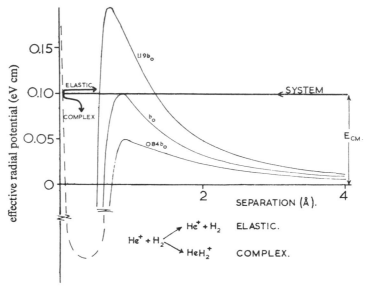

FIG. 8.—The effective radial potential (electrostatic plus centrifugal), in eV (c.m.), as a function of the inter-particle separation, at various impact parameters for the system $He^+ + H_2$ at constant energy E = 0·10 eV (c.m.).   (The well minimum and repulsive wall, shown dashed, are hypothetical). Also shown is the trajectory of the system for both elastic scattering and complex formation at impact parameters less than b_0, the critical one for orbiting collision.

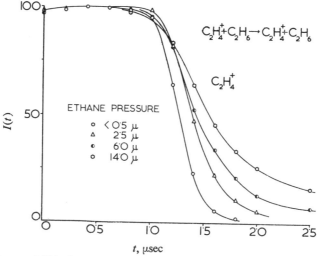

FIG. 9.—Dependence of $I(t)$, the normalized relative intensity of $C_2H_4^+$ ions with residence times greater than t, against t, the delay time in μsec, as a function of the ethane pressure in microns in the ionization chamber.   Primary $C_2H_4^+$ ion energy = 1·5 eV (lab.) (data obtained with source B).

complex would be much more facile. Reaction (13) may well involve chemical reaction yielding chemically identical products by the well-established H_2^- transfer reaction [22] : only the relevant isotopic experiments will reveal this. Why the complex follows exit channels leading to the original reactants and forsakes exothermic channels such as (14)

$$C_2H_4^+ + C_2H_6 \rightarrow C_4H_9^+ + H + 0.70 \text{ eV}, \tag{14}$$

remains unanswered.

We thank Prof. Richard Wolfgang for a characteristically illuminating discussion. Financial support from the United Kingdom Science Research Council and the Petroleum Research Fund of the American Chemical Society is gratefully acknowledged.

[1] A. Henglein and G. A. Muccini, *Z. Naturforsch.*, 1962, **17a**, 452 ; 1963, **18a**, 753 ; A. Henglein, *Ion Molecule Reactions in the Gas Phase* (American Chemical Society, Washington, D.C., 1966), *Adv. Chem. Series*, **58**, 63.

[2] G. K. Lavrovskaya, M. I. Markin and V. L. Talrose, *Kinetika i Kataliz*, 1961, **2**, 21.

[3] C. F. Giese and W. B. Maier, *J. Chem. Physics*, 1963, **39**, 739.

[4] For a recent survey, see J. Ross, ed., *Molecular Reams, Adv. Chem. Physics*, (Interscience, New York, 1966), vol. 10.

[5] Z. Herman, T. L. Rose, J. D. Kerstetter and R. Wolfgang, *J. Chem. Physics*, 1967, **46**, 2844.

[6] L. Matus, D. J. Hyatt and M. J. Henchman, *J. Chem. Physics*, 1967, **46**, 2439.

[7] K. Birkinshaw, D. Hyatt, L. Matus, A. J. Masson, I. Opauszky and M. J. Henchman, *Advances in Mass Spectrometry*, (Institute of Petroleum, London, 1968), vol. 4 in press.

[8] F. P. Abramson and J. H. Futrell, *J. Chem. Physics*, 1966, **45**, 1925.

[9] F. P. Abramson and J. H. Futrell, *J. Chem. Physics*, 1967, **46**, 3265.

[10] V. L. Talrose, *Pure Appl. Chem.*, 1962, **5**, 455.

[11] J. Durup and M. Durup, *J. Chim. Physique*, 1967, **64**, 386.

[12] L. D. Doverspike and R. L. Champion, *Bull. Amer. Physic. Soc.*, 1967, **12**, 236.

[13] J. H. Futrell and F. P. Abramson, *Ion-Molecule Reactions in the Gas Phase* (American Chemical Society, Washington, D.C., 1966) ; *Adv. Chem. Series.*, **58**, 107.

[14] L. Friedman and T. F. Moran, *J. Chem. Physics*, 1965, **42**, 2624.

[15] F. C. Fehsenfeld, A. L. Schmeltekopf, P. D. Goldan, H. I. Schiff and E. E. Ferguson, *J. Chem. Physics*, 1966, **44**, 4087.

[16] G. Gioumousis and D. P. Stevenson, *J. Chem. Physics*, 1958, **29**, 294.

[17] L. Friedman, *Ion-Molecule Reactions in the Gas Phase* (American Chemical Society, Washington D.C., 1966) ; *Adv. Chem. Series.*, **58**, 87.

[18] F. A. Mies and M. Krauss, *J. Chem. Physics*, 1965, **42**, 2703.

[19] J. C. Light, *J. Chem. Physics*, 1964, **40**, 3221.

[20] M. J. Henchman, *Disc. Faraday Soc.*, 1965, **39**, 63.

[21] M. S. B. Munson, J. L. Franklin and F. H. Field, *J. Physic. Chem.*, 1964, **68**, 3098.

[22] R. D. Doepker and P. Ausloos, *J. Chem. Physics*, 1966, **44**, 1951.

Copyright © 1977 by the American Chemical Society

Reprinted from Am. Chem. Soc. J. **99**:2943–2948 (1977)

Ion–Molecule Reactions in Monosilane–Benzene Mixtures. Long-Lived Collision Complexes[1]

W. N. Allen and F. W. Lampe*

Contribution from the Davey Laboratory, Department of Chemistry, The Pennsylvania State University, University Park, Pennsylvania 16802. Received July 19, 1976

On the basis of a previous study in this laboratory of ion-molecule reactions in SiH_4–C_6H_6 mixtures, it was reported[2] that internally excited $C_6H_6^+$ ions underwent a number of reactions with SiH_4, but that reactions of Si^+, SiH^+, SiH_2^+, and SiH_3^+ with C_6H_6 could not be detected. This previous study[2] was conducted in a mass spectrometer at ion-source pressures below 5×10^{-3} Torr, with reaction identification relying principally on the measurement of appearance potentials of the product ions. Since more recent reports[3-6] of ion-molecule reactions in similar systems have demonstrated the inadequacy of such apparatus and techniques for the study of complex systems involving more than one neutral reactant, a reinvestigation of the SiH_4–C_6H_6 was felt to be warranted. Accordingly, we have carried out a new study using tandem mass spectrometric techniques which avoid many of the ambiguities inherent in the previous work.[2]

The present results are in qualitative agreement with the earlier study[2] only with regard to the identification of the major products of ionized SiH_4–C_6H_6 mixtures. The earlier conclusions that Si-containing ions are unreactive toward C_6H_6 and that all ion-molecule reactions between SiH_4 and C_6H_6 are to be attributed solely to reaction of internally excited $C_6H_6^+$ ions with SiH_4 are incorrect. This paper constitutes a report of our reinvestigation of ion-molecule reactions in this system.

Experimental Section

The chemical reactions taking place when ions derived from electron impact on C_6H_6 collide with SiH_4 and ions from SiH_4 collide wih C_6H_6 were studied in a tandem mass spectrometer. This apparatus, which has been described previously,[5] consists of two quadrupole mass filters mounted in an "in-line" configuration and separated from each other by a collision chamber and ion lenses. It permits the injection of mass-selected ions, having kinetic energies variable down to about 1 eV, into a collision chamber containing the reactant molecule. Product ions scattered in the forward direction are collected with an acceptance angle of ~10° and mass analyzed. Ion intensity measurements were made with a Bendix Model 4700 continuous-dynode electron multiplier.

The energy of the ionizing electron beam in the ion source was 85 eV and trap currents used were in the range of 10-20 μA. The pressure of target gas in the collision chamber was measured with a CGS Barocel capacitance manometer. The reactant ion beam entering the collision chamber was found by retarding field measurements to have an energy spread of about 1 eV (full width at half-maximum).

Reaction cross sections were calculated from the relationship given in the equation

$$\sigma = \gamma I_s / I_p n l \qquad (1)$$

in which I_p and I_s are the multiplier currents due to primary and secondary ions, respectively, n is the number density of gas molecules in the collision chamber, l is the length of the collision chamber (1.0 cm), and γ is an instrumental calibration factor which takes into account different detection efficiencies of primary and secondary ions. For reactions in which very little momentum may be transferred from the primary ions to the product ions, such as exothermic hydride ion transfer,[7] γ (at 1 eV) was determined from the reaction ($\sigma = 47 \pm 14$ Å^2)[3,4]

$$CH_3^+ + SiH_4 \rightarrow SiH_3^+ + CH_4 \qquad (2)$$

to be 1.4. For endoergic reactions and those of sufficient complexity to require that considerable momentum be transferred to product ions, γ was determined from the reaction ($\sigma = 33 \pm 4$ Å^2)[8-12]

$$CH_3^+ + CH_4 \rightarrow C_2H_5^+ + H_2 \qquad (3)$$

to be 1.2. We believe the reaction cross sections reported to be accurate to within ±30%.

Monosilane was purchased from the J. T. Baker Chemical Co. It was subjected to several freeze-pump-thaw cycles, stored, and checked mass spectrometrically for satisfactory purity before use. Monosilane-d_4 was prepared from silicon tetrachloride (Peninsular Chem-Research, Inc.) by reduction with an excess of lithium aluminum deuteride.[13] It too was checked mass spectrometrically for satisfactory purity before use. The benzene used was Baker Analyzed Reagent, spectrophotometric grade. Except for degassing by freeze-pump-thaw cycles on the vacuum line it was used as received.

Results

Cross sections for formation of the various product ions in collisions of Si^+, SiH^+, SiH_2^+, and SiH_3^+ (at 1 eV of kinetic energy in the laboratory frame of reference) with C_6H_6 at ambient temperature (20 °C) are shown in Table I. Similarly, cross sections for formation of product ions in collisions of 1-eV $C_4H_2^+$, $C_4H_3^+$, $C_4H_4^+$, $C_6H_5^+$, and $C_6H_6^+$ ions with SiH_4 are shown in Table II. The cross sections in these tables have been corrected for the naturally occurring isotopes of ^{13}C (1.1%), ^{29}Si (4.7%), and ^{30}Si (3.1%).

It is apparent from Tables I and II that the major ionic products of ionized SiH_4–C_6H_6 mixtures that do not appear in the respective ionized pure compounds are those having m/e values in the range of 105-109, i.e., $SiC_6H_x^+$ ($x = 5$-9). This observation is in agreement with the previous report.[2] However, the product ions in the 105-109 m/e range are formed by reactions of Si-containing ions with C_6H_6 and this fact is a direct contradiction of the earlier reaction identification.[2] It must be

Table I. Reactions of Silicon-Containing Ions[a] with C_6H_6[b]

Reactant ion	Cross section in Å² for formation of product ion of m/e								
	78	79	81	83	105	106	107	108	109
Si^+	4.6				3.8	7.4			
SiH^+	14	2.2	3.6		6.6	1.4	6.7		
SiH_2^+	75	6.0			2.6	33	18		
SiH_3^+	8.0	1.5	2.8	1.0				8.6	18

[a] Reactant-ion energy of 1.0 eV in the laboratory frame of reference. [b] Collision chamber pressure of 1.0×10^{-3} Torr.

Table II. Reactions of Ions Derived from C_6H_6[a] with SiH_4[b]

Reactant ion	Cross section in Å² for formation of product ion of m/e											
	31	33	39	42	43	52	53	54	55	56	57	67
$C_4H_2^+$	8.9			1.6	2.0		2.8	3.9	6.5	2.2	0.2	3.4
$C_4H_3^+$	5.6	0.6			1.1			• 0.1	2.3		2.1	0.2
$C_4H_4^+$	1.4		0.6		1.2				1.2	2.3	0.5	0.5
$C_6H_5^+$	9.6				0.3							
$C_6H_6^+$	3.8		0.2			0.5						

[a] Reactant-ion energy of 1.0 eV in the laboratory frame of reference. [b] Collision chamber pressure of 1.0×10^{-3} Torr.

Table III. Reactions of Deuterated Si Ions[a] with C_6H_6[b]

Reactant ion	Cross section in Å² for formation of product ion of m/e										
	78	79	80	105	106	107	108	109	110	111	112
SiD^+	c	c		3.9	4.8	2.7	5.5				
SiD_2^+	c	c		1.1	12	17	23	17			
SiD_3^+	9.2	2.8	1.6			1.1	3.2	26	6.3	5.2	27

[a] Reactant-ion energy of 1.0 eV in the laboratory frame of reference. [b] Collision chamber pressure of 1.0×10^{-3} Torr. c Not measured.

concluded that the reaction identification of the earlier study[2] was incorrect and represents another example[3-6] of the unreliability of product-ion appearance potentials for the identification of ion–molecule reactions in systems containing two or more neutral reactants.

The most striking feature of the results shown in Table I is the detection of the ions of m/e 106, 107, and 109 in collisions of Si^+, SiH^+, and SiH_3^+, respectively, with C_6H_6. Since these ions represent $SiC_6H_6^+$, $SiC_6H_7^+$, and $SiC_6H_9^+$, respectively, they are simply adducts containing all the atoms of the colliding particles. Such direct detection of collision adducts or complexes is rare in beam studies, there being to our knowledge only two prior reports in the literature.[5,14] In addition to adduct formation, charge exchange to yield $C_6H_6^+$ (m/e 78) is another probable reaction of Si-containing ions with C_6H_6.

Significantly less reaction occurs when ions derived from C_6H_6 are injected into SiH_4 (Table II). The predominance of SiH_3^+ (m/e 31) as a product ion in these collisions is as expected, though, on the basis of previous reports of the tendency of SiH_4 toward H^- transfer to attacking positive ions.[3,4,15]

Table III shows the cross sections for formation of various product ions in collisions of SiD^+, SiD_2^+, and SiD_3^+ with C_6H_6. As with the data in Tables I and II these cross sections have been corrected for the naturally occurring isotopes of carbon and silicon. In calculating these corrected cross sections from the primary data it has been assumed that the fraction of total reaction of a given SiD_x^+ ion that yields products in the range of m/e 105–112 is the same as that for the corresponding SiH_x^+ ion. While this assumption is surely a reasonable one, its use must result in a greater uncertainty being assigned to the cross sections in Table III as compared with the cross sections in Tables I and II. A conservative estimate is that the cross sections in Table III are accurate with ±50%.

The effects of pressure and reactant-ion energy on the distributions of major products of collisions of Si-containing ions with C_6H_6 are shown in Figures 1–4. Reaction cross sections as functions of the energy of the reactant ions are shown in Figure 5. In view of the significantly lower reaction probabilities, detailed measurements of the pressure and energy dependencies of the reactions of ions derived from C_6H_6 with SiH_4 were not carried out. A few measurements of the energy dependence of the cross section for hydride transfer from SiH_4 to $C_6H_6^+$ showed typical exothermic behavior.[4,16]

Discussion

A. Nature of the Elementary Reactions. While some of the general features of the results have been remarked upon earlier, in this section we discuss the nature of the individual elementary reactions that take place when Si-containing ions collide with benzene.

1. $SiH_3^+ + C_6H_6$. As shown in Table I the major products of the reaction of SiH_3^+ ions of 1-eV kinetic energy with C_6H_6 are ions having m/e values of 78, 107, and 109, and the processes involved are clearly those shown by the reactions

$$SiH_3^+ + C_6H_6 \rightarrow C_6H_6^+ + SiH_3 \qquad (4)$$

$$SiH_3^+ + C_6H_6 \rightarrow SiC_6H_7^+ + H_2 \qquad (5)$$

$$SiH_3^+ + C_6H_6 \rightarrow SiC_6H_9^+ \qquad (6)$$

respectively. The charge-transfer reaction (eq 4) is endothermic[17,18] for ground-state ions by 19 kcal. As shown by Figures 1b and 5a the energy dependence of the cross section of eq 4 is in accord with the process being endothermic. The absence of an energy threshold for $C_6H_6^+$ formation in Figure 1b is to be attributed to contributions from the fast exothermic charge

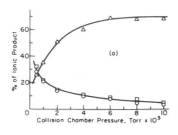

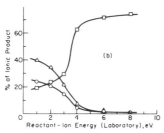

Figure 1. $SiH_3^+ + C_6H_6$ reaction. Dependence of major product distribution on (a) collision chamber pressure and (b) kinetic energy of reactant ions: ⊡ $C_6H_6^+$; ⊙ $SiC_6H_7^+$; ▲ $SiC_6H_9^+$.

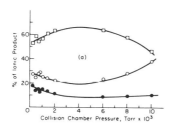

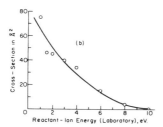

Figure 2. $SiH_2^+ + C_6H_6$ reaction. (a) Dependence of major product distribution of collision chamber pressure: ⊡ $C_6H_6^+$; ⊙ $SiC_6H_6^+$; ● $SiC_6H_7^+$. (b) Dependence of charge exchange cross section (eq 11) on kinetic energy of reactant ions.

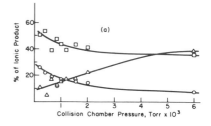

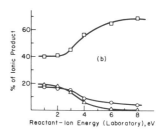

Figure 3. $SiH^+ + C_6H_6$ reaction. Dependence of major product distribution on (a) collision chamber pressure and (b) kinetic energy of reactant ions: ⊡ $C_6H_6^+$; ⊙ $SiC_6H_5^+$; ▲ $SiC_6H_7^+$.

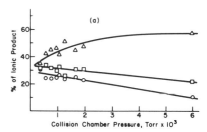

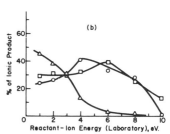

Figure 4. $Si^+ + C_6H_6$ reaction. Dependence of major product distribution on (a) collision chamber pressure and (b) kinetic energy of reactant ions: ⊡ $C_6H_6^+$; ⊙ $SiC_6H_5^+$; ▲ $SiC_6H_6^+$.

transfer from $^{29}SiH_2^+$ (for which corrections below 1 eV could not be made) and to the presence of internally excited SiH_3^+ in the beam. On the basis of the energy dependence of the respective reaction cross sections, reactions 5 and 6 are exothermic processes.

Minor products of the reaction are ions with m/e values of 79, 81, and 83, as may be seen in Table I. On the basis of the appearance of m/e 80 when SiD_3^+ is substituted for SiH_3^+ (Table III) we conclude that m/e 79 is $C_6H_7^+$ formed by the proton-transfer reaction depicted by the reaction

$$SiH_3^+ + C_6H_6 \rightarrow C_6H_7^+ + SiH_2 \tag{7}$$

The dependence of the cross section of eq 7 on kinetic energy is consistent with available thermochemical data[17-21] that indicate eq 7 to be about thermoneutral. The formation of the products having m/e 81 and 83 are best described by the reactions

$$SiH_3^+ + C_6H_6 \rightarrow SiC_4H_5^+ + C_2H_4 \tag{8}$$

$$SiH_3^+ + C_6H_6 \rightarrow SiC_4H_7^+ + C_2H_2 \tag{9}$$

respectively. The dependence of the cross sections on kinetic energy indicate that reactions 8 and 9 are both exothermic processes.

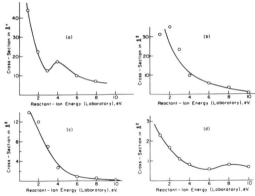

Figure 5. Dependence of reaction cross sections on kinetic energy of reactant ions: (a) $SiH_3^+ + C_6H_6 \rightarrow$ products; (b) $SiH^+ + C_6H_6 \rightarrow$ products; (c) $Si^+ + C_6H_6 \rightarrow$ products; (d) $C_6H_6^+ + SiH_4 \rightarrow SiH_3^+ + C_6H_7$.

The direct observation of $SiC_6H_9^+$ (m/e 109) at the detector is of major interest because it means that this adduct, or complex, of the reactants survives the 2×10^{-5}-s flight from the collision chamber. As shown in Figure 1a, increasing the pressure of C_6H_6 in the collision chamber for 1 eV SiH_3^+ ions increases the abundance of the $SiC_6H_9^+$ ion to 70% of the total product. Clearly this pressure effect is to be attributed to collisional stabilization of the energy-rich adducts, or complexes, initially formed. In the absence of such stabilization the $SiC_6H_9^+$ ions must contain all the energy of reaction and the relative kinetic energy of reactants as internal energy, and an appreciable fraction decomposes during the flight to the detector. The data in Figure 1a suggest (and it has been confirmed in a number of other pressure dependence studies of this system) that in the extrapolated limit of zero pressure, i.e., no collisional stabilization, an appreciable fraction of the energy-rich complexes reaches the detector and therefore has lifetimes in excess of 2×10^{-5} s. The increase in relative abundance of $SiC_6H_9^+$ due to pressure stabilization is accompanied by a concomitant decrease in the other ionic products, as shown in Figure 1a for $C_6H_6^+$ (m/e 107).

The effect on product distribution of increasing the relative kinetic energy of the reactants (i.e., internal energy of the adduct) for a C_6H_6 pressure of 1.0×10^{-3} Torr is shown in Figure 1b. In accord with the conclusion that $SiC_6H_9^+$ (m/e 109) is an energy-rich complex, the abundance of this ionic product falls rapidly with increasing energy. The relative abundance of the decomposition product $SiC_6H_7^+$ (m/e 107) falls less rapidly, while the relative abundance of $C_6H_6^+$ (m/e 78) increases sharply to over 70% of the product. This behavior is completely consistent with the following mechanistic scheme,

$$SiH_3^+ + C_6H_6 \rightarrow SiC_6H_9^{+*} \quad (6a)$$

$$SiC_6H_9^{+*} \rightarrow C_6H_6^+ + SiH_3 \quad (4a)$$

$$SiC_6H_9^{+*} \rightarrow SiC_6H_7^+ + H_2 \quad (5a)$$

$$SiC_6H_9^{+*} + C_6H_6 \rightarrow SiC_6H_9^+ + C_6H_6 \quad (10)$$

in which all ions shown, including both energy-rich and collisionally stabilized $SiC_6H_9^+$ ions, are registered at the detector. Other ionic products not shown in Figure 1 also arise from unimolecular decomposition of $SiC_6H_9^{+*}$.

Further insight into the reaction mechanism can be obtained by considering the product distribution of the reaction of SiD_3^+ ions with C_6H_6. These data, in terms of cross sections, are

shown in Table III for 1.0-eV SiD_3^+ ions and 1.0×10^{-3} Torr of C_6H_6. The ion at m/e 80 is accounted for entirely by $C_6H_6D^+$ formed by the D^+ transfer process analogous to reaction 7. This means that the benzene ion product (m/e 78 and 79) is comprised of 77% $C_6H_6^+$ and 23% $C_6H_5D^+$, respectively. The ionic product formed by loss of molecular hydrogen from the energy-rich complex, i.e., the ionic product of reaction 5, appears in this case at m/e 108, 109, and 110, corresponding to loss of D_2, HD, and H_2. Neglecting any possible contribution to m/e 110 by loss of a D atom, the data in Table III indicate that at least 73% of the product appears at m/e 109, i.e., corresponds to loss of HD from the energy-rich complex. These facts suggest that the SiD_3^+ ion retains to a very large extent its chemical identity in the complex and, further, that very little exchange of deuterium with the hydrogen atoms of the benzene ring takes place. Simple rupture of the Si–C bond leads to charge transfer (eq 4), while a 1,2-molecular hydrogen elimination forms the phenylsilyl ion. The reaction therefore appears to occur in a completely analogous manner to the classical addition of carbonium ions to aromatics,[22] and the mechanism proposed for major product formation is shown by eq 6b, 4b, and 5b. The energy-rich adduct is thus indicated to be a dienyl-type cation[22] characteristic of such ionic additions in liquid-phase organic chemistry. The benzyl-type structure of the product ion of reaction 5b has been previously proposed in studies of the electron-impact ionization of phenylsilane.[23]

$$SiD_3^+ + \text{(benzene)} \longrightarrow \left[\begin{array}{c} D\text{–}Si\text{–}D \\ D \quad H \\ + \end{array} \right]^* \quad (6b)$$

$$\left[\begin{array}{c} D\text{–}Si\text{–}D \\ D \quad H \\ + \end{array} \right]^* \longrightarrow \text{(benzene)}^+ + SiD_3 \quad (4b)$$

$$\longrightarrow \text{(phenyl)}\,SiD_2^+ + HD \quad (5b)$$

Recent evidence[24-29] showing the existence of silicon–carbon π bonds in compounds of the type $(CH_3)_2Si=CHR$ raises the possibility that the structure of the $SiC_6H_7^+$ might be analogous to that of the tropylium ion,[30,31] i.e.,

However, the formation of such a structure from the $SiC_6H_6D_3^{+*}$ formed in reaction 6b would be expected to lead to extensive scrambling and nearly statistical loss of hydrogen and deuterium, as was observed in studies of the formation of $C_7H_5D_2^+$ and $C_7H_4D_3^+$ ions by electron-impact ionization of $C_6H_5CD_3$.[32,33] The predominance of $C_6H_6^+$ formation in reaction 4b and of $SiC_6H_5D_2^+$ formation in reaction 5b strongly suggests that such scrambling does not occur. We conclude, therefore, that $SiC_6H_4^+$ ions are better described by the benzyl-type structure, as shown in eq 5b, and as suggested earlier.[23]

2. $SiH_2^+ + C_6H_6$. The major reaction of 1-eV SiH_2^+ ions with C_6H_6 is charge transfer, a reaction that is exothermic for ground-state ions and that for 1-eV ions proceeds with a very large cross section (Table I). Other major products formed with large cross sections are $SiC_6H_7^+$ (m/e 107) and $SiC_6H_6^+$ (m/e 106), while $SiC_6H_5^+$ (m/e 105) and $C_6H_7^+$ (m/e 79) arise from considerably less probable processes. It is noteworthy that no product corresponding to an adduct of the reactants was detected until the pressure of C_6H_6 in the collision chamber exceeded 6×10^{-3} Torr (i.e., only collisionally stabilized ions of m/e 108 are detected). This means no energy-rich collision complexes, i.e., $SiC_6H_8^{+*}$, have sufficiently long lifetimes to reach the detector and this is probably a consequence of the existence of the exothermic charge-transfer channel. As already discussed an exothermic charge-transfer channel does not exist for collisions of SiH_3^+ with C_6H_6, nor for collisions of SiH^+ and Si^+ and in all three of these systems energy-rich collision complexes are observed.

The reactions observed in this system for 1-eV ions and the enthalpy changes of the reactions, calculated where possible, for ground-state ions[17,19-21,34] are shown by

$$SiH_2^+ + C_6H_6 \rightarrow C_6H_6^+ + SiH_2;\ \Delta H° = -11\ kcal \quad (11)$$

$$SiH_2^+ + C_6H_6 \rightarrow C_6H_7^+ + SiH;\ \Delta H° = -14\ kcal \quad (12)$$

$$SiH_2^+ + C_6H_6 \rightarrow SiC_6H_5^+ + H_2 + H \quad (13)$$

$$SiH_2^+ + C_6H_6 \rightarrow SiC_6H_6^+ + H_2 \quad (14)$$

$$SiH_2^+ + C_6H_6 \rightarrow SiC_6H_7^+ + H \quad (15)$$

The dependence of the product distribution of this reaction on pressure is shown in Figure 2a and the dependence of the charge-transfer cross section on the kinetic energy of SiH_2^+ is shown in Figure 2b. The cross section of reaction 11 is seen to exhibit exothermic behavior which is in accord with the calculation from thermochemical data[16-20] of $\Delta H° = -11$ kcal. Detailed studies of the energy dependencies of the cross sections for reactions 12-15 were not carried out, but the magnitudes of the cross sections at 1 eV (Table I) suggest strongly that these reactions are also exothermic. Indeed, thermochemical data[17-21,34] confirm that reaction 12 is exothermic.

When SiD_2^+ ions of 1-eV kinetic energy collide with C_6H_6 the silicon-containing product ions are distributed over the m/e 105-109 range as shown in Table III. This rather broad distribution is in contrast to that observed in SiD_3^+-C_6H_6 collisions and suggests that appreciable exchange of hydrogen and deuterium atoms occur between the incoming SiD_2^+ and the benzene ring. However, the relative probabilities of formation of $SiC_6H_5D_2^+$ and $SiC_6H_6D^+$ ions by reaction 15 are not close to the 3:1 ratio predicted by randomization of H and D. These results are therefore also consistent with a benzyl-type structure for $SiC_6H_7^+$.

3. $SiH^+ + C_6H_6$. The data in Table I show that SiH^+ ions react with C_6H_6 to form ionic products having m/e values of 78, 79, 81, 105, 106, and 107. The respective reactions involved and, where possible, calculated enthalpy changes are shown in the reactions below.

$$SiH^+ + C_6H_6 \rightarrow C_6H_6^+ + SiH;\ \Delta H° = 29\ kcal \quad (16)$$

$$SiH^+ + C_6H_6 \rightarrow C_6H_7^+ + Si;\ \Delta H° = 20\ kcal \quad (17)$$

$$SiH^+ + C_6H_6 \rightarrow SiC_4H_5^+ + C_2H_2 \quad (18)$$

$$SiH^+ + C_6H_6 \rightarrow SiC_6H_5^+ + H_2 \quad (19)$$

$$SiH^+ + C_6H_6 \rightarrow SiC_6H_6^+ + H \quad (20)$$

$$SiH^+ + C_6H_6 \rightarrow SiC_6H_7^{+*} \quad (21)$$

As indicated by Table I and as confirmed by Figures 3a and 3b, energy-rich complexes with sufficient lifetime to be observed at the collector are formed in reaction 21. It is also to

be noted from the calculated enthalpy change and from the behavior of the cross section with energy (cf. Figures 3b and 5b) that the charge-transfer channel (eq 21) is endothermic. The amount of charge transfer that does occur below threshold, i.e., in the relative kinetic energy range of 0.73-1.5 eV (17-34 kcal/mol), must be attributed to the presence of internally excited SiH^+ ions in the reactant-ion beam. The presence of appreciable internal excitation in SiH^+ formed by electron impact on SiH_4 has been noted previously.[35,36] Reaction 17 is also endothermic and hence the extent of formation of $C_6H_7^+$ shown in Table I is also indicative of the presence of internally excited SiH^+ ions. The cross sections of reactions 18-21 exhibit the energy dependence that is typical of exothermic processes, as may be seen in Figures 3b and 5b for the specific cases of reactions 19 and 21.

If SiH^+ ions are replaced with 1-eV SiD^+ ions the ionic products containing six carbon atoms and one silicon atom are distributed as shown in Table III. This distribution suggests that the deuterium atom on the silicon is exchanged rather extensively with the ring hydrogens. In view of this complication and the absence of any classical organic chemical analogues we are unable to say more concerning the detailed mechanism.

4. $Si^+ + C_6H_6$. As shown in Table I the reaction of 1-eV Si^+ ions with C_6H_6 results in the formation of $C_6H_6^+$ (m/e 78), $SiC_6H_5^+$ (m/e 105), and $SiC_6H_6^{+*}$ (m/e 106). The pressure dependence and energy dependence of the distribution of these three products is shown in Figures 4a and 4b and the energy dependence of the total cross section is shown in Figure 5c. It is apparent from these figures that an energy-rich collision complex $SiC_6H_6^{+*}$, which has sufficient lifetime to be registered at the detector, is formed and that this energy-rich ion can be stabilized by collision so that at 6×10^{-3} Torr it counts for ~57% of the ionic product. Decomposition of this collision complex leads to the products $C_6H_6^+$ (m/e 78) and $SiC_6H_5^+$ (m/e 105). The suggested mechanism is depicted by the sequence of reactions 22-25, in which all ions shown are observed at the detector.

$$Si^+ + C_6H_6 \rightarrow SiC_6H_6^{+*} \quad (22)$$

$$SiC_6H_6^{+*} \rightarrow C_6H_6^+ + Si \quad (23)$$

$$SiC_6H_6^{+*} \rightarrow SiC_6H_5^+ + H \quad (24)$$

$$SiC_6H_6^{+*} + C_6H_6 \rightarrow SiC_6H_6^+ + C_6H_6 \quad (25)$$

We note again that in this system, in which an energy-rich collision complex is detected, the charge-transfer channel is endothermic by 24 kcal.[17] The appreciable amounts of $C_6H_6^+$ seen in Figure 4b and 5c at energies below the calculated threshold of 1.4 eV (laboratory) must be due to the presence of electronically excited Si^+ in the beam. This has been confirmed by ion-beam attenuation[37] experiments in CO_2 and by study of the endothermic reactions of Si^+ with D_2.[36] The results of these experiments lead to the conclusion that under our conditions the Si^+ ion beam consists of 62% Si^+ (^2P), which is the ground state, and 38% Si^+ (^4P) which lies 5.5 eV above the ground state.[38] Certainly the presence of this amount of excited state in the beam will account for the abundance of $C_6H_6^+$ at energies below the threshold for ground-state ions and for the shape of the cross section vs. energy curve. Further, it is tempting to interpret the apparent limiting abundance of pressure-stabilized collision complex of about 57%, seen in Figure 4a, as meaning that essentially all of the collision complexes of Si^+ (^2P) with C_6H_6 can be stabilized, but that very little of the collision complex of Si^+ (^4P) with C_6H_6, for which the charge-transfer channel is exothermic, can be stabilized at these pressures.

B. Phenomenological Rate Constants. It is not possible in beam experiments, such as described here, to obtain true re-

Table IV. Phenomenological Rate Constants[a] in SiH_4-C_6H_6 System for 1-eV Ions

Reaction	$k \times 10^{10}$, cm^3/s	$k \times 10^{10}$, (Langevin), cm^3/s
$Si^+ + C_6H_6 \rightarrow$ products	4.2	16.6
$SiH^+ + C_6H_6 \rightarrow$ products	8.9	16.4
$SiH_2^+ + C_6H_6 \rightarrow$ products	34	16.2
$SiH_3^+ + C_6H_6 \rightarrow$ products	10	16.0
$C_6H_5^+ + SiH_4 \rightarrow$ products	1.6	11.5
$C_6H_6^+ + SiH_4 \rightarrow$ products	0.7	11.5
$C_4H_2^+ + SiH_4 \rightarrow$ products	6.2	12.4
$C_4H_3^+ + SiH_4 \rightarrow$ products	2.3	12.3
$C_4H_4^+ + SiH_4 \rightarrow$ products	1.5	12.3

[a] Uncertainty in k is assumed to be ±30%, the same as in the cross sections.

action rate constants because the kinetic energy distributions of the two reactants are different. Nonetheless, from the viewpoint of the chemist it is useful to define a phenomenological rate constant in terms of the total reaction cross section as in the equation

$$k = \sigma V_r \qquad (26)$$

where σ is the experimental or phenomenological cross section for total reaction of a given ion and V_r is the relative velocity of reactants. The branching ratios, which are more sensitive to energy than are the total rate constants, may be obtained at the energy in question from the cross sections for formation of the individual products. Such phenomenological rate constants, as defined by eq 26, for the reactant pairs in this study are shown in Table IV. Also shown in Table IV are the rate constants for orbiting collisions according to the Langevin theory.[39] All the phenomenological rate constants are less than that predicted by Langevin theory with the exception of that for the reaction of SiH_2^+ with C_6H_6 and this exceeds the Langevin value by a factor of about 2. Over half of the total reaction of SiH_2^+ with C_6H_6 is simple electron exchange and reaction of SiH_2^+ with C_6H_6 is simple electron exchange and this strongly exothermic process that need not involve heavy-particle motion may well proceed by mechanisms other than Langevin orbiting collisions and complex formation.

C. Lifetimes of Energy-Rich Collision Complexes. It has been pointed out that energy-rich complexes of $SiC_6H_9^+$, $SiC_6H_7^+$, and $SiC_6H_6^+$ are registered directly at the detector. However, it is clear from Figures 1a, 3a, and 4a that some collisional stabilization does occur even at 10^{-3} Torr. The fractional contribution of the energy-rich complex to the total observed ionic product may be obtained as the intercept of an extrapolation of data such as in Figures 1a, 3a, and 4a to zero pressure. The limiting fraction of productions, which we denote by $F^0_{complex}$, clearly depends on the collision energy, or equivalently is a function of the natural lifetime, τ. If we assume that the fraction of collisions that do not revert back to reactants is given by the ratio of the experimental rate constant to the Langevin rate constant for orbiting (Table IV), it may be shown readily that the fraction $F^0_{complex}$ is given by the expression

$$F^0_{complex}(\tau) = \frac{e^{-t/\tau}}{k/k_L} \qquad (27)$$

In eq 27 τ is the lifetime of the energy-rich complex, t, the flight time to the detector is 2×10^{-5} s, k is the phenomenological rate constant, and k_L is the Langevin rate constant for orbiting collisions. Using eq 3, the data in Table IV, and extrapolated values of $F^0_{complex}$, we obtain lifetime estimates, of 8.2, 7.5, and

8.9 μs for $SiC_6H_6^{+*}$, $SiC_6H_7^{+*}$, and $SiC_6H_9^{+*}$, respectively. According to simple RRK theory,[40] the average lifetime of an energy-rich molecule with only one decomposition channel is given by

$$\tau \sim 10^{-13}(1) - (E_c/E))^{1-S} \qquad (28)$$

where S is the number of vibrational degress of freedom, E_c is the potential barrier separating the ground state of the energy-rich molecule from that of the product, and E is the total energy of the colliding reactants. The above experimental lifetimes of 8-9 μs are in accord with eq 28 if the ratio E_c/E is of the order of 0.4. This is certainly a reasonable number, but, unfortunately, the energetics of the system are not sufficiently well known at present to warrant more refined calculations using RRKM theory.[40]

Acknowledgment. This work was supported by the U.S. Energy Research and Development Administration under Contract No. ER(11-1)3416. We also wish to thank the Education Committee of the Gulf Oil Corporation for a grant that assisted in the construction of the tandem mass spectrometer.

References and Notes

(1) U.S. Energy Research and Development Administration Document No. COO-3416-28.
(2) D. P. Beggs and F. W. Lampe, J. Phys. Chem., 73, 4194 (1969).
(3) G. W. Stewart, J. M. S. Henis, and P. P. Gaspar, J. Chem. Phys., 57, 1990 (1972).
(4) T. M. H. Cheng, T. Y. Yu, and F. W. Lampe, J. Phys. Chem., 77, 2587 (1973).
(5) T. M. Mayer and F. W. Lampe, J. Phys. Chem., 78, 2433 (1974).
(6) T. M. Mayer and F. W. Lampe, J. Phys. Chem., 78, 2645 (1974).
(7) T. M. Mayer and F. W. Lampe, J. Phys. Chem., 78, 2195 (1974).
(8) R. A. Fluegge, J. Chem. Phys., 50, 4373 (1969).
(9) A. A. Herod and A. G. Harrison, Int. J. Mass Spectrom. Ion Phys., 4, 415 (1970).
(10) J. H. Futrell, J. Chem. Phys., 59, 4061 (1973).
(11) W. T. Huntress and R. F. Pinizzotto, J. Chem. Phys., 59, 4742 (1973).
(12) K. R. Ryan and P. W. Harland, Int. J. Mass Spectrom. Ion Phys., 15, 197 (1974).
(13) G. G. Hess and F. W. Lampe, J. Chem. Phys., 44, 2257 (1966).
(14) A. G. Urena, R. B. Bernstein, and R. G. Phillips, J. Chem. Phys., 62, 1818 (1975).
(15) T. M. H. Cheng and F. W. Lampe, J. Phys. Chem., 77, 2841 (1973).
(16) T. M. H. Cheng and F. W. Lampe, Chem. Phys. Lett., 19, 532 (1973).
(17) J. L. Franklin, J. G. Dillard, H. M. Rosenstock, J. T. Herron, K. Draxl, and F. H. Field, Natl. Stand. Ref. Data Ser., Natl. Bur. Stand., No. 26, 1 (1969).
(18) P. Potzinger, A. Ritter, and J. R. Krause, Z. Naturforsch. A, 30, 347 (1975).
(19) P. John and J. H. Purnell, J. Organomet. Chem., 29, 233 (1971).
(20) P. John and J. H. Purnell, J. Chem. Soc., Faraday Trans. 1, 69, 1455 (1973).
(21) P. Kebarle, R. Yamdagni, K. Hiraska, and T. B. McMahon, Int. J. Mass Spectrom. Ion Phys., 19, 71 (1976).
(22) T. S. Sorenson, "Carbonium Ions", G. A. Olah and P. von R. Schleyer, Ed., Wiley, New York, N.Y., 1970, Chapter 19.
(23) M. E. Freeburger, B. M. Hughes, G. R. Buell, T. O. Tiernan, and L. Spialter, J. Org. Chem., 36, 933 (1971).
(24) L. E. Gusel'nikov and M. C. Flowers, Chem. Commun., 864 (1967).
(25) M. C. Flowers and L. E. Gusel'nikov, J. Chem. Soc. B, 419 (1968).
(26) D. N. Roark and L. H. Sommer, J. Chem. Soc., Chem. Commun., 167 (1973).
(27) R. D. Bush, C. M. Golina, and L. H. Sommer, J. Am. Chem. Soc., 96, 7105 (1974).
(28) C. M. Golina, R. D. Bush, P. Orr, and L. H. Sommer, J. Am. Chem. Soc., 97, 1957 (1975).
(29) L. E. Gusel'nikov, N. S. Nametikin, and V. M. Vdorin, Acc. Chem. Res., 8, (1975).
(30) E. Hückel, Z. Phys., 70, 204 (1931).
(31) W. von E. Doering and L. H. Knox, J. Am. Chem. Soc., 76, 3203 (1954).
(32) J. B. Farmer, I. H. S. Henderson, C. A. McDowell, and F. P. Lossing, J. Chem. Phys., 22, 1948 (1954).
(33) P. N. Rylander, S. Meyerson, and H. M. Grubb, J. Am. Chem. Soc., 79, 842 (1957).
(34) A. E. Douglas, Can. J. Phys., 35, 76 (1957).
(35) P. Potzinger and F. W. Lampe, J. Phys. Chem., 73, 3912 (1969).
(36) W. N. Allen and F. W. Lampe, J. Chem. Phys., in press.
(37) B. R. Turner, J. A. Rutherford, and D. M. J. Compton, J. Chem. Phys., 48, 1602 (1968).
(38) C. E. Moore, Natl. Bur. Stand. (U.S.) Circ., No. 467, 1 (1949).
(39) G. Gioumousis and D. P. Stevenson, J. Chem. Phys., 29, 294 (1958).
(40) P. J. Robinson and K. A. Holbrook, "Unimolecular Reactions", Wiley-Interscience, New York, N.Y., 1972.

Editor's Comments
on Papers 21 Through 28

Although the formation of a long-lived complex is known to occur as an intermediate in many ion-molecule reactions, the energy distribution in the products of some reactions is incompatible with the occurrence of such a complex. In some reactions, even at low collision energies, the projectile ion appears to react with only part of the target molecule, leaving the other part unaffected. In such reactions, the energy of the leaving ion is the same as that of the projectile and that of the neutral fragment is approximately its proportional part of the energy of the original neutral. This assumes that there is no energy of reaction. If there is, it must be taken into account in the energy distribution of the products.

The earliest evidence for such stripping or pick-up reactions was reported by Henglein and Muccini (Paper 21). The investigators were studying ion-molecule reactions employing the Čermák mode,[1] by which secondary ions are formed in the ionization chamber by reaction with a beam of primary ions formed in the trap region and drifting back through the ionization chamber. If a secondary ion is formed by stripping a proton from a primary ion or by hydride ion transfer from a neutral to the primary ion, the resulting secondary will have approximately thermal energy initially and thus will be collected in the same manner as the primary ion. If the secondary ion is formed by abstraction of a hydrogen atom from a neutral molecule, its momentum will carry it in a direction parallel to the electron beam and it will not be collected. If, on the other hand, a complex is formed, secondary ions will leave the complex in all directions and some will be collectible. The cross sections for stripping processes are likely to be small, whereas those for reactions going through intermediate complexes may be large. The authors found several reactions that appeared to occur by a stripping process even at relatively small collision energies. Others appeared to proceed through a long-lived complex at low energies but to go over to a stripping mechanism at higher energies.

The remaining experimental papers in this section present results obtained by molecular beam methods. In Papers 22 and 23, a beam of primary ions is impacted on molecules in a collision chamber and the energies of the product ions are measured. Doverspike and Champion (Paper 23) also were able to measure the angular distributions of the products. The remaining experimental papers describe results obtained by crossed-beam methods. Papers 23, 24, and 25 describe reactions that appear to involve stripping mechanisms at the higher energies. Paper 23 shows that several channels exist for the reaction of D_2^+ with D_2 and with H_2. Doverspike, Champion, and Bailey[2] found that the reactions of Ar^+ and N_2^+ with D_2 approached the spectator stripping model. However, some back scattering was observed, and it became greater as the collision energy was reduced. This back scattering was attributed in part to head-on collisions and in part to the decomposition of a complex. Studies of the same systems reported in Papers 24 and 25 show a departure from the stripping model at low energies, but the nature of the energy and angular distributions was not in accord with a complex mechanism and so was attributed to a variety of direct mechanisms.

Paper 26, in which the neutral was employed either in a collision chamber or as a molecular beam, gives a detailed development of the change of the reaction from a long-lived complex to a stripping mechanism as the collision energy is increased. It is of especial interest that the transition from complex to stripping mechanism occurs at just the internal energy of the product ion that corresponds to the ion's dissociation energy. Thus a different mechanism must predominate if the same products are to be formed at higher energies. The results are also interesting because of the persistence of the complex up to a barycentric collision energy of 4.4 eV. Paper 27 also shows the transition of the reaction

$$C_2H_4^+ + C_2H_4 \rightarrow C_3H_5^+ + CH_3$$

from complex to stripping mechanism with increasing energy. It also shows that at sufficiently high energy some $C_3H_3^+$ is formed either by direct decay of the excited $C_4H_8^+$ complex or by decomposition of the excited $C_3H_5^+$ ion.

In an attempt to explain the energy distributions of the products of the reactions of Ar^+ and N_2^+ with D_2, Herman et al. (Paper 25) proposed that the difference in polarization energy between reactants and products be taken into account. The results obtained with this polarization model agreed well with experiment. Paper 28 extends the polarization model to the prediction of angular distributions of product ions with good results.

REFERENCES

1. Cermák, V., and Z. Herman, *Nucleon.* **19**: 106 (1961).
2. Doverspike, I. D., R. L. Champion, and T. L. Bailey, *J. Chem. Phys.* **45**: 4385 (1966).

21

Copyright © 1962 by Zeitschrift für Naturforschung

Reprinted from *Z. Naturforsch.* **17a**:452–460 (1962)

Mass Spectrometric Observation of Electron and Proton Transfer Reactions between Positive Ions and Neutral Molecules*

By A. Henglein and G. A. Muccini

Radiation Research Laboratories, Mellon Institute, Pittsburgh, Pa., U.S.A.
and Hahn-Meitner-Institut für Kernforschung, Berlin-Wannsee

(Z. Naturforschg. 17 a, 452—460 [1962] ; eingegangen am 16. März 1962)

The method of Cermak and Herman has been applied to mass spectrometric studies of symmetrical electron and proton transfer processes. The characteristics of the ion source used have been investigated both experimentally and theoretically. A new type of ionization efficiency curve is obtained if the current of a secondary ion is plotted as a function of the voltage between ionization chamber and electron trap at constant low voltage between the filament and the chamber. Essentially complete discrimination of primary ions has been achieved.

Electron transfer occurs with rather low cross section in methane but increases with molecular size and with increasing unsaturation. Large cross sections were observed in sulfur and iodine containing compounds. Double charge transfer reactions such as

$$NO^{++} + NO \rightarrow NO + NO^{++}, \qquad Xe^{++} + Xe \rightarrow Xe + Xe^{++}$$

have also been observed. Proton transfer reactions have been observed in several simple molecules. Some experimental results are presented which indicate that proton transfer may occur via a complex (at low kinetic energies) or as a stripping process (at higher energies).

Fragment ions have also been observed in the secondary mass spectra of several compounds. While part of these may result from the scattering of primary fragment ions, in some cases additional processes have to be postulated such as hydride ion transfer and dissociative charge transfer from vibrationally excited ions.

A simple new method for mass spectrometric studies of the interactions between ions and neutral molecules has recently been described by Cermak and Herman [1]. The electron accelerating voltage between the filament and the ionization chamber of a conventional ion source is kept below the ionization potential of the gas. The electrons traverse the chamber without causing any ionization and are then further accelerated by an electric field between the ionization chamber and the electron trap. The primary ions are accelerated in the direction opposite to the electron beam by this field before entering the ionization chamber. These primary ions are not able to pass the slit system of the mass spectrometer because of a kinetic energy component perpendicular to the direction of analysis. However, secondary ions produced by collisions with gas molecules in the chamber can be extracted into the analyzing section of the instrument if they are formed with negligible amounts of kinetic energy. Cermak and Herman demonstrated this in studies of dissociative charge transfer reactions in cases in which the transfer of mass and therefore of kinetic energy is extremely small.

The methodology of Cermak and Herman has been applied in studies carried out with a Consolidated Electrodynamics Corporation Model 21-103 C mass spectrometer. The sensitivity of the instrument was increased by using a Model 31 Cary (Vibrating Reed) Electrometer for the measurement of the ion currents. Studies of the characteristics of the ion source showed that essentially complete discrimination between primary and secondary ions is obtained: As a result it has been possible to investigate a number of typical resonant charge transfer reactions. In addition, the mass spectra of secondary ions of several simple compounds have been studied. It has been found that these secondary mass spectra contain not only the parent ions (formed by resonant charge transfer) but also protonated molecules as well as ions of lower masses resulting from ion-molecule reactions. It seems noteworthy to emphasize that the high degree of discrimination of primary ions makes it possible to detect certain secondary ions which cannot be observed in the conventional operation of the ion source.

* This work was supported, in part, by the U.S. Atomic Energy Commission.

[1] V. Cermak and Z. Herman, Nucleonics **19**, No. 9, 106 [1961].

Characteristics of the ion source

a) Experimental

The normal operation of the ion source is demonstrated in Fig. 1 for methane (curve 1). The current of the parent ion is given as a function of the electron accelerating voltage at constant trap voltage. A small current which decreases rapidly with decreasing electron voltage can still be observed below the ionization potential of methane (13.0 volts). Between 11.0 and 13.0 volts this current is attributed to the energy spread of the electron beam. At 11.0 volts the slope of curve 1 changes discontinuously and at lower voltages becomes nearly independent on the electron accelerating voltage. Fig. 2 shows the dependence of the CH_4^+-current on the pressure in the gas inlet system. Proportionality exists if the ion source is operated in the conventional way, i. e. with incident electron energies above the ionization potential of the methane (curve 1). The current increases with the square of the pres-

sure if the electron accelerating voltage is kept below 11.0 volts (curve 2). In this range only secondary CH_4^+ ions which result from some interaction of primary ions formed between the chamber and trap with gas molecules in the chamber are observed.

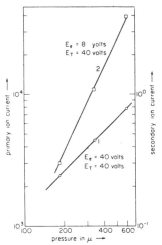

Fig. 2. Dependence of the CH_4^+ current on the pressure of methane in the gas inlet system (Curve 1: left ordinate scale, curve 2: right scale).

The formation of these secondary CH_4^+ ions is described in a more detailed manner by curve 3 in Fig. 1. The electron accelerating voltage E_e has been kept constant at 8.0 volts and the CH_4^+ current has been studied as a function of the trap voltage E_T Curve 3 represents an "ionization efficiency curve" for the secondary ion. The "appearance potential" here amounts to 5.0 volts. This corresponds exactly to the ionization potential of 13.0 volts of methane since the energy of the electrons is equal to

$$E_e + E_T = 13.0$$

when they reach the electron trap. It can therefore be concluded that the precursor of the secondary CH_4^+ ion is the primary CH_4^+ ion which transfers its charge in a collision with a methane molecule. "Secondary ionization efficiency curves" are therefore helpful in investigations of the nature of the primary ion. However the meaning of such secondary ionization efficiency curves is somewhat different from that obtained in more conventional ion sources. This will be discussed in detail in the following theoretical part.

The description of the characteristics of the ion source is completed by curve 2 in Fig. 1 where the

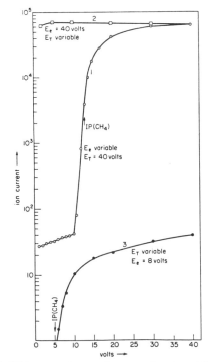

Fig. 1. CH_4^+ current from methane as a function of E_e or E_T, respectively. (E_e: voltage between filament and ionization chamber. E_T: voltage between ionization chamber and electron trap. Methane pressure in the gas inlet system: 600 μ. Repeller field: 3.8 volts/cm.)

ion current is plotted versus the trap voltage at constant accelerating voltage above the ionization potential of the gas. As it is well known from conventional operation the ion current is practically independent on E_T over a wide range.

b) Theoretical

The secondary ions cannot reach the collector if they have excessive kinetic energy either parallel to the long axis of the slits (i. e. in the direction of the primary ions) or perpendicular to this axis and to the direction of analysis. Only a beam within the divergence angles α and β (perpendicular to and in the plane of analysis, respectively) will pass through the whole slit system. The angle α is determined by the length l_1 of the exit slit of the ionization chamber and l_2 of the entrance slit of the collector system as well as the distance a between the two slits. The angle β is determined by the widths d_1 and d_2 of the exit slit of the ionization chamber and the exit slit of the ion accelerating system as well as their distance b. In the mass spectrometer employed here l_1, l_2 and a were 1.0, 1.26 and 40 cm, and d_1, d_2 and b were 0.15, 0.15 and 7.2 mm, respectively. The values of α and β are calculated to be equal to 0.056 and 0.0415 radians from these data. The maximum kinetic energy components U_α and U_β parallel and perpendicular to the direction of the primary ion beam which will allow analysis are given by

$$U_\alpha = \alpha^2 V, \qquad U_\beta = \tfrac{1}{4} \beta^2 V \qquad (1), (2)$$

where V is the ion accelerating high voltage of the ion source. In this work V was equal to 800 volts. U_α and U_β are found to amount to

$$U_\alpha = 2.5 \text{ eV}, \qquad U_\beta = 0.34 \text{ eV}. \qquad (3), (4)$$

Let x_0 be the distance between the ionization chamber and electron trap, x the distance between the chamber and a point between these two electrodes. The total kinetic energy of an electron which ionizes a molecule at this point is equal to

$$E_{tot} = E_e + U(x)$$

where U is the potential difference between the chamber and this point. If the field gradient between chamber and trap is linear $U = E_T x/x_0$. The primary ion formed at the distance x is accelerated by the potential U and enters the chamber with the kinetic energy $e U(x)$. At the appearance potential, AP, of the secondary ionization efficiency curve, all

ionizations take place immediately in front of the collector, i. e. $x = x_0$ and $E_{tot} = E_e + E_T$, and all primary ions entering the ionization chamber have the kinetic energy $e E_T$. However, at higher values of E_T ionization can occur between x_0 and a minimum distance x_1 which is given by the condition $E_e + U(x_1) = AP$. The primary ion beam therefore will have a distribution in kinetic energy between $e U(x_1)$ and $e E_T$. Since x_1 decreases with increasing E_T this distribution will become broader and broader.

The number of primary ions which are formed between x and $x + dx$ (or U and $U + dU$) and which will therefore obtain the kinetic energy $e U$ is equal to

$$N(U) \, dU \propto c \, \sigma(U) \, dx \qquad (5)$$

where c is the concentration of gas molecules in the ion sources and $\sigma(U)$ the cross section for ionization at the distance x, i. e. at total electron energy $E_e + U$. If $U = E_T (x/x_0)$

$$N(U) \, dU \propto c (x_0/E_T) \, \sigma(U) \, dU. \qquad (6)$$

The total number of primary ions formed between x_0 and x_1 will amount to

$$N_{tot} \propto c (x_0/E_T) \int_{AP-E_e}^{E_T} \sigma(U) \, dU, \qquad (7)$$

$\sigma(U)$ is easily derived from the conventional ionization efficiency curve. For example, curve 1 in Fig. 3 represents $\sigma(U)$ for CH_4^+ if E_e is equal to 8 volts.

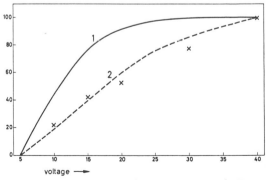

voltage ⟶

Fig. 3. Curve 1: Ionization efficiency curve of CH_4^+ (E_T constant at 40 volts. E_e variable. Abscissa: $E_e - 8$ volts. Curve 1 is normaziled at $E_e - 8 = 40$ volts). Curve 2: Total primary CH_4^+ current as function of E_T at $E_e =$ constant at 8 volts. (Curve 2 is calculated from curve 1 according to Eq. (7). Normalization of curve 2 at $E_T = 40$ volts). X: Observed secondary CH_4^+ current at various values of E_T (E_e constant at 8 volts. Normalization at $E_T = 40$ volts).

This curve is the measured primary ionization efficiency curve 1 in Fig. 1 (the scale of the abscissa is just shifted by 8 volts). Curve 2 in Fig. 3 is the integral

$$(1/E_T) \cdot \int_{AP-8}^{E_T} \sigma(U) \, dU$$

of curve 1 and represents the total primary CH_4^+ current at different voltages E_T. Both curves are normalized at 40 volts.

The number of secondary ions which are produced by reactions of the primary ions in the ionization chamber is proportional to

$$N'_{tot} \propto c \int_{AP-E_e}^{E_T} N(U) \, \sigma'(U) \, dU \qquad (8)$$

where $\sigma'(U)$ is the cross section of the ion-molecule reaction. By combining Eq. (6) and (8)

$$N'_{tot} \propto c^2 (x_0/E_T) \int_{AP-E_e}^{E_T} \sigma(U) \, \sigma'(U) \, dU \qquad (9)$$

is obtained. If σ' is independent of the kinetic energy of the primary ion, N'_{tot} becomes proportional to N_{tot}, i. e. the shape of the secondary ionization efficiency curve will be identical to that calculated from Eq. (7) (Fig. 3). In these considerations it has been assumed that all secondary ions reach the collector. However, if the secondary ions are formed with kinetic energies perpendicular to the direction of flight only a fraction, f, will be collected. As the kinetic energy of the primary ion increases f will decrease and N' will be described by the relation

$$N'_{tot}(E_T) \propto c^2 (x_0/E_T) \int_{AP-E_e}^{E_T} \sigma(U) \, \sigma'(U) \, f(U) \, dU \qquad (10)$$

Symmetrical charge transfer reactions

Symmetrical charge transfer processes have been studied by a number of authors [2-5]. These investigations have mainly been restricted to the noble gases. The reaction $H_2^+ + H_2 \rightarrow H_2 + H_2^+$ seems to be the only process studied in which molecular species are involved. The cross sections of such resonance processes are expected and have been found to be higher than gas collision cross sections. This arises because the resonance introduces a long range interaction

which would otherwise not occur. Relatively little variation of the cross section with the kinetic energy of the ion has been found. At energies above 200 eV in all cases the cross section observed falls very slowly and steadily as the relative kinetic energy of the collision partners increases. At lower kinetic energies, however, a small maximum has been observed in argon and in neon [6] and a very pronounced one in hydrogen [4]. These anomalies have been attributed to the occurrence of non-resonant processes in the noble gases due to the spin multiplicity of the lowest state of these ions. In the case of H_2 a side reaction in which a change of vibrational energy is involved has been assumed [4]. Scattering of ions in the case of exact resonance occurs primarily at small angles, the scattering intensity at 90° being practically zero [7]. It can therefore be assumed that all secondary parent ions formed in our ion source exclusively result from symmetrical charge transfer.

Fig. 3 contains a few points from curve 3 in Fig. 1. These points fit curve 2 in Fig. 3 fairly well. This curve is calculated on the assumption that both $f(U)$ and $\sigma'(U)$ in Eq. (10) are constant over the range from $5-40$ volts. The agreement indicates, as mentioned above, that the transfer of kinetic energy is very small and that the cross section of the observed process is not significantly dependent on the kinetic energy.

Fig. 4 shows the ionization efficiency curves for the secondary ions observed in methane and Figs. $5-7$ show similar data for some other simple molecules. The appearance potentials of the parent ions are always identical with those of the primary parent ions. All observed symmetrical charge transfer processes are listed in Table 1. The lowest cross section for transfer of a single electron has been found in methane. This reaction has been selected for reference in Table 1. In order to obtain relative cross sections, the ratio

$$\frac{\text{current of secondary ion at } E_e=8 \text{ and } E_T=40 \text{ volts}}{\text{current of primary ion at } E_e=40 \text{ and } E_T=40 \text{ volts}} \qquad (11)$$

has been measured relative to the similar ratio for methane. This procedure assumes that the primary ion current in the CERMAK–HERMAN operation of the ion source ($E_e=8$, $E_T=40$) is proportional to the

[2] H. S. W. MASSEY and E. H. S. BURHOP, Electronic and Ionic Impact Phenomena, Clarendon Press, Oxford 1952, p. 525.
[3] J. B. HASTED, Proc. Roy. Soc., Lond. A **205**, 421 [1951].
[4] H. B. GILBODY and J. B. HASTED, Proc. Roy. Soc., Lond. A **238**, 334 [1956].

[5] J. B. HASTED, Adv. in Electronics and Electron Phys. **13**, 1 [1960].
[6] A. ROSTAGNI, Nuovo Cim. **12**, 134 [1935].
[7] Reference [2], p. 443.

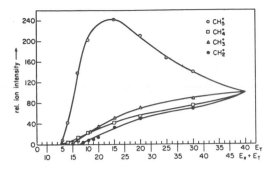

Fig. 4. Ionization efficiency curves of secondary ions in methane (E_e=constant at 8 volts. E_T variable. All curves normalized at E_T=40 volts).

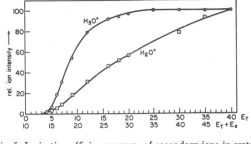

Fig. 5. Ionization efficiency curves of secondary ions in water (E_e=constant at 10 volts. E_T variable).

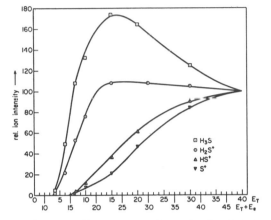

Fig. 6. Ionization efficiency curves of secondary ions in hydrogen sulfide (E_e=8 volts, E_T variable).

Fig. 7. Ionization efficiency curves of secondary ions in hydrogen chloride and ammonia (E_e=8 volts, E_T variable).

ion	relative cross section	ion	relative cross section
CH_4^{+a}	$(1.0)^a$	HCl^+	7.0
$C_2H_6^+$	3.3	NH_3^+	7.8
$C_2H_4^+$	10	H_2S^+	15
$C_2H_2^+$	14	CS_2^+	19
$c\text{-}C_6H_{12}^+$	5.0	I_2^+	27
$c\text{-}C_6H_{10}^+$	29	Ne^+	3.7
$C_6H_6^+$	13	Ar^+	9.3
NO^+	4.1	Kr^+	15
H_2O^+	4.2	Xe^+	23
CO_2^+	4.3		
Ar^{++}	1.1		
Kr^{++}	1.8		
Xe^{++}	3.0		
NO^{++}	0.3		

[a] Reference reaction: $CH_4^+ + CH_4 \rightarrow CH_4 + CH_4^+$

Table 1. Relative cross section of symmetrical charge transfer reactions.

ion current in the more conventional operation of the source (i. e. $E_e = 40$ volts). Since the primary ion current contained ions of kinetic energies between about 5 and 40 volts, the value of the cross section obtained is an average over this range. An additional complication arises since the primary molecular ions formed by electron impact will have various amounts of vibrational energy. The data obtained by the present method of measuring cross sections can be compared with literature values in the case of the noble gases. Our ratio of the cross sections in argon and neon amounts to $9.3/3.7 = 2.5$ which agrees with the ratio of $2.5 - 3.0$ calculated from the measurements of ROSTAGNI[6]. Absolute cross sections may be calculated from the data in Table 1 by using the known absolute cross section of the transfer process in argon (38×10^{-16} cm^2 at 20 eV).

The cross section depends significantly on the nature of the compound. In molecules of similar size (such as ethane, ethylene and acetylene) the cross section increases with increasing unsaturation. A similar increase is observed by going from cyclohexane to cyclohexene. High cross sections have been found in the sulfur containing compounds and in iodine. Table 1 also contains some examples of symmetrical double charge transfer. In these experiments, 110 volts were used instead of 40 to carry out the measurements and Eq. (11) adjusted accordingly. The process

$$NO^{++} + NO \rightarrow NO + NO^{++} \qquad (10)$$

was the only one found for the transfer of two charges in a molecular system.

Proton transfer reactions

The secondary mass spectra of some simple molecules are listed in Table 2. The ionization efficiency curves of these secondary ions are shown by Figs. 4 – 7. Ions of the form $H_{n+1}X^+$ from parent molecules H_nX have been observed in all cases. The secondary ionization efficiency curves of these ions begin at the same appearance potentials as those of the parent ions H_nX^+. Primary ions H_nX^+ must there-

Substance[a]	Ion	ionization potential (volts)[c]	relative intensity	
			primary spectrum[b]	secondary spectrum[b]
water	H_3O^+		2.4	29
	H_2O^+	12.61	100	100
	OH^+	~ 12.8	20	—
	O^+	13.61	0.7	—
hydrogen sulfide	H_3S^+		<0.04	2
	H_2S^+	10.47	100	100
	HS^+		38	4
	S^+	10.36	41	3
hydrogen chloride	H_2Cl^+		1.2	7
	HCl^+	12.90	100	100
	Cl^+	13.01	15	0.7
ammonia	NH_4^+		0.6	15
	NH_3^+	10.52 – 11.3	100	100
	NH_2^+		67	2
	NH^+		3.5	—
	N^+	14.54	7.8	—
methane	CH_5^+		2.8	20
	CH_4^+	13.1	100	100
	CH_3^+	9.9	78	120
	CH_2^+	11.9	12	6
	CH^+	11.13	5	—
	C^+		1	—
ethane	$C_2H_7^+$		not detectable	0.2
	$C_2H_6^+$	11.6	100	100
	$C_2H_5^+$	8.7	82	122
	$C_2H_4^+$	10.51	410	93
	$C_2H_3^+$		133	17
	$C_2H_2^+$	11.41	74	3
	C_2H^+		6	—
	C_2^+		0.8	—
ethylene	$C_2H_5^+$		not detectable	2
	$C_2H_4^+$	10.51	100	100
	$C_2H_3^+$		54	9
	$C_2H_2^+$	11.41	52	5
	C_2H^+		8	—
	C_2^+		1	—

[a] Pressure of the gas inlet system: 600 μ. Repeller field: 3.84 volts/cm.
[b] Primary spectra: $E_e = 40$ volts, $E_T = 40$ volts. Secondary spectra: $E_e = 8$ volts, $E_T = 40$ volts.
[c] Data taken from F. H. Field and J. L. Franklin, Electron Impact Phenomena, Academic Press Inc., New York 1957.

Table 2. Primary and secondary mass spectra of simple molecules.

fore be the precursors of the protonated species as well as the secondary parent ions.

$$H_nX^+ + H_nX \begin{cases} H_nX + H_nX^+ & (13) \\ H_{n-1}X + H_{n+1}X^+ & (14) \end{cases}$$

In order to compare the competing processes of electron and proton transfer the ratio of the currents of the secondary ions $H_{n+1}X^+$ and H_nX^+ is plotted in Fig. 8 versus the voltage E_T.

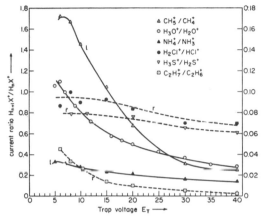

Fig. 8. Current ratio of protonated molecules and secondary parent ions as a function of E_T ($E_e = 8$ volts. l and r: left and rigth ordinate scale, respectively).

The general shape of the secondary ionization efficiency curves of the ions $H_{n+1}X^+$ differs markedly from that of the ions X_nX^+. A maximum at $10-15$ volts above the appearance potential can usually be observed (Figs. $4-7$). The decrease in the ratio $H_{n+1}X^+/H_nX^+$ in Fig. 8 indicates that electron transfer predominates more and more at higher kinetic energies. The shape of the $H_{n+1}X^+$ curves in Figs. $4-7$ may be explained by the inverse dependencies of σ, σ' and f on the kinetic energy eU of the primary ions [Eq. (10)]. The increase in $\sigma(U)$ at rather low kinetic energies determines the main features of the shape of the curve while the decrease in $\sigma'(U)$ and in $f(U)$ becomes predominant at higher kinetic energies. The decrease of σ' is well known from conventional studies on ion-molecule reactions [8, 9]. It may be described by the relation

$$\sigma' \propto U^a \qquad (15)$$

[8] D. P. STEVENSON and D. O. SCHISSLER, J. Chem. Phys. 29, 282 [1958].
[9] F. H. FIELD, J. L. FRANKLIN and F. W. LAMPE, J. Amer. Chem. Soc. 79, 2419 [1957].

over a certatin range of U. Values for a of -0.5 to -1.4 have been observed for different reactions [10].

The collection efficiency, $f(U)$, can no longer be assumed to be constant as in the electron transfer reactions since the transfer of the mass of the proton will be accompanied by the transfer of kinetic energy. This term is therefore expected to decrease above a certain value of U, but the decrease should depend strongly on the nature of the collision. The reaction may occur via an activated complex which dissociates into the final products after a lifetime much longer than the time of a molecular vibration. The existence of such complexes has been proven indirectly [9] and directly [11, 12] in several ion-molecule reactions. Complexes are probably preferentially formed at low kinetic energies. At higher kinetic energies the lifetime of the complex will become shorter than the time required for distribution of the excitation energy in the various degrees of freedom in the complex, i. e. there is practically no real complex formation. Reactions which are observed at higher kinetic energies are more likely to occur as stripping processes. Essentially the cross sections of such processes are not expected to exceed gas kinetic cross sections. In the case of complex formation the intermediate complex will move with half the original kinetic energy (eU) in the direction of the primary ion, the rest of the kinetic energy appearing as internal energy. The reaction product $H_{n+1}X^+$ will have the kinetic energy

$$\tfrac{1}{2}(eU)[(A+1)/(2A)] \sim \tfrac{1}{4}(eU) \qquad (16)$$

in the direction of the primary ion (where A is the mass of molecule H_nX). It will have an additional component of kinetic energy directed at random if part of the excitation energy of the complex and of the exothermicity of the reaction appears as kinetic energy of the final products. At values of U above 10 volts the maximum energy component U_a, at which collection is allowed, will have been reached. The collection efficiency is expected to fall significantly as the kinetic energy of the primary ion exceeds a few electron volts.

Where the secondary ion results from stripping of a proton from the primary ion the protonated molecule will be formed with the kinetic energy

[10] A. HENGLEIN, Z. Naturforschg. 17 a, 37 [1962].
[11] R. F. POTTIE and W. H. HAMILL, J. Phys. Chem. 63, 877 [1959].
[12] A. HENGLEIN, Z. Naturforschg. 17 a, 44 [1962].

$A^{-1}(A+1)^{-1}(eU)$ in the direction of the primary ion. This amount is much less than in the case of complex formation. There $f(U)$ is expected to depend only slightly on the kinetic energy in the range of $5-40$ volts. The ratio $H_{n+1}X^+/H_nX^+$ in Fig. 8 is only slightly dependent on E_T in the cases of hydrogen chloride, hydrogen sulfide and ammonia. If we can again assume that σ' and f of the electron transfer are nearly independent on energy it must be concluded that σ' and f of the proton transfer show the same behavior. The stripping model would therefore be more adequate to describe these reactions than the activated complex model (at least for kinetic energies above 5 eV). The very strong decrease in the current ratio CH_5^+/CH_4^+ in Fig. 8 indicates that this reaction occurs via a complex at low kinetic energies while a stripping reaction predominates at higher energies. This may also explain some observations of FIELD et al. [9] who studied the reaction $CH_4^+ + CH_4 \to CH_5^+ + CH_3$ by operating the ion source in the conventional way. They found the cross section to decrease at repeller field strengths between $10-100$ volts/cm but to become constant at higher field strengths.

The $C_2H_7^+$ ion which could not be detected in conventional studies on ion-molecule reactions in ethane [13] has been observed (Table 2, Fig. 8) with low intensity. Since the appearance potentials of $C_2H_6^+$, $C_2H_5^+$ and $C_2H_4^+$ from ethane no not differ very much, it is difficult to attribute the secondary $C_2H_7^+$ to one of these primary ions. A rough estimate shows that the cross section of the formation of $C_2H_7^+$ in ethane must be 100 times smaller than that of the proton transfer in water. This low cross section explains the failure to detect $C_2H_7^+$ in the conventional operation of an ion source since $C_2H_7^+$ is here masked by the C^{13} isotopic peak of the $C_2H_6^+$ ion.

Fragment ions in the secondary mass spectra

The secondary mass spectra in Table 2 contain a number of ions of lower mass numbers. Their secondary appearance potentials are identical with the known appearance potentials of these ions when formed by electron impact. It cannot be ruled out that primary ions are not scattered and pass through the slit system of the mass spectrometer. The collection efficiency of scattered ions is expected to be very small since most will have components of kinetic energy perpendicular to the direction of analysis. Furthermore, the scattering intensity at 90° is very low [14]. This would explain the rather low relative intensities of most of the fragment ions in Table 2. The table, however, contains a few examples which strongly indicate that there must be additional processes of formation of secondary fragment ions.

The ion CH_3^+ is the most abundant in the secondary mass spectrum of methane. Its intensity is even higher than that of CH_4^+ (the abundant ion in the primary mass spectrum). Furthermore, the secondary ionization efficiency curve of CH_3^+ always runs above that of CH_4^+ except for the immediate vicinity of the appearance potential of CH_3^+ (Fig. 4). This is in contrast to the behaviour of the primary ionization efficiency curves of these ions [15]. It must be concluded that CH_3^+ ions are formed by some ion-molecule reactions such as H^- transfer from methane

$$CH_3^+ + CH_4 \to CH_4 + CH_3^+ \qquad (17)$$

or dissociative electron transfer

$$CH_4^{+*} + CH_4 \to CH_4 + CH_3^+ + H \qquad (18)$$

If reaction (18) is initiated by a CH_4^+ ion in its ground state the energy deficit $D(CH_3^+ - H)$ has to be taken from the kinetic energy of the CH_4^+ ion. The cross section of this process would be very small since the collision occurs adiabatically in the energy range studied. However, if the CH_4^+ ion is formed with an amount of vibrational energy only about one-tenth of an electron volt smaller than $D(CH_3^+ - H)$ the rest of the energy deficit may easily be delivered by the kinetic energy. It is at present not possible to distinguish between reactions (17) and (18). Similarly the $C_2H_5^+$ ion occurs with abnormally high intensity in the secondary mass spectrum of ethane. It is therefore attributed to the analogous reactions

$$C_2H_5^+ + C_2H_6 \to C_2H_6 + C_2H_5^+ \qquad (19)$$

or $$C_2H_6^{+*} + C_2H_6 \to C_2H_6 + C_2H_5^+ + H. \qquad (20)$$

The OH^+ ion could not be detected in the secondary spectrum of water although its intensity as primary ion is very high. The absence of this ion as a

[13] F. W. LAMPE and F. H. FIELD, J. Amer. Chem. Soc. **81**, 3242 [1959].

[14] Reference [2], p. 496−497.

[15] A. HENGLEIN and G. A. MUCCINI, Z. Naturforschg. **15 a**, 584 [1960].

secondary ion would seem to corroborate the above ideas that the high intensities observed for CH_3^+ and $C_2H_5^+$ cannot be due to scattering of the primary ions. It also indicates that hydride ion transfer

$$OH^+ + H_2O \rightarrow H_2O + OH^+ \qquad (21)$$

does not occur. The ionization potential of OH seems to be slightly higher than that of water while that of CH_3 is much lower than that of methane (Table 2). In the case of water electron transfer

$$OH^+ + H_2O \rightarrow OH + H_2O^+ \qquad (22)$$

is expected to compete with reaction (21). Reaction (22) is of interest in considerations of the radiation chemistry of water. It explains the fact that there is no chemical evidence of OH^+ although the mass spectrum of water indicates that OH^+ is formed in high yield by high energy radiation. In the case of hydrogen chloride, the fragment Cl also has a slightly higher ionization potential than the molecule. The reaction $Cl^+ + HCl \rightarrow Cl + HCl^+$ may therefore partly be responsible for the very low relative intensity of Cl^+ in the secondary mass spectrum of hydrogen chloride.

Zum Stoßmechanismus bimolekularer Reaktionen

Teil II*):
Der Abstreifmechanismus der Reaktionen Ar+ + H2(D2) → ArH+ + H (ArD+ + D) bei Energien > 20 eV

Von *K. LACMANN* und *A. HENGLEIN*

Aus dem Hahn-Meitner-Institut für Kernforschung Berlin, Sektor Strahlenchemie, Berlin-Wannsee

(Eingegangen am 28. Dezember 1964)

Im Geschwindigkeitsspektrum der Ionen aus der Wechselwirkung (chemische Reaktion und Streuung) von Ar+-Ionen mit H_2 und D_2 wurde die Bande des ArH+- bzw. ArD+-Ions gefunden. Die Banden liegen praktisch an den Stellen, die nach dem Abstreifmodell zu erwarten sind. Hieraus wird geschlossen, daß das Ar+-Ion nicht völlig unelastisch mit der gesamten Wasserstoffmolekel zusammenstößt wie bei kleinen kinetischen Energien, bei denen die Voraussetzungen der Polarisationstheorie für Ion-Molekel-Reaktionen erfüllt sind; vielmehr tritt das Ar+-Ion im wesentlichen nur mit dem eingefangenen Wasserstoffatom in Wechselwirkung, ohne Impuls an das zweite Atom zu übertragen. Der Wirkungsquerschnitt nimmt stärker mit der kinetischen Energie im Schwerpunktsystem als nach der $1/\sqrt{E_s}$-Beziehung ab. Die Funktion $\sigma_{(E_s)}$ ist für die Übertragung eines H-Atoms und eines D-Atoms die gleiche. Die kinetischen Energien des einfallenden A+-Ions, oberhalb derer die Wirkungsquerschnitte der beiden Reaktionen gleich Null sind, unterscheiden sich um etwa den Faktor 2 und stimmen mit den nach dem Abstreifmodell errechneten überein. Auf einige kleine Abweichungen vom Abstreifmodell wird ebenfalls hingewiesen.

The bands of ArH+ and ArD+ ions have been observed in the velocity spectra of the ions resulting from the interaction (chemical reaction and scattering) of Ar+ ions with H_2 and D_2, respectively. The bands are practically located as it is expected from the stripping model. It is concluded that the Ar+ ion does not collide with the whole hydrogen molecule in a completely inelastic manner as it is the case at low kinetic energies, where the polarization theory of ion-molecule reactions is applicable. The Ar+ ion interacts only with the captured hydrogen atom without transfer of momentum to the second atom. The cross section stronger decreases with the relative kinetic energy of the collision partners than by the $1/\sqrt{E_s}$ law. The same function $\sigma_{(E_s)}$ was obtained for the transfer of an H and of a D atom. The critical energies of the incident Ar+ ions, above which the cross sections of the two reactions are zero, differ by a factor of about 2 and agree with the energies calculated from the stripping model. Some minor deviations from the stripping model are mentioned, too.

Theorie

a) Polarisationstheorie

Die Wirkungsquerschnitte der Ion-Molekel-Reaktionen

$$Ar^+ + H_2 \rightarrow ArH^+ + H \tag{1}$$

und

$$Ar^+ + D_2 \rightarrow ArD^+ + D \tag{2}$$

sind von Stevenson und Schissler [1] für verschiedene Energien des einfallenden Ar+-Ions im Bereich von etwa 0,3–5 eV gemessen worden. Sie sind bei kleinen kinetischen Energien von der Größenordnung

*) Teil I: A. Henglein, K. Lacmann und G. Jacobs, Ber. Bunsenges. physik. Chem. *69*, 279 (1965).

100 Å² [2], d.h. wesentlich größer als die gaskinetischen Stoßquerschnitte. Diese Erscheinung wird durch die Polarisationstheorie der Ion-Molekel-Reaktionen erklärt, nach der das Ion in der Molekel einen Dipol induziert, so daß es zu einer weitreichenden elektrostatischen Anziehung zwischen beiden kommt. Die gemessenen Wirkungsquerschnitte stimmten mit der nach der Formel

$$\sigma_p = 2\pi \frac{e}{V_0} \left(\frac{\alpha}{\mu}\right)^{1/2} = \pi e \left(\frac{2\alpha M}{\mu E_0}\right)^{1/2} = \pi e \left(\frac{2\alpha}{E_s}\right)^{1/2} \tag{3}$$

berechneten überein. Dabei bedeutet a die Polarisierbarkeit der Molekel, E_0, M, e und V_0 die kinetische Energie, Masse, Ladung und Geschwindigkeit des einfallenden Ions; mit μ und E_s werden die reduzierte Masse und kinetische Energie im Schwerpunktsystem der Stoß-partner bezeichnet $\left(E_s = \frac{1}{2}\mu V_0^2 = \frac{\mu}{M}E_0\right)$.

Zum Verständnis der folgenden Versuche und ihrer Diskussion erscheint es zweckmäßig, hier kurz auf die Grundlagen der Polarisationstheorie einzugehen [2]: Im Schwerpunktsystem bewegen sich die beiden Stoßpartner mit der Energie E_s aufeinander zu (im Laborsystem fällt das Ion mit der Energie E_0 ein, während die Molekel als praktisch ruhend betrachtet wird). Das Potential zwischen dem punktförmig gedachten Ion und dem induzierten Dipol der Molekel beträgt beim Abstand R der Stoßpartner:

$$E_{pot} = \frac{-\alpha e^2}{2R^4}. \tag{4}$$

Die anziehenden Kräfte auf die Stoßpartner sind immer zum Schwerpunkt des Systems gerichtet; unter dem Einfluß dieser Zentralkräfte beschreiben beide Teilchen eine rotationsartige Annäherungsbewegung (falls das Ion nicht zentral einfällt), für die der Drehimpuls zu jedem Zeitpunkt gleich groß ist. Fällt das Ion mit dem Stoßparamater b exzentrisch ein, beträgt der Drehimpuls:

$$L = \mu V_0 b. \tag{5}$$

Zu jedem Zeitpunkt der Annäherungsbewegung lassen sich die Geschwindigkeiten der Partner in je eine radiale Komponente (in Richtung auf den Schwerpunkt) und eine senkrecht hierzu charakterisieren. Die Summen der entsprechenden Energien stellen die kinetische Energie E_{kin}, mit der sich die Partner aufeinander zu bewegen, bzw. die Rotationsenergie E_{rot} des Systems dar. Alle diese Energien, deren Summe gleich ist der kinetischen Energie im Schwerpunktsystem

$$E_s = E_{pot} + E_{rot} + E_{kin} \tag{6}$$

hängen vom Abstand R der Partner ab. E_{rot} ist gleich $\frac{L^2}{2\mu R^2} = E_s \frac{b^2}{R^2}$. Das effektive radiale Potential

$$E_{eff} = E_{pot} + E_{rot} = -\frac{\alpha e^2}{2R^4} + \frac{E_s b^2}{R^2} \tag{7}$$

ist in Abb. 1 als Funktion von R bei konstantem Stoßparameter b für verschiedene Werte von E_s aufgetragen. Man erkennt, daß zur Annäherung der Stoßpartner bei einem exzentrischen Stoß ein Potentialwall, der durch das Zentrifugalpotential E_{rot} zustande kommt, zu überwinden ist. Die Höhe E'_{eff} des Walls und der Abstand R', bei dem er liegt,

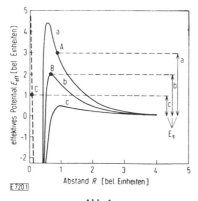

Abb. 1
Effektives Potential als Funktion des Abstands zwischen Ion und Molekel für verschiedene kinetische Energien im Schwerpunktsystem bei gleichem Stoßparameter

werden um so größer bzw. kleiner, je größer die Energie E_s des Systems ist. Aus $dE_{eff}/dR = 0$ erhält man

$$R' = \frac{e}{b}\left(\frac{\alpha}{E_s}\right)^{1/2} \tag{8}$$

und hieraus zusammen mit Gleichung (7) für die Höhe des Walls:

$$E'_{eff} = \frac{E_s^2 b^4}{2\alpha e^2}. \tag{9}$$

E_{kin} ist gleich der Differenz $E_s - E_{eff}$. Da der Potentialwall quadratisch mit E_s anwächst, ist er bei großem E_s höher als E_s selbst (Beispiel a), so daß sich die Teilchen nicht näherkommen können, als dem Abszissenwert des Punktes A in Abb. 1 entspricht (hier ist $E_{kin} = 0$). Dann beschreiben die Teilchen hyperbelähnliche Bahnen um einen gemeinsamen Schwerpunkt, ohne sich nahe genug zu kommen, um chemisch zu reagieren. Beispiel b in Abb. 1 beschreibt den Fall, in dem E_s gerade so groß ist wie der Potentialwall. Die Partikelchen nähern sich bis auf den Abstand R', d. h. den Abszissenwert des Punktes B in Abb. 1, an und rotieren nun auf Kreisbahnen. Bei kleinen Werten von E_s haben die Stoßpartner nach Überkommen des Potentialwalls noch radialen Impuls ($E_{kin} > 0$), so daß sie sich im Verlauf einer spiralartigen Rotationsbewegung weiter annähern (Beispiel c in Abb. 1). Bei sehr kleinen Abständen bestimmen anziehende Austauschkräfte und schließlich die abstoßenden Kräfte zwischen den Elektronenhüllen oder Atomkernen der Partner die potentielle Energie, welche bei kleinen Werten von R wieder ansteigt (gestrichelt in Abb. 1), weshalb es bei großer Annäherung (z. B. in Punkt C) zu einer rückläufigen Bewegung der Stoßpartner (oder ihrer Reaktionsprodukte) kommt. Wenn die Stoßpartner bei kleinem Abstand nach Überkommen des Walls reagieren, ist der Wirkungsquerschnitt der Reaktion gleich πb_{max}^2, wobei b_{max} denjenigen Stoßparameter bei gegebener Energie E_s bedeutet, bei der die Höhe des Potentialwalls gerade gleich E_s ist. Durch Gleichsetzen von E'_{eff} mit E_s in Gleichung (9) erhält man $b_{max}^2 = e\sqrt{\dfrac{2\alpha}{E_s}}$ und somit den Wirkungsquerschnitt von $\pi e\sqrt{\dfrac{2\alpha}{E_s}}$ nach Gleichung (3).

Die Polarisationstheorie enthält einige vereinfachende Annahmen: Der Wirkungsquerschnitt σ_p nach Gleichung (3) stellt einen Stoßwirkungsquerschnitt dar, der nur dann dem Wirkungsquerschnitt der chemischen Reaktion gleich ist, wenn diese bei jedem Stoß eintritt. Ist dies nicht der Fall, wird der Reaktionswirkungsquerschnitt:

$$\sigma = \eta \cdot \sigma_p, \tag{11}$$

wobei η ein Häufigkeitsfaktor ≤ 1 ist, der unter Umständen selbst von E_0 bzw. E_s abhängt. In dem von Stevenson und Schissler untersuchten Energiebereich ist η offenbar gleich 1,0. Das Ion darf nur dann als Punktladung betrachtet werden, wenn seine Dimensionen kleiner als der Stoßparameter b_{max} sind, was bei kleiner Ionenenergie weitgehend der Fall ist. Bei kleinen Werten von σ_p nach Gleichung (3) (bzw. großer Energie) wird das Potential nach Gleichung (4) nicht mehr allein die Annäherungsbewegung der Partner bestimmen, sondern chemische Bindungskräfte und abstoßende Kräfte zwischen den Elektronenhüllen maßgebend mitwirken. In dem extremen Fall, in dem der nach Gleichung (3) errechnete Wirkungsquerschnitt kleiner als der gaskinetische Stoßwirkungsquerschnitt wird, erscheint die Anwendung der Gleichung (3) überhaupt nicht mehr sinnvoll. Dies gilt für die Reaktion nach Gleichung (1) oberhalb $E_0 = 8$ eV, für die nach Gleichung (2) oberhalb 6 eV [3].

Zusammenstöße mit großem Wirkungsquerschnitt sind nach der Polarisationstheorie völlig unelastisch in bezug auf die Massen aller Atome in den Stoßpartnern. Welche Anteile der kinetischen Energie im Schwerpunktsystem den Produkten in Form von kinetischer bzw. von innerer Energie mitgegeben werden, ist nicht bekannt. Je nachdem, ob das eine oder andere Extrem vorliegt, sind die Grenzfälle des Komplexmodells, nämlich der elastische bzw. völlig unelastische Stoß mit Atomumlagerung (vgl. Teil I) verwirklicht. Bei einer Lebensdauer des Stoßkomplexes von mehr als der Zeit für eine Rotation sollte das Produkt-Ion im Geschwindigkeitsspektrum eine mehr oder weniger breite Bande hervorrufen, deren Maximum bei $\dfrac{M}{M + 2m} V_0$ liegt. Leider gelingt es gegenwärtig noch nicht, genügend energiehomogene Ar$^+$-Ionen-Strahlen von wenigen eV kinetischer Energie herzustellen, um das Geschwindigkeitsspektrum des Argoniums unter Reaktionsbedingungen aufzunehmen, unter denen die Voraussetzungen für die Anwendbarkeit der Polarisationstheorie erfüllt sind.

Nach Gleichung (3) haben die Reaktionen zwischen Ar$^+$ und H$_2$ bzw. D$_2$ denselben Wirkungsquerschnitt, wenn man sie bei der gleichen kinetischen Energie E_s im Schwerpunktsystem ablaufen läßt. Dies ist nicht nur bei Gültigkeit der Polarisationstheorie zu erwarten, sondern allgemein für jede Form der Potentialfunktion E_{pot}, solange diese für die Übertragung eines H- und D-Atoms denselben Verlauf hat. Denn immer hat E_{rot}, die zweite Komponente des effektiven Potentials, bei gleichem E_s denselben Wert. Zum Beispiel sollte der Wirkungsquerschnitt bei gleicher kinetischer Energie im Schwerpunktsystem auch dann denselben Wert für H- und D-Übertragung haben, wenn die Reaktion als Abstreifprozeß erfolgt, d. h. das einfallende Ar$^+$-Ion nur mit einem Atom der getroffenen Wasserstoffmolekel in Wechselwirkung tritt (als kinetische Energie im Schwerpunktsystem gilt dann allerdings $\dfrac{m}{M + m} E_0$).

b) Theoretische Geschwindigkeitsspektren, kritische Energien und Streuwinkel

Der maximale Streuwinkel, die Lage der Bande des Argoniums im Geschwindigkeitsspektrum und die kritische Energie E_c des einfallenden Ar$^+$-Ions sind in Tab. 1 für die verschiedenen Stoßmodelle zusammengestellt. In der Tabelle wird auch auf die Formeln im ersten Teil dieser Arbeit hingewiesen, nach denen die Berechnungen durchgeführt wurden. Die stärkste Streuung der Produkt-Ionen ist zu erwarten, wenn diese keine innere Energie mit sich führen (elastischer Stoß mit Atomumlagerung). Für die Reaktion Ar$^+$ + H$_2$ beträgt Θ_{max} unter diesen Umständen 0,038 Radiant. Weil die Winkel α und β, um die das Produkt-Ion

gerade noch gestreut werden darf, um nachgewiesen zu werden, nicht kleiner sind (vgl. Teil I), werden alle gebildeten ArH$^+$-Ionen unter unseren experimentellen Bedingungen selbst bei der stärksten Streuung, die auf Grund der Erhaltung des Impulses und der Energie denkbar ist, registriert. Wenn die Reaktion nach dem Abstreifmodell erfolgt, bewegen sich die Produkt-Ionen exakt in der Vorwärtsrichtung. Die Tabelle gibt auch Aufschluß über die maximalen Ablenkwinkel und die Breite der Bande, wenn die Reaktion prinzipiell zwar nach dem Abstreifmodell erfolgt, aber ein kleiner Energiebetrag von der Größe der Wärmetönung W in Form von kinetischer Energie beider Produkte frei wird. Dazu wurde ein Verhältnis $W/E_0 = 0{,}01$ angenommen. Die Wärmetönung der Reaktionen nach Gleichungen (1) und (2) schätzen wir auf 0,5 eV/ Elementarprozeß. (Das ArH$^+$-Ion dürfte isoelektronisch mit der HCl-Molekel sein. Der Reaktion Ar$^+$ + H$_2$ → ArH$^+$ + H entspricht dann die Reaktion Cl + H$_2$ → HCl + H. Weil D(H$_2$) und D(HCl) praktisch gleich groß sind, nämlich 4,5 eV [4], andererseits das Ar$^+$-Ion einen kleineren effektiven Radius als das Cl-Atom hat, wird die Reaktion zwischen Ar$^+$ und H$_2$ etwas exotherm sein, während die Reaktion Cl + H$_2$ praktisch thermoneutral ist. Die Wärmetönung von Ar$^+$ + H$_2$ ist dann gleich dem Überschuß der Bindungsenergie des ArH$^+$-Ions gegenüber der HCl-Molekel, der auf 0,5 eV geschätzt wird.) Mit $W = 0{,}5$ eV und $W/E = 0{,}01$ gelten die in Tab. 1 gegebenen Θ_{max}-Werte für $E_0 = 50$ eV; bei höheren Energien des einfallenden Ar$^+$ Ions sollte Θ_{max} noch kleiner werden. Im Falle der Reaktion Ar$^+$ + D$_2$ errechnet sich für Θ_{max} nach dem Grenzfall des elastischen Stoßes mit Atomumlagerung ein Wert, der die Winkel α und β übertrifft, so daß unter diesen Umständen nicht alle erzeugten ArD$^+$-Ionen in unserem Apparat nachgewiesen würden. Da aber, wie unten gezeigt werden wird, die Reaktion Ar$^+$ + H$_2$ (bei der auf jeden Fall alle Produkt-Ionen registriert werden) nach dem Abstreifmodell abläuft und wohl kaum ein anderes Stoßmodell für die Reaktion Ar$^+$ + D$_2$ anzunehmen ist, erscheint es sehr wahrscheinlich, daß auch alle gebildeten ArD$^+$-Ionen durch das Wiensche Filter bis zum Ionenauffänger gelangen.

Zur Errechnung der kritischen Energie des einfallenden Ions ist es notwendig, die Dissoziationsenergie des Argoniumions zu erkennen. D(Ar$^+$ — H) beträgt, wie oben abgeschätzt, 5,0 eV. Das Argoniumion hat aber eine niedriger liegende Dissoziationsgrenze D(Ar — H$^+$), und zwar ist die Differenz zwischen beiden gleich dem Unterschied der Ionisationsenergien des Argon- und Wasserstoffatoms. Diese Differenz beträgt 2,0 eV, so daß sich für D(Ar — H$^+$) ein Wert von 3,0 eV ergibt. Bei einer theoretischen Behandlung des Argoniumions ist kürzlich eine Dissoziationsenergie von 3,03 eV erhalten worden [5]. In einer früheren Mitteilung [6] haben wir diese Energie zur Errechnung der kritischen Energie des Ar$^+$-Ions benutzt. Jedoch glauben wir jetzt, daß es sinnvoller ist, D(Ar$^+$ — H) zu

Bd. 69, Nr. 4
1965

K. Lacmann u. A. Henglein: Zum Stoßmechanismus bimolekularer Reaktionen. II

289

verwenden, weil ja das Argoniumion aus $Ar^+ + H$ in dem Abstreifprozeß gebildet wird. Sollte die innere Energie des Argoniumions bei hohen Energien des einfallenden Ar^+-Ions 3,03 eV übertreffen, wird es prädissoziieren, sofern ausreichende Wahrscheinlichkeit für strahlungslose Übergänge zwischen den Potentialkurven für $(Ar^+ - H)$ und $(Ar - H^+)$ vorhanden ist. Sein geladenes Bruchstück wird sich mit derselben Lineargeschwindigkeit bewegen wie das ArH^+-Ion

selbst; weil unsere experimentelle Anordnung die Ionen hinsichtlich ihrer Geschwindigkeit und nicht ihrer Masse analysiert, wird das Bruchstück zur gleichen Stelle des Geschwindigkeitsspektrums gelangen wie ein stabiles Produkt-Ion. Den in Tab. 1 angegebenen Werten der kritischen Energie wurde deshalb eine Dissoziationsenergie des Argoniumions von 5,0 eV zugrunde gelegt.

Experimentelle Ergebnisse

Abb. 2 zeigt typische Geschwindigkeitsspektren für die Reaktionen des Ar^+-Ions mit H_2 bzw. D_2. Die relativ enge Bande bei der Geschwindigkeit V_0 rührt von Ar^+-Ionen her, die den Stoßraum ohne chemische Reaktion durchflogen haben. Diese Bande ist nach kleineren Geschwindigkeiten hin stark verbreitert, was auf die Streuung von Ar^+-Ionen an Wasserstoffmolekeln zurückzuführen ist. Aus diesem Untergrund hebt sich die Bande des ArH^+- bzw. ArD^+-Ions heraus, und zwar erscheint die ArH^+-Bande weitgehend symmetrisch um den Schwerpunkt bei 0,979 V_0, während das Zentrum der ArD^+-Bande bei 0,955 V_0 liegt. Beide Banden sind etwas breiter als die Bande des Primärions.

Ein Vergleich mit der Lage des Bandenschwerpunkts im theoretischen Geschwindigkeitsspektrum nach Tab. 1 lehrt, daß die Banden der Produkt-Ionen im Spektrum erscheinen, wie es auf Grund des Abstreifmodells zu erwarten war. In Tab. 2 ist die Geschwindigkeit des Bandenzentrums für verschiedene Energien des einfallenden Ar^+-Ions angegeben; ferner enthält die Tabelle die relative Abweichung in Prozent von der nach dem Abstreifmodell errechneten Bandenschwerpunktlage [angegeben wird hierbei $(V_e - V_t)/V_t \cdot 100$, wobei V_e und V_t die experimentell bzw. theoretisch bestimmten Geschwindigkeiten bedeuten]. Wie man erkennt, er-

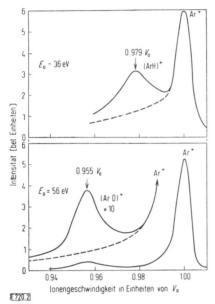

Abb. 2
Typische Geschwindigkeitsspektren für die Ionen aus der
Reaktion des Argonions mit H_2 und D_2

Tabelle 1
Theoretische Werte einiger charakteristischer Größen für verschiedene Stoßmodelle

Maximaler Streuwinkel, Lage der Argoniumion-Bande und kritische Energie des Ar^+-Ions	nach Gleichung in Teil I	Reaktion	
		$Ar^+ + H_2$ $\rightarrow ArH^+ + H$	$Ar^+ + D_2$ $\rightarrow ArD^+ + D$
Θ_{max} in Radiant	**Komplexmodell** Grenzfall: Völlig unelastischer Stoß mit Atomumlagerung		0	0
	Grenzfall: Elastischer Stoß mit Atomumlagerung	26	0,038	0,072
	Abstreifmodell			
	Grenzfall: Ideales Abstreifmodell		0	0
	Grenzfall: Freiwerden der Wärmetönung als kinetische Energie	34	0,016	0,022
Lage des Bandenzentrums im Geschwindigkeitsspektrum in Einheiten von V_0	**Komplexmodell**	24	0,9524	0,9091
	Grenzfall: Völlig unelastischer Stoß mit Atomumlagerung Grenzfall: Elastischer Stoß mit Atomumlagerung	27	0,915–0,989	0,843–0,975
	Abstreifmodell			
	Grenzfall: Ideales Abstreifmodell	29	0,9756	0,9524
	Grenzfall: Freiwerden der Wärmetönung als kinetische Energie	33	0,959–0,991	0,930–0,974
Kritische Energie E_c in eV	**Komplexmodell**	28	94	50
	Grenzfall: Völlig unelastischer Stoß mit Atomumlagerung Grenzfall: Elastischer Stoß mit Atomumlagerung		∞	∞
	Abstreifmodell			
	Grenzfall: Ideales Abstreifmodell	31	184	94

Tabelle 2

Geschwindigkeit des Bandenzentrums bei verschiedenen Energien und relative Abweichung vom theoretischen Wert (nach dem Abstreifmodell)
(ArH^+: $V_t = 0,9756\ V_0$. ArD^+: $V_t = 0,9524\ V_0$)

Energie des Ar^+ [eV]	ArH^+		ArD^+	
	V_e	$(V_e-V_t)/V_t \cdot 100$	V_e	$(V_e-V_t)/V_t \cdot 100$
	In Einheiten von V_0		In Einheiten von V_0	
22	0,9778	0,23	0,9603	0,83
41	0,9790	0,35	0,9565	0,43
50	0,9783	0,28	0,9562	0,40
60	0,9777	0,22	0,9562	0,40
70	0,9781	0,27	0,9586	0,65
75	0,9770	0,14	–	–
80	–	–	0,9608	0,86
84	–	–	0,9610	0,90
90	0,9770	0,14	–	–
110	0,9777	0,22	–	–
130	0,9790	0,35	–	–

V_t: theoretische Geschwindigkeit
V_e: experimentell gefundene Geschwindigkeit

scheint die Bande des Produkt-Ions nicht genau an der erwarteten Stelle, sondern ist um 0,2–0,8% der Geschwindigkeit nach höheren Geschwindigkeiten hin verschoben. Diese Abweichungen sind kaum größer als der Meßfehler, weshalb eine eingehende Diskussion dieses Details des Geschwindigkeitsspektrums nicht sinnvoll erscheint. Es sei aber erwähnt, daß die Spektren gut reproduzierbar waren, weshalb zumindest das Vorzeichen der Abweichung reell sein dürfte. In Teil III dieser Arbeit werden Reaktionen beschrieben werden, bei denen jene Abweichung viel ausgeprägter ist, weshalb wir die weitere Diskussion dort führen werden. Hier sei nochmals festgestellt, daß das Abstreifmodell in guter Näherung die Geschwindigkeitsspektren verstehen läßt; bei einem völlig unelastischen Stoß mit Atomumlagerung sollte die Bande des Produkt-Ions bei viel kleinerer Geschwindigkeit erscheinen. Da wegen der Gültigkeit der Polarisationstheorie anzunehmen ist, daß bei kleinen Ionenenergien ein unelastischer Stoßkomplex gebildet wird, muß im Bereich von einigen eV des einfallenden Ar^+-Ions ein Übergang vom Abstreif- zum Komplexmodell existieren; aus den bereits früher

erwähnten Gründen sind Messungen in diesem Bereich gegenwärtig noch nicht möglich.

Die Verbreiterung der Banden der Produkt-Ionen kann man nur so erklären, daß auch das neutrale Reaktionsprodukt in Bewegung gesetzt wird, d. h. das Abstreifmodell nicht exakt erfüllt ist. Als Maß der Verbreiterung mag die Differenz zwischen den Halbwertsbreiten der Banden des Produkt- und Ar^+-Ions gelten. In Abb. 3 ist diese Differenz in Prozenten der Geschwindigkeit V_0 als Funktion der Energie des einfallenden Ar^+-Ions aufgetragen. Wie man sieht, werden die Banden des ArH^+- bzw. ArD^+-Ions mit steigender Energie schmaler. Weil die Banden symmetrische Form haben und das Zentrum bei etwas höherer Geschwindigkeit als theoretisch erwartet erscheint, kann es sich bei der Impulsübertragung an das neutrale Reaktionsprodukt nicht um Energie handeln, die der kinetischen Energie des Ar^+-Ions entnommen wird, denn dann müßte die Bande als Ganzes bei kleineren Energien erscheinen. Daher ist anzunehmen, daß entweder die Wärmetönung der Reaktion in ungerichteter Weise als kinetische Energie frei wird oder daß innere Energie des Produkt-Ions (nach dem Abstreifmodell beträgt diese $E_0 \dfrac{m}{M+m}$) teilweise in kinetische Energie der beiden Produkte umgewandelt wird. Diese Details des Reaktionsmechanismus werden in Teil III an Hand der dort beschriebenen Reaktionen weiter besprochen werden, weil diese Effekte dort ausgeprägter sind.

Die Wirkungsquerschnitte der beiden Reaktionen sind in Abb. 4 in Abhängigkeit von der kinetischen Energie im Schwerpunktsystem dargestellt. In Übereinstimmung mit den im theoretischen Teil angestellten Überlegungen ist die Funktion $\sigma_{(E_a)}$ für beide Reaktionen die gleiche. Bei gleicher kinetischer Energie des ein-

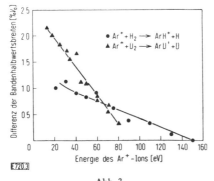

Abb. 3
Differenz der Halbwertsbreiten der Banden des sekundären und primären Ions als Funktion der Ar^+-Energie

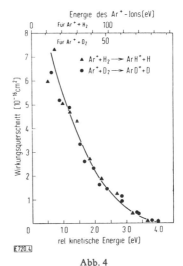

Abb. 4
Der Wirkungsquerschnitt als Funktion der kinetischen Energie im Schwerpunktsystem

Bd. 69, Nr. 4
1965

K. Lacmann u. A. Henglein: Zum Stoßmechanismus bimolekularer Reaktionen. II   291

fallenden Ar^+-Ions ist der Wirkungsquerschnitt zur D-Übertragung also immer kleiner als zur H-Übertragung. Dies stimmt überein mit früheren Untersuchungen über Reaktionen bei hoher Energie, bei denen ein Proton bzw. ein Deuteron vom einfallenden Ion auf den neutralen Reaktionspartner übertragen wird [7]. Der Wirkungsquerschnitt nimmt in Abb. 4 rasch ab, und zwar schneller als nach der Beziehung $\sigma \propto (E_s)^{-1/2}$, die bei kleinen Ionenenergien gilt [vgl. Gleichung (3)]. Oberhalb etwa 4 eV wird keine Reaktion mehr beobachtet. Die entsprechenden kritischen Energien von 164 eV (für ArH^+) und 84 eV (für ArD^+) stimmen recht gut mit den für das Abstreifmodell nach Tab. 1 errechneten überein.

Bei einer kinetischen Energie im Schwerpunktsystem von 1 eV (entsprechende Ar^+-Energie 41 bzw. 21 eV) ist der Wirkungsquerschnitt nach Abb. 4 etwa viermal kleiner als der gaskinetische Stoßwirkungsquerschnitt von $21 \cdot 10^{-16}$ cm². Nun wird man erwarten dürfen, daß eine Reaktion, die innerhalb einer Stoßzeit von der Größenordnung einer Atomschwingung eintreten muß, nicht bei jedem Stoß erfolgt. Wir möchten daher für den Reaktionswirkungsquerschnitt schreiben:

$$\sigma = \eta \cdot \sigma_K, \qquad (12)$$

wobei $\eta < 1$ die Wahrscheinlichkeit der Reaktion beim Stoß ist und σ_K den gaskinetischen Stoßwirkungsquerschnitt bedeutet, den man als annähernd konstant im untersuchten Energiebereich betrachten kann. Die Kurve in Abb. 4 gibt dann praktisch den funktionalen Zusammenhang zwischen dem Häufigkeitsfaktor η und der kinetischen Energie im Schwerpunktsystem wieder. Weil nicht-zentrale Zusammenstöße um so wahrscheinlicher sind, je größer der Stoßparameter ist, müssen ziemlich exzentrische Stöße – falls η wesentlich kleiner als 1 ist – den Hauptanteil an der Reaktion haben, denn andernfalls wären die relativ hohen Wirkungsquerschnitte von einigen 10^{-16} cm² nach Abb. 4 nicht zu verstehen. Die ArH^+- bzw. ArD^+-Ionen dürften daher bevorzugt in höheren Rotationsniveaus gebildet werden. Diese Vorstellungen stehen teilweise im Widerspruch zu einer Hypothese von Boelrijk und Hamill [3], wonach der Wirkungsquerschnitt einer Ion-Molekel-Reaktion bei hoher Energie gleich sein soll dem Produkt $p_K \sigma_K$, wobei p_K einen von der Energie unabhängigen Häufigkeitsfaktor bedeutet und auch σ_K (wie bei unserer Annahme) energieunabhängig ist. Nach diesen Autoren soll der Reaktionswirkungsquerschnitt also unabhängig von der Energie sein und oberhalb einer kritischen Energie sprunghaft auf Null fallen. Die Ergebnisse nach Abb. 4 zeigen, daß dies zumindest im

Fall der Reaktion des Ar^+-Ions mit H_2 und D_2 nicht zutrifft.

Giese und Maier [8] haben den Wirkungsquerschnitt der Reaktion $Ar^+ + D_2$ bis zu einer Ionenenergie von 25 eV gemessen und fanden größere Wirkungsquerschnitte, als nach der Polarisationstheorie [Gleichung (3)] zu erwarten waren. Unter ihren Versuchsbedingungen, unter denen die sekundären Ionen nachbeschleunigt wurden, gelangten aber nicht alle Produkt-Ionen zum Nachweis; sie führten deshalb einen Faktor ein, um für den Verlust an sekundären Ionen durch Streuung bei der Reaktion zu korrigieren. Zur Berechnung dieses Faktors war ein Stoßmodell anzunehmen, und zwar wählten jene Autoren das Modell eines unelastischen Zwischenkomplexes, durch dessen Zerfall den Produkten kinetische Energie isotrop im Schwerpunktsystem mitgegeben wird. Da aber für höhere Ionenenergien der Stoß weitgehend nach dem Abstreifmechanismus erfolgt (bevorzugte Vorwärtsstreuung), ist anzunehmen, daß jener Korrekturfaktor zu groß ausgefallen war und dementsprechend die von Giese und Maier bestimmten Wirkungsquerschnitte bei hohen Ionenenergien zu hoch sind. Homer et al. [9] berichteten kürzlich über wesentlich niedrigere Wirkungsquerschnitte der Reaktion $Ar^+ + D_2$. Die Wirkungsquerschnitte nach Abb. 4 sind um etwa 40% niedriger als die von Homer et al. angegebenen. Da die von uns gemessenen Wirkungsquerschnitte aus den in Teil I erwähnten Gründen eher zu niedrig als zu hoch sind, erscheint die Übereinstimmung mit den letztgenannten Autoren befriedigend.

Literatur

[1] D. P. Stevenson und D. O. Schissler, J. chem. Physics *29*, 282 (1958).

[2] Vgl. auch G. Gioumousis und D. P. Stevenson, J. chem. Physics *29*, 294 (1958).

[3] N. Boelrijk und W. H. Hamill, J. Amer. chem. Soc. *84*, 730 (1962).

[4] T. L. Cottrell, The Strengths of Chemical Bonds, Butterworth Scientific Publications, London 1958.

[5] T. F. Moran und L. Friedman, J. chem. Physics *40*, 860 (1964).

[6] A. Henglein und K. Lacmann, Institute of Petroleum/ASTM, Mass Spectrometric Symposium, Paris, Sep. 1964.

[7] A. Henglein und G. A. Muccini, Z. Naturforsch. *18 a*, 753 (1963).

[8] C. F. Giese und W. B. Maier II, J. chem. Physics *39*, 739 (1963).

[9] J. B. Homer, R. S. Lehrle, J. C. Robb und D. W. Thomas, Institute of Petroleum/ASTM, Mass Spectrometric Symposium, Paris 1964.

E 720

23

Reprinted from J. Chem. Phys. **46**:4718–4725 (1967)

Experimental Investigations of Ion–Molecule Reactions of D_2^+ with D_2 and H_2*

L. D. Doverspike and R. L. Champion

Department of Physics, University of Florida, Gainesville, Florida

INTRODUCTION

THE ion–molecule reactions

$$D_2^+ + D_2 \rightarrow D_3^+ + D \tag{1}$$

and

$$D_2^+ + H_2 \rightarrow D_2H^+ + H \tag{2a}$$

$$\rightarrow H_2D^+ + D \tag{2b}$$

have been studied in considerable detail. The purpose of these studies was to obtain detailed information concerning the kinematics and dynamics of the above reactions. Kinetic-energy distributions of the product ions have been measured for primary beam laboratory energies in the range 2–15 eV. Angular distributions for reaction products have been obtained at several collision energies above 6 eV. The apparatus and experimental procedure used in this investigation were identical to that employed in previous work performed in this laboratory,[1,2] and are described in Ref. 1.

Briefly, the experimental method consists of directing a mass-analyzed and velocity-selected ion beam into a collision chamber containing target gas at low pressure. The product ions are velocity analyzed with a 127° electrostatic velocity selector and mass-analyzed in a quadrupole field rf mass filter. The product-ion analysis and detection system rotates about the center of the scattering region, so that the angular distributions of the product ions can be obtained. The kinetic-energy distributions of the product ions provide detailed information concerning the kinematics of the reactions. Relative differential scattering cross sections are obtained from the product-ion angular distributions.

Several investigators using various experimental techniques have reported total-cross-section measurements for various isotopic forms of Reactions (1) and (2).[3–7] In some of the investigations attempts were made to determine the velocity distributions of the product ions for collision energies in the thermal energy range.[7–9] Also, various isotope effects in these reactions have been reported.[7,8,10,11] Total-cross-section measurements have established that reactions such as (1) and (2) have cross sections which are large ($\sim$80 Å²) at very low collision energies, but which fall rapidly to less than 1 Å² at primary-ion energies above approximately 15 eV. This total cross-section behavior restricted the present investigations to laboratory collision energies less than 15 eV while losses in primary-ion current limited the measurements to laboratory collision energies above 2 eV.

EXPERIMENTAL RESULTS

D_2^+–D_2 System

The kinetic-energy distributions of the product D_3^+ ions in reaction (1) (henceforth called the D_3^+ reaction) have been measured at 17 different primary-ion energies in the range 2–10 eV.[12] Four of the measured kinetic-energy distributions, which are typical of

* This research is supported by the U.S. Air Force Office of Scientific Research and by the National Aeronautics and Space Agency.

[1] R. L. Champion, L. D. Doverspike, and T. L. Bailey, J. Chem. Phys. **45**, 4377 (1966).
[2] L. D. Doverspike, R. L. Champion, and T. L. Bailey, J. Chem. Phys. **45**, 4385 (1966).

[3] H. Gutbier, Z. Naturforsch. **12a**, 499 (1957).
[4] D. P. Stevenson and D. O. Schissler, J. Chem. Phys. **29**, 282 (1958).
[5] C. F. Giese and W. B. Maier II, J. Chem. Phys. **39**, 739 (1963).
[6] D. W. Vance and T. L. Bailey, J. Chem. Phys. **44**, 486 (1966).
[7] B. G. Reuben and L. Friedman, J. Chem. Phys. **37**, 1636 (1962).
[8] V. L. Talroze, Izvest. Akad. Nauk SSSR Ser. Fiz. **24**, 1001 (1960).
[9] J. Durup and M. Durup, J. Chim. Phys. **64**, 386 (1967).
[10] C. F. Giese, Bull. Am. Phys. Soc. **9**, 189 (1964).
[11] J. H. Futrell and F. P. Abramson, in *Ion-Molecule Reactions in the Gas Phase* (American Chemical Society, Washington, D.C., 1966), Chap. 8.
[12] Unless otherwise stated, all kinetic energies refer to the laboratory coordinate system.

all the measurements, are shown in Figs. 1 and 2, along with the primary-ion kinetic-energy profiles. All these measurements were taken at a laboratory scattering angle of $\theta = 0°$. Two well-resolved maxima in the kinetic-energy profiles of D_3^+ can be seen in Figs. 2 and 1(b) along with the suggestion of an unresolved second maximum in Fig. 1(a). Two such resolved maxima were found at all primary-ion energies above 3 eV. An extensive target gas pressure study showed the intensities of both product-ion peaks varied linearly with the pressure at constant primary-ion intensity. The observed product-ion kinetic energies, E_3, corresponding

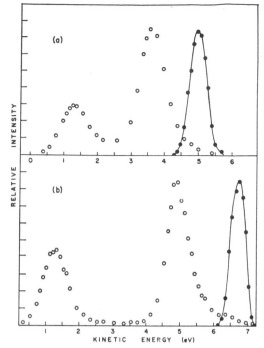

Fig. 2. D_3^+ product-ion kinetic-energy distribution (open circles) and primary-ion kinetic-energy distribution (solid circles). (a) $E_1 = 5.0$ eV. (b) $E_1 = 6.75$ eV.

is the laboratory scattering angle. The laboratory kinetic energies E_1 and E_3, determined from the peak values of the measured kinetic profiles, were used in Eq. (3) to calculate the Q's corresponding to the two maxima in the D_3^+ kinetic-energy distributions. These

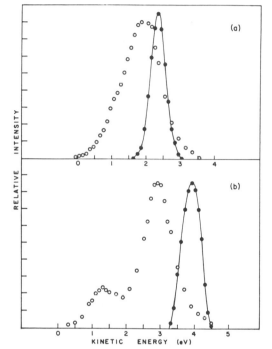

Fig. 1. D_3^+ product-ion kinetic-energy distribution (open circles) and primary-ion kinetic-energy distribution (solid circles). (a) $E_1 = 2.35$ eV. (b) $E_1 = 3.89$ eV.

to the peaks of the two maxima are presented in Fig. 3 as a function of primary beam energy, E_1.

Energy-momentum conservation can be used to show that the Q (the energy transformed from internal to translation) for reactions such as (1) and (2) is given by

$$Q = \left(1 + \frac{M_3}{M_4}\right) E_3 - \left(1 - \frac{M_1}{M_4}\right) E_1$$

$$- \frac{2}{M_4} (M_1 M_3 E_1 E_3)^{1/2} \cos\theta. \quad (3)$$

The subscripts 1–4 refer to the incident, target, detected, and unobserved particles, respectively, and θ

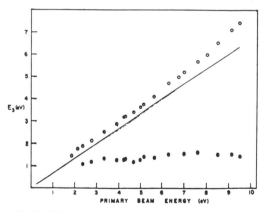

Fig. 3. D_3^+ product-ion kinetic energies (E_3) corresponding to intensity maxima observed in plots such as Figs. 1 and 2 for various collision energies, E_1. The open circles represent the higher-energy D_3^+ intensity maxima and the solid circles represent the lower-energy D_3^+ intensity maxima. The straight line represents $E_3 = (2/3) E_1$ which is the prediction of the pickup model.

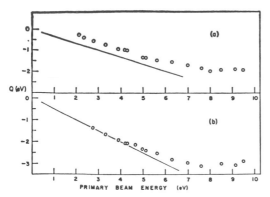

FIG. 4. Computed values of Q for the D_3^+ reaction, from Eq. (3) and each value of E_3, E_1 on Fig. 3. (a) Q vs E_1 for the higher-energy D_3^+ intensity maximum. The straight line represents $Q = -(1/3)E_1$ which is the prediction of the pickup model; (b) Q vs E_1 for the lower-energy D_3^+ intensity maximum. The straight line represents a completely inelastic collision, for which $Q = -E_1/2$.

results are plotted in Fig. 4 as a function of primary beam energy.

A double-peaked structure in the kinetic-energy distribution of the D_3^+ ions could arise from a reaction having a unique value of Q, if E_{3H} and E_{3L} (the high- and low-kinetic-energy maxima) correspond to c.m. scattering angles of $\chi = 0°$ and $\chi = 180°$, respectively. However, examination of the results displayed in Fig. 4 shows that this is not, in general, the case since for any given primary-ion energy E_1, $Q(E_1, E_{3H}) \neq Q(E_1, E_{3L})$. We therefore conclude that the two principal peaks in the D_3^+ KE distributions arise from two distinct reaction channels. Figure 4 also shows that the Q values depend markedly on collision energy and are negative throughout the collision energy range of this experiment. The Q's become more endothermic with increasing collision energy and finally reach endothermic limits at approximately 8-eV collision energy.

The energetic behavior of the high-energy group of D_3^+ is in fairly good agreement with the predictions of a simple pickup model, which has been discussed previously in connection with some earlier work on ion-molecule reactions by other investigators,[13] and in this laboratory.[2] This simple model assumes that the ncident ion picks up an atom from the target with negligible transfer of momentum to the residual target atom. Consequently, the product ion moves in the same direction as the incident ion with kinetic energy $E_3 = (M_1/M_3)E_1$ and the fraction $(M_3 - M_1)/M_3$ of the incident energy E_1 is converted into internal energy of the product ion. For the D_3^+ reaction this simple pickup model predicts that $E_3 = \frac{2}{3}E_1$ and $Q = -\frac{1}{3}E_1$. These straight lines are shown in Figs. 3 and 4(a) with the experimental values of E_3 and Q corresponding to the high-energy group of D_3^+. For primary-ion energies

[13] A. Henglein, Ref. 11, Chap. 5.

below approximately 6 eV, the slopes of the experimental curves are in rather good agreement with those predicted by the pickup model. However, significant deviation from this model is apparent at collision energies above 6 eV. This type of high-energy behavior, which will be discussed in more detail later, has been observed in the investigation of other ion–molecule reactions.[2]

If D_3^+ ions are formed by a completely inelastic process in which all the c.m. collision energy ($E_1/2$) appears as internal excitation energy of the reaction products, then the measured Q for such a collisional process is given by $Q = -E_1/2$. This straight line is shown in Fig. 4(b), along with the Q's calculated from the low-energy group of D_3^+. It is apparent that the low-energy group of D_3^+ corresponds to a completely inelastic collision at primary-ion energies below about 4.5 eV. The low-energy group of D_3^+ ions does not conform to any simple stripping or pickup model. Since a similar process has been observed in Reactions (2) (and tends to clarify the nature of the above reaction process), a detailed discussion of these results is deferred until the KE distributions of the D_2^+–H_2 system have been discussed.

D_2^+–H_2 System

Since it was found that D_3^+ formation proceeds by what appeared to be two distinct reaction channels, it was decided to investigate the isotopic reactions (2) (henceforth called the D_2H^+ reaction and H_2D^+ reaction) in the hope of gaining more information about the observed D_3^+ reaction channels.

Product-ion kinetic-energy distributions, which are typical of those found for the D_2H^+ and H_2D^+ reactions, are shown in Fig. 5. The kinetic-energy distribution of the D_2H^+ product ions in Fig. 5 shows a high-energy maximum with an indication of a small secondary maximum at a lower energy, but a low-energy group of D_2H^+ ions is notably absent. However, in addition to the D_2H^+ product ions, a low-energy group of H_2D^+ ions was observed. The energy resolution of the apparatus enabled the low-intensity peak of mass-4 product ions to be separated from the primary D_2^+ ions. The degree of separation between the mass-4 product and the primary ions can be seen in Fig. 5 where the sharp rise (at high energy) in the mass-4 profiles is due to primary D_2^+ ions. This primary-ion contamination does not vary appreciably as the target gas pressure is varied, while the intensity of the lower peak is found to be proportional to the target gas pressure. These low-energy mass-4 product ions are thought to be H_2D^+ ions. Justification of this assumption is given in the conclusions. Both product ions (D_2H^+, H_2D^+) were observed at all collision energies.

The experimental results for the D_2^+–H_2 system, when compared to those observed for the D_2^+–D_2 system, immediately suggest that the low-energy group of D_3^+ which was observed corresponds to deuteron transfer to the target molecule, rather than a deuterium atom transfer to the D_2^+ ion as postulated by the pickup model.

In Fig. 6, the observed kinetic energies corresponding to the high-energy maxima of the D_2H^+ profiles, and to the single maximum (although other maxima might be masked by the primary D_2^+ tail) of the H_2D^+ profiles, are plotted as functions of primary-ion energy. The curve $E_3 = \frac{4}{5}E_1$ in Fig. 6 is the result that would be expected if the D_2H^+ reactions followed the pickup model. The data conform to this model strikingly well except at the higher collision energies. The variation

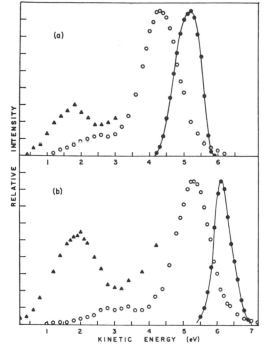

FIG. 5. D_2H^+ (open circles) and H_2D^+ (triangles) product-ion kinetic-energy distributions, and primary-ion kinetic-energy distribution (solid circles). (a) $E_1 = 5.15$ eV. (b) $E_1 = 6.2$ eV. The intensity scales for the H_2D^+ ions and the D_2H^+ ions are not exactly the same, the H_2D^+ intensity being somewhat exaggerated.

of Q (for D_2H^+ and H_2D^+ formation) with collision energy is given in Fig. 7 and its behavior is seen to be quite similar to that which was observed for the D_3^+ ions.

Since the pickup model satisfactorily describes the energetic behavior of the D_2H^+ reaction at moderate collision energies, one might expect the energetic behavior of the H_2D^+ reaction to be explainable in terms of a similar "stripping" process. In the formation of H_2D^+ this analogous process would be one in which a deuteron is transferred to the target molecule (H_2) with the residual D atom continuing on with energy $E_4 = E_1/2$. Under these conditions the KE distribution

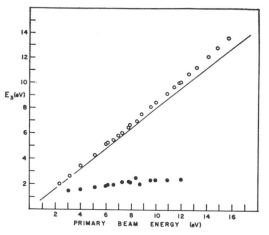

FIG. 6. Product-ion kinetic energies (E_3) corresponding to intensity maxima observed in plots such as Fig. 5 for various collision energies, E_1. The open circles represent the D_2H^+ product ion intensity maxima and the solid circles represent the H_2D^+ product-ion intensity maxima. The straight line represents $E_3 = (4/5) E_1$ which is the prediction of the pickup model.

of H_2D^+ should exhibit a maximum at $E_3 = E_1/4$ and the Q for this process should be given by $Q = -E_1/4$. Inspection of Fig. 6 shows that the H_2D^+ reaction does not follow this type of behavior over any extended range, nor does the low-energy group of D_3^+ ions observed from the D_3^+ reaction. Instead, at collision energies below about 5 eV, the H_2D^+ reaction proceeds via a completely inelastic process in which all the center-of-mass collision energy appears as internal excitation energy of the product. As in the D_3^+ reaction, the D_2H^+ and H_2D^+ reactions yield Q values which reach endothermic limits at the higher collision energies.

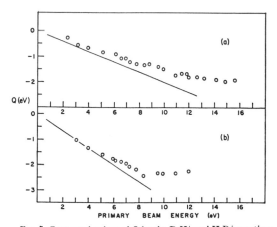

FIG. 7. Computed values of Q for the D_2H^+ and H_2D^+ reactions, from Eq. (3) and each value of E_3, E_1 on Fig. 6. (a) Q vs E_1 for the D_2H^+ reaction. The straight line represents $Q = -(1/5) E_1$ which is the prediction of the pickup model. (b) Q vs E_1 for the H_2D^+ reaction. The straight line represents a completely inelastic collision, for which $Q = -E_1/3$.

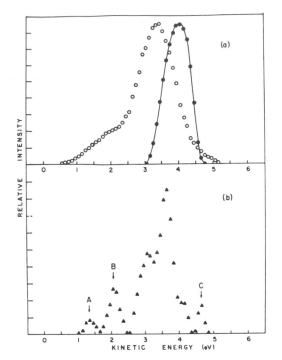

FIG. 8. (a) D_2H^+ product-ion kinetic-energy distribution (open circles), and primary-ion kinetic-energy distribution (solid circles) for $E_1 = 3.95$ eV. (b) Deconvoluted[14] results for the D_2H^+ profile in (a).

As mentioned before, there is an indication of secondary maxima occurring in the D_2H^+ profiles at energies below that of the main D_2H^+ peak. In Fig. 5, these secondary maxima are nearly resolved, but for the lower collision energies such as in Fig. 8(a) these secondary effects appear to be masked by the tailing of the main D_2H^+ peak. For curves such as Fig. 8(a) [and Fig. 1(a)], the product-ion KE profiles were deconvoluted.[14] This deconvolution process, the purpose of which is to correct for the finite spread in kinetic energy of the primary ions, permitted an estimate to be made concerning the location of any observed secondary maxima. Such a deconvoluted KE profile is shown in Fig. 8(b). If these deconvoluted results are indeed correct, then it can be seen in Fig. 8(b) that the D_2H^+ KE profile may possess fine structure which is unresolvable in the present apparatus. From the decon-

voluted result, it appears that the secondary maxima (at kinetic energies below that of the main peak) are due to two processes. The first, which is observed only at the lowest primary-ion energies, is one in which D_2H^+ is formed such that $Q = -E_{c.m.}$, i.e., a completely inelastic collision. This is denoted on Fig. 8(b) as Peak "B". In a second process, D_2H^+ appears to be formed with a resultant kinetic energy that yields a Q essentially the same as that associated with the D_2H^+ formed via the pickup model [Peak "A" on Fig. 8(b)], but with a kinetic energy, E_3, which indicates backscattering in the center of mass. In the pickup model the product which is forward-scattered in the c.m. leaves behind an H atom which is at rest in the lab system, and therefore has a velocity $-V_{c.m.}$ in the c.m. system. For the back-scattered D_2H^+ product, the H atom moves with a velocity $2V_{c.m.}$ in the lab system, or $+V_{c.m.}$ in the c.m. system. This type of behavior has been observed in ArD^+ and N_2D^+ formation[2] and, as in the present experiment, the relative cross section for this process is small compared to that for the forward-scattering pickup process.

Figure 8(a) also indicates that some D_2H^+ ions are formed with kinetic energies such that $Q > 0$. Such ions were seen only at very low collision energies and were observed for the D_3^+ reaction as well as the D_2H^+ reaction [e.g., see Fig. 1(a)]. An exothermic peak is indicated by "C" in Fig. 8(b). It should also be noted that the deconvolution process tends to divide the main trunk of the D_2H^+ distribution into two separate features. This behavior is not clearly understood. The centroid of the main deconvoluted peak [Fig. 8(b)] is essentially the same as the energy, E_3, assigned to the main peak of the unaltered data [Fig. 8(a)]. Another interesting feature of the deconvolution process is the appearance of a shoulder in the product-ion KE profile, occurring at the primary beam energy. This is due to some primary ions being transmitted through the rf mass filter even when it is tuned to pass the mass of the product ions.

ANGULAR MEASUREMENTS

For the three reactions investigated in these experiments, the D_3^+, D_2H^+, and H_2D^+ product ions are constrained by kinematics to have maximum laboratory scattering angles. However, all these maximum scattering angles are large in comparison with the angular resolution of the apparatus. For example, D_2H^+ has a maximum scattering angle of 26.5°, assuming $Q = 0$.

The angular distributions of the high-energy groups of D_3^+ and D_2H^+ were measured at primary beam energies between 7 and 11 eV. Some results obtained for the D_2H^+ reaction are shown in Fig. 9. No significant differences were found in the angular distributions of D_3^+ and D_2H^+, consequently the results in Fig. 9 are representative of both reactions. The angular distributions were deconvoluted[14] and the results of these

---

[14] This deconvolution process has been developed by G. E. Ioup of this laboratory. Details of the method will be published in a separate article in which the merits of the process will be discussed. It is to be emphasized that the conclusions concerning the present experiment would not be appreciably altered if this deconvolution were not employed. However, it is believed that the deconvolution results are of such interest that they should not be omitted from the present paper. As for the physical validity of the deconvolution process, the results which it has given so far are quite plausible, and furthermore the deconvolution of kinetic-energy profiles, other than Fig. 8(a), have yielded results which are physically consistent with those shown in Fig. 8(b).

deconvolutions are shown as dashed curves in Fig. 9. In this case, the deconvolution process corrects for the finite angular divergence of the primary beam and the finite acceptance angle of the product-ion detection system, but it does not correct for the angular broadening of the products which is caused by the thermal motion of the target gas. In the present experiment, angular broadening due to the thermal motion of the target gas can be quite large.

The energetic studies have shown that the high-energy group of D_3^+ and the D_2H^+ reaction proceed essentially via a simple pickup process, which assumes that the product ion continues in the same direction as the incident ion. The angular results do show that the product ions are much more forward collimated than is required kinematically, and are reasonably consistent with the assumptions of the pickup process. The angular distributions exhibit none of the basic features which are predicted by either the reactive glory or isotropic "hard-sphere" reaction models.

It was found that the angular profile of the primary beam became very broad for energies below 6 eV and, for this reason, meaningful angular measurements for the D_3^+ and D_2H^+ reactions could only be made above this energy. Attempts to obtain angular distributions for H_2D^+ or for the low-energy group of D_3^+ were unsuccessful because of relatively small product-ion intensities, and also because these product ions have very small kinetic energies and may diverge appreciably due to space-charge effects.

CONCLUSIONS

The experimental results indicate that many reaction channels exist for Reactions (1) and (2), and that the relative cross sections for these reaction channels depend strongly on collision energy. The relative importance of each reaction channel undoubtedly depends on the orientations of the colliding molecules, as well as on their relative velocities. It may well be that all energetically possible endothermic and exothermic reaction channels proceed with finite cross sections, some channels—those observed in this experiment—having much larger cross sections than others. In all cases where the product-ion kinetic-energy distributions exhibited two well-resolved maxima, E_{3L} and E_{3H}, some product ions of the same species were always found at energies E_3 such that $E_{3L} < E_3 < E_{3H}$. This may be due in part to the finite-energy spread of the primary ions, but it may also be true that reactive collisions involving a unique relative velocity would yield product ions with an energy distribution that would span a large range in energy. Any conclusions based on the experimental results are further complicated by the fact that the population distribution over the internal states of the primary ions is not known. This population distribution remains fixed, however, since the ion-source conditions are not altered as the collision energy is

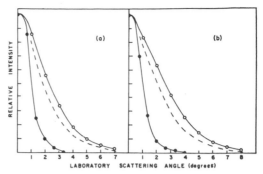

FIG. 9. Observed angular distributions for the D_2H^+ reaction (open circles) and primary beam angular profiles (solid circles). The dashed lines are the results of deconvolution. (a) $E_1 = 11$ eV. (b) $E_1 = 7$ eV.

varied. (The maximum electron energy for D_2^+ production is maintained at 55 eV.) Any population distribution of the primary ions predicted by theoretical calculations which take into account only ionization by electron impact could not be expected to yield too accurate an estimate of the actual distribution in the collision region, since such processes as charge transfer in the ion source and surface collisions might appreciably alter the population distribution due to electron impact alone. In spite of these complicating factors, a number of general remarks may be made as follows:

(A) The most probable reaction channels for both the D_3^+ and D_2H^+ reactions obey the predictions of a simple pickup model for which $E_3 = (M_1/M_3)E_1$ and $Q = (M_1 - M_3/M_3)E_1$. The data fit this model from the lowest laboratory collision energies (2 eV) up to moderate collision energies where the highly excited D_3^+ or D_2H^+ ions "prefer" not to store any more internal energy. Both the D_3^+ and D_2H^+ reactions begin to deviate from the pickup-model prediction at c.m. collision energies of about 4 eV. Above this collision energy the reactions proceed in such a way that the Q's do not appear to become more endothermic. Remarks concerning the values of these observed endothermic limits are made later.

(B) The H_2D^+ reaction (deuteron transfer to the target H_2) was observed, and this reaction becomes more endothermic as the collision energy is increased until an endothermic limit is attained. The results obtained for the H_2D^+ reaction are strikingly similar to those observed for the low-energy group of D_3^+, which strongly suggests that in the latter case the low-energy group of D_3^+ ions also arises from D^+ transfer. No simple collision model seems to describe these reactions adequately. Justification of the identification of mass-4 product ions as H_2D^+ rather than D_2^+ is, of course, necessary since no distinction can be made between the two ions in the present apparatus. It is possible that the mass-4 product ions are low-energy D_2^+ ions produced by inelastic collisions which leave the projectile ion and/or the target molecule in an excited state. However, this seems highly unlikely, since: (1) no low-energy D_2^+ ions were observed in $D_2^+ + D_2$ collisions; (2) the cross section for collisional excitation processes should be small at the low collision energies employed in the present experiment and furthermore the relative cross section should not decrease with increasing collision energy for $2 < E_1 < 15$ eV; (3) the kinematic behavior of the mass-4 product ions for the $D_2^+ + H_2$ reactants is consistent with that observed for the low-energy group of the mass-6 product ions for the $D_2^+ + D_2$ reactants.

(C) At low collision energies (≤ 2 eV in the c.m.), all three product ions are observed with intensity maxima at energies which give $Q = -E_{c.m.}$, corresponding to a completely inelastic collision. It is possible that these channels are due to formation of the excited complex ions $(D_4^+)^*$ and $(D_2H_2^+)^*$, which then dissociate into the observed reaction products, the relative velocity of separation of the dissociated products being fairly small. It should be emphasized, however, that this experiment neither proves nor disproves the existence of a long-lived ($\sim 10^{-12}$ sec) collision complex; rather, it suggests the possibility of its existence at low collision energies.

(D) For the D_3^+ and D_2H^+ reactions, groups of product ions were observed which had, at a given collision energy, the same Q's as those observed for the pickup process discussed in (A), but which correspond to back-scattering of D_3^+ and D_2H^+ in the c.m. system. The cross sections for this process are small with respect to the cross section for the pickup process.

(E) The dominant processes in Reactions (1) and (2) become more endothermic with increasing collision energy and the observed Q for each process finally attains an endothermic limit. This limiting behavior of Q versus collision energy is to be expected, since there is an upper limit to the amount of excitation energy the product ions can acquire without dissociating. Similar behavior has been observed in the ion–molecule reactions[2] $Ar^+ + D_2 \rightarrow ArD^+ + D$ and $N_2^+ + D_2 \rightarrow N_2D^+ + D$. In these cases, it was possible to establish that the N_2D^+ and ArD^+ product ions produced in the most endothermic reactions possessed excitation energies which were equal to the lowest ground-state dissociation energies of the product ions.

From the usual definition of Q, as the difference in the internal energies of the reactants and the products ($Q = U_1 + U_2 - U_3 - U_4$), and the assumption that the target molecule is in the ground state, one can readily show that for the D_3^+ reaction

$$Q = D(D_3^+) + U_1 - U_3 - D(D_2^+), \quad (4)$$

where U_1 is the excitation energy of the incident D_2^+; $D(D_2^+)$ is the dissociation energy of D_2^+ which is 2.69 eV; U_3 is the excitation energy of D_3^+; and $D(D_3^+)$

is the dissociation energy of D_3^+, with respect to dissociation into a ground-state D_2 molecule and a D^+ ion.

The observed endothermic limit of the low-energy group of D_3^+ in Reaction (1) is $Q = -3$ eV and when substituted into Eq. (4) gives

$$U_3(\text{max}) = D(D_3^+) + U_1 + 0.3 \text{ eV} \quad (5)$$

as the maximum observed excitation energy of the D_3^+ product ions. In the case of H_2D^+ formation

$$Q = D(H_2D^+) + U_1 - U_3 + D(HD) - D(D_2) - D(H_2^+),$$

$$(6)$$

where $D(H_2D^+)$ corresponds to dissociation into H^+ and ground-state HD and the other terms have their usual meaning. If the observed endothermic maximum for the H_2D^+ reaction ($Q = -2.5$ eV) is used in Eq. (6) along with the known values of the dissociation energies, then

$$U_3(\text{max}) \simeq D(H_2D^+) + U_1. \quad (7)$$

The results of Eqs. (5) and (7) clearly establish that both the H_2D^+ and the low-energy group of D_3^+ product ions possess (at the endothermic limit) sufficient excitation energy for the product ions to dissociate along the channels $D_3^+ \rightarrow D_2 + D^+$ or $H_2D^+ \rightarrow HD + H^+$.

The observed endothermic limit of the D_2H^+ reaction and the high-energy group of D_3^+ in reaction (1) is $Q \simeq -2$ eV. The same arguments when applied to these two reactions give

$$U_3(\text{max}) \simeq D(D_2H^+) + U_1 - 0.5 \text{ eV} \quad (8)$$

and

$$U_3(\text{max}) \simeq D(D_3^+) + U_1 - 0.7 \text{ eV} \quad (9)$$

as the maximum observed excitation energies of the D_2H^+ and the high-energy group of D_3^+ product ions, respectively. From the results expressed in Eqs. (8) and (9), it is not possible to definitely conclude that high-energy group of D_3^+ and D_2H^+ possess (at their endothermic limits) excitation energies which are sufficient for dissociation. If it is assumed that the primary ions which react are predominantly in the ground state (i.e., $U_1 = 0$), then the results indicate that the cross sections corresponding to reaction channels which give product ions with U_3 equal to the lowest ground-state dissociation energy of the product ions have small cross sections. It is also possible that the maximum excitation energies, $U_3(\text{max})$, observed in the two processes are equal to the ground-state dissociation energies of the product ions. This assumption, together with Eqs. (8) and (9), means that $U_1 \neq 0$, which implies that, at the endothermic limit,

228

the most likely reaction channels involve excited D_2^+ ions. On the basis of calculated Franck–Condon factors for D_2^+ production,[15] it is possible that an appreciable fraction (perhaps even a majority) of the primary D_2^+ ions are formed in vibrational levels such that $U_1 \geq 0.5$ eV. Just how the cross sections of such reactions depend upon the excitation state of the incident ion is unknown.

(F) According to the preceding discussion, the D_3^+, D_2H^+, and H_2D^+ product ions may be formed with internal energies in excess of their dissociation limits. Such ions can undergo "post-reaction dissociation," giving rise to a proton or deuteron and the remaining diatomic molecule. For example, if $(D_2H^+)^*$ is formed in Reaction (2a), with kinetic energy E_3^*, then the observed kinetic energy of a deuteron produced in dissociation of $(D_2H^+)^*$ should be roughly twice that for a proton. In fact, if the $(D_2H^+)^*$ dissociates isotropically in the c.m. system, D^+ and H^+ KE distributions should be generated which are symmetric (in the laboratory system) about $\frac{2}{3}E_3^*$ and $\frac{1}{3}E_3^*$ for D^+ and H^+, respectively. In this case, E_3^* is given by Eq. (3) with $Q \simeq -3$ eV, and may be approximated by $E_3^* \simeq E_1 - 3$ eV. Protons and deuterons have been observed for the $D_2^+ + H_2$ reactants and have intensity maxima in their KE distributions which agree quite well with that predicted by the above considerations. D^+ ions have been observed for the $D_2^+ + D_2$ reactants with D^+ kinetic energies around $\frac{1}{3}E_3^*$. Product ions which are thought to arise from post-reaction dissociation have been energy-analyzed for primary beam energies between 7 and 14 eV.

It may be argued that the observed D^+ and H^+ can

arise from reactions other than post-reaction dissociation, e.g., dissociative charge transfer to give H^+ for the $D_2^+ + H_2$ reactants and D^+ for the $D_2^+ + D_2$ reactants. However, energy distributions for D^+ and H^+ from dissociative charge transfer should not exhibit the observed dependence on E_1. Another process which yields D^+ ions is collision-induced dissociation, in which case the D^+ ions have energies $E_3 \simeq \frac{1}{2}E_1$.[1] Ions due to this process were observed in these experiments but were well removed in energy from those which are believed to be due to post-reaction dissociation. For the $D_2^+ + H_2$ reactants, it is necessary to justify the assumption that the observed mass-2 product ion is D^+ and not H_2^+ which arises from charge transfer. At all collision energies, large intensities of mass 2 were observed which, within the accuracy of the apparatus, had nearly thermal energies. This intensity maximum was attributed to charge transfer. Likewise, for the $D_2^+ + D_2$ reactants, very slow D_2^+ was observed.

Although it is not possible in the present apparatus to assign a value to the cross section associated with the post-reaction dissociation process, these results certainly indicate that for some charge-transfer measurements (where the total cross section is measured by collecting slow reaction products) great care should be taken to ensure that this kind of "contamination" process does not introduce sizable errors.

ACKNOWLEDGMENTS

The authors are very grateful to Professor T. L. Bailey for many valuable discussions as well as his extensive reading and criticizing the manuscript. We also wish to thank G. E. Ioup for his work and discussions regarding the deconvolution process.

---

[15] G. H. Dunn, J. Chem. Phys. **44**, 2592 (1966).

24

Reprinted from *J. Chem. Phys.* **49**:3058–3070 (1968)

Dynamics of the Reaction of N_2^+ with H_2, D_2, and HD

W. R. Gentry,* E. A. Gislason, B. H. Mahan, and Chi-Wing Tsao

*Inorganic Materials Research Division of the Lawrence Radiation Laboratory and Department of Chemistry
University of California, Berkeley, California*

(Received 12 April 1968)

Product-velocity-vector distributions have been determined for the reactive and inelastic scattering of N_2^+ by H_2, D_2, and HD. These distributions show that the reaction proceeds by a direct short-lived interaction rather than by a long-lived collision complex. Most products are scattered in the original direction of the N_2^+ projectile at a speed somewhat greater than calculated from the ideal stripping model. The internal excitation of N_2D^+ and N_2H^+ is very high and decreases somewhat with increasing scattering angle. For HD there is an isotope effect that favors N_2H^+ by large factors at small scattering angles, and N_2D^+ by smaller factors at large angles. The N_2^+ scattered from D_2 shows very little elastic component, but does reveal an inelastic process which is probably the collisional dissociation of D_2.

In an earlier paper[1] we reported measurements of the energy and angular distribution of products from the reaction of N_2^+ with H_2, D_2, and HD. Using intensity contour maps that show the complete product-velocity-vector distribution, we were able to demonstrate that the most probable reactive process is one in which the N_2H^+ or N_2D^+ is scattered forward, that is, in the original direction of the N_2^+ projectile. Over a considerable range of projectile energies, the most probable velocity of the forward scattered product is close to, but slightly greater than, that predicted by using the ideal stripping model for the reaction, as has been noted by other investigators.[2–5] However, we also observed product scattered through angles as large as 180° in the center-of-mass system, and this large-angle scattering became relatively more important as the projectile energy was increased. Thus, in addition to the stripping process, "rebound" reactive scattering occurs. In this paper, we report further observations of the reactive scattering and the first description of the nonreactive scattering of N_2^+ by isotopic hydrogen molecules.

EXPERIMENTAL

The instrument used in this work[6] consists of a magnetic mass spectrometer for preparation of a collimated beam of primary ions of known energy, a scattering cell to contain the target gas, and an ion-detection train made up of an electrostatic energy analyzer, a quadrupole mass filter, and an ion counter. These major components are described below in more detail.

PRIMARY-ION SPECTROMETER

Ions were formed in an electron bombardment source of the type described by Carlson and Magnuson.[7] In this source electrons from a tungsten filament

* Present address: Department of Chemistry, Massachusetts Institute of Technology, Cambridge, Mass.

[1] W. R. Gentry, E. A. Gislason, Y. Lee, B. H. Mahan, and C. Tsao, Disucssions Faraday Soc. **44**, 137 (1967).

[2] K. Lacmann and A. Henglein, Ber. Bunsenges Physik. Chem. **69**, 279 (1965).

[3] B. R. Turner, M. A. Fineman, and R. F. Stebbings, J. Chem. Phys. **42**, 4088 (1965).

[4] L. Doverspike, R. Champion, and T. Bailey, J. Chem. Phys. **45**, 4385 (1966).

[5] Z. Herman, J. Kerstetter, T. Rose, and R. Wolfgang, J Chem. Phys. **46**, 2844 (1967); Discussions Faraday Soc. **44**, 123 (1967).

[6] W. R. Gentry, University of California Lawrence Radiation Laboratory Rept. UCRL-17691.

[7] C. E. Carlson and G. D. Magnuson, Rev. Sci. Instr. **33**, 905 (1962).

oscillate along the axial direction of a cylindrical collision chamber which is surrounded by a solenoidal magnet. Ions are extracted axially through a circular hole, and pass through aperture lenses which produce an approximately parallel ion beam of circular cross section. At this point the ions can be accelerated or retarded to an energy suitable for magnetic momentum analysis. A quadrupole lens pair then focuses this beam onto the entrance slit of a magnetic momentum analyzer. The analyzer magnet was designed with the aid of the instrument described by Walton[8] to give high-order focusing of the ion beam. The object and image distances are 12 and 24 cm, slitwidths 2 mm, and the ion deflection angle 66°.

From the exit slit of the mass spectrometer the ions pass through a quadrupole lens pair which makes the beam parallel and restores it to a nearly circular cross section. After the quadrupole lens, ions are retarded or accelerated to their final energy and pass through an einzel lens and exit aperture before impinging on the entrance of the collision cell. The energy spread of the analyzed beam was no greater than 3% FWHM of the analysis energy, which was 25 eV.

VACUUM CHAMBER AND COLLISION CELL

The collision cell is located in the center of a large vacuum chamber which is evacuated by two 6-in. oil diffusion pumps equipped with liquid-nitrogen-cooled baffles. Inside the chamber is a large cylindrical copper cold shield that can be cooled by liquid nitrogen. On the top of the chamber is a 20-in.-diam rotatable lid mounted on ball bearings and made vacuum tight by a differentially pumped double Tec-Ring seal.[9] The exit aperture of the collision cell and the entire detection train are mounted on this lid and rotate with it.

The collision cell consists of two concentric cylinders. The inner cylinder is held stationary and contains the ion-beam entrance aperture (2×2 mm) while the outer cylinder contains the product-ion exit aperture (2-mm diam) and rotates with the detection train. This exit aperture can be positioned in a range of ±55° from the projectile-beam direction. The conductance of the apertures is small enough and the seals between the cylinders good enough that it is possible to maintain a pressure in the main vacuum chamber which is a factor of 10^3 smaller than the collision cell pressure. The distances of the entrance and exit apertures from the center of the scattering cell are 1.60 and 2.24 cm, respectively.

DETECTION TRAIN

Ions leaving the collision cell pass into a 90° spherical electrostatic energy analyzer. The exit aperture

of the collision cell and entrance aperture of the analyzer provide an angular resolution of 2.5° geometric full width, while the entrance and exit apertures of the analyzer give an energy resolution of 3% FWHM of the analysis energy.

The 90° deflection produced by the analyzer directs the ions into a vertical trajectory through a series of cylindrical lenses which focus the ions into a quadrupole mass filter. The axis of the mass filter can be floated at a dc level up to several hundred volts, which makes it possible to mass-analyze the ions at an axial kinetic energy of 15 V. Optimum focusing voltages for the experimental range of ion energies were determined and tabulated for use. The transmission of the lens–mass-analyzer system was found to be nearly constant in this range of ion energies.

After leaving the mass filter, ions strike a highly polished aluminum surface which is maintained at approximately −25 kV. The secondary electrons released by the ion impact then impinge at 25 keV energy on a lithium-drifted silicon wafer which is the sensing element of the counting system. The semiconductor detector, FET preamplifier, and linear amplifier have been described in detail by Goulding[10] and Goulding and Landis.[11] A Hamner Model NT-16 timer, two NS-11 10-Mc scalers, a NE-11 scanner, a NR-10 ratemeter, and a Model 33TC teletypewriter complete the counting system.

DATA ACQUISITION

Experiments were performed by selecting primary ions of the desired energy, adjusting the focusing to give a beam of maximum intensity, stability, and of optimum energy and angular shape. With the quadrupole mass spectrometer set for the appropriate mass, scans of intensity as a function of product energy were made by determining the counting rate at a series of energy selector and focusing potentials. Scans of the angular distribution at constant energy required no adjustment of the focusing potentials and were made simply by rotating the lid on which the detection train was mounted. The primary beam intensity was checked at intervals of 10–30 min and linear interpolation of any drift was used to calculate the beam intensity corresponding to each data point taken in the time interval.

Because of the favorable ratio of scattering-cell to background pressure, the number of product ions formed outside the scattering volume which reached the detector was negligible except when the detector was set at very small angles with respect to the primary beam. This background contribution was determined by noting the pressure in the main chamber with gas in the scattering cell, then evacuating

[8] E. T. S. Walton, Proc. Roy. Irish Acad. **57**, 1 (1954).
[9] D. Armstrong and N. Blais, Rev. Sci. Instr. **34**, 440 (1963).

[10] F. S. Goulding, Nucl. Instr. Methods **43**, 1 (1966).
[11] F. S. Goulding and D. Landis, Natl. Acad. Sci.–Natl. Res. Council Publ. 1184.

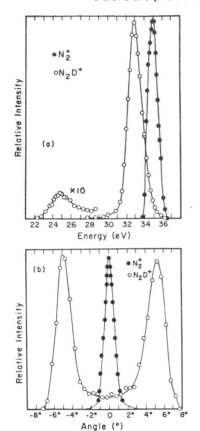

FIG. 1. Profiles of the N_2^+ beam and N_2D^+ product distributions. (a) A scan along the direction of the projectile beam ($\Theta = 0°$). The intense, high-energy product peak represents forward scattering, and the weaker low-energy peak is product scattered through $\theta = 180°$. (b) The N_2^+ and N_2D^+ intensities as a function of the laboratory angle Θ. The projectile laboratory energy was 75 eV. The scan of the product was made at 61 eV, which corresponds to the center-of-mass velocity, so the two peaks represent product scattered through $\pm 90°$ in the center-of-mass system. In both plots the N_2^+ beam intensity has been greatly reduced to make it comparable to product intensities.

the scattering cell and leaking target gas into the main vacuum chamber until the background pressure was restored, and then measuring the counting rate.

RESULTS AND DISCUSSION

Figure 1 shows typical primary data taken in energy and angular scans of the beam and product distribution. The energy scan was made with the detector aligned with the primary N_2^+ beam direction ($\Theta = 0°$ laboratory) and the intensity maxima at high and low energy correspond to scattering through angles of $\theta = 0°$ and $180°$, respectively, in the center-of-mass coordinate system. The angular scan of product intensity shown in Fig. 1(b) was made at an energy that corresponds to the center-of-mass velocity of the

N_2^+–D_2 system. Thus the two peaks observed at $\pm 5°$ in the laboratory coordinate system correspond to scattering through $\pm 90°$ in the center-of-mass system. Contour maps of scattered intensity were constructed by reading off the energies and angles corresponding to particular intensities from many such intensity profiles.

The maps here and in our earlier publication[1] show contours of constant intensity per unit velocity space volume. It can easily be shown that this intensity is independent of the coordinate system used to describe the scattering. Particle flux into a differential volume of velocity space must be conserved; thus,

$$I_{lab}(\Theta, \Phi, v) v^2 dv \sin\Theta d\Theta d\Phi = I_{c.m.}(\theta, \phi, u) u^2 du \sin\theta d\theta d\phi,$$

where (Θ, Φ) and v are the scattering angles and speed in the laboratory coordinate system and θ, ϕ, and u are the corresponding quantities in the center-of-mass system. Since the volume elements in the two systems are the same, I_{lab} and $I_{c.m.}$ are equal. They are related to the differential cross section per unit speed, I_{lab}' and $I_{c.m.}'$, used by Herschbach et al.[12] by the following equations:

$$I_{lab}'(\Theta, \Phi, v) = v^2 I_{lab}(\Theta, \Phi, v),$$

$$I_{c.m.}'(\theta, \phi, u) = u^2 I_{c.m.}(\theta, \phi, u).$$

On the other hand, theoretical treatments of scattering generally yield $I_n(\theta, \phi)$ defined as the number of particles in the state n reaching the detector per unit time per steradian. The relationship between $I_n(\theta, \phi)$ and $I_{c.m.}(\theta, \phi, u)$ is

$$I_n(\theta, \phi)(dn/du) = I_{c.m.}(\theta, \phi, u) u^2,$$

where dn/du represents the number of internal states of the product per unit speed in the center-of-mass system.

The actual intensity plotted in the maps is not $I_{c.m.}(\theta, \phi, u)$, of course, but this quantity averaged over the detector volume. The number of counts per second at the detector C were converted to normalized *relative* intensities $\bar{I}$ using the expression

$$\bar{I} = \bar{I}_{c.m.}(\theta, \phi, u) = \bar{I}_{lab}(\Theta, \Phi, v)$$

$$= 10^7 C/[i_0 P g(\Theta)(0.44 E_f)^{3/2}],$$

where i_0 is the peak incident beam intensity in units of 10^{-12} A, P is the scattering gas pressure in units of 10^{-4} torr, $g(\Theta)$ is the fraction of the scattering volume subtended by the detector at the laboratory scattering angle Θ, and E_f is the final laboratory energy of the ion. The factor $E_f^{3/2}$ normalizes the intensity to the detection volume in velocity space, which increases as $E^{1/2}$ due to the transmission band of the energy analyzer and as E_f due to the v^2 factor in the velocity volume element.

[12] W. B. Miller, S. A. Safron, and D. R. Herschbach, Discussions Faraday Soc. **44**, 108 (1967).

In our experiments $\bar{I}$ is the most convenient representation of our data for a number of reasons. It is better than the raw counts per second since it is normalized to standard values of several widely varying experimental parameters such as beam intensity, scattering gas pressure, and detector passband. Yet it is still a straightforward presentation of the intensities as measured in the laboratory. Provided the spread in initial relative velocities is small and $I_{c.m.}(\theta, \phi, u)$ is slowly varying over the detector volume, $\bar{I}$ is also a very good approximation to $I_{c.m.}(\theta, \phi, u)$. This is an advantage over plotting I_{lab}' (which is not equal to $I_{c.m.}'$), since we are ultimately interested in intensity distributions in the center-of-mass system. When the above conditions are not met, however, the experimental data will not directly yield intensities in the center-of-mass system. Thus any attempt to plot $I_{c.m.}'$ will require assumptions about $I_{c.m.}(\theta, \phi, u)$ or the initial relative velocity spread which are not directly available from the data. So neither $I_{c.m.}'$ nor I_{lab}' is satisfactory under all conditions, while $\bar{I}$ is, and we have used this cross section in all our contour maps.

From $\bar{I}$ it is possible to calculate the angular distribution function or differential cross section in the center-of-mass system $I(\theta)$ by

$$I(\theta) = \int_0^\infty \bar{I}(\theta, \phi, u) u^2 du. \qquad (1)$$

This differential cross section is still dependent on

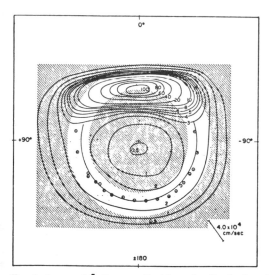

FIG. 2. A map of $\bar{I}$, the normalized intensity distribution of N_2H^+ in the center-of-mass coordinate system. The relative energy of N_2^+ and H_2 reactants was 5.6 eV. The outer shaded area represents values of Q greater than $+1.0$ eV, and in the inner shaded area Q is less than -2.5 eV. The circled points represent the actual maxima in the intensity which were located in the energy and angular scans. The circle which passes through many of these points corresponds to $Q = -1.7$ eV.

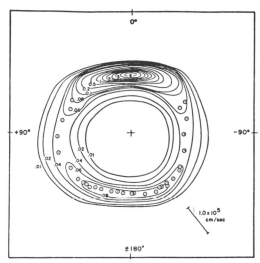

FIG. 3. A map of the normalized intensity $\bar{I}$ in the center-of-mass coordinate system for N_2D^+ from the N_2^+-D_2 reaction at 11.2 eV relative energy.

the initial kinetic energy, as is the total reaction cross section σ given by

$$\sigma = 2\pi \int_0^\pi I(\theta) \sin\theta d\theta. \qquad (2)$$

We have calculated both these quantities numerically from $\bar{I}(\theta, u)$.

Figures 2 and 3 show product intensity distribution $\bar{I}$ of N_2H^+ from the N_2^+-H_2 reaction at 5.6 eV relative energy, and of N_2D^+ from the N_2^+-D_2 reaction at 11.2 eV relative energy. These distributions and the ones reported previously[1] show a very pronounced peaking of intensity at zero angle in the center-of-mass system. This peaking was also observed at lower projectile energies by Wolfgang et al.[5] and by Bailey and co-workers[4] at energies comparable to those used by ourselves. The asymmetry of the distributions about $\theta = \pm 90°$ shows that the reaction proceeds by a direct or short-lived interaction, rather than by an intermediate complex that lives as long as a few rotational periods. The predominance of the forward scattering indicates that a projectile N_2^+ can with high probability pick up a hydrogen atom and continue on a nearly straight-line path, imparting little if any transverse momentum to the free hydrogen atom.

While small-angle scattering predominates, reaction products are observed at all angles in the center-of-mass system. Consequently, "rebound" processes occur in which the N_2H^+ product acquires a velocity component in the center-of-mass system opposite to the direction of approach of the N_2^+ projectile, and the free hydrogen atom receives considerable momentum. For large relative energies, a slight peaking in the angular distribution of product intensity appears at $\theta = 180°$. This backward peak is not evident for the $N_2^+-H_2$ reaction at 5.6 eV relative energy, as Fig. 2 shows, but is obvious for the same reaction at 8.1 eV relative energy,[1] as well as for the $N_2^+-D_2$ reaction[2] at 8.1 eV, and, as Fig. 3 shows, at 11.2 eV.

The significance of the product intensity distributions becomes clearer if we introduce the translational exothermicity of the reaction Q, which is defined by

$$Q = \tfrac{1}{2}(\mu' g'^2) - \tfrac{1}{2}(\mu g^2).$$

Here μ and g are, respectively, the reduced mass and relative speed of the reactants, and the primed quantities refer to the products. By assuming the reactants are in their ground states we can also write

$$Q = -\Delta E_0^0 - U,$$

$$Q(\text{eV}) = -2.5 + D(N_2-H^+) - U,$$

where ΔE_0^0 is the energy change for the reaction, U is the internal excitation energy of the products, and $D(N_2-H^+)$ is the unknown dissociation energy of N_2H^+ into a proton and a nitrogen molecule. The range of possible values of Q is limited by the value of ΔE_0^0 and the expectation that the products are unstable and undetectable if $U > D$. Thus,

$$-2.5 \leq Q \leq -\Delta E_0^0.$$

The lower limit for Q is rigorously specified except for the possibility that metastable N_2H^+ might live long enough (30 μsec) to reach the detector. The upper limit can only be estimated since ΔE_0^0 is unknown. An upper limit of zero for ΔE_0^0 can be deduced from the fact that the reaction between N_2^+ and H_2 is rapid even at thermal energies, and thus is very probably not endothermic. In addition, analogy to the isoelectronic molecule IICN suggests that the dissociation energy of N_2H^+ to N_2^+ and a hydrogen atom could be as high as 5.5 eV, which would correspond to a dissociation energy to N_2 and H^+ of 3.5 eV, and a value of -1 eV for ΔE_0^0. This estimate gives us a reasonable, if very approximate, estimate of 1 eV for the upper limit of Q.

The velocities of N_2H^+ relative to the center of mass that correspond to values of Q outside the allowed range are indicated by the shaded areas in Fig. 2. Such excluded regions of velocity space could have been indicated in Fig. 3, but due to high projectile energy and consequent very low product intensity, the data in Fig. 3 show enough asymmetry about the direction of the initial relative velocity vector to make comparison of the distribution to the allowed values of Q relatively uniformative.

From Fig. 2 we see that a certain amount of product intensity appears at values of Q greater than 1 eV, a region nominally excluded by energy conservation. This apparent discrepancy may be a result of an underestimate of $D(N_2-H^+)$, or may indicate that

an important fraction of reactant ions are in excited vibrational and electronic states. However, Moran and Friedman[13] estimate from overlap integral calculations that 90% of the N_2^+ ions formed by electron impact in the $X\,^2\Sigma_g^+$ state are in the ground vibrational level. It has also been suggested that as much as 10% of the ions may be in the $A\,^2\Pi_u$ state (excitation energy 1.12 eV) as long as 1–10 μsec after formation by electron impact. In our apparatus the 30-μsec flight time from the extraction aperture to the collision cell, plus the undetermined time spent in the ionization chamber before extraction, should reduce the fraction of ions in the $A\,^2\Pi_u$ state below 10%. Thus while it is not possible to rule out the presence of the excited electronic state, it would appear to be of minor importance.

It is likely the appearance of products in excluded regions of velocity space is a consequence of finite instrumental resolution combined with the motion of the target hydrogen molecules. Although the angular resolution in the laboratory coordinate system is better than 2°, it is no better than 35° in the center-of-mass system for $N_2^+-H_2$ collisions. This factor could be responsible for the intensity in regions around $|\theta| = 45°$ and $Q > 1$ eV in Fig. 2.

Even if the initial and final ion velocities could be measured exactly, the values of Q deduced are somewhat uncertain due to target motion. The error in Q arising from target motion is found to be

$$\delta Q = 2v_{II}[Mv_i - (M+m)v_f] + mv_{II}^2,$$

where M and m are the masses of the nitrogen molecule and abstracted hydrogen atom, v_i and v_f are the initial projectile and final product velocities, and v_{II} is the thermal velocity of the abstracted hydrogen atom. It is of interest to note that the uncertainty in Q vanishes to first order in the target velocity for the ideal stripping process, since in this case v_f equals $Mv_i/(M+m)$. For other combinations of v_i and v_f, the uncertainty in Q may be of the order of 0.2–0.3 eV. Finite energy resolution introduces an additional uncertainty of 0.2 eV. The combination of finite energy and angular resolution together with target motion seem sufficient to rationalize the presence of some scattered intensity in regions of velocity space corresponding to $Q > 1$ eV or $Q < -2.5$ eV.

Despite the uncertainties imposed by finite resolution and target motion, it is possible to make some general observations concerning the amount of the internal energy in the ion product. Figure 2 shows that for N_2H^+ scattered in the forward direction, the highest intensity appears at $Q = -2.5$ eV, which corresponds to products excited internally to levels very near the dissociation limit. For products scattered through angles of 90° or greater, the maximum intensity appears at an appreciably larger value of Q

[13] T. F. Moran and L. Friedman, J. Chem. Phys. **42**, 239 (1965).

approximately -1.7 eV. Such a Q value indicates that the back-scattered ions are internally excited, but to a noticeably smaller degree than products scattered forward. Apparently the recoil naturally associated with backscattering can lower the internal excitation of the ions somewhat, whereas recoil and product stabilization is accomplished less easily in the grazing reactive collisions associated with forward scattering.

In the analysis of the reaction of potassium with bromine, Bernstein and co-workers[14] assumed that the distribution of product internal energy was independent of scattering angle θ. While this assumption appears to be valid for the data available for the $K+Br_2$ system, it is clearly not valid for the $N_2^++H_2$ reaction. It must be emphasized, however, that the systems are quite different, since in our experiments the relative kinetic energy is greater than the exothermicity of reaction, whereas the opposite is true for the $K+Br_2$ system.

Several groups[2,4,5] as well as ourselves[1] have discussed the significance of the velocity of the forward scattered product. Here we only comment that for relative collision energies greater than about 5 eV the most probable laboratory velocity for forward scattered products is greater than the value predicted by the ideal stripping model. The reason for this is apparent from Fig. 2. Even at 5.1 eV relative energy the maximum in forward intensity lies at $Q=-2.5$ eV, the most negative value consistent with stable products. At higher relative energies, forward recoil and concommitant deviations from the ideal stripping model are necessary if stable products are to be formed.

The slight intensity peaking at $\theta=180°$ that is evident in Fig. 3 as well as the maps from other high-energy experiments requires comment. In the case of elastic scattering of atoms, such backward glory scattering implies that orbiting collisions occur in which the particles remain close for at least one rotational period. The high relative energies employed in the present experiment eliminate orbiting as a possible cause for backward peaking. However, it is not necessary to postulate such long-lived intermediates to explain a backward glory in molecular reactive scattering. Peaking at $\theta=180°$ can occur whenever it is possible for collisions with a nonzero impact parameter to lead to backward reactive scattering. One of several types of collision that could produce backward peaking is the ideal knockout process. In this case the N_2^+ would collide impulsively with one hydrogen atom so as to eject it from the molecule, and then pick up the remaining hydrogen atom to form N_2H^+. Such a process would produce backscattered ions and would occur with greatest probability when the axis of the hydrogen molecule was oriented perpendicular to the projectile trajectory and

the impact parameter was approximately equal to half the bond distance.

The kinematics of the ideal knockout process lead to the prediction that for 180° scattering the ratio of the product laboratory velocity to the velocity of the projectile is given by

$$v_f/v_i = M(M-m_2)/[(M+m_1)(M+m_2)],$$

where M is the projectile mass, m_2 is the mass of the ejected target atom, and m_1 is the mass of the target atom in the product. Thus the velocity ratio should depend only on the masses of the atoms, according to this model.

It is also possible to calculate the internal excitation energy of the product. According to the knockout model, this should be the sum of the exothermicity of the reaction and the kinetic energy of the projectile relative to the abstracted atom after the projectile has made an elastic collision with the ejected atom. The result is

$$U = -\Delta E_0^0$$
$$+[Mm_1/(M+m_1)][(M-m_2)/(M+m_2)]^2 \tfrac{1}{2}(v_i^2). \quad (3)$$

If U is greater than D, the smallest bond dissociation energy of the product, the product ions will be unstable, and none will reach the detector. Thus, after substituting D for U, Eq. (3) can be solved for a critical projectile velocity above which no back-scattered product should be detected. This critical velocity is higher the smaller the mass of the abstracted atom, and the closer the masses of the projectile and ejected atom. In their discussion of the kinematics of the knockout process, Light and Horrocks[15] omit the important term ΔE_0^0 in Eq. (3) and are led to an overestimate of the critical velocity for exothermic reactions.

Figure 4 shows a comparison between the experimental velocity ratios v_f/v_i for product scattered through 180°, and the predictions of the ideal knockout model. Also shown are the velocity ratios that correspond to the limiting Q values of $+1.0$ and -2.5 eV. The relative energy at which the predicted velocity ratio line intersects the $Q=-2.5$-eV curve is the critical energy above which the product is unstable, according to the knockout process.

From Fig. 4 it is clear that product is observed from all four isotopic reactions at relative energies well above the maximum values allowed by the knockout process. In particular, for the formation of N_2D^+ from HD, the model suggests no product should have been detected in any of the experiments performed except one. The model gives particularly high product excitation for this reaction because the projectile loses little energy upon elastic collision with

[14] T. T. Warnock, R. B. Bernstein, and A. E. Grosser, J. Chem. Phys. **46**, 1685 (1967).

[15] J. C. Light and J. Horrocks, Proc. Phys. Soc. (London) **84**, 527 (1964).

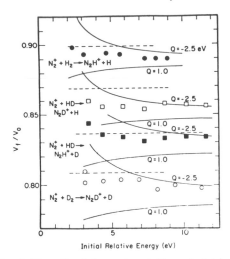

FIG. 4. The ratio of the most probable product laboratory velocity to the projectile velocity as a function of initial relative kinetic energy for products scattered through 180° in the center-of-mass system. The predictions of the ideal knockout model for each reaction are shown as horizontal dashed lines. The velocity ratios for the limiting Q values of $+1.0$ and -2.5 eV are also shown as solid lines.

the light hydrogen atom, and therefore moves with considerable energy relative to the heavy deuterium atom. The opposite is true for the formation of N_2H^+ from HD, and Fig. 4 shows that for this case the experimentally observed ratios are close to the values predicted from the knockout model at the lower relative energies.

In general, the experimental laboratory velocity ratios and, equivalently, internal excitation of the product are smaller than predicted from the ideal knockout model. Moreover, there is no evidence of an abrupt product intensity loss above the critical projectile energies derived from the knockout model. These considerations suggest that the ideal knockout model does not provide a good description of the backward scattering. Indeed, it is difficult to imagine the ejection and pickup of the two hydrogen atoms as separate events, given the sizes and strong long-range interaction of N_2^+ and H_2. We know of no other simple model which is successful in predicting final velocities for backscattered products from all four isotopic reactions.

Both the angular distribution of reaction products $I(\theta)$ and the total reaction cross section σ were calculated from the measured $\bar{I}(\theta, u)$ using Eqs. (1) and (2). Because of imprecise pressure calibration and the constant but uncertain transmission factors, only the relative values of $\bar{I}$ have meaning, and in order to obtain an absolute reaction cross section it was necessary to scale our calculation to some absolute measurement. We normalized our results for the total cross section of the N_2^+–D_2 reaction at 11.2 eV

relative energy to the measurement of Turner et al.[3] at this energy. The two other N_2^+–D_2 experiments for which complete maps of $\bar{I}$ are available then give total cross sections in good agreement with Turner's values.[3]

The angular distribution $I(\theta)$ of total product is shown in Fig. 5 for reactions with H_2 and D_2 calculated from the contour maps in this paper and our earlier work.[1] The angular distributions for the two reactions are very similar at the same relative energy. As the relative energy increases the intensity at all angles decreases, but the peaking at small angles becomes more pronounced. The general shape of these curves is similar to that found by Birely et al.[16] for the reactions of alkali metals with bromine and iodine but is even more sharply peaked at $\theta = 0°$.

The angular distribution curves of Fig. 5 do not show the intensity peak at 180° which is evident in Fig. 3 and the previously published intensity maps.[1] Inspection of the maps shows that the intensity distribution tends to narrow in velocity as well as peak in angle near $\theta = 180°$. Thus it is not surprising that the intergrated intensity $I(\theta)$ does not have a maximum at $\theta = 180°$. The projection of the detector

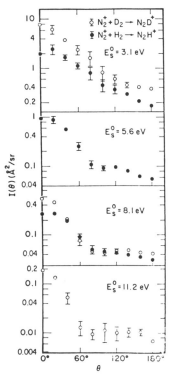

FIG. 5. The differential reaction cross section $I(\theta)$ as a function of center-of-mass scattering angle θ for the reactions of N_2^+ with H_2 and D_2 at several initial relative energies.

[16] J. Birely, R. Herm, K. Wilson, and D. Herschbach, J. Chem. Phys. 37, 993 (1967).

TABLE I. Relative intensities[a] and differential cross sections.

Expt. no.	E_a^0	$\bar{I}(0°)$ $\times 10^{-2}$	$\bar{I}(90°)$ $\times 10^{-2}$	$\bar{I}(180°)$ $\times 10^{-2}$	$I(0°)$ $\times 10^2$ ($Å^2$/sr)	$I(90°)$ $\times 10^2$ ($Å^2$/sr)	$I(180°)$ $\times 10^2$ ($Å^2$/sr)
			$N_2^+ + D_2 \rightarrow N_2D^+ + D$				
186	8.13	18.7	3.8	7.0	51.4	4.26	3.88
187	3.12	305	64.0	...	308	14.4	5.58
188	4.38	297	41.5	...	396	13.9	5.48
189	9.38	9.8	2.1	3.7	42.6	2.2	2.2
190	3.12	515	...	...	752	104	36.6
191	5.62	212	30.7	47.0	360	24.4	19.3
192	6.86	58.7	17.9	21.3	139	15.0	10.2
193	11.23	4.5	50	0.97	17.9	0.936	0.652
			$N_2^+ + H_2 \rightarrow N_2H^+ + H$				
194	2.33	1220	...	...	440	79.2	36.6
195	5.62	114	25.0	25.3	88.4	9.66	5.26
196	4.67	132	44.0	34.5	79.2	10.8	6.62
197	7.35	29.5	8.20	9.50	33.4	4.32	2.64
198	8.92	11.2	3.35	5.90	17.9	2.30	1.99
199	3.13	450	...	130	189	55.8	16.4
200	8.11	19.8	6.80	9.40	25.2	4.28	2.90
			$N_2^+ + HD \rightarrow N_2D$				
207a	4.34	111	75.3	69.0	55.8	11.1	10.4
208a	5.79	36.5	17.5	42.0	28.2	4.32	7.54
210a	7.25	7.53	3.58	11.0	8.19	1.42	2.60
211a	8.70	2.48	1.00	5.20	4.08	0.636	1.18
212a	10.2	0.72	0.42	1.88	1.81	0.374	0.604
213a	11.6	0.27	0.16	0.98	0.926	0.444	0.352
214a	3.39	160	...	...	54.0	24.6	12.1
			$N_2^+ + HD \rightarrow N_2H^+ + D$				
207b	4.34	513	13.6	15.7	756	5.70	5.28
208b	5.79	323	5.38	7.20	584	3.18	3.32
210b	7.25	174	2.70	3.30	396	2.60	1.97
211b	8.70	53.5	1.88	1.83	160	2.74	1.35
212b	10.2	14.9	0.98	0.87	60.4	1.75	0.670
213b	11.6	8.0	0.58	0.50	36.4	1.30	0.484
214b	3.39	413	...	...	480	12.9	6.36

[a] Arbitrary units.

profile on the contour maps is a rectangle, wider than it is long, since our laboratory speed resolution is better than our angular resolution. Given the character of the product distribution, it seems possible that the backward peaking evident on the intensity maps may result from the fact that the detector intercepts a larger fraction of the craterlike distribution at $\theta = 180°$ than at $\theta = \pm 90°$. In any case, failure to observe peaking at $180°$ in the product angular distribution $I(\theta)$ completely eliminates the possibility that the reaction proceeds through a long-lived collision complex.

No product intensity maps were obtained for the N_2^+–HD reactions, so it is not possible to compute the angular distributions at all angles and total cross sections for these cases. However, in Table I we give our measured values of $\bar{I}$ (peak value) and the computed $I(\theta)$ for $\theta = 0°$, $90°$, and $180°$. It can be seen that the intensity of N_2H^+ at $\theta = 0°$ greatly exceeds

that of N_2D^+ for all initial relative energies, regardless of whether $\bar{I}$ or I are compared. This very large isotope effect favoring hydrogen over deuterium in the forward-scattered products largely disappears if $\bar{I}$ or I is plotted as a function of E_a^0, the energy of the projectile relative to the *atom* abstracted.

Isotope effects similar to these have been observed and discussed in terms of the ideal stripping model by Henglein and co-workers.[17] This model predicts that the internal excitation of the forward-scattered product is

$$U = -\Delta E_0^0 + [m_1/(M + m_1)]E^0 \qquad (4a)$$

$$= -\Delta E_0^0 + E_a^0, \qquad (4b)$$

where M and m_1 are the masses of the projectile and abstracted atom and E^0 and E_a^0 are the projectile

---

[17] A. Henglein, K. Lacman, 5nd B. Knoll, J. Chem. Phys. 43, 1048 (1965).

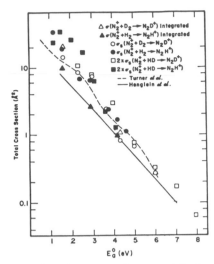

FIG. 6. Total reaction cross sections computed using Eq. (2) (triangles) and Eq. (5) (circles and squares). The results of Turner et al.[3] and Henglein[2] are indicated by the dashed and solid lines, respectively.

energies in the laboratory and relative to the abstracted atom. According to these expressions the internal energy of N_2D^+ will be greater than that of N_2H^+ for any given projectile laboratory energy. An infinite isotope effect is expected for projectile energies above 37.5 eV, for then U exceeds the dissociation energy of the product for deuterium but not for hydrogen abstraction. Our failure to observe such an infinite isotope effect, together with the observed magnitudes of the product velocities, is a clear demonstration that forward recoil and concomitant product stabilization occur and that the stripping model in its simplest form does not describe the forward scattering quantitatively.

It seems likely, however, that the origin of much of the small-angle isotope effects lies in Eq. (4b). In the energy region in which we have done our experiments, the intensity of products seems to be chiefly dependent on the amount of energy which must be disposed of. Our experimental observations that most of the products appear sharply peaked at $\theta = 0°$ and that only the minimum momentum necessary for product stability is transferred to the nonreacting atom strongly suggest that the N_2^+ projectile interacts almost exclusively with the abstracted atom. In this case the pertinent relative kinetic energy is E_a^0, the energy relative to the abstracted atoms. If $U > D$, then $U - D$ must be disposed of by recoil of the free atom. This becomes more difficult as E_a^0, and thus U, grows and therefore the intensity falls sharply.

Henglein's data[17] suggests that an H atom and a D atom are equally effective in disposing of a given energy excess $U - D$. An examination of our intensity

data confirms this over the entire energy range studied by us. We must be cautious, however, since our contour maps indicate that the scattering near $\theta = 0°$ is more strongly peaked than the passband of our detector in both velocity and angle. If this is true then $\bar{I}(\theta, u)$ and our computed $I(\theta)$ will be smaller than their true values. However, it can then easily be shown that

$$\sigma_s = I(\theta = 0°)(v_{max}/u_{max})^2 \qquad (5)$$

is, apart from a normalizing constant, the total amount of scattering in this region. The factor $(v_{max}/u_{max})^2$ is the square of the ratio of the laboratory velocity of the product at the peak to the center-of-mass velocity at the peak. It basically converts $I(\theta = 0°)$ to $I(\theta = 0°)$, so we can make use of the fact that the laboratory angular resolution (but not the center-of-mass resolution) of the detector is the same for all hydrogen isotopes. The normalizing constant mentioned above will then include the (constant) FWHM laboratory angular resolution. The cross section σ_s can be thought of as the total cross section for stripping. In fact, since our computed $I(\theta = 0°)$, though too small, is still about 10 times the values at $\theta = 90°$ and $180°$, we feel that the normalized σ_s is within experimental error equal to the true total cross section. Once again we have normalized our N_2^+–D result at 11.2 eV relative energy to Turner's result (0.27 Å²) at the same energy. Figure 6 shows σ_s plotted versus E_a^0 for the four possible reactions; the values for HD have been multiplied by 2 since it contains only one H atom compared to two in H_2 and one D atom compared to two in D_2. There is remarkably good agreement among the four reactions. The total cross section for N_2^+–D_2 measured by Turner[3] have also been plotted along with the curve derived by Henglein[17] from his N_2^+–H_2 and N_2^+–D_2 data, which also gives a reasonably good fit to his HD data. The best-fit straight line to our data has a slope $\partial \log\sigma_s/\partial E_a^0 = -0.42$, in good agreement with Henglein's value of -0.35 and our estimated value of -0.40 for Turner's data. Figure 6 also gives our total cross sections σ, computed from Eq. (2) as described earlier. The values of σ and σ_s are in satisfactory agreement at all energies.

Product which appears at larger angles in the center of mass involves substantial recoil of the nonreacting H or D atom. We have seen earlier that product at $\theta = 90°$ and $180°$ is noticeably less excited internally than product at $\theta = 0°$. Both of these facts suggest that the picture of an N_2^+ projectile interacting only with the abstracted atom which is successful at small angles breaks down for products at large angles. Thus it is not surprising that plots of $I(90°)$ or $I(180°)$ versus E_a^0 are no longer the same for the four possible reactions. The relative energy of the projectile and the entire molecule E_a^0 now becomes the significant dynamical parameter. The total amount of product $I(\theta)$ produced at $\theta=90°$ and $\theta=180°$ is shown as a function of E_s^0 in Fig. 7. For HD this involves summing the intensities for N_2H^+ and N_2D^+. All three molecules give about the same exponential dependence of product intensity on E_s^0 at both angles, which is in marked contrast to the behavior found at $\theta=0°$. The fact that intensity decreases as energy increases may be due to an increase in difficulty of stabilizing product molecules at the higher energies. However, the observation[1] that the most probable product internal excitation energy is below the dissociation limit for large-angle scattering suggests other factors may also be important. The most obvious possibility is that for small-impact-parameter collisions, charge transfer and dissociative charge transfer[18] start to compete significantly with chemical reaction as energy increases.

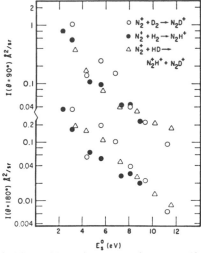

FIG. 7. Differential reactive cross sections at $\theta=90°$ and 180° as a function of energy of the projectile relative to the molecule. For HD, both isotopic products are included. Note the broken scale.

It is of interest to compare the relative probability of forming N_2H^+ and N_2D^+ from HD at the larger scattering angles. The experimental isotope effect is shown in Fig. 8 for $\theta=90°$ and 180°. In addition, the ratio $\sigma_s(N_2H^+)/\sigma_s(N_2D^+)$ for HD is shown in order to give an indication of the isotope effect at $\theta=0°$. It is obvious that for high projectile energies there is an enormous isotope effect favoring N_2H^+ at $\theta=0°$, a similar small effect at $\theta=180°$, and an intermediate effect at $\theta=90°$. The decrease in the isotope effect with increasing angle evidently indicates that the energy of the projectile relative to the abstracted atom, or equivalently the internal energy of the incipient N_2H^+ or N_2D^+, becomes less critical as the impulse imparted to the freed atom increases.

18 D. W. Vance and T. L. Bailey, J. Chem. Phys. 44, 486 (1966).

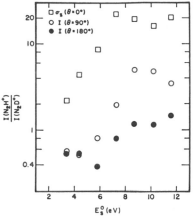

FIG. 8. Isotope effects in the differential and total cross sections for the N_2^+–HD reaction as a function of relative collision energy and scattering angle.

In contrast to the situation that holds for stripping or grazing collisions, in the intimate collisions associated with backscattering the atoms are strongly coupled, and the product internal energy and stability are not closely correlated with isotopic composition. It should be noted that at large angles and low energies there is a relatively small isotope effect that favors N_2D^+ over N_2H^+ from HD. The same phenomena appears when one compares N_2D^+ and

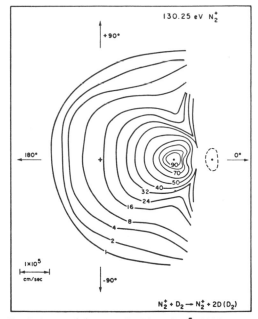

FIG. 9. A map of the normalized intensity I of N_2^+ scattered from D_2 at 16.3 eV relative energy in the center-of-mass coordinate system. The dashed ellipse at $\chi=0$ indicates the profile of the incident beam at 20% of its maximum intensity.

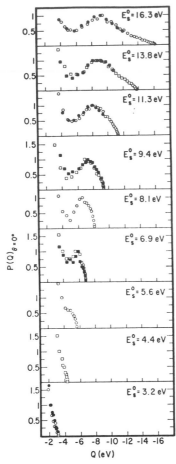

FIG. 10. The relative probability $P(Q)$ of finding a relative energy loss Q as a function of Q for N_2^+ scattered from D_2 at $\theta = 0°$. The two sets of data at the same energy refer to two separate experiments. Whenever possible, the scale of the abscissa for each experiment was determined by setting $P(Q) = 1$ at the inelastic maximum; for the low-energy experiments a convenient point is arbitrarily designated as $P(Q) = 1$.

N_2H^+ from D_2 and H_2. We have no explanation for this inverse isotope effect. It can be regarded as further evidence against the knockout model, which would predict that the product be dominantly N_2H^+ over the experimental energy range. A similar failure of the knockout model in hot-atom reactions has been noted by Cross and Wolfgang.[19]

The nonreactive scattering of N_2^+ by D_2 was investigated using projectile laboratory energies from 25–130 eV. In one of these experiments enough energy and angular scans were made to generate the contour

map shown in Fig. 9. Three important features are apparent. First, there is no evidence of any elastic scattering at large center-of-mass angles. In an investigation[20] of collisions of N_2^+ with He, we were able to detect N_2^+ scattered elastically and with some inelastic loss at angles as large as $\theta = 180°$. Consequently, the failure to detect large-angle elastic scattering in the N_2^+–D_2 system is not due to lack of sensitivity. We can conclude that in the N_2^+–D_2 system, collisions with small impact parameters lead virtually exclusively to reaction, charge transfer, dissociative charge transfer, or very inelastic nonreactive scattering.

The second feature to be noted from Fig. 9 is the relatively small total intensity of scattered N_2^+ except for the elastic scattering at very small center-of-mass angles. Together with the low intensity of N_2D^+ found for such high projectile energies, this suggests that most small-impact-parameter collisions produce neutral nitrogen molecules and D_2^+ or its dissociation products. This is consistent with the observations of Vance and Bailey[18] that the cross sections for charge transfer with and without dissociation increase with increasing projectile energy, and are of the order of 10 Å² for 100-eV ions.

The third feature of importance is the peak in the scattered N_2^+ intensity at $\theta = 0°$ and a Q value of -9.5 eV. To find the process responsible for this peak, we determined its intensity profile along $\theta = 0°$ at several projectile energies. The results are displayed in Fig. 10 as plots of $P(Q)$, the relative probability of finding a given value of Q, as a function of the relative energy loss Q. Apart from a scale factor, $P(Q)$ is given by

$$P(Q) = [I(\theta, u) u^3]/[2E_s I(\theta)],$$

which follows from the definition

$$P(Q) dQ = I(\theta, u) u^2 du$$

and the fact that

$$(\partial Q/\partial u)_{E_s^0} = 2E_s/u,$$

where E_s is the relative translational energy of the product. The inelastic feature first becomes discernable when the relative energy is 5.6 eV. As the relative energy increases, the Q that corresponds to the maximum intensity becomes more negative.

We have evaluated $I(\theta = 0°)$ for this inelastic scattering using Eq. (1). In this case the upper limit for the integral was the value of u for which Q was -4.5 eV. This was a somewhat arbitrary attempt to

[19] R. J. Cross and R. Wolfgang, J. Chem. Phys. 35, 2002 (1961).

[20] W. R. Gentry, E. A. Gislason, B. H. Mahan, and C. W. Tsao J. Chem. Phys. 47, 1856 (1967).

eliminate the contribution of the small Q scattering which comes from vibrational and rotational excitation. Since the inelastic feature is so sharply peaked in angle, we again used Eq. (5) to estimate the total cross section σ_s as a function of relative energy. This is shown in Fig. 11. We also computed the integrated total cross section σ using Eq. (1) and (2) from the data used in Fig. 9. The good agreement between σ and σ_s for this experiment suggests that as was true for reactive scattering, σ_s is a good approximation to the total cross section. The cross sections in Fig. 11 show an inelastic threshold near 4.5 eV, and increase from 0.01 Å^2 at 6 eV relative energy to 0.6 Å^2 at 16 eV.

There are three processes which might be responsible for this inelastic peak. One is a stripping collision that produces N_2D^+ with internal excitation sufficiently large so that dissociation to $N_2^+ + D$ occurs. The relative energy threshold for this process is 8.4 eV, however, well above the experimentally observed threshold. Moreover, stripping followed by isotropic dissociation would force the ratio of the final laboratory velocity of N_2^+ to the projectile velocity to be $\frac{28}{30} = 0.9333$ at all projectile energies. Experimentally, we find for this ratio

$$v_f/v_0 = 0.9462 + 0.000418(E^0 - 100)$$

over the range of energies studied. Stripping followed by product dissociation may contribute to the inelastic process at high projectile energies, but does not do so at the lower energies.

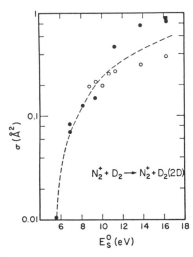

FIG. 11. The total cross section as a function of relative energy for inelastic scattering with energy loss greater than 4.5 eV for N_2^+ on D_2. Open and closed circles correspond to σ_s from experiments in which the beam energy spread was 1.2 and 0.96 eV FWHM, respectively. The square point is the fully integrated σ calculated from data of Fig. 9.

It is possible that collisional electronic excitation of N_2^+ is responsible for the inelastic peak. Excitation of the $B^2 \Sigma_u^+$ state of N_2^+ by collisions at energies well below 1 keV with H_2 and the rare gases has been observed spectroscopically.[21] However, if this were the principal source of inelastic loss, it is not clear why the threshold for excitation should be above 3.2 eV, the excitation energy of N_2^+ $B^2 \Sigma_u^+$. In addition, it is difficult to see why there should be a peak in the energy-loss spectrum which shifts smoothly to larger energy losses as the projectile energy increases. For example, at 8.1 eV relative energy, the $P(Q)$ curve peaks at $Q = -6.5$ eV, which does not correspond to any Franck–Condon favored transition to a known state of N_2^+. The fact that we did not observe any inelastic peaks at $\theta = 0°$ in our investigation of N_2^+–He collisions[20] suggests that the peak observed in the N_2^+–D_2 system does not arise from electronic excitation of N_2^+, but is connected with excitation of D_2.

Since the inelastic peak has a threshold near 4.5 eV, we feel that it may be associated with the collisional dissociation of D_2. The phenomenon of a stripping reaction followed by dissociation of N_2D^+ accomplishes this, but as mentioned above, this process has a threshold too high to be consistent with the data at low projectile energy. At relative energies above the critical value for product stability, the ideal knockout process can also lead to collisional dissociation of D_2. However, the critical initial relative energy at which N_2D^+ formed in the knockout process becomes unstable with respect to N_2^+ and D is 9.6 eV, well above the observed threshold. Thus neither the dissociative stripping nor the knockout process can be fully responsible for the inelastic peak.

Another mechanism for dissociation would be a curve-crossing phenomenon in which the reactants, moving on a surface that goes asymptotically to N_2^+ $X\,^2\Sigma_g^+$ and D_2 $^1\Sigma_g^+$, are transferred to the surface which has N_2^+ $X\,^2\Sigma_g^+$ and D_2 $^3\Sigma_u^+$ as its asymptotic states. Such a transition is allowed by spin conservation rules and could have a threshold at 4.6 eV, the dissociation energy of D_2. The process might also be pictured as a nearly vertical excitation to the $^3\Sigma_u^+$ state of a D_2 molecule which had been distorted by some degree to large internuclear separation. At low projectile energies the elongation of D_2 might be great, and transitions to the low-energy part of the $^3\Sigma_u^+$ curve of D_2 could occur and lead to Q values not much more negative than -4.6 eV. As the projectile energy increased, the time to distort D_2 would decrease, and the over-all process would become more vertical with respect to the D_2 internuclear separation. This would account for the

[21] C. Liu and H. P. Broida, Bull. Am. Phys. Soc. 13, 448 (1968).

increase in cross section and increase in excitation energy as the projectile energy increases. The Franck–Condon factor for excitation from the vibrational ground state of D_2 to the $^3\Sigma_u^+$ state shows a maximum at 9.5 eV, the observed limiting energy loss, and a shape which roughly approximates the inelastic peak observed in the highest-energy experiment. Thus collisional dissociation of D_2 by excitation to its lowest triplet state does provide an adequate, if not proven, explanation of the inelastic peak observed for N_2^+.

One other point concerning the inelastic scattering of N_2^+ should be recognized. The ratio of the most probable velocity of the inelastically scattered N_2^+ to the projectile velocity increases linearly with increasing projectile energy. Over much of the projectile energy range, this ratio is greater than the product (N_2D^+)-to-projectile velocity ratio expected for ideal stripping and is near or equal to the experimentally observed velocity ratio for N_2D^+. Also in this same range of high projectile energies, the inelastic cross section is comparable to or greater than the cross section for formation of N_2D^+. Therefore, in Henglein's experiments[2] in which velocity analysis of the scattered ions was carried out without mass analysis, some or even the major part of the apparent N_2D^+ signal was in fact inelastically scattered N_2^+. This possibility was recognized by Henglein.[2] Since the velocity of scattered N_2^+ and N_2D^+ are similar in the energy range where the inelastic and reactive cross sections are comparable, the ambiguity has rather little effect on the velocity ratios determined by Henglein. However, because of the increasing contributions of inelastic scattered N_2^+, Henglein's apparent total reaction cross section decreases with increasing projectile energy somewhat more slowly than the cross section found by ourselves and by Turner et al.[3] In fact, by combining the reactive and inelastic cross sections from Figs. 6 and 10 we predict that the logarithm of the cross section measured by Henglein should decrease linearly with a slope $\partial(\log\sigma_s)/\partial E_u^0 = -0.33$, which is in close agreement to his measured value of -0.35.

SUMMARY

We have used several representations of measured reactive and inelastic scattering to study the quali-

tative features of the N_2^+–H_2 reaction dynamics. From the velocity vector distributions $\bar{I}(\theta, \phi, u; E^0)$ we learned that the reaction proceeds by a direct interaction mechanism, that the internal excitation energy of the products is a function of their scattering angle, and that most products are scattered in the original direction of the N_2^+ projectile at a speed somewhat greater than that predicted from the ideal stripping model. Velocity ratio plots showed that the ideal knockout mechanism fails to give a quantitative explanation of the velocity or occurrence of backscattered products. From the differential reaction cross section $I(\theta; E^0)$ we drew further support for the direct interaction mechanism and the predominance of forward scattering. The isotopic variation of this differential cross section at three angles and its dependence on projectile energy showed that the degree to which the reaction product can be stabilized through recoil may in large measure account for observed isotope effects. The observations that the total reaction cross section falls with increasing projectile energy and is nearly the same for different isotopes at the same energy relative to the abstracted atom further reflect the importance of product stabilization through recoil at high collision energies. From the studies of nonreactive scattering we found an inelastic process which increases in importance with increasing projectile energy. Together with charge transfer and dissociative charge transfer, collisional dissociation is an important channel which competes with chemical reaction at high collision energies. We expect that more detailed conclusions can be drawn about the reaction dynamics when these data are compared to calculated molecular trajectories obtained using trial potential-energy surfaces.

ACKNOWLEDGMENTS

This work was supported by the U.S. Atomic Energy Commission. We would like to thank Mr. Arthur Werner for his assistance with some of the experiments. W.R.G. acknowledges the support received from an NSF predoctoral fellowship; E.A.G. acknowledges a postdoctoral fellowship from the National Center for Air Pollution Control of the Public Health Service.

25

Reprinted from *Faraday Soc. Discuss.* 44:123–136 (1967)

Crossed-Beam Studies of Ion-Molecule Reaction Mechanisms

By Z. Herman,* J. Kerstetter, T. Rose and Richard Wolfgang

Dept. of Chemistry, Yale University, New Haven, Connecticut, U.S.A.

Techniques have been developed for the crossed-beam study of certain simple reactions as a function of energy. An approximately monoenergetic ion beam is crossed with a thermal molecular beam and ionic products are characterized as to their nature, angular distribution, translational energy and, by difference, internal energy. Such measurements may be made as a function of energy available in the centre-of-mass system in the range 0·1-25 eV.

Initial studies using this method have been made on the reactions $Ar^+ + D_2 \rightarrow ArD^+ + D$ and $N_2^+ + D_2 \rightarrow N_2D^+ + D$. Contrary to earlier suggestions, their mechanisms are not dominated by an intermediate of relatively long life. Instead, even at the lowest energies, these reactions appear to be " direct " in that the collision complex has a lifetime comparable to, or less than, one molecular rotation. The results suggest a new general model for direct reactions which postulates dominance by long-range forces. This predicts product energies and angular distributions well in accord with experimental findings. At lowest reactant energies, part of the energy of reaction appears as translation, but as the incident velocity increases, the situation reverses and there is conversion of translational to internal energy. At highest reactant energies, spectator stripping is approached as a limiting case. Interpretation of the data by this model provides reasonable estimates for the critical internuclear distance in the intermediate state.

Over the past decade, there has been intensive exploration of the new field of chemical reactions above threshold or thermal energies, an area often called " hot " chemistry. Studies of energetic ion-molecule processes have been made using modified mass spectrometers,[1] while interactions of hot atoms have been investigated principally by use of recoil techniques.[2] Such work has served to uncover the richness of the field but has left some fundamental questions of mechanism unanswered. In this article, we report on first results of a study in which crossed beams, one of them accelerated, were used in an attempt to answer some of these problems.[3]

The apparatus[4] generates monoenergetic ions of variable energy which are crossed with a beam of thermal neutrals. The angular distributions and kinetic energies of product as well as reactant ions are measured by a combined mass spectrometer, stopping-potential analyzer. The energy of the reactant ions can be varied from 25 to 0·7 eV. Depending on the system used, this means that relative energies of collision as low as 0·1 eV may be achieved. Thus these studies can provide an effective overlap with standard " thermal " experiments.

Results on the simple reactions

$$Ar^+ + D_2 \rightarrow ArD^+ + D, \qquad \Delta H \cong -1\cdot2 \text{ eV,}^{[5]}$$
$$N_2^+ + D_2 \rightarrow N_2D^+ + D, \qquad \Delta H \cong -0\cdot3 \text{ eV.}^{[6]}$$

are reported here. These systems, which have been extensively studied in the past, proved suitable for an investigation of the fundamental question of whether reaction proceeded via a long-lived collision complex, or whether the mechanism was " direct " (i.e.), reaction occurs on the time-scale of a molecular vibration). Henglein and

* present address: Institute of Physical Chemistry, Czechoslovak Academy of Science, Prague, Czechoslovakia.

collaborators have provided convincing evidence that at high energies (> 30 eV) reaction proceeds in a direct process, postulating what is now known as the " spectator stripping " mechanism.[7] Other studies confirm this,[8] but also suggest—on a more circumstantial basis—that at low energies a long-lived complex type of mechanism becomes dominant.[6, 8, 9] Our data are not consistent with this latter conclusion. Instead, it appears that the same mechanism is operative over the whole energy range measured, and that at high energies this mechanism approaches spectator stripping as a limiting case. In this modified stripping process, long-range ion induced-dipole, rather than short range " chemical " forces, appear to determine the distribution of internal and translational energy in the products.

EXPERIMENTAL

A schematic diagram of the apparatus EVA (Evatron) is shown in fig. 1. The ion beam intersects with the molecular beam at a fixed 90° angle. Ions from the collision zone pass through an energy analyzer, a 60°-sector mass spectrometer and are detected by an electron multiplier. The beam sources are mounted on the rotatable lid of the scattering chamber and can thus pivot about the collision centre. The overall dimensions of the apparatus are small in order to obtain maximum particle densities.

The individual components of EVA are shown in fig. 2 and will be briefly described. A full description will appear elsewhere.[4]

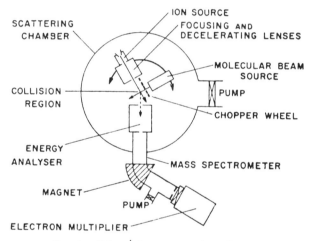

FIG. 1.—Schematic representation of EVA.

ION GUN.—This is a critical part of the design as a resonably intense low-energy beam of small cross-section, having a small energy and angular spread, is required. While this is not a difficult problem at higher energies, it had not previously been solved for the region around 1 eV. Reagent gas is ionized by a simple electron bombardment source, and ions are extracted at approximately 70 V through a circular aperture. The beam is then formed, decelerated and focused by a series of 14 circular and semi-circular elements. This lens assembly is based on a design of Gustafsson and Lindholm[10] suitably modified to make the lowest energy range accessible. The final beam emerges through a slit 1·0 × 0·5 mm.

MOLECULE SOURCE.—This is a Zacharias source constructed from fluted foil forming channels 1 cm long and each 1·4 × 10⁻⁴ cm² in area. The beam is collimated by a slit 2·5 mm vertical by 1 mm horizontal. It is pulsed by a rotating chopper wheel at a frequency of about 200 Hz. The source is operated at 55°C and is assumed to provide a corresponding Boltzmann distribution of particles.[11]

COLLISION REGION AND ENERGY ANALYZER.—The beams collide in a carefully screened field free region. Ions may pass through a detector slit 1 mm vertical by 0·2 mm horizontal into an energy analyzer. This corresponds to a resolution of approximately 0·9° horizontal and 3·3° vertical. The energy analyzer consists of a fine grid (90 lines/in.) at the potential of the last ion lens, followed by a finer screen (375 lines/in.) of variable potential which stops all ions below a given energy. The beam which passes is carefully refocused and accelerated to approximately 500 eV by a series of lenses.

MASS SPECTROMETER.—This is a simple 60°-sector instrument (7·5 cm radius) with an electron multiplier detector supplied by E.T.H. (Zürich).

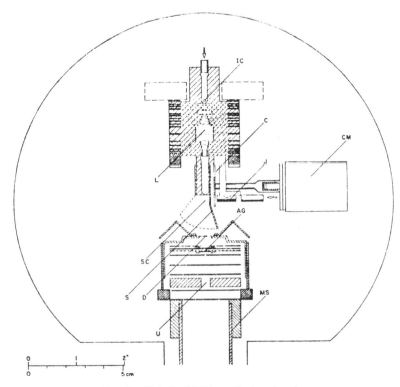

FIG. 2.—Detail of EVA scattering chamber.

IC, Ionization chamber; L, deceleration lenses; J, molecular jet; SC, scattering centre; S, screen with collimating slit; C, chopper; CM, chopper motor; D, detection slit; AG, energy analyzer grid; U, unipotential lens; MS, mass spectrometer.

AUXILIARY EQUIPMENT.—The scattering chamber is pumped at a rate of 300 l./sec. Background pressure ($\sim 10^{-5}$ torr) is sufficiently high that a significant amount of product is formed outside the beam intersection zone along the path of the ion beam. To discriminate against this noise, the output of the electron multiplier is fed into a lock-in amplifier operated in phase with the frequency of the chopper which modulates the beam.

PROCEDURE.—After appropriate tuning, the angular distributions of both reactant and product ions of all energies are measured. At lowest energies, a small correction is applied to compensate for scattering of the background (D.C.) product ions by the modulated D_2 beam. The energy spectra of both reactant and product ions are then measured at several angles.

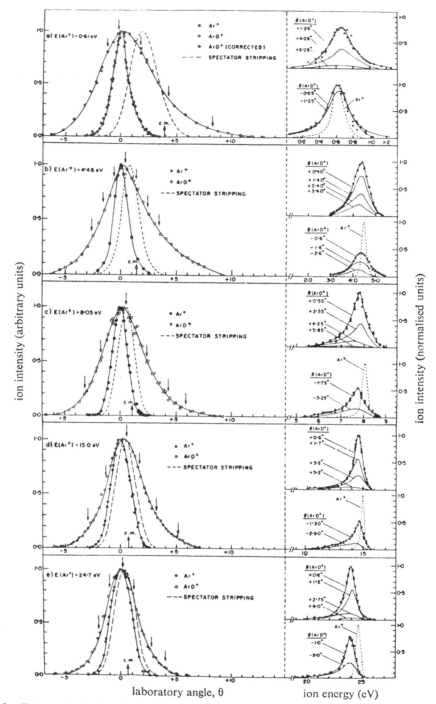

FIG. 3.—Data on $Ar^+ + D_2 \rightarrow ArD^+ + D$ reaction at several LAB energies of Ar^+. On the left are shown angular distributions. Calculated values for spectator stripping are dashed and the laboratory angles of centre-of-mass motion are indicated (cm). On the right, energy distributions, taken at angles indicated by arrows, are shown. Angular plots are normalized to unity for both product and reactant. Areas under energy spectra correspond to total relative intensities at the given angles.

RESULTS

Data for the reaction $Ar^+ + D_2 \rightarrow ArD^+ + D$ at each of several energies appear in fig. 3. Fig. 4 presents the corresponding information for $N_2^+ + D_2 \rightarrow N_2D^+ + D$ at two energies. The left section of each of these figures gives the total angular distributions

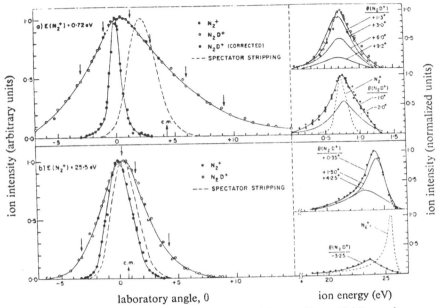

FIG. 4.—Data on $N_2^+ + D_2 \rightarrow N_2D^+ + D$ reaction at the highest and lowest LAB energies measured.

of reactant and product at specified ion beam energies. For comparison, a theoretical angular distribution of product ions as calculated from the spectator stripping model is indicated. At angles indicated by arrows, energy spectra were measured. These are at the right section of each figure.

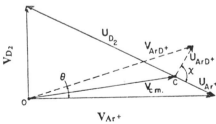

FIG. 5.—Construction and terminology used in Newton diagrams. The velocity of D_2, $v(D_2)$, is the most probable velocity α.

The data are summarized in the form of velocity vector Newton diagrams (fig. 6 and 7), constructed as indicated in fig. 5.[13] These show the most probable laboratory and centre-of-mass velocities of the reagents (v_{X^+}, v_{D_2} and u_{X^+}, u_{D_2} respectively) and the velocity of the centre-of-mass itself.[11] (The diagrams are so constructed that they show reactant velocities as they would be if the particles leave the collision centre [O in the LAB and C in the CM system] having had no interaction.)

Contour diagrams of the intensity of ionic products are then constructed as follows. The energy spectra at each laboratory angle θ are converted to velocity spectra by

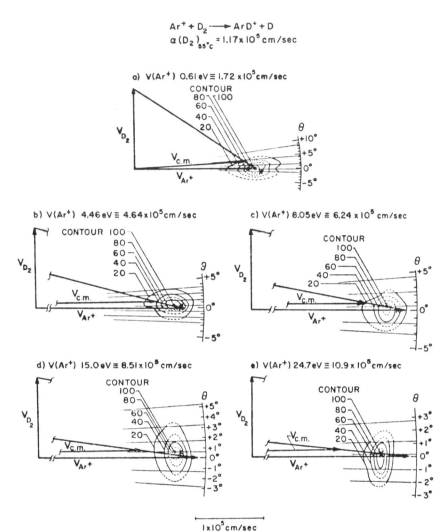

$$Ar^+ + D_2 \longrightarrow ArD^+ + D$$
$$\alpha (D_2)_{55°C} = 1.17 \times 10^5 \, cm/sec$$

a) $V(Ar^+)$ 0.61 eV $\equiv 1.72 \times 10^5$ cm/sec

b) $V(Ar^+)$ 4.46 eV $\equiv 4.64 \times 10^5$ cm/sec

c) $V(Ar^+)$ 8.05 eV $\equiv 6.24 \times 10^5$ cm/sec

d) $V(Ar^+)$ 15.0 eV $\equiv 8.51 \times 10^5$ cm/sec

e) $V(Ar^+)$ 24.7 eV $\equiv 10.9 \times 10^5$ cm/sec

1×10^5 cm/sec

FIG. 6.—Newton diagrams constructed from data in fig. 3. Relative intensities of product for given LAB ordinates v_{ArD^+}, θ and corresponding CM ordinates u_{ArD^+} are shown as contours. These intensities are relative to the LAB system, i.e., they are intensities as seen by a detector subtending a given solid angle with respect to the LAB origin, and sensitive to particles with velocities between v and $v + dv$. (Crosses $\times$ indicate estimated positions of peak intensity relative to the CM system. In the CM system intensities are as seen by a detector subtending a constant solid angle with respect to the CM origin and sensitive to CM velocities between u and $u + du$.)

multiplying every point by the corresponding velocity v. This accords with the transformation

$$I(E)dE = I(v)dv,$$

where I is intensity. The area under each such curve is then normalized to the corresponding total intensity at the angle θ, with this intensity at the maximum in the angular distribution being arbitrarily set equal to unity. The points are plotted on the Newton diagram at the appropriate laboratory velocities v_{XD^+} and angles θ, and appropriate contours are drawn.   There results a plot of intensities as seen by a detector subtending a constant small solid angle and sensitive to velocities between

TABLE 1.—SUMMARY OF RESULTS ON $Ar^+ + D_2$ REACTION

energy reactants		ang. displacement			vel. ArD⁺ CM	CM vel. prod.	transl.
LAB E_{Ar^+} (eV)	CM T_1 (eV)	observed $\Delta\theta$	spec. strip. $\Delta\theta$ (ss)	ang. width $\theta_{\frac{1}{2}}$	u_{ArD^+} ($\times 10^4$ cm/sec)	CM vel. react. u_{ArD^+}/u_{Ar^+}	exoergic. $Q_{expt.}$ (eV)
0·61	0·081	0·3	1·9	4·1°	1·7	0·91	+0·06
0·71	0·090	0·6	1·7	5·8	2·4	1·20	+0·18
2·23	0·228	0·2	1·0	3·5	2·5	0·79	+0·07
4·38	0·425	0·3	0·8	2·9	2·7	0·62	−0·09
4·46	0·431	0·3	0·8	3·0	2·6	0·61	−0·10
8·05	0·758	0·3	0·6	2·7	3·2	0·56	−0·25
15·0	1·391	0·3	0·4	1·8	4·4	0·57	−0·45
24·6	2·27	0·2	0·4	1·2	5·7	0·57	−0·71

E_{Ar^+}, LAB energy of Ar^+ beam; T_1, initial CM energy of reactants; $\Delta\theta$, observed displacement of peak of ArD^+ angular distribution with respect to that of Ar^+(LAB); $\Delta\theta(ss)$, peak displacement of ArD^+ as calculated from spectator stripping model; $\theta_{\frac{1}{2}}$, full width at half-maximum of ArD^+ angular distribution, minus full width at half-maximum of Ar^+ angular distribution; u_{ArD^+}, u_{Ar^+}, CM velocities of ArD^+ and Ar^+ respectively; $Q_{expt.}$ translational exoergicity calculated from data.

v and $v+dv$. (It should be noted that the corresponding volume elements on the Newton diagram are not constant but increase as v^2.)

Tables 1 and 2 present a partial summary of our results.   The laboratory angle of the peak of product intensity is given, as is the half-width of this distribution.   From

TABLE 2.—SUMMARY OF RESULTS ON $N_2^+ + D_2$ REACTION

energy reactants		ang. displacement			vel. N₂D⁺ CM	CM vel. prod.	transl.
LAB $E_{N_2^+}$ (eV)	CM T_1 (eV)	observed $\Delta\theta$	spec. strip. $\Delta\theta$ (ss)	ang. width $\theta_{\frac{1}{2}}$	$u_{N_2D^+}$ ($\times 10^4$ cm/sec)	CM vel. react. $u_{N_2D^+}/u_{N_2^+}$	exoergic. Q_{expt} (eV)
0·72	0·115	0·5°	2·2°	7·1°	2·8	0·89	+0·08
0·78	0·122	0·3	2·1	7·3	3·1	0·96	+0·12
1·04	0·155	0·5	1·9	5·5	2·9	0·78	+0·05
1·64	0·230	0·5	1·5	5·8	3·6	0·81	+0·09
3·06	0·407	0·5	1·0	3·8	3·9	0·66	−0·04
4·55	0·594	0·5	0·9	4·2	4·3	0·61	−0·13
5·35	0·693	0·3	0·8	4·7	4·6	0·59	−0·19
7·92	1·02	0·5	0·7	4·3	5·2	0·56	−0·34
11·0	1·40	0·3	0·6	2·9	6·5	0·59	−0·36
15·7	1·98	0·2	0·5	2·3	6·0	0·46	−1·2
25·5	3·21	0·5	0·4	2·0	8·2	0·49	−1·6

$E_{N_2^+}$, LAB energy of N_2^+ beam; T_1, initial CM energy of reactants; $\Delta\theta$, observed displacement of peak of N_2D^+ angular distribution with respect to that of N_2^+(LAB); $\Delta\theta(ss)$, peak displacement of N_2D^+ as calculated from spectator stripping model; $\theta_{\frac{1}{2}}$, full width at half-maximum of N_2D^+ angular distribution, minus full width at half-maximum of N_2^+ angular distribution; $u_{N_2D^+}, u_{N_2^+}$ CM velocities of N_2D^+ and N_2^+ respectively; $Q_{expt.}$ translational exoergicity calculated from data.

the Newton diagrams is taken the CM velocity u_{XD^+} for products appearing at the position of peak intensity. This quantity is used to calculate the translational exoergicity Q (see below).

In evaluating these results, it should be kept in mind that they are taken from Newton diagrams in which intensity contours are still relative to the LAB rather than the CM system.[14] (However, the estimated positions of peak intensity in

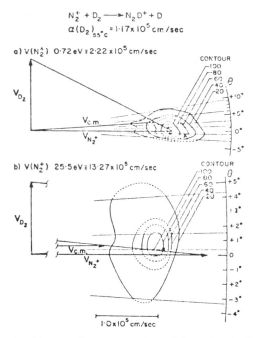

$$N_2^+ + D_2 \longrightarrow N_2D^+ + D$$
$$\alpha(D_2)_{55°C} = 1.17 \times 10^5 \text{ cm/sec}$$

a) $V(N_2^+)$ 0.72 eV = 2.22×10^5 cm/sec

b) $V(N_2^+)$ 25.5 eV = 13.27×10^5 cm/sec

1.0×10^5 cm/sec

FIG. 7.—Newton diagrams constructed from data in fig. 4.

the CM system are indicated.[13]) Furthermore, velocities in the centre-of-mass system are arbitrarily referred to a single centre-of-mass C corresponding to the most probable velocity α. Actually there is a distribution of centre-of-masses corresponding to the one-dimensional Boltzmann distribution of the D_2 beam.

DISCUSSION

DIRECT MECHANISM AGAINST LONG-LIVED INTERMEDIATE MODEL

Previous work on the reactions discussed here, particularly by Henglein,[7] Bailey and collaborators,[8] has shown that at higher energies (>4 eV LAB), they proceed by a direct mechanism, i.e., the interaction takes place on a time scale comparable to one vibration. At lower energies, however, it has been widely thought that ion-molecule reactions in general and these processes in particular, occur via an intermediate complex of sufficiently long lifetime to undergo normal unimolecular decomposition, i.e., having a lifetime of many vibrations.[6, 8, 9] The evidence for such a belief is, however, circumstantial, being based largely on a proposed interpretation of observed isotope effects.[6]

Our results give direct evidence that, even at the lowest energy measured (only ~0.1 eV in the CM system), the contribution of any long-lived intermediate is small and that reaction is dominated by a direct mechanism. If a complex that rotated many times were formed, the distribution of products would be symmetric about the CM. Inspection of the angular distributions (fig. 3 and 4) show that this cannot be the case. Furthermore, the energy spectra of the ionic products peak at too high an energy to be consistent with formation of an intermediate complex in which the momentum of the incident ion is shared between both products.[15] These observations are summarized in the Newton diagrams which clearly do not show the symmetry about the centre-of-mass (C) required by long-lived complex models.[16]

SPECTATOR STRIPPING MODEL

At the highest energy (25 eV) it appears that the spectator stripping model provides a reasonable representation of the reactions studied. In this mechanism, the freed D atom proceeds with a velocity vector unchanged from the D_2 reactant, i.e., the D atom has been a spectator to the reaction. The product ion will consequently carry the momentum of both the ionic reactant and half of that of the D_2 reactant. It will appear at a laboratory angle θ about half that of the CM motion of the system. The angular distribution calculated on the basis of spectator stripping, taking into account the Maxwellian distribution of the D_2 beam and the angular spread of the ion beam is shown by dashed lines in fig. 3 and 4. At 25 eV, although the experimental distributions are slightly broader, they have a peak near the calculated values. Similarly, the energy of the ionic products should be $(28/30)E_{N_2^+}$ and $(40/42)E_{Ar^+}$ according to spectator stripping, and the peaks of the actual energy spectra of N_2D^+ and ArD^+ approximate these values quite well.

Even at this highest energy, however, there are noticeable deviations and as the reactant ion energy decreases, these increase sharply. The direction of these deviations is the opposite from that expected were there a transition to a long-lived intermediate complex, i.e., with decreasing reaction energy, most product ions appear at angles increasingly smaller than calculated for spectator stripping, while their energies become *larger* than expected from spectator stripping. In terms of the Newton diagrams, this means that the peak in the distribution shifts even further forward with respect to the centre-of-mass, C.

A NEW MODEL OF DIRECT REACTIONS

The failure of the spectator stripping model presumably stems from its essential feature, the ignoring of all intermolecular forces, except insofar as they lead to the required transfer process. Any model of direct reactions must, in order to be more realistic, take into account these forces. The difficulty is that the short-range potentials involved are not well understood. However, the longer-range intermolecular potentials have been characterized. It therefore seems reasonable to construct a model which quantitatively takes into account these latter forces, but assumes that while short-range interactions serve to cause the actual transfer, they have no other effect on reactants or products. The success of such a model then gives us a measure of the extent to which long-range forces alone determine energy and momentum transfer in the reaction.

Reactions between ions and polarizable molecules would seem *a priori* to present a favourable system for this model since the potentials involved are both fairly long-range (r^{-4}) and moderately strong (~ 1 eV). We consider the reaction $X^+ + YZ \rightarrow XY^+ + Z$. As X^+ and YZ approach, a dipole is induced in YZ and the species are accelerated toward each other in accordance [17] with the approximate potential $V_{reactants} = -\alpha_{YZ}e^2/2r^4_{(reactants)}$. Here α_{YZ} is the polarizability of YZ, e is the unit of electric charge and $r_{(reactants)}$ is the separation $X^+ - YZ$. Transfer then occurs at some distance of closest approach, r_{X-YZ}, yielding products separated by a corresponding distance r_{XY-Z}. As a simple first approximation we assume that the freed particle Z initially has the same speed as does YZ just before reaction, although its direction may be changed.[18]   The products XY^+ and Z are then decelerated according to the potential $V_{products} = -\alpha_z e^2/2r^4_{(products)}$.   The net result will generally be that XY^+ will have a larger velocity than expected from spectator stripping, although this need not always be the case.   A schematic representation of this model is shown in fig. 8.

In addition to specifically considering only the long-range attractive forces, the model, as applied here to ion-molecule interactions, then assumes the following approximation I :   The ion-induced dipole potential depends on r^{-4} over all distances of separation.   (This approximation is strictly correct only for $r \to \infty$).

Let T_i and $T_{i,M}$ be the total kinetic energy and kinetic energy of species M respectively, in the centre-of-mass (CM) system, $i = 1,2,3,$ or 4, where 1 refers to reactants at infinite separation, 2 refers to reactants just prior to transfer, 3 refers to products just following transfer, 4 refers to products at infinite separation.

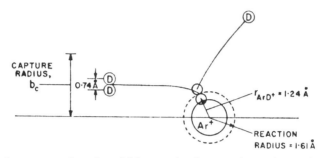

FIG. 8—Schematic representation of model for reaction between Ar^+ and D_2. D_2 approaching Ar^+ with impact parameter smaller than the classical Langevin [17] value $b_c = (2\alpha e^2/T_1)^{0.25}$ is captured. It is accelerated towards the ion by ion-induced dipole forces.   When the reaction radius is reached, one D atom is detached and the other freed to proceed with a speed initially unchanged from that of D_2 at moment of reaction.   In receding, it is decelerated by ion-induced dipole forces acting between ArD^+ and D.   Value of reaction radius indicated is that determined experimentally for peak in laboratory intensity on Newton diagrams.

Let X, Y, Z represent masses. r_{X-YZ} (and r_{XY-Z}) is the separation between the centre of charge of X^+ (or XY^+) and the electrical centre of YZ (or Z). T_1 is specified in the experiment.   We wish to compute the final energies T_4 and $T_{4,XY}$ and the final velocity in the CM system u_{XY^+} in order to compare this with the measured velocity v_{XY^+} in the LAB system.   Then

$$T_2 = T_1 + \alpha_{YZ} e^2/2r^4_{X-YZ}.$$

By conservation of momentum

$$T_{2,YZ} = XT_2/(X+Y+Z).$$

According to the model in its simplest form there is no change in the velocity of the freed atom Z at the moment of transfer.[18]   Therefore

$$T_{3,Z} = ZT_{2,YZ}/(Y+Z).$$

Hence

$$T_{3,XY} = ZT_{3,Z}/(X+Y).$$

Then

$$T_4 = T_3 - \alpha_Z e^2/2r^4_{XY-Z}, \text{ where } T_3 = T_{3,XY} + T_{3,Z},$$

$$T_{4,XY} = ZT_4/(X+Y+Z),$$

and

$$u_{XY} = (2T_{4,XY}/(X+Y))^{0.5}$$

Combining these equations, we get for the CM energy and velocity of the ionic product:

$$T_{4,XY} = \frac{Z}{X+Y+Z}\left[\frac{ZX}{(X+Y)(Y+Z)}\left(T_1 + \frac{\alpha_{YZ}e^2}{2r_{X-YZ}^4}\right) - \frac{\alpha_Z e^2}{2r_{XY-Z}^4}\right], \tag{1}$$

$$u_{XY} = \left\{\frac{2Z}{(X+Y)(X+Y+Z)}\left[\frac{ZX}{(X+Y)(Y+Z)}\left(T_1 + \frac{\alpha_{YZ}e^2}{2r_{X-YZ}^4}\right) - \frac{\alpha_Z e^2}{2r_{XY-Z}^4}\right]\right\}^{0.5}. \tag{2}$$

The model determines the partition of energy among the products. We define Q as the translational exoergicity (net conversion of internal to translational energy),

$$Q = T_4 - T_1 \tag{3}$$
$$= Q_0 - BT_1, \tag{4}$$

where

$$Q_0 = \frac{ZX}{(X+Y)(Y+Z)}\frac{\alpha_{YZ}e^2}{2r_{X-YZ}^4} - \frac{\alpha_Z e^2}{2r_{XY-Z}^4}, \qquad B = 1 - \frac{ZX}{(X+Y)(Y+Z)}.$$

The above equations yield the corresponding quantities for the spectator stripping model, $T_{4,XY}(SS)$, $u_{XY}(SS)$ and $Q(SS)$, if the terms containing α are set equal to zero. In any case, as the initial translational energy T_1 becomes large, these terms become relatively negligible and the model asymptotically approaches spectator stripping. This is, of course, just what has been experimentally observed.

COMPARISON WITH EXPERIMENT

To apply this model to actual reactions, three further working assumptions or approximations are made. II: The $X^+ - YZ$ complex is linear at the moment when Y is transferred. This assumption also implies that the parallel polarizability of

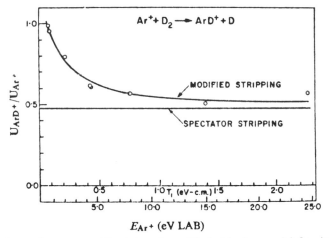

FIG. 9.—Predictions of spectator stripping and modified stripping model for $Ar^+ + D_2$ reaction compared with experiment (open circles). The ratio of calculated or experimental values of the CM velocity of product u_{ArD^+} to the CM velocity of reactant u_{Ar^+} are plotted against the energy of the Ar^+.

YZ, $\alpha_\parallel$, should be used. III: The charge is localized at the centre of X^+ throughout. IV: The YZ distance remains unchanged until reaction occurs.

We now normalize the model to the data at the lowest energy. This involves determining the parameters r_{X-YZ} and r_{XY-Z}. These are related: if Y and Z are identical or isotopic, r_{XY-Z} will be greater than r_{X-YZ} by half the Y—Z bond distance. The model is then tested by determining whether the reaction distance is reasonable and by using the same parameters to predict results at higher energies.

For the Ar^+ reaction, we use $\alpha_{\parallel} = 0.93 \times 10^{-24}$ cm³ [19] and 0·74 Å as the D—D bond distance.[20] u_{ArD+} is taken from the point of peak intensity in the Newton diagram (see table 1). r_{Ar-D_2} is then determined as 1·61 Å This corresponds to an Ar^+—D distance at the moment of reaction of 1·24 Å. This may be compared to the bond distance of 1·27 Å in the isoelectronic molecule DCl. Evidently the magnitude obtained is reasonable; indeed, the precision of the agreement must be somewhat fortuitous.[21]

The predictions of the model at higher energies, calculated using this value of r_{Ar-D_2}, are shown by the solid line in fig. 9. This represents the CM velocity u_{ArD+} (for the position of peak intensity in the LAB system) relative to the initial CM velocity of the Ar^+, u_{Ar+}. The data points are in agreement within experimental error.[21]

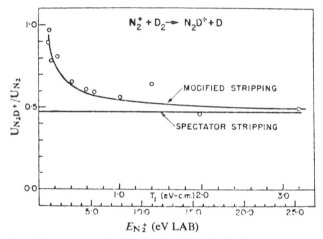

FIG. 10.—Predictions of spectator stripping and modified stripping model for $N_2^+ + D_2$ reaction compared with experiment.

A similar plot for the predictions of the model for the N_2^+ reaction is given in fig. 10. Again, agreement between calculated values and experimental results is good.[21] The model gives the N_2^+—D distance at the moment of reaction as 1·23 Å. This is less than the distance from the centre of the C—N bond to D of 1·65 Å in the iso-electronic molecule DCN, whereas with Ar^+, agreement between the analogous values was almost exact. Such a finding is reasonable. The centre of charge in N_2^+ may be significantly displaced from the centre of the bond and furthermore, the N_2^+ may not be axially aligned with the D_2 in the reaction configuration. Both effects would tend to reduce the reaction distance to less than the value indicated by the bonding in DCN.

ENERGY PARTITION

As given in eqn. (4), the translational exoergicity Q is maximal when the reagents are at rest. It then decreases linearly with reactant energy, becoming negative when $T_1 > Q_0/B$. Our findings are consistent with these expectations (see tables 1 and 2).

The internal energy of product is correspondingly

$$\mathscr{E}_{int} = -\Delta H - Q = BT_1 - \Delta H - Q_0 \tag{5}$$

For the reactions studied here, the translational exoergicity at zero reactant energy Q_0 is about 0·15 eV.

Conservation of energy requires that if $T_1 = 0$, $Q_0 \leqslant -\Delta H$. It is an interesting consequence that reactions exoergic by less than $-Q_0$ should not proceed by this mechanism at lowest energies.

SHAPE OF THE ANGULAR DISTRIBUTION

According to spectator stripping, the product ion should recoil directly forward. The present model, however, implies that its direction should depend on the impact parameter. Indeed, while this will generally be forward, backward recoil is not excluded.[15] The broadening of the angular distribution, as the energy decreases and spectator stripping becomes a poorer model, is qualitatively reasonable. A quantitative comparison is now being attempted.

The distribution of N_2D^+ is systematically broader than that of ArD^+. This is plausible. With N_2^+, there are two extra degrees of freedom: the angle between the N_2^+ and the D_2 axes at the time of reaction, and the extra sink for energy presented by the N_2^+ bond. A further broadening factor may be internal excitation of the N_2^+, particularly in the vibrational mode.

SUMMARY AND IMPLICATIONS

It has been shown that the reactions of Ar^+ and N_2^+ with D_2 are dominated by a direct or impact mechanism. This is true even at incident CM energies of only 0·1 eV—small compared to the additional kinetic energy generated as the particles are attracted into each other. Such a finding corrects earlier suggestions that a long-lived complex may be dominant. It also shows to be unjustified the common prejudice that in all chemical reactions where there is strong attraction between the reactants, a complex is necessarily formed if only the incident energy is low enough.

A model for direct reactions is proposed in which long-range forces dominate all observable aspects of the process: cross-section, angular distribution and energy partition. In the case of Ar^+ and N_2^+ reaction with hydrogen, previous work has already indicated that the cross-section approximately corresponds to the Langevin cross-section for close collision, as calculated using ion-induced dipole forces.[9] Our studies show that these same forces account remarkably well for the observed angular distribution and energy partition. However, the very success of the model implies that ion-molecule reactions may be a poor tool for probing the nature of shorter range intermolecular forces.

The model implies, and experiment confirms, that the translational exoergicity of ion-molecule reactions proceeding by a direct mechanism changes sign as a function of incident energy. At low reactant velocities, there is a net conversion of internal to translational energy while at higher velocities the converse is true.

The model proposed here, when modified to take into account the appropriate long-range forces, should also be applicable to direct reactions between species other than ions and non-polar molecules.

This work was supported by the U.S. Atomic Energy Commission and the National Aeronautics and Space Administration. We thank Dr. R. James Cross and Dr. Peter Hierl and Mr. Alfred Lee for their assistance and comments. T. Rose acknowledges support by an NIH Fellowship.

[1] See, e.g., C. F. Giese, *Adv. Chem. Physics*, 1966, **10**, 247 ; M. Henchman, *Ann. Reports*, 1965, **62**, 39, which give references to earlier reviews.

[2] R. Wolfgang, *Ann. Rev. Physic. Chem.*, 1965, **16**, 15.

[3] A preliminary communication has appeared : Z. Herman, J. Kerstetter, T. Rose and R. Wolfgang, *J. Chem. Physics*, 1967, **46**, 2844.

[4] To be described fully elsewhere.

[5] F. S. Klein and L. Friedman, *J. Chem. Physics*, 1964, **41**, 1789.

[6] T. F. Moran and L. Friedman, *J. Chem. Physics*, 1965, **42**, 2391.

[7] A. Henglein, K. Lacmann and G. Jacobs, *Ber. Bunsenges, Physik. Chem.*, 1965, **69**, 279.

[8] L. D. Doverspike, R. L. Champion and T. L. Bailey, *J. Chem. Physics*, 1966, **45**, 4385.

[9] B. R. Turner, M. A. Fineman and R. F. Stebbings, *J. Chem. Physics*, 1965, **42**, 4088.

[10] *Atomic and Molecular Processes* ed. by D. R. Bates (Academic Press, London, 1962), p. 705.

[11] The normal Maxwell-Boltzmann distribution in a bulk gas is proportional to (v^2/α^3) exp $(-v^2/\alpha^2)$, where α is the most probable velocity. The velocity distribution in the beam is proportional to the same expression multiplied by v.[12] However, because the time spent in the reaction zone is inversely proportional to the velocity, the distribution of interacting particles (at any given CM energy) is again given by the normal Maxwell-Boltzmann law. We thus use for our most probable velocity of D_2 in the reaction volume the usual expression $\alpha = (2kT/m)^{0.5}$, where T is the temperature of the source. The most probable velocity of the ions is measured directly.

[12] N. F. Ramsey, *Molecular Beams* (Oxford, 1956), p. 20.

[13] D. R. Herschbach, *Adv. Chem. Physics*, 1966, **10**, 332.

[14] Conversion of intensities from the laboratory to the centre-of-mass system is a complex matter since a distribution rather than a single centre-of-mass is involved. However, some idea of the effects of such a transformation can be obtained if a single centre-of-mass, that defined by the most probable D_2 velocity α, is assumed. The transformation is made by multiplying each laboratory intensity by the Jacobian factor u^2/v^2. This, in effect, changes the intensity from what it appears when viewed from the laboratory origin 0 to what it appears when viewed from the centre-of-mass origin C (see fig. 5). (The Jacobian (u^2/v^2) cos δ is appropriate to the case of elastic scattering (F. A. Morse and R. B. Bernstein, *J. Chem. Physics.*, 1962, **37**, 2019). However, it has been extensively used for transformations in thermal reaction systems in which product velocity is measured, and this application appears incorrect.) The results of such a transformation will be meaningless in the vicinity of C but should be increasingly valid as u increases. The positions of CM peak intensity estimated in this manner are indicated in fig. 6 and 7. (Note that in both systems intensities are given per unit volume of configuration space on the Newton diagram ; but these volumes are not uniform, increasing in size as v^2 in the LAB and u^2 in the CM systems.)

[15] Doverspike, Champion and Bailey [8] in measurements of energy distribution of N_2D^+ and ArD^+ formed at relatively high energies ($\sim$40 eV) find a small peak " backward " of the CM. They interpret this as indicating a contribution by a long-lived complex. However, it seems plausible that such a peak represents " rebound " processes as must necessarily occur if reaction takes place on collision at very small impact parameters and consequent " head-on " collisions.

[16] Such symmetry can be expected only if the intensities have been properly converted to the CM system.[13] In the Newton diagrams shown, intensities are plotted in the LAB system. However, even upon making the appropriate Jacobian transformation, the intensity distributions would remain qualitatively similar and quite unsymmetric with respect to the centre-of-mass.

[17] E. W. McDaniel, *Collision Phenomena in Ionized Gases* (John Wiley, 1964), p. 71.

[18] The assumption that the velocity of the freed atom remains unchanged at the moment of transfer is essentially carried over from the spectator stripping model. (That model also implies that there is no change in direction, which we do not assume here.) This is a convenient approximation but one which is likely to become inadequate at low energies. Further elaboration of the present modified stripping model is likely to require that this assumption be replaced by one which is physically more meaningful. Related work is in progress and will be published elsewhere.

[19] Landolt-Bornstein, *Zahlenwerte and Funktionen* (Springer, Berlin, 1951), vol. I, pt. 3, p. 510 ; Moelwyn-Hughes, *Physical Chemistry* (Pergamon Press, London, 1957), p. 373, 385.

[20] Bond distances are taken from : *Spec. Publ.*, no. 11 (Chemical Society, 1958).

[21] Calculations of this type may also be carried out using the estimated position of peak intensity in the CM system (see footnote 14 and fig. 6 and 7). This leads to reaction radii smaller by about 0·3 Å in the Ar^+ system and 0·4 Å in the N_2^+ system. Plots analogous to fig. 9 and 10, using these radii to provide theoretical estimates of the positions of CM peak intensity at higher energies, show equally good agreement between theory and experiment.

26

Reprinted from J. Chem. Phys. 56:851–854 (1972)

Kinematics of the C+(D₂, D)CD+ Reaction

C. R. IDEN, R. LIARDON, AND W. S. KOSKI

Department of Chemistry, The Johns Hopkins University, Baltimore, Maryland 21218

INTRODUCTION

The $C^+(D_2, D)CD^+$ reaction has been studied previously by Maier.[1] In the barycentric energy range 0.4–2.5 eV, the expected endothermic nature of the reaction was verified by measurement of the total cross section. However, the threshold energy was not accurately measured. Statistical phase-space theory has been used by Truhlar to determine theoretically the cross section as a function of energy,[2] and good agreement was obtained with the experimental results. The question arose whether the apparent success of the theory was related to the possible formation of a relatively long lived complex. In view of the fact that the reaction is endothermic and that the CD_2^+ complex may be bonded strongly relative to the energy of $C^+ + D_2$, the system may form a potential well of significant depth. Indeed, if the CD_2^+ complex had the same heat of formation as CH_2^+ formed from the electron bombardment of CH_3, the depth of the well would exceed 4 eV. Under such conditions one would expect, on the basis of simple RRK theory,[3] to have a high probability of persistent complex formation at low relative energies,[4,5] provided the accessibility of the state representing the deep well is not restricted for other reasons.[6] Therefore, it was considered worthwhile to study this reaction in detail, and this publication gives the results of the measurements of the angular and energy distributions of the CD^+ product ion. These distributions show that at low relative energies the reaction proceeded through a persistent complex as an intermediate, and at the higher energies studied the CD^+ ions were produced by a direct mechanism.

EXPERIMENTAL

The apparatus used in this study is shown in Fig. 1. The instrument has been described in detail elsewhere[7,8]; only a brief description shall be given here. Ions are formed in an electron bombardment source, extracted, and focused into a quadrupole mass spectrometer. After mass selection, the ion energy is fixed, and the ions are refocused by a lens system into a well collimated beam. The neutral reactant is introduced

by one of two possible techniques. A molecular beam, utilizing a fused glass capillary array, may be mounted at a fixed angle of 90° to the ion beam. For reactions with lower cross sections, a reaction chamber is used. The chamber is constructed in such a way that the ionic products may exit unimpeded at large laboratory angles. The detector consists of a retarding potential grid energy analyzer, a second quadrupole mass spectrometer, and a Bendix continuous channel electron multiplier. This entire assembly may scan

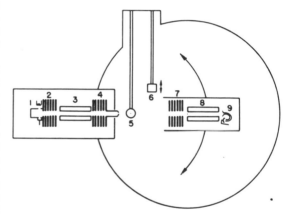

FIG. 1. Diagram of the instrument. The component parts of the apparatus are: 1, ion source; 2, lens system; 3, quadrupole mass filter; 4, lens system; 5, reaction chamber; 6, attenuation chamber; 7, retarding potential energy analyzer; 8, quadrupole mass filter; and 9, Channeltron electron multiplier.

around the reaction center over a large laboratory angle. Ion counting techniques are used in obtaining the angular and energy distributions of both reactant and product ions.

The C^+ ion beam was prepared by bombarding CO with 75 eV electrons. Conditions were maintained in the source such that only the ground state (^2P) was produced.[9] The mass analyzed beam ($\sim 5 \times 10^{-11}$ A) was directed into the reaction chamber containing 10–30 mtorr of deuterium. Operating pressure in the scattering region was maintained at 5×10^{-7} torr.

257

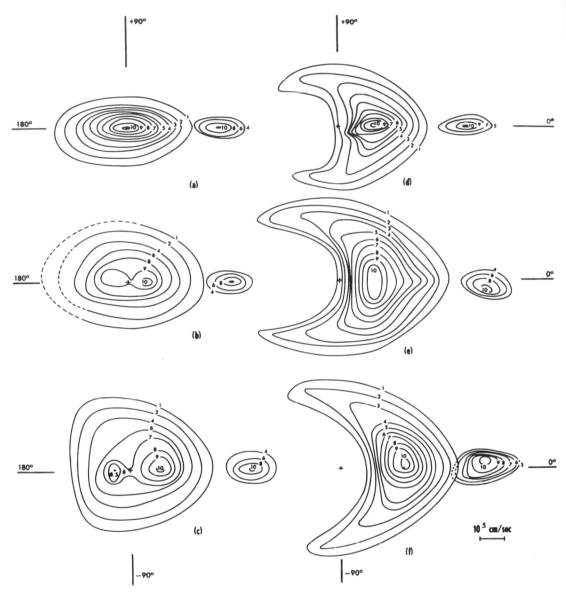

FIG. 2. The C[+] and CD[+] ion velocity distributions are shown in Cartesian coordinates in the center of mass system for the barycentric energies (a) 3.47 eV, (b) 4.40 eV, (c) 5.98 eV, (d) 7.04 eV, (e) 8.48 eV, and (f) 9.14 eV. In each case the C[+] distribution appears to the right of the associated CD[+] distribution. The small cross denotes the tip of the center of mass velocity vector.

RESULTS AND DISCUSSION

The energy and angular distributions of CD[+] formed by the reaction $C^+(D_2, D)CD^+$ were measured for barycentric energies between 3.47 eV and 9.14 eV. The data are given in Fig. 2 where the displays are intensity contour maps of particle flux in velocity space using Cartesian coordinates in the center of mass system.[10]

Examination of the contour maps reveals the main mechanistic features of the reaction. Figure 2(a) gives a product ion distribution which is nearly isotropic and symmetrical within experimental error about the barycentric angle ±90°. This symmetry[11] is consistent with persistent complex formation with a lifetime exceeding the rotational period of the system $(10^{-12}$ sec$)$. In Fig. 2(b) a small asymmetry about $\pm 90°$ begins to develop and indicates the onset of a more direct type of mechanism. The relative energy is 4.4 eV or 17.6 eV in the laboratory system. Simple application of the conservation of energy and momentum shows that at this projectile energy the internal energy of the CD⁺ ion is beginning to exceed its bond energy for the complex mechanism. As one proceeds to Fig. 2(c), the asymmetry continues to increase indicating an increasing fraction of the CD⁺ is formed by a direct mechanism. Figure 2(d) shows the peak of the CD⁺ distribution well forward of the center of mass represented by the small cross in the plot. Figures 2(e) and 2(f) give the distributions as the projectile energy is further increased. This sequence of contour maps shows the change from a completely complex mechanism to one that is completely direct.

In addition to experiments which led to the contour maps in which both angular and energy distributions were measured, a number of experiments were performed for which the product energy distribution was measured at an angle of zero degrees. The results of such experiments are given in Fig. 3, where the translational exoergicity, the difference between the relative translational energy of the reactants and products, is plotted against the C⁺ ion laboratory kinetic energy. For the reaction in question, the translational exoergicity, Q, was obtained from the most probable values of the velocity of the C⁺(V_1) and CD⁺(V_3) by the equation

$$Q = [30.00V_1^2 + 56.00V_3^2 - 84.00V_1V_3]1.0359 \times 10^{-12} \text{ eV}.$$

We will now consider the variation of Q as a function of the projectile energy. Conservation of energy requires

$$T_R + U_R = T_P + U_P + \Delta E_0.$$

T and U represent the translational and internal energy of the reactants (R) and products (P); ΔE_0 is the heat of the reaction. Since $Q = T_P - T_R$ then

$$Q = U_R - U_P - \Delta E_0.$$

In our case the C⁺ ion is in its ground state, and if one assumes the D₂ is also in its ground state, $U_R = 0$. Upper and lower bounds now can be set for Q. If CD⁺ is in its ground electronic state and if the ion contains no vibrational or rotational energy, then $U_P = 0$ and $Q = -0.40$ eV. A minimum value for Q is found when $U_P = 4.10$ eV, the bond dissociation energy of the ground electronic state. This minimum value of Q is -4.50 eV. Figure 3 gives the plot of Q vs primary ion energy. The solid lines give the calculated values for complex formation and for spectator stripping. The measured values over the first portion of

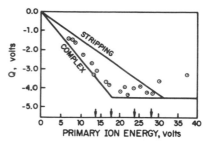

FIG. 3. The translational exoergicity is shown as a function of the C⁺ ion laboratory energy. Calculated values of Q are shown for complex formation and the spectator stripping model.

the curve closely approximate those expected from complex formation. The value of Q decreases until the bond dissociation limit of CD⁺ is reached. The value of Q then tends to remain constant and finally rises at the high energy end of the curve. The four arrows on the abscissa of Fig. 3 show the energies at which the first four contour maps of Fig. 2 were made. The behavior of Q with increasing C⁺ kinetic energy correlates with the mechanistic features obtained from contour maps. The first evidence of a direct mechanism appears at 17.6 eV which is just the C⁺ ion kinetic energy at which the internal energy of CD⁺ exceeds its bond energy for complex formation. Hence, the reaction dynamics must change if continued production of stable CD⁺ is to be realized.

Recently, Tulley et al.[12] have studied the dynamics of the reaction N⁺(O₂, O)NO⁺, and they reported a similar variation of Q with projectile kinetic energy. The values of Q were always somewhat more positive than the calculated values for spectator stripping, and the bond dissociation for Q was never reached exactly. They made a careful kinematic analysis and concluded that the experimental deviations from the ideal curves result from a finite energy and angular resolution of the apparatus. In view of the similarity of our apparatus to theirs, it is probable that the same explanation is applicable to our results.

In view of the fact that the reaction proceeds through a long lived intermediate, it is of interest to examine the results to see what information can be obtained on the nature of the complex. Recent ab initio calculations[13] on H–C–H⁺ show that the ground state is a $^2A_1$ state with a H–C–H angle of 141° (C_{2v} symmetry). The first excited state ($^2\Pi$) is linear at 0.13 eV above the ground state. There is, of course, no assurance that either of these states represent the configuration of the reaction intermediate. Recent preliminary calculations[14] in our laboratory indicate that an attractive potential is present as the C⁺ ion approaches the D₂ molecule irrespective of relative orientation, and in the C–H–H⁺ colinear configuration the $^2\Pi$ state may be lower in energy than the $^2\Sigma^+$ state. It appears, therefore, that the reaction may proceed through transition state intermediates represented by a variety of molecular configurations. An additional conclusion can be drawn about the geometry of the intermediate from the contour maps in Fig. 1. At low relative energies the distribution of the CD^+ velocity vectors is nearly isotropic. Herschbach and co-workers have shown that an isotropic distribution is to be expected if a large component of the rotational angular momentum is found in the products.[11] Furthermore, it was shown that a linear intermediate will not give an isotropic product distribution. The inference is that very probably the reaction can proceed through a transition state that is represented by a linear or bent geometrical configuration; however, the configuration of the long lived intermediate that gives rise to the isotropic distribution of CD^+ is bent.

Somewhat more information is available on the electronic states of CD^+. Douglas and Herzberg have reported two states for this ion, the ground state $^1\Sigma^+$ [15] and $^1\Pi$ [16] state. Two further band systems in the molecular spectra are ascribed to a $^1\Delta \rightarrow {}^1\Pi$ transition and a $^3\Sigma \rightarrow {}^3\Pi$ transition.[17] Moore et al.[18] have performed limited CI calculations on CH^+ and find the first four states are $X\,^1\Sigma^+$, $^3\Pi$, $^1\Pi$, and $^2\Sigma^+$ with energies 0.00, 0.64, 4.01, and 11.49 eV, respectively. It should be noted that all of these states dissociate into the same atomic species, $C^+(^2P) + D(^2S)$.

At present, it is not possible to distinguish experimentally between the reaction leading to the ground state of CD^+ and the reactions leading to any of the first three excited states of the ion. As shown by the values of Q, a large amount of energy goes into the internal modes of the product. Because the amount of energy is not large enough to attain the first excited electronic state of D, the energy must go into the CD^+ molecular ion. Since the first four states of CD^+ dissociate into the ground states of C^+ and D, the maximum value of Q only reflects this fact and does not permit one to infer directly the channels involved.

Moore et al.,[18] in the calculation of the low lying states of CH^+, found the $^3\Sigma^+$ state strongly repulsive.

It would then appear that products formed via this channel would dissociate to $C^+(^2P) + D(^2S) + D(^2S)$. If a large amount of rotational energy is present in the CD^+, the channel involving the $^1\Pi$ state would be unlikely because this state possesses a very shallow well.[16,17,18] Thus, the more attractive states in our energy domain for the CD^+ product are the $^1\Sigma^+$ and the $^3\Pi$, and the reaction can be represented by the equation

$$C^+(^2P) + D_2(^1\Sigma_g{}^+) \rightarrow CD^+(^1\Sigma^+, {}^3\Pi) + D(^2S).$$

The threshold and cross section for this reaction have been measured in an effort to determine the electronic state of the CD^+ at low energies and the results are given in a separate article in this Journal.

ACKNOWLEDGMENT

This work has been supported by the U.S. Army Research Office—Durham.

[1] W. B. Maier II, J. Chem. Phys. **46**, 4991 (1967).
[2] D. G. Truhlar, J. Chem. Phys. **51**, 4617 (1969).
[3] H. S. Johnston, *Gas Phase Reaction Rate Theory* (Ronald, New York, 1966).
[4] R. Wolfgang, Accounts Chem. Res. **3**, 48 (1970).
[5] C. R. Iden, R. Liardon, and W. S. Koski, J. Chem. Phys. **54**, 2757 (1971).
[6] E. A. Gislason, Bruce H. Mahan, Chi-wing Tsao, and Arthur S. Werner, J. Chem. Phys. **54**, 3897 (1971).
[7] D. F. Munro, Ph.D. dissertation, The Johns Hopkins University, 1969.
[8] C. R. Iden, Ph.D. dissertation, The Johns Hopkins University, 1971.
[9] P. S. Wilson, R. W. Rozett, and W. S. Koski, J. Chem. Phys. **52**, 5321 (1970).
[10] R. Wolfgang and R. J. Cross, Jr., J. Phys. Chem. **73**, 743 (1969).
[11] W. B. Miller, S. A. Safron, and D. Herschbach, Discussions Faraday Soc. **44**, 108 (1967).
[12] J. C. Tully, Z. Herman, and R. Wolfgang, J. Chem. Phys. **54**, 1730 (1971).
[13] C. F. Bender and H. F. Schaeffer, III, J. Mol. Spectry. **37**, 423 (1971).
[14] A. Pipano, K. C. Lin, M. M. Heaton, and Joyce J. Kaufman (unpublished).
[15] A. E. Douglas and G. Herzberg, Can. J. Res. **A20**, 71 (1942).
[16] A. E. Douglas and J. R. Morton, Astrophys. J. **131**, 1 (1960).
[17] M. Carre, Physica **41**, 63 (1969).
[18] P. L. Moore, J. C. Browne, and F. A. Matsen, J. Chem. Phys. **43**, 903 (1965).

27

Reprinted from *J. Chem. Phys.* **51**:452–454 (1969)

Crossed-Beam Studies of Energy Dependence of Intermediate Complex Formation in an Ion–Molecule Reaction

Z. Herman,* A. Lee,† and R. Wolfgang†

Department of Chemistry, University of Colorado, Boulder, Colorado 80302

(Received 31 March 1969)

The question of whether reaction involves an intermediate persistent complex has always been central to chemical kinetics, but conclusions on this point have usually been circumstantially based on the composition of the products. Recently, more immediate evidence has become available: for ion–molecule reactions through product velocity measurements by Durup,[1] Doverspike,[2] Henchman,[3] Henglein,[4] and co-workers; and for thermal neutral–neutral reactions through crossed-beam angular distribution measurements by Herschbach,[5] Kinsey,[6] and collaborators. We report here initial crossed-beam experiments on complex formation where both velocity and angular spectra were measured, and in which the collision energy was varied over the relevant range (~1–10 eV lab). The resulting data facilitate understanding of the energy-dependent competition between various modes of persistent-complex and direct reactions. Results on the following processes are presented:

$$C_2H_4^+ + C_2H_4 \rightarrow \begin{array}{c} [C_4H_8^+] \\ CH_3 + C_3H_5^+ \end{array} \rightarrow C_3H_3^+ + H_2 \; .$$

Here $C_4H_8^+$ is the possible intermediate[7] and $C_3H_5^+$ and $C_3H_3^+$ are the observed products, with the latter being favored at higher energies.

The crossed-beam apparatus EVA was used.[8,9] The $C_2H_4^+$ produced by electron bombardment[10] was mass-analyzed, formed into a beam with energy variable between 1 and 10 eV, and crossed with a thermal C_2H_4 beam. Results on measurements of angular and energy spectra of $C_3H_5^+$ and $C_3H_3^+$ are illustrated in Fig. 1 for typical energies. These diagrams give the velocity and angular distributions of products with respect to the center of mass and the relative collision axis. Contours show relative intensities, in the plane of measurement, using Cartesian velocity phase space.[11]

At lower energies the distribution of products is, within experimental error, symmetric with respect to a plane passing through the center of mass and normal to the collision axis. This is what would be expected if a complex persisting for at least several rotations (and which has thus "forgotten" the direction of the reactants) were formed ($\gtrsim 10^{-12}$ sec).[12] The forward and backward peaking is a consequence of the angular

momentum of the system.[5] These data are in striking contrast to the very asymmetric distribution found, even at c.m. energies of 0.1 eV, for certain other reactions.[9] At higher energies, however, $C_3H_5^+$ does show an increasingly strong forward peak (relative to the

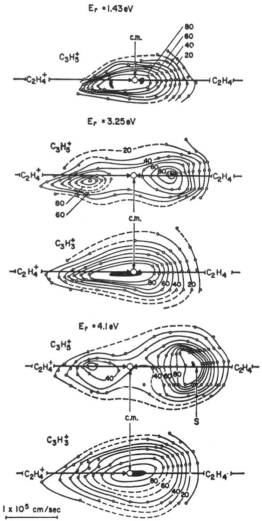

FIG. 1. Velocity-vector diagrams in the center-of-mass system, of angular and velocity distributions of $C_3H_5^+$ and $C_3H_3^+$ at three relative energies of colliding $C_2H_4^+$ and C_2H_4. Arrows represent center-of-mass velocity vectors of reactants. A vector from the center of mass indicates the velocity and the direction, with respect to the $C_2H_4^+$ vector, of product shown at any given point. Contours show product intensities normalized to 100 for each diagram. Total cross-section ratios $C_3H_3^+/C_3H_5^+$ are 0.15, 0.80, and 1.8 [T. Tiernan and J. H. Futrell, J. Phys. Chem. **72**, 3080 (1968)] at the three energies. (No plot for $C_3H_3^+$ is given at the lowest energy because of low intensity.) S indicates the product velocity for "spectator stripping."

incident ion). This indicates that the lifetime of $C_4H_8^+$ becomes shorter so that much of it breaks up before it has undergone even one rotation. The reaction thus makes a continuous transition to a direct mechanism as the available energy increases.

The data for $C_3H_3^+$ provide an interesting elaboration of the model. This product would result from further decomposition of highly excited $C_3H_5^+$. Such internally excited $C_3H_5^+$ will form by decay of $C_4H_8^+$ in which relatively little energy goes into translation. As expected on this basis, Fig. 1 shows that, at low collision energies, $C_3H_3^+$ is produced with lower average velocity than the surviving $C_3H_5^+$.[13]

At higher initial energies $C_3H_3^+$ has a more symmetric distribution than does $C_3H_5^+$. Apparently, there is then still a substantial probability of forming a persistent $C_4H_8^+$, but this either decays directly to $C_3H_3^+$, or, more likely, yields $C_3H_5^+$ so excited that it usually decomposes further to $C_3H_3^+$. A lesser quantity of $C_3H_3^+$ may form directly on initial collision, or by decay of directly formed $C_3H_5^+$.

It is appropriate to mention that such studies are potentially of importance to our understanding of unimolecular processes. The data in Fig. 1 (lower energies) can yield spectra of translational energies of products from the decay process. Such information may be as valuable as lifetime measurements in testing the theory of unimolecular decay.

This work was supported by the U.S. Atomic Energy Commission and by NASA. Valuable suggestions and assistance by Dr. Peter Hierl and Mr. John Krenos are gratefully acknowledged.

* Visiting Professor, on leave from Institute of Physical Chemistry, Czechoslovak Academy of Science, Prague.

† Present address: Sterling Chemistry Laboratory, Yale University, New Haven, Conn. 06520.

[1] J. Durup and M. Durup, J. Chim. Phys. **64**, 386 (1967).

[2] L. D. Doverspike and R. L. Champion, J. Chem. Phys. **46**, 4718 (1967).

[3] L. Matus, I. Opauszky, D. Hyatt, A. J. Masson, K. Birkinshaw, and M. J. Henchman, Discussions Faraday Soc. **44**, 146 (1967).

[4] A. Ding, A. Henglein, and K. Lacmann, Z. Naturforsch. **232**, 779 (1968).

[5] W. B. Miller, S. A. Safron, and D. R. Herschbach, Discussions Faraday Soc. **44**, 108 (1967).

[6] (a) D. O. Ham, J. L. Kinsey, and F. S. Klein, Discussions Faraday Soc. **44**, 174 (1967). (b) D. O. Ham and J. L. Kinsey, J. Chem. Phys. **48**, 939 (1968).

[7] Earlier indications of the formation of such a complex are given in F. H. Field, J. L. Franklin, and F. W. Lampe, J. Am. Chem. Soc. **79**, 2419 (1957), and J. Durup, M. Durup, *Advances in Mass Spectrometry*, Proceedings of Berlin Conference, 1967 (to be published).

[8] Z. Herman, J. D. Kerstetter, T. L. Rose, and R. Wolfgang, Rev. Sci. Instr. **40**, 538 (1969).

[9] Z. Herman, J. D. Kerstetter, T. L. Rose, and R. Wolfgang, Discussions Faraday Soc. **44**, 123 (1967).

[10] The internal energy of the $C_2H_4^+$ so produced is estimated as ranging from 0–1.2 eV, with peak intensity close to 0 eV.

[11] R. Wolfgang and R. J. Cross, Jr., J. Phys. Chem. **73**, 743 (1969).

[12] Such symmetry is in itself a necessary but not sufficient condition of persistent-complex formation. Direct reactions which could give such a distribution are conceivable though unlikely. One possibility in the present case is that electron transfer can occur rapidly, compared to the time of collision. Thus $C_2H_4^+$ could in effect approach the center of mass from either direction. However, the fact that the distribution is quite unsymmetric at higher energies shows that electron transfer is not fast compared to the collision period. Furthermore, in the system $C_2D_4^+ + C_2H_4$, products of different isotopic composition all show symmetric distributions, a finding which is implausible if the results are a consequence of direct reaction between colliding $C_2D_4^+ + C_2H_4$ and $C_2D_4^+ + C_2H_4^+$.

[13] In decomposition of $C_4H_8^+$ most of the kinetic energy is carried by H_2. Thus the velocity vector of the $C_3H_3^+$ should be quite similar to that of $C_3H_5^+$ parent.

28

Reprinted from *J. Chem. Phys.* **52**:5687–5694 (1970)

Impulsive Reaction Model of Ion–Molecule Reactions: Angular Distributions*

Daniel T. Chang† and John C. Light‡

The Department of Chemistry and The James Franck Institute, The University of Chicago, Chicago, Illinois 60637

(Received 18 December 1969)

The direct reaction model of ion-molecule reactions proposed by Wolfgang to explain the velocity dependence of products of exothermic reactions is refined and extended to yield the angular distribution of products as well. The model consists of consecutive "two-body" trajectories for reactants and products with "reflection" of radial relative velocities at a reaction radius. It is found that a single value of the reaction radius yields velocity and angular distributions in substantial agreement with experiment for the $Ar^+ + D_2$ reaction over a substantial energy range. Some predictions of the model and some possible modifications are also discussed.

I. INTRODUCTION

The use of crossed beams with angular and velocity analysis of the product ions[1–5] has recently shown that many simple ion–molecule reactions, like many neutral reactions, proceed by a direct mechanism without the formation of an intermediate complex. In particular, the asymmetry of the angular distributions with respect to reflection about 90° in the center-of-mass (c.m.) system persists over a wide energy range (from 0.1–2.3 eV in the c.m. system for $Ar^+ + D_2$[1]), indicating that the complex mechanism for this reaction, if it occurs at all, must occur at very low energies.

The simplest model of direct reactions, the spectator stripping model,[6,7] appears to yield a satisfactory description of the process at higher energies, but gives significant errors in the energy and angular distributions at lower energies.[1,2] In 1967 Wolfgang and co-workers[1] modified this model to take into account the effect of the polarization potentials on the energy distribution of the product ions, and found remarkably good agreement when one parameter, the "reaction radius," was fixed.

In this brief paper we extend the direct reaction model of Wolfgang to calculate both the energy and angular distribution of the products of the reaction $Ar^+ + D_2 \rightarrow ArD^+ + D$. To be more specific, we make an assumption that the forces are impulsive at the reaction radius and for this reason we shall refer to it as the impulsive

reaction model. Again the theory contains only the reaction radius as a parameter, and we now find that a single value of this parameter allows us to fit both the energy and angular distribution over the measured energy range with considerable success. The most serious discrepancy remaining between the theory and this experiment now lies in the value of the total reactive cross section.

The impulsive reaction model is in a real sense a classical physical model of the dynamics of the reactive collision event. It consists very simply of three parts: an incoming trajectory governed by the long-range forces with no coupling between the translational and internal motions, an impulsive reactive event in which the relative radial velocities of the particles are reversed and in which the products become bound, and finally, an outgoing trajectory of the products similar to the incoming trajectory. This model is classical and deterministic in that each initial trajectory yields a single final trajectory resulting in a specific velocity and scattering angle. Although there appears to be no *a priori* reason for this model not to apply to neutral reactions, we have confined ourselves, in this article, to the application to ion–molecule reactions.

In the next section we define the model and give the equations which determine the scattering in the center-of-mass system. In Sec. III we present the results from the model of the $Ar^+ + D_2$ reaction, transform to the laboratory coordinates, and perform the averaging which

263

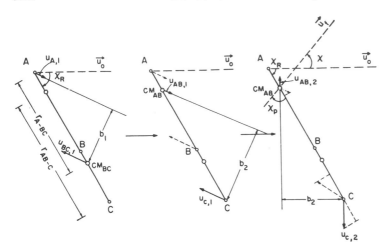

FIG. 1. Configuration and velocity decomposition at the reaction radius. (a) Initial velocities in initial coordinates. (b) Initial velocities in final coordinates. (c) Velocities after reversal of radial components.

allows direct comparison with experiment.[1] The last section is a discussion of the validity of the model and possible modifications.

II. IMPULSIVE REACTION MODEL

As mentioned previously, the impulsive reaction model consists of two trajectories connected by an assumption. In this section we shall only present the model, leaving possible justifications and comments until the last section. In the following derivations we make use of classical mechanics only and the attendant conservation laws of energy and angular momentum. We work in the center-of-mass system in which the total momentum of the system is zero. We assume the reactants A and BC are in their ground electronic states and neglect the internal vibrational and rotational energy that the initial molecule BC may have.

The initial trajectory is determined by the initial impact parameter b_0 of A with respect to the center of mass of BC, the initial relative kinetic energy T_0, and the long-range spherically symmetric potential $V_{A-BC}(r)$ which acts between the centers of mass of A and BC. The trajectory of A and BC then lies in the plane determined by the initial velocity vector, U_0 and r. The angle between U_0 and r at a distance r_{A-BC} is given by[8]

$$\chi_R = \int_{r_{A-BC}}^{\infty} \frac{b_0}{r^2} \left(1 - \frac{V_{A-BC}(r)}{T_0} - \frac{b_0^2}{r^2}\right)^{-1/2} dr, \quad (1)$$

where r_{A-BC} is such that the radical in the integrand is always real between r_{A-BC} and infinity. The relative kinetic energy at this point is

$$T_1 = T_0 - V_{A-BC}(r_{A-BC}). \quad (2)$$

Conservation of angular momentum is used to define the impact parameter at r_{A-BC}, b_1 (see Fig. 1),

$$b_1 = J/(2\mu_R T_1)^{1/2} = b_0 (T_0/T_1)^{1/2}, \quad (3)$$

where J is the initial angular momentum of the system, and μ_R is the reduced mass of the reactants, $\mu_R = M_A(M_B + M_C)/(M_A + M_B + M_C)$.

We now assume that when the particles reach the reaction radius, r_{A-BC}, reaction will take place instantaneously and the products will begin to separate. In order to determine the final trajectory and thus the velocity and scattering angle, we must fix the initial conditions of the outgoing particles. To do this we make the following assumptions:

(1) The configuration of the three particles is collinear at this point.
(2) The BC distance has not changed from that of the isolated molecule.
(3) Velocities are assigned to the particles such that the magnitudes are unaltered, but the *radial* components are reversed in direction, i.e., the particles are now separating (see Fig. 1).

The first two assumptions fix the configuration of the system to give the starting distances for the outgoing trajectory, and the third assumption, coupled with the conservation equation, fixes the initial outgoing relative velocities and the partitioning of angular momentum between the rotation of the product molecule and the orbital angular momentum.

The mathematical prescription to do this is somewhat tedious, but straightforward. The distance between the center of mass of the new molecule AB and C is given by

$$r_{AB-C} = [M_A/(M_A + M_B)]r_{A-B} + r_{BC}, \quad (4)$$

where r_{A-B} is determined by the reaction distance r_{A-BC} and r_{BC},

$$r_{A-B} = r_{A-BC} - [M_C/(M_B + M_C)]r_{BC}. \quad (5)$$

Thus the reaction distance in terms of the product molecule AB and C is determined.

The velocity assignment before reaction is that B and C each have the velocities of the center of mass of BC with magnitudes

$$U_{C,I} = U_{B,I} = U_{BC,I} = (M_A/M)(2T_1/\mu_R)^{1/2}, \quad (6)$$

where the I denotes the fact that this is before reaction and

$$M = M_A + M_B + M_C.$$

The magnitude of the momentum of the center of mass of the AB molecule is the vector sum of the momenta of A and B. Since these momenta have opposite direction, the velocity is

$$U_{AB,I} = (M_A U_{A,I} - M_B U_{B,I})/(M_A + M_B)$$
$$= [M_A M_C/M(M_A + M_B)](2T_1/\mu_R)^{1/2}. \quad (7)$$

The orbital angular momentum of AB with respect to C is

$$J_I = [(M_A + M_B)U_{AB,I} + M_C U_{C,I}]b_2, \quad (8)$$

where

$$b_2 = b_1\left[1 + \frac{M_B}{r_{A-BC}}\left(\frac{r_{BC}}{M_B + M_C} - \frac{r_{A-B}}{M_A + M_B}\right)\right]. \quad (9)$$

By conservation, the rotational angular momentum of AB is given by

$$J_{AB} = J - J_I. \quad (10)$$

The outgoing trajectory starts after we reverse the radial components of $U_{AB,I}$ and $U_{C,I}$ such that the product system is now separating. This gives the vectors shown in Fig. 1(c). The magnitude of the velocities are unchanged, but the relative translational kinetic energy is changed because of the change in the relative masses. The relative kinetic energy of the products at r_{AB-C} is

$$T_2 = [M_A M_C/(M_A + M_B)(M_B + M_C)]T_1$$
$$= \tfrac{1}{2}\mu_P(U_{AB,2} + U_{C,2})^2, \quad (11)$$

where μ_P is the reduced mass of the products,

$$\mu_P = (M_A + M_B)M_C/M. \quad (12)$$

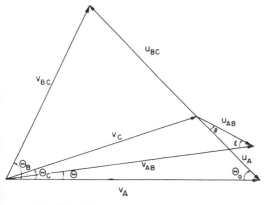

FIG. 2. Velocity diagram for in-plane scattering.

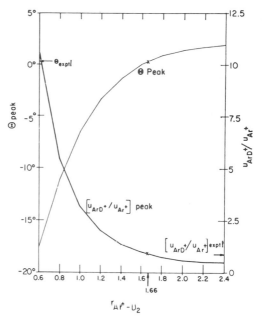

FIG. 3. $u_{ArD}{}^+/u_{Ar}{}^+$ and Θ_{peak} vs $r_{Ar}{}^+_{-D_2}$. $T_0 = 0.081$ eV.

The remainder of the energy, including the heat of reaction, goes into the rotation and vibration of the product molecule. The orbital angular momentum of the product is unchanged by the change in the radial components of the velocity.

The final impact parameter of the products, b_f, as they separate is given by the final translational energy, T_f, and the final angular momentum, J_I,

$$b_f = J_I/(2\mu_P T_f)^{1/2} = b_2(T_2/T_f)^{1/2}. \quad (13)$$

Finally, if the spherically symmetric potential determining the product trajectory is $V_{AB-C}(r)$, then the final kinetic energy is

$$T_f = T_2 + V_{AB-C}(r_{AB-C}), \quad (14)$$

and the deflection angle for the outgoing trajectory is

$$\chi_P = \int_{r_{AB-C}}^{\infty} \frac{b_f}{r^2}\left(1 - \frac{V_{AB-C}(r)}{T_f} - \frac{b_f^2}{r^2}\right)^{1/2} dr. \quad (15)$$

It may happen that the radical under the integrand becomes imaginary for $r_{AB-C} < r < \infty$. This indicates that reaction cannot occur at this impact parameter with this assumed partitioning of energy. Such a case must then be counted as an inelastic nonreactive collision or else an indication that a complex is formed. In this case the simple model breaks down for this range of energies and impact parameters, and an alternative treatment must be devised. However, in the cases studied so far, this has not been a problem.

It should be noted that the order in which the operations of decomposing the velocities in the new

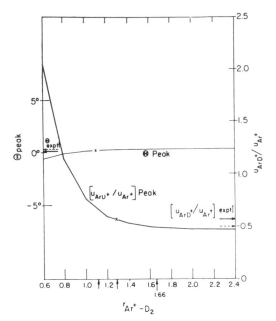

FIG. 4. u_{ArD^+}/u_{Ar^+} and Θ_{peak} vs $r_{Ar^+-D_2}$. $T_0 = 0.758$ eV.

center-of-mass system and reversal of the radial components are performed is immaterial—the same equations result. This is a result of the "instantaneous" nature of the operations and is essential for the preservation of microscopic reversibility.

This then completes the model of a reactive collision. For a given initial kinetic energy and impact parameter for which the critical radius r_{A-BC} is reached, the initial deflection is found from Eq. (1). Using Eqs. (2)–(14) the parameter of the outgoing trajectory and the rotational angular momentum of the product are determined. The final deflection is then found from Eq. (15) to yield the total deflection angle χ,

$$\chi = \pi - \chi_R - \chi_P. \tag{16}$$

Since χ may range from π to $-\infty$, while the measured scattering angle θ is always positive, we assign the values to θ as follows[5]:

(i) $\chi > 0$; $\theta - \chi$,

(ii) $\chi < 0$; $\Delta\chi = \chi - 2\pi k \; k$ integer,

    (a) $\pi \geq \Delta\chi \geq 0$; $\theta = \Delta\chi$,

    (b) $2\pi > \Delta\chi > \pi$; $\theta = 2\pi - \Delta\chi$.   (17)

The c.m. differential cross section per unit solid angle, $I(\theta, \phi)$, is then given by

$$I(\theta, \phi) \; \sin\theta d\theta d\phi = \sum_{b_i} P(b_i) b_i db_i d\phi, \tag{18}$$

where $P(b_i)$ is the probability of reaction (to be discussed later) and the b_i are those values of the impact

parameter which correspond to scattering with the same θ. Since there is cylindrical symmetry, $I(\theta, \phi)$ is independent of ϕ, and we have

$$I(\theta) = (\sin\theta)^{-1} \sum_{b_i} P(b_i) b_i \mid db_i/d\chi \mid_{b_i}. \tag{19}$$

Since this model gives only a single value of the final relative velocity for each T_0, the c.m. differential cross section per unit velocity space is

$$I(\theta, U) = I(\theta) U^{-2} \delta(U - U_{AB}), \tag{20}$$

where U_{AB} is the final velocity predicted by the model.

The translational exoergicity of the reaction, Q, is sometimes measured. It is defined by

$$Q = T_f - T_0. \tag{21}$$

Note that this is a function of T_0 only, not the impact parameter. The total internal excitation of the product molecule is then given by

$$E_{AB} = -\Delta H - Q, \tag{22}$$

where ΔH is the heat of reaction and is negative for exothermic reactions. As in the spectator stripping model, at high enough values of the initial translational energy, E_{AB} may exceed the binding energy of AB and the product molecule will be unstable.

The final quantity of interest is the total reactive cross section, σ_R. This is obtained by integration of Eq. (18),

$$\sigma_R(T_0) = 2\pi \int_0^{b_m} P(b_0) b_0 db_0, \tag{23}$$

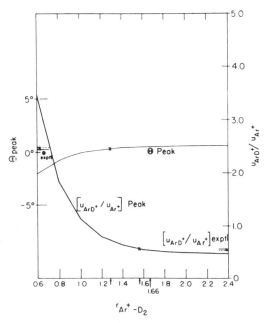

FIG. 5. u_{ArD^+}/u_{Ar^+} and Θ_{peak} vs $r_{Ar^+-D_2}$. $T_0 = 2.270$ eV.

where the maximum impact parameter is determined by the requirement that the initial trajectory must go to the reaction radius, $r_{\text{A-BC}}$. The value of b_m therefore depends on T_0 and the potentials.

III. Ar⁺+D₂: COMPARISON OF THEORY AND EXPERIMENT

Since the reaction

$$Ar^+ + D_2 \rightarrow ArD^+ + D$$

has been studied in detail over a range of experimental energies, this reaction was chosen to make detailed comparisons. In addition Wolfgang and co-workers[1] have shown that the velocity distribution of the products is well represented by a model of this type. The experimental information available consists, basically, of the angular and velocity distributions of the product ion and the total reaction cross section as a function of initial energy. Our aim is, of course, to fit all of these data as well as possible with a single value of our single parameter, $r_{\text{A-BC}}$.

There are several phases in this comparison. First, because there is a forward glory in the c.m. reactive scattering and because the final velocity depends only on T_0, we can determine the predicted positions of the peaks of the lab angular and velocity distributions

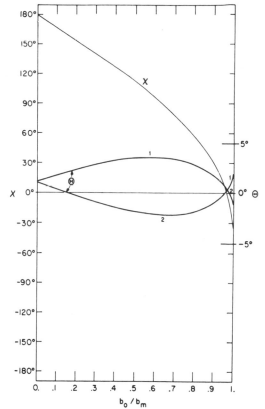

FIG. 7. χ and Θ vs b_0/b_m. $T_0 = 2.270$ eV.

without determining the entire angular distribution. This allows us to vary $r_{\text{A-BC}}$ and determine the variation of these peaks easily at the various experimental energies. Once the most appropriate value of $r_{\text{A-BC}}$ has been determined, we can then evaluate the c.m. angular distribution for a given T_0, confirming that a forward glory exists, and giving the desired distribution. Finally, to compare most directly with experiment, this distribution is transformed into laboratory coordinates, averaged over the acceptance angle of the detector, and averaged over the initial beam energy spread and angular resolution to yield an intensity distribution quite comparable to that which was observed. The only effect excluded is the role of out-of-plane scattering which enters in a most complicated manner.

The potential used in the calculation are the long-range attractive polarization potentials

$$V(r) = -\alpha e^2/2r^4, \qquad (24)$$

with the values of α given by

$$\alpha_{D_2} = 0.93 \text{ Å}^3, \qquad \alpha_D = 0.67 \text{ Å}^3.$$

The heat of reaction is $-\Delta H = 1.2$ eV, the dissociation energy of ArD⁺ is then taken to be 3.03 eV. The

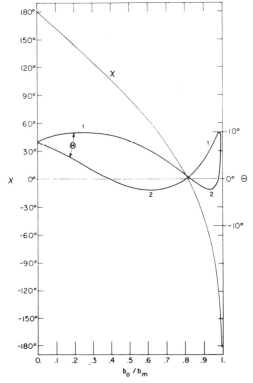

FIG. 6. χ and Θ vs b_0/b_m. $T_0 = 0.081$ eV.

TABLE I. Summary of results on $Ar^+ + D_2$ reaction.

E_{Ar^+} (eV)	T_0 (eV)	Θ_{peak}			u_{ArD+}/u_{Ar+}			Q (eV)		
		Exptl	IRM[a]	Spec. strip.	Exptl	IRM	Spec. strip.	Exptl	IRM	Spec. strip.
0.61	0.081	0.3°	0.35°	1.94°	0.91	0.90	0.48	0.06	0.06	−0.04
0.71	0.090	0.6	0.43	1.80	1.20	0.87	0.48	0.18	0.05	−0.05
2.23	0.228	0.2	0.64	1.01	0.79	0.66	0.48	0.07	−0.02	−0.12
4.38	0.425	0.3	0.59	0.74	0.62	0.58	0.48	−0.09	−0.12	−0.22
4.46	0.431	0.3	0.57	0.72	0.61	0.58	0.48	−0.10	−0.13	−0.23
8.05	0.758	0.3	0.47	0.54	0.56	0.54	0.48	−0.25	−0.30	−0.40
15.0	1.391	0.3	0.38	0.41	0.57	0.51	0.48	−0.45	−0.63	−0.73
24.6	2.27	0.2	0.34	0.35	0.57	0.50	0.48	−0.71	−1.09	−1.19

[a] These results were calculated from the present model, the impulsive reaction model.

laboratory angular distribution is related to the c.m. angular distribution by

$$I_L(\Theta) = I(\theta) V_{AB}^2 / U_{AB}^2 \, | \cos\xi |^{-1}, \quad (25)$$

where V_{AB} is the laboratory velocity of the product AB, Θ is the laboratory angle, and

$$\partial V / \partial U \,|_{V_{AB}} = | \cos\xi |^{-1}. \quad (26)$$

We may note from the Newton diagram, Fig. 2, that there may be 0, 1, or 2 values of θ corresponding to each value of Θ. For our present purposes we note that since $I(\theta)$ has a singularity at $\theta = 0$ (which will be shown later) that the position of the peak in the lab system depends only on U_{AB} for fixed T_0 and beam angles of 90°. The product c.m. velocity, U_{AB}, in turn depends only on T_0 and the reaction radius parameter r_{A-BC}. Thus to fix r_{A-BC} and to check the consistency of this interpretation, we first plot, for fixed T_0, the calculated values of (U_{ArD+}/U_{Ar+}) and $\Theta(peak)$ as a function of $r_{Ar^+-D_2}$. These are given in Figs. 3–5 for three values of T_0. The most probable experimental[1] values of these two quantities and the reaction radii at which they occur are also shown.

We may draw several conclusions from these graphs. First, at each energy, one value of $r_{Ar^+-D_2}$ gives both peak positions to high accuracy. This is a most comforting result since it implies that the reaction dynamics are governed by both a short-range repulsion and long-range attraction. It also confirms the fact that it is the forward scattering peak in the c.m. system which is entirely responsible for the peak in the lab angular distribution. Second, although as the energy increases, the best value of $r_{Ar^+-D_2}$ decreases, the single value of $r_{Ar^+-D_2} = 1.66$ Å yields very good results at all energies due to the fact that the peak positions become less sensitive to $r_{Ar^+-D_2}$ as the energy increases. We shall, therefore, use only this value of this parameter throughout the remainder of the paper. The information on the energy dependence of $\Theta(peak)$ and (U_{ArD+}/U_{Ar+}) is summarized in Table I. For comparison, the experi-

mental values, the predictions of this model with $r_{Ar^+-D_2} = 1.66$ Å, and the predictions of the spectator stripping model are given. It is easily seen that the present model is in good agreement with experiment, and is considerably superior to the spectator stripping model. Wolfgang used the value of $r_{Ar^+-D} = 1.61$ Å to explain the velocity distribution in this reaction, and we find the optimum value to be very close to this.

Having determined a value for r_{Ar^+-D}, we may now determine the angular distributions predicted by our model. It is here that this model differs most strongly from the spectator stripping model which predicts only forward scattering in the c.m. system. The maximum impact parameter which will allow reaction to occur is that above which the distance of closest approach of the reactants is larger than $r_{Ar^+-D_2}$. For low energies this is just determined by the orbiting collisions, at higher energies it is determined by the largest zero of the radial velocity. We have, for the polarization potential, b_m equal to the larger of

$$b_m = \max(2\alpha e^2 / T_0)^{1/4},$$

$$= \max r_{Ar^+-D_2} [1 + (\alpha e^2 / 2T_0 r_{Ar^+-D_2}^4)]^{1/2}. \quad (27)$$

In Figs. 6 and 7 we show θ and Θ plotted against the reduced impact parameter, b_0/b_m, for two energies. The two branches of Θ are physical, being due to the fact that the c.m. angular distribution has cylindrical symmetry about the relative velocity vector U_0. Thus the equal intensity in plane at θ and $-\theta$ contribute to the lab intensity at two different values of Θ. The branch of Θ labeled 1 comes from c.m. scattering outside (above) the triangle of initial velocity vectors in the Newton diagram.

In order to calculate $I(\theta)$ and $I_L(\Theta)$, we must make an assumption about the probability of reaction, $P(b)$, which enters Eqs. (18) and (19). We make only the simplest assumption in this paper, that it is unity,

$$P(b) = 1. \quad (28)$$

The c.m. intensity is uninteresting, merely showing the forward peak with some scattering at all angles. Of more interest are the laboratory angular distribution $I_L(\Theta)$ shown in Fig. 8 for $T_0 = 0.081$ eV. These show, in addition to the central peak due to forward scattering, the total width of the lab angular distribution. The side peaks occur in the calculated distribution because of the Jacobian factor in Eq. (25), the points at which dv/du becomes infinite. These are actually singular only because our model gives a fixed value of the final velocity, U_{AB} (a delta function distribution). Because of the finite resolution of the detector and the velocity spread of the initial beam, these peaks would not be observed even if the model were rigorous. They do, however, serve to define the total breadth of the angular distribution to be expected.

Finally, to make a direct comparison with the experimental angular distribution,[1] we must average our calculated $I_L(\Theta)$ distributions over the angular resolution of the detector (0.9° in Ref. 1) over the velocity spread of the thermal (55°C, Ref. 1) D_2 beam, and finally over the angular resolution of the initial beams (2° width of half-height, taken from Fig. 3, Ref. 1). The details of these averaging processes are given in Ref. 9; the final result, together with the experimental results are shown for $T_0 = 0.081$ eV ($E_{Ar}{}^+ = 0.61$ eV) in Fig. 9. As can be seen, there is substantial agreement on the peak position and the total width, but some disagreement on the width at half-height. The curves were normed to the same maximum. A more complete discussion of this disparity is given below.

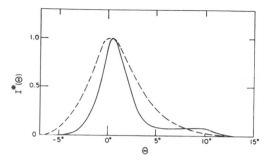

FIG. 9. $I_L{}^*(\Theta)$ vs Θ. $T_0 = 0.081$ eV $E_{Ar}{}^+ = 0.61$ eV. - - -, Wolfgang and co-workers.

IV. DISCUSSION

In the last section we saw that this simple one-parameter model gives substantial agreement with the experimentally observed angular and energy distribution for the reaction $Ar^+ + D_2 \rightarrow ArD^+ + D$. This model is comparable in spirit to two recent direct models of reaction put forth by Suplinskas and co-worker[10] and by Kuntz et al.[11] Suplinskas has followed the classical trajectories of three particles on a drastically simplified potential-energy surface consisting of long-range polarization forces and hard-core repulsions (two body). This is more realistic (and more complicated) than our model in that the two-body repulsions are used and the "reactive configuration" is not fixed. Excellent agreement with experiment has also been obtained.[10] The "direct interaction model with product repulsion" (DIPR) used by Kuntz has a soft two-body repulsive force between the product atom and one of the atoms of the diatomic. It was applied to neutral systems with no long-range attractive forces and was found to yield distributions in good agreement with computer trajectory studies on more realistic potential surfaces.

Both of these models yield a distribution of final translational energies in the c.m. system as well as angular distribution. This distribution is assured in the DIPR model by allowing nonlinear (but fixed) "reactive configurations." The impulsive reaction model proposed here is a good first approximation, but it is obviously deficient in two important respects. First, the predicted independence of the translational exoergicity and the angular distributions is not found experimentally.[2,3] Also the fact that our calculated lab angular distribution has a narrower half-width than is observed experimentally[1] is almost certainly due to the fact that the final c.m. translational energy is fixed to a single value for each initial energy. The quantization of internal energy alone implies this is wrong, and the real dynamics of the reactions must also broaden the distribution.

The model proposed in this paper can be modified to yield translational energy spreads in the c.m. system in two ways. First, the collinear assumption can be

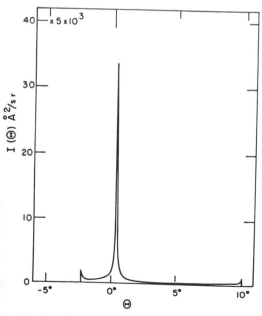

FIG. 8. $I_L(\Theta)$ vs Θ. $T_0 = 0.081$ eV and $E_{Ar}{}^+ = 0.61$ eV.

relaxed while still keeping the assumption of reversal of "radial" velocities. The second method would be to assume a given radial velocity spread instead of strict reversal of radial velocities while keeping the collinear configuration. This assumption would involve the introduction of an additional parameter and either one would destroy some of the striking simplicity of the model. The rather good agreement between experiment and our simple model seems to indicate that although the final c.m. velocity distribution is certainly not a delta function, it is rather sharply peaked about the value we fix with our reaction radius.

Taking the view that the model is valid enough as is to have predictive value, we may make two qualitative predictions on the basis of Figs. 6 and 7. They are: as the kinetic energy is decreased (below 0.081 eV) for this system, a *backward* glory should appear as the scattering angle goes through $-\pi$ at large impact parameter. This would have the effect of producing a roughly symmetric angular c.m. distribution which would be difficult to distinguish experimentally from the onset of complex formation[5] except by the relatively narrow velocity distribution which would persist.[12] At the other limit, as the energy is increased (above 2.27-eV c.m.), the cross section for product formation should decrease sharply when the predicted internal energy of the product exceeds the dissociation energy. This last effect is also predicted by the spectator stripping model (at different energies) and has some experimental support.[6] It should be noted, however, that Bailey and co-workers[5] have observed the ArD⁺ product at initial energies above the predicted critical energy. This also implies that the model should be modified to take into account a distribution of internal energies (or translational exoergicities). The impulsive reaction model reduces to a modified stripping model at high energies in that the predicted velocity distribution and translational exoergicity become very close to the spectator stripping values.

The only major assumption in the model for which no method of rational modification is apparent is the assumption of the probability of reaction as a function of b. We have assumed that $P(b)$ is a step function, being unity if the reaction radius is reached and zero otherwise. It is apparent from experiment that this is not valid for some reactions, but we know of no simple method of calculating or estimating $P(b)$.

Finally, this model appears to provide a simple explanation of the major characteristic of direct ion–molecule reactions. If the qualitative features predicted continue to stand the test of comparison with experimental data, further development toward a more realistic model along the lines suggested above will be warranted. The simplicity with which the peaks in the angular and velocity distributions and the total angular spread can be calculated will add to its utility.

ACKNOWLEDGMENTS

The authors gratefully acknowledge the general support of the James Franck Institute by the Advanced Research Projects Agency.

* This research was supported in part by a grant from the National Science Foundation, NSF-GP-8674.
† Present address: Department of Chemistry, Brandeis University, Waltham, Mass.
‡ Alfred P. Sloan Foundation Fellow.

[1] Z. Herman, J. Kerstetter, T. Rose, and R. Wolfgang, Discussions Faraday Soc. **44**, 123 (1967).
[2] W. R. Gentry, E. A. Gislason, B. H. Mahan, and C. Tsao, J. Chem. Phys. **49**, 3058 (1968).
[3] W. R. Gentry, E. A. Fislason, Y. Lee, B. H. Mahan, and C. Tsao, Discussions Faraday Soc. **44**, 137 (1967).
[4] A. Henglein, K. Lacman, and G. Jacobs, Ber. Bunsenges. Physik. Chem. **69**, 279 (1965).
[5] L. D. Doverspike, R. L. Champion, and T. L. Bailey, J. Chem. Phys. **45**, 4385 (1966).
[6] A. Henglein, Advan. Chem. Ser. **58**, 66 (1966); A. Henglein, K. Lacmann, and N. Knoll, J. Chem. Phys. **43**, 1048 (1965).
[7] R. E. Minturn, S. Datz, and R. L. Becker, J. Chem. Phys. **44**, 1149 (1966).
[8] H. Goldstein, *Classical Mechanics* (Addison–Wesley, Reading, Mass., 1950).
[9] D. Chang, Ph.D. thesis, "Two Problems in Molecular Dynamics," University of Chicago, Chicago, Ill., 1969.
[10] T. F. George and R. J. Suplinskas (private communication).
[11] P. J. Kuntz, M. H. Mok, and J. C. Polanyi, J. Chem. Phys. **50**, 4623 (1969); P. J. Kuntz, Ph.D. thesis, University of Toronto, 1968.
[12] *Note added in proof:* Dr. John Tully has correctly pointed out in a private communication that the magnitude of the backward glory will be smaller than that of the forward glory, even in the limit as $E \rightarrow 0$.

EFFECTS OF ENERGY IN
VARIOUS MODES

Editor's Comments
on Papers 29 Through 35

A reacting system that includes an energy barrier to reaction must have sufficient energy to surmount the barrier if reaction is to occur. Thus a reaction that is normally endoergic with reactants in their ground states may be caused to occur if enough energy is added to the system. In Papers 38 through 41 the effect of translational energy in causing reactions to occur is demonstrated and the energy threshold for such a reaction is shown to be the heat of the reaction. It would be expected that sufficient vibrational energy in one or both of the reactants will also help bring about an endoergic reaction. This was first demonstrated by von Koch and Friedman (Paper 29) for the reaction

$$H_2^+ + He \rightarrow HeH^+ + H \tag{1}$$

which is endoergic for ground state H_2^+ but proceeds at a rapid rate when H_2^+ is in the third or higher vibrational level. Chupka and Russell (Paper 30) have refined this experiment by employing photoionization. A similar study (Paper 31) of the effect of vibrational energy of $C_2H_2^+$ on the very slightly endoergic reaction

$$C_2H_2^+ + H_2 \rightarrow C_2H_3^+ + H \tag{2}$$

also shows the cross section to be very small for thermal ions but to increase rapidly with increasing vibrational energy.

Most studies of the effect of vibrational energy on reaction rate have employed vibrationally excited ions as the excited species. One study by Schmeltekopf et al. (Paper 32) has showed the effect of vibrational excitation of the neutral reactant.

For the reaction

$$O^+ + N_2 \rightarrow NO^+ + N \tag{3}$$

a strong dependence of the reaction rate on vibrational energy was found. Since reaction (3) is quite exoergic, it appears that it must involve a considerable energy of activation. Paper 33 presents a theoretial study of this reaction.

While reactions exhibiting an energy barrier increase in rate with increasing vibrational energy, reactions that have no energy barrier are either independent of, or exhibit a slow reduction in, rate with increasing vibrational excitation. Paper 34 gives a beautiful example of this, showing the steady decrease in rate for the reaction

$$NH_3^+ + NH_3 \rightarrow NH_4^+ + NH_2 \tag{4}$$

with increase in the vibrational quantum number of NH_3^+. Paper 35 also shows the effect of vibrational energy, this time of H_3^+, on its reaction with ethylene. The total rate of the reaction varies very little with vibrational energy, but as would be expected, the extent of decomposition of $C_2H_5^+$ to $C_2H_3^+$ increases with increasing vibrational energy. A much less elegant study of the rates jof reaction of methane with CH_4^+ and CH_3^+ showed that the various reactions were independent of the energy of the electrons employed in forming the reactant ions.[1] Since any excess energy in the CH_4^+ and the greater part of the excess energy in CH_3^+ had to be in vibration, these results give further credence to the fact athat vibrational energy has very little effect on the rates of reaction in the absence of an energy barrier.

REFERENCE

1. Field, F. H., J. L. Franklin, and M. S. B. Munson, *J. Am. Chem. Soc.* **85**: 3575 (1963).

Hydrogen–Helium Ion–Molecule Reactions*

Helge von Koch† and Lewis Friedman

Chemistry Department, Brookhaven National Laboratory, Upton, Long Island, New York

THE hydrogen–helium ion–molecule reactions,

$$H_2^+ + He \rightarrow HeH^+ + H, \qquad (a)$$

$$He^+ + H_2 \rightarrow HeH^+ + H, \qquad (b)$$

provide an interesting opportunity for the investigation of the energy distribution in reactants and products in elementary reactions. The molecule ions involved in these reactions have sufficiently few electrons and are structurally simple enough to permit accurate quantum mechanical computation of their respective dissociation energies. Consequently, their thermochemistry is probably better known than that of most other ion–molecule reactions. Reaction (a) is endothermic by 1.1 eV and (b) is exothermic by 8.3 eV.[1] The relatively large excess of energy available in (b) tends to make (a) the most probable source of HeH+. This is supported by experimental evidence later in the paper. The role of translational energy in supplying the heat of reaction (a) can be investigated by determining the reaction rate as a function of the velocity of H_2^+ ions in the mass spectrometer ion source. The role of internal energy on the endothermic reaction can be examined by comparing theoretical and experimental values of the reaction rate constant. If the experimental value is smaller than expected, we may assume that only a fraction of the total H_2^+ has sufficient internal energy for reaction. Since we can estimate the population of excited vibrational states in H_2^+ produced, by electron impact on H_2, we can then determine the minimum internal energy requirement.

One might expect that H_2^+ produced by electron impact on molecules other than H_2, CH_4 for example, would have a different internal energy distribution which should be reflected in the experimental values of the rate constant.

Gioumousis and Stevenson[2] have shown that the thermal reaction rate constant for an ion-molecule reaction can be related to the microscopic reaction cross section. They proved that the thermal rate constant derived by Eyring, Hirschfelder, and Taylor[3] for

$$H_2^+ + H_2 \rightarrow H_3^+ + H \qquad (c)$$

is directly obtained from the product of the velocity and the velocity dependent cross section, i.e.,

$$k = 2\pi (e^2\alpha/\mu)^{\frac{1}{2}} = g\sigma(g), \qquad (1)$$

where e is the charge on the ion, α the molecule polarizability, μ the reduced mass of the reacting system, and $\sigma(g)$ the velocity dependent microscopic reaction cross section. Eyring, Hirschfelder, and Taylor formulated the rate in terms of absolute reaction-rate theory and assumed that a very small activation energy arises from the rotational energy of the system as a whole. Their rate constant is practically a collision number, for unit concentrations of reactants, which very accurately predicts the rates of the exothermic isotopic hydrogen ion–molecule reactions.[4,5]

The reactions of He and H_2 in an ion source of a mass spectrometer have been investigated by Hertz-

---

* Research performed under the auspices of the U. S. Atomic Energy Commission.

† Permanent address: Royal Institute of Technology, Stockholm 70, Sweden.

[1] (a) E. A. Hylleraas, Z. Physik **71**, 739 (1931); (b) A. A. Evett, J. Chem. Phys. **24**, 150 (1956).

[2] G. Gioumousis and D. P. Stevenson, J. Chem. Phys. **29**, 294 (1958).

[3] H. Eyring, J. O. Hirschfelder, and H. S. Taylor, J. Chem. Phys. **4**, 479 (1936).

[4] D. P. Stevenson and D. O. Schissler, J. Chem. Phys. **29**, 282 (1958).

[5] B. G. Reuben and L. Friedman, J. Chem. Phys. **37**, 1636 (1962).

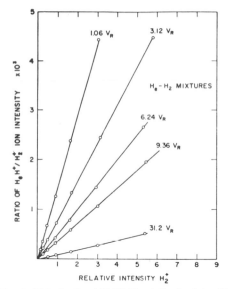

FIG. 1. Plot of ratio of HeH$^+$/H$_2$ relative ion intensities as a function of H$_2^+$ relative intensity at various repeller voltages.

berg *et al.*[6a] and a value of 0.07 ± 0.03 was reported for the ratio of cross sections of reactions (a)/(c). They remarked that the appearance potentials of HeH$^+$ and H$_2^+$ were identical establishing the importance of (a) as a pathway to HeH$^+$.[6b]

If the H$_2^+$–He reaction were exothermic, one would expect a ratio of rates proportional to the ratios of $(\alpha/\mu)^{\frac{1}{2}}$ for the two systems or a ratio of the order of 0.44 rather than 0.07. The brief communication of Hertzberg *et al.* indicated that their result was for ions with an average energy of above 10 eV. This energy is sufficient to permit the ions to surmount a rather high potential barrier. Since under these circumstances only a fraction of the H$_2^+$–He collisions were effective in producing HeH$^+$, the collision efficiency is probably more sensitive to the distribution of internal energy in the H$_2^+$ than its relative translational energy. Polanyi[7] has investigated the problem

[6] (a) M. Hertzberg, D. Rapp, I. B. Ortenburger, and D. D. Briglia, J. Chem. Phys. **34**, 343 (1961). (b) Kaul, Lauterbach, and Taubert [Z. Naturforsch. **16a**, 624 (1961)] in a similar investigation reported the appearance potential of HeH$^+$ to be 16.5 ± 0.3 eV and suggested that excited H$_2^+$ was the reactant ion. The threshold datum of Hertzberg *et al.* was reported with no error limit, so that it is difficult to tell whether a real discrepancy exists in the literature or whether both results agree within error limits. Several other ion–molecule investigations have included work on the H$_2$–He system [M. Pahl, Ergenbnisse der Exakten Naturwissenschaften **35**, 182 (1962); C. F. Geise and W. B. Maier III, J. Chem. Phys. **35**, 1913 (1961)]. The preliminary results of Geise and Maier are of particular interest in that they disagree with results of Herztberg *et al.* and our data. They have deferred a discussion of these differences pending a more detailed study of their own novel technique. We have no explanation for this discrepancy. The authors are grateful to Dr. D. P. Stevenson for calling the above pertinent references to our attention.

[7] J. C. Polanyi, J. Chem. Phys. **31**, 1338 (1959).

of energy distribution in elementary reactions and concluded that for endothermic reactions internal energy is of much greater importance than translational energy. In exothermic reactions Polanyi concluded that a large fraction of the heat of reaction would be found as internal energy. In an experiment with crossed molecular beams Herschbach, Kwei, and Norris[8] found that for the reactions

$$K+CH_3I \rightarrow KI+CH_3,$$

$$K+C_2H_5I \rightarrow KI+C_2H_5,$$

more than 80% of the energy of reaction went into the vibrational and rotational excitation of the products. This is consistent with results obtained in the study of the H$_2^+$–H$_2$ reaction.[5] Very little experimental evidence is available to test the validity of Polanyi's conclusions with respect to endothermic reactions. It is the purpose of this paper to report a study of the H$_2^+$–He reaction from this point of view.

EXPERIMENTAL

Experimental measurements were made with a Consolidated Engineering Corporation 180°, 5-in. radius mass spectrometer. Gas was admitted into the ion source through a glass pinhole leak with a radius of approximately 0.01 mm. The ion source was run with 2500-V ion accelerating voltage, 50-V electron accelerating voltage, and 10-μA ionizing electron current. Ions were accelerated from the region of the electron beam in the ion source by a variable positive potential, ranging from 0–30 V, applied to both electrodes of the split repeller in the ion source. Spectra were determined by varying the magnetic field to the approximate focus of a given peak and then scanning the peak with a very small variation of the ion accelerating voltage, i.e., coarse magnetic and very fine electrostatic scanning. Ion currents were amplified with a model 31 Applied Physics Corporation vibrating reed electrometer. Current sensitivity of 10^{-15} A at the collector were obtained using a 10^{11}-Ω input resistor. The following procedure for determining the phenomenological cross section for the ion-molecule reaction Q was used.

$$Q = I_s/I_p n l, \tag{2}$$

where I_s is the secondary or product ion current, I_p the reactant ion current, and n and l are, respectively, the molecule concentration and the ion path length in the ion source. For a given repeller voltage, appropriate source focus settings were made, and the respective primary and secondary ion abundances were determined. The concentration of neutral molecules was computed from measured values of corresponding ion yields and independently determined ionization

[8] D. R. Herschbach, G. H. Kwei, and J. A. Norris, J. Chem. Phys. **34**, 1842 (1961).

and collection efficiencies.[9] The measurements of ion yields were repeated frequently over a period of several hours while the sample was being pumped out of the sample reservoir through the mass spectrometer. Semilogarithmic plots of the data as a function of time gave data from which $Q_{H_3^+}$ and Q_{HeH^+} could be determined at a number of different pressures. No attempt was made to redetermine l, the ion path length in the source. This was obtained from the low repeller values of $Q_{H_3^+}$ and the theoretical value of $Q_{H_3^+}$.

Standard procedures were used to measure relative ionization efficiency curves.[5]

He⁺–H₂ REACTION

Since the secondary ions observed in the experiment are produced from a mixture of H₂ and He subjected to electron impact ionization and dissociation, we must first consider the relative contributions of He⁺–H₂ and H₂⁺–He reactions to the production of HeH⁺. The question of whether the HeH⁺ is produced by a bimolecular collision of either ion-molecule pair is not completely trivial. Data are presented in Fig. 1 which establish this mechanism for a number of repeller potentials. All plots are quite linear, but some indication of a reaction of higher order appears at the high repeller runs. This is discussed later in the paper.

The procedure for identification of the reactant ion in an ion-molecule reaction is comparison of the ionization efficiency curves of the product with possible reactant ionization efficiency curves. Data are presented in Fig. 2 for HeH⁺, H₂⁺, and He⁺, all normalized to the same relative ion intensity at 50 V ionizing electron energy. The value of ionizing electron energy which produces 50% of the maximum He⁺ brings about a very small change (of roughly 4%) in the HeH⁺/H₂⁺ ratio. Thus in this energy region, one would estimate an upper limit of 8% of the HeH⁺ formed by reaction of He⁺ with H₂.

The HeH⁺ and H₂⁺ ionization efficiency curves are nearly parallel in the region of linear increase with electron energy. The HeH⁺ is shifted by approximately 1 eV higher on the energy scale. The observation of 8% HeH⁺ from He⁺ reacting with H₂ may be in error because of an improper comparison of the HeH⁺ ionization efficiency curve with the corresponding curve for all the H₂⁺ produced by electron impact on H₂. The proper comparison would be with an ionization efficiency curve for H₂⁺ in the fifth or higher vibrational states which would have an ionization efficiency curve coincident with that of HeH⁺ in the linear rise region. Examination of the vibrational levels of H₂⁺

[9] The method of determining ion collection efficiency was that of D. P. Stevenson and D. O. Schissler, J. Chem. Phys. **29**, 282 (1958). The total ion yield of H₂⁺ from pure H₂ or rare-gas ions produced by electron bombardment of rare gases was measured at the repeller in the ion source, and this value was then compared with the respective collected ion currents. Ionization cross sections were taken from F. Lampe, J. L. Franklin, and F. H. Field, J. Am. Chem. Soc. **79**, 6129 (1957).

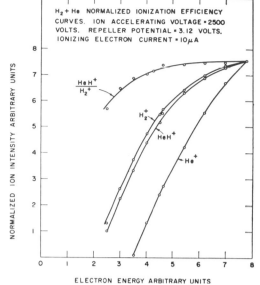

FIG. 2. Normalized ionization efficiency curves for H₂–He mixtures. The ratio of HeH⁺/H₂⁺ as a function of electron energy is plotted on the same energy axis.

indicates that the fifth state is the lowest with sufficient energy to provide the necessary amount for the heat of reaction. Consequently, we may conclude Fig. 2 shows that perhaps there is an upper limit of 8% of the reaction going via He⁺–H₂, but that it is more probable that this value is high and that for the purposes of this discussion the reaction may be considered as going exclusively via H₂⁺ and He.

The conclusion that He⁺ does not produce HeH⁺ on reaction with H₂ may be rationalized by considering the problem of dissipating the 8.3-eV heat of reaction without dissociating the product. If we accept Polanyi's argument that in exothermic atomic reactions of the type

$$A + BC \rightarrow AB + C,$$

the heat of reaction will be almost entirely concentrated in the vibration of the A–B bond formed, then it will be impossible to observe HeH⁺. We may be somewhat more quantitative and assume statistical equilibration of the excess energy among the internal degrees of freedom of a linear HeH₂⁺ intermediate. If we assume 0.25 eV or less for the average value of the vibrational quantum in HeH₂⁺ and then calculate the way in which n quanta are distributed among the oscillators such that less than 1.7 eV, the HeH⁺ dissociation energy, is deposited in the HeH⁺ bond, we conclude that only a few percent of the HeH⁺ could survive. The speculation above is presented only to further rationalize the difficulty in observing HeH⁺ formed at low pressures in a highly exothermic reaction.

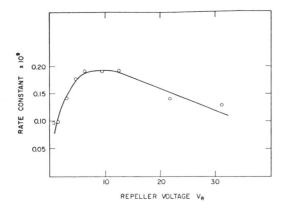

FIG. 3. Rate constant k_a for production of HeH$^+$ from H$_2$–He mixtures plotted as a function of repeller voltage. Values are computed from phenomenological cross sections using Eq. (3).

H$_2^+$–He REACTION

We have concluded that the bulk of the HeH$^+$ yield is formed via the H$_2^+$–He reaction. The experimental data obtained from the mass spectrometer are phenomenological cross sections Q taken as a function of repeller voltage in the ion source. Stevenson and Gioumousis have shown that for ions with kinetic energy large with respect to that of the reactant molecule, the thermal rate constant k is related to the phenomenological cross section Q by the relation

$$k = (eEl/2m_1)^{\frac{1}{2}}Q, \qquad (3)$$

where E is the electric field in the ion source, l the ion path length, and m_i the mass of the reactant ion. Figure 3 shows a plot of k as a function of the repeller voltage V_r. k was computed from experimental values of Q using Eq. (3). This relation gives k proportional to $QE^{\frac{1}{2}}$. We recall that Stevenson and Gioumousis derived the following relation for Q:

$$Q = 2\pi (2m_1\alpha e^2/\mu)^{\frac{1}{2}} \cdot (1/eEl)^{\frac{1}{2}}, \qquad (4)$$

so that k should be independent of E. Combination of Eqs. (3) and (4) gives Eq. (1), the Eyring, Hirschfelder, and Taylor equation for k. It is clear that k does not have the theoretically predicted independence of electric field. The rate constant for the H$_3^+$ reaction showed a similar decrease with electric field in the high range of repeller voltage, 10–30 V, but was well behaved in the lower range.

In spite of the fact that the theory does not seem to apply to the data presented over the complete range of ion translational energy in the experiment we have taken the position that the theory is basically correct but possibly incomplete when applied to endothermic reactions. We attempt to demonstrate that the Gioumousis–Stevenson model provides a useful basis for discussing these reactions.

We note that the thermal rate constants plotted in

Fig. 2 were derived for reactions with very small activation energy, reactions with unit collision efficiency. The valid assumption was made that for relatively low-energy impact processes the potential function for the system could be approximated by consideration of ion-induced dipole forces alone. If, however, collisions take place which bring the reactants sufficiently close together to bring into play short-range repulsive forces, then one can imagine scattering processes which do not lead to reaction. Thus collision efficiency and the value of the rate constant is reduced with increasing ion energy. This is perhaps more easily visualized by imagining the system moving across a valley or well on a potential energy surface with sufficient momentum to be reflected back out of the range of the ion induced dipole attraction. Hamill and co-workers[10,11] have considered this problem and modified the Gioumousis–Stevenson model to take into consideration short range repulsive interactions. A correction for the breakdown of the Gioumousis–Stevenson theory in the high-energy region can be made, for these reactions, by assuming that the probability of repulsive interactions between H$_2^+$ and He and H$_2^+$ and H$_2$ are very similar and that the ratio of rate constants would be predicted by the theory even though the individual rates may be in error. Plots of k_a/k_c as a function of repeller voltage are shown as the points in Fig. 4. The solid line is calculated and is discussed later. The decrease in k_a has been almost exactly compensated for by a decrease in k_c. The absolute value of the ratio is considerably lower than that predicted by the ratio rate constants computed according to Eq. (1). The low-energy region of Fig. 4 is essentially the same as Fig. 3.

The increase in rate or cross section with energy at relatively low repeller voltage clearly establishes a kinetic energy dependence of the H$_2^+$–He ion–molecule reaction. A similar dependence on kinetic energy has been obtained for the yield of KBr from K atoms and HBr in a crossed molecular beam experiment by Greene, Roberts, and Ross.[12] The average ion kinetic energy corresponds to one-fourth the repeller voltage plotted as the abscissa in Fig. 3. This arises from the fact that the ions originate halfway between the repeller and ion exit slit. The ion energy varies from a value of approximately kT in the electron beam to one-half eV$_r$ at the exit slit. We have ignored the effect of field penetration of the ion accelerating voltage and anode voltage into the ion source.

The effect of kinetic energy on the reaction saturates at a relatively low average ion energy, of the order of 1.5 eV. In the region in which the rate constant is insensitive to ion kinetic energy, the theoreti-

[10] D. A. Kubose and W. H. Hamill, J. Phys. Chem. **65**, 183 (1961).
[11] N. Boelrijk and W. H. Hamill, J. Am. Chem. Soc. **84**, 730 (1962).
[12] E. F. Greene, R. W. Roberts, and J. Ross, J. Chem. Phys. **32**, 940 (1960).

cal value of the ratio of rates $k_a/k_c/(0.44)$ is still more than three times the experimental ratio (~ 0.1). Reactant ion kinetic energy does not appear to be able to bring about the reaction of the bulk of the H_2^+ ions in the mass spectrometer ion source. The data in Figs. 3 and 4 suggest that the H_2^+–He reaction proceeds via excited H_2^+ ions with sufficient internal energy to overcome the reaction potential barrier and that a necessary condition for reaction is a minimum critical kinetic energy in the reactants.

Let us consider the effect of a threshold kinetic energy on the value of rate constants computed from experimentally determined phenomenological cross sections. If a minimum energy W_t is required for reaction of primary ions I_p, then there will be an initial fraction of the ion path length l_t over which no reaction can take place. In a uniform electric field E in the ion source $W_t/W = l_t/l$, where W is the kinetic energy the primary ions have acquired when they reach the exit slit of the ion source. We can then express the observed secondary ion yield in terms of the amount that would have been formed over the entire path l minus the fraction that would have been formed in the distance l_t if there were no kinetic energy threshold for reaction.

$$I_s = I_p n (Ql - Q_t l_t), \qquad (5)$$

Q and Q_t are the phenomenological cross sections for ions reacting with path lengths l and l_t, respectively. This expression gives the observed secondary ion yield in terms of two hypothetical yields. From Eq. (4) we see that for the same ion–molecule reaction over different path lengths in the same electric field

$$Q_t/Q = (l/l_t)^{\frac{1}{2}}. \qquad (6)$$

We can then simplify Eq. (5) to give

$$I_s = I_p n Q[l - l_t(l/l_t)^{\frac{1}{2}}]. \qquad (7)$$

The secondary ion current that we would expect for reaction everywhere in the path length l, I_{s_0} can now be introduced so that

$$I_s = I_{s_0}[1 - (l_t/l)^{\frac{1}{2}}]. \qquad (8)$$

If we now express our rate constant k in terms of the corresponding constant k_0, we have

$$k_0 = (eEl/2m_1)^{\frac{1}{2}}(I_{s0}/I_p n l) \qquad (9)$$

and

$$k = k_0[1 - (l_t/l)^{\frac{1}{2}}] \qquad (10a)$$

or

$$k = k_0[1 - (W_t/W)^{\frac{1}{2}}]. \qquad (10b)$$

The calculated curve in Fig. 4 shows that Eq. (10b) gives a good fit of the experimental data if we insert 0.27×10^{-9} cm³/molecule sec for k_0. This value is derived later in this paper from the theoretical expression of the rate constant. The equation gives a good correspondence to the experimental data even at very low

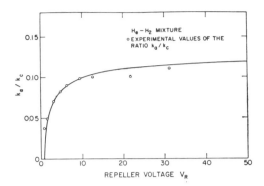

FIG. 4. Ratio of rate constants k_a/k_c plotted as points as a function of repeller voltage. The smooth curve is a ratio k_a/k_c with k_a calculated from Eq. (10b) using indicated values of W_t and k_0; the smooth curve is calculated for $W_t = 0.35$ eV, $k_0 = 0.27 \times 10^{-9}$ cm³/molecule sec.

repeller energy, where the assumption that we can neglect the effect of the thermal energy with respect to the ion kinetic energy may be questioned. The magnitude of the threshold kinetic energy used to obtain the fit of experimental data was 0.35 eV. A shift of 0.05 eV in kinetic energy gave a distinctly poorer fit of the experimental data. This is considerably less than the heat of reaction, 1.1 eV, and can hardly be considered as contributing significantly to the activation energy of the reaction. The activation energy or energy required to surmount the potential barrier in the reaction must be supplied largely by internal energy in the H_2^+. We can treat the H_2^+–He system as an "exothermic" reaction by considering only those H_2^+ ions formed with sufficient internal energy as available for reaction. Inspection of the vibrational energy levels in H_2^+ [1a] reveals that only those in the fifth vibrational state or higher can react if the HeH$^+$ dissociation energy recently computed by Evett[1b] is used to compute the thermochemistry of the reactions. The fraction of H_2^+ in the fifth or higher vibrational states is obtained from an estimate of electron impact transition probabilities.

Coolidge, James, and Present[13] obtained from consideration of the Franck–Condon principle the following relation for transition probabilities near the excitation threshold:

$$Fe, v, e', v' = K(V - V_v)M^2_{e,v,e',v'};$$

K is a constant, $(V - V_v)$ is the excess electron energy above the excitation threshold, and M the product of the integrals of the vibrational wavefunctions for the respective lower and upper states, i.e., the overlap integral. The excitation processes that we are concerned with are brought about by electrons with energy considerably above the excitation threshold so

[13] A. S. Coolidge, H. M. James, and R. D. Present, J. Chem. Phys. 4, 193 (1936).

TABLE I. Overlap values—$M^2_{v0,v'v'}$. $H_2\ {}^1\Sigma_g^+ \rightarrow (H_2^+\ {}^2\Sigma_g^+)_{v'}$

v	Krauss and Kropf	Our calculations	
		With δ function	With wavefunctions
0	8.0	20.6	8.96
1	15.6	18.0	16.11
2	18.0	15.6	17.79
3	16.3	12.6	15.53
4	13.0	10.4	12.29
5	9.2	6.4	9.03
6		4.0	6.31
7			4.46
8			3.04
9			2.19
10			1.47
11			0.96
12			0.65
13			0.46
14			0.30
15			0.23
16			0.13
17			0.07
18			0.02

that the $(V-V_v)$ term for the respective vibrational levels in H_2^+ is approximately the same and may be assumed to contribute very little to the relative transition probability. A rigorous treatment would require an evaluation of the matrix element of the perturbation function in the transition as a function of electron energy. We have not attempted this, but have assumed, for the excess energies involved, that we may obtain approximate ratios of transition probabilities to the respective vibrational states directly from ratios of the squares of the respective overlap integrals.

Krauss and Kropf[14] have published values for the squares of the overlap of the Morse function for the ground state of H_2 with Morse functions for the first six vibrational states of H_2^+. Recently, Cohen, Hiskes, and Riddell[15] have obtained vibrational wavefunctions for H_2^+ by computation of a variational approximation to the wavefunction of the three-body system. We have used these functions to compute the squares of the overlap integrals of the first 19 vibrational states in H_2^+ with the Morse function for the ground state of H_2. It is of interest to compare the values of M^2 obtained in this way with values of Krauss and Kropf and also data obtained by the assumption that one may replace the H_2^+ wavefunction by a constant times a δ function at the classical turning point of the potential energy curve. The latter appears to be a

rather poor approximation. The results given in Table I show rather good agreement between data obtained from Morse functions for H_2^+ and the wavefunctions of Cohen, Hiskes, and Riddell. Multiplying the value of the thermal rate constant [Eq. (1)] with 0.29, the fraction of H_2^+ found to be in the fifth or higher vibrational states, we get $k_0 = 0.27 \times 10^{-9}$ cm³/molecule sec, which has been inserted in Eq. (10b). As we probably have overestimated the population of the more highly excited states in H_2^+ because of the neglect of the energy dependence of transition probabilities, the value of 29% might be somewhat high. However, the results are in excellent agreement with the experimental values of k_a/k_c and provide strong support for the assertion that this reaction does proceed via vibrationally excited H_2^+.

REACTION OF CH₄-He MIXTURES

We remarked in the introduction that it might be possible to obtain a different distribution of excited H_2^+ ions by the electron impact ionization and dissociation of methane and in this way obtain a reaction cross section or rates that differed from the results with H_2-He mixtures. The possibilities of reaction giving HeH⁺ from CH₄-He mixtures are manifold:

$$CH_4^+ + He \rightarrow HeH^+ + CH_3 \qquad -3.2\ eV,$$
$$CH_3^+ + He \rightarrow HeH^+ + CH_2 \qquad -5.4\ eV,$$
$$CH_2^+ + He \rightarrow HeH^+ + CH \qquad -5.9\ eV,$$
$$CH^+ + He \rightarrow HeH^+ + C \qquad -4.2\ eV,$$
$$H_2^+ + He \rightarrow HeH^+ + H \qquad -1.1\ eV,$$

as well as the $He^+ - CH_4 \rightarrow HeH^+ + CH_3$ reaction. All of the above except the last reaction with He^+ ions are more endothermic than the H_2^+-He reaction. The respective heats of reaction computed from data of Field and Franklin[16] are tabulated along side each reaction equation. In order for the respective hydrocarbon ions to react they must have sufficient internal energy to overcome the reaction endothermicity.

The excited hydrocarbon ions produced by electron impact on methane may dissipate their excitation energy by unimolecular dissociation prior to collision with helium atoms leading to ion–molecule reaction. The CH_4^+ ion must be metastable with lifetime of the order of 10^{-5} sec and energy at least 3.19 eV in order to react with He and produce HeH⁺. If the excitation energy is rapidly redistributed among the various internal degrees of freedom in CH_4^+ dissociation into CH_3^+ and H would take place with a threshold energy of 1.3 eV, and in times very short compared with the CH_4^+ residence time in the ion source. There is much evidence pointing toward the validity of the assumption of rapid equilibration of excess energy in

[14] M. Krauss and A. Kropf, J. Chem. Phys. 26, 1776 (1957).
[15] S. Cohen, J. R. Hiskes, and R. J. Riddell, Jr., UCRL-8871.
[16] F. H. Field and J. L. Franklin, Electron Impact Phenomena (Academic Press Inc., New York, 1957).

polyatomic ions produced by electron impact.[17] This is the basic assumption of the statistical theory of mass spectra. This theory predicts lifetimes of the order of 10^{-9} sec and less for excited ions with energy a few volts above threshold decomposition energy so that such ions could not participate readily in ion–molecule reactions.

We can estimate threshold energies for the decomposition processes of CH_4^+ to C^+ by successive loss of H atoms by taking the difference in the respective ion heats of formation and the heat of formation of the hydrogen atoms. The ion heats of formation are,

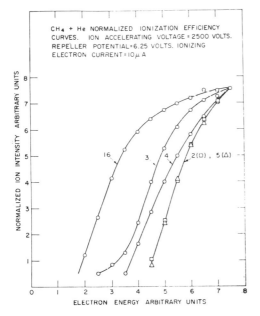

FIG. 5. Ionization efficiency curves of ions produced by electron impact on CH_4–He mixtures. H_2^+ and HeH^+ points fall on identical curve.

according to Field and Franklin, CH_3^+, 11.3 eV; CH_2^+, 14.4 eV; CH^+, 17.3 eV; and C^+, 18.7 eV. With 2.2 eV for the heat of formation of H we have a 5.3 eV threshold for decomposition of CH_3^+ into CH_2^+ and H. In this and the other decomposition reactions the threshold for decomposition is less than the heat of reaction with He so that the latter is disfavored. This makes the H_2^+–He reaction with H_2^+ formed from CH_4 by electron impact the most probable source of HeH^+ because in this case the dissociation threshold is greater than the heat of reaction. The conclusion is strongly supported by data given in Fig. 5 on the ionization efficiency curves. The normalized relative ion intensity of HeH^+ and H_2^+ plotted as a function

[17] H. M. Rosenstock, M. D. Wallenstein, A. L. Wahrhaftig, and H. Eyring, Proc. Natl. Acad. Sci. U. S. **38**, 667 (1952).

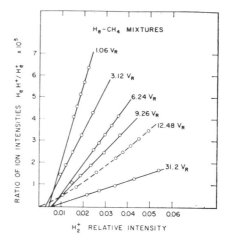

FIG. 6. Ratio of relative ion intensities HeH^+/He^+ as a function of H_2^+ relative intensity at various repeller voltages.

of electron energy coincide. This indicates that all of the H_2^+ formed from CH_4 has sufficient energy to react with He, or that we have a superposition of ionization efficiency curves with reactions involving He^+ and CH_4 as well as the H_2^+–He reaction.

The yields of HeH^+ are very much smaller than yields observed in mixtures of H_2 and He (of the order of 1% of the latter), so that there is some concern about the possibility of contributions of higher-order reactions or possibly wall reactions. The order of the reaction appeared, from plots of the ratio of HeH^+/He^+ against H_2^+ in Fig. 6, quite normal except for the fact that the straight lines did not intercept the origin in a way expected for a pure second-order

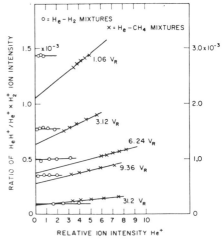

FIG. 7. Plots of $HeH^+/He^+ \times H_2^+$ relative intensities as a function of He^+ relative intensity. Left ordinate refers to He–H_2 mixtures and right ordinate to He–CH_4 mixtures.

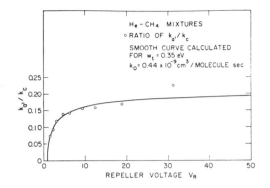

FIG. 8. Plots of k_a/k_c for He–CH_4 mixtures. Values of phenomenological cross sections for HeH^+ production were taken from intercepts of plots in Fig. 7 to separate third-order component in the reaction.

bimolecular process. Careful study of the data for 12-V repeller indicated distinct curvature of the plots near the origin, which suggests a departure from the bimolecular mechanism. If we assume that the reaction is a mixture of second- and third-order processes with

$$d[HeH^+]/dt = k[H_2^+][He] + k'[H_2^+][He]^2.$$

The third-order process may be assumed to be arising from a charge-transfer excitation and ionization process

$$He^+ + CH_4 \rightarrow CH_4^+ + He,$$

$$CH_4^+ \rightarrow H_2^+ + CH_2,$$

$$H_2^+ + He \rightarrow HeH^+ + H.$$

Then a plot of the $HeH^+/[H_2^+][He]$ ratio as a function of He concentration should give straight lines with slopes proportional to k', the third-order rate constant, and intercepts proportional to k, the second-order rate constant. This is done in Fig. 7 along with similar plots of data in the H_2–He system. The latter give for the most part the straight lines parallel to the He axis expected for bimolecular processes. There is some evidence that with higher values of repeller voltage a very small third-order reaction entered. The data from CH_4–He mixtures all give lines much larger with positive slope, indicating a rather significant third-order reaction. Independent analysis of the order of the reaction by plotting the logarithm of the HeH^+ product as a function of the logarithm of the respective reactants separately showed an approximate second-order dependence on He (slightly lower than a

slope of 2) and first-order dependence on H_2^+ at medium and higher values of repeller voltage. A similar analysis of the order of reaction with H_2–He mixtures showed the reaction to be well behaved, first order both in H_2 and He. The dependence on H_2^+ may be somewhat misleading in that its pressure is proportional to the CH_4 neutral concentration, as well as the concentrations of all ions derived from CH_4 by electron impact. At very low repeller voltages the approximate second-order dependence on He was maintained but a fractional order in H_2^+ noted. We have not been able to satisfactorily explain these results.

If we assume that we can obtain a good second-order phenomenological cross section from the extrapolated intercepts of the plots in Fig. 7, then we obtain from these Q values, ratios of rate constants plotted as a function of energy in Fig. 8. The same behavior with respect to kinetic energy noted in the H_2–He reaction is observed here. The threshold kinetic energy used for the calculated curve through the experimental points is 0.35 eV, and the asymptotic value of the ratio of rates of k_a/k_c at infinite kinetic energy is somewhat higher than the value obtained in the H_2–He system (0.44 compared to 0.27). We would conclude that the H_2^+ produced from CH_4 by charge transfer or electron impact processes had a somewhat larger population of vibrationally excited species in fifth or higher states than H_2^+ obtained from electron impact on H_2, approximately 1.5 times the fraction produced by electron impact on H_2.

The mechanism of production of HeH^+ from mixtures of CH_4 and He is indeed complicated. The results are consistent however with those obtained from the H_2–He study. Only a fraction of the total H_2^+ can react, and a necessary condition for this reaction is kinetic energy of the order of 0.35 eV in the ion reactant. The third-order reaction was somewhat surprising. It turns out that we probably have observed a reaction usually concealed as a very small fraction of a much more abundant process which was observable here because of the relatively small yield of H_2^+ produced by electron impact on CH_4.

ACKNOWLEDGMENTS

The authors wish to thank Mr. A. P. Irsa and Mr. Theodore Middleton for their assistance in carrying out some of the experiments. We are indebted to Dr. Max Wolfsberg and Dr. F. S. Klein for reading and discussing this manuscript prior to publication.

30

Reprinted from J. Chem. Phys. **49**:5426–5437 (1968)

Photoionization Study of Ion–Molecule Reactions in Mixtures of Hydrogen and Rare Gases*

W. A. Chupka and Morley E. Russell†

Argonne National Laboratory, Argonne, Illinois 60439

(Received 16 October 1967)

The reactions producing HeH^+, NeH^+, and ArH^+ in mixtures of hydrogen with rare gases were studied by photoionization mass spectrometry. The HeH^+ and NeH^+ ions are produced by vibrationally excited H_2^+ ions; the thresholds for reaction were found to be very near the $v=3$ and $v=2$ states of H_2^+, respectively. Above the vibrational threshold the reaction cross section increases with vibrational quantum number and there is no evidence for a kinetic-energy threshold for these states. When accelerated to sufficient kinetic energy, H_2^+ ions in vibrational states below threshold react with a cross section which increases as the deficit in vibrational energy decreases. The rate constant for reaction of those vibrational states that are near or about the reaction threshold and that contribute most to the observed reaction becomes nearly constant at low repeller voltages. The threshold for the reaction producing HeH^+ can be made consistent with a theoretical value for the dissociation energy $D_0(HeH^+) = 1.835$ eV. The difference $D_0(NeH^+) - D_0(HeH^+)$ is found to be 0.25 ± 0.03 eV. This yields $D_0(NeH^+) = 2.085$ eV. In argon–hydrogen mixtures, the ArH^+ is formed by an exothermic reaction of H_2^+ in all vibrational states with Ar and also by reaction of Ar^+ ions in the $^2P_{3/2}$ ground state and (with $\sim 30\%$ larger cross section) of Ar^+ in the $^2P_{1/2}$ excited state with H_2. The cross section has little or no dependence on vibrational energy. Chemi-ionization processes leading to the formation of ArH^+ by excited H^* and Ar^* atoms were observed. The dissociation energy of ArH^+ was found to be certainly greater than 2.647 eV and, from a tentative interpretation of the chemi-ionization processes, probably greater than 3.397 eV.

INTRODUCTION

The ion–molecule reactions occurring in mixtures of hydrogen with rare gases have been investigated in detail by earlier workers,[1-3] who used electron-impact ionization. The reactions

$$H_2^+ + He \rightarrow HeH^+ + H, \qquad (1)$$

$$H_2^+ + Ne \rightarrow NeH^+ + H, \qquad (2)$$

are of particular interest since it has been shown that neither reaction occurs with thermal H_2^+ ions in the vibrational ground state, but each does occur with H_2^+ ions with vibrational energies above a certain critical value (specific to each reaction). These critical values were determined from the electron-impact appearance potential of the ion produced in the reaction, but the accuracy was not sufficient for a positive determination of the minimum vibrational quantum number required for reaction.

Friedman and co-workers[2,3] made a very thorough and careful study of these reactions and, from a detailed and imaginative analysis of their data, reached several very interesting conclusions regarding the existence of kinetic-energy thresholds for the reactions, the relative efficacy of kinetic and internal energy in promoting reaction, the extent of vibrational de-excitation, and the applicability of the

* Work performed under the auspices of the U.S. Atomic Energy Commission.

† Permanent address: Department of Chemistry, Northern Illinois University, DeKalb, Ill.

[1] V. W. Kaul, V. Lauterbach, and R. Taubert, Z. Naturforsch. **16a**, 624 (1961).

[2] (a) H. von Koch and L. Friedman, J. Chem. Phys. **38**, 1115 (1963); (b) T. F. Moran and L. Friedman, J. Chem. Phys. **39**, 2491 (1963).

[3] (a) F. S. Klein and L. Friedman, J. Chem. Phys. **41**, 1789 (1964); (b) Lewis Friedman, Advan. Chem. Ser. **58**, 87 (1966).

Gioumousis–Stevenson theory of ion–molecule reactions. However, the limitations of electron-impact techniques make it desirable to reinvestigate these reactions by photoionization, with its far greater energy resolution.

In the case of hydrogen–argon mixtures, two reactions must be considered even at low ionizing-electron energies, namely,

$$H_2^+ + Ar \rightarrow ArH^+ + H, \qquad (3)$$

$$Ar^+ + H_2 \rightarrow ArH^+ + H. \qquad (4)$$

Since the ionization potentials of argon and hydrogen are so nearly the same, Kaul et al.[1] were unable to decide which reaction occurs or whether both do. From their relatively inaccurate appearance potentials, they decided that, if Reaction (3) occurs, the H_2^+ ions must again be vibrationally excited, and if Reaction (4) occurs, the reacting Ar^+ ions must be in the electrically excited $^2P_{1/2}$ state. However, these conclusions were not certain.[4] Again, the high resolution of the photoionization technique can easily provide unambiguous answers to these questions.

Finally, accurate determination of thresholds for the above reactions can lead to the determination of dissociation energies for ions of the rare-gas hydrides.

EXPERIMENTAL

The apparatus and techniques used in these experiments are practically the same as those used in previous studies of exothermic ion–molecule reactions.[5] Monochromator slitwidths of $300\,\mu$ and $50\,\mu$ were used, corresponding to photon bandwidths of 2.5 Å (~ 0.05 eV) and 0.4 Å (~ 0.008 eV), respectively. The Hopfield continuum was used as a photon source and all data were taken by counting ions for measured time intervals at varied photon wavelengths.

The gases were premixed just before being introduced into the ionization chamber; the flow rate of each gas was independently variable. There was no provision for accurate measurement of gas pressure inside the ionization chamber since this was not important for the main purpose of this investigation.

EXPERIMENTAL RESULTS

A. H₂–He, H₂–Ne Mixtures at Low Repeller Voltage

In the case of mixtures of hydrogen with helium or neon, the concentration of rare gas was about four times greater than that of hydrogen. This ratio was determined by ionization-gauge readings, which were corrected for differences in the ionization cross sections of the gases. The ionization gauge was located in the source region of the mass spectrometer but

outside the ionization chamber. The total pressure of gas in the ionization chamber was adjusted so that with 17.0-eV photons the intensity of rare-gas hydride ions was about 10% of that of H_2^+. The repeller voltage was nominally set to zero, at which value the hydride ion intensity at high photon energy was not far below its maximum. Under these conditions, there was still a small repeller voltage (probably caused by contact potentials) which resulted in fairly efficient ejection of ions from the ionization chamber. The data were taken with monochromator slits $50\,\mu$ wide so that the photon bandwidth was 0.4 Å (~ 0.008 eV). The ion intensities and counting times were such that above the ionization thresholds (about 16.20 and 15.97 eV for HeH^+ and NeH^+, respectively) the random statistical error is less than 4% for all points. The results are shown in Fig. 1. Also indicated on Fig. 1 are the thresholds for formation of the excited vibrational states of H_2^+ from the H_2 molecule in its vibrational and rotational ground state. The ionization potential of H_2 is taken[6] to be 15.425 eV, slightly smaller than the previously accepted value; the vibrational energies of H_2^+ were taken from the calculation of Wind.[7] As is readily apparent from Fig. 1, neither reaction occurs with H_2^+ in its vibrational ground state. The minimum vibrational quantum number required for reaction to occur at zero kinetic energy appears to be $v \geq 3$ for HeH^+ formation and $v \geq 2$ for NeH^+ formation. (The very small amount of reaction products detected at lower vibrational quantum numbers is readily attributed to reaction just outside the ionization chamber where ions begin to be accelerated and possibly to small acceleration inside the ionization chamber as a result of contact potential differences. As will be shown later, ions with vibrational energy below threshold can react if their kinetic energies are high enough.)

Another conclusion which is readily evident from inspection of Fig. 1 is that the reaction cross section is not constant for the distributions of H_2^+ ions formed above the reaction threshold, but rather is seen to increase at the thresholds for production of the H_2^+ ion in successively higher vibrational states.

Determining how the reaction cross section depends on the vibrational quantum number necessitates a detailed consideration of the mechanism of photoionization of hydrogen in this range of photon energies. Most of the photoionization in this energy range occurs by autoionization. The relevant characteristics of this process have been discussed earlier.[5] Both theory and experiment support the hypothesis that autoionization tends to leave the resulting H_2^+ ion predominantly in the highest energetically possible vibrational state. With this assumption, one may

[4] For a useful review see M. Pahl, Ergeb. Exakt. Naturw. **34**, 182 (1962).

[5] W. A. Chupka, M. E. Russell, and K. Refaey, J. Chem. Phys. **48**, 1518 (1968).

[6] W. A. Chupka and J. Berkowitz, J. Chem. Phys. **48**, 5726 (1968).

[7] H. Wind, J. Chem. Phys. **43**, 2956 (1965).

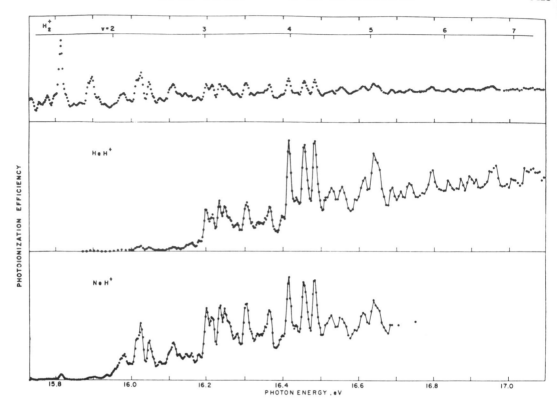

FIG. 1. Photoionization-efficiency curves for H_2^+, HEH$^+$, and NeH$^+$ formed in mixtures of hydrogen with rare gases. The photon bandwidth was 0.4 Å (~0.008 eV). The thresholds for the vibrational states of H_2^+ are indicated.

determine the relative cross section for reaction as a function of vibrational energy by a method described in detail earlier.[5] Briefly, it consists in measuring the relative heights of autoionization peaks above the apparent continuum for the hydride ion and the H_2^+ ion. The ratio of the heights of corresponding peaks is then proportional to the cross section for reaction of the H_2^+ ion in the highest vibrational state energetically possible at the particular photon energy.

Values of relative cross section so obtained from the more prominent autoionization features are shown in Table I. The relatively small variation of individual values for a particular value of v shows that the vibrational distribution is nearly the same for all the autoionization peaks within a vibrational interval. This is consistent with (but not necessarily a proof of) the hypothesis above. The greatest amount of scatter occurs for the formation of HeH$^+$ at $v=3$. However, the consistency for the formation of NeH$^+$ at $v=3$ indicates that a few of the helium data points may be somewhat in error. In Fig. 2 the average values of σ_r(NeH$^+$) and σ_r(HeH$^+$) are plotted as a function of the vibrational energy.

The threshold for Reaction (1) producing HeH$^+$ may be calculated from the very accurate theoretical value of the dissociation energy of HeH$^+$ calculated by Wolniewicz,[8] who found D_e(HeH$^+$) = 16448.3 cm^{-1} = 2.039 eV, the error probably being much less than 0.001 eV. Unfortunately, the zero-point energy of HeH$^+$ is less accurately known. The most accurate published value at present is that of Anex,[9] who calculates the zero-point energy to be 0.204 eV. However, Kołos and Wolniewicz[10] have recently made a more accurate calculation of the zero-point energy which they determine to be 0.195 eV, the uncertainty being less than ~0.001 eV. Using the latter value, one obtains D_0(HeH$^+$) = 1.844 eV. The value of ΔE for Reaction (1) is calculated to be $\Delta E(1) = D_0(H_2^+) - D_0$(HeH$^+$) = 2.647 − 1.844 = 0.803 eV. The value 0.803 eV is indicated by an arrow in Fig. 2. Since this calculated value of ΔE is very likely correct to within ~0.002 eV, it is seen that the H_2^+ ion with $v=3$ and no other rotational or translational energy cannot react with a helium atom of zero kinetic energy.

[8] L. Wolniewicz, J. Chem. Phys. 43, 1087 (1965).
[9] B. G. Anex, J. Chem. Phys. 38, 1651 (1963).
[10] W. Kołos and L. Wolniewicz (private communication).

In Fig. 2 (obtained at nominally zero repeller voltage) we see that the reaction between H_2^+ in the $v=3$ state and helium has a cross section equal to about a third of the apparent asymptotic value for large v. This requires that the kinetic (and possibly rotational) energy of H_2^+ be very effective in promoting Reaction (1), the cross section for $v=3$ rising from zero to about a third of its maximum value within the first several tenths of a volt above threshold. To verify this conclusion we investigated the kinetic-energy dependence of the reaction cross section by measuring the ion yield at different values of the repeller voltage.

Some of the kinetic energy needed for the reaction may be provided by the contact potential differences, as mentioned earlier. The importance of this contribution was investigated by estimating the actual repeller voltage at the nominally zero value by use of the ion–molecule reaction

$$H_2^+ + H_2 \rightarrow H_3^+ + H. \qquad (5)$$

The cross section for this reaction is known[11] to behave at lower repeller voltages according to the Gioumousis–Stevenson theory,[12] which predicts $Q \propto V^{-1/2}$, where Q is the phenomenological cross section and V the repeller field. From measurements of the H_3^+/H_2^+ intensity ratio at 10- and 0-V nominal repeller voltage, the repeller voltage at 0-V nominal value was calculated to be about 0.4 V. Thus the

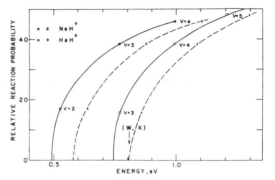

FIG. 2. Relative reaction probability for formation of HeH$^+$ and NeH$^+$ as a function of energy. The solid lines show the dependence on vibrational energy of H_2^+ alone. The dashed lines show the dependence on the average total internal energy of the collision complex as explained in the text. The threshold marked (W, K) is the theoretical value calculated from the data of Wolniewicz and Kolos as explained in the text.

kinetic energy of H_2^+ ions will range from thermal to about 0.2 eV in the ionization chamber, and this is certainly sufficient to cause some reaction of H_2^+ ($v=3$). However, as noted above, a very rapid increase of cross section immediately above threshold is required.

The next step is to see whether the apparent disagreement between experimental and theoretical thresholds can be rationalized on the basis of possible ion kinetic energy. For this purpose we wish to plot the values of σ_r as a function of the average total energy of the collision pair. This involves augmenting the vibrational energy with a small additional amount due to rotational and translational energy. This correction, which is most significant at the lowest energy, can be approximated as follows. The most probable rotational quantum number of H_2 (the number of $\sim 66\%$ of all molecules) in equilibrium at room temperature is $J=1$, and because of the operation of selection rules this tends to be preserved in the autoionization process as well as in direct ionization. Thus, the large autoionization peaks correspond mostly to production of H_2^+ ions with E (rotation) $=2B\approx$ 60 cm^{-1} $=0.008$ eV. The average thermal translational energy of H_2 (and hence of H_2^+) is $\frac{3}{2}kT$ but, because the Gioumousis–Stevenson collision cross section varies as $E^{-1/2}$, the average energy of those making collisions is calculated to be $kT=0.026$ eV. The average rotational energy of the H_2 collision partner is ~ 0.022 eV. The average total c.m. energy of a collision pair due to the repeller field of ~ 0.4 V/cm is calculated as follows. The average kinetic energy of H_2^+ due to the repeller field is $\frac{1}{4}eV_r$ (since ionization occurs about halfway between the repeller plate and the exit hole). However, if the kinetic energy is weighted by the Gioumousis–Stevenson cross section, this average becomes $\frac{1}{6}eV_r$. Furthermore, the kinetic energy in the center-of-mass system is two-thirds of this, so that the desired quantity is $\frac{1}{9}eV_r=0.04$ eV. Thus

TABLE I. Relative reaction cross sections $\sigma_r(\text{ion}) = \sigma(\text{ion})/\sigma(H_2^+)$ for formation of hydride ions by reactions of H_2^+ ions with vibrational quantum number v. Both the individual ratios of the heights of prominent autoionization peaks and the average ratio for each value of v are listed.

v	$\sigma(\text{HeH}^+)/\sigma(H_2^+)$	$\sigma(\text{NeH}^+)/\sigma(H_2^+)$
2		1.33
		1.12
		1.11
		1.11
		Av $=1.17$
3	1.27	2.76
	1.04	2.38
	1.74	2.84
	1.86	2.96
	1.72	2.62
	1.70	2.39
	Av $=1.56$	.Av $=2.66$
4	3.66	3.05
	4.00	3.35
	3.86	3.07
	Av $=3.84$	Av $=3.16$
5	4.4	
	3.9	
	5.2	
	4.8	
	Av $=4.6$	

[11] H. Gutbier, Z. Naturforsch. 12a, 499 (1957).
[12] G. Gioumousis and D. P. Stevenson, J. Chem. Phys. 29, 294 (1958).

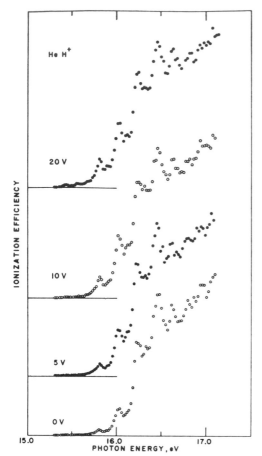

FIG. 3. Photoionization-efficiency curves for formation of HeH⁺ in a helium–hydrogen mixture. The repeller voltages were 0, 5, 10, and 20 V; the photon bandwidth was 2.5 Å ($\sim$0.05 eV).

a rough estimate of the average internal energy (in addition to vibration) of a collision pair is taken to be $E = 0.10$ eV. Values of σ_r plotted as a function of the average total internal energy of the collision pair are also given in Fig. 2. It is seen that the data can thus be brought into agreement with the theoretical threshold.

B. Effect of Repeller Voltage V_r

The photoionization efficiency curves of HeH⁺ and NeH⁺ were measured for helium–hydrogen and for neon–hydrogen mixtures at nominal repeller voltages of 0, 5, 10, and 20 V. Since the distance from the repeller plate to the ion exit hole was 1 cm, the field strength in V/cm is also given by these numbers. The monochromator slits in these measurements were 300 μ wide so the photon bandwidth was 2.5 Å ($\sim$0.05 eV). The conditions of gas pressure and rel-

ative concentrations were practically the same as those of the measurements shown in Fig. 1. The data, taken point by point at intervals of 2 Å, are shown in Figs. 3 and 4 for helium and neon, respectively. For comparison, the photoionization efficiency curve for H₂⁺ (the shape of which is invariant with repeller voltage) taken with 300-μ slits is shown in Fig. 5. The scales of the ordinates for all the plots within one figure has been chosen so that the values of the ordinates for the data taken at 725 Å (17.1 eV) are approximately equal.

It is immediately obvious from these figures that kinetic energy of H₂⁺ is quite effective in causing reactions that are otherwise endoergic. At $V_r = 20$ V, even H₂⁺ ($v = 0$) reacts with helium with a significant cross section, even though the reaction is endoergic by about 0.8 eV.

In order to display these data more graphically and to make a comparison with some results of von Koch and Friedman,[2a] we have followed the procedure of the latter workers and plotted the radios k_a/k_c as a function of repeller voltage in Figs. 6 and 7, where k_a is the rate constant for Reaction (1) or (2) and k_c is the rate constant for Reaction (5). The values of k_c were measured later in a separate ex-

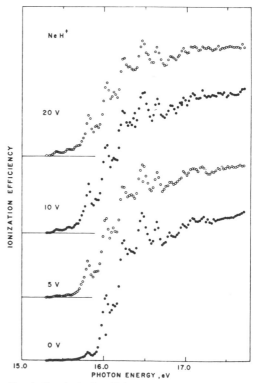

FIG. 4. Photoionization-efficiency curves for formation of NeH⁺ in a neon–hydrogen mixture. The repeller voltages were 0, 5, 10, and 20 V. The photon bandwidth was 2.5 Å ($\sim$0.05 eV).

periment but under conditions apparently identical to those employed in the measurements of Figs. 3 and 4. In addition, the ratio k_a/k_c was measured separately with photons of 725-Å wavelength and with a repeller voltage which could be varied from −1.5 to +40 V. The data shown in Figs. 6 and 7 for 725 Å were so obtained. As the nominal repeller voltage was made negative, the ion intensities dropped rapidly but the values of k_a and k_c as well as the ratio k_a/k_c became and remained constant. The voltage at which k_a/k_c became constant was taken as the true zero of repeller voltage and corresponded to a nominal negative voltage of about one tenth of a volt.

The ratio k_a/k_c is also equal to the ratio Q_a/Q_c of the phenomenological cross sections; and since k_c is nearly independent of repeller voltage in this region,[2,11] this ratio is nearly proportional to k_a. The values of k_a were calculated from the data plotted in Figs. 3 and 4 for H_2^+ ions formed at photon energies corresponding to some of the more prominent peaks of these figures (except for $\lambda = 725$ Å). These peaks, indicated in Fig. 5, are seen to correspond to the production of H_2^+ ions with maximum vibrational quantum numbers of 0, 1, 2, 3, 4, and 7— as noted in Figs. 6 and 7. It should be noted that the total ion intensity (rather than the height of the peak above the adjacent apparent continuum) at a particular photon energy was used in calculating k_a. Thus the rate constants are those for a distribution of vibrational quantum numbers from 0 to the maximum. This distribution may not be far from that given by the Franck–Condon factors[13,14] applicable to direct ionization, truncated at v_{max}, except that it will probably have an abnormal abundance of v_{max}. The wavelength 725 Å does not correspond to a pronounced peak in the ionization efficiency of H_2 but rather to nearly pure directionization. Hence the vibrational distribution produced at 725 Å should be given fairly well by the calculated Franck–Condon factors truncated at $v = 7$. (These factors are small for $v > 7$ and can be neglected here.) This vibrational distribution should most closely approximate the one produced by electron impact at the 50 V used by von Koch and Friedman.[2a] (However this latter distribution will also have very significant contributions from autoionization, the effect of which will be to enhance the relative abundance of the lowest vibrational states—particularly $v = 0$.) Because of the large uncertainty in the relative concentrations of H_2 and rare gas in the ionization chamber, the probable error of the absolute magnitude of k_a/k_c is ∼30%. However, the relative values of k_a/k_c are much more accurate, the probable error being about 5% or less.

It will be noted that there are no clear thresholds in Figs. 6 and 7—not even for some distribution (e.g., $v \leq 2$, 1, and 0 in Fig. 6) that definitely cannot react at the lowest apparent kinetic energy used (corresponding to 0.56-V repeller voltage). As has been mentioned earlier, indefiniteness in the threshold is due to reaction in the acceleration region immediately outside the ionization chamber. No correction was made for this contribution since it could not be estimated reliably from the present data. Partly because of this effect, the present data do not permit an unequivocal decision whether the ratio of k_a/k_c approaches zero as the repeller voltage approaches zero for H_2^+ ions with zero initial kinetic energy but with sufficient vibrational energy to react (e.g., for much of the H_2^+ formed at $\lambda = 725$ Å).

Both for HeH$^+$ and NeH$^+$ formed by reaction of H_2^+ ions produced by impact of 50-V electrons, the ratio k_a/k_c found by Friedman and co-workers[2] decreased somewhat more rapidly at the lowest repeller voltages than does the ratio determined from our present data for $\lambda = 725$ Å. This behavior is probably due to the greater relative abundance of the lower vibrational states of H_2^+ in the experiments of Friedman and co-workers. They interpreted their results as showing the existence of a kinetic-energy threshold for reaction, even when the H_2^+ ions had sufficient vibrational energy to react. However, their interpretation was based on several assumptions which, while at that time tenable and consistent with their lower resolution electron-impact data, are now disproved by the results of the present photoionization study.

Friedman and co-workers[2] assumed (1) that H_2^+ ions with vibrational energy insufficient for reaction did not react at any kinetic energy and (2) that H_2^+ ions with sufficient vibrational energy to react had a rate constant k_a for reaction which was (a) zero below a threshold kinetic energy, (b) a constant k_0 above this threshold energy, and (c) the same for all vibrational states above the critical energy for reaction. They then derived an equation for the variation of k_a/k_c with repeller voltage. By adjusting the value of k_0, they were able to obtain a good fit

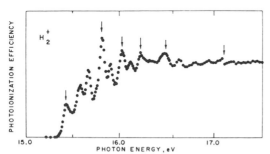

FIG. 5. Photoionization-efficiency curve for H_2^+ The photon bandwidth was 2.5 Å (∼0.05 eV). The arrows indicate the photon energies at which the data of Figs. 6 and 7 were taken.

[13] See, for instance, Table I of Ref. 2.
[14] M. E. Wacks, J. Res. Natl. Bur. Std. (U.S.) 68A, 631 (1964).

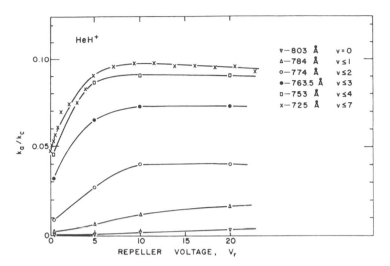

FIG. 6. Ratio of rate constants k_a/k_c as a function of repeller voltage V_r for formation of HeH[+] and H$_3$[+] by H$_2$[+] produced by photons of the indicated wavelength.

to their experimental data for both HeH[+] and NeH[+]. The required value of k_0 agreed well with the Gioumousis–Stevenson value multiplied by the fraction of H$_2$[+] ions with vibrational quantum numbers above the threshold for reaction, this fraction being calculated from Franck–Condon factors for direct ionization. The threshold quantum numbers, based on data then available, were taken to be $v \geq 5$ for helium and $v \geq 2$ for neon. The fact that their derived equation and value of k_0 agreed well with experiment was taken as support for their assumption. It is now seen that these assumptions are not correct and the good agreement must be fortuitous.

In a more recent work Klein and Friedman[3a] observe a variation with ion kinetic energy of the isotope effect for these reactions which "appears to conflict with results obtained from studies of the

H$_2$[+]–He and H$_2$[+]–Ne reactions, which show very little kinetic-energy transfer." Still more recently Friedman[3b] has modified assumption (1) by stating that "the cross-sections for reaction of internally excited species are assumed to be an order of magnitude or more larger than processes which require direct energy transfer." The present data prove conclusively that the original assumptions (1) and (2) are incorrect and show quantitatively the variation of reaction cross section with both internal and kinetic energy.

That assumption (1) above is incorrect is evident from Figs. 3, 4, 6, and 7 where it can be seen that all vibrational states react when the kinetic energy is sufficiently high although, at any given kinetic energy, those states whose reaction is less endoergic react with higher cross section.

There is no evidence in our data for assumption

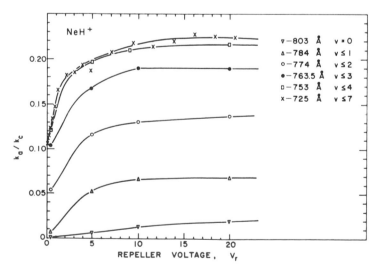

FIG. 7. Ratio of rate constants k_a/k_c as a function of repeller voltage V_r for formation of NeH[+] and H$_3$[+] by H$_2$[+] produced by photons of the indicated wavelength.

(2a), particularly in the case of helium whose threshold energy was taken by Friedman *et al.*[2] to be 0.35 eV. Our data, particularly those taken at 725 Å and with nominal negative repeller voltages, show that if such a threshold exists it must be of the magnitude of thermal energies or still lower.

There is now no reason for assumption (2b) to hold. Since the rate constant for H_2^+ ions with vibrational energy below the critical value varies with kinetic energy, the simplest and most reasonable explanation of the other data of Figs. 6 and 7 is that k_a varies with kinetic energy especially rapidly at low kinetic energies. Moran and Friedman[2b] considered such an explanation for their observations, but discarded it mainly because of the good agreement of those observations with their theoretical calculations. In these calculations they assumed that there was no conversion from translational to internal energy; such energy conversion was felt to be improbable by analogy with some data available for collisions between neutral gas molecules and because k_a/k_c does not increase much for repeller voltages above 5 V. The agreement referred to is seen now to be fortuitous: the data on neutral gases are for weakly interacting systems and are not applicable to this case, and the observation regarding the behavior of k_a/k_c merely means that k_a does not change much with kinetic energy above about 3 eV ($\sim\frac{1}{4}eV_r$, as discussed by von Koch and Friedman[2a]) for those states of H_2^+ that contribute most of the reaction—and this is quite reasonable. The data of Figs. 6 and 7 show that endoergic states with less vibrational energy have values of k_a that are still increasing at higher values of V_r, but these values of k_a are always much less than those for higher vibrational states. Thus, in contrast to the step function that Friedman and co-workers assumed for the curve of k_a vs kinetic energy for all vibrational states above the critical value, the present data support smoother functions that reach an approximately constant value at repeller voltages in the range 5–10 V for the vibrational states of H_2^+ lying just below the critical value for reaction (e.g., for $v=2$ and $v=3$ for reaction with helium). The value of V_r at which k_a becomes nearly constant increases as the reaction becomes more endoergic. This behavior is very reasonable and is qualitatively what would be expected, e.g., on the basis of the statistical "phase-space" theory of ion–molecule reactions.[15] Thus, the increase of k_a at repeller voltages higher than 5 or 10 V (expected by Friedman and co-workers if conversion from translational to internal energy was effective) actually does occur for the most endoergic reactions (of the lowest vibrational states of H_2^+), but the cross sections for these reactions are always much less than those for the exothermic (and near exothermic) reactions. This is

not unreasonable; in fact, it is qualitatively predicted by the "phase-space" theory.

Assumption (2c) is inconsistent with the data shown in Fig. 2. However, the plot of Fig. 2 is based on the assumption that autoionization produced predominantly the H_2^+ ion in the highest energetically allowed vibrational state. This assumption has been experimentally verified[16] recently by photoelectron energy analysis for six of the most intense autoionization lines of H_2.

The good agreement between the value of k_0 which Friedman and co-workers calculated on the basis of the Gioumousis–Stevenson theory and the value required to fit their data is now seen to be fortuitous. This is especially evident in the case of helium since they took $v=5$ as the vibrational threshold for reaction, while Figs. 1 and 2 show it to be very near $v=3$. Even if it is actually just above $v=3$ (so that the critical value is $v=4$) their agreement is still fortuitous since it is seen from Fig. 6 that k_a for $v=3$ and 2 (as well as for $v=4$) must contribute very significantly to their value for k_0, which is their limiting value of k as $V_r\rightarrow\infty$.

In the case of neon, Friedman and co-workers considered the reaction to be energetically allowed for H_2^+ with $v\geq2$. Although this is in accord with the data of Fig. 1, it is seen from Fig. 7 that $v=1$ also contributes significantly to the reaction at higher repeller voltages. Moreover, by comparison with the data for HeH^+ in Fig. 2, it is very likely that H_2^+ with $v=2$ and zero kinetic and rotational energy cannot react.

The present data are too crude to allow accurate determination of the microscopic reaction cross section as a function of kinetic energy for individual vibrational states. This will require data with resolution such as that of Fig. 1 and taken carefully at small intervals of repeller voltage. This is now quite feasible.

Moran and Friedman[2b] suggest that collisional deactivation of vibrationally excited states occurs at very low kinetic energies and may account for their decreasing values of k_a/k_c at low repeller voltage and kinetic-energy threshold. Our data show even more conclusively that many collisions are nonreactive not only at low kinetic energies but at the higher kinetic energies also. The latter follows from the fact that, especially in helium, there is a very appreciable contribution to the reaction from H_2^+ ions in states that von Koch and Friedman considered to be unreactive. Their observed low value of k for the reaction can then be explained only if a large fraction of all collisions (given by the Gioumousis–Stevenson collision-rate expression, which gives at least a lower limit to the collision rate) are unreactive, even for H_2^+ ions in at least some of the vibrational states above the reaction threshold. None of these observations

---

[15] J. C. Light and J. Lin, J. Chem. Phys. **43**, 3209 (1965).

[16] J. Berkowitz and W. A. Chupka (unpublished data).

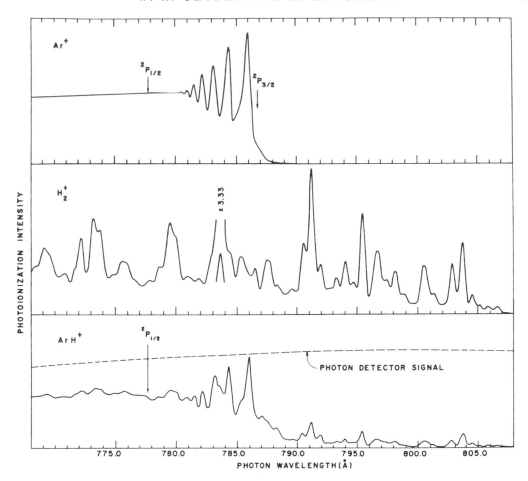

FIG. 8. Smoothed recorder traces of the photoionization intensity of Ar⁺, H₂⁺, and ArH⁺ as a function of photon wavelength. The gas was an argon–hydrogen mixture; the photon bandwidth was 0.4 Å (~0.008 eV). Also shown is the signal from the photon detector.

offer conclusive proof that the unreactive collisions result in vibrational deactivation of H_2^+ ions. However, the rapid decrease of k with decreasing repeller voltage observed by Friedman and co-workers strongly suggests such a competing deactivation process.

Moran and Friedman[2b] considered that such vibrational deactivation could not occur by conversion of vibrational to translational energy since, as they were erroneously led to conclude from their data, the reverse process did not seem to occur. (However, this conclusion was later modified by Friedman.[3]) This objection is now removed since we have seen from Figs. 6 and 7 that kinetic energy is quite effective in causing reaction of H_2^+ ions with vibrational energy below the threshold for reaction. As an alternative, Moran and Friedman proposed the mechanism of collision-induced dipole radiation. However, they estimated the radiative lifetime to be of the order of 10^{-8} sec, which is typical of a strong electronic transition with $f \approx 1.0$ and ΔE equal to several volts. In the present case of emission of vibrational quanta, the radiative lifetime would be larger than the 10^{-5}–10^{-6} sec the ions spend in the ionization chamber. Such a long-lived collision complex for a system with so few degrees of freedom is extremely unlikely (but if it did exist it would be detectable with the mass spectrometer). Thus, any vibrational de-excitation of the H_2^+ ions must occur predominantly by conversion of vibrational to translational (and possibly also rotational) energy.

Light and Lin[15] have applied the "phase-space" theory of ion–molecule reactions to the data of von Koch and Friedman[2a] on Reaction (1). Their calculated rate constant as a function of kinetic energy for the assumed Franck–Condon vibrational distribution of H_2^+ is in excellent agreement with the data of von Koch and Friedman (see Fig. 6 of Ref. 15). Only at the lowest repeller voltage are our data for a roughly similar vibrational distribution (725-Å data) in significant disagreement with those of von Koch and Friedman. Light and Lin used the value $D_0(\text{HeH}^+) = 1.727$ eV obtained from a less accurate calculation than that of Wolniewicz, whereas the correct value is $D_0(\text{HeH}^+) = 1.844$ eV. Use of this slightly higher value for $D_0(\text{HeH}^+)$ will shift the calculated values of k upward slightly, possibly giving even better agreement with the data of von Koch and Friedman at higher kinetic energies, but going slightly higher (in the direction indicated by our data) at low kinetic energies. In any case, within experimental error, there is no disagreement between our data and the calculations of Light and Lin. Furthermore our experimental variation of k_a/k_c with vibrational distributions and with repeller voltage is qualitatively as expected from the "phase-space" theory.

C. Dissociation Energy of NeH+

From the curves (Fig. 2) of reaction probability as a function of vibrational energy for formation of NeH+ and HeH+, we determine the difference in threshold energies of the reactions to be $\Delta = 0.25 \pm 0.03$ eV. This value is based on the reasonable assumption that the shapes of the curves are similar for these very similar reactions. This difference is just the difference between the dissociation energies of the two hydride ions, i.e.,

$$\Delta = D_0{}^0(\text{NeH}^+) - D_0{}^0(\text{HeH}^+) = 0.24 \text{ eV}.$$

If we accept the value of $D_0{}^0(\text{HeH}^+) = 1.844$ eV, then we obtain the value $D_0{}^0(\text{NeH}^+) = 2.08 \pm 0.03$ eV. This value is in excellent agreement with the much less precise value 2.15 ± 0.3 eV obtained by Kaul et al.[1] by electron impact.

D. Ion–Molecule Reactions in H₂–Ar Mixtures

Approximately equimolar amounts of hydrogen and argon were introduced into the ionization chamber at a total pressure such that reaction products accounted for no more than about 15% of the total ionization. The monochromator slits were both 50 μ wide so that the photon band width was 0.4 Å ($\sim$0.008 eV) and the repeller voltage was nominally zero (actual voltage $\sim$0.6 V). For this gas mixture, no data were taken at higher repeller voltages. For the purpose of the present investigation, it was sufficient to scan with the monochromator and to obtain a strip-chart record of ion intensity vs photon wavelength for each ion of interest.

Figure 8 shows the smoothed traces of such scans for H₂+, Ar+, and ArH+ ions, as well as the current from the photon monitor. (It should be noted that these curves are not accurately equivalent to relative photoionization efficiency curves, since the ion intensity at every point should be divided by the photon monitor current corrected for the wavelength dependence of the photoelectric efficiency of the detector. It is not necessary to do this for the present purpose.) Several conclusions are obvious by inspection of Fig. 8:

(a) Both Reactions (3) and (4) occur in this system.

(b) The H₂+ ion in its vibrational ground state (and higher states) is capable of reacting at nominally zero kinetic energy. Thus Reaction (3) is exothermic.

(c) The cross section for Reaction (3) shows little or no variation with vibrational energy of the H₂+ reactant ion. This is similar to the behavior found for other exothermic reactions[5] and is in strong contrast to that found for Reaction (1) and (2) which are endothermic for H₂+ ions in the ground state.

(d) Reaction (4) occurs for both the $^2P_{3/2}$ ground state and $^2P_{1/2}$ excited state of the Ar+ ion.

(e) The reaction cross section for reaction of $^2P_{1/2}$ excited Ar+ ions is somewhat higher than that of the $^2P_{3/2}$ ground-state ions. On the assumption that immediately above threshold the abundances of these states of the Ar+ ions are in the ratio $^2P_{3/2}/^2P_{1/2} = 2/1$ given by the statistical weight, the ratio of the two reaction cross sections may be estimated from the magnitude of the small (but measurable) rise at the $^2P_{1/2}$ threshold[17] in the ArH+ curve of Fig. 8. The resulting ratio of cross sections is $\sigma(^2P_{1/2})/\sigma(^2P_{3/2}) = 1.3$.

It is easily seen that the shape of the curve for the ArH+ ion in Fig. 8 should be independent of pressure (so long as only small amounts of reactant ions are consumed) and the relative amounts of argon and hydrogen, since the intensity of ArH+ ions depends on the product of pressures $P(\text{H}_2) \cdot P(\text{Ar})$ for both Reactions (3) and (4). It can easily be shown that

$$\frac{\text{ArH}^+ [\text{Reaction (4)}]}{\text{ArH}^+ [\text{Reaction (3)}]} = \frac{\sigma_i(\text{Ar})Q(4)}{\sigma_i(\text{H}_2)Q(3)},$$

where σ_i is the photoionization cross section and Q the phenomenological cross section for the indicated reaction. By comparing the curve of Fig. 8 and using

[17] It should be noted that there is no similar rise (nor in fact any feature) at the $^2P_{1/2}$ threshold in the Ar+ curve in Fig. 8. This has been confirmed in several measurements, including a careful point-by-point measurement of the ionization efficiency of argon from the ionization threshold to well past the $^2P_{1/2}$ threshold with a resolution width of 0.04 Å ($\sim$0.8 mV) [J. Berkowitz and W. A. Chupka (unpublished data)]. With higher resolution, many more autoionized peaks are resolved, but they merge smoothly into an apparent continuum below the $^2P_{1/2}$ threshold and there is no structure at this threshold. This is in contrast to the findings of other workers [e.g., R. E. Huffman, Y. Tanaka, and J. C. Larrabee, J. Chem. Phys. 39, 902 (1963)] but in accord with theoretical predictions [U. Fano (private communication)]. Nevertheless, above the $^2P_{1/2}$ threshold Ar+ ions are formed in both $^2P_{1/2}$ and $^2P_{3/2}$ states in relative amounts given approximately by the relative statistical weights.

values of photoionization cross sections obtained from the graphs of Cook and Ching,[18] we obtain

$$Q(4)/Q(3) = 1.4.$$

From the theory of Gioumousis and Stevenson,[12] we calculate

$$\frac{Q(4)}{Q(3)} = \left[\frac{m(Ar^+)\alpha(H_2)}{m(H_2^+)\alpha(Ar)}\right]^{1/2} = 3.1.$$

Thus there is a discrepancy of about a factor of two.

Stevenson and Schissler[19] have measured $k(4)$, the rate constant for Reaction (4), and obtained the value $k(4) = 1.68\times10^{-9}$ cm³ molecule⁻¹·sec⁻¹; the theoretical value[12] was calculated to be 1.50×10^{-9} cm³ molecule⁻¹·sec⁻¹. Stevenson and Schissler concluded that Reaction (3) did not occur, while we estimate from our data that it actually contributed roughly about 20% to their measured ArH⁺ intensity. Their corrected value for $k(4)$ would be approximately 1.35×10^{-9} cm³ molecule⁻¹·sec⁻¹, which is only slightly below the theoretical value. Thus, we conclude from our measurement of the ratio $Q(4)/Q(3)$ that $Q(3)$ is roughly twice the theoretical value. Similar discrepancies have been noted for other proton-transfer processes (e.g., $H_2^+ + O_2 \rightarrow HO_2^+ + H$).[12]

E. Chemi-ionization and Reactions of Ar_2^+

In argon–hydrogen mixtures, small amounts of ArH⁺ were found to be formed at photon wavelengths below the ionization threshold for H_2 (and for Ar). To get sufficient intensity for a measurement of the photoionization-efficiency curve for the formation of ArH⁺ in this wavelength region, the photon resolution width was increased to 2.5 Å. The result is shown in Fig. 9.

A similar process producing H_3^+ was found to occur in hydrogen alone.[5] It was shown to be due to the reaction of excited metastable 2^2S hydrogen atoms and of excited H_2 molecules with H_2 molecules in their ground state. The photoionization-efficiency curve for H_3^+ in pure hydrogen is also shown in Fig. 9 for comparison. The most prominent feature of the (dashed) curve for H_3^+ in this region is the peak at 14.78 eV, which has a low-energy shoulder with a threshold at about 14.66 eV. The peak has been shown[5] to correspond to the production of the metastable H* atoms by a strong predissociation of the D $^1\Pi_u$ $v = 3$ state of H_2 located at 14.78 eV, while the shoulder corresponds to the production of H* in the dissociation continuum which has its threshold at

14.66 eV. The relative intensities of the two processes are not known accurately but, at comparable pressures, the ArH⁺ ions are only about a fifth to a tenth as intense as the H_3^+ ions. It is seen that, within experimental error, the major peak of H_3^+ has its counterpart at about 14.79 eV on the ArH⁺ curve. Thus, this peak may correspond to the production of ArH⁺ by the reaction

$$H^*(2^2S) + Ar \rightarrow ArH^+ + e^-. \tag{6}$$

However, although there is a good correspondence between the positions of some of the other peaks, the relative intensities of ArH⁺ are much higher.

In a recent study of chemi-ionization in argon by photon absorption, Huffman and Katayama[20] found that Ar_2^+ ions were formed by the process

$$Ar^* + Ar \rightarrow Ar_2^+ + e^-, \tag{7}$$

where Ar* could be any one of many optically allowed excited states of argon above the threshold level at 842.81 Å (14.71 eV). The positions of these states and their relative intensities (the latter measured roughly from Fig. 2 of Huffman and Katayama) are also shown by the bars of Fig. 9. There is an excellent correspondence in position and a fairly good correspondence in intensity between the bars and the structure of the ArH⁺ curve. Therefore, the curve very probably indicates the occurrence of either or

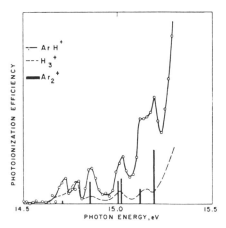

FIG. 9. Photoionization-efficiency curve (solid line) of ArH⁺ below the ionization threshold of H_2^+. The peaks indicate formation of ArH⁺ by excited H* and Ar* atoms. The dashed line is the photoionization-efficiency curve for H_2^+ formed in pure hydrogen. The vertical bars indicate the positions and approximate relative intensities of peaks in the photoionization-efficiency curve for Ar_2^+ produced by chemi-ionization in pure argon (taken from the data of Ref. 20).

[18] G. R. Cook and B. K. Ching, "Absorption, Photoionization, and Fluorescence of Some Gases of Importance in the Study of the Upper Atmosphere," Aerospace Corporation Report No. TDR-469(9260-01)-4, January 1965.

[19] D. P. Stevenson and D. O. Schissler, J. Chem. Phys. 29, 282 (1958).

[20] R. E. Huffman and D. H. Katayama, J. Chem. Phys. 45, 138 (1966).

both of the processes

$$Ar^* + H_2 \rightarrow ArH^+ + (H + e^-), \tag{8}$$

$$Ar^* + Ar \rightarrow Ar_2^+ + e^-,$$

$$Ar_2^+ + H_2 \rightarrow ArH^+ + Ar + H. \tag{9}$$

It should be noted that the relative intensities of ArH$^+$ produced by Reaction (8) need not be the same for the various states of Ar* as in the case of the first step of Reaction (9) as measured by Huffman and Katayama since the relative reaction cross sections may be different.

Accurate measurements of the dependence of ArH$^+$ intensity on argon pressure and on repeller voltage should distinguish between these two possibilities. Some preliminary measurements show that Reaction (8) very likely occurs and is predominant at higher repeller voltages which suppress the second step of Reaction (9). However, there is some evidence that Reaction (9) also occurs appreciably at the lower repeller voltages.

If Reaction (8) occurs, there is a possibility that it can occur even with the Ar in lower excited states than are required for production of Ar$_2^+$. However, measurements have not yet been extended to longer wavelengths. Also, it is not yet established whether or not H$^-$ is produced together with ArH$^+$. Consequently it is not known whether the other product of Reaction (8) is H$^-$ or H$+e^-$.

It is seen from Fig. 9 that Reaction (6) is necessary to account for the peak at 14.79 eV and the high intensity underlying the peak at 14.71 eV. Argon does not absorb in the neighborhood of 14.79 eV. Similarly, either Reaction (8) or (9) is necessary to account for the peak at 14.71 eV. The other peaks could be attributed to either Ar* or H*, although the relative intensities strongly suggest the former.

Lower limits to $D_0^0(ArH^+)$ may be calculated on the assumption that the Reactions (6), (7), and (8) are exothermic.

F. Dissociation Energy of ArH$^+$

The fact that Reaction (3) occurs readily for H$_2^+$ in its vibrational ground state sets the definite lower limit

$$D_0^0(ArH^+) \geq D_0^0(H_2^+) = 2.647 \text{ eV}.$$

If, as seems probable, Reaction (6) occurs with thermal H$^*(2^2S)$ atoms, then one may calculate the limit $D_0^0(ArH^+) \geq 3.397$ eV. However, there is still

a possibility that Reaction (6) occurs only for H* atoms with kinetic energy. It has been shown[8] that the H* atoms produced by the absorption peak at 14.78 eV have an average kinetic energy of $\sim$0.07 eV with a rather large spread (the width at half maximum being $\sim$0.10) due to the original thermal velocity of H$_2$. The absorption peak at 15.01 eV produces H* atoms with an average kinetic energy of 0.18 eV and a width at half maximum of $\sim$0.16 eV. It may be argued that the peak observed at $\sim$15.02 eV is due to such fast Ar* atoms and that there is a kinetic-energy threshold for Reaction (8) at $\sim$0.25 eV. This interpretation would lead to the limit $D_0^0(ArH^*) \geq \sim$3.15 eV.

If, as seems very probable, the peak at $\sim$14.72 eV corresponds to reaction of Ar* in the state at 14.71 eV, designated as[?] $4d[\frac{1}{2}]_1^0$ by Huffman and Katayama, then other lower limits to $D_0^0(ArH^+)$ may be calculated. If the process occuring is either Reaction (9) or Reaction (8) with production of H$+e^-$, then the lower limit may be calculated to be $D_0^0(ArH^+) \geq$ 3.366 eV. If in Reaction (8) the H$^-$ ion is produced, the limit must be reduced by the electron affinity of the H atom (0.747 eV) to $D_0^0(ArH^+) \geq 2.619$ eV.

The reaction Ar$_2^+ + H_2 \rightarrow ArH^+ + Ar + H$ has been reported[21] to occur. If this reaction is then considered to be exothermic and the lower limit $D_0^0(Ar_2^+) \geq$ 1.049 eV obtained by Huffman and Katayama[20] is used, one obtains the same limit as above, namely $D_0^0(ArH^+) \geq 3.366$ eV. However, it should be noted that such conclusions based on electron-impact data are not to be considered very precise since the reactant Ar$_2^+$ ions may be formed with broad vibrational-energy distributions and the reaction may not be exothermic for the ground state of Ar$_2^+$. If Reaction (9) is occurring by absorption at $\sim$14.72 eV, then the calculated limit is precise and the reacting Ar$_2^+$ is in the vibrational state (or states) produced at the chemi-ionization threshold of Huffman and Katayama.

We conclude that the lower limit to $D_0^0(ArH^+)$ is certainly as high as 2.647 eV and probably 3.397 eV.

An upper limit to $D_0^0(ArH^+)$ can be calculated from the observation of the reaction

$$H_2 + ArH^+ \rightarrow H_3^+ + Ar.$$

The upper limit is the proton affinity of H$_2$, which is somewhat uncertain but probably[5,22,23] is $\sim$4.5 eV.

---

[21] Unpublished results reported in Ref. 4.
[22] R. E. Christoffersen, J. Chem. Phys. **41**, 960 (1964).
[23] H. Conroy, J. Chem. Phys. **41**, 1341 (1964).

31

Copyright © 1974 by the American Institute of Physics

Reprinted from J. Chem. Phys. 61:2122–2128 (1974)

Photoionization and ion cyclotron resonance studies of the reaction of vibrationally excited $C_2H_2^+$ ions with H_2 *

S. E. Buttrill Jr.[†]

Department of Chemistry, University of Minnesota, Minneapolis, Minnesota 55455

J. K. Kim[‡] and W. T. Huntress Jr.

Jet Propulsion Laboratory, California Institute of Technology, Pasadena, California 91103

P. LeBreton and A. Williamson

Department of Chemistry, California Institute of Technology, Pasadena, California 91103

I. INTRODUCTION

The reaction

$$C_2H_2^+ + H_2 \rightarrow C_2H_3^+ + H \tag{1}$$

is slightly endothermic for ground state $C_2H_2^+$ ions at thermal kinetic energies. Using the values $\Delta H_f(C_2H_2^+)$ $= 317.2^1$ and $\Delta H_f(C_2H_3^+) = 266$ kcal/mole,[2] the endothermicity of Reaction (1) is $\Delta H \cong +1$ kcal/mole. This endothermicity is sufficiently small that it may be considered within the error limits for the determination of the thermodynamic value for $\Delta H_f(C_2H_3^+)$. The heat of formation of the $C_2H_3^+$ ion has proven to be a difficult quantity to measure; mainly because the appearance thresholds for the ion are poorly defined even in photoionization spectra. Since the observation of Reaction (1) would result in a good upper limit to $\Delta H_f(C_2H_3^+)$ for comparison with photoionization values,[2,3] this reaction was studied using both ion cyclotron resonance (ICR) and photoionization mass spectrometric (PIMS) techniques.

Photoionization mass spectrometric studies were conducted near threshold in order to examine the dependence of Reaction (1) on $C_2H_2^+$ ion vibrational energy in the first few vibrational levels of the ground electronic state. Ion cyclotron resonance studies were used to examine the dependence of Reaction (1) on pressure and time. From the combined use of both ICR and PIMS methods, it becomes possible to study individual ion–molecule processes in greater detail than is possible by using either method alone. Ambiguities present in data from either PIMS or ICR experiments can be resolved by appropriately designed experiments using the complementary instrument. In this study, the dependence of Reaction (1) on internal energy is obtained from photoionization experiments, and the rate constant and mechanism for Reaction (1) is obtained from ICR experiments. The ICR experiments are also used to determine the reactivity of the totally unexcited $C_2H_2^+$ ions remaining

at high H_2 pressures and long reaction times; data which cannot be obtained from the PIMS instrument.

II. PHOTOIONIZATION STUDIES

A. Experimental

The JPL–CIT photoionization mass spectrometer used in this study has been described in detail elsewhere.[4] Briefly, it consists of a McPherson model 225 monochromator with a standard Hinteregger lamp. Ions are mass selected with an Extranuclear 324-9 quadrupole mass filter and detected with a Channetron multiplier using conventional ion counting techniques. A cooled photomultiplier with a coating of sodium salicylate monitors the intensity of the light passing through the ion source.

For these experiments, the analog signal from the photomultiplier was fed into a voltage-to-frequency converter whose output was used as the time base for a digital counter. The ion pulses were fed into this counter so that the reading on the counter was proportional to the ratio of ion count rate to light intensity. Since the light intensity provided the time base, the net effect of this arrangement was that at wavelengths where the H_2 lamp spectrum was weak, ions were counted for longer times and conversely for wavelengths where the light intensity was strong. The counter design was such that the numerical reading was equal to the total number of ions counted during the measurement. A simple stepping-switch programmer was built to control the mass setting of the quadrupole mass filter so that up to ten different masses could be monitored during a run. As each measurement of (ion count)/(integrated light intensity) was completed, it was automatically printed on a Hewlett Packard printer, and the programs advanced to the next preselected mass. At the end of each cycle of the programmer, the monochromator wave-

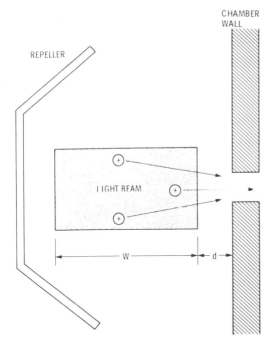

CHAMBER WALL

REPELLER

LIGHT BEAM

W

d

FIG. 1. Schematic drawing of the ion source showing the shaped repeller and the orientation of the photoionizing light beam. The distances w, the width of the beam, and d, the distance from the beam to the ion exit hole, are defined as shown in the drawing.

length was manually reset, and the entire process repeated. This arrangement made it possible to collect and record data of uniform precision with the maximum possible efficiency obtainable without complete computer control of the instrument.

The experimental conditions for the present study were as follows: monochromator resolution measured as 0.7 Å full-width at half-maximum; acetylene pressure constant at 2.5×10^{-4} torr as measured by an MKS Baratron capacitance monometer; and ion source repeller voltage essentially zero. The H_2 pressure was varied from zero up to 2.0×10^{-3} torr. At each wavelength, the $C_2H_2^+$ and $C_2H_3^+$ signals were recorded at several different H_2 pressures.

B. Kinetic analysis

The differential equations describing the variation of the primary ion current I_p and secondary ion current I_{si} as a function of position within the source are

$$dI_p/dx = I_p \Sigma_i \, \sigma_i[N_i], \tag{A}$$

$$dI_{si}/dx = I_p \sigma_i[N_i], \tag{B}$$

where σ_i is the cross section for the formation of the ith secondary ion and $[N_i]$ is the number density of the neutral involved. The appropriate solutions are

$$I_p = I_p^0 \exp(-\Sigma x), \tag{C}$$

$$I_{si} = (I_p^0 \sigma_i[N_i]/\Sigma)[1 - \Sigma \exp(-\Sigma x)], \tag{D}$$

where $\Sigma = \Sigma_i \sigma_i[N_i]$ and I_p^0 is the initial primary ion current.

However, the ions are not all formed at the same distance from the source exit hole since the plane of the light beam is parallel to the nominal path followed by the ions. A schematic drawing of the source is shown in Fig. 1, which defines the parameters, w, the width of the light beam, and d, the distance from the light beam to the source exit. In order to obtain expressions for the experimentally observed ion current, it is necessary to average Eqs. (C) and (D) over the width of the light beam.

$$\bar{I}_p = \frac{1}{w} \int_d^{d+w} I_p(x) \, dx = \frac{I_p^0}{\Sigma w}(1 - e^{-\Sigma w}), \tag{E}$$

$$\bar{I}_{si} = \frac{I_p^0 \sigma_i[N_i]}{\Sigma^2 w}(\Sigma w - e^{-\Sigma d}[1 - e^{-\Sigma w}]). \tag{F}$$

Because of the complicated form of Eqs. (E) and (F), it is most convenient to obtain relative cross sections for Reaction (1) from the ion current ratio $C_2H_3^+/C_2H_2^+$ as a function of H_2 pressure.

The ratio of Eq. (F) to Eq. (E) yields after some rearrangement

$$\frac{\bar{I}_{si}}{\bar{I}_p} = \frac{\sigma_i[N_i]}{\Sigma}\left[\frac{\Sigma w}{e^{-\Sigma d} - e^{-\Sigma(d+w)}} - 1\right]. \tag{G}$$

Expanding the exponentials to second order in Σ leads to the approximation

$$\frac{\bar{I}_{si}}{\bar{I}_p} = \sigma_i[N_i]\frac{d + \frac{1}{2}w}{1 - \Sigma(d + \frac{1}{2}w)}$$

and it follows that a plot of $\bar{I}_{si}/\bar{I}_p$ vs N_i will have a slope proportional to σ_i provided that the sum of all the terms in Σ involving that particular N_i are negligible compared to $1/(d + \frac{1}{2}w)$. This condition is met in the present case since the ratio $C_2H_3^+/C_2H_2^+$ never exceeded 0.08 at the highest pressure of H_2 used.

C. Results and discussion

Typical plots of $C_2H_3^+/C_2H_2^+$ vs H_2 pressure at three different wavelengths are shown in Fig. 2. The slope of each plot is proportional to the cross section for Reaction (1). Since the exact values of d and w are not well defined and since neither the light intensity nor the pressure in the ion source are completely uniform, no attempt was made to calculate absolute cross sections. The relative cross section for Reaction (1) obtained from plots such as those in Fig. 2 is shown in Fig. 3 as a function of the wavelength of the ionizing photons. A clear step is present at 1067 Å and a less distinct break appears at 1046 Å. These features coincide with the onsets found by Botter, Dibeler, Walker, and Rosenstock[5] for the formation of $C_2H_2^+$ with one and two quanta of excitation in ν_2, the carbon–carbon stretching mode. From recent photoionization data, Dibeler and Walker[6] have determined relative Franck–Condon factors of $1.00:0.33:0.11$ for the formation of acetylene ions with

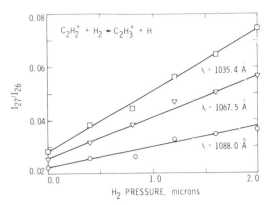

FIG. 2. Plots of $C_2H_3^+/C_2H_2^+$ vs hydrogen gas pressure in the source. The slope of each plot is proportional to the total cross section at that wavelength.

$\nu_2 = 0, 1$, and 2, respectively. These values are in good agreement with those obtained from the photoelectron spectrum of C_2H_2.[7]

For the purpose of the following discussion, we divide the wavelength range covered by Fig. 3 into three regions: (a) 1090–1067 Å, (b) 1067–1046 Å, and (c) 1046–1030 Å. In region (a), all of the acetylene ions are in their electronic and vibrational ground state; in region (b), some acetylene ions are in the $\nu_2 = 1$ state while the rest are in the ground state; and in region (c), ions are formed in the ground state, in the $\nu_2 = 1$ state and in the $\nu_2 = 2$ state. If we assume that the Franck–Condon factors give the relative probability of forming acetylene ions in the $\nu_2 = 0$, 1, and 2 states (that is, we assume that autoionization processes do not greatly change the distribution of vibrational states produced by direct ion-

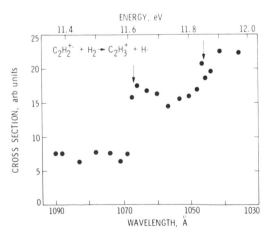

FIG. 3. Experimental relative cross section for Reaction (1) as a function of photoionizing wavelength. Each point is obtained from the slope of a plot such as those in Fig. 2. A clear step is apparent at 1067 Å and a smaller and less distinct step is present at 1046 Å.

TABLE I. Relative reaction cross section for the reaction $C_2H_2^+ + H_2 \rightarrow C_2H_3^+ + H$ as a function of the number of quanta of excitation in the acetylene ion carbon–carbon stretching mode ν_2.

ν_2	0	1	2
Relative section	7.5 ± 1.0	45 ± 5	90 ± 15

ization), then it is possible to express the average observed relative cross section in each region in terms of the relative cross sections for Reaction (1) for acetylene ions with $\nu_2 = 0$, 1, and 2.

$$\sigma_a = 1.00\ \sigma(\nu_2 = 0)\,, \tag{H}$$

$$\sigma_b = 1.00\ \sigma(\nu_2 = 0) + 0.33\ \sigma(\nu_2 = 1)\,, \tag{I}$$

$$\sigma_c = 1.00\ \sigma(\nu_2 = 0) + 0.33\ \sigma(\nu_2 = 1) + 0.11\ \sigma(\nu_2 = 2)\,. \tag{J}$$

Solving Eqs. (H)–(J) for the relative cross sections gives the numerical values shown in Table I.

The rapid increase in the cross section for Reaction (1) with increasing vibrational excitation is as expected for an endothermic reaction. Using $\Delta H_f (C_2H_3^+) = 266$ kcal/mole, Reaction (1) is endothermic by 1.1 kcal/mole, and if all of the reactants are in the ground state, the reaction cross section should be zero. Table I, however, shows that the relative cross section for $\nu_2 = 0$ is about one-sixth of that for $\nu_2 = 1$. At 300 K approxi-

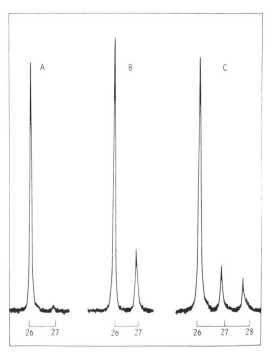

FIG. 4. Conventional ICR spectra at 15 eV of the ions in (a) 2×10^{-6} torr C_2H_2, (b) 2×10^{-6} torr C_2H_2 plus 3.5×10^{-5} torr H_2, (c) 2×10^{-6} torr C_2H_2 plus 1.7×10^{-5} torr D_2.

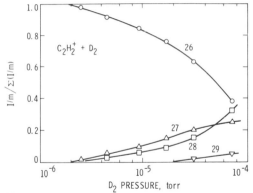

FIG. 5. Pressure dependence of the mass-normalized ion intensities in C_2H_2–D_2 mixtures at constant C_2H_2 pressure. Electron energy 15 eV.

mately 14% of the neutral acetylene molecules have one quantum of excitation in either of the two doubly degenerate bending modes ν_4 or ν_5 corresponding to 1.7 or 2.1 kcal/mole of excitation, respectively. Dibeler, Walker, and McCulloh[1] have observed these hot bands in the photoionization mass spectrum of acetylene below the IP of the parent ion. If the ionization of acetylene near threshold is adiabatic (i.e., if all of the vibrational quantum numbers remain unchanged), then almost all of the observed reactivity of $C_2H_2^+$ ions with $\nu_2 = 0$ can be accounted for by the presence of ions with either $\nu_4 = 1$ or $\nu_5 = 1$.

III. ION CYCLOTRON RESONANCE STUDIES

A. Intensities vs pressure

Figure 4 shows the results of conventional continuous-drift ICR experiments in C_2H_2, C_2H_2–H_2 and in C_2H_2–D_2 mixtures at 15 eV electron impact energy (below the ionization potential of hydrogen), and at constant C_2H_2 pressure. At zero hydrogen (or D_2) pressure, only the acetylene parent ion at $M/e = 26$ is observed. The C_2H_2 pressure is sufficiently low that no secondary ions due to reactions of $C_2H_2^+$ with C_2H_2 are observed. Addition of H_2 results in the appearance of the $M/e = 27$ ion, which identifies the occurrence of Reaction (1). Addition of D_2 results in the appearance of two ions at $M/e = 27$ and $M/e = 28$. Double resonance ejection experiments at low D_2 pressures confirm that these new ions are due entirely to reaction of $C_2H_2^+$ ions with D_2.

$$C_2H_2^+ + D_2 \rightarrow C_2HD^+ + HD \qquad (2)$$
$$ C_2H_2D^+ + D \ . \qquad (3)$$

The results of these experiments yield $k_2/k_3 = 1.6 \pm 0.2$. At higher D_2 pressures, further exchange and reaction occurs which results in a decrease in the $M/e = 27$ to $M/e = 28$ ratio and in the appearance of an ion at $M/e = 29$ (Fig. 5). A portion of the intensity at $M/e = 28$ at

low D_2 pressures may also be due to the exchange reaction if exchange occurs with equilibration of the hydrogen atoms in a long-lived $[C_2H_2D_2^+]^*$ intermediate complex. In this case, the ratio of $M/e = 26$, 27, 28 ions expected from the nonreactive exchange process is $1:4:1$. By subtracting one-fourth of the intensity at $M/e = 27$ from the intensity at $M/e = 28$, an upper limit of $k_2/k_3 = 2.7$ is obtained. The abstraction process, Reaction (3), most likely occurs via a direct process, and not via an intermediate $[C_2H_2D_2^+]^*$ intermediate complex, since the latter process would result in the production of nearly equal intensities of $M/e = 28$ and $M/e = 29$ ions. No ions are observed at $M/e = 29$ at low D_2 pressures.

B. Intensities vs time

The JPL ICR spectrometer was used in the McMahon-Beauchamp ion trapping mode[8,9] to obtain mass spectra of ions as a function of reaction time. Figure 6 shows a semilogarithmic plot of the normalized ion intensities vs storage time observed in C_2H_2 and in a C_2H_2–H_2 mixture. The electron energy used was 15 eV, below the IP of H_2 (15.4 eV). Similar results were also obtained at an electron impact energy of 40 eV, where the H_2^+ ion was ejected from the cell by an rf pulse at the cyclotron frequency of H_2^+ applied concurrently with the ion formation pulse. No significant differences were noted between the data at 15 and at 40 eV.

The curve labelled $(C_2H_2^+)_0$ in Fig. 6 is the decay curve for $C_2H_2^+$ ions in C_2H_2 alone. No $C_2H_3^+$ ions were observed without H_2 present in the mixture. On addition of H_2 the $C_2H_3^+$ ion appears, and the decay rate for the $C_2H_2^+$ ions increases at short times. At long times, however, the decay rate for $C_2H_2^+$ ions is slowed and the decay eventually becomes equal to the decay rate in

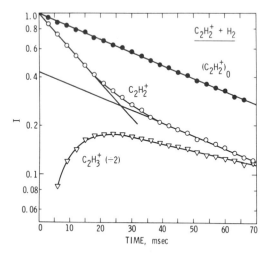

FIG. 6. Temporal behavior of the ions in C_2H_2 and in a C_2H_2–H_2 mixture at 15 eV. C_2H_2 pressure 3.66×10^{-7} torr, added H_2 pressure 2.16×10^{-5} torr. Solid dots correspond to ion intensities without added hydrogen, open circles are the observed ion intensities with added hydrogen. For the sake of clarity, the $C_4H_2^+$ and $C_4H_3^+$ ions are not shown.

C_2H_2 alone. This change in slope in the C_2H_2–H_2 mixture most likely indicates the point at which the excited $C_2H_2^*$ ions have disappeared by reaction with H_2. The subsequent decay rate at long times in the mixture is equal to the decay rate without added hydrogen, corresponding to the reaction of ground state $C_2H_2^+$ ions with C_2H_2 neutrals.

The above analysis requires that the rate constant for the reaction of excited $C_2H_2^*$ ions with C_2H_2 is equal to the rate constant for the reaction of ground state $C_2H_2^+$ ions with C_2H_2. The available evidence indicates that this is indeed the case. The reaction of $C_2H_2^*$ ions with C_2H_2

$$C_2H_2^+ + C_2H_2 \rightarrow C_4H_3^+ + H \tag{4}$$
$$\rightarrow C_4H_2^+ + H_2 \tag{5}$$

has been examined[10] using the photoionization mass spectrometer over the region from 1088 to 1035 Å; from the threshold for formation of the $C_2H_2^*$ ion to wavelengths where $C_2H_2^*$ ions in the $\nu_2 = 2$ level of the ground electronic state are formed. Although a small change in the product distribution for the reaction was noted with decreasing wavelength, the total cross section for the reaction remained constant to within ±5% over this range.[10] The total rate constant for the reaction, measured in this work by the ICR ion storage technique, remains constant for electron impact energies from 12 to 40 eV, and does not change over the C_2H_2 pressure range from 2×10^{-7} to 2×10^{-6} torr. No curvature was noted in any of the decay curves used to measure the total rate constant out to times approaching 100 msec. The total rate constant, $k_4 + k_5$, measured in these experiments is given in Table II.

At long times, the $C_2H_3^+$ ion in Fig. 6 also shows a linear decay with time. This decay rate is independent of hydrogen pressure, but changes with C_2H_2 pressure. Figure 7 shows the mass spectra of the ions present at 10 and at 90 msec in a C_2H_2–H_2 mixture where most of the ions present originate from ionization of H_2. The electron energy used was 30 eV. The major ion present at 10 msec is the $C_2H_3^+$ ion formed by the reaction of H_3^+ with C_2H_2. The mass spectrum of the ions present at

TABLE II. Rate constants.

Reaction	Rate constant, 10^{-9} cm^3/sec	Theoretical collision rate constants[a]
$(C_2H_2^+)^* + H_2 \rightarrow C_2H_3^+ + H$	0.063 ± 0.018[b]	1.53–1.83
$C_2H_2^+ + H_2 \rightarrow C_2H_3^+ + H$	< 0.002	
$\begin{array}{l} + C_2H_2 \rightarrow C_4H_3^+ + H \\ \rightarrow C_4H_2^+ + H_2 \end{array}$	1.52 ± 0.10	1.19–1.62
$C_2H_3^+ + H_2 \rightarrow C_2H_4^+ + H$	< 0.001	1.53–1.83
$C_2H_3^+ + C_2H_2 \rightarrow C_4H_3^+ + H_2$	0.71 ± 0.08	1.18–1.61

[a]The minimum value is for the angle-averaged polarizability of the neutral, and excludes grazing collisions. The maximum value includes grazing collisions and uses the maximum component of the polarizability tensor.
[b]For the distribution of vibrational energies produced by electron impact at 15–40 eV.

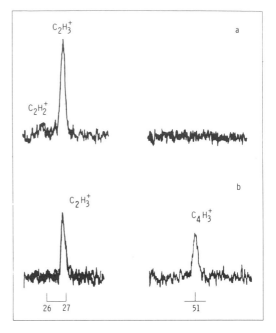

FIG. 7. ICR mass spectra at 30 eV of the ions present in a mixture of 3.0×10^{-7} torr C_2H_2 and 8.61×10^{-5} torr H_2 at (a) 10 msec, (b) 90 msec. Continuous ejection of the $C_2H_3^+$ ion results in nearly complete disappearance of the $C_4H_3^+$ ion.

90 msec show extensive conversion of the $M/e = 27$ ions into $C_4H_3^+$ ions at $M/e = 51$. Continuous cyclotron ejection of the $C_2H_3^+$ ions in this mixture results in almost complete disappearance of the ions at $M/e = 51$ at 90 msec. This identifies the reaction

$$C_2H_3^+ + C_2H_2 \rightarrow C_4H_3^+ + H_2 . \tag{6}$$

The decay of $C_2H_3^+$ ions was measured over a wide range of C_2H_2 and H_2 pressures, consistently yielding the result $k_6 = 7.0 \pm 0.9 \times 10^{-10}$ cm^3/sec.

C. Kinetic analysis

In order to describe the temporal behavior of the ions in Fig. 6, the simplest possible model consistent with the general features of the observations is used:

$$(C_2H_2^+)^* + H_2 \xrightarrow{k_1} C_2H_3^+ + H$$
$$+ C_2H_2 \xrightarrow{k_0^*} \text{products}$$
$$C_2H_2^+ + C_2H_2 \xrightarrow{k_0} \text{products} .$$

The asterisk designates the initial distribution of excited ions, and the rate constant k_1 is an average over the distribution of the rate constants for the reaction with H_2. Acetylene ions in the ground vibronic state do not react with H_2, and $k_0^* = k_0$. With these assumptions the observed $C_2H_2^+$ ion intensity is

$$I(C_2H_2^+) = e^{-k_0 [C_2H_2] t} \left[P_0 + P_0^* e^{-k_1 [H_2] t} \right], \tag{K}$$

where P_0 is the fraction of ions in the ground vibronic state at $t = 0$ and P_0^* is the fraction of excited ions at $t = 0$ $(P_0 + P_0^* = 1)$. Equation (K) reproduces the behavior of $C_2H_2^+$ ions observed in Fig. 6.

In order to determine the parameters P_0^*/P_0 and k_1 from experiment for this model, two methods were used. In the first, the difference in the slopes of the semilogarithmic plots, with and without added H_2, were taken at short times in order to calculate k_1. The slope of the $C_2H_2^+$ curve in Fig. 6 is given by

$$\frac{d}{dt} \ln I = -k_0 [C_2H_2] - \frac{P_0^* k_1 [H_2] e^{-k_1 [H_2] t}}{P_0 + P_0^* e^{-k_1 [H_2] t}} . \tag{L}$$

For $k_1 [H_2] t \ll 1$, Eq. (L) becomes

$$\frac{d}{dt} \ln I = -k_0 [C_2H_2] - P_0^* k_1 [H_2] .$$

The difference in slopes at short times is therefore simply

$$\Delta S \simeq - P_0^* k_1 [H_2] . \tag{M}$$

At long times, Eq. (K) yields

$$I_\infty(t) = P_0 e^{-k_0 [C_2H_2] t}$$

$$= P_0 I_0(t) , \tag{N}$$

where $I_0(t)$ is the observed $C_2H_2^+$ intensity without added hydrogen. From the extrapolated value of $I_\infty(t)$ at $t = 0$:

$$I(0)/I_\infty(0) = 1 + (P_0^*/P_0) . \tag{O}$$

From Eqs. (M) and (O), consistent values of $k_1 = 6.3 \pm 1.8 \times 10^{-11}$ cm^3/sec and $P_0^*/P_0 = 1.3 \pm 0.3$ are obtained for H_2 pressures in the range from 4×10^{-6} to 2×10^{-5} torr.

In the second method, the function

$$\ln \left[\frac{I(t)}{I_\infty(t)} - 1 \right] = \ln \left(\frac{P_0^*}{P_0} \right) - k_1 [H_2] t \tag{P}$$

is plotted vs time. Equation (P) is obtained from Eqs. (K) and (N). The rate constant k_1 is obtained from the slope and P_0^*/P_0 from the logarithm of the intercept at $t = 0$. This method has the advantage that the short time assumption is not required. Figure 8 shows a plot of this function for the data in Fig. 6. This method also yields values $k_1 = 6.3 \pm 1.8 \times 10^{-11}$ cm^3/sec and $P_0^*/P_0 = 1.3 \pm 0.3$, which at least demonstrates the consistency of the model.

The experimental quantity $P_0^*/P_0 = 1.3$ is much larger than would be predicted from the experimental Franck-Condon factors[6,7] for formation of vibrationally excited $C_2H_2^+$ ions. The electron energy used, 15 eV, is well below the AP of the first electronically excited state of the ion (16.3 eV[7]). The photoionization mass spectrum,[5] however, shows very broad and intense autoionization structure near 13 and 15 eV. It is quite likely, therefore, that autoionization contributes extensively to the electron impact data and that the fraction of excited ions is consequently much larger than would be predicted from the Franck-Condon distribution. Unfortunately, since this distribution of excited ionic states is unknown, it is therefore not possible to determine rate constants as a function of vibrational energy from the electron im-

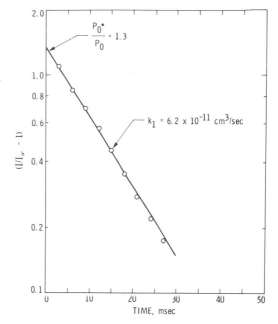

FIG. 8. Semilogarithmic plot of the function $[I(t)/I_\infty(t) - 1]$ for $C_2H_2^+$ ions vs time from the data for the C_2H_2–H_2 mixture in Fig. 6 (see text).

pact data for comparison with the relative cross sections determined from the photoionization measurements. Until this distribution can be well established, more detailed kinetic modeling than performed here is unwarranted.

IV. CONCLUSIONS

Vibrationally excited $C_2H_2^+$ ions react with H_2 to produce $C_2H_3^+$ ions. Photoionization experiments show that $C_2H_2^+$ ions in the $\nu_2 = 1$ and $\nu_2 = 2$ state definitely react with H_2. The small cross section for the reaction measured for $\nu_2 = 0$ ions is largely due to the presence of thermally excited ions and neutrals. Although experiments at lower temperatures are necessary to determine conclusively from photoionization experiments if the reaction occurs for totally unexcited acetylene ions, the ICR ion storage data at long reaction times indicates that the rate constant for totally unexcited $C_2H_2^+$ ions in the ground vibronic state is less than about 3% of the rate constant for the initial distribution of ions formed by electron impact at 15 eV. At long times in the ICR experiment, the only remaining unreacted ions must be those with vibrational and translational energies less than required for occurrence of the reaction.

The amount of energy contained in one quantum of vibrational energy in the ν_2 stretching mode of the acetylene ion is 5.2 kcal/mole; in the ν_4 and ν_5 bending modes of the neutral, 1.7 and 2.1 kcal/mole, respectively. The data indicates that ions in the ground vibronic state do not react, but that ions formed from neutrals with one quantum of excitation in the ν_4 or ν_5

mode do react. These results suggest that the endo-thermicity of the reaction is less than 2 kcal/mole. The value $\Delta H_f(C_2H_2^+) = 317.2 \pm 0.2$ kcal/mole is well established,[1] which yields an upper limit of $\Delta H_f(C_2H_3^+) \leq 267$ kcal/mole from the photoionization studies of Reaction (1) in this work. This number is in good agreement with the value $\Delta H_f(C_2H_3^+) = 266$ kcal/mole reported by Chupka, Berkowitz, and Refaey. The value $\Delta H_f(C_2H_3^+) = 269$ kcal/mole obtained by Brehm appears to be too large. This latter value yields a reaction endothermicity of 4 kcal/mole, which could not be overcome by one quantum of bending mode vibrational energy.

*This paper represents results of one phase of research carried out at the Jet Propulsion Laboratory, California Institute of Technology, under contract No. NAS7-100, sponsored by the National Aeronautics and Space Administration.

†NASA Research Affiliate, Jet Propulsion Laboratory, April–August 1973.

‡NASA Resident Research Associate.

[1] V. H. Dibeler, J. A. Walker, and K. E. McCulloh, J. Chem. Phys. **59**, 2264 (1973).

[2] W. A. Chupka, J. Berkowitz, and K. M. A. Refaey, J. Chem. Phys. **50**, 1938 (1969).

[3] B. Brehm, Z. Naturforsch. A **21**, 196 (1966).

[4] P. LeBreton, A. Williamson, J. L. Beauchamp, W. T. Huntress, Jr., and A. Lane (unpublished).

[5] R. Botter, V. H. Dibeler, J. A. Walker, and H. M. Rosenstock, J. Chem. Phys. **44**, 1271 (1966).

[6] V. H. Dibeler and J. A. Walker, Int. J. Mass Spectrom. Ion Phys. **11**, 49 (1973).

[7] D. W. Turner, C. Baker, A. D. Baker, and C. R. Brundle, *Molecular Photoelectron Spectroscopy* (Wiley-Interscience, New York, 1970), pp. 190–191.

[8] T. B. McMahon and J. L. Beauchamp, Rev. Sci. Instrum. **43**, 509 (1972).

[9] W. T. Huntress, Jr., and R. F. Pinizzotto, Jr., J. Chem. Phys. **59**, 4742 (1973).

[10] S. E. Buttrill, Jr. (unpublished results).

32

Reprinted from *J. Chem Phys.* **48**:2966–2973 (1968)

Afterglow Studies of the Reactions He+, He(2³S), and O+ with Vibrationally Excited N₂

A. L. Schmeltekopf, E. E. Ferguson, and F. C. Fehsenfeld

Aeronomy Laboratory, Environmental Science Service Administration, Boulder, Colorado

INTRODUCTION

In the past several years the following reactions have attracted considerable interest, both as regards their reaction mechanisms and because of certain geophysical applications:

$$He^+ + N_2 \rightarrow N^+ + N + He \qquad (1a)$$

$$\rightarrow N_2^+ + He, \qquad (1b)$$

$$He(2^3S) + N_2 \rightarrow N_2^+ + He + e, \qquad (2)$$

$$O^+ + N_2 \rightarrow NO^+ + N. \qquad (3)$$

Reaction (1) has been studied by several groups[1-6] with good agreement on the over-all rate constant. Only Warneck[4] and Heimerl et al.[6] have determined the branching ratio, (1a) to (1b). Reaction (2), a Penning ionization, has the interesting property that it excites a vertical transition in the N_2, i.e., the Franck–Condon principle is obeyed. Reaction (3) has also been measured by several groups[7-9] but with less satisfactory agreement. Wide discrepancies in earlier published measurements of Reaction (3) were apparently due to experimental difficulties associated with the large vibrational effect which is reported here. In the present investigation the above reactions have been studied as a function of vibrational temperature of the N_2, the first such laboratory measurements carried out. The study and understanding of Reactions (1) and (2)

were necessary in order to obtain the measurements for Reaction (3) by the present experimental techniques.

EXPERIMENTAL

Figure 1 shows a sketch of the apparatus used for measuring the He+, He(2³S), and O+ reactions. This apparatus has been described in detail elsewhere[3] and the only changes made for this experiment are in the N_2 inlet and in the addition of the electron gun shown in the sketch. The details of the detection apparatus, the discharge, and the gas-handling system are given in previous papers.[3,10] Basically the system is as follows: a fast pump (500 liter/sec) is used to give the gas a bulk-flow velocity of 8×10^3 cm/sec at a pressure of 0.38 torr in a stainless-steel tube 8 cm in diameter and 1 m long. The gas (He) is ionized and excited by a dc discharge using an aluminum cylindrical cathode and small wire anode. The excited states decay in the flowing weak plasma for ~2 msec before reaching the point where O_2 may be added to the He afterglow for the O+ studies. The helium afterglow contains the species He+, He₂+, and He(2³S). The O_2, when added, reacts[2] with the He+ to form O+, and the He(2³S) reacts[11] to form both O+ and O₂+. These ions then flow for ~2 msec in a plasma of cool electrons. Any O+ which is formed in an excited state will be de-excited by super-elastic collisions with electrons to the ground state.[7] At the N_2 inlet point there are either He+ and He₂+ ions and He(2³S) metastables or, when O_2 is added, O+ and O₂+ ions, flowing in a He background gas such that the neutral He concentration exceeds the ion concentrations by over 5 orders of magnitude.

The ion signals are detected at the end of the tube by a quadrupole mass spectrometer which samples through a 0.5-mm hole in a 0.25-mm-thick piece of

[1] J. Sayers and D. Smith, Discussions Faraday Soc. **37**, 167 (1964).

[2] E. E. Ferguson, F. C. Fehsenfeld, A. L. Schmeltekopf, and H. I. Schiff, Planetary Space Sci. **12**, 1169 (1964).

[3] F. C. Fehsenfeld, A. L. Schmeltekopf, P. D. Goldan, H. I. Schiff, and E. E. Ferguson, J. Chem. Phys. **44**, 4087 (1966).

[4] P. Warneck, Planetary Space Sci. (to be published).

[5] T. F. Moran and L. Friedman, J. Chem. Phys. **45**, 3837 (1966).

[6] J. Heimerl, R. Johnsen, and M. A. Biondi, 20th Annual Gaseous Electronics Conference, San Francisco, Calif., 18 October 1967.

[7] F. C. Fehsenfeld, A. L. Schmeltekopf, and E. E. Ferguson, Planetary Space Sci. **13**, 219 (1965).

[8] M. J. Copsey, D. Smith, and J. Sayers, Planetary Space Sci. **14**, 1047 (1966).

[9] P. Warneck, Planetary Space Sci. **15**, 1349 (1967).

[10] P. D. Goldan, A. L. Schmeltekopf, F. C. Fehsenfeld, H. I. Schiff, and E. E. Ferguson, J. Chem. Phys. **44**, 4095 (1966).

[11] E. E. Muschlitz and M. J. Weiss, *Atomic Collision Processes*, M. R. C. McDowell, Ed. (North-Holland Publ. Co., Amsterdam, 1964), p. 1073.

301

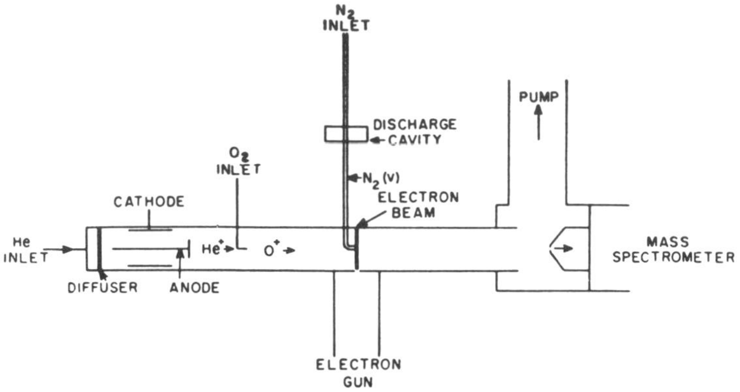

FIG. 1. Schematic sketch of the experimental apparatus.

molybdenum. For rate-constant measurements the signal of interest (O⁺ for example) is observed as a function of reactant N_2 gas flow. The rate is calculated from the decrease of the primary-ion signal, as described previously.[3]

The new and somewhat difficult measurement in this particular experiment, which has not previously been described, is the measurement of the vibrational temperature (T_v) of the N_2. The N_2 was vibrationally excited by passing the N_2 through a microwave discharge. The mechanism of vibrational excitation appears to be that the energetic electrons in the discharge produce a transitory N_2^- which quickly autoionizes to leave the N_2 vibrationally excited.[12] The N_2 vibrations quickly come into a Boltzmann distribution by mutual collisions, as expected from theoretical cross sections,[13] thus establishing a proper vibrational temperature. The N_2 vibrational temperature is lowered by wall collisions, but these are much less efficient[14] than the mutual-exchange collisions, so that as the N_2 vibrationally cools while passing down the tube it retains a Boltzmann distribution of vibrational states.

The microwave discharge takes place in a quartz side tube as shown in Fig. 1. This side tube is ≈ 50 cm long and 1 cm diameter. In order to vary the vibrational temperature at the exit of the tube two methods have been used: either the cavity may be moved further from the tube exit, thus allowing vibrational relaxation on the walls of the tube, or the discharge power can be reduced. Both of these methods were used in the experiment, but the first was preferred because the discharge stability was better at high power.

DETAILS OF VIBRATIONAL-TEMPERATURE MEASUREMENT

One basic method was used to measure T_v, namely excitation (and simultaneous ionization) of the $N_2(v)$

in a vertical electronic transition, governed by known Franck–Condon factors, to a radiation state and observation of the light emitted. With the use of known transition probabilities, this light intensity is then related to the vibrational distribution. The first and primary method of excitation was the so-called Penning ionization reaction,

$$\mathrm{He}(2^3S) + N_2 X\,^1\Sigma_g(v) \rightarrow N_2^+ B\,^2\Sigma_u^+(v') + \mathrm{He} + e.$$

$$(2)$$

This reaction might, *a priori*, not be expected to ionize the N_2 vertically; however, observation of the light emitted from the $B\,^2\Sigma_u^+$ state of N_2^+ ($B\,^2\Sigma_u^+ \rightarrow X\,^2\Sigma_g$ 1st-negative system) showed that the vibrational levels in the $B\,^2\Sigma_u^+$ state were populated according to the Franck–Condon factors connecting the $N_2 X\,^1\Sigma_g$ state to the $N_2^+ B\,^2\Sigma_u^+$ state. When the N_2 was not vibrationally excited this observation was unambiguous. Robertson[15] has previously made the same observation. When the N_2 was vibrationally excited the analysis always led to a Boltzmann vibrational distribution for the $N_2 X\,^1\Sigma_g$ state, as expected theoretically. In order to test further the rather surprising result that the Penning reaction excited a vertical electronic transition from ground-state N_2 to the $N_2^+ B$ state, the electron gun shown in Fig. 1 was added. The electron beam passes through the vibrationally excited N_2 just as it enters the reaction vessel, and the light emitted due to the electron ionization and excitation is observed spectroscopically from the side. The electron gun yielded a current of 10 mA of 5000-V electrons which ionized the N_2, partially into the $N_2^+ B\,^2\Sigma_u^+$ state. Electrons of 5000-V energy would be expected to ionize and excite the N_2 in a vertical transition so that by directly comparing the N_2^+ spectrum excited by the electron beam with that produced by Reaction (2), a check of our interpretation for Reaction (2) can be made. The spectra were identical, thus establishing that the Penning ionization (2) does indeed obey the Franck–Condon

[12] G. J. Schulz, Phys. Rev. **125**, 229 (1962).
[13] D. Rapp, J. Chem. Phys. **43**, 316 (1965).
[14] J. E. Morgan, L. F. Phillips, and H. I. Schiff, Discussions Faraday Soc. **33**, 118 (1962).

[15] W. W. Robertson, J. Chem. Phys. **44**, 2456 (1966).

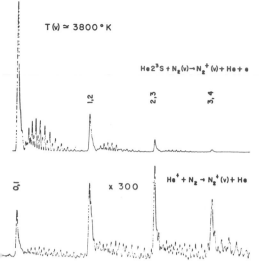

$T(v) \simeq 3800°K$

$He \cdot 2^3S + N_2(v) \rightarrow N_2^+(v) + He + e$

1,2   2,3   3,4

$\times 300$

$He^+ + N_2 \rightarrow N_2^+(v) + He$

FIG. 2. A portion of the $N_2^+B\ ^2\Sigma_u^+ \rightarrow X\ ^2\Sigma_g^+$ 1st negative-system spectrum produced by reacting $He(2^3S)$ (upper) and He^+ (lower) with N_2.

rinciple. A typical spectrum produced by $He(2^3S)$ is nown in the upper part of Fig. 2.

In order to interpret the N_2^+ 1st-negative-system ntensities in terms of a vibrational-population distribution for the N_2 ground state, accurate Franck–Condon factors are required for $N_2X\ ^1\Sigma_g \rightarrow N_2^+B\ ^2\Sigma_u^+$, nd relative radiative-transition probabilities for the st-negative system $N_2^+B\ ^2\Sigma_u^+ \rightarrow N_2^+X\ ^2\Sigma_g^+$ must be nown. Calculations of these were made using the ydberg–Klein–Rees calculational method.[16,17]

The Einstein transition probability for the 1st-egative system is given by[18]

$$A_{B_v,I_{v'}} = (64\Pi^4/3hd_B)\nu_{BI}^3 q_{B_v,I_{v'}} R_e^2(B, I), \quad (4)$$

here B refers to N_2^+B and I to N_2^+X. These are not all nown, but if $R_e^2(B, I)$ is independent of r, the inter-uclear separation, (as the computations indicate it is) nen by using the lifetime measurements of Bennett nd Dalby,[19] one finds that

$$A_{B_v,I_{v'}} = 8.13 \times 10^{-5} \nu_{BI}^3 q_{B_v,I_{v'}}, \quad (5)$$

r the populations of the various vibrational levels v f the B state are given by

$$N_{B_v} = 7.78 \times 10^{18} \frac{I_{B_v,I_{v'}}(\text{erg/cm}^2 \cdot \text{sec} \cdot \text{sr})}{(\nu_{BI}^4 q_{B_v,I_{v'}})}, \quad (6)$$

here $I_{B_v,I_{v'}}$ is the intensity of the v, v' band of the 1st-

[16] R. N. Zare, Univ. of Calif. Radiation Laboratory Rept. CRL 10925, November 1963.
[17] R. N. Zare, A. L. Schmeltekopf, D. L. Albritton and W. J. arrop (unpublished).
[18] G. Herzberg, *Spectra of Diatomic Molecules* (D. Van Nostrand o., Inc., New York, 1950).
[19] R. G. Bennett and F. W. Dalby, J. Chem. Phys. **31**, 434 1959).

negative system, ν_{BI} is the wavenumber, and $q_{B_v,I_{v'}}$ is the Franck–Condon factor for the transition. The basic assumption of the method is that the ground state $N_2X_{v''}$ is excited in such a way that the transition is a vertical one, so that the distribution of $N_2^+B_v$ vibrational states produced by the excitation is predictable from the Franck–Condon factors connecting X to B. Thus

$$N_{B_v} \propto \sum_{X=0}^{\infty} q_{B_v,X_{v''}} N_{X_{v''}}. \quad (7)$$

We know N_{B_v} experimentally and need to solve for $N_{X_{v''}}$.

To calculate the vibrational temperature, defined in the usual way by the Boltzmann equation,

$$N_{X_{v''}} = N_{X_0} \exp(-E_{X_{v''}}/kT_v), \quad (8)$$

the N_{B_v} obtained from spectroscopic data, using Eq. (6), is compared to the predicted N_{B_v} by use of Eqs. (7) and (8) for various temperatures. This is done on a computer so that the temperature at which the square of the difference between the predicted N_{B_v} and the measured N_{B_v} is minimized is obtained. A number of the bands of the 1st-negative system were measured (the bands $\Delta v = +3, +2, +1, 0, -1$). The agreement between temperatures measured from the various bands was good, and the data indicated reasonably good Boltzmann distributions. Plots of this type are identical for $He(2^3S)$ excitation and for high-velocity electrons. We thus feel confident that the distribution is Boltzmann at all temperatures and that the temperature determinations are reliable.

From the above, an obvious way to get the temperature more easily is available. Calculating, by use of Eqs. (7) and (8), the N_{B_v} to be expected at a number of temperatures, a plot as shown in Fig. 3 is obtained. (Note that the scale for $v=0$ is compressed by 2 compared to the others.) On this plot the expected popula-

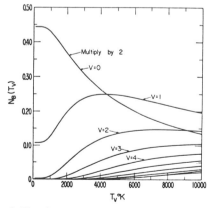

FIG. 3. Plot of the predicted population N_B of vibrational levels of the $B\ ^2\Sigma_u^+$ state of N_2^+ produced by Franck–Condon ionization of $N_2X\ ^1\Sigma_g^+$ vs $N_2X\ ^1\Sigma_g^+$ vibrational temperature T_v.

tion distribution for any temperature is given. At 300°K, 11% of the B-state population should be in $v=1$. The population measured for *room* temperature N_2 ionization by both $He(2^3S)$ and electrons shows precisely this population distribution, thus supporting the method. By obtaining the ratio of the intensity of the 0, 0 band, for example, with N_2 vibrationally excited to that for N_2 at room temperature, one can (by use of the curve for $v=0$ in Fig. 3) read directly the vibrational temperature. Thus, obtaining the vibrational temperature is a matter of reading an experimental intensity ratio and looking up the temperature on Fig. 3.

The principal experimental usefulness of using only the light from the $v=0$ level of the B state is the fact that we no longer need to obtain a detailed population distribution for the temperature measurement. Thus only a small $He(2^3S)$ concentration yields enough light to make a reliable temperature measurement. Then, when measuring Reaction (3), enough O_2 can be added to the He afterglow to remove almost all of the $He(2^3S)$ before the N_2 is added, thus removing the possibility of having extended sources of O^+, while still having enough $He(2^3S)$ to make the vibrational-temperature measurements of the N_2 simultaneously with the rate measurement. This improves the stability and reproducibility. It took some time ($\sim$5 min) for the temperature of the N_2 coming out of the side tube to reach equilibrium. This was monitored by watching the intensity of the 0, 0 band on the spectrometer until equilibrium had been reached, at which time the rate measurements were taken with the mass spectrometer.

If the He afterglow, containing principally He^+, He_2^+, and $He(2^3S)$, is impinged on the N_2 flow and the spectrum of the 1st-negative system of N_2^+ is observed, an apparent non-Boltzmann distribution of N_2 vibrational states is formed. However, if the He^+ and He_2^+ are first eliminated by adding a small amount of O_2, [or some other gas which reacts with the He^+ ions faster than with the $He(2^3S)$], one finds that the spectrum changes and the deduced N_2 vibrational population becomes Boltzmann. Thus the spectrum excited by He^+ (directly or indirectly, see below) is not the same as that excited by Reaction (2), and this leads to confusion if they are observed simultaneously. This effect was also observed by Robertson.[15] The top half of Fig. 2 shows the spectra excited by Reaction (2), where the N_2 has been vibrationally excited to 3800°K and the He^+ and He_2^+ have been removed by adding a small amount of O_2 upstream. The lower half of the figure shows the spectra resulting from Reaction (1b). In this case the $He(2^3S)$ has been removed by adding H_2 upstream. The reaction of He^+ with H_2 is unobservably slow,[3] while the reaction $He(2^3S)+H_2\rightarrow H_2^+ + He+e$ has been measured to have a rate constant $\approx 10^{-10}$ cm³/sec.[20] If both of the reactions (1b) and (2)

---

[20] W. P. Sholette and E. E. Muschlitz, J. Chem. Phys. **36**, 3368 (1962).

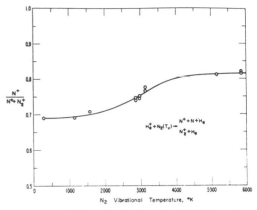

FIG. 4. Fractional N^+ production from the $He^+ + N_2$ reaction as a function of N_2 vibrational temperature.

are allowed to go on simultaneously, then the sum of the two spectra is of course observed, giving a relatively high intensity for lines from high vibrational states.

RESULTS

$He^+ + N_2$

Reaction (1) has been measured by a number of workers and there is a substantial agreement among all of these results, five independent experiments[1-6] yielding values of k_1 from 1.3 to 1.7×10^{-9} cm³/sec. Most measurements have not separated the two channels of reaction however. Warneck[4] finds in photoionization-mass-spectrometry studies that $k_{1a}=0.71$ and $k_{1b}=0.76\times 10^{-9}$ cm³/sec, or $k_{1a}\approx k_{1b}$. Heimerl *et al.*[6] find $k_{1a}=0.52\ k_1$ and $k_{1b}=0.48\ k_1$. The present result differs somewhat with $k_{1a}=0.7k_1$ and $k_{1b}=0.3k_1$. The dependence of this branching ratio on N_2 vibrational temperature has been measured in the present study and is shown in Fig. 4.

The higher ratio of k_{1a} to k_{1b} obtained in the present experiments may reflect N_2^+ loss by electron recombination in the flowing afterglow tube. This would make k_{1a}/k_{1b} appear larger than it really is. This would not affect our total k_1 rate measurement which is based on He^+ loss.

Very recently, Inn[21] has examined the spectrum from the reaction

$$He^+(^2S)+N_2(X\ ^1\Sigma_g^+)\rightarrow He(^1S)+N_2^+(C\ ^2\Sigma_u^+)$$

(1b')

followed by radiation,

$$N_2^+(C\ ^2\Sigma_u^+)\rightarrow N_2(X\ ^2\Sigma_g^+)+h\nu,$$

(9)

the second-negative system of N_2^+. From the intensity of the second-negative system, Inn deduces that a

---

[21] E. C. Y. Inn, Planetary Space Sci. **15**, 19 (1967).

substantial fraction ($\geq 5\%$) of the charge transfers must go via Reaction (1b′), follwed by radiation.

It has been suggested[22] that Reaction (1a) is also a charge transfer into a near-resonant vibrational level of the N_2^+C state (1b′), followed by the known predissociation,[23]

$$N_2(C\,^2\Sigma_u^+)\rightarrow N(^4S)+N^+(^3P). \quad (10)$$

Since Reactions (9) and (10) occur at comparable rates, if this mechanism is correct it follows that the N_2^+C state populated by charge transfer from He^+ has about the same lifetime against radiation as it does against predissociation. If this picture is true, one might expect to change the ratio of N^+ to N_2^+ in Reaction (10) by vibrationally exciting the ground-state N_2 so that for the N_2 molecules with $v>0$ the resonance now is with the $v=5$ or higher vibrational level of the C state (as contrasted to $v=4$ for $N_2X\,^1\Sigma_g v=0$), thus allowing the N_2^+ to be stabilized by a radiative transition to the X state of N_2^+, rather than predissociating into N^++N.

To test the above hypothesis careful measurements of the products of Reaction (1) were made under conditions where all the $He(2^3S)$ had been removed from the He stream by reaction with k_v.[20,24] The only source left for N_2^+ then was from Reaction (1b). Figure 4 shows the results of this measurement as a function of the N_2 vibrational temperature. As can be seen, there is only a moderate change in the branching ratio between 300°–6000°K vibrational temperature of the N_2. The ratio of (1a) to (1b) increases slightly with N_2 vibrational temperature. Reaction (1) was carefully checked for a change in rate constant with N_2 vibrational temperature. No change in the over-all rate constant was observable over the range of vibrational temperature 300–6000°K. The fraction of He^+ charge transfer which ultimately populates the $B\,^2\Sigma_u^+$ state is estimated crudely from photon intensities to be of the order of a percent or so (probably from 0.4%–1%). The resulting spectrum is shown in Fig. 2b. The spectrum appears to be essentially the same at $T_v=300°$ and 3800°K. The origin of $N_2^+B\,^2\Sigma_u^+$-state population has not been determined. It may be a consequence of $N_2^+C\,^2\Sigma_u^+$ charge transfer with ground-state N_2, i.e.,

$$N_2^+C\,^2\Sigma_u^++N_2X\,^1\Sigma_g^+\rightarrow N_2X\,^1\Sigma_g^++N_2B\,^2\Sigma_u^+,$$
$$\quad (11)$$

or of direct He^+ charge transfer,

$$He^++N_2X\,^1\Sigma_g^+\rightarrow N_2^+B\,^2\Sigma_u^++He, \quad (1b′)$$

or of an as yet unidentified process. (Experiments in another apparatus in which the pressure can be varied are planned in order to attempt to resolve this question

in the future.) It is also important to note that the spectrum of Fig. 2(b) may not accurately reflect the vibrational population of the B state as produced, since an unknown amount of vibrational relaxation by collision may occur before radiation.

$He(2^3S)+N_2$

The cross section for Reaction (2) has been measured by Benton, et al.[24] to be 6.4 Å^2, and by Sholette and Muschlitz[20] to be 7.0 Å^2. It is quite surprising to us that this reaction obeys the Franck–Condon principle, or in other words, that the population of vibrational levels produced in the N_2^+B state of $He(2^3S)$ ionization is the same as that produced by 5-keV electron ionization, since the interaction of the helium metastable and N_2 molecule results in a considerably longer-duration collision than the ionization by a fast electron interacting by means of its electric field. From the point of view of electron rearrangement, Reaction (2) is rather complicated. The usual picture of a spin–flip process $He(2^3S)+e\rightarrow He(1^1S)+e$ involves an actual exchange of electrons, i.e., ejection of an electron originally on the $He(2^3S)$ and capture of an external electron, in this case from N_2. It would seem that a rather substantial interaction of the $He(2^3S)$ and the N_2 is involved, in so much as $He(2^3S)$ internal electronic energy is utilized to eject an N_2 electron. One might thus have expected a somewhat more complicated distribution of N_2^+ vibrational states to result than in the case of fast electrons. Reaction (2) was checked for a rate-constant dependence of T_v by integrating the total population density of N_2^+ obtained by Reaction (2) from the integrated intensity of the 1st-negative system of N_2^+, measured as a function of the N_2 vibrational temperature. The total population density, although changed in distribution, remained constant. The above measurements were made with absolute intensity calibration so that the absolute number of radiations from the B state could be determined. Also absolute density measurements of the $He(2^3S)$ were made by absorption techniques, so that the efficiency of conversion of $He(2^3S)$ to N_2^+B could be made. It was found that at least 75% of the $He(2^3S)$ reacting with N_2 at any vibrational temperature yielded a 1st-negative photon.

Evidence for Franck–Condon ionization of N_2 by the metastables has previously been suggested by Cermak[25] on the basis of retarding potential experiments on the electron ejected by Reaction (2). Cermak's experiments, while not quantitative on this point, also suggested that a sizeable fraction of Reaction (2) populates the N_2^+B state, with a comparable N_2^+X-state population and a lower $N_2^+A\,^2\Pi_u$-state population. Cermak's experiments suggested Franck–Condon ionization [by both the $He(2^3S)$ and the $He(2^1S)$ metastable] of N_2, CO, COS, and CO_2, but not NO, nor O_2. Robertson[15] also found the N_2^+B-state populations,

[22] R. F. Stebbings, A. C. H. Smith, and H. Ehrhardt, J. Chem. Phys. **39**, 968 (1963).

[23] F. R. Gilmore, J. Quant. Spectry. Radiative Transfer **5**, 369 (1965).

[24] E. E. Benton, E. E. Ferguson, F. A. Matsen, and W. W. Robertson, Phys. Rev. **128**, 206 (1962).

[25] V. Čermák, J. Chem. Phys. **44**, 3781 (1966).

produced by He(2^3S) from ground-vibrational-state N₂, to follow Franck–Condon factors, but found marked deviations from Franck–Condon population in the case of the $O_2^+ A\ ^2\Pi_u$ state.

$O^+ + N_2$

Figure 5 shows the rate constant for the O^+ reaction with vibrationally excited N₂. The rate constant is seen to rise sharply, from its value of 1.3×10^{-12} cm³/sec at room temperature, above 1200°K. At 1200°K $v=1$ is appreciably populated ($\approx 6\%$), so that the rate constant for $v=1$ must not be much larger than that for $v=0$. Above 1200°K, $v=2$ is starting to be populated, and the increase in rate constant approximately parallels the increase in $v=2$ as a function of temperature. Thus $v=2$ must have a much larger rate constant than either $v=0$ or $v=1$. This kind of analysis can be used to invert the data on Fig. 5 to give the rate constant for the various vibrational levels of N₂ with O^+. This has been done with the help of a computer to yield the result shown in Fig. 6.

In measuring the rate constant as a function of vibrational temperature, the following procedure was used. A particular N₂ flow was established in the side-arm tube. This reduced the O^+ signal, due to the reaction between O^+ and vibrationally unexcited nitrogen. A microwave discharge was then initiated in the side tube, producing the vibrationally excited N₂. The vibrational temperature of the N₂ was determined in a region near the entry of the N₂ into the flow tube and was assumed to be constant at that temperature throughout the reaction region.

Thus, during the course of an experiment, the N₂ flow was held constant while the N₂ vibrational tem-

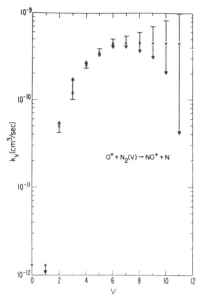

F1G. 6. Reaction rate constant k_r as a function of vibrational level of N₂ for the reaction $O^+ + N_2 \rightarrow NO^+ + N$. This is produced from the data in Fig. 5. If the values indicated by $\times$ are used the line shown in Fig. 5 is obtained. The ↑ indicate that if these points are used, the upper edge of the points in Fig. 5 are fitted, and if ⊥ is used, the lower edge is fitted.

perature was varied. The variation of the O^+ signal as a function of N₂ vibrational temperature was the raw data from which the rate constants were calculated. Runs were made at N₂ flows of 0.5, 1.0, 1.5, and 2.0 atm·cc/sec with satisfactory agreement. The decline of the O^+ signal as a function of the rate of addition of the vibrationally excited N₂ is expressed directly in terms of a rate constant in the form

$$k = [\tilde{K} A \bar{v}^2 / Q(N_2) d] \log_{10}([O_0^+]/[O^+]), \quad (12)$$

where k is the reaction-rate constant in cubic centimeters per second, A is the cross-sectional area of the flow tube in square centimeters, $\bar{v}$ is the average flow velocity in centimeters per second, $Q(N_2)$ is the rate of introduction of N₂ in particles per second, d the length of the reaction zone in centimeters, $[O_0^+]$ the concentration of O^+ at the end of the reaction zone before the introduction of the N₂, and $[O^+]$ the concentration of O^+ at the end of the reaction region in the presence of a fixed quantity of N₂. The quantity $\tilde{K}$ is a dimensionless parameter which is arrived at by a machine solution of the transport equations using a standard explicit-finite-difference method. In the present experiment the following set of assumptions were used to compute $\tilde{K}$:

(1) The flow was assumed to be laminar with a slip of about 10%.

(2) Account was taken of the small pressure gradient in the reaction zone.

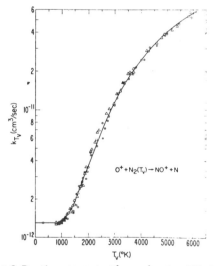

F1G. 5. Reaction rate constant k_{T_v} as a function of N₂ vibrational temperature for the reaction $O^+ + N_2 \rightarrow NO^+ + N$. The points are data and the line is a smooth fit used to derive Fig. 6.

(3) The reactant gas was assumed to issue uniformly from a 1-cm-diam tube, coaxially situated in the reaction vessel.

(4) The initial distribution of O^+ ions downstream at the reactant-gas inlet had reached the fundamental coaxial-diffusion mode.

(5) Axial as well as radial diffusion was considered.

It was further assumed that the transport properties of this flowing system were changed by the discharge in N_2 only in the alteration of the distribution of O^+.

Given the pertinent experimental parameters, d, A, $\bar{v}$, Q, the neutral and ambipolar diffusion coefficient, the flow rate of the helium buffer gas, and the pressure at some point in the tube, the machine supplies a table of values for $\bar{K}$ associated with a given logarithmic decrease in O^+. For the present set of assumptions, $\bar{K}$ is a slightly varying function of O^+ decrease. For example, for a change from $\log_{10}[O_0^+]/[O^+]=0.1$ to $\log_{10}[O_0^+]/[O^+]=4$, $\bar{K}$ would increase by 14%, varying about a value of $\bar{K}=3.35$ for $\log_{10}[O_0^+]/[O^+]=1$.

To gain a better perspective into this solution, let us imagine a simple situation in which the loss of O^+ is expressed by the rate equation

$$d[O^+]/dt = -k[O^+][N_2], \tag{13}$$

which has the solution

$$k = (1/[N_2]\tau)\ln([O_0^+]/[O^+]), \tag{14}$$

where the concentration of $[N_2]=Q(N_2)/A\bar{v}$ and the reaction time $\tau=d/\bar{v}$. Expressed in terms of Eq. (12), $\bar{K}$ would be equal to 2.30 and would be independent of the change in O^+.

The reason that $\bar{K}$ is larger in the former case is due chiefly to the parabolic velocity profile of the flowing gas, coupled with an ion distribution which tends to be peaked on axis. The reason $\bar{K}$ is not constant in the former case is twofold. In the first place, end effects arise because the ions react with large quantities of reactants introduced on, and thus initially located on, the tube axis. In the second place, axial diffusion effectively increases the transport velocity of the O^+ ions down the tube, since the attenuation of the ions by reaction creates a concentration gradient down the tube.

It is necessary to exclude the possibility that impurities could have caused the increased reaction rate for Reaction (3) when the N_2 is discharged. The maximum observed increase in rate constant due to the N_2 excitation brings the effective rate constant up to over 5×10^{-11} cm³/sec, a factor 40 larger than for unexcited N_2. Thus, assuming that no rate constant for excited states of N_2 may be greater than the orbiting-collision rate constant of 1.2×10^{-9} cm³/sec, a concentration of excited states of $\approx5\%$ of the N_2 flow would be required. There is no possibility of creating more than 1 part in 10^4 of ions or metastables[26] in the tube we used, es-

pecially in view of the fact that this concentration must exist 10 cm or more downstream of the discharge in a tube 1 cm in diameter. Thus, metastables or ions could not give an effect as large as was seen.

The N atoms can be produced in concentrations of $\approx1\%$ in some cases; however, the N atoms were removed by titrating the N_2 afterglow with NO, where the reaction

$$N+NO \rightarrow N_2+O \tag{15}$$

(which has a visible end point) exchanges the free N atoms for free O atoms, and no effect on the signal strength was observed. The N_2 produced in the NO titration is known to be vibrationally excited,[14] but this does not measurably effect the O^+ signal, presumably because the vibrational excitation added this way is small compared to that caused by the microwave discharge. It is not surprising that the N atoms gave no effect on the O^+ signal in view of the fact that the charge transfer of O^+ with N atoms is endothermic and therefore not possible at the thermal energy of the experiment.

The role of chemical impurities produced by discharging the nitrogen can be discounted since no possible impurity could have a sufficient product of concentration and rate constant to explain the result. The rate constant for $O^++NO \rightarrow NO^++O$ is so low[10] that over a 200% NO impurity would be required for example. NO is a likely impurity in an N_2 discharge and also the one which would be most deceptive, since it could lead to the same product ion NO^+. Additionally, NO and most other impurities would have reacted with the O_2^+, and this was not observed. Also, the impurities produced in the discharge would not decay by rapid wall relaxation with the known vibrational-deactivation coefficient, as discussed below. The presence of impurities other than NO would in all probability have been manifested in the mass spectrum which was observed between 1 and 100 amu.

The enhanced reaction between O^+ and discharged N_2 was reproducible with different sidearm tubes, N_2 bottles, and N_2 flows, in experiments carried out over a 2-year period. The conclusion that the effect is indeed a vibrational one appears to be inescapable.

The measured rate constant, $1.3\pm0.2\times10^{-12}$ cm³/sec at room temperature, corresponding to a reaction efficiency $\eta \sim 10^{-3}$ for Reaction (3), is remarkably low, a fact which has been commented upon by Giese[27] and others. The reason for this is not clear, but the following possibility seems plausible. At thermal energy (no excess kinetic or vibrational energy of the reactants) it seems likely that the reaction proceeds via the ground-state N_2O^+ intermediate molecular ion, i.e.,

$$O^+(^4S)+N_2(^1\Sigma) \rightarrow N_2O^+(X\ ^2\Pi) \rightarrow NO^+(^1\Sigma)+N(^4S). \tag{16}$$

[26] W. L. Starr, J. Chem. Phys. **43**, 73 (1965).

[27] C. F. Giese, Advan. Chem. Ser. **58**, 20 (1966).

The relative inefficiency of the reaction ($\sim$1 reaction per 10^3 collisions) may then be attributed to the spin-forbidden nature of the N$_2$O$^+$-formation step. The reaction probability $\sim$10^{-3} is perhaps reasonable for the spin hindrance; Wigner[28] and Herzberg[29] have remarked that the unimolecular dissociation of N$_2$O is expected to proceed only about 10^{-3} as fast as the normally expected rate because of the multiplicity change in the decomposition.

With increased vibrational energy, the rate constant (and cross section) for the reaction increases markedly, as shown in Fig. 5. The present speculation is that perhaps at a certain vibrational level of the N$_2$ the lowest quartet state of N$_2$O$^+$ becomes energetically accessible, and the reaction then proceeds via

$$O^+(^4S) + N_2(^1\Sigma) \rightarrow N_2O^+(^4\Lambda) \rightarrow NO^+(^1\Sigma) + N(^4S),$$

$$(17)$$

where Λ, the unknown orbital-angular-momentum state, is most likely Σ. In this case, the N$_2$O$^+$-complex formation is not spin forbidden and might be expected to occur with much greater efficiency. Figure 6 shows how the rate-constant data as a function of vibrational temperature are unfolded to yield rate constant vs N$_2$ vibrational-quantum number. The steep rise in rate constant occurs at $v=2$. According to the above-postulated mechanism, this would locate the threshold for Reaction (17) to be exothermic between $v=1$ and $v=2$. This in turn would locate the position of the N$_2$O$^+$ quartet state at $\sim$2.8 eV above the N$_2$O$^+$ ground state. Unfortunately, no independent evidence on the location of N$_2$O$^+$ quartet states is presently available.

The steep onset in rate constant for an endothermic reaction with added reactant kinetic energy has been demonstrated by Giese and Maier[30] in several cases. Reaction (3) is known to have a sharp increase in cross section with O$^+$ kinetic energy,[27,31] which is consistent with the above picture. This increase occurs below $\sim$0.5 eV[27] but above 1400°K according to recent results of Warneck.[9] It will be interesting to see if the translational-energy onset is the same as the vibrational, i.e., $\sim$1500°K ($\sim$1200°K in addition to the normal 300°K available).

The importance of Reaction (3) in the earth's ionosphere, where it plays a major role in determining the maximum electron density, by virtue of converting the major ion produced into a molecular ion which can efficiently capture electrons, is well known.[32] The likely importance of the increase in rate constant with N$_2$ vibrational excitation for decreasing the ionospheric electron density under disturbed atmospheric conditions has also been discussed.[33] The discovery of the large effect of vibrational excitation on the rate constant plays an important role in rationalizing differences in various early laboratory measurements of Reaction (3). The first measurements of Reaction (3) were made in stationary afterglows[34,35] in which the N$_2$ was discharged in the same breakdown pulse used to produce the O$^+$. This necessarily vibrationally excited the N$_2$ and led to higher values for the rate constant than is appropriate to ground-vibrational state N$_2$. It is indeed possible to correlate the enhancement of the rate constant with the average breakdown-discharge power applied in these experiments, which were repetitively pulsed. Copsey et al.[8] have largely (perhaps completely) resolved this problem by single-shot discharge afterglow studies and arrive at a value $k = 2.4 \times 10^{-12}$ cm^3/sec, in reasonably good agreement with the present flowing afterglow result. It is not apparent at this time whether the remaining disagreement is due to a small remaining vibrational effect from the single discharge, from a contribution to O$^+$ loss by excited ions, probably O$^+(^2D)$, in the stationary afterglow, or whether the present flowing-afterglow measurement is now too low for reasons which have escaped our notice. The fact that the most recent flowing-afterglow results on the reaction

$$O^+ + O_2 \rightarrow O_2^+ + O \qquad (18)$$

agree well with those of Copsey et al.[8] ($k_{18} = 2.0 \times 10^{-11}$ cm^3/sec) encourages us to believe that the two experimental approaches are reasonably correctly analyzed and that the deviation in k_3 is due to some difference such as reactant states. It is found experimentally that Reaction (18) is much less sensitive to discharge-excitation conditions than Reaction (3).

ACKNOWLEDGMENTS

This work has been supported in part by the U.S. Defense Atomic Support Agency. The authors would like to express their appreciation to H. I. Schiff and D. L. Albritton for helpful discussions of this work.

[28] E. P. Wigner, Nachr. Ges. Wiss. Gottingen, 375 (1927).
[29] G. Herzberg, Z. Physik. Chem. (Leipzig) **B17**, 68 (1932).
[30] C. F. Giese and W. B. Maier, J. Chem. Phys. **39**, 197 (1963).
[31] R. F. Stebbings, B. R. Turner, and J. A. Rutherford, J. Geophys. Res. **71**, 771 (1966).

[32] E. E. Ferguson, F. C. Fehsenfeld, P. D. Goldan, and A. L. Schmeltekopf, J. Geophys. Res. **70**, 4323 (1965).
[33] A. L. Schmeltekopf, F. C. Fehsenfeld, G. I. Gilman, and E. E. Ferguson, Planetary Space Sci. **15**, 401 (1967).
[34] J. Sayers and D. Smith, *Atomic Collision Processes* (North-Holland Publ. Co., Amsterdam, 1964), p. 871.
[35] G. F. O. Langstroth and J. B. Hasted, Discussions Faraday Soc. **33**, 298 (1962).

33

Ab Initio Calculation of Potential Energy Curves for the Ion Molecule Reaction
$O^+ + N_2 \rightarrow NO^+ + N$

A. Pipano* and Joyce J. Kaufman

Department of Chemistry, The Johns Hopkins University, Baltimore, Maryland 21218

INTRODUCTION

It is generally observed experimentally that cross sections for exothermic ion–molecule reactions decrease with increasing ion kinetic energy. An important exception to this behavior is exhibited by the exothermic reaction $O^+ + N_2 \rightarrow NO^+ + N$ (ground state reactants to ground state products), the cross section of which goes through a broad maximum as a function of O^+ kinetic energies.[1] A theoretical justification for this apparently anomalous low energy behavior was given in a paper by Kaufman and Koski.[1] From the rules for the possible symmetry and spin permitted electronic states of the intermediate N_2O^+ formed from the various states of reactants and products combined with experimental spectroscopic and thermochemical data, schematic potential energy curves were drawn. Those curves were essentially two cuts through the potential surface for the reaction, presumably close to the reaction path. A three-dimensional plot is sufficient in this case, as the intermediate N_2O^+ is assumed to be linear in accordance with Walsh's rules.[2] The spin and symmetry rules indicated that both ground state reactants $O^+(^4S_u) + N_2(^1\Sigma_g^+)$ and ground state products $NO^+(^1\Sigma^+) + N(^4S_u)$ are permitted to combine uniquely only to a $^4\Sigma^-$ state of the intermediate N_2O^+. These $^4\Sigma^-$ repulsive curves cross the attractive potential curves of the ground and some of the excited states of the intermediate NNO^+. It is both spin and symmetry forbidden to go directly from ground state reactants to the $^2\Pi_i$ ground state of N_2O^+. Thus it was suggested that the observed increase of the cross section for production of NO^+ with energy, starting in the electron volt region, could be interpreted as arising from crossing of the ground state reactant repulsive $^4\Sigma^-$ curve with ground and higher lying attractive states of the intermediate opening up additional channels for reaction. Thus a more comprehensive interpretation of this reaction would depend upon the availability of more detailed and more accurate potential curves.

CALCULATIONS AND RESULTS

The present investigation is an *ab initio* calculation of the potential energy curves pertinent to the understanding of the above mentioned ion–molecule reaction. The only possibility within the limit of computational facilities (core size and time) available when we initiated this research using an SCF–CI program based on Slater orbitals was to use a minimal basis of Slater orbitals. In an earlier paper by the authors[3] it was shown that this basis gives quite correct ionization potentials for the N_2O molecule both in the SCF and CI methods, which means that the energy distances between curves near the minimum of the ground state neutral molecules are reasonably accurate. The exponents for the basis functions were obtained by using Slater's rules and are identical with Set 1 of exponents in Ref. 3. All nonzero, unique one- and two-electron integrals were computed at each internuclear distance to about six decimal place accuracy, using a modified version of Stevens' molecular integral program. A configuration interaction calculation (including all single and double excitations) was carried out at each point and for each state so that hopefully the potential energy curves for both reactants and products would dissociate correctly. No attempt was made to calculate the exact ground state SCF molecular orbitals for the neutral molecule, as it was found, in accordance with results obtained by Buenker and Peyerimhoff,[4] that they are quite inadequate for limited CI calculation of the ionic states. Molecular orbitals, with approximately the correct charge distribution, were obtained at each internuclear distance by using the CI programs

309

TABLE I. The effect of bending[a] the linear N–N–O⁺ molecule on the CI[b] energy of its first two Π states.

Electronic state of linear molecule ($C_{\infty v}$)	Electronic state after bending (C_S)	Total energy (linear) a.u.	Total energy (bent) a.u.	energy increase eV
$^2\Pi_x$	$^2A'$	−182.526242	−182.525188	0.029
$^2\Pi_y$	$^2A''$	−182.526242	−182.525041	0.033
$^4\Pi_x$	$^4A'$	−182.358775	−183.357627	0.031
$^4\Pi_y$	$^4A''$	−182.358775	−183.357509	0.034

[a] Bending of 5° relative to the linear molecule and for $R(N–N) = 2.11$ a.u., $R(N–O) = 2.24$ a.u. in both cases.   [b] CI wavefunctions include the corresponding state and all single and double excitations relative to it.

and iterating on the MO's, used for a previously calculated distance, until the largest first-order contribution of a singly excited state to the energy of the ground state neutral molecule was less than 5×10^{-4} a.u. (thus approximately satisfying the Brillouin theorem[5]). The MO's for the first internuclear distance were obtained by using the SCF MO's, calculated for Set 2 of exponents in Ref. 3, as first approximation. For each electronic symmetry state of NNO⁺ the ground state configuration together with all single and double excitations relative to it were included in the CI calculations. Higher roots of the same symmetry and spin were obtained by using a program by Bender[6] which is based on a modified Nesbet algorithm.[7] It should be noted that the computational error in those

higher roots is larger than the error in the corresponding lowest root.

Using Set 4 of exponents in Ref. 3, the effect of bending the molecule was tested for the ²Π and ⁴Π states. Those CI calculations were done for the experimental equilibrium internuclear distance [$R(N–N) = 2.11$ a.u., $R(N–O) = 2.24$ a.u.] and an angle of bending of 5° relative to the axis of the linear molecule. The results are given in Table I and show that the energies of both Π states increase with bending in accordance with Walsh's rules.[2] The effect of bending on the Σ states was not tested, as it is well known that the energy of those states increases when a molecule is distorted to a lower symmetry.[8]

The calculated energies as a function of internuclear

TABLE II. CI[a] energies for various states[b] of the linear N–N–O system as a function of N–O distance, keeping the N–N distance fixed ($R_{N-N} = 2.11$ a.u.).

N–O	N–N–O	N–N–O⁺				
Distance a.u.	$^1\Sigma^+$ (332)[a]	$^2\Pi_1$ (123)	$^2\Sigma^+$ (177)	$^2\Pi_2$[e] (123)	$^4\Pi$ (1379)	$^4\Sigma^-$ (1310)
2.24	−1.209659	−0.710993	−0.558676		−0.581197	
2.40	1.219706	−0.741338	−0.558477		−0.600833	−0.520827
2.60	−1.208882	−0.755140	−0.546839		−0.592100(4)	−0.626167
2.80	−1.188289	−0.758642	−0.529706(8)[b]		−0.578470(4)	−0.698936
3.00	−1.171751	−0.744861	−0.553693(8)		−0.561241(4)	−0.740657
3.30	−1.155334	−0.725676	−0.546014(8)	−0.452975(7)	−0.602549(2)	−0.778816
3.60	−1.149548	−0.699661	−0.558425(8)	−0.485308(7)	−0.628183(2)	−0.793480
3.90	−1.149032	−0.670348	−0.575354(8)	−0.504046(7)	−0.640183(2)	0.789969
4.20	−1.147813	−0.647528	−0.575180(8)	−0.523981(7)	−0.648225(2)	−0.788489
4.50	−1.145698	−0.631256	−0.573146(8)	−0.543506(7)	−0.654297(2)	−0.790682
5.00	−1.145652	−0.606659	−0.571218(8)	−0.555118(7)	−0.655778(2)	−0.780909
5.50	−1.144881	−0.593620	−0.569305(8)	−0.564017(7)	−0.656871(2)	−0.777352
6.00	−1.144400	−0.586440	−0.567905(8)	−0.568420(7)	−0.657298(2)	−0.774825
7.00	−1.143924	−0.580377	−0.566312(8)	−0.572203(7)	−0.657636(2)	−0.771580

Total energy relative to −182.000000 a.u.

[a] The number of configurations involved in each case is given in parenthesis.
[b] The serial number of the configuration, which has the largest weight on the wavefunction is given in parenthesis whenever it is different from 1. The corresponding configurations are given in the text.
[e] This state was obtained as the second root of the Hamiltonian matrix for the ²Π₁ state.

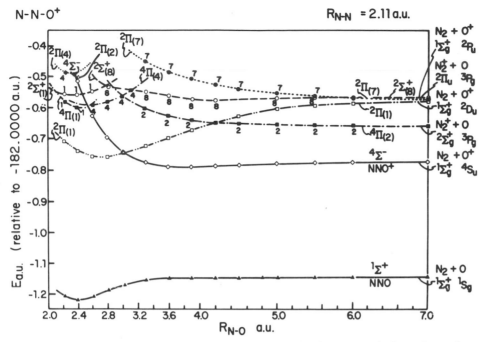

FIG. 1. CI energies of NNO$^+$ vs R_{N-O} a.u. [O^z approaching N$_2^y$ linearly, $(x+y=+1)$, $R_{N-N}=2.11$ a.u.].

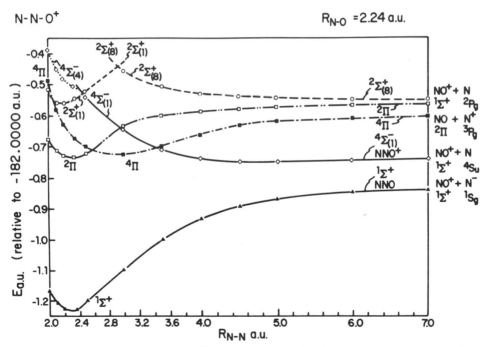

FIG. 2. CI energies of NNO$^+$ vs R_{N-N} a.u. [N^z receding from NO^y linearly, $(x+y=+1)$, $R_{N-O}=2.24$ a.u.].

TABLE III. CI [a] energies for various states[b] of the linear N–N–O system as a function of N–N distance, keeping the N–O distance fixed ($R_{N-O} = 2.24$ a.u.).

N–N Distance a.u.	Total energy relative to -182.000000 a.u.				
	N–N–O		N–N–O$^+$		
	$^1\Sigma^+$ (332)[a]	$^2\Pi_1$ (123)	$^2\Sigma^+$ (177)	$^4\Pi$ (1379)	$^4\Sigma^-$ (1310)
2.00	-1.164989	-0.676273	-0.512171	-0.499195	$-0.390589(4)$
2.11	-1.209659	-0.710993	-0.558676	-0.581197	$-0.465724(4)$
2.20	-1.225371	-0.730695	-0.555223	-0.625414	$-0.481524(4)$
2.35	-1.225856	-0.731073	-0.543656	-0.673958	$-0.507093(4)$
2.50	-1.198860	-0.719867	-0.517925	-0.699617	-0.540086
3.00	-1.100591	-0.633389	$-0.456609(8)$ [b]	-0.725705	-0.643321
3.50	-0.997418	-0.603779	$-0.509692(8)$	-0.697302	-0.710011
4.00	-0.931875	-0.590265	$-0.528568(8)$	-0.660865	-0.738326
4.50	-0.893330	-0.581858	$-0.537133(8)$	-0.634102	-0.746763
5.00	-0.870616	-0.576119	$-0.542745(8)$	-0.620233	-0.747446
6.00	-0.848280	-0.570403	$-0.550261(8)$	-0.614998	-0.744812
7.00	-0.841157	-0.569274	$-0.552890(8)$	-0.606222	-0.743439
8.00	-0.836200	-0.568972	$-0.553400(8)$	-0.605882	-0.742783

[a] The number of configurations involved in each case is given in parenthesis.

[b] The serial number of the configuration, which has the largest weight in the wavefunction is given in parenthesis whenever it is different from 1. The corresponding configurations are given in the text.

distance for various states of the linear N–N–O system are listed in Table II and Table III together with the number of configurations involved in each case. As the CI programs used were written for nonlinear polyatomic molecules,[9] only a subgroup of the full linear symmetry of the molecule was used and the number of configurations involved in all cases is somewhat larger than one would expect. The resulting potential energy curves are plotted in Fig. 1 (keeping the N–N distance fixed) and Fig. 2 (keeping the N–O distance fixed). The configurations which have the largest weight in the resulting wave functions are listed below according to their serial numbers (here all the σ orbitals are listed first followed by the π orbitals; they are not listed in order of increasing energies since this order changes as a function of internuclear distance):

$^1\Sigma^+$ (1) $(1\sigma)^2(2\sigma)^2(3\sigma)^2(4\sigma)^2(5\sigma)^2(6\sigma)^2(7\sigma)^2(1\pi)^4(2\pi)^4$,

$^2\Pi$ (1) $(1\sigma)^2(2\sigma)^2(3\sigma)^2(4\sigma)^2(5\sigma)^2(6\sigma)^2(7\sigma)^2(1\pi)^4(2\pi)^3$,

(7) $(1\sigma)^2(2\sigma)^2(3\sigma)^2(4\sigma)^2(5\sigma)^2(6\sigma)^2(7\sigma)^1(8\sigma)^1(1\pi)^4(2\pi)^3$,

$^2\Sigma^+$ (1) $(1\sigma)^2(2\sigma)^2(3\sigma)^2(4\sigma)^2(5\sigma)^2(6\sigma)^2(7\sigma)^1(1\pi)^4(2\pi)^4$,

(8) $(1\sigma)^2(2\sigma)^2(3\sigma)^2(4\sigma)^2(5\sigma)^2(6\sigma)^2(7\sigma)^2(8\sigma)^1(1\pi)^4(2\pi)^2$,

$^4\Pi$ (1) $(1\sigma)^2(2\sigma)^2(3\sigma)^2(4\sigma)^2(5\sigma)^2(6\sigma)^2(7\sigma)^2(1\pi)^4(2\pi)^2(3\pi)^1$,

(2) $(1\sigma)^2(2\sigma)^2(3\sigma)^2(4\sigma)^2(5\sigma)^2(6\sigma)^2(7\sigma)^1(8\sigma)^1(1\pi)^4(2\pi)^3$,

(4) $(1\sigma)^2(2\sigma)^2(3\sigma)^2(4\sigma)^2(5\sigma)^2(6\sigma)^2(7\sigma)^2(1\pi)^3(2\pi)^3(3\pi)^1$,

$^4\Sigma^-$ (1) $(1\sigma)^2(2\sigma)^2(3\sigma)^2(4\sigma)^2(5\sigma)^2(6\sigma)^2(7\sigma)^2(8\sigma)^1(1\pi)^4(2\pi)^2$.

In order to ascertain into which fragments each state of the intermediate separates, the corresponding CI first order density matrices were calculated at the largest N–O and N–N distances ($R_{N-N} = 2.11$ a.u., $R_{N-O} = 7.00$ a.u., and $R_{N-N} = 8.00$ a.u., $R_{N-O} = 2.24$ a.u.). Transformations were then performed on the resulting density matrices, so that they would be expressed in terms of "fragmental" MO's (or AO's). The spin and symmetry of each of the fragments was then determined by a straight forward analysis of the occupancies of its MO's (or AO's). The results for reactants and products are given in Tables IV and V, respectively together with the energies of the corresponding states at the largest separations calculated. The energies are relative to the $^4\Sigma^-$ state at the same internuclear distance and are compared to the corresponding experimental values for the energies of the fragments relative to ground ionic state reactants

TABLE IV. Energies of fragments into which the calculated electronic states of N-N-O+ separate as a result of increasing the N-O distance (reactants).

Electronic state of intermediate[a]	Fragments	Energy eV		
		Calculated[b] at $R_{N-O} = 7.00$ a.u.	Experimental[c]	
$^2\Sigma^+$ (8)	$N_2(^1\Sigma_g^+) + O^+(^2P_u)$	5.585	5.017	
$^2\Pi$ (7)	$N_2^+(^2\Pi_u) + O(^3P_g)$	5.425	3.082	
$^2\Pi$ (1)	$N_2(^1\Sigma_g^+) + O^+(^2D_u)$	5.203	3.325	
$^4\Pi$ (2)	$N_2^+(^2\Sigma_g^+) + O(^3P_g)$	3.100	1.962	
$^4\Sigma^-$ (1)	$N_2(^1\Sigma_g^+) + O^+(^4S_u)$	0.000	0.000	

[a] The numbers in parenthesis refer to the serial number of the dominant configuration in the wavefunction.
[b] Relative to the energy of the $^4\Sigma^-$ state at the same internuclear distance.
[c] Relative to the energy of the ground state reactants [$N_2(^1\Sigma_g^+) + O^+(^4S_u)$].

or products. The results in Table IV indicate that the energy distances between curves near the separation limit for the reactants are also in reasonable agreement with experiment. The two apparent discrepancies in case of the two $^2\Pi$ states of reactants [$^2\Pi(1)$ and $^2\Pi(7)$] can be accounted for by looking at the corresponding curves in Fig. 1. In the case of reactants, the energy of the $^2\Pi(7)$ state decreases and the energy of the $^2\Pi(1)$ state increases while increasing the N-O distance, so that it is quite possible that the corresponding curves cross at larger N-O separations. The results in Table V show not quite as good agreement for the calculated energy differences as compared to the experimental values. The apparent discrepancies may be rationalized by examination of the corresponding curves in Fig. 2. The energy of the $^2\Sigma^+(8)$ is decreasing and that of the $^2\Pi(1)$ state increasing (both of which separate to the same products) while increasing the N-N distance. Therefore it is quite possible that the corresponding curves cross at larger N-N separations. The energy of the $^4\Pi$ state is also decreasing as a function of N-N distance and thus may eventually cross some of the higher curves at larger N-N separations. As no attempt was made simultaneously to optimize the N-O distance, while changing the N-N distance or vice versa, the resulting lower ground state energy of reactants, compared to the ground state energy of products, near the dissociation limits has no significance. Moreover, due to a steeper energy drop along the $^4\Sigma^-$ curve, when the N-O distance is increased, (compared to the effect of increasing the N-N distance) it is anticipated that such a simultaneous optimization would yield the correct results. Although the energy separations between potential curves are in reasonable agreement with experimental data, the results inevitably reflect the deficiencies of a minimal basis set. Both equi-

librium and dissociation calculated equilibrium distances are larger than one should expect, due to an insufficient number of polarization functions (p, d, and f type functions) in the basis. In this particular case, the minimal basis does not even provide a reasonably sufficient number of MO's for the CI expansion. The basis set yields fifteen MO's, eleven of which are doubly occupied in the single determinant wavefunction for the $^1\Sigma^+$ state of the neutral N_2O molecule. This means that only four unoccupied MO's remain for excitations, two of which belong to σ symmetry, the other two forming a degenerate pair of π orbitals. It is obvious that this small number of unoccupied MO's is incapable of yielding, in the CI expansions, all the configurations needed for correct dissociation along all calculated potential curves. For this reason the $^2\Pi(1)$ state dissociates to higher energy fragments both when the N-N or N-O distances are increased. When the N-O distance is increased the $^2\Pi(1)$ state dissociates to $N_2(^1\Sigma_g^+) + O^+(^2D_u)$ instead of $N_2^+(^2\Sigma_g^+) + O(^3P_u)$. Similarly when the N-N distance is increased, it separates to $NO^+(^1\Sigma^+) + N(^2P_u)$ instead of $NO^+(^1\Sigma^+) + N(^2D_u)$. Another possible result of the inflexibility of the basis is that in some cases the configuration which has the largest weight in the wavefunction changes from point to point along the potential curve. This is the reason that in Table II some points are missing for the second root of the $^2\Pi$ state and in Table III the corresponding points are entirely omitted.

DISCUSSION

Almost concomitant with the publication of the original paper on the theoretical justification of the apparently anomalous behavior of the $O^+ + N_2$ reaction[1] a paper appeared on the photoelectron spectroscopy of NNO.[10] The ionization potential data in that paper enabled Kaufman and Koski to enhance

TABLE V. Energies of fragments into which the calculated electronic states of N-N-O+ separate as a result of increasing the N-N distance (products).

Electronic state of intermediate[a]	Fragments	Energy (eV)		
		Calculated[b] at $R_{NN} = 8.00$ a.u.	Experimental[c]	
$^2\Sigma^+$ (8)	$NO^+(^1\Sigma^+) + N(^2P_u)$	5.153	3.576	
$^2\Pi$ (1)	$NO^+(^1\Sigma^+) + N(^2P_u)$	4.729	3.576	
$^4\Pi$ (1)	$NO(^2\Pi) + N^+(^3P_g)$	3.725	5.265	
$^4\Sigma^-$ (1)	$NO^+(^1\Sigma^+) + N(^4S_u)$	0.000	0.000	

[a] The numbers in parenthesis refer to the serial number of the dominant configuration in the wavefunction.
[b] Relative to the energy of the $^4\Sigma^-$ state at the same internuclear distance.
[c] Relative to the energy of the ground state products [$NO^+(^1\Sigma^+) + N(^4S_u)$].

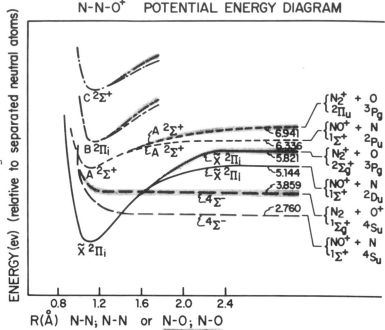

FIG. 3. NNO⁺ potential energy diagram (schematic).

the detail in their schematic potential curve for the reaction. This curve, Fig. 3, is the one which they had used this past two years for guidance in interpretation of experimental ion–molecule results.[11] It is gratifying to see that the *ab initio* CI computations performed in the present research confirmed the role of the $^4\Sigma^-$ state as the original path for the reactants. The positions of the other electronic states as indicated in the schematic curve and their proper dissociated products are also quite well substantiated. The calculated curves show an even greater wealth of structure than would have been anticipated prior to such a detailed calculation.

No broadening of the vibrational peaks in the photoelectron spectrum was observed by Brundle and Turner[10] indicating that mass spectroscopic appearance potentials of NO⁺ and N₂⁺ were not occurring by potential energy surface crossing out of the directly populated electronic states of the ion. Our potential energy surfaces confirm this behavior since the repulsive curves which would lead to the first appearance potential of NO⁺($^4\Sigma^-$) or to N₂⁺($^4\Pi$) cross the lower ionic states of NNO⁺ to the right. Our potential energy curves also would seem to lend some credence to the controversial observation of Natalis and Collin[12] that in the photoelectron spectrum of N₂O there is a new ionization process occurring around 14.3 eV which appears as a broad diffuse band showing no structure and extending over almost 2 eV. They postulated the ionic state would have the configuration $(1\pi)^4(7\sigma)^2$ $(2\pi)^2(\pi)^*$ and might be a $^2\Pi$, $^2\phi$ or $^4\Pi$ state where

they predicted the $^4\Pi$ state to be the most probable due to spin coupling energy. Our calculations do show a low lying $^4\Pi$ state with a wealth of interesting structure; in addition, parts of the region may also be crossed by the repulsive $^4\Sigma^-$ curve.

By following the higher roots of states of the same symmetry, we are able to calculate both the adiabatic curves and the diabatic curves as defined by Lichten.[13] This gives insight both about low energy reactions which most often proceed along adiabatic surfaces and about higher energy reactions which have sometimes been observed to proceed along diabatic surfaces.

Inspection of Fig. 1 (O⁺ approaching N₂ keeping the N–N distance constant) indicates a possible basis for the observed increase in rate constant by a factor of 40 in a room temperature flowing afterglow experiment when two quanta of vibrational energy are put into the N₂.[14] The original reactant $^4\Sigma^-$ NNO⁺ curve crosses the $\tilde{X}\,^2\Pi_i$ NNO⁺ at a point somewhat above the $v=0$ state. Increasing the vibrational energy of the N₂ could make the probability of transition from the $^4\Sigma^-$ state to the $^2\Pi_i$ state much larger.

A recent paper by Leventhal[15] indicates that for O⁺+N₂ at as low as 3 eV relative energy there is still no evidence of complex formation. The question arises whether going to zero relative energy would lead to observable complex formation. As O⁺+N₂ come in on their original $^4\Sigma^-$ curve at a cross point A they get into an N₂O⁺ state (for example the $\tilde{X}\,^2\Pi_i$ state) and at a cross point B there is a certain probability of dissociating to ground state $NO^+ + N$ along its unique $^4\Sigma^-$ curve. Now if $O^+ + N_2$ have zero relative kinetic energy they will fall into the well represented by $\tilde{X}\ ^2\Pi_i\ N_2O^+$ and oscillate with a certain vibrational frequency characteristic of N_2O^+. This means that during one rotational period of N_2O^+, the $NO^+ + N$ ($^4\Sigma^-$) will cross point B many times (10–20). If the transition probability (time) is right relative to the rotational period the NNO^+ will dissociate into products $NO^+ + N$ irrespective of how small the relative velocity is since point A is already above point B.[16] This type of reasoning is thus applicable to reactions which have similar potential surfaces; namely where the crossing point A of the reactant curve to an intermediate complex curve of a different symmetry is higher than the crossing point B of the product curve. It is also applicable if the reaction proceeds along a single surface where there is a trough in the three dimensional reaction surface such that the reactants may proceed downward in energy to the products.

Our calculations indicate that the two $^2\Pi$ curves for the reactants $O^+(^2D_u) + N_2(^1\Sigma_g^+)\ ^2\Pi(1)$ and $O(^3P_g) + N_2^+(^2\Pi_u)\ ^2\Pi(7)$ must cross or pseudocross at extremely large N–O internuclear distances. This explains the experimental observation that metastable $O^+(^2D)$ ions react with N_2 in the low energy range to form principally $N_2^+(^2\Pi_u)$ while the ion–molecule reaction to form NO^+ has a very small probability.[18] The $^2\Pi(1)$ curve $[O^+(^2D_u) + N_2(^1\Sigma_g^+)]$ is near resonant with $^2\Pi(7)$ curve $[O(^3P_g) + N_2^+(^2\Pi_u)]$ for N_2^+ in the $A\ ^2\Pi_u$, $v=1$ state. Thus charge exchange takes place at very large distances and the $O(^3P_g) + N_2^+(^2\Pi_u)$ are formed where they then dissociate along a slightly repulsive curve.

CONCLUSIONS

The most significant conclusion of the present research is the unambiguous confirmation of the overriding role that spin and symmetry restrictions play in any type of collisional process. The original paper[1] states that these rules govern the course of all ion-molecule reactions as well as all reactive and nonreactive collisions. It does not matter if the reactants ever form even a transiently bound intermediate complex or not, the reactants and products must proceed along the symmetry and spin permitted paths, mitigated possibly by curve crossings, which are themselves also fixed by the potential energy surfaces. However, it should be stressed that it is not sufficient merely that the reactants have the proper spin and symmetry to permit them to combine to a particular electronic state of the intermediate, they must be the exact pair which does combine to this state. The same holds true for the products. There is an absolute uniqueness in the correspondence of an intermediate state to a pair of separated fragments. This uniqueness has permitted the derivation of a corresponding comprehensive general theory for molecular decompositions based on symmetry and spin restrictions.[19]

ACKNOWLEDGMENTS

The majority of the computations were performed on the CDC 6600 computer at New York University. The authors wish to thank Mr. R. Malchie of the NYU Computing Center for his cooperation. The preliminary computations were carried out on the CDC 6600 at the CDC Washington Data Center. We should like to thank B. Feely of CDC who graciously arranged our use of the computer and R. Balthrop and especially J. V. Caron of the CDC Data Center for their cooperation. Programs made available by Professor J. W. Moskowitz, Dr. R. M. Stevens, and Dr. C. F. Bender are also gratefully acknowledged as is the guidance of Professor I. Shavitt in developing and writing the CI programs.

This research was supported in part by the AEC and in part by BRL under Contract No. DAAD05-70-0027.

One of us (J.J.K.) should also like to thank the U.S. Air Force Office of Scientific Research, Office of Aerospace Research, Propulsion Division (under Contract AF49(638)1530 and F44620-69-C-0110) for their long time support of her theoretical research which led to the understanding necessary to formulate this problem.

* Present address: Israel Aircraft Industries Ltd., LOD Airport, Israel.
[1] Joyce J. Kaufman and W. S. Koski, J. Chem. Phys. **50**, 1942 (1969) and references therein.
[2] A. D. Walsh, J. Chem. Soc. **1953**, 2266.
[3] A. Pipano and Joyce J. Kaufman, Chem. Phys. Letters **7**, 99 (1970).
[4] R. J. Buenker and S. D. Peyerimhoff, J. Chem. Phys. **53**, 1368 (1970).
[5] L. Brillouin, *Les Champs Self Consistents de Hartree et de Fock*, (Herman, Paris, 1934), page 19.
[6] C. F. Bender, I. Shavitt, and A. Pipano (unpublished).
[7] R. K. Nesbet, J. Chem. Phys. **43**, 311 (1965).
[8] A. H. Jahn and E. Teller, Proc. Roy. Soc. **A161**, 220 (1937).
[9] A. Pipano and I. Shavitt (unpublished).
[10] C. Brundle and D. W. Turner, J. Mass Spectrometry and Ion Phys. **2**, 195 (1969).
[11] Joyce J. Kaufman, speech presented at Wright-Patterson Air Force Base, June, 1970.
[12] P. Natalis and J. E. Collin, J. Mass Spectrometry and Ion Phys. **2**, 221 (1969).
[13] W. Lichten, Advan. Chem. Phys. **13**, 41 (1968) and prior references cited therein.
[14] A. L. Schmeltkopf, E. E. Ferguson, and F. C. Fehsenfeld, J. Chem. Phys. **48**, 2966 (1968).
[15] J. J. Leventhal, J. Chem. Phys. **54**, 5102 (1971).
[16] In the thermal energy reaction of $O^+ + N_2$ it could be hypothesized following the arguments of Polyani[17] that there is a barrier in the exit channel $NO^+ + N$ because using target N_2 in the $v=2$ vibrational state increases the rate constant by a factor of 40. (See Ref. 14).
[17] J. C. Polanyi, *Proceedings of the Conference on Potential Energy Surfaces in Chemistry*, edited by W. A. Lester, Jr., (IBM Research Lab., San Jose, Cal., 1971), p. 10.
[18] J. A. Rutherford and D. A. Vroom (private communication, August 1971).
[19] Joyce J. Kaufman, Ellen Kerman, and W. S. Koski, Intern. J. Quantum. Chem. **4S**, 391 (1971).

Reprinted from *J. Chem. Phys.* **48**:1527–1533 (1968)

Ion–Molecule Reactions of NH_3^+ by Photoionization*

W. A. Chupka and M. E. Russell

Argonne National Laboratory, Argonne, Illinois

(Received 4 October 1967)

Photoionization-efficiency curves were measured for NH_3^+, NH_4^+, and H_3O^+ ions in NH_3 gas and in gaseous mixtures of NH_3 and H_2O The results are interpreted to determine how the relative reaction cross section depends on the vibrational energy of the NH_3^+ ion for the reactions

$$NH_3^+ + NH_3 \rightarrow NH_3^+ + NH_2, \tag{1}$$

$$NH_3^+ + H_2O \rightarrow NH_4^+ + OH. \tag{2}$$

The cross section for Reaction (1) decreases with increasing vibrational energy while that for Reaction (2) is nearly independent of vibrational energy. The results are discussed in terms of the formation of an intermediate complex and in terms of stripping models The upper limit of the proton affinity was determined to be ≥ 8.54 eV for NH_3 and ≥ 7.0 eV for H_2O.

The use of photoionization for the study of ion–molecule reactions has made it possible to investigate how the vibrational and rotational energy of the reacting ion affects the reaction cross section.[1] The study of an exothermic reaction producing H_3^+ in hydrogen showed that, at very low repeller voltage, increasing vibrational energy of the ion led to a decrease in reaction cross section and that, at higher repeller voltage, an opposite dependence was found. Because the photoionization of the hydrogen molecule near threshold is dominated by autoionization processes, the vibrational-energy distribution of the H_2^+ ions formed at a particular photon energy is not readily apparent from the photoionization efficiency curve. The photoionization of NH_3 on the other hand, occurs almost entirely by direct ionization in the threshold region and the step-like character of the ionization-efficiency curve gives a clear display of the vibrational excitation of the ion. Furthermore, NH_3^+ ions are produced with vibrational energies varying over a range of about 2 eV, and this feature makes this a very convenient system for a test of the effect of vibrational energy on reaction rate.

Ion–molecule reactions occurring in ammonia have been studied earlier by electron-impact methods.[2–4] The major reaction detected was

$$NH_3^+ + NH_3 \rightarrow NH_4^+ + NH_2, \tag{1}$$

which was found to be exothermic. No information regarding the effect of vibrational energy could be obtained from this work.

EXPERIMENTAL RESULTS AND DISCUSSION

The apparatus and techniques used in this experiment were the same as those used in the study of ion–molecule reactions in hydrogen,[1] with a few small changes. The present data were taken with monochromator slits $300\ \mu$ wide which gave a photon bandwidth of 2.4 Å. The source of light was the many-line spectrum of hydrogen.

---

* Work performed under the auspices of the U.S. Atomic Energy Commission.

[1] W. A. Chupka, M. E. Russell, and K. Rafaey, J. Chem. Phys. **48**, 1518 (1968), preceding article.

[2] V. L. Tal'rose and E. L. Frankevich, "Investigation of the Ion-Molecular Elementary Act of a Radiation Chemical Process," in *Proceedings of the All-Union Conference on Radiation Chemistry Moscow, 1957* (Consultants Bureau, Inc., New York, 1959) pp. 11–16.

[3] L. M. Dorfman and P. C. Noble, J. Phys. Chem. **63**, 908 (1959).

[4] G. A. W. Derwish, A. Galli, A. Giardini-Guidoni, and G. G. Volpi, J. Chem. Phys. 39, 1599 (1963).

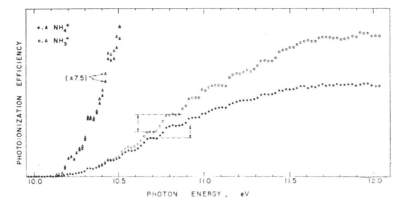

FIG. 1. Photoionization-efficiency curves for NH_3^+ and NH_4^+ formed in NH_3 gas. The ordinate scale of the data is adjusted so that the data points of the first plateau at ~10.2 eV coincide on the average. The dashed lines indicate how step heights were measured. The ratio of corresponding step heights (NH_4^+/NH_3^+) is proportional to the reaction cross section for NH_3^+ ions with the corresponding vibrational excitation.

A. Reaction (1) in NH₃

The photoionization-efficiency curves of NH_3^+ and NH_4^+ were measured at four different repeller voltages with pressures of NH_3 gas such that the amount of reaction was less than 20% for a repeller voltage of 0.5 V/cm and less than 10% for all other repeller voltages. The curves for a repeller voltage of 2.7 V/cm are shown in Fig. 1, where the scale factor for the NH_4^+ curve is adjusted so that the data points on the first plateau at about 10.20 eV are made to coincide (on the average) with those for NH_3^+, as can be seen from the magnified portion of Fig. 1. More than 10 000 ions were counted at each selected wavelength above the ionization threshold, and the total probable error at any point above the ionization threshold is about 1% or less. This probable error is only that due to statistical fluctuations and does not include any systematic error such as that involved in the correction of the photon-intensity measurement for the variation of detector response with photon wavelength. Such systematic errors should be the same for the NH_3^+ and NH_4^+ measurements, which were carried out within minutes of each other under identical conditions. Since the main purpose of this work is to measure the relative intensities of NH_4^+ and NH_3^+ as a function of wavelength, no great attempts to minimize such systematic errors was made, other than the procedures described in the earlier work.[1]

The curve of Fig. 1 for NH_3^+ is in good agreement with those of Watanabe and coworkers.[5] The step-like character of the curve is well understood.[5,6] The ionization in this energy region is very predominantly direct ionization, even though the presence of very strong absorption lines in this region indicates formation of "superexcited states."[7] Most of these strong absorption bands just above the ionization potential are members of Rydberg weries converging to vibrationally excited NH_3^+ ions in the electronic ground state. Such super-

excited states can autoionize only by conversion of vibrational to electronic energy. The situation is therefore exactly analogous to that of molecular hydrogen. (This situation should exist in general whenever the 0→0 transition is not highly probable in direct ionization.) However, while these states in hydrogen autoionize with rates comparable to that of the competing predissociation, in ammonia autoionization is not detectable and hence is much less probable than predissociation. The difference in behavior may be attributed to much faster predissociation in ammonia than to hydrogen, or to a much slower autoionization rate in ammonia, or to both.

Duncan,[8] who has obtained high-resolution absorption spectra in this region, remarks that all bands at wavelengths shorter than 1675 Å are sharp. He inferred (incorrectly) that the ionization potential is higher than 87 483 cm⁻¹ because of the existence of a discrete band at this position. The existence of this discrete band shows that the predissociation lifetime for such states of NH_3 in this region of the spectrum cannot be more than about one or two orders of magnitude shorter than for most of the relevant states of hydrogen. It follows then that the autoionization rates for NH_3 cannot be much more than, and may even be much less than, those of H_2 in spite of the fact that the particular vibrational mode predominantly excited in the NH_3 core has associated with it an oscillating electric dipole. The autoionization theory of Berry[9] implies that this circumstance would result in a much increased autoionization rate. On the other hand, the lower vibrational frequency found in the present experiment tends in the opposite direction. Measurement of the ionization-efficiency curve for NH_3 at higher resolution, with the purpose of setting limits to the amounts of autoionization, might provide an informative test of such a theory.

The vibrational mode that is predominantly excited both in direct ionization and in the Rydberg series con-

[5] K. Watanabe and J. R. Mottl, J. Chem. Phys. 26, 1773 (1957); K. Watanabe and S. P. Sood, Sci. Light (Tokyo) 14, 36 (1965).
[6] W. A. Chupka, J. Chem. Phys. 30, 191 (1959), especially p. 203.
[7] R. L. Platzman, Vortex 23, 372 (1962).

[8] A. B. F. Duncan, Phys. Rev. 47, 822 (1935); 49, 211 (1936); 50, 700 (1936).
[9] R. S. Berry, J. Chem. Phys. 45, 1228 (1966).

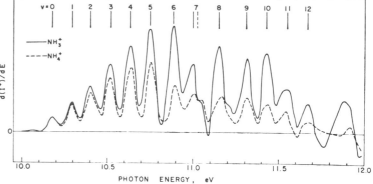

FIG. 2. The first derivative of the curves of Fig. 1, with vibrational quantum number assignment for ν_2. The assignments for $\nu > 6$ are not reliable.

verging to the electronic ground state of the ion is the ν_2 out-of-plane bending mode of the planar ion (or the planar Rydberg states). This is readily understood on the basis of the Franck–Condon principle since the NH_3 molecule in the electronic ground state is pyramidal while the ion (and higher Rydberg states) are planar.[5]

The vibrational structure of the curve of photoionization cross section is more graphically displayed by taking its first derivative. Such a derivative, obtained by computer after a slight smoothing of the data, is shown in Fig. 2 for both NH_3^+ and NH_4^+. Again the curves for NH_3^+ and NH_4^+ are adjusted so that the peaks corresponding to the $0 \rightarrow 0$ transition coincide in height.

It can be seen from Fig. 2 that the $0 \rightarrow 0$ transition occurs at 10.17 eV, in good agreement with the ionization potential of 10.166 eV obtained by Watanabe and Sood.[5] The small peak at about 10.06 eV is a "hot band," i.e., the $0 \leftarrow 1$ transition also seen by Watanabe and Mottle.[5] The ν_2 out-of-plane-bending frequency was found[5] to be about 960 cm⁻¹ (0.119 eV) for the highest Rydberg states. This agrees very well with the spacing

of the most prominent peaks of Fig. 2. Also the $6 \leftarrow 0$ bands are the strongest in the highest Rydberg states.[5]

As can be seen from Fig. 2, on which the ν_2 vibrational quantum numbers of the ion are indicated, the $6 \leftarrow 0$ transition is also the most probable in the direct ionization (as would be expected). However, there is definite evidence for either some excitation of other vibrational modes or some slight traces of autoionization superposed on the direct ionization. (This is somewhat more easily seen in Fig. 1 since some of these effects have been minimized by the smoothing that preceded taking the derivative plotted in Fig. 2.) Above the $6 \leftarrow 0$ peak of Fig. 2, the simple vibrational progression is not as clearly evident.

The variation of the cross section for Reaction (1) as a function of the vibrational energy of the NH_3^+ ion was determined by the following procedure. The data for all repeller voltages were plotted as shown in Fig. 1. Then the ratio of the heights of the vibrational steps for NH_4^+ and for NH_3^+ was measured for each vibrational quantum number ν_2 up to about 9 or 10, where the steps become rather indistinct. An example of the step

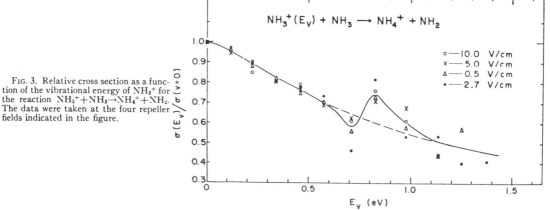

FIG. 3. Relative cross section as a function of the vibrational energy of NH_3^+ for the reaction $NH_3^+ + NH_3 \rightarrow NH_4^+ + NH_2$. The data were taken at the four repeller fields indicated in the figure.

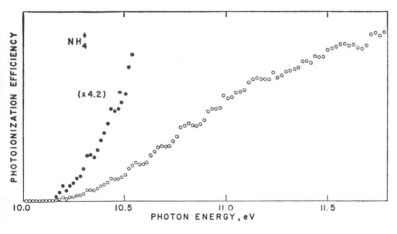

Fig. 4. Photoionization-efficiency curve for NH_4^+ formed by the reaction $NH_3^+ + H_2O \rightarrow NH_4^+ + OH$. The repeller field was ~ 1.0 V/cm.

heights whose ratio was used to determine the relative reaction cross section of NH_3^+ with $v=5$ is shown in Fig. 1. The occurrence of other minor structure on the steps was ignored, but the measurement of the step heights in the NH_3^+ and NH_4^+ curves included it in a consistent manner. The relative reaction cross sections so obtained are plotted in Fig. 3 as a function of the vibrational energy of NH_3^+.

The general behavior of the data shown in Fig. 3 is similar to that found[1] at low repeller voltages for the reaction of H_2^+ with H_2 to form H_3^+. Thus, for this exothermic reaction, the relative cross section decreases as the vibrational energy of the reactant NH_3^+ increases. This behavior is expected if the reaction takes place by the initial formation of a relatively long-lived collision complex which then decomposes along all energetically allowed reaction paths with relative probabilities determined by the same type of statistical considerations as have been applied to unimolecular ions excited by other means. The "phase-space" theory of Light[10] is one such form of treatment which has been applied to some simple ion-molecule reactions with some success. The qualitative description of this reaction according to such a statistical theory is as follows. The collision complex formed by NH_3^+ ($v=0$) has just enough energy to decompose to form the reactants (i.e., this is a thermoneutral reaction) while the reaction to form NH_4^+ is strongly exothermic. Hence, for $v=0$ practically all collision complexes decompose to form NH_4^+. As the internal energy of the collision complex is increased by increasing the vibrational energy of the reactant ion, the probability of the backward reaction increases.

The trend of the reaction cross section as a function of the vibrational energy of NH_3^+ is the same (within experimental error) at all repeller voltages. This trend is in sharp contrast to that for the formation of H_3^+

[10] J. C. Light, J. Chem. Phys. **40**, 3221 (1964); J. C. Light and J. Lin, J. Chem. Phys. **43**, 3209 (1965).

in hydrogen,[1] for which the reaction cross section measured at the higher repeller voltages began to increase with increasing vibrational energy of the reactant ion This increasing cross section was attributed to a proton-transfer mechanism that did not involve formation of an intermediate complex. For the formation of NH_4^+, the absence of such reverse behavior at higher repeller voltages does not exclude the possibility of the occurrence of such a mechanism of proton (or H atom) transfer. One can only conclude either (1) that even at the higher repeller voltages such a process is much less probable than the complex-formation mechanism assumed to operate at the lower repeller voltages or (2) that the probability of proton (H atom) transfer is nearly independent of, or varies inversely with, the vibrational energy of NH_3^+ in the particular vibrational mode predominantly excited. That the latter explanation is reasonable can be justified as follows. The vibrational mode predominantly excited in this case is a bending mode rather than a stretching mode as for H_2^+. To a first approximation, this would not be expected to be effective in changing the probability of a proton

Table I. Comparison of the apparent ionization cross section of NH_3 for (a) nearly complete reaction and (b) very little reaction, in order to demonstrate depletion of low vibrational states for Case (a).

Photon wavelength (Å)	Apparent photoionization cross section (arbitrary units)		$\sigma(95\%)/\sigma(7\%)$
	(a) 94% Reaction	(b) 7% Reaction	
1205	7 380	7 050	1.05
1180	37 100	31 200	1.19
1155	93 500	73 300	1.28
1130	164 500	114 000	1.44
1105	226 000	145 000	1.57
1080	266 000	176 000	1.51
1055	284 000	187 000	1.52

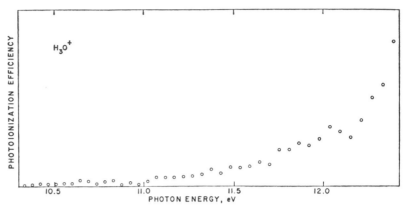

FIG. 5. Photoionization-efficiency curve for H_3O^+ formed by the reaction $NH_3^+ + H_2O \rightarrow H_3O^+ + NH_2$. The sudden rise above ~12.2 eV is due to the production of H_2O^+ and its subsequent reaction with H_2O.

either surmounting or penetrating a potential barrier in a proton–transfer process. If the reaction occurs by H-atom transfer from NH_3 to NH_3^+, again the effect of vibrational excitation of the NH_3^+ is expected to be small and uncertain in direction.

The smooth downward trend of the curve in Fig. 3 is apparently interrupted in the region $v=6$ and 7. Since the relative cross section for $v=6$ seems abnormally low while that of the following $v=7$ seems abnormally high, one might be tempted to attribute this behavior to an inadvertent departure from consistency in the measurement of the step heights. A careful check indicated that this did not seem to be the cause. If this behavior is not some kind of artifact but has physical significance, it may be explained in several ways.

One possibility is that a part of the step-like feature in the neighborhood of $v=7$ on Fig. 1 may be due to excitation of a stretching mode that might have a larger reaction cross section as explained above. As can be seen from Fig. 2, the peak at $v=7$ appears to be a doublet and the positions of the components suggest a Fermi resonance of the $v=7$ overtone with some combination involving one or more other normal modes. However, the two parts of the doublet seem to have very different reaction cross sections. Such a difference would imply that the two vibrational eigenfunctions are quite different (i.e., that little mixing of the zero-approximation eigenfunctions occurs) and hence the resonance cannot be strong.

Another possible explanation is that part of the feature at about $v=7$ is due to autoionization, which might produce NH_3^+ in vibrational states other than $v=7$ of the v_2 mode, e.g., in v_2 vibrations with $v<7$ or in modes other than v_2. Similarly, one cannot exclude the possibility that direct ionization may produce the NH_3^+ ion with vibrational excitation wholly or partly in modes other than v_2. Similar features at other photon energies may be unresolved. Measurement of the photoionization-efficiency curves at sufficiently high resolution and with sufficiently small wavelength increments should resolve this question.

B. Nearly Complete Reaction in NH₃

Reaction (1) is calculated to be exothermic by about 1.4 eV. This figure is based on fairly well-established thermochemical data,[11] the most uncertain being the proton affinity of NH_3 (taken to be 214 kcal/mole = 9.3 eV). A fraction of this energy (a rather small one if the energy distribution is determined statistically) may appear as kinetic energy of the reaction products, with nearly half the kinetic energy going into the NH_4^+ product ion. It is conceivable that putting more excitation energy into the NH_3^+ reactant may possibly result in higher kinetic energy for the NH_4^+ product ion; and this in turn could decrease the collection efficiency of the mass spectrometer and thereby produce the apparent drop in reaction cross section shown in Fig. 3. This situation may be more serious than in the case of H_3^+ formation. In the latter reaction, the exothermicity is approximately the same but only one-fourth of the kinetic energy released appears in the H_3^+ ion, and several experiments indicate that the H_3^+ product ions have little kinetic energy.[12] Apparently no such data exist for NH_4^+ ions produced by Reaction (1).

The variation of collection efficiency with kinetic energy was minimized in these NH_4^+ measurements by opening the defining slits of the mass spectrometer until no further increase in transmission occurred. As an additional test, however, the pressure of NH_3 was raised and the repeller voltage was dropped until Reaction (1) was found to be approximately 94% complete. The ionization efficiency of the remaining NH_3^+ was then measured at several wavelengths and compared with the same measurement carried out on NH_3^+ under conditions of slight reaction in order to observe the preferential depletion of the lower vibrational levels of NH_3^+ in the case of nearly complete reaction as expected on

[11] V. I. Vedeneyev, L. V. Gurvich, V. N. Dondrat'yev, V. A. Medvedev, and Ye. L. Frankevich, *Bond Energies, Ionization Potentials, and Electron Affinities* (Edward Arnold and Co., London, 1966).
[12] B. G. Reuben and L. Friedman, J. Chem. Phys. **37**, 1636 (1962).

the basis of the results shown in Fig. 3. From these results, shown in Table I, it is seen that the expected depletion of low vibrational states occurs for the case of nearly complete reaction.

C. Reactions in Gaseous Mixtures of NH₃ and H₂O

NH_3 and H_2O gases were premixed just before being introduced into the ionization chamber, and the pressures of the two gases were adjusted so that at a repeller voltage of about 1 V/cm, about 15% of the NH_3^+ ions reacted to form NH_4^+ and so that about 75% of the NH_4^+ was produced by the reaction

$$NH_3^+ + H_2O \rightarrow NH_4^+ + OH. \qquad (2)$$

The amount of NH_4^+ produced by Reaction (1) was readily measured by shutting off the H_2O gas flow while keeping the flow of NH_3 constant.

Ionization-efficiency curves of NH_4^+ were taken with both NH_3 and the large amount of H_2O and then with the H_2O flow cut off. The latter curve was subtracted from the former to yield the ionization-efficiency curve of NH_4^+ formed by Reaction (2). This curve is shown in Fig. 4. The data are not quite as precise as those of Fig. 1. A careful comparison of the curve of Fig. 4 with the curve for NH_3^+ of Fig. 1 shows that, within experimental error, vibrational energy apparently has no effect on the reaction cross section. When the two curves are normalized at the second plateau (~ 10.32 eV) whose height is more accurately measured than that of the first plateau, the curve of Fig. 4 is about 10% higher than that of NH_3^+ of Fig. 1 at 11.5 eV. Thus vibrational energy of NH_3^+ has much less effect on the cross section for Reaction (2) than on that for Reaction (1) and appears to act in the opposite direction. Within the greater uncertainty of the measurement for Reaction (2), there may be no effect at all.

D. Proton Affinities of NH₃ and H₂O

Since Reaction (2) is observed to occur with NH_3^+ ions in the vibrational ground state, it must be exothermic. A lower limit to the proton affinity (P.A.) of NH_3 may then be calculated by use of well-established thermochemical data for NH_3 and H_2O.[11] The limit so calculated is P.A. $(NH_3) \geq 8.54$ eV $= 196$ kcal/mole. This is in agreement with the accepted values[11] of 211–214 kcal/mole obtained by calculation.

The value of the proton affinity of H_2O is not well established. A recent tabulation[11] lists values ranging from 7.3 to 8.2 eV for this quantity. More recently Van Raalte and Harrison[13] measured the appearance potential of H_3O^+ from several alcohols and esters and obtain the value P.A. $(H_2O) = 151 \pm 3$ kcal/mole $= 6.55$ eV.

It may be seen from Fig. 1 that NH_3^+ ions with

11 D. Van Raalte and A. G. Harrison, Can. J. Chem. **41**, 3118 (1963).

vibrational energies ranging from 0 to about 1.5 eV can be made by photoionization. This circumstance offers the possibility of determining the proton affinity of H_2O by attempting to observe the reaction

$$NH_3^+ + H_2O \rightarrow NH_2 + H_3O^+ \qquad (3)$$

as a function of the vibrational energy of the NH_3^+ ion. Thermochemical calculations show that Reaction (3) will be energetically possible for NH_3^+ ions in their ground state if P.A. $(H_2O) \geq 7.85$ eV and for NH_3^+ ions with 1.5-eV vibrational energy if P.A. $(H_2O) \geq 6.35$ eV. The tabulated values of P.A. (H_2O) are well above 6.35 eV and hence Reaction (3) should be energetically possible within the wavelength range covered in this experiment.

Reaction (3) was investigated under approximately the same conditions as those used to study Reaction (2). The observed H_3O^+ ion intensities were corrected for the formation of H_3O^+ from H_2O^+ which was produced by scattered light. The latter contribution was readily measured by removing the NH_3 gas. The corrected photoionization-efficiency curve of H_3O^+ is shown in Fig. 5. At all wavelengths the intensity of H_3O^+ was much less than that of NH_4^+. At 11.5 eV the NH_4^+ ion from Reaction (2) was about 100 times as abundant as the H_3O^+ ion from Reaction (3). Because the probability of Reaction (3) is much less than that of Reaction (2) and because of the relatively large scattered-light correction, the data of Fig. 5 show more scatter than those of Figs. 1 and 4 and a clear threshold for Reaction (3) is not apparent. If Reaction (3) occurs at all for ground-state NH_3^+ ions, its cross section is less than 0.01 of that of the competing Reaction (2). The data indicate that Reaction (3) has a threshold in the neighborhood of 11.0 eV. The rather sudden rise at energies above about 12.2 eV is due to the photoionization of H_2O which has a threshold at 12.6 eV; but because of "hot bands" and the photon bandwidth, the rise extends to lower energies. Above the threshold at 12.6 eV, the intensity of H_3O^+ is several orders of magnitude greater than those in Fig. 5.

If the threshold for Reaction (3) is at 11.0 eV, this leads to the value P.A. $(H_2O) = 7.0$ eV $= 161$ kcal/mole. This value of proton affinity should perhaps be considered a lower limit since the threshold may lie at still lower energies but, because of the severe competition of Reaction (2), may be undetectable in the present experiment.

The value P.A. $(H_2O) = 151$ kcal/mole obtained by Van Raalte and Harrison is probably too low, being obtained from appearance potentials which are probably too high. All but one of the processes investigated by these workers involve a sequence of two decompositions producing a total of three fragments. The appearance potential for such processes is almost always considerably higher than the thermochemical threshold. The one process which apparently involves only a single fragmentation step is the production of H_3O^+ from ethanol. This process has been studied by photoionization in this laboratory[14] and its threshold has been tentatively determined to be ≤ 11.5 eV. Since the photoionization-efficiency curve for this process has considerable curvature at threshold, further measurements must be made before the threshold can be accurately determined. However, the upper limit of 11.5 eV corresponds to the limit P.A. $(H_2O) \geq 7.0$ eV, which agrees with the value obtained from Reaction (3) and with the tabulated values.[11]

It should be noted that, as mentioned earlier, Reaction (3) should be energetically possible within the wavelength range covered, it has very low probability and is barely detectable, presumably because of competition by the more exothermic Reaction (2). This serves to emphasize again that the nonobservation of an ion–molecule reaction is not a good criterion of endothermicity, especially in cases in which other reactions are known to occur. It would be much more reliable, though still not rigorous, to use nonobservation of an ion–molecule reaction as a criterion that the reaction not observed is less exothermic than reactions observed to occur for the reactant in question.

E. Discussion

It is of interest to consider why the cross section for Reaction (1) decreases with increasing vibrational energy of the NH_3^+ reactant ion while the cross section for Reaction (2) is nearly insensitive to vibrational energy and may even have a slight opposite dependence. As was already mentioned in Sec. A, the behavior of Reaction (1) is at least qualitatively in accord with formation of an intermediate complex and the "phase-space theory" except for one anomalous feature. How-

ever, Reaction (2) behaves differently. One possible explanation for such a difference might be the difference in the exothermicity of the reactions. If Reaction (2) were more exothermic than Reaction (1) and other factors were equal, the backward reaction of the intermediate complex for Reaction (2) would have a lower rate relative to the forward reaction, especially at low internal energies of the complex. However, the opposite situation is the case: Reaction (2) is less exothermic than Reaction (1) by about 14 kcal/mole, the difference between the bond-dissociation energies $D(HO–H)$ and $D(H_2N–H)$.[11] The other relevant factors of the phase-space theory are the densities of states of products and reactants. Detailed calculations will be required for any definite conclusion. However, the numbers and types of degrees of freedom are so similar for both reactions that it seems unlikely the observed difference in behavior can be explained by these factors.

If Reaction (2) occurs predominantly by hydrogen-atom transfer without formation of a complex, the insensitivity of the cross section to the vibrational energy of NH_3^+ is readily explained since the NH_3^+ ion is likely to be a relatively passive acceptor. Vibrational energy in the stretching modes of the H_2O molecule would presumably have much more effect.

In order to explain the observed effects of vibrational energy on reaction cross section, it will probably be necessary first to perform independent experiments of the type already being done[15] in order to determine whether the reactions proceed by complex formation or by some type of stripping mechanism. No firm conclusions regarding this question can be drawn from the present data. However, it seems very likely that complex formation and the phase-space theory can explain all the observations.

---

[1] W. A. Chupka and M. E. Russell (unpublished data).

[15] Z. Herman, J. D. Kerstetter, T. L. Rose, and R. Wolfgang, J. Chem. Phys. **46**, 2844 (1967).

35

Internal Energy Effects on the Reaction of H_3^+ with Ethylene

Andre S. Fiaux, David L. Smith, and Jean H. Futrell*

*Contribution from the Department of Chemistry, University of Utah,
Salt Lake City, Utah 84112. Received June 19, 1975*

Several ion-molecule reactions of H_3^+ were originally studied using moderately high pressure, single source mass spectrometers.[1] Although such studies provided information on reaction pathways and, in some cases, reaction rate constants, the experimental conditions were such that one learned very little about the actual dynamics of the reactions. Since these experiments were carried out in a single ionization chamber, there was considerable uncertainty over which reactants led to which products. This is a particularly severe limitation in the case where multiple ion-neutral collisions occur. The ion-molecule product spectrum obtained in these experiments did show, however, that H_3^+ ions which are typically formed with considerable vibrational excitation energy are probably deactivated through collisions with H_2. Interpretation of such results is further complicated by the fact that raising the H_2 pressure may also change the product spectrum by changing the probability of collisional stabilization of highly excited product ions, which, in the absence of collisions, would undergo dissociation. Details of the collisional relaxation of H_3^+ are totally obscured in such experiments because of an inherent broad distribution in the number of H_3^+-H_2 collisions an ion undergoes before reaction with an additive molecule.

More recently ion-molecule reactions of H_3^+ have been studied by ion cyclotron resonance (ICR) mass spectrometry.[2] These results are more informative because the techniques of double resonance can often be used to relate reactants and products; however, this is still a single reaction vessel experiment and the possibility of sequential reactions and, to a certain extent, collisional stabilization of highly excited ions remains. Furthermore, since the probability of H_3^+ reacting with the additive substance is approximately constant throughout the reaction time, there is still a very broad distribution in the number of H_3^+-H_2 collisions an ion undergoes before reacting. Thus neither the single source mass spectrometer nor the conventional ion cyclotron resonance mass spectrometer is capable of measuring unambiguously the product spectrum resulting from reaction of H_3^+ as an explicit function of the number of H_3^+-H_2 collisions.

This paper reports results on the reactions of H_3^+ with ethylene obtained using a tandem mass spectrometer. H_3^+ ions are produced in a moderately high pressure source, mass analyzed, and injected into an ICR cell where reaction with ethylene follows. Because the collisional relaxation reactions of H_3^+ with H_2 and the reactions of H_3^+ with C_2H_4 occur in physically separated parts of the instrument, the product distribution characteristic of the reactions of H_3^+ with ethylene may be measured directly as a function of the average number of H_3^+-H_2 collisions occurring prior to reaction with C_2H_4.

It will be shown that a variety of different kinds of information may be obtained by detailed analysis of the results for the reaction of H_3^+ with C_2H_4. One question to be answered focuses on how the excess energy of an exothermic reaction is divided among the vibrational and translational degrees of freedom of the ion and neutral products. This information permits evaluation of the rate of deactivation of excited H_3^+ ions as they undergo collisions with H_2, and, in a broader sense, provides a basis for formulation of models which are useful in specifying parameters necessary for ion-molecule equilibrium studies. Results obtained using deuterium isotopes are used to elucidate the reaction mechanism and provide information on isotope effects in the unimolecular decomposition reactions of superexcited ions.

Experimental Section

The tandem mass spectrometer used in this work (illustrated in Figure 1) has been described in detail elsewhere[3] and will be discussed only briefly here. The first stage of the instrument is a 180°, 5.7 cm radius mass spectrometer equipped with a source or primary ionization chamber capable of efficient operation at pressures up to approximately 100 μm. The magnetically collimated electron beam (electron energy of 50 eV) produces a sheath of ions within the source approximately midway between the split repellers and the ion exit slit. By choosing specific repeller fields and source pressures, the average number of ion-neutral collisions in the source may be changed from 0 to a maximum of approximately 30. Ions extracted from the source are mass analyzed, decelerated to less than 0.1 eV translational energy, and subsequently injected into the cavity of an ion cyclotron resonance mass spectrometer.

The ICR cell used in this apparatus contains two sections of equal length and assumes the dual role of a second reaction chamber with a mass analyzer. Typical drift fields and trapping plate voltages used in this work were 0.5 V/cm and +0.5 V, respectively. Relative ion power absorptions are determined using a conventional oscillator-detector which has been calibrated at different frequencies using ion beams whose absolute intensities are measured using the "total ion current" collector. The relative abundance of different ions is obtained by multiplying the relative ion power absorption by the ion mass.

Results

Because the present experiments are performed using two reaction chambers which, to a good approximation, are completely isolated, the reactions may be divided into two groups.

A. S. Fiaux, D. L. Smith, and J. H. Futrell

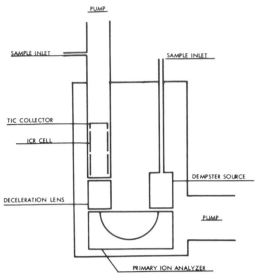

Figure 1. Block diagram of tandem mass spectrometer used in these studies.

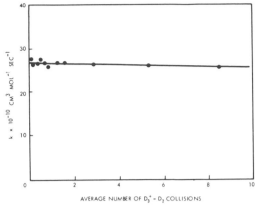

Figure 2. The total rate of reaction of D_3^+ with C_2H_4 as a function of the average number of $D_3^+-D_2$ collisions.

In the source of the first stage of the instrument, H_2^+ ions formed by electron impact may react as follows:

$$H_2^{+*} + H_2 \rightarrow H_3^{+**} + H \qquad \Delta H = -1.7 \text{ eV} \quad (1)$$

$$H_3^{+**} + H_2 \rightarrow H_3^{+*} + H_2^* \qquad (2)$$

The H_2^+ ions are formed with a Franck-Condon distribution of vibrational states and will usually be converted into H_3^+ ions upon their first collision with a neutral hydrogen molecule.[4] Although the H_3^+ excitation energy is not known exactly, simple statistical calculations suggest that there should be a broad distribution with an average around 1.5 eV.[5] Since only a small fraction of the H_2^+ ions are produced in the ground vibrational state,[6] the exoergicity[7] given for (1) constitutes only an upper limit. If the source pressure is sufficiently high, energy relaxing collisions illustrated by (2) occur, giving H_3^+ ions that are at least partially deactivated.

H_3^+ ions extracted from the source are mass analyzed, decelerated, and injected into the second chamber (ICR cell) where the following reactions occur:

$$H_3^+ + C_2H_4 \rightarrow C_2H_4^+ + H_2 + H \quad \Delta H = +1.1 \text{ eV} \quad (3)$$

$$H_3^+ + C_2H_4 \rightarrow C_2H_5^+ + H_2 \quad \Delta H = -2.7 \text{ eV} \quad (4)$$

$$H_3^+ + C_2H_4 \rightarrow C_2H_3^+ + 2H_2 \quad \Delta H = -0.5 \text{ eV} \quad (5)$$

The energetics given here assume that all reactants and products are in their ground state. In the case of (3), the heat of reaction is only a lower limit since, as will be discussed later, the products may always be formed with considerable translational energy.

The rate constant for reaction of D_3^+ with C_2H_4 is given in Figure 2 as a function of the average number of $D_3^+-D_2$ collisions which is varied from 0 to 10. This rate constant was measured by monitoring the decrease in the D_3^+ resonance signal as the C_2H_4 pressure is raised in the ICR cell. From application of pseudo-first-order kinetics we can write

$$\ln [D_3^+]^0/[D_3^+] = k\tau[C_2H_4] \qquad (I)$$

where $[D_3^+]^0$ and $[D_3^+]$ are the relative concentrations of D_3^+ in the ICR cell when the C_2H_4 pressure is zero and P, respec-

tively, τ is the time ions spend in the ICR cell (the reaction time), and $[C_2H_4]$ is the C_2H_4 concentration in the cell for pressure P. The relative D_3^+ concentrations are obtained from the corresponding ICR signal, τ is measured by a pulsing technique,[8] and $[C_2H_4]$ is calculated from the C_2H_4 pressure which is measured using a calibrated ionization gauge. This same method has been used to measure the rate of reaction of CH_4^+ with CH_4 and gives results which compare well with previous measurements.[3]

Product distributions resulting from reaction of H_3^+ with C_2D_4 as well as from reaction of H_3^+ with C_2H_4 are given as a function of the average number of $H_3^+-H_2$ collisions occurring prior to reaction in Figures 3 and 4, respectively. The reaction chamber pressure and reaction time were chosen such that less than 20% of the H_3^+ ions actually underwent reaction with ethylene. Since the rate of this reaction is approximately collision limited, it follows that subsequent reaction of ion products may be neglected; that is, the reaction of H_3^+ with ethylene occurs under single-collision conditions. The product distributions given in Figures 3 and 4 may be interpreted as relative rate constants characteristic of the corresponding reaction channels.

The average number of $H_3^+-H_2$ collisions may be derived from the average number of ion-neutral collisions in the source, $\bar{n}$, which is given by $k\tau[H_2]$ where k is the ion-neutral collision rate constant, τ is the source residence time, and $[H_2]$ is the neutral density in the source. In the present work, the ion-neutral collision rate constant is replaced by the $H_2^+(H_2,H)H_3^+$ reaction rate constant. This simplification introduces only a small error since this reaction is known to occur on nearly every collision.[4] Due to the reduced mass effect, the $H_2^+-H_2$ collision rate constant is slightly larger than the $H_3^+-H_2$ collision rate constant and may be corrected for by multiplying $k_1\tau[H_2]$ by 0.9. The quantity $k_1\tau[H_2]$ is determined directly by measuring the H_2^+ and the H_3^+ ion current coming from the source and applying eq II.

$$\bar{n} = 0.9k_1\tau[H_2] = 0.9 \ln ([H_2^+] + [H_3^+])/[H_2^+]) \quad (II)$$

The H_2^+ and H_3^+ ion currents are measured by increasing the magnetic field until the beams strike a special collector located at a radius of curvature slightly less than that required for the ions to pass into the deceleration lens. This procedure eliminates possible mass and energy discrimination in the deceleration lens. Under high pressure conditions the average number of ion-neutral collisions is assumed proportional to the pressure in the main vacuum chamber; this pressure is measured using an ionization gauge.

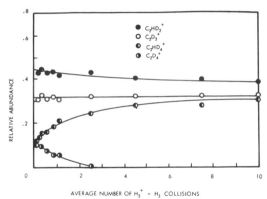

Figure 3. Distribution of products for the reaction of H_3^+ with C_2D_4 as a function of the average number of H_3^+–H_2 collisions in the source.

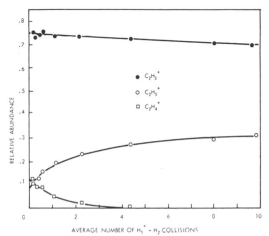

Figure 4. Distribution of products for the reaction of H_3^+ with C_2H_4 as a function of the average number of H_3^+–H_2 collisions in the source.

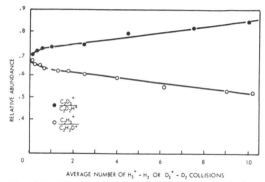

Figure 5. The $C_2H_3^+/C_2H_2D^+$ and $C_2D_3^+/C_2D_2H^+$ ratios resulting from the reaction of D_3^+ with C_2H_4 and H_3^+ with C_2D_4 as a function of the corresponding average number of D_3^+–D_2 or H_3^+–H_2 collisions in the source.

Table I. Summary of Rate Constants for the Rate of Reaction of D_3^+ (H_3^+) with Ethylene

Method	$k \times 10^9$ cm³ mol⁻¹ s⁻¹
Flowing afterglow[10] (H_3^+)	2.02
High pressure MS[11] (H_3)	0.99
Trapped ion ICR[12] (H_3^+)	3.6
Present (D_3^+)	2.8 ± 0.6
Langevin[13] (D_3^+)	2.4
(H_3^+)	2.9

Except for the limiting case of no collisions, there will always be a substantial fraction of the ions which have experienced either more or fewer collisions than the average given by (II). We have considered this problem in detail in another publication[9] and have shown that the probability, P, of an ion undergoing exactly n collisions when the *average number* of ion–neutral collisions is $\bar{n}$ is given by

$$P(n,\bar{n}) = \frac{\bar{n}^n}{n!}e^{-\bar{n}} \qquad (III)$$

which is the Poisson Distribution Law. Thus any interpretation of changes in the product distributions given in Figures 3 and 4 with variation of the number of ion–neutral collisions occurring in the source must take into account this distribution. While (III) shows that there is a relatively broad distribution in the number of ion–neutral collisions occurring prior to reaction in the present apparatus, experiments performed with single chamber instruments[1,2] have a much broader distribution than that given by (III) because these experiments also involve a distribution of reaction time.

The present results are given as a function of the average number of H_3^+–H_2 collisions which can be obtained from the average number of ion–neutral collisions. Assuming that reaction of H_2^+ with H_2 occurs on the first collision, and using

(III), it can be shown[9] that the average number of H_3^+–H_2 collisions, $\bar{n}'$, is given by (IV)

$$\bar{n}' = 1 - P(0) \sum_{n=1}^{\infty} (n-1)P(n) \qquad (IV)$$

where $P(n)$ is given by (II) and $P(0)$ is the probability of having no collisions. Thus each data point given in Figures 3 and 4 represents measurements of both the product distribution as well as the H_2^+–H_3^+ ion currents coming from the ion source. These latter quantities are used with (II) to give the average number of ion–neutral collisions, $\bar{n}$. The average number of H_3^+–H_2 collisions is then calculated using eq III and IV.

Results are given in Figure 5 for the $C_2H_3^+/C_2H_2D^+$ and $C_2D_3^+/C_2D_2H^+$ ratios for the reactions of D_3^+ with C_2H_4 and for H_3^+ with C_2D_4 as a function of the average number of D_3^+–D_2 or H_3^+–H_2 collisions occurring in the source. A small correction for contribution of the $C_2H_4^+$ ion to the m/e 28 ion in the case of reaction of D_3^+ with C_2H_4 has been made by using the relative abundance of the $C_2D_4^+$ observed for reaction of H_3^+ with C_2D_4.

Discussion

Rate of Reaction. Rate constants for the total rate of reaction of $H_3^+(D_3^+)$ with ethylene obtained previously[10,11,12] and in the present study are given in Table I. While the agreement is not particularly good, there is a general consensus that the reaction proceeds at a rate approaching the ion–neutral collision rate. If the reduced mass correction factor is taken into account, the present value is in very good agreement with the trapped ion ICR result. Because of inherent uncertainties in estimation of pressure and reaction time, the rather low value obtained using a high pressure mass spectrometry should be discounted in comparison with the other, more precise measurements.

There appears to be a serious discrepancy between the trapped ion ICR and the flowing afterglow measurements. Both instruments have previously been used with considerable success to measure absolute rates of ion–molecule reactions. While both methods are generally capable of precise measurement, the flowing afterglow apparatus has the advantage of having completely equilibrated reactants. Thus one might suspect that the discrepancy between these two measurements results from the presence of internally excited H_3^+ ions which react faster in the case of the trapped ion ICR measurements. However, the present results given in Figure 2 show that this rate remains constant as the average number of D_3^+–D_2 collisions occurring prior to reaction with ethylene is varied from 0 to 10. It will be shown that these collisions lower the average H_3^+ excitation energy by nearly 3 eV. From this we conclude that the discrepancy in the rate constants obtained using the trapped ion ICR method and the flowing afterglow technique is not due to presence of internally excited H_3^+ ions. The present results support the conclusion that the rate constant is very nearly that calculated from the Langevin collision limit.[13]

Reaction Mechanism. Product distributions for reaction of D_3^+ with C_2H_4 and H_3^+ with C_2D_4 were measured to gain information on the reaction mechanism. Results are given in Figure 3 as a function of the average number of H_3^+–H_2 (D_3^+–D_2) collisions occurring prior to reaction with C_2D_4. Since product ions containing more than one H, such as $C_2D_2H^+$, were not observed (that is, they accounted for less than 1% of all products) we conclude that reaction does not proceed through formation of a $C_2D_4H_3^+$ intermediate which then randomly eliminates a hydrogen molecule.

Information on the likelihood of random elimination of an X_2 molecule (where X may be either H or D) by a $C_2X_7^+$ ion may be obtained from work by Abramson and Futrell[14] who carried out a detailed study of (6)

$$CX_3^+ + CX_4 \rightarrow [C_2X_7^+] \rightarrow C_2X_5^+ + X_2 \quad (6)$$
$$\Delta H = -1.0 \text{ eV}$$

using several isotopically labeled reactants. In addition, Huntress and Bowers[2] as well as Aquilanti, Galli, and Volpi[15] have studied the reaction

$$X_3^+ + C_2X_6 \rightarrow [C_2X_7^+] \rightarrow C_2X_5^+ + X_2 \quad (7)$$
$$\Delta H = -1.1 \text{ eV}$$

While these studies do not provide a consistent view of the hydrogen molecule elimination process, they do clearly show that all combinations of X_2 occur when $C_2X_7^+$ is involved as an intermediate. The contrast between these previous studies and the present results permits us to conclude that H_3^+ does not react with C_2H_4 through formation of a $C_2X_7^+$ intermediate.

The results in Figure 3 show that $C_2D_3^+$ and $C_2D_2H^+$ are the only $C_2X_3^+$ products observed for the reaction of H_3^+ with C_2D_4. Similarly, $C_2H_3^+$ and $C_2H_2D^+$ are the only dissociation products observed for reaction of D_3^+ with C_2H_4. The $C_2D_3^+/C_2D_2H^+$ and $C_2H_3^+/C_2H_2D^+$ ratios for these two reactions are given in Figure 5 and provide further information on the reaction mechanism. It should be emphasized that changing the number of collisions prior to reaction changes only the X_3^+ excitation energy and that all conditions in the reaction chamber analyzer remain constant.

These results show that the two ratios have a value of approximately 0.69 in the case of reaction of highly excited X_3^+ ions. If a $C_2D_4H^+$($C_2H_4D^+$) intermediate were to undergo completely random elimination of HD and D_2 (HD and H_2), a $C_2D_3^+/C_2D_2H^+$ or ($C_2H_3^+/C_2H_2D^+$) ratio of 0.66 would be observed. Since this is relatively close to the observed value, we tentatively conclude that X_3^+ reacts with C_2X_4 to give

$C_2X_5^+$ which then randomly eliminates an X_2 molecule.

This interpretation is valid only if isotopic randomization within the $C_2X_5^+$ ion is fast compared to the elimination reaction. Voracheck, Meisels, Geanangel, and Emmel[16] have shown that the maximum activation energy for hydrogen atom migration in the $C_2H_5^+$ ion is 0.08 eV. Their results support the ab initio calculations of Hariharan, Lathan, and Pople[17] which predict the bridged form of the ethyl ion to be only 0.04 eV more stable than the classical form. Since the activation energy for hydrogen atom migration is over an order of magnitude smaller than the activation energy for hydrogen elimination, randomization prior to elimination is highly probable.

If reaction were to occur rapidly without randomization of energy among the $C_2X_5^+$ vibrational degrees of freedom, the $C_2D_3^+$ or $C_2H_3^+$ products are expected to dominate over the $C_2D_2H^+$ and $C_2H_2D^+$ products, respectively. An extreme case of this picture is X^- transfer to give the $C_2D_3^+$ and $C_2H_3^+$ products exclusively. In such a case, the $C_2D_3^+/C_2D_2H^+$ and $C_2H_3^+/C_2H_2D^+$ ratios would both be larger than the 0.66 statistical value. In addition, both ratios would be expected to decrease with decreasing X_3^+ energy. Since one ratio increases while the other decreases as the X_3^+ energy is decreased we conclude that any model which consistently predicts a preferential loss of the X that was originally part of the X_3^+ can be dismissed.

The behavior of the $C_2D_3^+/C_2D_2H^+$ and $C_2H_3^+/C_2H_2D^+$ ratios in Figure 5 may, however, be rationalized in part by expected isotope effects. Table II is a representative summary of observed isotope effects on the elimination of X_2 by CX_4^+, $C_2X_4^+$, and $C_2X_5^+$. These results have been chosen to illustrate the present state of our understanding of isotope effects for such reactions of "similar" ions. Very good data exist for these systems and illustrate the range of isotope effects observed as the method of ion preparation (structure, energetics) and observation time are varied.

The isotope effects given in Table II represent the experimentally measured ratio of probabilities for elimination of H_2 and HD or D_2 divided by the ratio for completely statistical elimination by the parent ion. Thus isotope effects greater than 1.0 indicate a preference for eliminating H_2 over HD or for HD over D_2. A ratio less than 1.0 corresponds to the opposite preference and an isotope effect of 1.0 indicates statistical elimination. The results for elimination by $CH_2D_2^+$ were obtained using a double focusing spectrometer to positively identify the ions and clearly show a strong preference for elimination of the lighter particle.[18] Similar results were obtained for other substituted methanes and also for the hydrogen atom elimination reaction. A recent study[19] of the $CHDCHD^+$ ion shows the same trend. In addition, this study shows that lowering the electron energy in the immediate vicinity of the threshold enhances the isotope effect. A similar observation was recently made in the collision induced dissociation spectra of substituted methanes.[20]

In the case of metastable ions[21-23] rather dramatic isotope effects may be observed, as shown in Table II. In this study the tandem ICR mass spectrometer has been used in a novel way to study the metastable ion decompositions. Using a deuterium/ethylene mixture in the source, $C_2H_4D^+$ ions produced by (4) are extracted, mass analyzed, and injected into the ICR cell where subsequent unimolecular decomposition products may be observed. The time interval between ion formation and injection into the ICR cell is estimated to be about 3×10^{-6} s, the time that the ion remains in the ICR cell is approximately 3×10^{-3} s. Details of this study will be published elsewhere.[23]

Because energy distributions of reacting species are not precisely defined in the present experiments, isotope effects cannot be calculated reliably from current theories of isotope effects. Qualitatively the data can be rationalized in terms of different frequencies in the reaction coordinate for H and D bond cleavage.[18,19] Thus the increased probability of elimination of H over D is caused by a larger zero-point energy in the case of H and because of a higher vibrational frequency of the lighter particles. The larger zero-point energy effectively lowers the activation energy barrier. Since the rate of reaction is strongly dependent on the excitation energy when it is in the vicinity of the threshold, it follows that isotope effects will become much larger in the case of parent ions with very small excitation energies. Thus large isotope effects are observed in the case of reactions occurring at their thresholds as illustrated by studies of metastable ions.

Table II shows that for the case of reaction of highly excited X_3^+, the resulting $C_2X_5^+$ eliminates X_2 with essentially no isotope effect. As will be discussed later, X_3^+ is initially formed with an average excitation energy of nearly 2 eV. Since (5) is exothermic by 0.5 eV for ground state X_3^+ ions, it follows that the $C_2X_5^+$ ions probably have around 2 eV of energy in excess of the dissociation energy. Under these conditions, a minimal isotope effect is expected. Increasing the number of X_3^+-X_2 collisions decreases the excitation energy in the X_3^+ ion as well as in the $C_2X_5^+$ ion. Under high collision conditions the X_3^+ ion is completely deactivated and the $C_2X_5^+$ ion can have no more than 0.5 eV excitation energy. Under these conditions an isotope effect of 1.3 is observed. Thus the present results are qualitatively consistent with our simplistic understanding of isotope effects in unimolecular reactions.

The present results show only modest isotope effects while the results of Lohle and Ottinger[22] as well as the metastable studies using the tandem ICR[23] show that hydrogen elimination by the $C_2X_5^+$ ion is characterized by large isotope effects when the reaction occurs in the threshold energy range. The rather small isotope effect observed in the present study suggests that dissociation of the $C_2X_5^+$ ions primarily occurs at energies substantially above the threshold energy of the reaction. This is discussed further later.

H_3^+ Excitation Energy. Information on the amount of excitation energy with which H_3^+ is initially formed as well as the rate at which this ion is relaxed through collisions with H_2 may be deduced from product distributions observed for reaction of H_3^+ as the number of H_3^+/H_2 collisions in the ion source of the tandem ICR is varied. Limits on the initial H_3^+ excitation energy may be obtained from the observation of endothermic reaction channels such as (3), as shown in Figure 4. Since $C_2H_4^+$ accounts for 14% of the product distribution in the zero collision limit, we may conclude that at least 14% of the H_3^+ are formed with a minimum of 1.1 eV of excitation energy.

The value of 1.1 eV for the endothermicity of (3) is obtained by assuming that all products are formed with negligible internal and translational energy. This model would be appropriate if the "charge-transfer" reaction occurred through protonation followed by adiabatic, random elimination of an H or D atom. However, since $C_2H_4^+$ and $C_2D_4^+$ are by far the major isotopic products observed for reaction of D_3^+ or H_3^+ with C_2H_4 and C_2D_4, respectively, we conclude that charge transfer does not proceed through this mechanism. Rather, the isotopic results suggest that charge transfer occurs as the ion and neutral are being drawn together by the attractive ion-neutral interaction potential but before an intimate collision actually occurs. This same mechanism has been proposed and discussed in more detail elsewhere for the reaction of H_3^+ with ammonia[24] and methanol.[9]

If, as discussed above, the neutral product of charge transfer is H_3 rather than H_2 and HX, the endothermicity of the charge-transfer channel is likely substantially greater than 1.1 eV. Porter, Stephens, and Karplus[25] and Conroy[26] have shown that the potential energy surface for H_3 is repulsive and that

Table II. Representative Summary of Isotope Effects on the Elimination of H_2, HD, and D_2 by CX_4^+, $C_2X_4^+$, and $C_2X_5^+$

Parent ion	Neutral products	Isotope effect	Method
$CH_2D_2^+$	H_2/HD	5.1	Electron impact (70 eV)[18]
$CH_2D_2^+$	HD/D_2	2.3	Electron impact (70 eV)
$CHDCHD^+$	H_2/HD	1.8	Electron impact (20 eV)[19]
$CHDCHD^+$	HD/D_2	1.4	Electron impact (20 eV)
$CHDCHD^+$	HD/D_2	1.8	Electron impact (13.5 eV)
$CHDCHD^+$	H_2/HD	33	Metastable ions, DFMS[21]
$CHDCHD^+$	HD/D_2	23	Metastable ions, DFMS
$C_2H_2D_3^+$	H_2/HD	5.4	Metastable ions, DFMS[22]
$C_2H_2D_3^+$	HD/D_2	31	Metastable ions, DFMS
$C_2H_4D^+$	H_2/HD	6.3	Metastable ions, tandem ICR[23]
$C_2H_4D^+$	H_2/HD	1.1	Present (0 D_3^+-D_2 collisions)
$C_2H_4D^+$	H_2/HD	1.3	Present (10 D_3^+-D_2 collisions)
$C_2D_4H^+$	HD/D_2	1.0	Present (0 H_3^+-H_2 collisions)
$C_2D_4H^+$	HD/D_2	1.3	Present (10 H_3^+-H_2 collisions)

Table III. Calculated Distribution of Vibrational Excitation Energy in H_3^+

H_3^+ vibrational quantum number	Probability	Vibrational excitation[a] energy, eV
0	0.031	0
1	0.085	0.4
2	0.147	0.7
3	0.204	1.1
4	0.203	1.5
5	0.147	1.9
6	0.082	2.2
7	0.039	2.6
8	0.013	3.0
9	0.001	3.4

[a] Each H_3^+ quantum of vibrational energy is worth 0.37 eV.

this species should immediately dissociate into H_2 and H. Depending upon the geometry (triangular or linear) and internuclear separation of the H_3, the H_2 and H products are expected to have between 0.5 and 1.0 eV of translation energy. Adding this to the intrinsic endothermicity of the reaction implies that production of $C_2H_2^+$ may be as much as 2 eV endothermic.

The total amount of excitation energy which could appear in principle as excitation energy in H_3^+ is given by

$$E_{(excess)} = \epsilon^P_{int} + \epsilon^P_{trans} = -\Delta H + \epsilon^R_{int} + \epsilon^R_{trans} \quad (V)$$

where ϵ^P_{int} and ϵ^P_{trans} are the internal excitation energy and center-of-mass translational energy of the products; ΔH is the heat of reaction calculated from the standard heats of formation of ground state reactants and products; and ϵ^R_{int} and ϵ^R_{trans} are the internal excitation energy and center-of-mass translational energy of the reactants.

The participation of $E_{(excess)}$ among the internal and translational degrees of freedom has been estimated[5] by assuming that the energy is deposited randomly among all possible vibrational and translational states of the products. Since details of the calculation are published, only the results will be given here for comparison with the present experimental results. The probabilities of H_3^+ being formed in various vibrational levels and having the corresponding vibrational excitation energies are given in Table III. If this simplistic model is even approximately correct, H_3^+ ions are expected to have excitation energies from 0 eV up to over 3 eV. The calculated average H_3^+ excitation energy is approximately 1.5 eV and compares well with an experimental measurement of 2.0 eV made by Durup and Durup.[27]

The results in Table III show that, if the excess energy in the reaction of H_2^+ with H_2 is distributed randomly among the vibrational and translational states of the products, 73% of the H_3^+ are expected to have 1.1 eV or more excitation energy. If, on the other hand, the charge-transfer reaction channel requires 2.0 eV excitation energy, Table III suggests that only 18% of the initially formed H_3^+ ions could undergo this reaction. Until the kinematics of this reaction are better understood, a more detailed analysis of the present data is not possible. However, the experimental results appear consistent with the notion that H_3^+ ions are formed with a broad distribution of excitation energy. This conclusion is supported by similar results obtained for reaction of D_3^+ with methane,[5] ammonia,[24] and methanol.[9]

As the number of X_3^+–X_2 collisions is increased from zero, the endothermic charge-transfer reaction channel closes and the relative abundancies of the other products, $C_2X_3^+$ and $C_2X_5^+$, remain constant, as shown in Figures 3 and 4. Further, the product distributions remain constant as the number of X_3^+–X_2 collisions is increased from 10 to over 20. As will be discussed later, the relative abundance of $C_2X_3^+$ is expected to depend only on the amount of excitation energy in the initially formed $C_2X_5^+$ ion. Consequently we shall assume that, when the $C_2X_5^+$ excitation energy remains constant, the X_3^+ excitation energy is also constant. Thus the present results, as well as results obtained for reaction of H_3^+ with methane,[5] ammonia,[24] and methanol,[9] indicate that H_3^+ ions are probably completely deactivated after 5 to 10 collisions. This conclusion, coupled with a recent measurement of the D_3^+–H_2 rate of reaction,[28] shows that 2 to 3 reactive collisions are necessary to deactivate X_3^+ completely.

$C_2H_5^+$ Lifetime. The duration of time that a $C_2H_5^+$ ion, which is made through reaction of H_3^+ with ethylene, exists before undergoing dissociation to $C_2H_3^+$ and H_2 is fundamental to our understanding of the partitioning of excess energy among the various degrees of freedom of the $C_2H_5^+$ and H_2. The relative abundancies of $C_2H_5^+$ and $C_2H_3^+$ observed in the flowing afterglow instrument,[10] in a high pressure mass spectrometer,[11] in a conventional ICR apparatus[12] and in the present instrument are given in Table IV and provide definite limits on the $C_2H_5^+$ lifetime. All values have been chosen such that they are probably representative of reaction of very low energy H_3^+ ions.

The reaction of low energy H_3^+ ions with ethylene occurs according to the following

$$H_3^+ + C_2H_4 \longrightarrow \begin{cases} C_2H_5^+ + H_2^* & (8) \\ C_2H_5^{+*} + H_2 + 2.5\,eV & (9) \\ \xrightarrow{M} C_2H_5^+ + M^* & (10) \\ C_2H_3^+ + H_2 & (11) \end{cases}$$

which can be used to rationalize the different product distributions given in Table IV. In the flowing afterglow apparatus, at a pressure of 0.5 Torr, the $C_2H_5^+$ accounts for 0.94 of the products. At the same pressure (same ion–neutral collision frequency) the fraction of products represented by $C_2H_5^+$ in the high pressure mass spectrometer is 0.7. Moreover, the qualitative effect of increasing the pressure increases the probability of observing the $C_2H_5^+$ product. Observation of much more $C_2H_5^+$ in the flowing afterglow apparatus results from the much longer reaction times (time between ion creation and observation of products) associated with this experimental method. This comparison suggests that reaction 12 also contributes to the product distribution.

Since hydrogen was the bath gas in both experiments and since the total number of collisions was not varied at fixed pressures,

$$C_2H_3^+ + H_2 \rightleftharpoons C_2H_5^{+*} \qquad (12)$$

$$\xrightarrow{M} C_2H_5^+ + M^* + 2.2\,eV \qquad (13)$$

contributions from (12) and (9) followed by (10) cannot be evaluated explicitly. In the high pressure mass spectrometry studies, source residence times were not measured and there is only the qualitative deduction that some of the ions must have lifetimes in the range of collision frequencies sampled by these experiments (4×10^{-7}–9×10^{-6} s).

It is interesting to note that (9) or (12) have very nearly the same excitation energy (2.5 or 2.2 eV, respectively), and thus should have approximately the same lifetimes. Consequently, independent of whether (9) or (12) dominates the product distribution, these experiments involve collisional stabilization of $C_2H_5^+$ ions, some of which must have lifetimes that are comparable to the ion–neutral collision time. Assuming that Langevin collision frequencies apply, competition between (9) and (10) and between (12) and (13) sets the limit that some $C_2H_5^+$ ions have lifetimes of this magnitude.

As discussed above, the tandem ICR instrument has been used to study the metastable decomposition of $C_2H_4D^+$ ions produced through the reaction of D_3^+, with ethylene; this experiment observes directly the unimolecular decomposition in the time interval from 3×10^{-6} s after formation to 3×10^{-3} s. Since the daughter ion ($C_2H_3^+$, $C_2H_2D^+$) relative abundance was 3 to 4% of the parent ion ($C_2H_4D^+$) relative abundance, it follows that, for those $C_2X_5^+$ ions which live as long as 3×10^{-6} s, the majority (96 to 97%) actually have lifetimes greater than 10^{-3} s. This result is consistent with the work of Kim, Theard, and Huntress[12] who used a pressure of the order of 10^{-5} Torr, intermediate between the high pressure flowing-afterglow results discussed above and the present low pressure results. Since they observed slightly more $C_2H_5^+$ than is observed in the present study, and since the ion–neutral collision time in their experiments is of the order of 10^{-4} s, it follows that some $C_2H_5^{+*}$ lifetimes fall in the range of 10^{-4} s. Increasing further the ion–neutral collision time to 10^{-2} using the present instrument causes only a small decrease in the relative abundance of the $C_2H_5^+$ ion.

From these several experiments, we may conclude that $C_2H_5^+$ ions produced by the reaction of collisionally relaxed H_3^+ with C_2H_4 exhibit a range of lifetimes from $<10^{-7}$ to $>10^{-3}$ s. This is readily rationalized in terms of a distribution of energy deposited in $C_2H_5^{+*}$ and a strong energy dependence of the ethyl ion decomposition reaction 11. The present experimental results show that approximately 30% have lifetimes greater than 10^{-3} s and 3–4% decay in the time interval from 3 μs to 3 ms and require that most of the $C_2H_5^{+*}$ initially formed ($>60\%$) have lifetimes shorter than microseconds.

This conclusion is consistent with statistical theories of unimolecular dissociation such as the quasi equilibrium theory of mass spectra. In fact, for $C_2H_5^+$, one calculation[29] suggests that the rate of dissociation will change by six orders of magnitude as the internal energy is increased by 0.1 eV. Since both the internal energy and the translational energy of the reactants (H_3^+ and C_2H_4) are well defined, the manner in which the heat of reaction is partitioned among the degrees of freedom of the products depends on the kinematics of the reaction.

Energy Partitioning. In the present experiment, the reaction time is about 10^{-3} s and the average time between collisions is of the order of 10^{-2} s. If we assume that those $C_2H_5^+$ ions which live longer than 10^{-3} s do not have enough energy to dissociate, we may obtain information about the partitioning of the exothermicity in the reaction of H_3^+ with ethylene. Figure 4 shows that reaction of the lowest energy H_3^+ ions gives 0.31 $C_2H_5^+$. Since formation of $C_2H_3^+$ is 0.5 eV exothermic, we conclude that 0.31 of the collisions result in the hydrogen molecule carrying away at least 0.5 eV of energy either as internal or translational energy.

The energy limit of 0.5 eV assumes that the activation energy barrier is given by the standard heats of formation of the reactants and products. Since the reverse of such reactions may have energy barriers of several tenths of an electron volt, the energy of the hydrogen molecule could be much less than 0.5 eV. Elwood and Cooks[30] have measured the kinetic energy release for this reaction and found that the barrier to the reverse reaction is 0.019 ± 0.002 eV. Consequently we conclude that 0.31 of the hydrogen molecules produced in the reaction of H_3^+ with ethylene do carry away at least 0.5 eV of excitation energy.

While the idea that reaction exothermicity is divided statistically among the products has often been invoked, it cannot explain the present results. The ion, $C_2H_5^+$, has a large number of degrees of freedom and, furthermore, the quanta within these degrees of freedom are very small when compared to the vibrational quantum for the hydrogen molecule and the density of ion states is correspondingly much higher. If the reaction exothermicity were randomly divided among the possible states of the products, essentially all would appear in the $C_2H_5^+$ ion.

It has been shown that the hydrogen molecule resulting from the initial protonation reaction typically contains large amounts of energy. To show whether this energy is relative translational energy of the products or is vibrational excitation energy in H_2, the reaction of KrH^+ with ethylene was studied. This is a unique substitution since Kr and H_2 have identical proton affinities; thus the reaction exothermicities remain unchanged. While these results will be discussed elsewhere,[31] it is noted that reaction of stabilized KrH^+ gives essentially the same product distribution as is observed for reaction of stabilized H_3^+ ions. It follows that 30% of the krypton atoms must also carry away a minimum of 0.5 eV. If both KrH^+ and H_3^+ react with ethylene through the same mechanism, we may surmise that the hydrogen molecule carries away the excess energy as translation and not as vibration.

Herman, Kerstetter, Rose, and Wolfgang[32] have studied the conversion of internal energy into translational energy in ion–neutral beam studies and explained their results at low relative energy by invoking a "modified stripping" model. This model, later generalized and now known as the "polarization reflection"[33] or "impulsive reaction"[34] model, considers the effects of the long range polarization forces in both the approach and retreat trajectories on the kinetic energy of the products. It is appropriate to consider this model in conjunction with the present study performed at very low kinetic energy such that long-range attractive forces dominate. Moreover, in the case of H_3^+, isotopic studies show that the reaction intermediate is best characterized as a $C_2H_5^+ \cdot H_2$ species and not a $C_2H_7^+$ ion, consistent with proton transfer occurring via an impulsive mechanism.

According to this model, reaction commences as the ion and the neutral are drawn together by the long-range ion-induced dipole force. While less than 0.1 eV initially, the relative kinetic energy at the point of closest approach is of the order of 1 eV. At this point, the H^+–H_2 bond breaks, a H^+–C_2H_4 bond is formed simultaneously, and the neutral H_2 is elastically scattered off the repulsive barrier of the $C_2H_5^+$. As the H_2 leaves the collision center, its kinetic energy decreases due to the polarization forces. Since the basic assumptions of this model have been adequately described,[33,34] only the results will be given here.

Using the same notation as Hierl, Herman, and Wolfgang, the final energy of the products of the center-of-mass system, T_4, is given by

$$T_4 = [(X)(Z)P/(X + Y)(Y + Z)] - P' \qquad (VI)$$

Table IV. Product Distributions for the Reaction of H_3^+ with Ethylene as Observed Using High Pressure Mass Spectrometry,[11] Flowing Afterglow,[10] ICR,[12] and the Present Tandem Instrument

Method	Pressure, Torr	Product distribution	
		$C_2H_5^+$	$C_2H_3^+$
Flowing afterglow[10]	0.1–0.3	0.94	0.06
High pressure ms[11]	0.05	0.51	0.49
	0.10	0.56	0.44
	0.20	0.67	0.33
	0.30	0.68	0.32
ICR[12]	3×10^{-4}	0.40	0.60
Present results ICR	10^{-6}	0.31	0.69
High pressure MS	0.5	0.7	0.3

where Y is the mass of the transferred particle, X and Z are the masses of the other two particles, P is the potential energy which is converted into kinetic energy as the reactants are brought from infinite separation to the distance of closest approach, and P' is the translational energy of the products which is converted into potential energy as the products separate from the point of closest approach to infinity. The quantities P and P' are assumed to be given by

$$P = e^2\alpha(C_2H_4)/2r^4$$
$$P' = e^2\alpha(H_2)/2r^4 \qquad (VII)$$

where α is the polarizability of ethylene and hydrogen, respectively, e is unit charge, and r is the distance of closest approach. If an arbitrary but reasonable value of r of 2 au (2.1 Å) is chosen, (VI) predicts a relative kinetic energy of the products of 1.2 eV which is in accord with the present experimental results.

A more severe test of the applicability of this model to the present results lies in comparison of the H_3^+ and KrH^+ results since the relative kinetic energies of the products, according to the polarization reflection model, are functions of the mass and polarizabilities of hydrogen and krypton. Without actually specifying r, the final kinetic energies of the products, $T_4(KrH^+)$ and $T_4(H_3^+)$, are $T_4(KrH^+) = 11.0/r^4$ eV Å⁴ and $T_4(H_3^+) = 14.0/r^4$ eV Å⁴. Thus, if the distance of closest approach is the same for both reactions, the model predicts that the products of both reactions will have approximately the same kinetic energy. This prediction is required and is in agreement with the present experimental results which show that $C_2H_5^+$ is produced with the same internal energy in both reactions. This correlation constitutes a rather severe test of the applicability of the model because H_2 and Kr have very different polarizabilities ($\alpha(H_2) = 5.4$ au, $\alpha(Kr) = 16.7$ au) and masses.

Since KrH^+ has an internuclear separation which is about twice as large as the internuclear separation in H_3^+, the present results suggest that the effective distance of closest approach depends primarily on the $C_2H_5^+$ and not on the incoming ion geometry. If one breaks r up into two parts, one part representing the distance of closest approach between the proton and the ethylene molecule and another part representing the distance of closest approach between the proton and the outgoing neutral product (Kr or H_2), the difference in the bond lengths in KrH^+ and H_3^+ effectively decreases the H_2 kinetic energy relative to that of the Kr atom. This prediction, while certainly speculative, is consistent with the present results. Note, however, that Chang and Light[34] found the best fit by using the same r for both reactants and products.

Conclusions

Both the rate of reaction and the product ion distribution

A. S. Fiaux, D. L. Smith, and J. H. Futrell

were measured as an explicit function of the average number of H_3^+-H_2 collisions occurring prior to reaction with ethylene. The present results indicate that the reaction is very fast; within experimental error it proceeds at the ion-neutral collision rate as calculated using the Langevin ion-induced dipole theory. The present results also show that the rate of reaction is independent of the H_3^+ vibrational excitation energy.

The observed product distribution shows that at least 14% of the H_3^+ ions are formed initially with a minimum of 1.1 eV of excitation energy and that collisions with hydrogen molecules rapidly deactivate the excited ions. Coupled with other experimental and theoretical studies from this laboratory,[5,9,24,28] we conclude that H_3^+ ions are initially formed with a rather broad distribution of excitation energies and with an average energy of 1.5 to 2.0 eV. While 5 to 10 collisions with H_2 are required to completely deactivate the H_3^+ ion, only 2 or 3 of these are reactive collisions.

Results obtained using isotopically substituted reactants show that H_3^+ reacts under the present low energy and low pressure conditions by transferring a proton to the ethylene which, if energetically possible, will randomly eliminate a hydrogen molecule. This reaction scheme is illustrated by (8), (9), and (10). The present results, coupled with previous studies[2,14,15] of the $C_2H_7^+$ ion, strongly suggest that the intermediate in the first step of the reaction is best described as $C_2H_5^+\cdot H_2$ and not as a $C_2H_7^+$ ion.

A critical analysis of past and present results on the reaction of H_3^+ with ethylene shows that those $C_2H_5^+$ ions which have enough energy to decompose to $C_2H_3^+$ and H_2 do so with lifetimes in the time interval from 10^{-8} to 10^{-6} s. Isotope effects for elimination of H_2, HD, and D_2 by the $C_2X_5^+$ ions are much less than the isotope effects observed for the same reaction in the special case where the $C_2X_5^+$ ions live at least 10^{-6} s. This result further supports our contention that the majority of the $C_2H_3^+$ are produced from $C_2H_5^+$ ions which have lifetimes less than 10^{-6} s.

One especially interesting result of the present study is the formation of stable ethyl ions in a two-body process which do not require stabilization by a third body. This situation is possible only if the hydrogen molecule carries away a substantial portion of the exothermicity of the reaction. Thus the many previous studies which have assumed that all of the reaction exothermicity is deposited in the particle with the higher density of states (in this case, $C_2H_5^+$) should be reconsidered. Parallel studies performed in this laboratory on reactions of H_3^+ with methanol[9] as well as of KrH$^+$ with ethylene[30] support the hypothesis that at least 31% of the H_2 neutrals produced in the reaction of vibrationally relaxed H_3^+ with ethylene are produced with at least 0.5 eV of translational energy.

The "polarization stripping" model as described by Hierl, Herman, and Wolfgang[33] has been applied to the reaction of H_3^+ and KrH$^+$ with ethylene to predict the relative translational energy of the products of the reactions. This model predicts that the products will have approximately the same translation energy in both reactions. This result correlates very well with the present results which show that the $C_2H_5^+$ ions formed with the same excitation energy and thus the kinetic energy of the products is the same in both reactions. Since Kr

and H_2 have very different polarizabilities and masses, and since the KrH$^+$ and the H_3^+ are geometrically very different, we feel that this correlation between theory and experiment strongly supports the applicability of the "polarization stripping" model to these reactions.

Ion-neutral crossed beam studies, presently in progress in our laboratory, are measuring angular and energy distributions of the products of the reactions as a function of the translational energy of the reactants. These results, coupled with the present thermal energy results, will provide further insight to the dynamics of these reactions.

Acknowledgment. The authors are grateful to the National Science Foundation for support of this research through Grant GP 33870X. A. Fiaux acknowledges support from the Suiss National Foundation for the Encouragement of Research.

References and Notes

(1) For examples see S. Wexler, *J. Am. Chem. Soc.*, **85**, 272 (1963); M. S. B. Munson, J. L. Franklin, and F. H. Field, *J. Am. Chem. Soc.*, **85**, 3584 (1963); V. Aquilanti and G. G. Volpi, *J. Chem. Phys.*, **44**, 2307 (1966).
(2) W. T. Huntress, Jr., and M. T. Bowers, *Int. J. Mass Spectrom. Ion Phys.*, **12**, 1 (1973), and references cited therein.
(3) D. L. Smith and J. H. Futrell, *Int. J. Mass Spectrom. Ion Phys.*, **14**, 171 (1974).
(4) For a recent measurement of this rate see: L. P. Theard and W. T. Huntress, Jr., *J. Chem. Phys.*, **60**, 2840 (1974).
(5) D. L. Smith and J. H. Futrell, *J. Phys. B*, **8**, 803 (1975).
(6) F. von Busch and G. H. Dunn, *Phys. Rev. A*, **5**, 1726 (1972).
(7) Energetics given in the present paper are calculated using the following standard heats of formation: $\Delta H_f(H_3^+)$ = 266, $\Delta H_f(C_2H_5^+)$ = 220, $\Delta H_f(C_2H_4^+)$ = 253, $\Delta H_f(C_2H_3^+)$ = 282, and $\Delta H_f(KrH^+)$ = 266.
(8) D. L. Smith and J. H. Futrell, *Int. J. Mass Spectrom. Ion Phys.*, **10**, 405 (1973).
(9) A. Fiaux, D. L. Smith, and J. H. Futrell, *Int. J. Mass Spectrom. Ion Phys.*, accepted for publication.
(10) J. A. Burt, J. L. Dunn, M. J. McEwan, M. M. Sutton, A. E. Roche, and H. I. Schiff, *J. Chem. Phys.*, **52**, 6062 (1970).
(11) V. Aquilanti and G. G. Volpi, *J. Chem. Phys.*, **44**, 3574 (1966).
(12) J. K. Kim, L. P. Theard, and W. T. Huntress, Jr., *J. Chem. Phys.*, **62**, 45 (1975).
(13) G. Gioumousis and D. P. Stevenson, *J. Chem. Phys.*, **29**, 294 (1958).
(14) F. P. Abramson and J. H. Futrell, *J. Chem. Phys.*, **45**, 1925 (1966).
(15) V. Aquilanti, A. Galli, and G. G. Volpi, *J. Chem. Phys.*, **47**, 831 (1967).
(16) J. H. Vorachek, G. F. Meisels, R. A. Geanangel, and R. H. Emmel, *J. Am. Chem. Soc.*, **95**, 4078 (1973).
(17) P. C. Hariharan, W. A. Lathan, and J. A. Pople, *Chem. Phys. Lett.*, **14**, 385 (1972).
(18) L. P. Hills, M. Vestal, and J. H. Futrell, *J. Chem. Phys.*, **54**, 3834 (1971).
(19) S. M. Gordon, G. J. Krige, and N. W. Reid, *Int. J. Mass Spectrom. Ion Phys.*, **14**, 109 (1974).
(20) E. Constatin, *Int. J. Mass Spectrom. Ion Phys.*, **13**, 343 (1974).
(21) I. Baumel, R. Hagemann, and R. Botter, 19th Annual Conference on Mass Spectrometry and Allied Topics, Atlanta, Ga., May 1971.
(22) U. Lohle and Ch. Ottinger, *J. Chem. Phys.*, **51**, 3097 (1969).
(23) R. D. Smith and J. H. Futrell, results to be published.
(24) D. L. Smith and J. H. Futrell, *Chem. Phys. Lett.*, **24**, 611 (1974).
(25) R. N. Porter, R. M. Stephens, and M. Karplus, *J. Chem. Phys.*, **49**, 5163 (1968).
(26) H. Conroy, *J. Chem. Phys.*, **47**, 912 (1967).
(27) J. Durup and M. Durup, *J. Chim. Phys. Phys.-Chim. Biol.*, **64**, 386 (1967).
(28) D. L. Smith and J. H. Futrell, *Chem. Phys. Lett.*, accepted for publication.
(29) M. L. Vestal, *J. Chem. Phys.*, **43**, 1356 (1965), and by private communication.
(30) Private communication from T. A. Elwood and R. G. Cooks.
(31) R. D. Smith and J. H. Futrell, results to be published.
(32) Z. Herman, J. Kerstetter, T. Rose, and R. Wolfgang, *J. Chem. Phys.*, **46**, 2844 (1967); *Discuss. Faraday Soc.*, **44**, 123 (1967).
(33) P. M. Hierl, Z. Herman, and R. Wolfgang, *J. Chem. Phys.*, **53**, 660 (1970).
(34) D. T. Chang and J. C. Light, *J. Chem. Phys.*, **52**, 5687 (1970).

Editor's Comments
on Papers 36 Through 37

36 ČERMÁK and HERMAN
Réactions des Ions Excités dans N_2, O_2, CO, SO_2, COS et CS_2

37 HENGLEIN
Mass Spectrometric Observations on Reactions of Excited Ions in Carbon Disulfide and Some Aromatic Compounds

It would be expected that ions in excited electronic states might undergo reactions that would be endoergic and hence impossible for ions in the ground state. A number of such reactions have been observed, perhaps the first being those reported by Čermák and Herman (Paper 36). The states of the reacting ions were determined by comparing the appearance potential of the product with known electronic energies of the primary ion. Thus Čermák and Herman attributed the formation of N_3^+ to the $B^2\Sigma_u^+$ state of N_2^+ (at about 18.8 eV) and that of O_3^+ to the $^2\Pi_u$ state of O_2^+. Several other reactions were also observed. In these instances, the detection of the reaction of the electronically excited primary was possible because reactions of lower energy states of the ion were endoergic and hence forbidden. Henglein (Paper 37) observes several similar reactions. In later studies, Munson et al.[1] and Bowers et al.[2] found the appearance potential of N_3^+ to be 21 to 22 and 22.6 eV, respectively, and thus corresponding to a more highly excited state than that proposed by Čermák and Herman. In fact, Bowers et al. found evidence that two excited states of N_2^+ reacted to form N_3^+.

Although it is possible to determine the energy level of the excited reactant ions, the relative abundance of these ions will usually be small compared with that of the ground state ions and so will have to be measured by an independent method. Cress et al.[3] have shown that by using a pulsed ion source with variable reaction time, it is possible to determine both the unimolecular decay rate and the ion-molecule reaction rate constant, and have obtained values of approvimately $1.4 \times 10^{-5} s^{-1}$ and 5.8×10^{-11} cc/molecule s respectively for these rate constants in nitrogen. Relatively few such measurements have been made, however.

A single second-order ion-molecule reaction gives a linear semilog plot of the decay of the reactant ion against either time or neutral concentration (the other being kept constant). If two states of the ion having different reaction rate constants are present, the semilog plot will be curved, but it will become linear when the more reactive ion has disappeared. It is then possible to determine the fractions of the two states in the original beam from the slope of the curve for the slower reaction to obtain that rate constant and by extrapolation. From this and the composite curve, it is possible to obtain the rate constant for the faster reaction. Turner et al.[4] originally developed this method and used it to determine the fractions of the excited ions in O^+ and O_2^+ beams prepared by electron impact on O_2. Lao et al.[5] also used this to determine the proportion of $4P(C^+)$ in a C^+ beam from CO.

REFERENCES

1. Munson, M. S. B., F. H. Field, and J. L. Franklin, *J. Chem. Phys.* **37**: 1790 (1962).
2. Bowers, M. T., P. R. Kemper, and J. B. Laudenslager, *J. Chem. Phys.* **61**: 4394 (1974).
3. Cress, M. C., P. M. Becker, and F. W. Lampe, *J. Chem. Phys.* **44**: 2212 (1966).
4. Turner, B. R., J. A. Rutherford, and D. M. J. Compton, *J. Chem. Phys.* **48**: 1602 (1968).
5. Lao, R. C. C., R. W. Rozett, and W. S. Koski, *J. Chem. Phys.* **49**. 4202 (1968).

36

Reprinted from *J. Chim. Phys.* 57:717–719 (1960)

RÉACTIONS DES IONS EXCITÉS DANS N_2, O_2, CO, SO_2, COS et CS_2

V. Čermák and Z. Herman
Institut de Chimie Physique de l'Académie Tchécoslovaque des Sciences

Les résultats de nombreux travaux ont permis d'établir que les réactions molécules-ions devaient être athermiques ou exothermiques pour être observables. Il est alors très important de savoir si l'énergie cinétique ou l'énergie d'excitation de l'un des réactants peut être utilisée pour rendre exothermique une réaction endothermique donc interdite.

Jusqu'à présent on ne connaît expérimentalement qu'un seul exemple de réaction endothermique rendue possible par l'apport d'énergie d'excitation plutôt que cinétique de l'un des réactants. Il s'agit de la réaction

$$N_2{}^+ + N_2 \rightarrow N_3{}^+ + N$$

étudiée par SAPOROSCHENKO (¹) sous pression très élevée dans un spectromètre de masse d'un type spécial. L'intervention possible d'états excités dans les réactions molécules-ions a été discutée dans deux autres cas par HERRON et SCHIFF (²) et MEISELS (³).

Au cours d'un travail (⁴), effectué à l'aide d'un spectromètre de masse du type NIER (⁵), sur les réactions molécules-ions produites dans les gaz minéraux, nous avons observé un certain nombre de réactions nouvelles indiquées dans le tableau I. Les sections efficaces phénomènologiques, également mentionnées dans ce tableau, sont calculées d'après la relation (*) :

$$Q_f = \frac{i_s}{i_p . M . l} \tag{1}$$

dans laquelle i_s est le courant des ions secondaires, i_p le courant total des ions primaires, M le nombre de molécules par cm³ dans la source d'ions et l la distance entre le faisceau des électrons et la fente de sortie de la chambre d'ionisation (⁶).

Nous avons constaté que le potentiel d'apparition des ions secondaires N_3^+, O_3^+ et CS_3^+ corres

(*) On a supposé que l'efficacité d'extraction et de collection des ions primaires et secondaires était la même. Elle peut être différente, si les ions secondaires acquièrent, lors de leur formation, une certaine énergie cinétique.

pondait à celui des ions primaires N_2^+ dans l'état $B^2\Sigma_u^+$, des ions primaires O_2^+ dans l'état $^2\Pi_u$ et des ions primaires CS_2^+ dans leur premier état excité (⁷). Le potentiel d'apparition des autres ions secondaires était difficile à mesurer d'une façon précise par suite de la très faible intensité des courants ioniques, mais il semblait être lui aussi plus élevé que celui des ions primaires dans leur état fondamental.

Le caractère nouveau de ces réactions est dû au fait que les ions primaires réagissent dans un état dont la durée de vie moyenne (⁸) τ est de l'ordre de 10^{-7} à 10^{-8} s. Le temps de résidence de la plupart des ions excités dans la source étant très court à cause de l'émission radiative, la réaction secondaire se produit pratiquement dans le faisceau électronique et la vitesse thermique des ions primaires est comparable à celle acquise dans le champ électrique.

Si l'on considère une réaction secondaire à partir d'une seule espèce d'ions primaires excités formés par impact électronique on peut admettre, pour la section efficace la valeur moyenne $\bar{Q}$ donnée par la relation :

$$n_s = n_{p_0} . M . \bar{s} . \bar{Q} \tag{2}$$

dans laquelle n_s est le nombre d'ions secondaires formés par unité de temps, et $\bar{s}$ est la distance moyenne parcourue pendant la durée de vie moyenne des ions excités et n_{p_0} le nombre d'ions primaires excités. Lorsque les ions ont une vitesse thermique moyenne $\bar{v} = \sqrt{\dfrac{8\,kT}{\pi m}}$ on peut calculer $\bar{s}$ dans un champ électrique homogène d'intensité E :

$$\bar{s} = \frac{1}{4\pi} \int_0^\tau \int_0^\pi s . 2\pi \sin\delta\, d\delta\, dt \tag{3}$$

où δ représente l'angle entre les vecteurs vitesse thermique initiale et champ électrique. Posons $a = \dfrac{e\mathrm{E}}{m}$ (m étant la masse de l'ion), pour $a\tau \lesssim \bar{v}$

la relation (3) devient

$$\bar{s} = \bar{v}\,\tau + \frac{1}{9}\frac{a^2\tau^3}{\bar{v}} \tag{4}$$

et, pour $a\tau > \bar{v}$:

$$s = \frac{\bar{v}^2}{a}\left(\frac{11}{18} + \frac{6}{18}\,ln\,\frac{a\tau}{v}\right) + \frac{1}{2}\,a\tau^2 \tag{5}$$

Si nous considérons la relation (1), on peut en déduire la section efficace moyenne $\bar{Q}$ donnée par la relation

$$\bar{Q} = \frac{i_s}{i_{p_0}Ms} = Q_f\frac{i_p l}{i_{p_0}\bar{s}} = Q_f\frac{Q_i l}{Q_e\bar{s}} \tag{6}$$

dans laquelle Q_i est la section efficace d'ionisation de tous les ions primaires et Q_e la section efficace d'excitation des états réagissant. La durée de vie moyenne de l'état $B^2\Sigma_u^+$ étant connue [9] ($6,6\ 10^{-8}$ s), on peut calculer le produit $\bar{Q}\frac{Q_e}{Q_i}$ pour la réaction dans l'azote en supposant que seul cet état excité réagit.

$$\bar{Q}\frac{Q_e}{Q_i} = 1,7\ 10^{-17}\frac{1,9\ 10^{-1}}{3,8\ 10^{-3}} = 8,5\ 10^{-16}\ cm^2$$

Le rapport $\frac{Q_e}{Q_i}$, correspondant aux conditions de la réaction, n'est pas connu. Mais nous pouvons essayer de calculer une valeur approximative de $\bar{Q}$ en prenant la valeur minimum mesurée par STEWART [10] de la section efficace totale pour le niveau $v = 0$ de l'état $B^2\Sigma_u^+$($Q_e = 9,2\ 10^{-18}$ cm²).

$$\bar{Q} = 8,5\ 10^{-16}\frac{2,8\ 10^{-16}}{9,2\ 10^{-18}} = 2,6\ 10^{-14}\ cm^2.$$

Si le nombre d'ions primaires excités réagissant diminue selon la relation

$$N_t = N_0 e^{-\frac{t}{\tau}} \tag{7}$$

dans laquelle N_t est le nombre d'ions au temps t, N_0 le nombre d'ions au temps $t = 0$ et si $k =$ la constante de vitesse de la réaction molécules-ions, le nombre d'ions secondaires n_s sera donné par :

$$n_s = n_{p_0}M\int_0^{t_r} k\,e^{-\frac{t}{\tau}}\,dt \tag{8}$$

soit :

$$n_s = n_{p_0}M\bar{k}\,\tau\left(1 - e^{-\frac{t_r}{\tau}}\right) \tag{9}$$

t_r étant le temps de résidence des ions primaires dans la source d'ions et $\bar{k}$ la constante de vitesse moyenne entre le temps $t = 0$ et $t = t_r$. En remplaçant n_s et n_{p_0} par i_s et i_{p_0}, on peut calculer d'après

les données expérimentales le produit $\bar{k}\frac{Q_e}{Q_i}$.

$$\bar{k}\frac{Q_e}{Q_i} = \frac{i_s}{i_p M\tau\left(1 - e^{-\frac{t_r}{\tau}}\right)} \tag{10}$$

Dans le cas particulier de l'azote on obtient pour

$$E = 10\ V/cm\ et\ t_r = \sqrt{\frac{2\,m\,l}{eE}} = 10^{-6}\,s.$$

$$\bar{k}\frac{Q_e}{Q_i} = \frac{3,3\ 10^{-18}}{6,6\ 10^{-8}} = 5\ 10^{-11}\ cm^3\ mol^{-1}s^{-1}.$$

et en admettant pour Q_e la valeur de STEWART

$$\bar{k} = 5\ 10^{-11}\frac{2,8\ 10^{-16}}{9,2\ 10^{-18}} = 1,5\ 10^{-9}\ cm^3\ mol^{-1}s^{-1}.$$

La valeur de k ainsi évaluée est plus grande que la valeur théorique de k pour la réaction des ions nonexcités de vitesse thermique [12]

$$k = 2\pi e\left(\frac{\alpha}{\mu}\right)^{1/2} = 8,26\ 10^{-10}\ cm^3\ mol^{-1}s^{-1}. \tag{11}$$

où α est la polarisibilité de la molécule et μ la masse réduite. La divergence entre ces valeurs pourrait signifier par exemple, que la valeur de Q_e serait en réalité plus grande à cause de la participation de l'état $C^2\Sigma^+$ à la réaction. Cependant la valeur théorique de la constante de vitesse de la réaction entre ions excités et molécules, l'intensité du champ électrique, les divers facteurs influençant l'extraction des ions ne sont pas connus avec une précision suffisante pour conclure plus nettement.

TABLEAU I

*Sections efficaces phénoménologiques
des réactions secondaires.*

Tensions d'extraction 160 V, intensité électronique 80 μA,
énergie des électrons 80 eV.

Gaz (mélange)	Ions secondaires	$Q_f.10^{16}$ cm²
A — H₂	AH⁺	110
N₂	N₃⁺	0,17
O₂	O₃⁺	1,2
CO	C₂O⁺	0,13
N₂ — CO	NCO⁺	0,12
SO₂	S₂O₃⁺	0,05
CO₂	CO₃⁺	0,003
	C₂O₂⁺	0,006
	C₂O₃⁺	0,03
COS	CS₂O⁺	0,11
CS₂	CS₃⁺	4,0
	C₂S₂⁺	0,63
	C₂S₃⁺	0,13
	C₂S⁺	0,23

BIBLIOGRAPHIE

(1) M. SAPOROSCHENKO. — *Phys. Rev.*, 1958, **111**, 1550.
(2) J. P. HERRON et H. I. SCHIFF. — *Canad. J. Chem.*, 1958, **36**, 1159.
(3) G. G. MEISELS. — *J. chem. Phys.*, 1959, **31**, 284.
(4) V. ČERMAK et Z. HERMAN. — *Coll. czech. chem. Comm.*, à paraître.
(5) A. O. NIER. — *Rev. sci. Instr.*, 1947, **18**, 398.
(6) D. P. STEVENSON et D. O. SCHISSLER. — *J. chem. Phys.*, 1958, **26**, 282.

(7) J. COLLIN. — *J. chim. Phys.*, sera publié.
(8) G. HERZBERG. — *Molecular spectra and molecular structure.* I. D. van NOSTRAND. — *Comp.* New-York, 1957.
(9) R. G. BENNETT et F. W. DALBY. — *J. chem. Phys.*, 1959, **31**, 434.
(10) D. T. STEWART. — *Proc. Phys. Soc.*, 1956, **69** A, 437.
(11) F. H. FIELD, J. L. FRANKLIN et F. W. LAMPE. — *J. amer. chem. Soc.*, 1957, **79**, 2419.
(12) H. EYRING, J. O. HIRSCHFELDER et H. S. TAYLOR. — *J. chem. Phys.*, 1936, **4**, 479.

335

37

Reprinted from Z. Naturforsch. **17a**:37–43 (1962)

Mass Spectrometric Observations on Reactions of Excited Ions in Carbon Disulfide and some Aromatic Compounds*

By Arnim Henglein

Radiation Research Laboratories, Mellon Institute, Pittsburgh, Pa., USA
and Hahn–Meitner-Institut für Kernforschung, Berlin-Wannsee, Germany

(Z. Naturforschg. 17 a, 37—43 [1962] ; eingegangen am 24. Oktober 1961)

Reactions have been observed between excited ions of carbon disulfide and carbon disulfide, water, iodine and ethylene. The ions CS_3^+, $C_2S_2^+$, $C_2S_3^+$, H_2OS^+, CSI^+, SI^+, $C_2H_3S^+$ and $C_2H_4S^+$ are formed in these reactions. An additional contribution to the $C_2S_2^+$ ion results from reaction of CS^+ with CS_2. The reaction of I_2^+ with CS_2 leads to the ion CS_2I^+. Reactions of excited molecular ions have also been found in benzonitrile, chlorobenzene, bromobenzene and iodobenzene. Ions of the C_{12} series such as $C_{12}H_9CN^+$, $C_{12}H_{10}^+$ and $C_{12}H_9^+$ are formed in these processes. In all cases, the appearance potential of the secondary ion is lower than the first dissociation limit of the molecular ion. It is therefore concluded that the excited ions are stable towards unimolecular decomposition. Repeller field studies show that they must be metastable towards decay by photon emission. In the case of aromatic ions this is explained by fast internal conversion of electronically excited ions into high vibrational levels of their electronic ground states. The form of the ionization efficiency curves of the secondary ions at high electron energies corresponds to what is expected if the excited precursors are formed in optically allowed transitions.

Most of the fundamental work on ion-molecule reactions during recent years has been carried out on processes which are initiated by ions in their ground states. Since the only reactions which can generally be observed in a mass spectrometer occur with high cross section and therefore with little activation energy only exothermic or thermo-neutral reactions have been expected to occur. However, as has recently been shown, endothermic reactions may be found if they are induced by ions carrying some excess energy. Several reactions induced by I_2^+ carrying 1.3 eV of excitation energy have been detected [1]. Cermak and Herman [2] have reported reactions of excited ions formed from simple molecules containing double or triple bonds such as N_2, O_2, CS_2, CO, CO_2 and SO_2. In the present paper several reactions of excited CS_2^+ ions with water and iodine are described as well as a number of reactions of excited molecular ions of some aromatic compounds. In addition, some considerations about the lifetime and nature of the excited ions will be presented.

All experiments have been carried out with a Consolidated Electrodynamics Corporation Model 21-103 C mass spectrometer. The experimental details have already been reported [1].

Theoretical Considerations

Four standard criteria have been generally applied in the investigation of ion-molecule reactions [3, 4].

1. Secondary ions are recognized by the proportionality of their intensity to the square of the pressure in the ionization chamber.

2. Comparisons of the appearance potential of the secondary ions with those of the primary ions lead to the identification of the precursor ion.

3. The ratio of secondary to primary ion current (i_s/i_p) is found to decrease with increasing repeller field strength.

4. Considerations of thermochemical data allow certain reactions which are endothermic in nature to be excluded from further consideration.

In the case of excited ions some complications are involved in criteria 2, 3 and 4 which we now regard in some detail.

Ionization efficiency curves and appearance potentials

The ionization efficiency curve of a molecular ion is schematically shown by curve 1 in Fig. 1. The

* This work is supported, in part, by the U.S. Atomic Energy Commission.

[1] A. Henglein and G. A. Muccini, Z. Naturforschg. 15 a, 584 [1960].

[2] V. Cermak and Z. Herman, J. Chim. Phys. 57, 717 [1960].

[3] D. P. Stevenson and D. O. Schissler, J. Chem. Phys. 29, 282 [1958].

[4] F. H. Field, J. L. Franklin and F. W. Lampe, J. Amer. Chem. Soc. 79, 2419 [1957].

sharp change in slope at point a above the appearance potential AP(A) is due to the beginning of the excitation of a higher electronic state B of the ion. The observed curve 1 is therefore the sum of

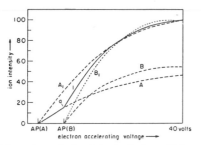

Fig. 1. Schematic analysis of the ionization efficiency curve of a molecular ion.

the corresponding curves A and B of the ground state and the excited state. In order to compare ionization efficiency curves we normalize them at 40 volts of electron accelerating voltage. If a reaction is induced by the excited state B only, the secondary ion will appear at AP(B) and its intensity will follow the ionization efficiency curve B_1 (B_1 is the normalized curve B). However, if only the ion in its ground state is able to initiate a reaction, the secondary ion will appear at AP(A) but its ionization efficiency curve A_1 will deviate from curve 1 especially at voltages only slightly above the appearance potential (A_1 is the normalized curve A). The secondary ions described in this paper show ionization efficiency curves of the type represented by curve B_1. In the following paper a reaction in acrylonitrile will be reported where the precursor is only the molecular ion in its ground state.

The form of the ionization efficiency curve of a secondary ion at high electron accelerating voltages may give some information about the nature of the formation of the excited precursor by electron impact. The cross section Q depends on the electron energy T according to equation (1), if the transition from the ground state of the neutral molecule to the excited molecule ion is optically allowed.

$$Q = \frac{4\,\pi\,a_0^{\,2}\,R^2}{T}\,\frac{f}{E}\,\ln c \cdot T \tag{1}$$

(a_0: BOHR-radius; R: RYDBERG constant; f: oscillator strength of transition; E: energy of excited

state; c: constant). The plot of $i_s \cdot T$ versus $\log T$ will yield a straight line. Optically forbidden transitions will show a much stronger decrease in Q with increasing T than given by equation (1).

Dependence of i_s/i_p on the repeller field strength

The activated complex which is formed in the reaction of a slow primary ion P^+ with a molecule M may often have different ways of decomposition to yield various secondary ion S^+:

$$P^+ + M \to (PM)^{+*} \nearrow \begin{matrix} P^+ + M \\ S_1^+ + M_1 \\ S_2^+ + M_2 \end{matrix} \tag{2}$$

Each of these paths of dissociation is characterized by a unimolecular rate constant [5,6]

$$k_i = \nu\,(1 - E_i/E)^{\,\alpha-1} \tag{3}$$

where E is the total excitation energy of the complex and E_i the energy required for dissociation along path i, ν is about 10^{-13} sec^{-1}, α the number of degrees of vibrational freedom. According to the polarization theory [7] of ion-molecule reactions, the cross section of the formation of the complex decreases with the reciprocal square root of the kinetic energy of the ion P^+ or repeller field strength ε, respectively.

$$\sigma \propto \varepsilon^{-0.5}. \tag{4}$$

In fact, a number of ion-molecule reactions have been observed in simple molecules where the total cross section σ_t was proportional [3] to $\varepsilon^{-0.5}$. However, if several modes of dissociation of the intermediate complex can compete, the total cross section may depend on the repeller field strength according to

$$\sigma_t \propto \varepsilon^a \tag{5}$$

where a is different from -0.5. Part of the kinetic energy of the primary ion will appear as excitation energy of the complex. E in equation (3) will therefore increase with increasing repeller field strength. As a result, the relative frequencies of the various paths of decomposition of the complex may change. The cross section of the formation of a particular secondary ion will decrease more strongly with increasing repeller field strength than expected from the $\sigma \sim \varepsilon^{-0.5}$ relation, if another path of decomposi-

[5] M. BURTON and J. L. MAGEE, J. Phys. Chem. **56**, 842 [1952].
[6] R. F. POTTIE and W. H. HAMILL, J. Phys. Chem. **63**, 877 [1959].

[7] G. GIOUMOUSIS and D. P. STEVENSON, J. Chem. Phys. **29**, 294 [1958].

tion becomes more frequent. Many ion-molecule reactions are already known, the cross sections of which depend on ε by an exponent a between -0.5 and -1.0 [3, 4, 8]. Our data presented in Fig. 10 show even higher negative values of a for some reactions studied. At higher kinetic energies of the primary ion the most frequent product ion of the collision is probably the ion P^+ itself or its dissociation products. This is well known from studies of collision induced dissociation of ions [9-11]. At these higher ion energies the lifetime of the complex is shorter than the time required for distribution of the excitation energy in the various degrees of freedom of the complex; i. e. there is practically no real complex formation. Perhaps, this process is already competing with many ion-molecule reactions at lower ion energies *.

The current ratio i_s/i_p is given by the relation

$$i_s/i_p = \sigma_{t(\varepsilon)} \, l \, c \qquad (6)$$

where c is the concentration of gas molecules and l is the path length of the primary ion in the ionization chamber. This path length is constant and is independent of the repeller field strength, if the lifetime of the primary ion is much longer than the time of its residence in the ionization chamber; i. e. $10^{-7} - 10^{-6}$ seconds. However, if an excited primary ion is deactivated by either photon emission or dissociation within a much shorter time, it can initiate bi-molecular reactions only along a short part of its total path. The path length l_1 which it traverses in its excited state will be proportional to the repeller field strength. The current ratio now will become:

$$i_s/i_p = \sigma_{t(\varepsilon)} \, c \, l_{1(\varepsilon)} . \qquad (7)$$

Introduction of the dependences of σ and l_1 on ε leads to

$$i_s/i_p \sim \varepsilon^a \cdot \varepsilon^1 \sim \varepsilon^{1+a} . \qquad (8)$$

If the exponent a lies in the range of -0.5 to -1.0, the current ratio will now increase or at least be constant with increasing repeller field strength.

[8] F. W. Lampe, F. H. Field and J. L. Franklin, J. Amer. Chem. Soc. 79, 6132 [1957].
[9] J. Mattauch and H. Lichtblau, Phys. Z. 40, 16 [1939].
[10] A. Henglein, Z. Naturforschg. 7 a, 165 [1952].
[11] C. E. Melton and H. M. Rosenstock, J. Chem. Phys. 26, 568 [1957].
* If dissociation of the complex into $P^+ + M$ is negligible,

Experimental Results

The ion-molecule reactions observed in this work are listed in Table 1. Reaction I has been used as a standard reaction to calculate relative cross sections. The current ratios i_s/i_p of this table have been measured at a pressure of 500 μ in the reservoir of the gas inlet system. These observations were carried out at electron accelerating voltages of 40 volts and a repeller field strength of 3.84 volts/cm. The ion current of all ions reported in this paper as secondary ions have been found to be proportional to the square of the pressure.

Fig. $2-9$ show the ionization efficiency curves for the secondary ions and a number of primary ions. All curves have been normalized at 40 volts. The plots show the ionization efficiency curves in the range from 9 to 40 volts and in some cases also from 40 to 170 volts. All measurements of the repeller field dependence of the ratio i_s/i_p are compiled in Fig. 10.

Carbon disulfide

Cermak and Herman [2] have already interpreted the formation of the secondary ions CS_3^+, $C_2S_2^+$ and $C_2S_3^+$ as products of reactions of the excited carbon disulfide ion. The present studies confirm this conclusion. A comparison of the ionization efficiency curves of the secondary ions given in Fig. 3 with those of the primary ions in Fig. 2 clearly shows that

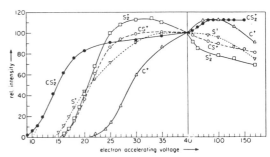

Fig. 2. Ionization efficiency curves of primary ions in carbon disulfide.

there should exist as many ion-molecule reactions of
$$|a| < 0.5$$
than reactions of $|a| > 0.5$. However, reactions for which the exponent a has a lower negative value than 0.5 have never been observed. It is, therefore, concluded that the formation of P^+ (or its dissociation products) is an important competing process the relative frequency of which much stronger increases with increasing ε than that of any other secondary ion.

No.	Secondary ion	Mode of Formation	i_s/i_p[a]	relative cross[b] section
		water		
I	H_3O^+	$H_2O^+ + H_2O \rightarrow H_3O^+ + OH$	$2.4 \cdot 10^{-2}$	1
		carbon disulfide		
II	CS_3^+	$CS_2^{+*} + CS_2 \rightarrow CS_3^+ + CS$	$1.1 \cdot 10^{-3}$	0.044
III	$C_2S_2^+$	$CS_2^{+*} + CS_2 \rightarrow C_2S_2^+ + S_2$	$6.0 \cdot 10^{-5}$	0.0025
IV	$C_2S_2^+$	$CS^+ + CS_2 \rightarrow C_2S_2^+ + S$	$7.2 \cdot 10^{-4}$	0.03
V	$C_2S_3^+$	$CS_2^{+*} + CS_2 \rightarrow C_2S_3^+ + S$	$6.0 \cdot 10^{-5}$	0.0025
		carbon disulfide-water[c]		
VI	H_2OS^+	$CS_2^{+*} + H_2O \rightarrow H_2OS^+ + CS$	$1.3 \cdot 10^{-4}$	0.011
		carbon disulfide-iodine[c]		
VII	CS_2I^+	$I_2^+ + CS_2 \rightarrow CS_2I^+ + I$	$5.7 \cdot 10^{-3}$	0.42
VIII	CSI^+	$CS_2^{+*} + I_2 \rightarrow CSI^+ + SI$	$4.2 \cdot 10^{-4}$	0.036
IX	SI^+	$CS_2^{+*} + I_2 \rightarrow SI^+ + CSI$	$1.2 \cdot 10^{-3}$	0.10
		benzonitrile		
X	$C_{12}H_9CN^+$	$C_6H_5CN^{+*} + C_6H_5CN \rightarrow C_{12}H_9CN^+ + HCN$	$3.2 \cdot 10^{-3}$	0.13
		chlorobenzene		
XI	$C_{12}H_9^+$	$C_6H_5Cl^{+*} + C_6H_5Cl \rightarrow C_{12}H_9^+ + HCl + Cl$	$2.7 \cdot 10^{-3}$	0.11
		bromobenzene		
XII	$C_{12}H_{10}^+$	$C_6H_5Br^{+*} + C_6H_5Br \rightarrow C_{12}H_{10}^+ + Br_2$	$6.3 \cdot 10^{-3}$	0.26
		iodobenzene		
XIII	$C_{12}H_{10}^+$	$C_6H_5I^{+*} + C_6H_5I \rightarrow C_{12}H_{10}^+ + I_2$	$1.6 \cdot 10^{-2}$	0.70

[a] at 500 μ of pressure in the reservoir of the gas inlet system.
[b] reaction I was chosen as reference reaction.
[c] each component had a pressure of 250 μ in the reservoir of the gas inlet system.

Table 1. List of ion-molecule reactions.

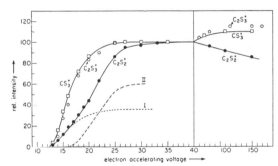

Fig. 3. Ionization efficiency curves of secondary ions in carbon disulfide.

only an excited parent ion appearing at 13.0 ± 0.3 volts can be the precursor of the secondary ions. The secondary ions have the same appearance potential and therefore the same precursor. The appearance potential of 13.3 volts agrees well with that of 13.60 ± 0.19 which has been attributed by COLLIN [12] to the $^2\Pi_{3/2\,\mu}$ state of the CS_2^+ ion.

Although the ionization efficiency curves for CS_3^+ and $C_2S_3^+$ are essentially identical, the curve of the $C_2S_2^+$ ion rises less steeply just above the

[12] J. COLLIN, J. Chim. Phys. 57, 424 [1960].

appearance potential. The slope of the $C_2S_2^+$ curve then increases at higher voltages. It is concluded that $C_2S_2^+$ is formed at higher voltages not only by reaction III but also by an additional process.

If it is assumed that $C_2S_2^+$ is exclusively formed by CS_2^{+*} between 13.3 and 15.0 volts this range of the $C_2S_2^+$ curve can be extrapolated to higher voltages using the form of the CS_3^+ curve. Curve 1 has been obtained this way. It gives the contribution of reaction III to the formation of $C_2S_2^+$. If curve 1 is subtracted from the $C_2S_2^+$ curve, curve II is obtained. This additional contribution has essentially the same form as the ionization efficiency curve of the primary CS^+ ion which is the only carbon containing ion with an appearance potential of about 16 volts. We therefore attribute this part of the formation of $C_2S_2^+$ to reaction IV (Table 1) of this ion.

The excited state of the CS_2^+ ion is able to undergo many reactions with other molecules. Fig. 4 and 5 show that the ions H_2OS^+, CSI^+ and SI^+ are formed in mixtures of carbondisulfide with water and iodine, respectively. The curves in these figures make clear that no other ion than CS_2^{+*} can be the precursor of those secondary ions. CS_2I^+ has been observed too. Since its ionization efficiency curve is

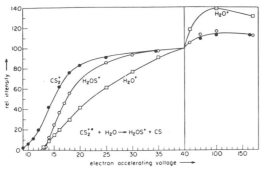

Fig. 4. Ionization efficiency curves of H_2OS^+ and H_2O^+ in 1 : 1 water-carbon disulfide.

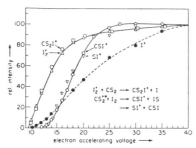

Fig. 5. Ionization efficiency curves of ions in 1 : 1 iodine-carbon disulfide.

identical with that of I_2^+, reaction VII in Table 1 has been formulated.

Some preliminary experiments have been carried out with mixtures of carbondisulfide and hydrocarbons. The secondary ions $C_2H_3S^+$ and $C_2H_4S^+$ have been observed in mixture with ethylene. The appearance potentials of 13.4 and 12.8 of these ions again are much higher than those of the parent ions in this mixture [$AP(CS_2^+)$: 10.1; $AP(C_2H_4)$: 10.5 volts]. Either an excited molecular ion of CS_2 or C_2H_4 must therefore be the precursor.

Aromatic compounds

A number of ion-molecule reactions have been found in benzonitrile and the phenyl halides and are listed in Table 1. That the secondary ions observed result from the reactions of excited molecular ions can be readily determined in these compounds for two reasons. First, the secondary ions have rather high intensities. Second, they belong to the C_{12} series such as $C_{12}H_{10}^+$, $C_{12}H_9^+$ and $C_{12}H_9CN^+$. The mass spectra of these aromatic compounds contain only a small number of primary C_6 ions which

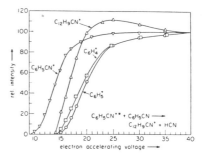

Fig. 6. Ionization efficiency curves of ions in benzonitrile.

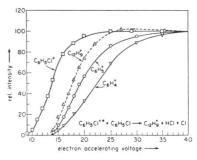

Fig. 7. Ionization efficiency curves of ions in chlorobenzene.

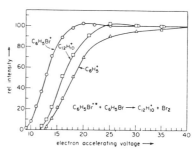

Fig. 8. Ionization efficiency curves of ions in bromobenzene.

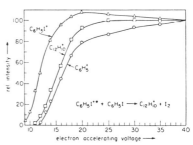

Fig. 9. Ionization efficiency curves of ions in iodobenzene.

have sufficient intensity to be possible precursor of secondary ions of the C_{12}-series. Only $C_6H_5^+$ and, in some cases, $C_6H_4^+$ have to be regarded besides the parent ion. The ionization efficiency curves of those ions are shown together with those of the secondary ions in Fig. 6 – 9. It can be seen that the secondary ion always appears below the first dissociation limit of the parent ion; i. e. below $AP(C_6H_5^+)$ and $AP(C_6H_4^+)$, but much higher than the AP of the parent ion. It is again concluded that excited states of the molecular ions are the precursors.

Nature and life time of the excited ions

Excited ions may be unstable towards either dissociation or photon emission. Spontaneous dissociations of metastable ions are well known in mass spectrometry [13, 14]. The daughter ions of such metastable transitions generally have the same appearance potentials as the same ions produced as primary ions in the ionization chamber. Metastable molecular ions of the aromatic compounds studied here have been observed in their mass spectra. However, it does not appear that metastable ions which are capable of unimolecular decomposition are also the initiators of our bimolecular reactions since our secondary ions always appear below the first dissociation limit of the parent ion. For example, in carbon disulfide CS^+ and S^+ appear at 15.7 and 14.3 volts, respectively; i. e. above the 13.3 volts of the excited state of CS_2^+. Similarly it was observed in the previous studies of reactions of excited I_2^+ ions that the secondary ions appear about 1 volt below the dissociation of I_2^+ into $I^+ + I$ (ref. [1]). It must therefore be concluded that the excited ions, which initiate bimolecular reactions, are stable with respect to dissociation.

The high relative cross sections in Table 1 already indicate that the excited ions do not decay by photon emission within about 10^{-8} sec., i. e. the time for optically allowed transitions. This is especially evident for the primary ions in the aromatic compounds where reaction cross sections as nearly as large as that of the water reaction have been observed. As already mentioned excited ions of short lifetime will pass only a very short path l_1 in the ionization chamber along which they have a chance to meet a gas molecule. The current ratio i_s/i_p of their reactions should therefore be smaller by about two orders of magnitude than that of the standard reaction. It must be remembered here that Table 1 gives only minimum values of i_s/i_p and of relative

cross sections of excited ion-reactions since only part of all primary parent ions formed by electron impact will have excess energy.

The ratio i_s/i_p of most of the reactions studied depends on the repeller field strength according to equation (5). The logarithmic plots in Fig. 10

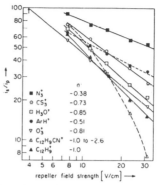

Fig. 10. Ratio i_s/i_p as function of repeller strength. The ratio at 3.84 volts/cm has been normalized to 100. a is the exponent in equation (5).

generally yield straight lines for repeller field strengths above 5 volts/cm. The data of the excited ion-molecule reactions $I_2^{+*} + H_2O \rightarrow H_2OI^+ + I$ [1], $N_2^{+*} + N_2 \rightarrow N_3^+ + N$ [2], $O_2^{+*} + O_2 \rightarrow O_3^+ + O$ [2] and of two reference reactions $H_2O^+ + H_2O \rightarrow H_3O^+ + OH$ and $A^+ + H_2 \rightarrow AH^+ + H$ are included in the figure. The slope of the latter reaction was found to be equal to -0.51 which agrees with the findings of STEVENSON [3]. A value of -0.85 was observed for the reaction in water which again agrees fairly well with the value of -0.73 calculated from the data of LAMPE et al. [8]. The exponent a varies over a wide range for the other reactions. It becomes even higher than -1.0 in some cases. Reaction X, for example, is extremely dependent on the repeller field strength. The corresponding curve in Fig. 10 shows an increase in slope with increasing repeller field strength. This is attributed to some competing reactions of the activated complex as mentioned above.

An increase of i_s/i_p with increasing ε (positive a) as expected for reactions of short living excited ions. has never been observed. The reaction in nitrogen showed a smaller negative exponent a than 0.5. The value of -0.38 found here is the only indication that there might be a short living precursor in this

[13] J. A. HIPPLE, Phys. Rev. 71, 594 [1947].
[14] See for complete reference F. H. FIELD and J. L. FRANKLIN, Electron Impact Phenomena, Academic Press Inc. Publ., New York 1957.

reaction. If there is not some hitherto unknown influence of the repeller field strength on the ratio i_s/i_p of excited ion-reactions, it must be concluded that the lifetimes of the precursors of our reactions are longer than 10^{-6} sec.

The stability of the excited aromatic ions may be explained by fast internal conversion from the electronically excited ion to high vibrational levels of its electronic ground state. Internal conversion is known to be very efficient in deactivating higher electronic states in polyatomic neutral molecules [15]. Because of the large energy difference between the lowest excited state and the ground state, internal conversion generally stops at the lowest excited state which further decays by photon emission. In the case of aromatic ions, however, the spacing between

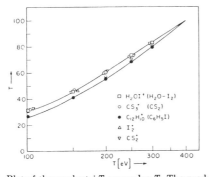

Fig. 11. Plot of the product iT versus $\log T$. The product iT at $T = 400$ eV has been made equal to 100. (i: ion current; T: kinetic energy of bombarding electrons).

the levels always seems to be much smaller. This may be concluded from the experiments of Fox and HICKAM [16] on ionized benzene, in which four excited states below the first dissociation limit were found. It seems therefore plausible that internal conversion of excited ions will lead to the ground state of the ion. According to this explanation, the ions do not enter their bimolecular reactions with neutral molecules in the same electronic states in which they were formed by electron impact. A similar explanation may be possible for the stability of the excited iodine ion [1]. Internal conversion again may be favored by the large number of excited states just above the appearance potential of the ion [17]. No general explanation can be given for the other ions consisting of a small number of atoms. Perhaps some of these excited ions decay by light emission to give a lower metastable state of long lifetime.

Some plots of the product iT versus $\log T$ are shown by Fig. 11. The diagram contains data for some secondary ions as well as for the primary ions I_2^+ and CS_2^+. Although equation (1) is only valid at electron energies of several hundred eV reasonably straight lines result from the data in Fig. 11. The deviations at the lower energies in Fig. 11 are due to the failure of the relationship in this range. Apparently all of the excited ions described are formed in optically allowed transitions.

The author wishes to thank Mr. G. K. BUZZARD for his assistance in carrying out the experiments reported here.

[15] M. KASHA and S. P. McGLYNN, Ann. Rev. Phys. Chem. 7, 403 [1956].
[16] R. E. FOX and W. M. HICKAM, J. Chem. Phys. 22, 2059 [1954].
[17] J. D. MORRISON, H. HURZELER, M. G. INGHRAM and H. E. STANTON, J. Chem. Phys. 33, 821 [1960].

Editor's Comments
on Papers 38 Through 41

38 GIESE and MAIER
Dissociative Ionization of CO by Ion Impact

39 CHUPKA, BERKOWITZ, and GUTMAN
Electron Affinities of Halogen Diatomic Molecules as Determined by Endoergic Charge Transfer

40 HUGHES, LIFSHITZ, and TIERNAN
Electron Affinities from Endothermic Negative-Ion Charge-Transfer Reactions. III. NO, NO_2, SO_2, CS_2, Cl_2 Br_2, I_2, and C_2H

41 BEAUCHAMP
Ion Cyclotron Resonance Studies of Endothermic Reactions of UF_6^- Generated by Surface Ionization

Most ion-molecule reactions that have been studied have been exoergic and have had little or no energy of activation. This has been primarily because of instrument limitations and not because there are no ion-molecule reactions that are endoergic or that have energies of activation. But a few studies have focused on endoergic processes, and some of them have proved very valuable. Probably the first such study, reported by Giese and Maier (Paper 38), involved the decomposition of CO upon collision with rare gas ions of various energies. The collision energy in the center of mass at which the product ion just formed was the (endothermic) heat of reaction and the experimental results agreed well with those calculated with the aid of known thermochemical values.

Probably the greatest value of endoergic ion-molecule reactions has been in studies of negative ions. With positive molecular ions, the ionization potential and hence the heat of formation of the molecular ion can be determined at least fairly well by direct ionization. This is not true for parent negative ions. Electron attachment to a molecule usually does not produce an ion of sufficient lifetime to be detected. Futher, if it is formed, it necessarily is in an excited state of unknown energy so the heat of formation cannot be deduced. On the other hand, if an endoergic charge exchange reaction to give the desired ion can be performed, the center of mass energy at onset will give the heat of reaction from which the electron affinity can be computed.

Negative ions of halogen molecules had been known to exist ever since Hogness and Harkness[1] had observed I_2^- and deduced that it was formed by charge exchange from I^-. Much later, Reese et al.[2] found the electron affinity of F_2^- to be 3.0 eV by determining the appearance potential of the ion from SO_2F_2. The existence of halogen molecular negative ions and the electron affinities of the halogen molecules were of considerable interest and were thus subjected to study in several laboratories in the early 1970s. Chupka, Berkowitz, and Gutman (Paper 39) devised a special ion source to permit controlling the velocity of ions in the source and to allow the primary ions to collide at controlled energy with the neutral molecules present. In this way, several kinds of negative ions whose heats of formation were known were impacted on halogen molecules at various energies. The onset for charge transfer, after correction for the thermal motion of the molecules, was the heat of reaction; it permitted computation of the electron affinities of the halogen molecules. Their results are in excellent agreement with those obtained by DeCorpo and Franklin,[3] who employed an entirely different method, and of Hughes, Lifshitz, and Tiernan (Paper 40), who employed tandem mass spectrometers to provide mass selected primary beams and mass analysis of the products.

Both Chupka et al. and Hughes et al. have used their techniqies to determine electron affinities of several other small molecules. In addition, Refaey and Franklin have applied a modification of the method of Chupka et al. to measure the thresholds for the formation of a number of ions formed by impacting I^- on several small molecules in a study of SO_2 by this method.[4]

Beauchamp (Paper 41) has applied ion-cyclotron resonance to the study of endoergic collision reactions of UF_6^- with UF_6, SF_6, and BF_3 and has deduced the heats of formation of several UF_x^- ions from these data. The study is interesting in that it shows still another application for the ion-cyclotron technique and also in that it gives further evidence for the very large electron affinities of the hexafluorides, especially, WF_6, UF_6, and PtF_6, which have electron affinities of 4.5, 4.9, and 6.8 eV, respectively.

In all these endoergic collision studies, it has been necessary to make corrections for the velocity of the target molecules. These will usually have a thermal energy distribution. Methods of calculating such corrections have been discussed by Berkling et al.[5], von Busch et al.[6], and Chantry[7]. Their methods have been applied in all the papers dealing with the accurate determination of threshold values.

REFERENCES

1. Hogness, T. R., and R. W. Harkness, *Phys. Rev.* **32**: 784 (1928).
2. Reese, R. M., V. H. Dibeler, and J. L. Franklin, *J. Chem. Phys.* **29**: 880 (1958).
3. DeCorpo, J. J., and J. L. Franklin, *J. Chem. Phys.* **54**: 1885 (1971).
4. Refaey, K. M. A., and J. L. Franklin, *J. Chem. Phys.* **65**: 1994 (1976).
5. Berkling, K., R. Helbring, K. Kramer, H. Pauly, C. Schlier, and P. Toschek, *Z. Physik.*, **166**: 406 (1962).
6. Busch, Fr. von, H. J. Strunk, and C. Schlier, *Z. Phys.* **199**: 518 (1967).
7. Chantry, P. J., *J. Chem. Phys.* **55**: 2746 (1971).

38

Reprinted from *J. Chem. Phys.* **39**:197–200 (1963)

Dissociative Ionization of CO by Ion Impact*

Clayton F. Giese and William B. Maier II

Department of Physics, University of Chicago, Chicago, Illinois

INTRODUCTION

THE bond-dissociation energy of CO and the related quantity, the heat of vaporization of carbon were, up to recent years, subjects of a long controversy.[1] After several decisive experiments,[2–4] there is now general agreement that $D(CO) = 11.11$ eV and $L(C) = 170$ kcal/mole. A few pieces of evidence remain, however, which are not in accord with the new measurements.

In particular, the work of Lindholm[5,6] on the dissociative ionization of CO by ion impact would seem to favor $D(CO) = 9.61$ eV. Lindholm uses an apparatus in which a primary ion beam selected by one mass spectrometer crosses a reaction chamber. Reaction products are extracted from the chamber at right angles to the primary ion direction, and analyzed by a second mass spectrometer. This second mass spectrometer is arranged to discriminate against secondary ions which acquire appreciable forward momentum in the interaction. Further, Lindholm believes it to be improbable that the kinetic energy of the low-energy primary ion enters into the energetics of reactions of the class he studies. He thus concludes that for any reaction which he detects with fairly large observed cross section, the energy imparted to the target molecule in the interaction equals the recombination energy of the primary ion used.

Lindholm observes the reaction $Ne^+ + CO \rightarrow C^+ + O + Ne$ and believes it to be induced by charge exchange. Thus, the semiquantitative theory of Massey and Burhop[7] should apply. Their predictions for nonresonant charge exchange are that the maximum cross section should occur for the ion velocity

$$v_m = a \mid \Delta E \mid /h, \qquad (1)$$

and that for $v \ll v_m$ the cross section should have the form:

$$\sigma \propto \exp(-a \mid \Delta E \mid /4hv), \qquad (2)$$

where a is a length found empirically to be about 7×10^{-8} cm in typical cases and ΔE is the recombination energy of the primary ion minus the energy required to produce the particular state of the secondary ion involved. Using well-known ionization potentials,[8] one may readily calculate that for $D(CO) = 9.61$ eV, $\Delta E = +0.68$ or $+0.78$ eV, while for $D(CO) = 11.11$ eV, $\Delta E = -0.82$ or -0.72 eV. The former number in each case is for recombination of the $^2P_{\frac{1}{2}}$ state, the latter for the $^2P_{\frac{3}{2}}$ state. Experimentally, Lindholm measures a cross section for this reaction which is fairly large and which increases with decreasing ion kinetic energy, down to 25 eV. If $\Delta E = -0.72$ eV, one would predict a cross section which is very small at low energies and which reaches a maximum at about 3000 eV. Lindholm thus concludes that ΔE must be positive; takes the charge exchange to be a resonance or near-resonance process to a state of CO^+ excited beyond the dissociation limit, the excess energy coming off as kinetic energy of the C^+ and O fragments; and concludes that $D(CO) = 9.61$ eV.

In Lindholm's later paper,[6] published after it had become clear that $D(CO) = 11.11$ eV, he finds that new measurements have not resolved the difficulty. He concludes that Eqs. (1) and (2) are not applicable to this case and that the cross section must exhibit a decrease below the minimum energy obtained in his studies. The measurements reported here were made in an attempt to resolve this difficulty.

MEASUREMENTS

The measurements in the present report were made with the double mass spectrometer which has been described.[9] Because C^+ ions were produced with a distribution of kinetic energies in the ion source of the source mass spectrometer, there was a rather large

---

* This research has been supported by grants from the National Science Foundation, the Louis Block Fund for Basic Research, and the Esso Education Foundation.

[1] A. G. Gaydon, *Dissociation Energies and Spectra of Diatomic Molecules* (Chapman and Hall Ltd., London, 1953); L. Brewer and A. Searcy, Ann. Rev. Phys. Chem. **7**, 259 (1956).
[2] L. M. Branscomb and S. J. Smith, Phys. Rev. **98**, 1127 (1955).
[3] C. R. Lagergren, Dissertation Abstr. **16**, 770 (1956).
[4] W. A. Chupka and M. G. Inghram, J. Chem. Phys. **59**, 100 (1955).
[5] E. Lindholm, Arkiv Fysik **8**, 433 (1954).
[6] E. Gustafsson and E. Lindholm, Arkiv Fysik **18**, 219 (1960).
[7] H. S. W. Massey and E. H. S. Burhop, *Electronic and Ionic Impact Phenomena* (Oxford University Press, New York, 1952).
[8] C. E. Moore, "Atomic Energy Levels" National Bureau of Standards Circ. 467, Vol. I (1949).
[9] C. F. Giese and W. B. Maier II "Energy Dependence of Cross Sections for Ion-Molecule Reactions. Transfer of Hydrogen Atoms and Hydrogen Ions" (to be published).

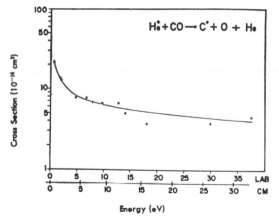

FIG. 1. Cross section for the reaction $He^++CO \rightarrow C^++O+He$ as a function of He^+ ion kinetic energy. Two energy scales are given, the upper scale for laboratory energy, the lower for energy in the center-of-mass system.

background of C^+ which leaked through this mass spectrometer when it was tuned to Ne^+ ions, particularly at low accelerating voltages. For this reason, it was necessary to obtain the Ne^+ ions of low kinetic energy by retardation from initial energies of 6 to 12 eV. Further, to obtain Ar^+ kinetic energies greater than 20 eV, it was necessary to accelerate the ions after analysis in the source mass spectrometer. In cases where it was possible to cross check a result by obtaining the same kinetic energy both with and without retardation and acceleration, the agreement was adequate.

DISCUSSION OF RESULTS

Measured cross sections for the reactions $He^++CO \rightarrow C^++O+He$, $Ne^++CO \rightarrow C^++O+Ne$, and $Ar^++CO \rightarrow C^++O+Ar$ are given as functions of primary ion

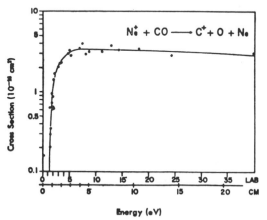

FIG. 2. Cross section for the reaction $Ne^++CO \rightarrow C^++O+Ne$ as a function of Ne^+ ion kinetic energy. Two energy scales are given, the upper scale for laboratory energy, the lower for energy in the center-of-mass system.

kinetic energy in Figs. 1–3. To exhibit best the interesting onsets of these cross sections, the results are presented in semilogarithmic plots.

The reaction $He^++CO \rightarrow C^++O+He$ is definitely exothermic, ΔE being $+2.21$ eV for $D(CO)=11.11$ eV. Thus, one expects no threshold behavior, and, indeed, the cross section (Fig. 1) increases as the ion kinetic energy is reduced, down to the lowest energy (0.6 eV) obtained.

The reaction $Ne^++CO \rightarrow C^++O+Ne$ is endothermic, with $\Delta E=-0.72$ eV. This means that *regardless of the detailed mechanism assumed for the reaction* the reaction

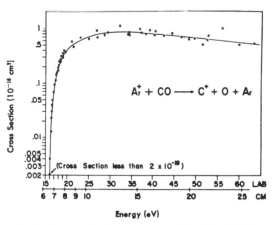

FIG. 3. Cross section for the reaction $Ar^++CO \rightarrow C^++O+Ar$ as a function of Ar^+ ion kinetic energy. Two energy scales are given, the upper scale for laboratory energy, the lower for energy in the center-of-mass system.

cannot occur below a laboratory ion kinetic energy of E_{lab}, where

$$E_{lab}=0.72 \text{ (total mass/target mass)}=1.23 \text{ eV.}$$

We see from Fig. 2 that the cross section does drop off sharply below an ion kinetic energy of 5 eV. Since, at present, we have no theory to provide a guide to the proper technique for establishing a threshold energy, it is dangerous to take the onset energy seriously. A vertical line through the steep portion of the curve cuts the energy axis at 1.3 ± 0.1 eV, in good agreement with the threshold energy above.

Another endothermic process is $Ar^++CO \rightarrow C^++O+Ar$, for which $\Delta E=-6.62$ eV for the $^2P_{\frac{1}{2}}$ state of Ar^+, and -6.44 eV for the $^2P_{\frac{3}{2}}$ state.[8] The laboratory threshold energy should be 6.44 $(68/28)=15.65$ eV. Figure 3 shows that the cross section for this reaction drops off at low energies. Again, one presently has no guide to the proper means for establishing a threshold, but the steep portion of the curve cuts the axis at 16.0 eV, and it appears that any reasonable criterion for picking a threshold will give a result within 0.5 eV of this value.

TABLE I. Comparison of present cross sections with cross sections of Lindholm.[a]

Reaction	Energy	Present results $(10^{-16}$ cm$^2)$	Lindholm $(10^{-16}$ cm$^2)$	Ratio	γ
He$^+$+CO→C$^+$+O+He	25 eV	4.7	1.7	0.36	0.379
Ne$^+$+CO→C$^+$+O+Ne	25 eV	3.0	0.8	0.26	0.806
Ar$^+$+CO→C$^+$+O+Ar	50 eV	0.65	0.06	0.093	1.04

[a] See reference 6.

The onset of the reaction involving Ar$^+$ is not as sharp in terms of laboratory energy as for the reaction with Ne$^+$; however, in terms of energy in the center-of-mass system, they are more comparable, the reaction reaching full development in about 2.5 eV in the case of the Ne$^+$ reaction and in about 5 eV in the case of the Ar$^+$ reaction. Part of the difference is certainly the spread in ion energy from the small mass spectrometer. The effect of this energy spread is impossible to assess quantitatively because the distribution of ions with energy is not well known, but we believe the size of the effect to be no more than a few tenths of an electron volt. A further cause for differences in the threshold behavior for these reactions lies in the fact that the ion beam is a mixture of two states separated by 0.097 eV for Ne$^+$ and 0.178 eV for Ar$^+$. These two causes account only for a small portion of the smearing out of the onset; the remainder must be attributed to the characteristics of the reaction. Very possibly the form of the cross-section curve just above threshold is governed by available phase space as a function of excess energy.

The detailed nature of the reactions studied here is not clear. It is not possible, at present, to tell whether the reaction proceeds by formation of an excited CO$^+$ ion through charge transfer, followed by dissociation, or by a direct interaction of the three atoms followed by a separation of the three particles. It may be possible to investigate the mechanism by which the reaction proceeds by studying the cross sections for the formation of CO$^+$ and O$^+$.

The discussion in Ref. 9 of the uncertainties involved in the results applies here, also, except that in calculating the present results, the K_2 of this earlier report was always set equal to unity. We have no reliable information as to the value which K_2 actually assumes, but rough estimates—see below, Fig. 3 in Ref. 9, and Table I—indicate that K_2 is probably not a strongly varying function of energy over the energy ranges for which we have data. For the endothermic reactions presented here, K_2 might be as large as two for the higher energies and somewhat smaller than unity for points nearest the thresholds. For the reaction He$^+$+CO→C$^+$+O+He, K_2 could conceivably be 4 or so and would increase at lower energies. Thus, the cross sections given here may be in error by a rather sizeable multiplicative factor, but the relative energy dependence should not

be greatly different from that of the true, total cross sections.

In comparing the present results with those of Lindholm,[6] one must keep in mind that the experiments are quite different. The present experiment differs from Lindholm's in that there is no discrimination against processes in which substantial momentum is transferred from the primary to the secondary ion. Obviously, if a reaction is endothermic and requires that the kinetic energy of the primary ion be utilized, the products, neutral and charged, must acquire momentum in the forward direction. Table I compares the present cross sections with those of Lindholm.[6] Lindholm's results correspond to the differential cross sections integrated over some effective solid angle centered around 90° to the primary ion direction in the laboratory system, while our values correspond to an integration over a fairly large solid angle centered around 0°. Thus, one expects that if the reactions studied are really the same, Lindholm's values should be smaller than the present results. At present, it is not possible to predict what the ratio of Lindholm's cross sections to those given here should be because one knows neither the solid angle of acceptance of Lindholm's apparatus nor how to treat the complicated three-body separation, but the following very crude analysis may be of some value. If we assume that the excess energy in the center-of-mass system is shared equally among the three particles, we can calculate the ratio γ of the speed of the center of mass in the laboratory to the speed of the C$^+$ ion in the barycentric system. Values of γ are given in Table I. The larger the value of γ, the more the C$^+$ ions tend to be distributed in the original direction of the primary ion, which explains, qualitatively, the variation in the ratio of Lindholm's cross sections to those presented in this paper.

CONCLUSIONS

As a result of the present experiment a number of conclusions can be drawn:

(1) The difficulty of relating Lindholm's[6] results to the well-accepted bond energy of CO is now resolved.

(2) Lindholm's apparatus does not appear to discriminate very effectively against processes in which momentum is transferred to the secondary ion.

(3) It would appear to be quite risky to assume, as has been the practice, that the total energy given to the target system equals the recombination energy of the projectile ion, independent of its kinetic energy.[10]

---

[10] This has also been suggested by Tal'roze on the basis of some comparisons of mass spectra obtained after charge transfer from primary ions of varying kinetic energy. V. L. Tal'roze, Izv. Akad. Nauk S.S.S.R. Ser. Fiz. 24, 1001 (1960).

(4) The efficiency of processes of the type studied here, in which kinetic energy of the projectile particle is utilized, is surprisingly high, and is worthy of some theoretical treatment.

(5) If the method used in the present report proves to have any generality, it appears that it is a technique capable of measuring dissociation energies with an accuracy, perhaps, of 0.1 eV.

39

Reprinted from *J. Chem. Phys.* **55**:2724–2733 (1971)

Electron Affinities of Halogen Diatomic Molecules as Determined by Endoergic Charge Transfer*

W. A. Chupka and J. Berkowitz

Argonne National Laboratory, Argonne, Illinois 60439

AND

David Gutman

Department of Chemistry, Illinois Institute of Technology, Chicago, Illinois 60616

(Received 4 August 1970)

Ion-pair formation by photon absorption at threshold wavelengths has been used to prepare I^-, Br^-, and F^- ions with approximately room-temperature thermal energies, as verified by retarding-potential measurements. The primary ions were accelerated and their reactions with halogen molecules studied at laboratory kinetic energies from 0.0 to about 4.0 eV. Thresholds were determined for endoergic reactions of the type $X^- + Y_2 \rightarrow X + Y_2^-$, where X may be the same as Y. At least two reactions were used in determining each electron affinity. The agreement was good in all cases. The values of electron affinity obtained are 3.08 ± 0.1 eV for F_2, 2.38 ± 0.1 eV for Cl_2, 2.51 ± 0.1 eV for Br_2, and 2.58 ± 0.1 eV for I_2. Interhalogen molecular ions such as IBr^- were also observed, and measurement of the threshold for formation gave the value 2.7 ± 0.2 eV for the electron affinity of IBr. The retarding-potential measurements of F^- from F_2 strongly support a value for the dissociation energy of F_2 in the neighborhood of 1.6 eV.

I. INTRODUCTION

The electron affinities of atoms and molecules are quantities of considerable importance in many areas of chemistry and physics. A wide variety of techniques have been used to measure these quantities. These techniques and their results have been described in several recent review articles.[1-3] The most reliable and accurate determinations of electron affinities have been made for atoms and a few radicals; there appears to be no stable molecule for which a very accurate and generally accepted value of the electron affinity has been determined.

Recently several techniques which are particularly suitable for the determination of the electron affinities of stable molecules have been developed or attempted. One such technique involves the study of reactions of negative ions with molecules. Thus Henglein and Muccini[4] observed charge transfer from negative ions to stable neutral molecules at moderate repeller fields in a single-chamber ion source. They concluded that the observed reactions must be exothermic and were thus able to establish the relative order of the electron affinities of several molecules. Similar studies were carried out by Kraus *et al.*[5] More recently the tandem-mass-spectrometer technique has been used by Vogt *et al.*,[6,7] by Paulson,[8] and by Bailey *et al.*[9] to study charge-transfer processes and ion–molecule reactions over a range of kinetic energies from the order of 1 eV up to several hundred electron volts. However, it has not been possible to decrease the ion kinetic energies down to the thermal range, and furthermore the reactant ions have a rather large spread in kinetic energy. For these reasons, accurate thresholds could not be determined for endoergic processes. Instead a monotonic increase of cross section with decreasing kinetic

energy is taken to indicate that the reaction is exoergic, while for an endoergic reaction the cross section must vanish at sufficiently low energy. With these criteria, molecules could be ordered according to their electron affinities. However, the inability to obtain good data down to thermal kinetic energies and the unknown states of vibrational excitation of molecular ions used as projectiles are serious limitations of this technique.

A related technique has been used by Helbing and Rothe[10] and by Baede *et al.*[11] These workers use velocity-selected beams of alkali atoms and measure the cross section for charge transfer to halogen diatomic molecules as a function of the velocity of the alkali atoms. They were able to determine thresholds for the charge-transfer processes and thereby obtain values for the electron affinity of the halogen molecule.

In the experiments described here, we prepare atomic halogen negative ions by the ion-pair formation process

$$X_2 + h\nu \rightarrow X^+ + X^-. \tag{1}$$

This process occurs at photon energies below the ionization potential of X_2 for all the halogens except Cl_2. The X^- ions are then accelerated and undergo endoergic reactions with the halogen molecule Y_2 (which may be the same as X_2) to produce Y_2^- and XY^-, i.e.,

$$X^- + Y_2 \rightarrow X + Y_2^-, \tag{2}$$

$$X^- + Y_2 \rightarrow XY^- + Y. \tag{3}$$

The kinetic energy thresholds for the reactions are measured, and these yield a value (or at least a lower limit) for the electron affinity of Y_2 and, in some cases, XY as well.

II. EXPERIMENTAL APPARATUS AND PROCEDURE

The apparatus used in the present experiment is a combination vacuum uv monochromator and mass spectrometer which has been described previously.[12,13] The major innovation was in the design of the ionization and collision chamber. Instead of a single ionization chamber such as was used in earlier studies of ion–molecule reactions,[14] the apparatus had a double chamber, shown schematically in Fig. 1. The gases X_2 and Y_2 are introduced into the chamber through separate inlets in order to minimize the formation of interhalogen compounds. The chamber is so open internally that no attempt is made to maintain any significant concentration difference within. The primary X^- ions are formed by photon absorption in the region between the repeller plate R and the first grid G_1. A very small potential difference (between 0.1 and 0.2 V) was maintained between R and G_1 in order to repel the X^- ions toward G_1. Of this potential drop, the portion across the region of ion formation was kept at about 0.03 eV or less in order to minimize the kinetic energy spread of the projectile ion. The projectile was then accelerated to the desired collision energy by a potential applied between the closely spaced grids G_1 and G_2. The region between G_2 and G_3 is a field-free region in which the reactions of interest occur. Those primary and reaction-product ions that go through grid G_3 are then accelerated and focused into a magnetic mass spectrometer as previously described.[12] The purpose of the present source is to measure the relative microscopic cross sections of endoergic reactions as a function of kinetic energy rather than to measure the integral phenomenological cross section. (The latter is the quantity that is usually measured directly in the typical single-chamber ion source.)

The accelerating potential between grids G_1 and G_2 was supplied as a staircase voltage that also controlled the address advance of a multichannel scaler into which the mass-analyzed ion signal of interest was fed. The accelerating potential was swept repetitively over a chosen range until the desired number of counts was accumulated. This electronic system has been described in detail in an earlier publication[15] dealing with its

TABLE I. Thresholds for ion-pair formation.

Molecule	Calc. threshold (Å)	Wavelength used (Å)	D_0 (eV)	E.A. (atom) (eV)
F_2	797	793	1.59	3.448
Cl_2	1044	1047	2.475	3.613
Br_2	1186	1190	1.971	3.363
I_2	1388	1388 (also 1395)	1.542	3.063

application to photoelectron energy analysis by a retarding-potential technique. In its present application, it served a similar purpose in part. In most experiments the accelerating potential was varied from slightly negative values to an appropriate maximum positive value. This provided a retarding-potential analysis of the kinetic energy of the projectile ion and a determination of the zero of the energy scale. The zero was placed at the position of the half-maximum of the total sharply rising part of the curve for the primary ion.

In the case of an endoergic reaction at threshold, the reaction products must move in the direction of motion of the center of mass of the system, i.e., very nearly in the direction of motion of the projectile ion. Thus at threshold, primary and product ions are collected with very nearly the same efficiency and the ratio of product to primary ion intensities will be proportional to the cross section for the reaction. (The mass spectrometer is operated with such large slits that the small difference in kinetic energy results in no significant difference in collection efficiency.) However, at higher collision energies, this ratio can give an underestimate of the cross section if the product ions are appreciably scattered away from the direction of the reactant-ion trajectory. The outer walls of the collision chamber as well as the plates containing grids G_2 and G_3 were insulated from one another and could be placed at different potentials so as to optimize the collection of reaction products. It was found that such adjustments led to only negligibly small increases in the ratio of product to primary ions at the energies encountered in the present experiments. This finding suggests that the angular distribution of product ions may be strongly peaked along the projectile trajectory. In the measurements of thresholds reported here, all walls of the collision chamber were kept at the same potential in order to provide a field-free collision region. In the final measurements, the voltage increment used was 0.02 eV.

All four atomic halogen negative ions were used as primary reactant ions in these experiments, although the most suitable and frequently used ones were I^- and Br^-. Table I gives the calculated thresholds for the ion-pair formation process and the experimental wavelengths that correspond to local maxima in the cross section for ion formation. In the calculation of thresh-

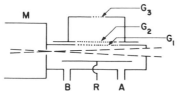

FIG. 1. Diagram of ionization-collision chamber. A and B are gas inlets, M is a monochromator with exit slit, and G_1, G_2, and G_3 are grid plates.

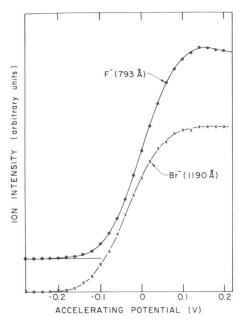

FIG. 2. Retarding-potential curves for kinetic energy analysis of Br⁻ ions produced from Br₂ at 1190 Å and of F⁻ ions produced from F₂ at 793 Å.

olds, the dissociation energies of the diatomic molecules (except for F_2) were taken from Herzberg[16] and the electron affinities from those given by Berry.[1]

The case of F_2 requires further discussion. A recent measurement by photoionization[17] has been interpreted to yield a value of $D_0^0(F_2) = 1.34 \pm 0.03$ eV $= 30.9$ kcal mole⁻¹, which is considerably lower than previously accepted values, e.g., 36.76 ± 0.06 kcal mole⁻¹ $- 1.59$ eV as given by Wagman et al.[18] Still more recently F_2 has been investigated by photoionization mass spectrometry[19] and the data were interpreted to give $D_0^0(F_2) = 1.59 \pm 0.01$ eV. Of interest here are the data on the process of ion-pair formation. The experimental data in Refs. 17 and 19 are practically identical. The threshold occurs in the neighborhood of 800 Å, though somewhat different values of the threshold wavelength were inferred because the two groups took account of thermal energy in different ways. However, the major difference is one of interpretation. Berkowitz et al.[19] find the threshold to be consistent with the dissociation energy determined by the dissociative-ionization process and conclude that the ions are formed at threshold with no excess kinetic energy. Dibeler et al.[17] are forced to conclude that the ion-pair formation process releases the ions with about 0.2 eV kinetic energy at threshold. Such excess kinetic energy (0.1 eV for the F⁻ ion) would be a significant source of error in the present work.

A retarding-potential curve was therefore determined for the F⁻ ion produced from F_2 at 793 Å and compared with a similar curve for Br⁻ formed from Br₂ at 1190 Å.

Both curves (Fig. 2) were taken under the same conditions (except for the wavelength of light used). The position of half-maximum occurs at a slightly more negative value for Br⁻ (indicating slightly more kinetic energy), but the difference (∼0.03 eV) is approximately equal to the experimental error. Since the Br⁻ ions have no excess kinetic energy (other than that due to the original thermal energy of the Br₂ molecule), it must be concluded that the F⁻ ions have none also. If the value $D_0^0(F_2) = 1.34$ eV for the dissociation energy of F_2 were correct, the F⁻ ions produced at 793 Å would have about 0.15 eV kinetic energy. This is clearly not the case and this observation provides further confirmation for the higher value[19] of $D_0^0(F_2)$.

The case of Cl_2 also requires some comment. Cl_2 is unique among the halogens in that the threshold for ion-pair formation occurs at energies above the ionization potential of the molecule. Thus Cl⁻ ions cannot be prepared in the absence of electrons. Also Cl⁻ ions have been observed[20] at energies much lower than the threshold calculated in Table I and were found to be produced by the process

$$Cl_2^* + Cl_2 \rightarrow Cl_2^+ + Cl^- + Cl. \qquad (4)$$

In the present experiment, it was found that the Cl⁻ ion was produced only in rather low abundance in the region of the calculated threshold for ion-pair formation, and a retarding-potential energy analysis indicated that while most of the Cl⁻ ions possessed kinetic energies in the thermal range, a small tail—possibly resulting from Process (4)—extended to higher energies. Hence Cl⁻ was not a very suitable reactant ion and was used only in collisions with Cl_2.

The ions Br⁻ and I⁻ were produced in good abundance at the calculated thresholds. In the case of I⁻ a number of experiments were also carried out at 1395 Å, which corresponds to a "hot band," since it seemed possible that some narrowing of the kinetic energy distribution of the ions might result. However, neither the retarding-potential measurement of the kinetic energy nor the thresholds for endoergic charge transfer was found to differ outside of experimental error from the results using 1388-Å photons. Therefore the thresholds reported here were measured by use of the more intense ion-formation process at 1388 Å.

The pressure of sample gas in the collision chamber was not accurately measured in these experiments. A rough estimate based on ionization-gauge readings indicated that the total pressure was of the magnitude of 10⁻³ torr. At higher pressures, the retarding-potential curve for the primary ion began to degrade. During the measurements reported here, the sample pressure was always kept at a value well below that at which significant degradation occurred.

The measurements were taken in the following manner. First the reactant-ion intensity was measured as a function of the accelerating voltage (between grids

G_1 and G_2). Conditions of pressure, repeller voltage, etc., were adjusted until no further improvement in the sharpness of the retarding potential curve could be obtained. Then the reactant-ion intensity was again measured as a function of the accelerating voltage, and the corresponding measurement for the product ion followed immediately. Each measurement was made at least twice. The product-ion intensity (usually very small) always rose at the same voltage as the reactant-ion intensity did. This initial rise was due to ions formed by collisions which occurred after the primary ions passed through grid G_3 and during their subsequent acceleration into the focusing system of the mass spectrometer. This background was not serious and was to some extent useful since at its threshold it had the same shape as the curve for the primary ion and served to monitor the apparent energy distribution of the primary ion in the course of the measurement of the product ion. The curve for the product ion was divided by that for the reactant ion to give the relative cross section for the reactant as a function of collision energy.

III. EXPERIMENTAL RESULTS

A. General

More than one reactant ion was used for each halogen gas. One of the reactant ions was always the atomic negative ion of the halogen gas being studied. When a mixture of two halogens was used (one to provide the projectile ion and the other the target molecule), there was an additional complication due to an exothermic (or nearly thermoneutral) ion–molecule reaction which occurred in the ionization region of the source. The reaction

$$X^- + Y_2 \rightarrow XY + Y^- \qquad (5)$$

is exothermic or only slightly endothermic for all pairs of halogens investigated here. Thus, in addition to the projectile ion X^- formed by photon absorption in X_2, a large amount of Y^- formed by the ion–molecule reaction was found present in the primary beam at large concentrations of Y_2. Since the ratio Y^-/X^- is proportional to the pressure of the target molecule Y_2, the latter pressure was decreased until the amount of Y^- in the primary beam was negligible for the purpose of this experiment. The rate of formation of Y_2^- by Reaction (2) was also decreased under these conditions, and therefore the experiments with mixed halogens required much longer measuring times than those with only a single gas.

B. Determination of Thresholds

The experimental relative cross-section curves, given by the ratio of product ions to reactant ions as a function of accelerating voltage, all show a significant amount of curvature at the threshold. This curvature extends over a range of energy consistent with the

thermal energy distribution of the target gas and the kinetic energy distribution of the projectile ion as given by its retarding-potential curve. Unfortunately the threshold behavior of the cross section for reactions of the type studied here is not known.

Phase-space arguments reviewed by Truhlar[21] suggest that the cross section σ near threshold for an ion–molecule reaction in which two particles collide to give two other particles may be given by

$$\sigma(E_{ex}) = CE_{ex}^n, \qquad (6)$$

where E_{ex} is the excess translational energy in the center of mass, C is a constant, and n is a number between 1 and 5/4. Thus the phase-space arguments suggest an abrupt threshold with an approximately linear energy dependence of the cross section for our processes. However, this dependence is expected to hold only over an energy interval of about 0.1 eV or less. Phase space calculations extended over larger energy ranges (Ref. 21 and references given therein) generally show a gradual departure from linearity with negative curvature, the cross section reaching a maximum and eventually decreasing with energy. Previous experimental data[22] show this qualitative behavior as do our experimental curves, the curvature at the threshold being of the magnitude expected from the thermal energy spread of the reactants.

The effect of thermal energy and especially of the thermal kinetic energy of the target gas can be very appreciable. Even for monoenergetic projectiles, the spread in relative translational energy of the collision partners can be very much more than the average thermal energy of the target gas. This can easily be shown by an example similar to one Chantry and Schulz[23] give to explain the large spread in kinetic energy of the fragments produced by dissociation of molecules with a random thermal kinetic energy distribution. Consider a beam of monoenergetic ions of mass m_0 and of velocity v_0 and target gas molecules of mass m_t with velocities either $+v_t$ or $-v_t$ along the direction of the ion trajectory. The relative velocity of collision will be $v_0 \pm v_t$ and the energy spread is given by

$$\tfrac{1}{2}\mu\{(v_0+v_t)^2 - (v_0-v_t)^2\} = \tfrac{1}{2}\mu(4v_0v_t), \qquad (7)$$

where μ is the reduced mass. This quantity can be much more than $\tfrac{1}{2}m_t v_t^2$ when $v_0 > v_t$.

If we assume that $\tfrac{1}{2}m_0 v_0^2 \gg kT$, which will be a sufficiently good approximation for present purposes, then for monoenergetic ion beam incident on a random thermal gas, the distribution of the relative collision energy will be nearly symmetric about the value for the motionless gas. In the special case in which the cross section for the reaction of interest is a linear function of the excess collision energy above threshold, it can be shown that convolution with a distribution function which is symmetric about the value for the motionless gas yields a curve which is linear at high energies and

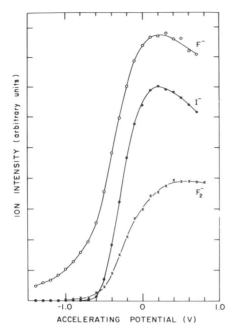

FIG. 3. Retarding-potential curves for F^-, F_2^-, and I^- ions formed at 1388 Å in a mixture of I_2 and F_2. The primary ion formed is I^-.

has a curved section in the threshold region. However, the linear portion will coincide with the original straight line and thus will extrapolate to the proper threshold. Thus, in this case while the curve may have a very extensive tail, extrapolation of the linear section will nevertheless yield the desired threshold value.

Methods for properly calculating the distribution of relative collision energy for various experimental situations have been given earlier.[24,25] These methods have recently been applied to the case of a beam of projectiles incident on a random thermal gas.[26,27] For monoenergetic projectiles incident on a random thermal gas at temperature T, Chantry[26] has shown that to a very good approximation $W_{1/2}$, the full-width at half-maximum of the distribution of relative collision energy is given by

$$W_{1/2} = (11.1\gamma kTE_0)^{1/2}, \qquad (8)$$

where $\gamma = m_0/(m_0 + m_t)$ and E_0 is the nominal kinetic energy in the center-of-mass system. This expression is very useful in estimating the amount of curvature to be expected at threshold and in determining the position of threshold.

For the reactions studied here, the values of $W_{1/2}$ in the region of the threshold are in the range 0.15–0.4 eV with most of them in the lower part of that range. For a cross section that varies linearly with excess energy, the distribution given by Chantry[26] yields a tail in the experimental curve but linear extrapolation yields a

threshold value which is shifted to lower energies only by the amount $3\gamma kT$ which is nearly negligible. The largest errors will result in the case of a step-function threshold. Chantry[26] has shown that convoluting the energy distribution in the center-of-mass system with such a step function yields a sigmoid curve and that the extrapolation of the nearly linear portion associated with the inflection point yields a threshold value which is too low by very nearly $0.6W_{1/2}$ where $W_{1/2}$ is evaluated at the threshold energy E_0. None of the present data for the reactions yielding Y_2^- are found to fit the assumption of a step-function threshold. However, the shapes of the experimental curves indicate that the cross-sections curves all depart from linearity in that they possess a small negative curvature. The result will be that the extrapolated threshold will be somewhat shifted to lower energies.

The effect of energy distribution is taken into account in the following approximate manner. The linear portion of the cross-section curve is extrapolated to zero. To the value of the collision energy so determined is added 0.025 eV, the average rotational energy of the target molecules (the vibrational energy is negligible) and $0.2W_{1/2}$ where $W_{1/2}$ is evaluated at E_0 by successive approximation. The total correction ranges from about 0.06 to about 0.1 eV and is the major source of error in determining the value of the threshold. The reproducibility of determining the intercept of the extrapolated linear portion is much better than this.

A more conservative treatment might have employed a correction of $0.3W_{1/2}$ with an estimated error of somewhat greater than that quantity in order to completely span the range of threshold behavior from linear to steplike. However, trial convolutions of the energy distribution with estimated cross section curves indicate that the latter have an energy dependence closer to linear than steplike. Hence the lesser correction employed above should be more accurate, although the difference is only a few hundredths of an electron volt.

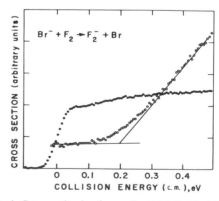

FIG. 4. Cross section for the reaction $Br^- + F_2 \rightarrow F_2^- + Br$ as a function of collision energy (open circles). Solid points show the primary ion current.

The above corrections are based on the assumption that all forms of internal and kinetic energy are equivalent in causing reaction to occur. This may not be true but the fraction of vibrationally excited molecules is very small and it is unlikely that rotational energy is much more effective than translational energy in causing reaction. Therefore it is unlikely that large errors result from failure of this assumption.

C. Fluorine

When I^- ions were formed in a mixture of F_2 and I_2 gases, it was found that large amounts of F_2^- ions were produced at the lowest accelerating voltages by the reaction

$$I^- + F_2 \rightarrow F_2^- + I. \qquad (9)$$

Therefore it was concluded that this reaction is either exothermic or approximately thermoneutral and hence that $E.A.(F_2) \gtrsim E.A.(I) = 3.063$ eV. It was also found

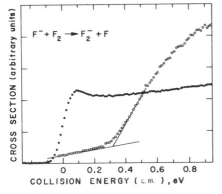

FIG. 5. Cross section for the reaction $F^- + F_2 \rightarrow F_2^- + F$ as a function of collision energy (open circles). Solid points show the primary ion current.

that the F^- ion was formed in even larger abundance than F_2^- by the exothermic reaction

$$I^- + F_2 \rightarrow IF + F^-. \qquad (10)$$

Retarding-potential curves were taken for all three ions observed and are shown in Fig. 3. It is seen that although the F_2^- ions have nearly the same kinetic energy distribution as the reactant I^- ions, the F^- ions have a high kinetic energy tail due to the exothermicity of Reaction (10). Division of the curve for F_2^- by that for I^- indicates that the cross section for Reaction (9) increases with energy for a short interval above zero. This might indicate that Reaction (9) is endothermic by a very small amount (comparable to kT). Alternatively, this behavior may be due to more favorable competition with the very exothermic reaction (10) at higher kinetic energies.

The reactions of Br^- and of F^- with F_2 to form F_2^- were found to be endoergic. The relative cross sections for these reactions as a function of collision energy are

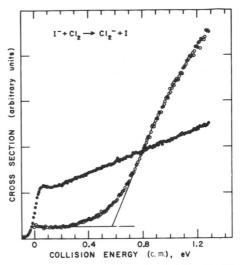

FIG. 6. Cross section for the reaction $I^- + Cl_2 \rightarrow Cl_2^- + I$ as a function of collision energy (open circles). Solid points show the primary ion current.

shown in Figs. 4 and 5. The thresholds were determined as previously described and the resulting values for the electron affinity of Br_2 are given in Table II.

D. Chlorine

The charge-transfer reactions of all halide ions with Cl_2 were found to be endoergic. Figures 6 and 7 show the data for I^- and Br^- reactant ions, respectively. The Cl^- ion was not a very suitable reactant ion as explained in Sec. II. Nevertheless it was used to determine a rough value for the threshold, which was in good agreement with those determined by using I^- and Br^-

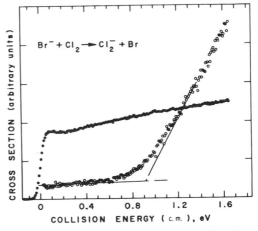

FIG. 7. Cross section for the reaction $Br^- + Cl_2 \rightarrow Cl_2^- + Br$ as a function of collision energy (open circles). Solid points show the primary ion current.

TABLE II. Measured thresholds and electron affinities.

Projectile (X⁻)	Target (Y₂)	Product ion	Threshold correction (eV, c.m.)	Corrected threshold (eV, c.m.)	E.A.(Y₂) (eV) Experimental values	E.A.(Y₂) (eV) Selected value
I^-	F_2	F_2^-		$\lesssim 0$	$\gtrsim 3.06$	
Br^-	F_2	F_2^-	0.07	0.27	3.10	3.08 ± 0.10
F^-	F_2	F_2^-	0.06	0.38	3.07	
I^-	Cl_2	Cl_2^-	0.09	0.66	2.41	
Cl^-	Cl_2	Cl_2^-		(1.31)	(2.30)	2.38 ± 0.10
Br^-	Cl_2	Cl_2^-	0.10	1.01	2.35	
I^-	Br_2	Br_2^-	0.08	0.59	2.48	2.51 ± 0.10
Br^-	Br_2	Br_2^-	0.08	0.84	2.53	
I^-	I_2	I_2^-	0.07	0.49	2.57	2.58 ± 0.10
Br^-	I_2	I_2^-	0.07	0.77	2.59	
						E.A. (IBr) 2.7 ± 0.2
I^-	Br_2	IBr	0.07	0.43	2.63	
Br^-	I_2	IBr^-	0.06	0.53	2.83	

reactant ions. The values of EA(Cl_2) so determined are given in Table II.

E. Bromine and Iodine, Mixed Halogens

Both Br^- and I^- ions were used in charge transfer with Br_2 and I_2. The experimental data are shown in Figs. 8–11 and the electron affinities given in Table II.

In addition to the process yielding the Y_2^- ion, the process forming the IBr^- ion was studied using $I^- + Br_2$ and $Br^- + I_2$ as reactants. The data are shown in Figs. 12 and 13 and the electron affinity of IBr is given in Table II. The mixed halogen ions XY^- were observed in a number of cases, but no systematic study was carried out except for IBr^-. When observed and measured, the XY^- intensity was appreciably less than that of the Y_2^- ion. Thus in the reaction of Br^- with I_2 at a laboratory collision energy of about 5 eV, the IBr^- intensity was only about a third of the I_2^- intensity.

IV. DISCUSSION

A. Errors

The major experimental uncertainty in the values of electron affinity determined in these experiments is that due to the effect of the thermal energy of the target gas. The good agreement between values determined by different projectiles for which the quantitative effect of this energy spread is different indicates that this error is probably not large.

A proper threshold determination for the reactions

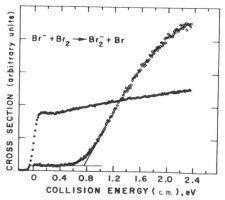

FIG. 8. Cross section for the reaction $Br^- + Br_2 \rightarrow Br_2^- + Br$ as a function of collision energy (open circles). Solid points show the primary ion current.

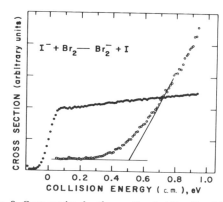

FIG. 9. Cross section for the reaction $I^- + Br_2 \rightarrow Br_2^- + I$ as a function of collision energy (open circles). Solid points show the primary ion current.

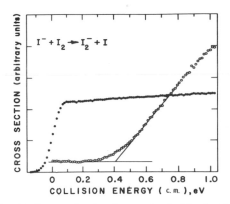

FIG. 10. Cross section for the reaction $I^-+I_2 \rightarrow I_2^-+I$ as a function of collision energy (open circles). Solid points show the primary ion current.

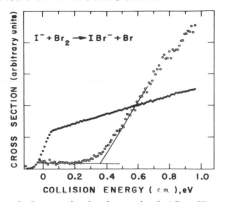

FIG. 12. Cross section for the reaction $I^-+Br_2 \rightarrow IBr^-+Br$ as a function of collision energy (open circles). Solid points show the primary ion current.

studied here yields rigorously only a lower limit to the electron affinity of the halogen molecules. However, the consistency among the various determinations again supports the assumption that the actual value of the electron affinity has been determined.

B. Comparison with Other Data

Table III lists the values of the electron affinities determined in this work as well as some of those determined by other workers. The only other experimental value for the electron affinity of F_2 is that determined by Reese et al.[28] from the appearance potential of F_2^- formed by dissociative electron capture by SO_2F_2. It is in good agreement with our value—well

within the error usually associated with electron-impact appearance potentials.

The electron affinity E.A.(X_2) of the X_2 halogen molecule may be estimated by making use of the suggestion of Mulliken[29] that

$$E.A.(X_2) = E.A.(X) - \tfrac{1}{2}D(X_2), \quad (11)$$

where E.A.(X) is the electron affinity of the X atom and $D(X_2)$ is the dissociation energy of X_2. Values so estimated are given in Table III and are found to be surprisingly close to our measurements.

Wahl et al.[30] have made extensive calculations of the dissociation energies of the negative ions F_2^- and Cl_2^-. By use of values of the electron affinity of the atom

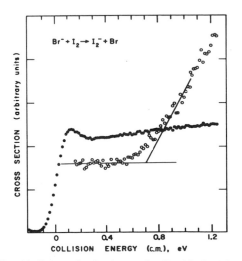

FIG. 11. Cross section for the reaction $Br^-+I_2 \rightarrow I_2^-+Br$ as a function of collision energy (open circles). Solid points show the primary ion current.

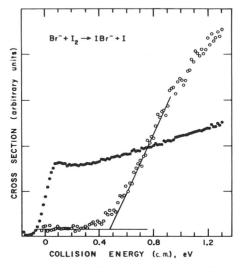

FIG. 13. Cross section for the reaction $Br^-+I_2 \rightarrow IBr^-+I$ as a function of collision energy (open circles). Solid points show the primary ion current.

TABLE III. Values of the electron affinity of halogen molecules.[a]

Molecule	E.A. (eV)	Reference	Method[b]
F_2	3.08 ± 0.1	This work	CTT
	≥3.0	28	AP
	2.7	Eq. (11)	E
	3.49	30	TC
Cl_2	2.38 ± 0.1	This work	CTT
	2.5 ± 0.3	31	CTS
	2.39	30	TC
	2.4	Eq. (11)	E
Br_2	2.51 ± 0.1	This work	CTT
	2.6 ± 0.3	31	CTS
	2.4	Eq. (11)	E
	2.23 ± 0.1	10	CTT
	2.8	11	CTT
I_2	2.58 ± 0.1	This work	CTT
	2.4 ± 0.3	31	CTS
	2.3	Eq. (11)	E

[a] Values quoted are assumed to be adiabatic.
[b] The methods denoted by these abbreviations are as follows: AP—appearance potential, CTS—charge-transfer spectra, CTT—charge-transfer threshold, E—estimate based on Eq. (11), and TC—theoretical calculation.

and the dissociation energy of the neutral diatomic molecule, the electron affinities of the molecules may be calculated and are also given in Table III. Their value for F_2 is higher than ours by 0.4 eV, but these authors estimate that their error may be nearly that large and in the direction to decrease the discrepancy. In the case of Cl_2 the agreement with our value is excellent.

Person[31] has estimated both vertical and adiabatic electron affinities for Cl_2, Br_2, and I_2. The adiabatic values are given in Table III and are in agreement with our values—well within Person's estimated error.

Helbing and Rothe[10] have measured the threshold energy for charge transfer from Cs to Br_2 and determined the electron affinity of Br_2 to be 2.23 ± 0.1 eV, which is somewhat lower than our value. Baede et al.[11] performed similar measurements for charge transfer from K, Na, and Li to Br_2 and obtained the value 2.8 eV for the electron affinity of Br_2. Since the measurements of Helbing and Rothe were performed with two molecular beams intersecting at right angles, the data will be relatively free of the effect of the thermal motion of the target gas as treated by Chantry.[26] This is discussed by Rothe and Fenstermaker.[27]

Baede et al.[11] used a collision chamber filled with halogen gas and furthermore chose their thresholds somewhat arbitrarily a little above the point at which the signal rose above the noise. Their cross sections show considerable curvature at threshold and it is very likely that taking proper account of the effect of the thermal energy of the target gas would yield significantly higher thresholds and thus a lower value for the electron affinity of Br_2.

C. Mechanism of Reactions Studied

Many of the reactions studied here obviously occur by charge transfer. However, when X and Y in Reaction (2) are identical, the reaction may also occur by exchange of atoms. That such an exchange reaction can proceed with significant cross section is suggested by the occurrence of Reaction (3), in which the mixed-halogen negative ion is produced. The present study does not distinguish between the two types of reaction in the case in which X and Y are identical. In isotopic studies of reactions between O^- and O_2 and between O^- and NO_2 to produce O_2^- and NO_2^-, respectively, Vogt[6] found that the probability of charge transfer without atom exchange was at least 10 times that of atom exchange at kinetic energies of 5 eV and higher.

In any case, the reactions studied here must involve crossing of potential surfaces of the fairly stable Y_3^- ion. The absolute values of the cross sections have not been accurately determined here but are roughly estimated to be of the order of 1–10 Å² at their maximum. Thus the crossing of potential surfaces must be quite probable.

* Work performed under the auspices of the U.S. Atomic Energy Commission.

[1] R. S. Berry, Chem. Rev. 69, 533 (1969).
[2] B. Moiseiwitsch, Advan. Atomic Mol. Phys. 1, 61 (1965).
[3] L. M. Branscomb, in Atomic and Molecular Processes, edited by D. R. Bates (Academic, New York, 1962), Chap. 4.
[4] A. Henglein and G. A. Muccini, J. Chem. Phys. 31, 1426 (1959).
[5] K. Kraus, W. Muller-Duysing, and H. Neuert, Z. Naturforsch. 16a, 1385 (1961).
[6] D. Vogt, Intern. J. Mass. Spectry. Ion Phys. 3, 81 (1969).
[7] D. Vogt, B. Hauffe, and H. Neuert, Z. Physik 232, 439 (1970).
[8] J. F. Paulson, J. Chem. Phys. 52, 958, 963 (1970).
[9] T. L. Bailey and P. Mahadevan, J. Chem. Phys. 52, 179 (1970).
[10] R. K. B. Helbing and E. W. Rothe, J. Chem. Phys. 51, 1607 (1969).
[11] A. P. M. Baede, A. M. C. Moutinho, A. E. DeVries, and J. Los, Chem. Phys. Letters 3, 530 (1969).
[12] J. Berkowitz and W. A. Chupka, J. Chem. Phys. 45, 1287 (1966).
[13] W. A. Chupka and J. Berkowitz, J. Chem. Phys. 47, 4320 (1967).
[14] W. A. Chupka, M. E. Russell, and K. Rafaey, J. Chem. Phys. 48, 1518 (1968).
[15] J. Berkowitz and W. A. Chupka, J. Chem. Phys. 51, 2341 (1969).
[16] G. Herzberg, Molecular Spectra and Molecular Structure. I. Spectra of Diatomic Molecules (Van Nostrand, New York, 1950).
[17] V. H. Dibeler, J. A. Walker, and K. E. McCulloh, J. Chem. Phys. 51, 4230 (1969).
[18] D. D. Wagman, W. H. Evans, V. B. Parker, I. Halow, S. M. Bailey, and R. H. Schumm, Natl. Bur. Std. (U.S.) Tech. Note 270-3 (1968).
[19] J. Berkowitz, W. A. Chupka, P. M. Guyon, J. Holloway, and R. Spohr, paper presented at the Eighteenth Annual Conference on Mass Spectrometry and Allied Topics, San Francisco, 14–19 June, 1970.
[20] J. Berkowitz and W. A. Chupka, unpublished data.
[21] D. G. Truhlar, J. Chem. Phys. 51, 4617 (1969).
[22] W. B. Maier II, J. Chem. Phys. 46, 4991 (1967).
[23] P. J. Chantry and G. J. Schulz, Phys. Rev. Letters 12, 449 (1964).
[24] K. Berkling, R. Helbing, K. Kramer, H. Pauly, Ch. Schlier, and P. Toschek, Z. Physik 166, 406 (1962).
[25] Fr. von Busch, H. J. Strunck, and Ch. Schlier, Z. Physik 199, 518 (1967).

[26] P. J. Chantry, J. Chem. Phys. **55**, 2746 (1971), this issue.

[27] E. W. Rothe and R. W. Fenstermaker, J. Chem. Phys. (unpublished).

[28] R. M. Reese, V. H. Dibeler, and J. L. Franklin, J. Chem. Phys. **29**, 880 (1958).

[29] R. S. Mulliken, J. Am. Chem. Soc. **72**, 600 (1950).

[30] A. C. Wahl, P. J. Bertoncini, G. Das, and T. L. Gilbert, Intern. J. Quantum Chem. **1s**, 123 (1967), and private communication.

[31] W. B. Person, J. Chem. Phys. **38**, 109 (1963).

40

Reprinted from *J. Chem. Phys.* **59**:3162–3181 (1973)

Electron affinities from endothermic negative-ion charge-transfer reactions. III. NO, NO$_2$, SO$_2$, CS$_2$, Cl$_2$, Br$_2$, I$_2$, and C$_2$H[*]

B. M. Hughes, C. Lifshitz[†], and T. O. Tiernan[‡]

Aerospace Research Laboratories, Chemistry Research Laboratory Wright-Patterson Air Force Base, Ohio 45433

(Received 26 March 1973)

Translational energy thresholds have been measured for electron transfer reactions from various atomic negative ions to NO, NO$_2$, SO$_2$, CS$_2$, Cl$_2$, Br$_2$, I$_2$, and C$_2$H$_2$ using a tandem mass spectrometer. These thresholds, corrected where necessary for thermal broadening, were utilized with known thermodynamic data to calculate electron affinities for the molecules indicated. These results were also shown to be consistent with electron affinity limits established by observations of certain exothermic reactions of the molecular negative ions. The isotope distributions in the products from the reactions of monoisotopic Cl$^-$ and Br$^-$ with natural Cl$_2$ and Br$_2$, respectively, suggest that these reactions proceed via a linear X$_3^-$ intermediate in which good "mixing" of the available energy occurs. The implications of this model with respect to the significance of the measured energy thresholds is discussed.

I. INTRODUCTION

Increasing recognition of the important role of negative ions in many reactive environments including the normal and perturbed upper atmosphere, flames, lasers and discharges, and biological systems, has prompted a renewed interest in electron affinities. The electron affinity is the most important energetic quantity associated with negative ion formation, since it provides a direct indication of negative ion stability. Electron affinities have been accurately determined for a number of atoms using photodetachment techniques.[1,2] The latter method is generally less definitive for molecules, however, because for these, the vertical detachment energy is not necessarily the same as the electron affinity. While a variety of other experimental methods have also been applied in an effort to measure molecular electron affinities (these are described or summarized in several recent articles[1–4]), the resultant values which have been reported are in considerable disagreement. Thus, the electron affinities for most molecules are not considered to be accurately known.

The existing uncertainties in our present knowledge of electron affinities are largely due to the limitations of the experimental methods previously applied for such determinations. Quite recently, however, two types of collision experiments have been developed which appear to be generally applicable for obtaining accurate data for molecular species. Both of these are beam experiments and are based on the measurement of energy thresholds for processes producing the molecular negative ion.

One such method employs energy-selected alkali atom beams as the ionizing agent and the threshold is determined for the process,

$$X + AB \rightarrow X^+ + AB^- , \qquad (1)$$

where X denotes the alkali atom and AB the molecule of interest. In some studies of such reactions, the positive ions have been detected, while in others, the negative ion products were monitored. This method has now been applied for various simple molecules in several laboratories.[5–10] In the second type of experiment, atomic negative ions are used as the ionizing reactant, and the threshold measured is that for the electron-transfer reaction,

$$X^- + AB \rightarrow X + AB^- . \qquad (2)$$

This technique has been employed in our own laboratory, as described in preceding papers of this series[11,12] and has also been adopted elsewhere[13] but using quite different instrumentation than that which we employ. The studies in our laboratory utilize a tandem mass spectrometer which is particularly well suited for these experiments because it is capable of producing a mass resolved ion beam with controlled translational energy, which is variable over the range from ~0.3 to 170 eV (laboratory energy). Furthermore, the high energy resolution achievable with the instrument, ±0.3 eV (lab) over the entire accessible range, ensures a considerable degree of accuracy in the threshold measurements. The application of this method was described in some detail in our previously reported study of O$_2$,[12] in which it was shown that a consistent value of the electron affinity could be obtained from experiments with several different atomic ions, provided that the threshold data were corrected for the effects of Doppler broadening.[14–17]

In the present study, we have extended the negative ion charge transfer experiments to several other small molecules, including some of the halogens, NO, SO_2, CS_2, and C_2H_2. Also, we have accomplished more extensive experiments with NO_2, which support the conclusions reported in our earlier preliminary note.[11] Since the completion of our experiments, electron affinities for some of these molecules have been reported by other workers.[9,10,13] It is also appropriate therefore, to compare our results with these findings from other laboratories.

II. EXPERIMENTAL

A. Apparatus and Techniques

The apparatus used in these experiments is a tandem mass spectrometer which has been described previously.[18,19] Briefly, the first stage or ion gun of the apparatus is a double-focusing mass spectrometer which produces a well-collimated mass-selected ion beam having a kinetic energy of ~170 eV in the laboratory scale, the half-width of the energy distribution at half-maximum being 0.3 eV. This beam passes through a thick-element field lens,[18,20] which retards the translational energy to the desired interaction energy without introducing any additional spread, and then into a shielded field-free collision chamber. Product ions which emerge from the collision chamber are reaccelerated by a Soa-type lens,[21,22] mass analyzed in a second sector mass spectrometer and detected using an electron multiplier operated in pulse-counting mode. The application of the instrument for measuring energy thresholds of endothermic ion-neutral processes and the methods employed to quantitatively define the energy scale have been discussed in previous papers from our laboratory.[12,23]

For the present experiments collision chamber pressures were kept sufficiently low (5–30 μ) to ensure that the observed products could be formed only by single collision events and are therefore truly bimolecular. The collision chamber temperature was maintained at 30 °C in all experiments. The intensity of the reactant ion beam was monitored throughout the experiments and appropriate corrections were made to the reaction cross sections for any observed changes. In general, the primary beam current was constant to within about 10%–15% for any given experiment in which the reaction cross section was determined as a function of energy. Rate coefficients reported in this study were determined on a relative basis as described in earlier papers,[11,12] using the O^-/NO_2 charge-transfer reaction as a reference standard, taking the value of $k = 1.2 \times 10^{-9}$ cm^3/molecule second for this reaction.[24]

Calculated cross-section curves to simulate the effects of Doppler broadening of the thresholds, (as described below), were plotted using a Hewlett-Packard Model 9100A Programmable Calculator and an H–P Model 9125A Plotter.

B. Threshold Behavior and Corrections for Doppler Broadening

As discussed in our earlier paper,[12] the cross sections plotted as a function of energy, for endothermic negative-ion charge-transfer reactions which we observe, often exhibit considerable tailing or curvature in the threshold region. This feature, which tends to obscure the true onsets in some instances, can be attributed in part to Doppler broadening of the effective reaction energy, owing to the thermal velocity distribution of the neutral target molecules in the collision chamber. The severity of this broadening and the magnitude of the error inherent in the uncorrected or linearly extrapolated threshold can be shown to depend on both the width of the effective energy distribution, (which can be calculated assuming a Boltzmann distribution of neutral particle velocities), and on the shape of the true threshold function. It is currently not possible to specify *a priori* the nature of the threshold law for endothermic ion-neutral collisions. The statistical phase-space theory of reactions predicts that the cross section for such a process should rise sharply from threshold with no delayed onset and that it should exhibit an exponential dependence on the excess translational energy available for the reaction.[25,26] Little experimental data has been obtained, however, to test the predictions of the statistical theory. In addition, it is not clear, even if the threshold law can be predicted, just what range of validity it has.[26] In principle, one could apply a deconvolution technique to unfold the effective energy distribution from the experimental data and thereby arrive at the residual or true threshold function (which is presumably that applicable to the reaction at 0 °K). Successful application of deconvolution methods usually requires preliminary smoothing of the data however, and even then there is some question as to the uniqueness of the solution. Therefore, we have employed a somewhat simpler method, already described in some detail,[12] to effect Doppler corrections where these are necessary. This procedure entails calculation of the effective energy distribution, folding this distribution into an assumed threshold function, and fitting the curve thus synthesized to the experimental data points. From the best fit achieved, the true threshold energy is thus deduced.

As described previously,[12] an approximate one-dimensional form of the effective energy distribution function, derived by Bethe,[27] can be employed with sufficient accuracy for the present results. This distribution function is given by

$$\omega(E_{max})dE_{max} = \tfrac{1}{2}(M/\pi\mu kTE_0)^{1/2}$$
$$\times \exp[(-M/4\mu kTE_0)(E_0 - E_{max})^2]dE_{max} ,$$
$$(3)$$

where E_{max} is the maximum energy available from the collision, M is the mass of the neutral target, μ is the reduced mass, and E_0 is the center-of-mass energy for stationary neutral targets, that is,

$$E_0 = (M/m + M)E_{lab} . \qquad (4)$$

The convolution integral which gives the calculated cross section may then be expressed as

$$f(E_0)dE_0 = \int_{-\infty}^{+\infty} f(E_{max})\omega(E_{max})dE_{max} , \qquad (5)$$

where $f(E_{max})$ is the assumed threshold function.

C. Sources of Error

For those reactions observed in the present study which involve the interaction of heavy ions such as I^- and Br^- with relatively light target molecules, Doppler broadening is the principle source of error in the measured energy thresholds. This error is substantially reduced by treating the data as described above. However, there is still a small residual error even in these cases, as a result of our application of an approximate form of the energy distribution function.[27] An exact treatment of the thermal effect has recently been presented by Chantry,[17] who also compared the more accurate calculations with the results from the one-dimensional derivation.[27] Such a comparison indicates that for the reactions reported here, the largest error arising from this source would be for the I^-/NO reaction, and even in this case, it is very small (~0.06 eV). The uncertainty in the effective reaction energy which is due to the translational energy spread in the reactant ion beam is obviously minor for heavy ion reactions; thus, for the I^-/NO process this is on the order of 0.05 eV. The cumulative uncertainties for this reaction, therefore, indicate error limits of about ±0.1 eV on the measured threshold and on the electron affinity thus determined. The reproducibility of the measured thresholds is considerably better than these limits indicate.

For the reactions of lighter mass ions such as O^- and S^-, the uncertainty in the threshold energy which originates from the translational energy distribution of the incident ion beam becomes more important than that due to Doppler smearing of the threshold. This is the case, of course, because the width of the effective energy distribution which results from the thermal motion of the target molecules is much smaller for reactions in which the ion and neutral are of more nearly equal mass. Also, as will be seen, the threshold functions which best describe many of the lighter ion reactions are linear functions, and as described in our earlier

paper, under these circumstances, the uncorrected linearly extrapolated threshold energy is essentially the true threshold energy. For these reactions therefore, the error limits on the thresholds (and on the electron affinities), are specified on the basis of the uncertainty in the effective energy arising from the reactant ion translational energy spread. We have not attempted to make corrections for the latter factor.

In recently reported charge transfer experiments of the same type as those described here, Chupka et al.[13] have assumed that the average rotational energy of the neutral target molecules is directly available for overcoming the endothermicity of the reaction and have corrected their threshold energies accordingly. As far as we are aware, there is no definitive experimental evidence to support the assumed equivalence of the rotational and translational energies in effecting such reactions. Moreover, statistical theory calculations by Truhlar,[26] for an endothermic ion-molecule process, suggest that the initial rotational energy has an extremely small effect on the reaction cross section, certainly less than the translational energy. Therefore, we have made no threshold corrections for the rotational energy in the present experiments. Such corrections, in any event, would be extremely small.

It is presumed for the charge transfer reactions to be described that the reactant ions are in the ground state, that the product ions are formed at threshold without excess internal or translational energy, and that there is no activation energy barrier.. Some indication that the reactant ions are indeed in the ground state is provided by our observations that for a given reaction, no differences in the cross section or threshold behavior can be detected when the reactant ion is generated from several different source molecules.[12] We are unable to determine the energy states of the product ions formed in the present experiments, however, which means that from a rigorous standpoint, the electron affinities derived are only lower limits. However, the fact that a consistent value of the electron affinity for a given molecule can be obtained from several different reactions lends support to the assumption that this is actually the true electron affinity. The possibility that there is an activation energy for the negative ion-molecule processes with which we are concerned cannot be entirely ruled out, but this seems unlikely in view of the strong attractive ion-induced dipole forces which characterize such reactions, and which should dominate any small repulsive forces.

Finally, it should be noted that the threshold behavior observed in our experiments could be affected by changes in the collection efficiency of the apparatus with increasing reactant ion energy, a problem which we have discussed previously.[12,23] In an earlier paper,[12] we reported measurements of the cross section for the O^-/O_2 endothermic charge transfer reaction, over the laboratory energy range from ~ 0.3 to 170 eV, and showed that our results were in very good agreement with other beam observations on this reaction by Rutherford and Turner[28] and by Bailey and Mahadevan.[29] Similar comparisons between our measured cross-sections as a function of energy and those reported from other laboratories have also been made for positive ion processes,[23] and consistency was also demonstrated in these cases. It is reasonable, therefore, to assume that discrimination effects are relatively minor in so far as these affect the actual thresholds determined. Such effects may influence the shape of the cross sections at energies above that at which the maximum cross section is observed, and may account for the skewing on the high energy side. However, we are unable to assess these effects in detail with our existing instrumentation.

III. RESULTS AND DISCUSSION

A. NO

Although there have been several rather recent determinations of the electron affinity of NO by various techniques, the true value is still a matter of some uncertainty. Page and co-workers[3] reported a value of 0.83 eV derived from experiments with the magnetron. This is in close agreement with the result of Williams and Hamill,[30] 0.85 eV, obtained from RPD measurements of the appearance potentials for the dissociative ionization and ion-pair processes in ethyl nitrite. Also consistent with these values is the lower limit of E. A. $(NO) \geqslant 0.65$ eV determined by Stockdale et al.[31] from the appearance potentials for dissociative electron attachment to NO_2. It was subsequently demonstrated by other workers[32] that such appearance potentials are strongly dependent upon the gas temperature and that the correct value must be obtained by effectively extrapolating the data measured at a given temperature to 0 °K. Spence and Schulz[33] estimated that by correcting the data of Stockdale et al.,[31] a value of E. A. $(NO) \geqslant 0.035$ eV would be obtained. The latter result is in much better agreement with the laser photoelectron spectroscopy value for the NO electron affinity, (0.021 eV), which was recently reported by Celotta et al.[34] Berkowitz et al.[35] have applied a technique similar to that employed in the present study, and concluded that 0.1 eV $\leqslant$ E.A. $(NO) \leqslant 0.5$ eV, the lower value being favored. The upper limit of 0.5 eV was specified largely on the basis of earlier ob-

servations of the reaction

$$NO^- + O_2 \rightarrow O_2^- + NO \qquad (6)$$

in flowing afterglow experiments.[36] It should also be mentioned that Paulson[37] has reported observation of the reaction

$$NO^- + N_2O \rightarrow N_2O^- + NO \qquad (7)$$

although he found that this reaction was apparently endothermic for low energy NO^- ions. However, since Paulson's data did not permit determination of the energy threshold and since, in any case, the electron affinity of N_2O is not well established, this yields no definitive information as to the NO electron affinity.

Measurements of the electron affinity of NO by the chemiionization method described earlier have been reported from two laboratories,[38,39] and these results indicate that the electron affinity of NO is on the order of 0.1 eV or less.

In the present study, we have examined the charge transfer reactions of I^- and O^- with NO. The cross sections measured as a function of translational energy for these reactions are plotted in Figs. 1 and 2, respectively. Calculated cross-section curves, obtained as described earlier assuming a step-function threshold dependence, are also shown in Fig. 1 for the I^-/NO reaction. These were plotted for three assumed threshold energies which correspond to values of 0, 0.1, and 0.2 eV, respectively for the NO electron affinity, if the value for the iodine atom electron affinity is taken as 3.063 eV.[40] In the region up to the maximum,

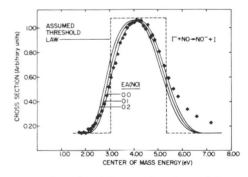

FIG. 1. Translational energy dependence of the cross section for the electron transfer reaction of I^- with NO. Points are experimental data; solid curves were obtained by convoluting the calculated effective energy distribution function for the reactants with the indicated delta threshold function (dotted lines). The three curves shown were computed for three different assumed threshold energies corresponding to E.A. $(NO) = 0, 0.1,$ and 0.2 eV, respectively (see text for details).

TABLE I. Thresholds measured for endothermic electron transfer reactions to selected molecules and molecular electron affinities calculated from these values.

Incident ion	Target molecule	Threshold function deduced	Corrected threshold (eV, c.m.)	E.A. of target molecule (eV)	
				Calculated from threshold[a]	Selected value
I^-	NO	Delta	3.05	0.015 ± 0.1	0.015 ± 0.1
O^-	NO	Linear	1.60	-0.14 ± 0.2[b]	
I^-	NO_2	Linear	0.86	2.21 ± 0.1	2.28 ± 0.1
Br^-	NO_2	Linear	1.02	2.34 ± 0.1	
F^-	NO_2	Linear	1.28	2.17 ± 0.2	
I^-	SO_2	Linear	2.06	1.00 ± 0.1	0.99 ± 0.1
S^-	SO_2	Linear	1.11	0.97 ± 0.1	
Cl^-	SO_2	Linear	3.34	0.27 ± 0.1[c]	
O^-	CS_2	Linear	1.00	0.47 ± 0.2	0.50 ± 0.2
Cl^-	CS_2	Linear	3.30	0.31 ± 0.2[c]	
S^-	CS_2	Linear	1.56	0.52 ± 0.2	
I^-	Cl_2	Step	0.75	2.32 ± 0.1	2.32 ± 0.1
Cl^-	Cl_2	Linear	1.34	2.27 ± 0.2	
I^-	Br_2	Linear	0.50	2.57 ± 0.2	2.62 ± 0.2
Br^-	Br_2	Linear	0.70	2.66 ± 0.2	
I^-	I_2	Linear	0.59	2.48 ± 0.2	2.42 ± 0.2
Br^-	I_2	Linear	1.00	2.36 ± 0.2	
Cl^-	C_2H_2	Linear	2.17	1.82 ± 0.4[d]	
S^-	C_2H_2	Linear	1.10	2.21 ± 0.4[d]	2.21 ± 0.4[d]

[a]The atomic electron affinities used for these calculations and the reference sources are respectively: E.A. (I) = 3.065 eV (2); E.A. (Br) = 3.363 eV (2); E.A. (Cl) = 3.613 eV (2); E.A. (F) = 3.448 eV (2); E.A. (O) = 1.465 eV (2); E.A. (S) – 2.077 eV [W.C. Lineberger and B.W. Woodward, Phys. Rev. Lett. 25, 424 (1970)].

[b]Refer to text.

[c]Values obtained from Cl^- reactions are in many cases anomalously low. See text.

[d]Value given is E.A. (C_2H); see text.

most of the data points are fitted quite accurately by the calculated curve corresponding to an electron affinity of 0 eV for NO, although there is some scatter in the data in the immediate threshold region. On the basis of this data, it may be concluded that

$$0 \text{ eV} \leqslant \text{E.A. (NO)} \leqslant 0.1 \text{ eV} .$$

The threshold obtained for the O^-/NO charge transfer reaction (Table II and Table I) also indicates that the electron affinity of NO is very near zero. In this case, the Doppler width is small and a linear threshold law is obviously applicable over an energy span of several volts. As we have demonstrated previously,[12] linear extrapolation will yield essentially the correct threshold for such a reaction. However, while Doppler broadening is negligible for the O^-/NO reaction, the smearing of the threshold which arises from the translational energy spread in the center of mass for this reactant pair becomes more significant. Therefore, the error limits on the electron affinity determined

from this reaction threshold are somewhat larger than for the I^-/NO process. Thus, the data plotted in Fig. 2 lead to a value of E.A. (NO) $\geqslant -0.14 \pm 0.2$ eV, which is still consistent with the value noted above.

We have also observed, in the present study, an electron transfer reaction between NO^- and CS_2, which occurs at very low incident ion energies, and which shows a decreasing cross section with increasing energy, (Fig. 3). This reaction therefore appears to be exothermic, and since the electron affinity of CS_2 is ~0.5 eV, as will be shown later in this paper, this result implies that E.A. (NO) $\leqslant 0.5$ eV. Again, this upper limit is in agreement with the value derived from endothermic reaction thresholds. This conclusion presumes, of course, that the NO^- reactant ions are not internally excited.

As described in our earlier preliminary report,[41] an O^- product which originates from the neutral NO target was also detected from the O^-/NO interac-

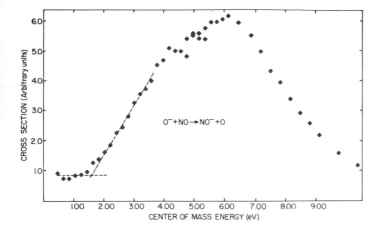

FIG. 2. Translational energy dependence of the cross section for the electron transfer reaction of O^- with NO.

tion. This product was observed by using isotopically labeled $N^{16}O$ and monitoring the $^{18}O^-$ product when $^{16}O^-$ was employed as the reactant ion. The data plotted in Fig. 4 shows that the cross section for this process is largest at very low kinetic energy and decreases rapidly at higher energies. This O^- product obviously is not formed by dissociative charge transfer since it is detected at energies well below the charge-transfer threshold. A possible explanation is that this product results from decomposition of a very short-lived intermediate collision complex, $(^{16}O--N--^{18}O)^-$ * which cannot be detected in our experiments. If such a mechanism does indeed apply, then the lifetime of the intermediate must be sufficiently long to permit some redistribution of the energy available from

the collision. A more detailed analysis of the kinematics of this reaction is necessary before definitive conclusions can be drawn as to the mechanism.

B. NO_2

In a previous communication,[11] we reported the results of preliminary charge transfer experiments with NO_2 which indicated that E. A. $(NO_2) \geqslant 2.30 \pm 0.15$ eV. The threshold measurements upon which this determination was based were not corrected for Doppler broadening, although it was noted in the earlier paper that such corrections would be small for the reactions considered. In an attempt to further substantiate our previously determined value for the NO_2 electron affinity, we have here studied reactions of several other nega-

TABLE II. Rate coefficients for various exothermic electron transfer reactions to NO_2 [a].

Reactant ion and source molecule	Electron affinity (eV) of corresponding atom, radical or molecule	Reaction rate coefficient, $k (\times 10^9$ cm^3/molecule sec)[b]
$O^-(N_2O)$	1.465 ± 0.005 [c]	1.2
$D^-(ND_3, D_2O, D_2S)$	0.776 [d]	1.2
$S^-(COS, CS_2)$	2.077 ± 0.0005 [e]	0.82
$SH^-(H_2S, C_3H_7SH)$	2.319 ± 0.010 [c]	0.78
$SO_2^-(SO_2)$	0.90 ± 0.1 [f]; 1.0 [g]	0.43
$CS_2^-(CS_2)$	0.44 ± 0.2 [f]	0.41
$CF_3^-(C_2F_6)$	2.0 ± 0.1 [h]	0.37
$C_2F_5^-(C_3F_8)$	2.3 ± 0.1 [h]	0.32

[a]Measured with reactant ions of ~ 0.3 eV laboratory energy.
[b]Determined by normalizing relative rates for all reactions to the absolute value cited for the O^-/NO_2 reaction, which is taken from Ref. 24.
[c]Reference 2.
[d]Reference 53.
[e]W. C. Lineberger and B. W. Woodward, Phys. Rev. Lett. 25, 424 (1970).
[f]Present work.
[g]Reference 54.
[h]C. Lifshitz and R. Grajower, Int. J. Mass Spectrom. Ion Phys. 3, 211 (1969).

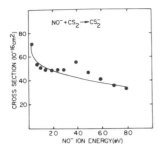

FIG. 3. Cross section as a function of energy for the electron transfer reaction of NO⁻ with CS₂.

tive ions with NO_2 and have repeated the experiment with Br⁻ which was described before.[11] Data obtained for the reaction of I⁻, Br⁻, and F⁻ with NO_2, as a function of energy, are plotted in Figs. 5, 6, and 7, respectively. Corrections for thermal broadening of the I⁻ and Br⁻ thresholds were accomplished by convoluting the appropriate thermal velocity distributions with various assumed threshold functions and fitting the calculated curves to the experimental data points, as already described. The solid curves indicated in Figs. 5 and 6 represent the "best fits" which could be achieved, and the corresponding linear threshold functions thus deduced are shown by dotted lines. From the thresholds obtained in this manner and the known electron affinities of I (3.063 eV [2,40]) and Br (3.363 eV [2,40]), calculated values of E. A. (NO_2) ≥ 2.21 ± 0.1 eV and ≥ 2.34 eV ± 0.1 eV are determined for the I⁻ and Br⁻ reactions, respectively (Table I). Doppler broadening is negligible for the F⁻ reaction threshold, but again the over-all error here is somewhat larger, owing to the greater effective reactant ion energy distribution for this case. As shown in Table I, the linearly extrapolated threshold indicated in Fig. 7 leads to E. A. (NO_2) ≥ 2.17 eV ± 0.2 eV. A consistent value of the NO_2 electron affinity is therefore obtained from all three of the reactions discussed. Also, this value is in good agreement with that reported in our preliminary note.[11]

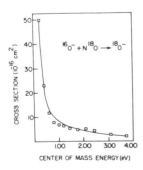

FIG. 4. Cross section as a function of energy for the switching reaction, $^{16}O^- + N^{18}O \rightarrow {}^{18}O^-$.

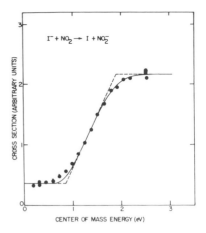

FIG. 5. Translational energy dependence of the cross section for the electron transfer reaction of I⁻ with NO_2. Points are experimental data; solid curve is the calculated "best fit" curve obtained by convoluting the effective energy distribution function with the linear threshold function indicated by dotted lines (see text).

It is apparent that the calculated cross section functions for the I⁻/NO_2 and Br⁻/NO_2 reactions shown in Figs. 5 and 6 do not fit the experimental data in the threshold region as well as the corresponding functions calculated for reactions with NO and O_2.[12] The longer tailing observed in the neighborhood of threshold for the NO_2 reactions may arise from other sources than the thermal smearing. This tailing was found to be quite reproducible and cannot be attributed to scatter in the data. Similar threshold behavior was observed for the I⁻/NO_2 electron transfer reaction by Berkowitz et al.[35] The latter authors suggested that this effect might be due to the geometry change required in the transition from neutral NO_2 to the NO_2^- ion, since the latter has a somewhat smaller bond angle Ferguson and co-workers have previously discussed such factors for other triatomic species.[42,43] This effect could result in lower transition probabilities in the threshold region, giving apparently larger thresholds for the electron transfer reaction.

It is interesting to note that the cross section for the Br⁻/NO_2 reaction (Fig. 6), which was measured at c. m. energies up to 5 eV, appears to exhibit some structure. After onset and the initial rise to a plateau, a second rise occurs in the neighborhood of about 3 eV. Virtually identical behavior was observed by Nalley et al.[9] in the collisional ionization of NO_2 with cesium, although the second rise in their cross section curve occurred at slightly higher energy. These workers noted that the second break in their excitation function occurred approximately at the onset energy expected for the $^3B_1$ state of NO_2^-. Obviously, excitation to a higher state of NO_2^- could also explain the energy dependence of the cross section observed in our experiments. While the apparent onset in the present case, (~2.1 eV above the initial break), does not correspond as closely to the recently estimated energy of the $^3B_1$ state, which is thought to lie ~2.6–2.8 eV above the ground state,[42] there is sufficient uncertainty to preclude any strict comparison. The formation of some excited state of NO_2^- at higher energies in the Br^- reaction seems to be the only reasonable explanation for the observed dependence since the splitting in the two thresholds is clearly too large to be explained on the basis of excitation of the bromine atom product.

There is also a hint of structure in the excitation function shown in Fig. 7. Here, fewer data points were taken in the threshold region; however, and the apparent break about one volt above threshold is most likely due simply to scatter in the data.

Electron-transfer reactions from several other negative ions to NO_2 were observed at near-thermal collision energies. Rate coefficients determined for these reactions are summarized in Table II. The cross section for the O^-/NO_2 reaction has previously been shown to decrease with increasing translational energy of the incident ion,[28,44,45] as is usually the case for exothermic processes. Similar energy dependences were obtained for the other reactions reported in Table II, and typical results are shown for the D^- and SH^- reactions in Figs. 8 and 9, respectively. This behavior and the large rate coefficients determined for the reactions shown in Table II indicate that all of the reactions

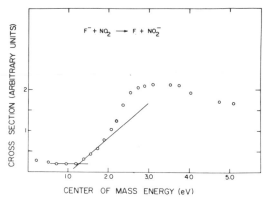

FIG. 7. Translational energy dependence of the cross section for the electron transfer reaction of F^- with NO_2. Linearly extrapolated threshold is shown.

are exothermic. Therefore, it follows that E.A.(NO_2) is greater than that of O, D, S, SH, SO_2, CS_2, CF_3, and C_2F_5. Of these species, SH has the highest electron affinity (Table II), and this data thus places a lower limit of 2.32 eV on the electron affinity of NO_2, which is quite consistent with the lower limit established by the threshold measurements described above. It will be noted that our observations regarding the SH^- reaction with NO_2, (which we also reported earlier[11]), are in direct contradiction to the results reported by Vogt,[46] which indicated that this reaction is endothermic. However, Ferguson and co-workers[47] have recently confirmed our findings for thermal SH^- ions in flowing afterglow experiments. The latter authors determined a rate coefficient of 3.6 $\times 10^{-10}$ cm^3 sec^{-1} for charge transfer from SH^- to

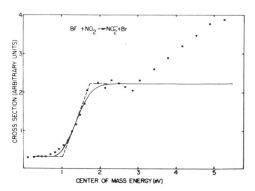

FIG. 6. Translational energy dependence of the cross section for the electron transfer reaction of Br^- with NO_2. Points are experimental data; solid curve is the calculated "best fit" to the data; dotted line is the threshold function used to calculate the excitation function.

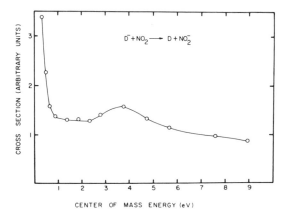

FIG. 8. Cross section as a function of energy for the electron transfer reaction of D^- with NO_2.

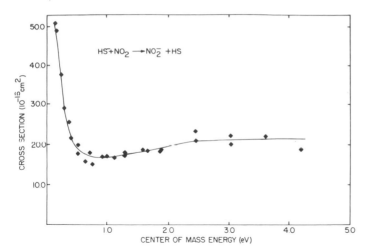

FIG. 9. Cross section
as a function of energy for
the electron transfer reaction of SH⁻ with NO₂.

NO_2, which is somewhat lower than our measured value of 7.8×10^{-10} cm³ molecule⁻¹ sec⁻¹ for 0.3 eV reactant ions, but which clearly indicates that the reaction is fast at low energies and is thus unquestionably exothermic.

In a later section of this paper, it will be shown that endothermic charge transfer experiments observed with chlorine indicate that E.A. $(Cl_2) \gtrsim 2.32 \pm 0.1$ eV. Since this value is very nearly the same as that which we find for NO_2 from similar threshold measurements, (within the limits of uncertainty), it was of interest to study the reactions of NO_2^- with Cl_2 and of Cl_2^- with NO_2 at low kinetic energy. When 0.3 eV NO_2^- ions were impacted on Cl_2, the charge transfer reaction

$$NO_2^- + Cl_2 \rightarrow Cl_2^- + NO_2 \qquad (8)$$

was observed to occur and a rate coefficient of 2.2×10^{-10} cm³/molecule second was determined for this reaction. At higher energies, as shown in Fig. 10, the cross section for this reaction drops sharply. Taken together, these facts indicate that Reaction (8) is exothermic, which means that E.A. $(Cl_2) \gtrsim$ E.A. (NO_2), and therefore E.A. $(NO_2) \lesssim 2.4$ eV, based upon our value for Cl_2 and the error limits cited above. An attempt was also made to observe the reverse of Reaction (8), but charge transfer from Cl_2^- to NO_2 could not be detected. This may be due in part to the fact that there is a favorably competing reaction channel for this interaction at low energy which yields Cl^- and NO_2Cl, as recently reported by Ferguson et al.[47] It was also determined in the present study that NO_2^- ions undergo charge transfer to I_2, the rate coefficient measured for this reaction being 1.7×10^{-10} cm³/molecule sec.

$$NO_2^- + I_2 \rightarrow I_2^- + NO_2 . \qquad (9)$$

This reaction is expected since our results to be discussed subsequently indicate that E.A. $(I_2) \gtrsim 2.42 \pm 0.2$ eV. Again, we were unable to detect the reverse of Reaction (9).

Our observations of the exothermic processes described above, in particular, Reaction (8) and the SH⁻ reaction with NO_2, bracket the electron affinity of NO_2 within a rather narrow range, that is,

2.32 eV < E.A. (NO_2) < 2.4 eV .

The result is in excellent agreement with the recent findings of Ferguson et al.,[47] but is slightly lower than most of the current estimates obtained by the chemiionization technique using alkali atom beams.[48-50] The latter measurements yield values for E.A. (NO_2) in the range from about 2.4 to 2.7 eV. It now seems clear that the earlier values obtained for NO_2 by photodetachment,[51] (3.1 eV), and from magnetron studies,[3] (3.9 eV), are definitely in error. It should also be noted that the

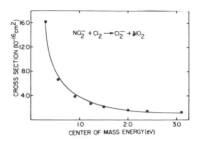

FIG. 10. Cross section as a function of energy for the electron transfer reaction of NO_2^- with Cl_2.

TABLE III. Summary of literature values for molecular electron affinities.

Molecule	Electron affinity (eV)	Method[a]	Reference
NO	0.015 ± 0.1	ECT, BG	Present data
	0.09	ECT, BG	35
	0.1 ± 0.1	CI, BG	9
	~0	CI, BG	7
	0.047 ± 0.005	PES	34
	0.024 + 0.010 − 0.005	PES	b
NO$_2$	2.28 ± 0.1	ECT, BG	Present data
	2.30 ± 0.15	ECT, BG	11
	2.04	ECT, BG	35
	2.70 ± 0.2	CI, BG	50
	2.46 ± 0.1	CI, BG	38
	2.45	CI, BB	7
	2.50 ± 0.1	CI, BB	48
	2.54 ± 0.05	CI, BB	51
	3.9	M	3
	< 3.9	P	c
	3.1	P	49
	2.38 ± 0.06	TBCT	47
SO$_2$	0.99 ± 0	ECT, BG	Present data
	1.0 ± 0.05	P	54
	1.0 < E.A. < 1.12	TBCT	52
CS$_2$	0.50 ± 0.2	ECT, BG	Present data
	1.0 < E.A. < 1.12	TBCT	52
Cl$_2$	2.32 ± 0.1	ECT, BG	Present data
	2.38 ± 0.1	ECT, BG	13
	2.5 ± 0.5	CI, BG	7
	2.3; 2.45; 2.5	CT, BG	48
	2.52 ± 0.17	DEC	4
	2.35 + 0.15	TBCT	47
Br$_2$	2.62 ± 0.2	ECT, BG	Present data
	2.51 ± 0.1	ECT, BG	13
	2.6; 2.55	CI, BG	48
	2.3 ± 0.1	CI, BB	5
	2.87 ± 0.14	DEC	4
I$_2$	2.42 ± 0.2	ECT, BG	Present data
	2.58 ± 0.1	ECT, BG	13
	2.5; 2.55; 2.3	CI, BG	48
	2.54	CI, BB	49
C$_2$H	2.21 ± 0.4	ECT, BG	Present data
	3.73 ± 0.05	P	54
	2.2	TBCT	62
	2.7	M	3

[a]ECT—Endothermic negative-ion charge transfer; CI—Chemionization; P—Photodetachment electron spectroscopy; DEC—Dissociative electron capture; M—Magnetron; TBCT—Thermal bracketing negative ion charge transfer or ion–molecule reaction; BG—Beam–Gas; BB—Beam–beam.

[b]M. W. Siegel, R. J. Celotta, J. L. Hall, J. Levine, and R. A. Bennett, Phys. Rev. A **6**, 607 (1972).

[c]D. E. Milligan, M. E. Jacox, and W. A. Guillory, J. Chem. Phys. **52**, 3864 (1970).

E.A. (NO$_2$) indicated by the present experiments is considerably larger than the lower limit of 2.04 eV reported by Berkowitz *et al*. from the same type of threshold determinations.[35] A summary of recently reported electron affinity data for NO$_2$ and the other molecules studied in this paper is presented in Table III.

C. SO$_2$

Kraus *et al*.[52] have previously reported the observation of the charge transfer reaction,

$$CS_2^- + SO_2 \rightarrow CS_2 + SO_2^- \qquad (10)$$

from experiments in which reactant ions of a "few eV" energy were employed. We have also observed this reaction with 0.3 eV CS_2^- ions and have determined a rate coefficient of 1.3×10^{-10} cm^3/molecule second for the process. Since we were unable to detect the reverse of Reaction (10) at this energy, the forward reaction is definitely exothermic, which indicates that E.A. (SO$_2$) ≥ E.A. (CS$_2$). From our value for CS$_2$ (to be discussed in Sec. III D), a lower limit of 0.50 ± 0.2 eV can thus be placed on the electron affinity of SO$_2$.

The present experiments also show that charge transfer from D$^-$ to SO$_2$ is exothermic for near-thermal incident ion energies. This establishes the order, E.A. (SO$_2$) ≥ E.A. (D) = 0.776 eV.[53] The cross section for this reaction, plotted in Fig. 11 as a function of energy, drops sharply with increasing energy up to about 1.8 eV. Such a decrease is usually characteristic of exothermic processes. Above this energy, however, the cross section rises again, suggestive of a threshold corresponding to the onset of another process. Conceivably, this is indicative of the formation of an

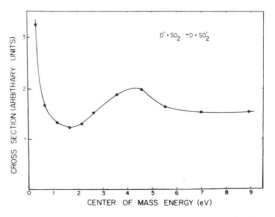

FIG. 11. Cross section as a function of energy for the electron transfer reaction of D$^-$ with SO$_2$.

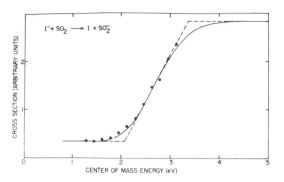

FIG. 12. Translational energy dependence of the cross section for the electron transfer reaction of I⁻ with SO_2. Points are experimental data; solid curve is calculated "best fit" excitation function derived from the linear threshold function indicated by dotted line.

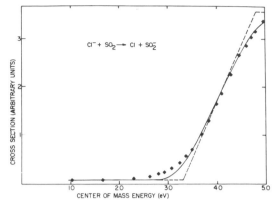

FIG. 14. Translational energy dependence of the cross section for the electron transfer reaction of Cl⁻ with SO_2. Points, experimental data; solid curve, calculated "best fit" excitation function; dotted line, linear threshold function.

excited SO_2^- state. In connection with this behavior it is interesting to note that the cross section for photodetachment of SO_2^- has been observed by Feldmann[54] to exhibit two energy thresholds, having a separation of about 1.6 eV. Unfortunately, there is no other information available on the negative ion states of SO_2, and it is therefore difficult to say whether or not postulation of such an excited state is reasonable. In SO_2 itself, the energy difference between the ground state and the first excited state is significantly larger than 1.6 eV. As Feldmann[54] has noted, some parallel can be drawn between SO_2^- and ClO_2 which is isoelectronic with this negative ion. SO_2 and ClO_2 have very nearly the same bond angles, (119 and 117°, respectively), and according

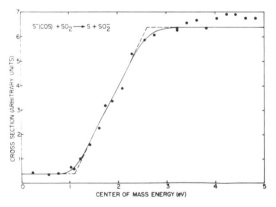

FIG. 13. Translational energy dependence of the cross section for the electron transfer reaction of S⁻ with SO_2. Points, experimental data; solid curve, calculated "best fit" excitation function; dotted line, linear threshold function.

to the Walsh diagram,[55] the addition of another electron to the (b_1) orbital of SO_2 should result in only a slight decrease in the O–S–O bond angle for SO_2^-. However, there appears to be no excited state of ClO_2 which lies 1.6 eV above the ground state.

The charge transfer reaction of I⁻, S⁻, and Cl⁻ with SO_2 were also examined in the present study, and were found to be endothermic in all three cases. The excitation functions measured for these reactions are shown in Figs. 12–14 along with the calculated functions derived by making appropriate corrections for Doppler broadening. The energy thresholds thus deduced and the values of E.A.(SO_2) calculated therefrom are summarized in Table III. The values of 1.00 ± 0.1 and 0.97 ± 0.1 eV for E.A. (SO_2), obtained from the I⁻ and S⁻ reactions, respectively, are in excellent agreement and are consistent with the photodetachment data of Feldmann.[54] The markedly lower value for E.A.(SO_2) which is indicated by the Cl⁻ reaction must be considered to be anomalous. Similarly inconsistent data was obtained from several Cl⁻ reactions with other molecules, as will be discussed in later sections. The behavior of Cl⁻ is not readily understood, but it is clearly different from the other halide reactant ions in that it usually yields energy thresholds which are too large to correspond to formation of the product negative ions in their lowest energy state.

D. CS_2

As mentioned above, our observation of Reaction (10) with low energy CS_2^- ions indicates that E.A.

$(CS_2) \leqslant$ E. A. (SO_2), and therefore from the data listed in Table III for SO_2, E. A. $(CS_2) \leqslant 0.99 \pm 0.1$ eV. As would be expected from this upper limit, the reactions of O^-, Cl^-, and S^- with CS_2 were all found to be endothermic with 0.3 eV reactant ions. Translational energy thresholds for electron transfer from these ions to CS_2 were measured, as shown by the data plotted in Figs. 15–17. Doppler broadening is negligible for these reactions and the threshold functions are clearly linear. Consequently, the threshold energies indicated in Table I were obtained by simple linear extrapolation of the excitation functions. The values for the electron affinity of CS_2 calculated from the thresholds (Table I) for the O^- and S^- reactions are in excellent agreement. Again, the Cl^- reaction gives an anomalously high threshold and thus, a lower value for the electron affinity of CS_2. The O^- and S^- experiments yield an average value of E. A. $(CS_2) \geqslant 0.50 \pm 0.2$ eV.

It should also be mentioned that in the case of the S^- reaction with CS_2 there is a prominent reaction channel other than electron transfer which produces S_2^-. As indicated by the excitation function for this process (Fig. 18), the threshold of S_2^- production is even lower than that for electron transfer. The cross section for S_2^- formation is larger by approximately a factor of six than the electron transfer cross section for the S^-/CS_2 interaction at 3 eV in the center-of-mass energy scale.

E. Cl_2, Br_2, I_2

As indicated by the data summarized in Table III, the electron affinities of the halogens have recently been measured by both the endothermic electron transfer and collisional ionization threshold methods. In still other experiments, the halogen electron affinities were obtained from appearance potentials for formation of the molecular negative

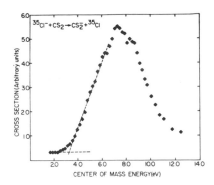

FIG. 16. Translational energy dependence of the cross section for the electron transfer reaction of Cl^- with CS_2. Linearly extrapolated threshold is shown.

ions by dissociative electron capture, in conjunction with measurements of the translational energies of the product ions.[4] The general consensus from these several determinations is that the electron affinities of Cl_2, Br_2, and I_2 lie in the range from ~ 2.3 to 2.9 eV. There is still sufficient uncertainty, even among experiments of the same type, however, to prevent a definitive ordering of the halogen electron affinities. As would be expected from these earlier estimates, the present study indicates that the reactions of Cl^-, Br^-, and I^- with the halogen molecules are all endothermic at low reactant ion energies. Again, translational energy thresholds for the electron transfer reactions are observed, as shown in Figs. 19–25.

Doppler corrections were accomplished where necessary, but thermal broadening of the thresholds

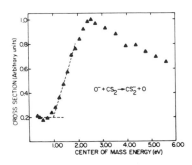

FIG. 15. Translational energy dependence of the cross section for the electron transfer reaction of O^- with CS_2. Linearly extrapolated threshold is shown.

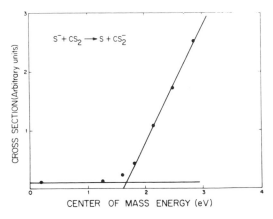

FIG. 17. Translational energy dependence of the cross section for the electron transfer reaction of S^- with CS_2. Linearly extrapolated threshold is shown.

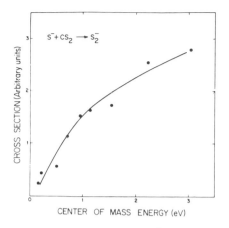

FIG. 18. Cross section as a function of energy for the S-atom transfer reaction of S⁻ with CS_2.

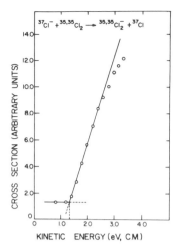

FIG. 20. Energy dependence of the cross section for the electron transfer reaction of Cl⁻ with Cl_2. Linearly extrapolated threshold is shown.

is negligible for the majority of these interactions and in such cases, the linearly extrapolated thresholds should be essentially correct. It will be noted in Fig. 25 that the reaction of Cl⁻ with I_2 yields not only I_2^- but also ClI⁻ and I⁻ as products, even at very low energies. Therefore the threshold for formation of I_2^- in this instance is difficult to determine with any precision since the cross section is extremely small. The threshold energies measured for the several reactions are tabulated in Table I, along with the values of the electron affinities deduced from these thresholds. The averaged data, also indicated in Table I, suggests that E. A. (Br_2) > E. A. (I_2) > E. A. (Cl_2), although in our experiments

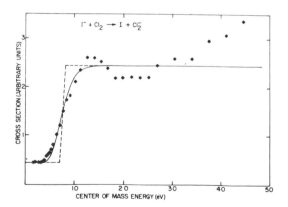

FIG. 19. Translational energy dependency of the cross section for the electron transfer reaction of I⁻ with Cl_2. Points, experimental data; solid curve, calculated "best fit" excitation function; dotted line, step threshold function.

also, the magnitude of the uncertainties in these values precludes an absolute ordering.

Several exothermic negative ion reactions with the halogens were also observed at 0.3 eV (lab) energy in the present study. These reactions, which are listed with their rate coefficients in Table IV, provide a further indication of the relative halogen electron affinities. Since electron transfer occurs from SH⁻ to both Cl_2 and I_2, at this low energy, this places a lower limit of 2.32 eV on the electron affinities of both of these molecules. Also, the fact that electron transfer takes place from Cl_2^- to Br_2, while the reverse reaction from Br_2^- to Cl_2 could not be observed, clearly indicates that E.A. (Br_2) > E.A. (Cl_2). As can be seen from Table IV, however, our experiments show that the corresponding reactions of Cl_2^- with I_2 and of I_2^- with Cl_2 both occur with 0.3 eV reactant ions, although the rate coefficient is substantially larger for electron transfer from Cl_2^- to I_2. Presumably, this indicates that the electron affinities of Cl_2 and I_2 are very nearly the same, that of I_2 being slightly larger. We also attempted to observe reactions between Br_2^- and I_2 and between I_2^- and Br_2. Unfortunately, the intensities of both the Br_2^- and I_2^- reactant ions which can be obtained in our instrument are quite low and the cross sections for the indicated reactions are apparently extremely small. Thus, we cannot draw any positive conclusions with respect to the relative electron affinities of I_2 and Br_2 from these results. In general, however, the exothermic reactions observed with the halogens are consistent with the electron affinity data obtained from the endothermic charge transfer thresholds.

The electron affinity of Cl_2 is definitely lower than that of both Br_2 and I_2.

It can also be seen from Table IV that the molecular halogen ion-halogen molecule interactions exhibit several prominent reaction channels other than electron transfer which lead to mixed halogen product ions. This suggests that at low energies, these interactions involve fairly long-lived intermediates in which some redistribution of the collision energy occurs. Similar conclusions can be drawn from isotopic labeling experiments for endothermic reactions with the halogens, as described below.

F. Isotope Distributions of Products from Endothermic Electron Transfer Reactions with Halogen Molecules

The fact that both chlorine and bromine have two isotopes with relatively large natural abundances facilitates the study of isotopic mixing in the products of mass selected Br^- and Cl^- with the respective diatomic molecules. As discussed in several previous papers from our laboratory,[56] the extent of isotopic mixing in the products from ion–molecule reactions can provide information on the mechanisms of such processes and some indication of the nature of intermediate collision complexes, if such are formed. Such information is, of course, not

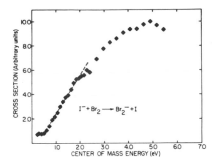

FIG. 21. Energy dependence of the cross section for the electron transfer reaction of I^- with Br_2. Linearly extrapolated threshold is shown.

as definitive as that obtained from detailed scattering experiments in which the product ion energies and angular distributions are measured. It is quite clear however, from the marked change in the isotopic composition of the halogen ion products with increasing reactant translational energy, that the reaction mechanism which is involved at energies near threshold is different from that which dominates in the energy regime several volts above threshold. Table V shows the observed relative intensities of $^{79}Br^{79}Br^-$, $^{79}Br^{81}Br^-$ and $^{81}Br^{81}Br^-$ from the charge-transfer reaction of both $^{79}Br^-$ and $^{81}Br^-$ with natural bromine, at various incident ion energies. The corresponding products from the reactions of $^{35}Cl^-$ and $^{37}Cl^-$ with natural chlorine are also listed in Table V. It can be seen that in the higher energy region where the cross section for electron transfer from Br^- to Br_2 is essentially

TABLE IV. Exothermic electron transfer and ion-molecule reactions involving Cl_2, Br_2, and I_2.[a]

Reaction	Rate coefficient $k (\times 10^{-10}$ cm^3/molecule sec)[b]
$SH^- + Cl_2 \rightarrow Cl_2^- + SH$	3.8
$\rightarrow Cl^- + (?)$	2.2
$SH^- + I_2 \rightarrow I_2^- + SH$	4.9
$\rightarrow I^- + (?)$	0.94
$Cl_2^- + {}^*Cl_2 \rightarrow {}^*Cl_2^- + Cl_2$	1.5
$\rightarrow Cl_3^- + Cl$	0.0084
$I_2^- + I_2 \rightarrow I_3^- + I$	1.1
$Cl_2^- + Br_2 \rightarrow Br_2^- + Cl_2$	1.8
$\rightarrow Br_2Cl^- + Cl$	0.38
$\rightarrow BrCl^- + (?)$	0.16
$\rightarrow BrCl_2^- + Br$	0.10
$Cl_2^- + I_2 \rightarrow I_2^- + Cl_2$	0.77
$\rightarrow ICl_2^- + I$	0.30
$\rightarrow ICl^- + (?)$	0.30
$\rightarrow I_2Cl^- + Cl$	0.55
$I_2^- + Cl_2 \rightarrow Cl_2^- + I_2$	0.20
$\rightarrow Cl I_2^- + Cl$	0.30
$\rightarrow Cl I^- + (?)$	0.24
$\rightarrow Cl_2 I^- + I$	0.28

[a]For reactant ions of ~0.3 eV laboratory energy.
[b]Determined by normalizing the relative rates for all reactions to the absolute value of 1.2×10^{-9} cm^3/molecule sec for the O^-/NO_2 electron transfer reaction, which is taken from Ref. 24.

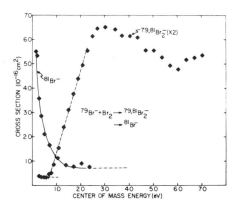

FIG. 22. Energy dependences of the cross sections for the electron transfer reaction of Br^- with Br_2 (showing linearly extrapolated threshold), and for the atom switching reaction.

TABLE V. Energy dependence of isotopic mixing in the products from endothermic charge transfer reactions of Br^- with Br_2 and Cl^- with Cl_2.

Reaction	Impacting ion	Ion energy (eV)	Product ions (%)		
			$^{79,79}Br_2^-$	$^{79,81}Br_2^-$	$^{81,81}Br_2^-$
$Br^- + Br_2 \rightarrow Br_2^- + Br$	$^{79}Br^-$	3.1	0.390	0.484	0.126
	$^{81}Br^-$	3.1	0.131	0.503	0.366
	$^{79}Br^-$	4.5	0.372	0.492	0.134
	$^{81}Br^-$	4.5	0.143	0.497	0.361
	$^{79}Br^-$	6.5	0.289	0.499	0.213
	$^{81}Br^-$	6.5	0.222	0.503	0.275
	$^{79}Br^-$	7.0	0.268	0.505	0.227
	$^{81}Br^-$	7.0	0.243	0.498	0.259
	$^{79}Br^-$	10.5	0.256	0.502	0.245
	$^{81}Br^-$	10.5	0.252	0.504	0.245
			$^{35,35}Cl_2^-$	$^{35,37}Cl_2^-$	$^{37,37}Cl_2^-$
$Cl^- + Cl_2 \rightarrow Cl_2^- + Cl$	$^{35}Cl^-$	2.0	0.643	0.319	0.038
	$^{37}Cl^-$	2.0	0.319	0.539	0.142
	$^{35}Cl^-$	3.4	0.658	0.311	0.032
	$^{37}Cl^-$	3.4	0.289	0.558	0.155
	$^{35}Cl^-$	12.0	0.569	0.373	0.053
	$^{37}Cl^-$	12.0	0.560	0.374	0.066

independent of energy (10.5 eV, lab, or 7.0 eV, c.m., see Fig. 22), the isotopic distribution of the Br_2^- product ions approaches that which is characteristic of natural bromine, (that is, the relative ratio of $^{79}Br^{79}Br/^{79}Br^{81}Br/^{81}Br^{81}Br$ in natural bromine is 0.255/0.499/0.244). Moreover, at these energies, the isotopic distribution of Br_2^- products is virtually identical from the reactions of both $^{79}Br^-$ and $^{81}Br^-$, indicating that the impacting ion is not incorporated in the products. At translational energies nearer threshold, however, the Br_2^- products exhibit an isotopic distribution which is quite different from the normal bromine distribution, and obviously some isotopic scrambling has occurred. Also, at lower energies, the isotopic distributions of the product ions are significantly different for the cases where $^{79}Br^-$ and $^{81}Br^-$ are the impacting ions. The variations with the energy of

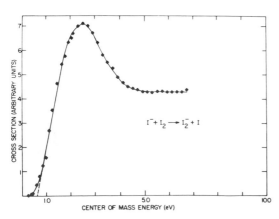

FIG. 23. Energy dependence of the cross section for the electron transfer reaction of I^- with I_2. Linearly extrapolated threshold is shown.

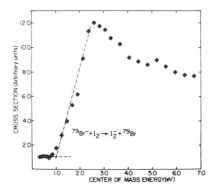

FIG. 24. Energy dependence of the cross section for the electron transfer reaction of Br^- with I_2. Linearly extrapolated threshold is shown.

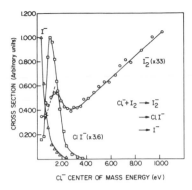

FIG. 25. Energy dependences of the cross sections for the electron transfer, atom transfer and atom switching reactions of Cl⁻ with I₂.

the isotope distributions in the products from the Cl^-/Cl_2 charge-transfer reaction are analogous to those discussed above for the bromine reaction, as indicated by the data in Table V, (the relative ratio of $^{35}Cl^{35}Cl/^{35}Cl^{37}Cl/^{37}Cl^{37}Cl$ in natural chlorine is 0.569/0.376/0.061).

It is instructive to consider possible mechanisms for the reactions just discussed and to calculate the isotopic distributions of products which would be expected on the basis of these models. One such mechanism is long-range electron transfer which involves no bonding between the incident negative ion and the neutral halogen molecule, but which results in formation of the negative molecular ion. In this case, the isotopic distributions of the product Br_2^- and Cl_2^- ions should be just that which corresponds to the natural isotopic abundances of the halogen molecules. Another possibility is an atom abstraction mechanism, in which the product molecular ion is formed by the creation of a new bond between the incident ion and one atom of the target molecule, with simultaneous rupture of the bond between the two atoms of the target molecule. In this case, it is assumed that there is equal probability of abstracting either atom of the target molecule, but the incident ion is always incorporated in the product ion. A third possibility is a mechanism which involves formation of a linear intermediate $[Br_3]^-$ or $[Cl_3]^-$ complex, in which the available collision energy is distributed equally among the vibrational degrees of freedom. Here, it is assumed that it is equally probable that either of the two "partial" bonds of the complex will be broken to yield the product molecular ion. A somewhat similar mechanism for the charge transfer process would entail a cyclic intermediate complex, having three "partial" bonds. Again, the col-

lision energy is assumed to be randomized and loss of any one of the three atoms to give the product ion is considered to be equally likely. The isotopic product distributions which are calculated for these four reaction models for the reactions of $^{79}Br^-$ and $^{81}Br^-$ with natural bromine molecules and for the reactions of $^{35}Cl^-$ and $^{37}Cl^-$ with natural chlorine molecules are shown in Tables VI and VII. Also given in these tables are the experimentally measured isotopic product distributions for these reactions at two reactant ion translational energies, one in the neighborhood of the reaction threshold and the other well above the energy at which the maximum cross section is observed.

It is evident, from the data in Tables VI and VII, that the isotopic distributions of the Br_2^- and Cl_2^- products, at translational energies in the threshold region, are in striking agreement with the distributions calculated for a reaction mechanism which involves a linear triatomic intermediate complex. At translational energies several volts above threshold however, the experimental product distributions are quite consistent with the isotopic distributions calculated for a long range electron transfer mechanism, where no isotopic scrambling occurs. As indicated by the more extensive experimental results in Table V, there is apparently a gradual transition from the complex mechanism to the long range mechanism as the energy increases above threshold. Probably, more than one mechanism is operative in the intermediate energy regime. In any event, this data provides strong evidence that intermediate $[Br_3]^-$ and $[Cl_3]^-$ collision complexes are indeed formed in the energy region near threshold. Since these complexes were never actually detected in our experiments, (at collision chamber pressures up to 30 μ, the maximum used), their lifetimes must be shorter than 20–30 μsec, which is approximately the transit time for these species from the collision chamber to the detector in our tandem instrument.

Linear symmetric structures of the polyhalide ions which correspond to the intermediate complexes discussed above were predicted from molecular orbital calculations,[57] which showed that the bonding in these ions could be described in terms of "p" orbitals without involving the higher "d" orbitals. Later measurements of the nuclear quadrupole coupling constants in the polyhalide ions confirmed this interpretation of the bonding.[58,59] The infrared and Raman spectra of the polyhalide ions determined by Person et al.[60] are also consistent with this structural picture. The symmetric structures which our isotopic data indicate for the $[Br_3]^-$ and $[Cl_3]^-$ intermediate complexes therefore appear to be quite reasonable.

TABLE VI. Isotope distributions in products from endothermic charge transfer reactions of Br^- with Br_2.

Mechanism	Impacting ion	Calculated product ion Distributions (%)		
		$^{79,79}Br_2^-$	$^{79,81}Br_2^-$	$^{81,81}Br_2^-$
I. $Br^- \to Br_2$ (long-range electron transfer)	$^{79}Br^-$	25.5	49.9	24.4
	$^{81}Br^-$	25.5	49.9	24.4
II. $Br^- \cdots Br \dashv Br$ (atom abstraction)	$^{79}Br^-$	50.6	49.4	0.0
	$^{81}Br^-$	0.0	50.6	49.4
III. $Br \cdots Br \cdots Br$ (linear complex)	$^{79}Br^-$	38.1	49.7	12.2
	$^{81}Br^-$	12.8	50.2	37.0
IV. cyclic complex	$^{79}Br^-$	42.2	49.6	8.2
	$^{81}Br^-$	8.5	50.4	41.7

Reaction	Ion energy (eV)	Impacting ion	Experimental product ion distributions (%)		
			$^{79,79}Br_2^-$	$^{79,81}Br_2^-$	$^{81,81}Br_2^-$
$Br^- + Br_2 \to Br_2^- + Br$	1.9	$^{79}Br^-$	38.0	50.5	11.9
	1.9	$^{81}Br^-$	12.1	50.5	37.4
	10.5	$^{79}Br^-$	25.6	50.2	24.5
	10.5	$^{81}Br^-$	25.2	50.4	24.5

G. C_2H_2

In previous studies in our laboratory,[41] we observed an exothermic reaction of O^- ions with C_2H_2 which occurs at near thermal energy and which yields C_2H^- as a product. This reaction was also observed in a more recent study by Stockdale *et al.*[61] Bohme and co-workers[62] recently reported equilibrium constants for reactions involving C_2H^- and other compounds which produce C_2H_2 and the corresponding negative ion of the compound. The acetylene molecular negative ion was not detected in any of these investigations.

In the present experiments we studied reactions of other negative ions with acetylene, but were unable to detect any such process which produces $C_2H_2^-$. The reactions of S^-, SH^-, and Cl^- were observed to yield C_2H^- at higher translational energies, although the cross section for the SH^- reaction was too small to permit a reliable threshold measurement. The excitation functions for the S^- and Cl^- reactions are shown in Figs. 26 and 27, respectively and the thresholds determined for these are listed in Table I. It is easily demonstrated that for reactions of the type,

$$X^- + C_2H_2 \to C_2H^- + XH \qquad (11)$$

the electron affinity of C_2H can be calculated from the relation:

$$E.A. (C_2H) = \Delta H_f(C_2H) + \Delta H_f(XH) - \Delta H_f(X)$$
$$- \Delta H_f(C_2H_2) + E.A.(X) - E_t, \qquad (12)$$

where ΔH_f is the heat of formation of the respective species and E_t is the threshold energy in the center-of-mass for the reaction. Using Eq. (12) with the measured energy thresholds and taking the appropriate thermodynamic data from the compilation by Franklin *et al.*,[63] the C_2H electron affinity values shown in Table I are obtained. As was observed to be the case with some of the other molecules already discussed, the Cl^- reaction with C_2H_2 gives an apparently lower value of the electron affinity for C_2H than do the other negative ion reagents. The value of E.A. $(C_2H) = 2.16 \pm 0.4$ eV obtained from the S^- reaction is therefore presumed to be the preferred value. The latter is in excellent agreement with the electron affinity derived by Bohme *et al.*,[62] (2.2 eV), from the flowing afterglow experiments mentioned above. Other values previously obtained for E.A. (C_2H) are listed in Table III.

H. "Symmetric" Electron Transfer Reactions of Molecular Negative Ions

Several of the molecular negative ions investigated in the present experiments were observed to undergo "symmetric" electron transfer reactions with their corresponding neutral species. Reactions (13)–(15), which were detected by monitoring the isotopic product ions indicated are typical of such processes:

$$^{14}NO_2^- + {}^{15}NO_2 \to {}^{14}NO_2 + {}^{15}NO_2^-, \qquad (13)$$

TABLE VII. Isotope distributions in products from endothermic charge transfer reactions of Cl^- with Cl_2.

Mechanism	Impacting ion	Calculated product ion distributions (%)		
		$^{35,35}Cl_2^-$	$^{35,37}Cl_2^-$	$^{37,37}Cl_2^-$
I. $Cl^- \to Cl_2$ (long-range electron transfer)	$^{35}Cl^-$	56.9	37.6	6.1
	$^{37}Cl^-$	56.9	37.6	6.1
II. $Cl^- \rightrightarrows Cl\!\!+\!\!Cl$ (atom abstraction)	$^{35}Cl^-$	75.4	24.6	0.0
	$^{37}Cl^-$	0.0	75.4	24.6
III. $Cl \rightrightarrows Cl \rightrightarrows Cl$ (linear complex)	$^{35}Cl^-$	66.0	31.0	3.0
	$^{37}Cl^-$	28.4	56.3	15.3
IV. (cyclic complex)	$^{35}Cl^-$	69.3	28.6	2.0
	$^{37}Cl^-$	19.0	62.5	18.5

Reaction	Ion energy (eV)	Impacting ion	Experimental product ion distributions (%)		
			$^{35,35}Cl_2^-$	$^{35,37}Cl_2^-$	$^{37,37}Cl_2^-$
$Cl^- + Cl_2 \to Cl_2^- + Cl$	3.4	$^{35}Cl^-$	65.8	31.1	3.2
	3.4	$^{37}Cl^-$	28.7	55.8	15.5
	12.0	$^{25}Cl^-$	56.9	37.3	5.6
	12.0	$^{37}Cl^-$	56.0	37.4	6.6

$$^{32}SO_2^- + {}^{34}SO_2 \to {}^{32}SO_2 + {}^{34}SO_2^- , \qquad (14)$$

$$C^{32,32}S_2^- + C^{32,34}S^- \to C^{32,32}S_2 + C^{32,34}S_2^- . \qquad (15)$$

Rate coefficients measured for these reactions at 0.3 eV (lab) translational energy are comparatively large, $k_{13} = 4.2 \times 10^{-10}$ cm³/molecule sec, $k_{14} = 6.7 \times 10^{-10}$ cm³/molecule sec; and $k_{15} = 1.5 \times 10^{-10}$ cm³/molecule sec. With increasing reactant ion trans-

lational energy, the cross sections fall off sharply and are approximately an order of magnitude smaller at 10 eV (Fig. 28). These "symmetric" electron transfer reactions seem to be fairly general for small molecular negative ions, but as will be seen in the following paper of this series (paper IV), these are not observed with more complex polyatomic negative ions such as SF_6^-.

Reactions of the type described are interesting because they could provide a means for relaxation

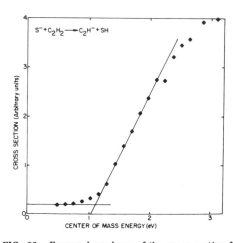

FIG. 26. Energy dependence of the cross section for the dissociative electron transfer reaction of S^- with C_2H_2 to give C_2H^-. Linearly extrapolated threshold is shown.

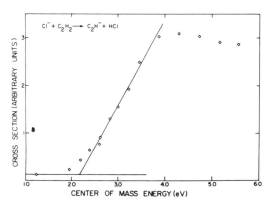

FIG. 27. Energy dependence of the cross section for the dissociative electron transfer reaction of Cl^- with C_2H_2. Linearly extrapolated threshold is shown.

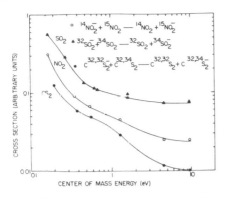

FIG. 28. "Symmetric" electron transfer reaction cross sections of NO_2^-, SO_2^-, and CS_2^- with their corresponding molecules as a function of energy.

of the internal excitation deposited in the negative ion by the electron attachment process, and might significantly increase the apparent lifetimes of these species, particularly under conditions where many collisions occur.

I. Comparison of the Negative Ion Charge Transfer and Alkali Atom Chemiionization Techniques

Some of the electron affinities which have been determined from negative ion charge transfer thresholds in the present study have also been recently measured in other laboratories using the alkali atom chemiionization technique. It is appropriate to compare these results and comment on the relative merits of the two techniques. A review of the data obtained by these two methods, (see summary in Table III), shows that in general, the electron affinities derived from chemiionization thresholds are somewhat larger than those calculated from negative ion charge transfer thresholds. This situation also holds for the case of O_2, for which the electron affinity is considered to be very accurately known from photodetachment electron spectroscopy. Our previously reported value for E. A. (O_2) from negative ion charge transfer experiments[12] is in excellent agreement with the latter measurement but several of the chemiionization[7,48] values are higher.

It has been suggested[50] that chemiionization experiments may give more accurate electron affinity data than negative ion charge transfer methods because in the latter, the energy necessary for reaction includes both the endothermicity of the process and that necessary to overcome the rotational barrier in the outgoing channel. No such barrier is expected for separation of the ion-pair formed by chemiionization. If a significant rotational barrier

is actually involved for the ion-molecule reactions in question, then the measured translational energy thresholds would be larger than the true endothermicity of the process and the electron affinities calculated would be too low.

Evidence obtained in the present investigation tends to refute the foregoing arguments. The best indication that there is not a high rotational barrier in the outgoing charge transfer channel comes from the isotope scrambling experiments with the halogens. As already discussed we have concluded that the Cl^-/Cl_2 and Br^-/Br_2 reactions proceed via an X_3^- intermediate (X=Cl, Br) in which the two X–X bonds are equivalent. This means that at threshold, the only collisions contributing to reaction are those having a very low impact parameter, or even head-on collisions $(b = 0)$. It follows therefore, that the orbital angular momentum for the reactants, $L = \mu v b$, is rather low. Conservation of angular momentum in the reaction demands that

$$L + I = L' + I' , \qquad (16)$$

where I is the rotational angular momentum and the primed quantities are for the products. Thus, since both L and I are small in this case, L' and I' must also be small and so a very low rotational barrier is expected for product formation. Only at higher translational energies, where long-range charge transfer becomes more prominent, do large impact parameter collisions contribute to reaction. In this instance, however, the available energy is already greater than the sum of the endothermicity and the rotational barrier. It seems unlikely that very long-range (high impact parameter) charge transfer collisions contribute at all to the reactions in question. None of the molecules discussed capture electrons directly, since the resulting negative ions which would be formed are unstable with respect to autodetachment. Thus, during charge transfer, the collision must be close enough to permit stabilization of the molecular negative ion by vibrational energy transfer. That such vibrational energy transfer does occur for collision at energies near threshold is also clearly shown by the halogen isotope scrambling results, since for these energies the two X–X bonds in the X_3^- intermediate are equivalent (that is, the energy is equally distributed between them). Good "mixing" of the internal energy therefore occurs in these cases. A similar situation is expected for the other negative-ion charge-transfer reactions studied here, and indeed, in separate experiments, the corresponding intermediates have been observed to be stable. (NO_2X^- and SO_2X^-, where X=Cl, F). From these considerations, we believe that the translational energy thresholds measured in the present experiments correspond to the true endothermicities of these negative ion charge transfer reactions.

*Presented in part at the 18th Annual Conference on Mass Spectrometry and Allied Topics, San Francisco, CA, June 1970, and at the 27th Symposium on Molecular Structure and Spectroscopy, Columbus, OH, June 1972 (Paper K-6).

†Ohio State University Research Foundation Visiting Research Associate under Contract F33615-67C-1758; from the Hebrew University, Jeruslam, Israel.

‡Author to whom reprint requests should be addressed.

[1]B. Moiseiwitsch, Adv. At. Mol. Phys. **1**, 61 (1965).

[2]R. S. Berry, Chem. Rev. **69**, 533 (1969).

[3]F. M. Page and G. C. Goode, *Negative Ions and the Magnetron* (Wiley, New York, 1969).

[4]J. J. DeCorpo and J. L. Franklin, J. Chem. Phys. **54**, 1885 (1971).

[5]R. K. B Helbing and E. W. Rothe, J. Chem. Phys. **51**, 1607 (1969).

[6]A. P. M. Baede, A. M. C. Moutinho, A. E. deVries, and J. Los, Chem. Phys. Lett. **3**, 530 (1969).

[7]K. Lacmann and D. R. Herschbach, Chem. Phys. Lett. **6**, 106 (1970).

[8]A. M. C. Moutinho, A. P. M. Baede, and J. Los, Physica (Utr.) **51**, 432 (1971).

[9]S. J. Nalley, R. N. Compton, H. C. Schweinler, and P. W. Reinhardt, Rept. No. ORNL-TM-2620, "Collisional Ionization Between Cesium and Selected Molecules," Oak Ridge National Laboratory, Oak Ridge, TN, April 1971.

[10]S. J. Nalley and R. N. Compton, Chem. Phys. Lett. **9**, 529 (1971).

[11]C. Lifshitz, B. M. Hughes, and T. O. Tiernan, Chem. Phys. Lett. **7**, 469 (1970).

[12]T. O. Tiernan, B. M. Hughes, and C. Lifshitz, J. Chem. Phys. **55**, 5692 (1971).

[13]W. A. Chupka, J. Berkowitz, and D. Gutman, J. Chem. Phys. **55**, 2724 (1971).

[14]K. Berkling, R. Helbing, K. Kramer, H. Pauly, Ch. Schlier, and P. Toschek, Z. Phys. **166**, 406 (1962).

[15]Fr. von Busch, H. J. Strunck, and Ch. Schlier, Z. Phys. **199**, 518 (1967).

[16]E. W. Rothe and R. W. Fenstermaker, J. Chem. Phys. **54**, 4520 (1971).

[17]P. J. Chantry, J. Chem. Phys. **55**, 2746 (1971).

[18]J. H. Futrell and C. D. Miller, Rev. Sci. Instrum. **37**, 1521 (1966).

[19]B. M. Hughes and T. O. Tiernan, J. Chem. Phys. **55**, 3419 (1971).

[20]K. R. Spangenberg, *Vacuum Tubes* (McGraw-Hill, New York, 1948), pp. 349 and 379.

[21]E. A. Soa, *Jenaer Jahrbuch 1959/1* (Zeiss, Jena, 1959).

[22]J. A. Simpson, Rev. Sci. Instrum. **32**, 1283 (1961).

[23]T. O. Tiernan and R. E. Marcotte, J. Chem. Phys. **53**, 2107 (1970).

[24]F. C. Fehsenfeld, A. L. Schmeltekopf, D. B. Dunkin, and E. E. Ferguson, ESSA Technical Report ERL +35-AL3, U.S. Dept. of Commerce, Washington, DC, 1969.

[25]P. Pechukas and J. C. Light, J. Chem. Phys. **42**, 3281 (1965).

[26]D. G. Truhlar, J. Chem. Phys. **51**, 4617 (1969).

[27]H. A. Bethe Rev. Mod. Phys. **9**, 69 (1937); H. A. Bethe and G. Placzek, Phys. Rev. **51**, 450 (1937).

[28]J. A. Rutherford and B. R. Turner, J. Geophys. Res. **72**, 3795 (1967).

[29]T. L. Bailey and P. Mahadevan, J. Chem. Phys. **52**, 179 (1970).

[30]J. M. Williams and W. H. Hamill, J. Chem. Phys. **49**, 4467 (1968).

[31]J. A. D. Stockdale, R. N. Compton, G. S. Hurst, and P. W. Reinhardt, J. Chem. Phys. **50**, 2176 (1969).

[32]D. Spence and G. J. Schulz, Phys. Rev. **188**, 280 (1969).

[33]D. Spence and G. J. Schulz, Phys. Rev. A **3**, 1968 (1971).

[34]R. Celotta, R. Bennett, J. Hall, M. W. Siegel, and J. Levine, Bull. Am. Phys. Soc. **15**, 1515 (1970). In the thesis which describes these results (M. W. Siegel, Univ. Colorado, Boulder, thesis, 1970; Univ. Microfilms Order No. 70-23754), a somewhat higher value of E.A. (NO) = 0.047 ± 0.005 eV is cited, which is obtained after compensating for the rotational energy distributions in the states involved in the transition.

[35]J. Berkowitz, W. A. Chupka, and D. Gutman, J. Chem. Phys. **55**, 2733 (1971).

[36]F. C. Fehsenfeld, E. E. Ferguson, and A. L. Schmeltekopf, J. Chem. Phys. **45**, 7844 (1966).

[37]J. F. Paulson, J. Chem. Phys. **52**, 959 (1970).

[38]S. J. Nally, R. N. Compton, H. C. Schweinler, and V.E. Anderson, J. Chem. Phys. (to be published)

[39]K. Lacmann and D. R. Herschbach, Chem. Phys. Lett. **6**, 106 (1970).

[40]R. S. Berry and C. W. Reimann, J. Chem. Phys. **38**, 1510 (1963).

[41]B. M. Hughes and T. O. Tiernan, Paper No. 15, 16th Annual Conference on Mass Spectrometry and Allied Topics, Pittsburgh, PA, May 1968.

[42]E. E. Ferguson, F. C. Fehsenfeld, and A. L. Schmeltekopf, J. Chem. Phys. **47**, 3085 (1967).

[43]E. E. Ferguson, *Advances in Electronics and Electron Physics*, edited by L. Marton (Academic, New York, 1968), Vol. 24, p. 1.

[44]F. A. Wolf and B. R. Turner, J. Chem. Phys. **48**, 4226 (1968).

[45]T. O. Tiernan and B. M. Hughes, Paper No. 13, 16th Annual Conference on Mass Spectrometry and Allied Topics, Pittsburgh, PA, May 1968.

[46]D. Vogt, Int. J. Mass Spectrom. Ion Phys. **3**, 81 (1969).

[47]D. B. Dunkin, F. C. Fehsenfeld, and E. E. Ferguson, Chem. Phys. Lett. **15**, 257 (1972); E. E. Ferguson (private communication).

[48]A. P. M. Baede, Physica (Utr.) **59**, 541 (1972).

[49]K. Lacmann and D. R. Herschbach, Bull. Am. Phys. Soc. **16**, 213 (1971).

[50]C. B. Leffert, W. M. Jackson, and E. W. Rothe, J. Chem. Phys. (to be published); E. W. Rothe (private communication).

[51]P. Warneck, Chem. Phys. Lett. **3**, 532 (1969).

[52]K. Kraus, W. Muller-Dusping, and H. Neuert, Z. Naturforsch. A **16**, 1385 (1961).

[53]D. Vogt, B. Hauffe, and H. Neuert, Z. Phys. **232**, 439 (1970).

[54]D. Feldmann, Z. Naturforsch. A **25**, 621 (1970).

[55]G. Herzberg, *Electronic Spectra of Polyatomic Molecules* (Van Nostrand, New York, 1966).

[56]J. H. Futrell and T. O. Tiernan, in *Ion-Molecule Reactions*, edited by J. H. Franklin (Plenum, New York, 1972), p. 477.

[57]G. C. Pimentel, J. Chem. Phys. **19**, 446 (1951).

[58]C. D. Cornwell and R. S. Yamasaki, J. Chem. Phys. **27**, 1060 (1957).

[59]R. S. Yamasaki and C. D. Cornwell, J. Chem. Phys. **30**, 1265 (1959).

[60]W. B. Person, G. R. Anderson, J. N. Fordemwalt, H. Stammreich, and R. Forneris, J. Chem. Phys. **35**, 908 (1962).

[61]J. A. D. Stockdale, R. N. Compton, and P. W. Reinhardt, Int. J. Mass Spectrom. Ion Phys. **4**, 401 (1970).

[62]D. K. Bohme, E. Lee-Ruff, and L. B. Young, J. Am. Chem. Soc. **94**, 5153 (1972).

[63]J. L. Franklin, J. G. Dillard, H. M. Rosenstock, J. T. Herron, and K. Draxl, Natl. Stand. Ref. Data Ser. **26** (1969).

41

Ion cyclotron resonance studies of endothermic reactions of UF$_6^-$ generated by surface ionization

J. L. Beauchamp

Contribution No. 5164 from the Arthur Amos Noyes Laboratory of Chemical Physics, California Institute of Technology, Pasadena, California 91125

I. INTRODUCTION

Ion cyclotron resonance spectroscopy (ICR) has previously been employed to investigate the formation and reactions of positive and negative ions derived from uranium hexafluoride,[1] considerably extending the earlier negative ion studies of Stockdale, Compton, and Schweinler.[2] With thermal electrons the attachment reaction

$$UF_6 + e \longrightarrow \begin{cases} UF_6^- , & (1) \\ UF_5^- + F , & (2) \end{cases}$$

leads to the formation of UF$_6^-$ with an unusually slow rate constant ($k \cong 5 \times 10^{-10}$ cm^3 molecule$^{-1} \cdot$ sec^{-1}) at low pressures ($< 10^{-5}$ torr).[1] At higher electron energies the the dissociative attachment Process (2) is observed, with a threshold of 0.9 ± 0.1 eV.[2] The species UF$_5^-$ reacts rapidly with UF$_6$ by electron transfer

$$UF_5^- + UF_6 \rightarrow UF_6^- + UF_5 ; \qquad (3)$$

electron transfer from other negative ions, including Cl$^-$ and SF$_6^-$

$$Cl^- + UF_6 \rightarrow UF_6^- + Cl , \qquad (4)$$

$$SF_6^- + UF_6 \rightarrow UF_6^- + SF_6 , \qquad (5)$$

appears to be a general process leading to the formation of UF$_6^-$.[1]

The electron transfer Reaction (4) indicates a high electron affinity for UF$_6$ (≥ 3.61 eV) and lattice energy calculations for the crystalline solid NO$^+$UF$_6^-$ suggest a value of 6.1 eV.[3] The surface ionization estimate of 2.9 eV by Page and Goode appears to be somewhat low.[4] The internal energy of UF$_6^-$ formed in the direct electron attachment Reaction (1) can be appreciable, and it was suggested that UF$_6^-$ containing excess internal excitation reacts to form the species UF$_7^-$ observed in UF$_6$ alone,[1]

$$(UF_6^-)^* + UF_6 \rightarrow UF_7^- + UF_5 . \qquad (6)$$

The majority of the UF$_6^-$ formed by electron attachment is unreactive with UF$_6$. The reaction of SF$_5^-$ with UF$_6$ indicates that the species UF$_7^-$ is reasonably stable,

$$SF_5^- + UF_6 \rightarrow UF_7^- + SF_4 , \qquad (7)$$

with $D(UF_6 - F^-) \geq 54 \pm 10$ kcal/mole.[1]

The present study was initiated to investigate endothermic reactions of UF$_6^-$ with translational energy in the range 0–40 eV in order to better characterize the thermochemical properties and processes which lead to formation of the various species UF$_n^-$ in UF$_6$ alone. In earlier studies the failure to observe reactions of UF$_6^-$ with other molecules at thermal ion energies was attributed to the thermochemical stability of the reactant ion.[1] In the present study endothermic reactions of UF$_6^-$ with BF$_3$ and SF$_6$ are also examined. In accomplishing these objectives new ICR techniques have been developed which promise to have wide application for the investigation of endothermic reactions of positive and negative ions which can be generated by surface ionization.

II. EXPERIMENTAL

The formation of negative ions on heated surfaces is a widely recognized phenomenon.[4,5] Although previous studies have suggested the possible formation of negative ions on the directly heated filament of an ICR spectrometer,[6] the conditions under which this occurs and the various manifestations of the process have not been well characterized. A cross sectional view of the ICR cell used in this study is shown in Fig. 1(a).[7] The spacing of the trapping and drift electrodes are 2.54 and 1.0 cm, respectively. Charged particles generated by thermionic emission from a rhenium ribbon filament (dimensions 0.001 × 0.032 in.) enter through a 2 mm circular aperture in the trapping electrode, traverse the cell in the direction of the magnetic field and are collected on an electrode used to monitor the emission current. The entrance and exit apertures are covered with a 85% open mesh screen to prevent field penetra-

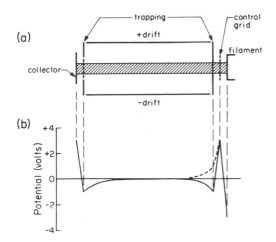

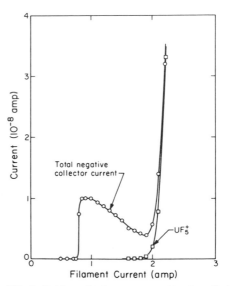

FIG. 1. (a) Cross sectional view of ICR cell. The shaded region indicates thermionically emitted particles which traverse the cell in the direction of the primary magnetic field. (b) Space potential along the ion beam. In the configuration shown negative ions have a kinetic energy of 3 eV in the cell, which confines scattered and product ions with a kinetic energy less than 1 eV (determined by the trapping voltage). The dashed portion of the curve indicates the change in space potential effected by modulating the voltage applied to the trapping plate adjacent to the filament.

tion into the cell. A control grid is mounted between the filament and the trapping electrode. In trapped ion experiments[8] this grid is used in gating the electron beam. In the present studies the grid is biased to assist in the extraction of electrons and negative ions from the ribbon filament. This considerably increases the usable currents at low energies.

The total negative collector current at 70 eV, recorded in the presence of 10^{-6} torr UF₆, is shown as a function of filament current in Fig. 2. The onset of thermionic emission of electrons in the absence of UF₆ occurs at ≈ 2 A. In the presence of UF₆ the onset for emission is 0.8 A (Fig. 2). The shape of the curve suggests that negative surface ionization is occurring at low filament currents, with the increase in current above 2 A being due to electron emission.

To verify that the UF₆ does not alter the electron emission characteristics of the filament, the yield of UF_5^+ formed by electron impact ionization was measured as a function of filament current (Fig. 2). The yield of UF_5^+ is directly proportional to the electron emission current. Above 2 A the collector current and UF_5^+ signal follow each other closely. Below 2 A the UF_5^+ signal vanishes, leading to the conclusion that thermionic emission at low filament currents is due entirely to negative ions. There is no visual access to the filament in the ICR apparatus used in these experiments. A rhenium ribbon filament of similar dimensions was mounted on a test stand and exposed to UF₆ at a pressure of 5×10^{-6} torr as measured by an ionization gauge. Negative ion emission was observed with a yield curve iden-

tical to that shown in Fig. 2. The onset for negative ion emission occurs at a filament temperature of 800 °C, determined by viewing the assembly through a quartz window with an optical pyrometer. At this temperature radiation from the filament is just barely visible.

The kinetic energy of ions formed by surface ionization in the ICR cell is determined by the bias applied to the filament. The variation of potential along the negative ion beam is shown in Fig. 1(b), with conditions appropriate for having negative ions at 3 eV in the ICR cell. The trapping well confines product ions with translational energy less than 1 eV in the direction of the ion beam. These ions drift from the source to the resonance region where they are observed using a marginal oscillator detector.[7] Lock-in detection and amplification of the signal is accomplished by modulating the voltage applied to the trapping plate adjacent to the filament.[9] A square wave voltage is applied symmetrically with respect to ground. The potential illustrated in Fig. 1(b) is appropriate for the trapping portion of the cycle. During the other half cycle the modification of the potential indicated by the dashed line does not allow for ion confinement.

A simple experiment permits observation of the ions being formed on the filament. By adjusting the ion energy so that it is less than the depth of the trapping well, negative ions will be reflected as they approach the trapping electrode. With trapping voltage modulation, however, ions traverse the cell and are reflected by the trapping plate adjacent to the collector. When the trapping plates come to the same potential, ions

FIG. 2. Total negative ion current reaching the collector as a function of the filament current. The accelerating voltage is 70 eV. Also included is the signal intensity of UF_5^+ resulting from electron impact ionization of UF₆. The two curves are normalized together at 2.3 A, where UF_6^- formation is negligible in comparison to electron emission.

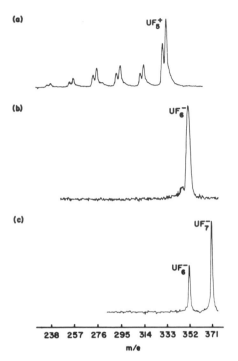

FIG. 3. ICR single resonance spectra of (a) positive ions formed at 70 eV in an enriched sample containing 40. 7% $^{235}UF_6$, (b) negative ions being formed on the filament, and (c) product ions observed from reaction of UF_6^- with UF_6 at a relative kinetic energy of 5.0 eV. The major species are identified in each instance and the m/e values correspond to the series $^{238}UF_n$, $n = 0-7$.

termine the ion energy distribution in the *center* of the ICR cell under actual operating conditions.

General descriptions of ICR techniques and apparatus are given elsewhere.[7-9] These experiments employed an instrument with a 15 in. magnet capable of a maximum field of 23.5 kG. The cell used is similar to that described by McMahon and Beauchamp,[8] and is equipped for operation either in drift or trapped ion modes. The best resolution which could be achieved in the drift mode with an oscillator frequency of 95 kHz at m/e 352 was a peak width of 1.3 amu (FWHM). This was sufficient to resolve $^{235}UF_6^-$ from $^{238}UF_6^-$ [Fig. 3(a)] and is limited by coherence times of ions in the radio frequency field of the detector.[7]

UF_6 was obtained from Varlacoid Chemical Co. An enriched sample containing 40. 7% $^{235}UF_6$ was made available by Dr. R. N. Compton of Oak Ridge National Laboratory. Other chemicals were readily available commercial samples and were used as supplied except for degassing at liquid nitrogen temperatures. All mixtures were prepared by admitting the components through separate leak values. As noted previously,[1] difficulties are encountered in measuring the pressure of UF_6. Consequently the reported values are only approximate.

III. RESULTS

A. Reactions in UF_6 alone

In the pressure range between 10^{-6} and 10^{-5} torr secondary products from reactions of thermionically generated UF_6^- and UF_6 are observed [Fig. 3(c)]. In the transverse geometry of the present experiment, only the slow products are detected. The yield of various UF_n^- ions in UF_6 alone are shown in Fig. 4, with UF_6^- energies between 0 and 40 eV in the center of mass.

present in the cell at that instant of time will be confined in the well and can be detected. The signal appears as a transient absorption at the beginning of the trapping cycle. Signal intensities are sufficient to use the phase-sensitive detector. A boxcar integrator gives a better signal to noise ratio, however. The 70 eV positive ion mass spectrum of UF_6 and the negative ion spectrum at 1 amp filament current are compared in Figs. 3(a) and 3(b). The only negative ion observed over the full range of filament currents is UF_6^-.

Exact calibration of the ion energy in the ICR cell is difficult. A correction for the electron energy scale can be generated by observing inelastic excitation spectra[10] and is generally less than 0.4 eV. The excitation spectra measurements give only an estimate of the peak in the distribution and not the complete distribution. Using the trapping electrodes as a stopping grid assembly leads to vanishing negative ion currents at the collector within 0.2 eV of the nominal value. Reported laboratory ion energies are estimated to be correct to within 0.3 eV. While this leaves much to be desired, the specific construction features and dimensions of the ICR cell provide constraints which would make it difficult to install a stopping grid analyzer which would de-

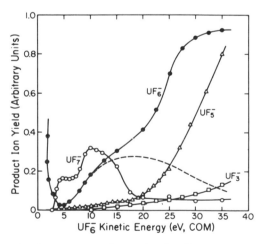

FIG. 4. Product ion yields as a function of ion energy in the center of mass (COM) from the reaction of UF_6^- with UF_6 at 3×10^{-6} torr. The yield for UF_6^- has been reduced by a scale factor of 3.0 to compare with the remaining products.

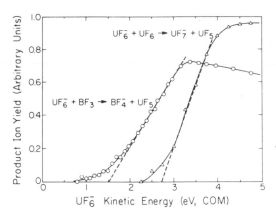

FIG. 5. Threshold region for reaction of UF_6^- with BF_3 and UF_6 to give BF_4^- and UF_7^-, respectively.

Product ion signal intensities are divided by UF_6^- current measured on the collector to determine relative product ion yields. A filament current of 1.0 A (corresponding to a temperature of 900 °C) and a trapping voltage of 1.0 volt were employed. All curves are directly comparable except for UF_6^- which has been reduced by a scale factor of 3.0.

The cross section for the formation of UF_7^- from UF_6^- in the reaction

$$UF_6^- + UF_6 \rightarrow UF_7^- + UF_5 , \tag{8}$$

increases rapidly with kinetic energy above 2.3 eV. The threshold region, considered in greater detail below, is shown in Fig. 5, recorded with a trapping well depth of 0.4 eV. The structure at 7.5 eV (Fig. 4) suggests a second process leading to the formation of UF_7^-. Above 10 eV the decrease in the yield of UF_7^- may be due to momentum transfer to the product ion, causing it to escape the trapping well. Alternatively, excess internal energy in the product ion may subsequently lead to unimolecular decomposition. An attempt was made to determine the maximum cross section for formation of UF_7^- at 10 eV relative kinetic energy. Assuming a small attenuation of the reactant beam, the product ion current $I(UF_7^-)$ is given by

$$I(UF_7^-) = I(UF_6^-)nQ_r l , \tag{9}$$

where $I(UF_6^-)$ is the UF_6^- beam current, n is the UF_6 number density, Q_r is the reaction cross section, and l is the path length. By increasing the filament current such that electron emission dominates, the current of UF_5^+ can be recorded at 70 eV and is given by

$$I(UF_5^+) = I_e n Q_i l , \tag{10}$$

where Q_i is the 70 eV cross section for formation of UF_5^+ from UF_6(UF_5^+ is 59% of the total ionization at 70 eV). By taking ratios, Q_r is given by

$$Q_r = \frac{(UF_7^-)I_e Q_i}{I(UF_6^-)I(UF_5^+)} . \tag{11}$$

ICR detection of UF_5^+ and UF_7^- yields the ratio $I(UF_7^-)/$

$I(UF_5^+)$ and the currents I_e and $I(UF_6^-)$ are directly measured at the collector. Repeated determinations gave $Q_r \cong Q_i$. Although the total ionization cross section of UF_6 is not known, it is estimated to be ~ 20 Å^2.[11] Hence $Q_r \cong 12$ Å^2, which is probably correct within a factor of 2. Working backward using Eq. (9), this value for Q_r yielded a pressure of $\sim 6 \times 10^{-6}$ torr, which is within the estimated range employed.

The yield of UF_6^- passes through a minimum at 4.0 eV and rises sharply at higher and lower energies. The low energy portion of the curve is attributed to trapping of elastically scattered UF_6^-. The higher energy portion of the curve is probably due to electron transfer as well as decomposition of internally excited UF_7^-. The yield of UF_6^- at relative kinetic energies above 10 eV depended markedly on trapping voltage. With a well depth of 0.4 V, the product ion yield indicated by the dashed line was observed. This suggests that UF_6^- formation is accompanied by momentum transfer, and probably occurs only in energetic encounters at small impact parameters. With isotopically enriched UF_6, the electron transfer reaction

$$^{235}UF_6^- + ^{238}UF_6 \rightleftharpoons ^{235}UF_6 + ^{238}UF_6^- , \tag{12}$$

could not be observed in either the drift or trapped ion mode using ion cyclotron double resonance and ion ejection techniques.[12] Analogous results have been reported for SF_6, namely symmetric electron transfer is not observed at low ion energies.[13,14]

The product ion UF_5^- becomes abundant at relative kinetic energies above 20 eV. The formation of slow UF_5^- at lower energy occurs with low probability, exhibiting an onset at ~ 5 eV. Possible processes for UF_5^- formation include

$$UF_6^- + UF_6 \begin{cases} \rightarrow UF_5^- + UF_7 , & (13) \\ \rightarrow UF_5^- + F + UF_6 , & (14) \\ \rightarrow UF_5^- + F_2 + UF_5 . & (15) \end{cases}$$

It has previously been shown that UF_7 is a stable species.[1] Thus, Reaction (13) is the lowest energy process leading to formation of UF_5^-. This process would appear to be competitive with Reaction (7); the predominance of UF_7^- suggests that E.A.$(UF_7) \gtrsim$ E.A.(UF_5), which is shown to be the case below. The minimum energy required for Process (14) is E.A.$(UF_6) + E_1$ $= 4.9 + 0.9 = 5.8$ eV, where E_1 is the threshold energy for Process (2) and the value for E.A.(UF_6) is derived below. This is close to the observed threshold of ~ 5 eV. Reaction (15) would be energetically less favorable than Process (14) since $D(F-F)$ is much less than $D(UF_5-F)$; Process (15) might result, however, from the decomposition of UF_7^-.

Interestingly, UF_4^- is not observed. The species UF_3^- is observed with a gradual onset at a relative kinetic energy of 10 eV. Most likely UF_3^- results from the decomposition of UF_5^- formed with excess internal energy.

B. Mixture of UF_6 with BF_3

In a 40:1 mixture of BF_3 and UF_6 at a total pressure of $\sim 4 \times 10^{-5}$ torr, the reaction

TABLE I. Summary of thermodynamic data for uranium compounds.[a]

Species	ΔH_f^{298} (kcal/mole)
UF_7	≤ -493 [b]
UF_6	-511
UF_5	-455
UF_4	-377
U	115 [c]
UF_7^-	-618 [c]
UF_6^-	624 [c]
UF_5^-	-509

[a]Except as noted, data is from the summary given in Ref. 1.
[b]Calculated assuming UF_7 is stable with respect to dissociation to give $UF_6 + F$ (see text).
[c]This work.

$$UF_6^- + BF_3 \rightarrow BF_4^- + UF_5 , \tag{16}$$

results in the formation of BF_4^-. The variation of reaction cross section with relative kinetic energy is shown in Fig. 5. The thermal motion of the target molecules leads to a significant doppler broadening in the interaction energy in experiments where a nominally monoenergetic beam interacts with gas molecules in a reaction chamber. This phenomenon has been treated in detail by Chantry.[15] At the relative energy E_0 the distribution of kinetic energies in the center of mass has a full width at half maximum, given by

$$W_{1/2} = (11.1 \gamma kTE_0)^{1/2} , \tag{17}$$

where $\gamma = m/(m + M)$, m and M being the projectile and target molecule masses.[15] Extrapolation of the rising linear portion of the cross section for Reaction (16) as indicated in Fig. 5 gives an onset of 1.5 eV in the center of mass. At $E_0 = 1.5$ eV, $W_{1/2} = 0.61$ eV. By considering model cross sections, Chantry has shown that extrapolation of the linear portion of a cross section which increases in proportion to the excess energy above threshold will lead to a threshold which is too low by $3\gamma kT$.[15] For Reaction (16) this amounts to 0.066 eV, giving an estimated threshold of 1.6 ± 0.2 eV. This is higher than the apparent onset by an amount approximately equal to $W_{1/2}$, again in agreement with Chantry's analysis. Similarly, a threshold energy of 2.7 ± 0.2 eV is estimated for Reaction (8). At this energy $W_{1/2} = 0.62$ eV.

Thermochemical data utilized in this study for compounds containing uranium are summarized in Table I. The bond dissociation energy $D(BF_3-F^-) = 71 \pm 5$ kcal/mole (3.1 eV) is known from other studies,[16] and with the measured threshold for Reaction (16) gives $D(UF_5-F^-) = 108 \pm 6$ kcal/mole (4.7 eV). Using the values $\Delta H_f(UF_6) = -511$ kcal/mole, $\Delta H_f(UF_5) = -455$ kcal/mole, and $\Delta H_f(F^-) = -61$ kcal/mole, this bond energy gives $\Delta H_f(UF_6^-) = -624$ kcal/mole and E.A.$(UF_6) = 4.9 \pm 0.5$ eV. This is low in comparison to the lattice energy estimate of 6.1 eV.[3] Assuming that the latter is correct, the measured threshold for Reaction (11) gives $\Delta H_f(UF_5) = -481$ kcal/mole, which corresponds to $D(UF_5-F)$

$= 49$ kcal/mole. This is low in comparison to the average bond energy in UF_6 of 123 kcal/mole.

The threshold of 2.7 eV for Reaction (8) can be combined with $D(UF_5-F^-) = 108$ kcal/mole to give $D(UF_6-F^-) = 46 \pm 10$ kcal/mole (2.0 eV) and $\Delta H_f(UF_7^-) = -618$ kcal/mole. The derived value for $D(UF_6-F^-)$ of 46 ± 10 kcal/mole is somewhat below the lower limit of 54 ± 10 kcal/mole provided by observation of Reaction (7). The uncertainties in both measurements are considerable, however.

If UF_7 is stable with respect to dissociation to give UF_6 and F,[1] then the derived heat of formation for UF_7^- gives E.A.$(UF_7) \leq 5.5$ eV. This is consistent with the derived value E.A.$(UF_6) = 4.9 \pm 0.5$ eV and the fact that UF_7^- does not undergo electron transfer to UF_6.

The dissociation of UF_7^- to UF_6^- and F is endothermic by only 0.4 eV. If Reaction (8) proceeds by an idealized stripping mechanism[17] at higher energies, then the internal excitation of UF_7^- will exceed 0.4 eV at a relative kinetic energy of 15.1 eV. At this energy the cross section for formation of UF_7^- drops off considerably with a rise in the cross section for UF_6^- (Fig. 4).

C. Mixture of UF_6 with SF_6

The electron transfer Reaction (5) is exothermic by E.A.(UF_6) – E.A.$(SF_6) = 4.9 - 0.6 = 4.3$ eV. In a 40:1 mixture of SF_6 and UF_6 at a total pressure of $\sim 4 \times 10^{-5}$ torr, the reverse reaction

$$UF_6^- + SF_6 \rightarrow SF_6^- + UF_6 , \tag{18}$$

is observed at relative kinetic energies above 7.5 eV. Variation of the reaction cross section with kinetic energy in the center of mass is displayed in Fig. 6. The gradual onset above 7.5 eV is considerably higher than the calculated threshold of 4.3 eV and yields only an upper limit of ~ 8.1 eV for E.A.(UF_6).

IV. SUMMARY AND CONCLUSION

The present work supports a high value for the electron affinity of UF_6, the threshold for Reaction (11) giving E.A.$(UF_6) = 4.9 \pm 0.5$ eV. This is compared with available data for other hexafluorides in Table II. The present results are not without the ambiguities which

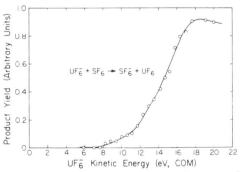

FIG. 6. Threshold region for endothermic electron transfer from UF_6^- to SF_6.

TABLE II. Electron affinities of selected hexafluoride molecules.

Species	Electron affinity (eV)	Reference
SF$_6$	0.6 ± 0.1	a, b
SeF$_6$	3.0 ± 0.2	c
TeF$_6$	3.2 ± 0.2	c
WF$_6$	4.5 ± 0.2	c
UF$_6$	4.9 ± 0.5	This work
PtF$_6$	6.8 ± 1.0	3

[a]C. Lifshitz, T. O. Tiernan, and B. M. Hughes, J. Chem. Phys. 59, 3182 (1973).
[b]R. N. Compton and C. D. Cooper, J. Chem. Phys. 59, 4140 (1973).
[c]C. D. Cooper and R. N. Compton, Bull. Am. Phys. Soc. 19, 1106 (1974).

accompany the measurement of thermodynamic properties from reaction thresholds. Formation of UF$_6^-$ by surface ionization undoubtedly leaves the reactant with considerable internal energy. Even at room temperature the average internal energy of UF$_6$ is high (~ 0.2 eV). The present experiments do not allow the effects of internal excitation on the thresholds for Reactions (8), (16) and (18) to be determined. The method of extrapolation to zero current assuming a linear threshold dependence for the cross section leaves much to be desired, particularly since discrimination against fast products above threshold may distort the true dependence of the cross section on ion energy. If the threshold for Reaction (18) had been used to determine E.A.(UF$_6$), it would be difficult to explain the observed threshold for Reaction (16), which would be *too low* by nearly 3 eV. The higher threshold for Reaction (18) could result, however, from a vertical rather than an adiabatic process, leaving both UF$_6$ and SF$_6^-$ vibrationally excited. Because of these uncertainties it would be of interest to redetermine the threshold for Reaction (16) in a crossed beam experiment utilizing UF$_6^-$ which has been collisionally relaxed in a high pressure ion source. Photodetachment experiments underway in our laboratory may also provide useful information relating to E.A.(UF$_6$).

The endothermicity of Reaction (8) is determined to be 2.7 ± 0.2 eV. This confirms the conjecture in the earlier trapped ion study of UF$_6$ that Reaction (6) involves UF$_6^-$ with excess internal excitation.[1] The formation of UF$_6^-$ by attachment of thermal electrons could yield an ion with up to 4.9 eV internal excitation. In the experiments designed to directly observe ions from the filament, UF$_6^-$ formed by surface ionization did not react to form UF$_7^-$ at near thermal ion energies.

The experimental methodology developed in the course of this work has extensive applications for studying endothermic reactions of both positive and negative ions formed by surface ionization. Investigations of endothermic reactions of alkali ions would be of particular interest.[18,19] It is important to note that the experimental arrangement discriminates against product ions which have appreciable kinetic energy in the direction of the reactant ion beam. For this reason

the processes observed tend to be those which occur at relatively large impact parameters with little momentum transfer. Processes amenable to study include electron transfer and stripping reactions in which atomic ion transfer to or from the neutral reactant leaves a slow product ion. Processes which take place at small impact parameter or involve complex formation are not good candidates for study except at kinetic energies comparable to the trapping well depth. It would be of interest to extend the results of the present study with an investigation of the energetic reactions of UF$_6^-$ with UF$_6$ using an in-line configuration to detect the fast products.

The ease with which UF$_6^-$ can be formed by negative surface ionization is due in part to the high electron affinity of UF$_6$. This deserves further study, particularly with metal surfaces of low work function. It may be possible to develop simple but intense sources of UF$_6^-$ for electromagnetic isotope separators and heavy ion accelerators. In addition the phenomenon provides the basis for the development of a very simple detector for the presence of UF$_6$ in vacuum systems at low pressures.

ACKNOWLEDGMENTS

This work was supported in part by the Energy Research and Development Administration under Grant No. AT(04-3)767-8 and the Ford Motor Company Fund for Energy Research administered by the California Institute of Technology. The instrument used in these studies was funded by the National Science Foundation under Grant No. NSF-GP-18383.

[1]J. L. Beauchamp, J. Chem. Phys. 64, 718 (1976).
[2]J. A. D. Stockdale, R. N. Compton, and H. C. Schweinler, J. Chem. Phys. 53, 1502 (1970).
[3]N. Bartlett, University of California, Berkeley (private communication); see also, N. Bartlett, Angew. Chem. Int. Ed. Engl. 7, 433 (1968).
[4]F. M. Page and G. C. Goode, *Negative Ions and the Magnetron* (Wiley-Interscience, London, 1967).
[5]R. O. Jenkins and W. G. Trodden, *Electron and Ion Emission from Solids* (Dover, New York, 1965).
[6]R. C. Dunbar, J. Am. Chem. Soc. 90, 5676 (1968).
[7]For a general discussion of ICR, see J. L. Beauchamp, Annu. Rev. Phys. Chem. 22, 527 (1971).
[8]T. B. McMahon and J. L. Beauchamp, Rev. Sci. Instrum. 43, 509 (1972).
[9]T. B. McMahon and J. L. Beauchamp, Rev. Sci. Instrum. 42, 1632 (1971).
[10]D. P. Ridge and J. L. Beauchamp, J. Chem. Phys. 51, 470 (1969).
[11]The ionization cross section of U is estimated to be ~15 Å^2 from the relative data of G. DeMaria, R. P. Burns, J. Drowart, and M. G. Inghram, J. Chem. Phys. 32, 1373 (1960). The range 20 ± 10 Å^2 for UF$_6$ likely encompasses the actual value, which has not been determined.
[12]J. L. Beauchamp, Adv. Mass Spectrom. 8, 717 (1973).
[13]M. S. Foster and J. L. Beauchamp, Chem. Phys. Lett. 31, 482 (1975).
[14]C. Lifshitz, T. O. Tiernan, and B. M. Hughes, J. Chem. Phys. 59, 917 (1973).
[15]P. J. Chentry, J. Chem. Phys. 55, 2746 (1971).
[16]J. C. Haartz and D. H. McDaniel, J. Am. Chem. Soc. 95,

8562 (1973).

[17] A. Henglein, Adv. Chem. Ser. 58, 63 (1966).

[18] R. D. Wieting, R. H. Staley, and J. L. Beauchamp, J. Am.

Chem. Soc. 97, 924 (1975).

[19] R. H. Staley and J. L. Beauchamp, J. Am. Chem. Soc. 97, 5920 (1975).

AUTHOR CITATION INDEX

I

Roberts, R. W., 277
Robertson, W. W., 302, 305
Robinson, C. F., 46
Robinson, P. J., 203
Roche, A. E., 330
Rose, T. L., 197, 230, 256, 262, 270, 322, 330
Rosenstock, H. M., 4, 203, 280, 338, 379
Ross, J., 197, 277
Rostagni, A., 210
Rothe, E. W., 180, 358, 379
Rozett, R. W., 260, 332
Rudolph, P. S., 72
Russell, M. E., 316, 358
Rutherford, J. A., 203, 308, 332, 379
Ryan, K. R., 141, 203
Rylander, P. N., 203

Safron, S. A., 232, 260, 262
Saporoschenko, M., 335
Sayers, J., 180, 301, 308
Schaeffer, H. F., III, 260
Schiff, H. I., 180, 197, 301, 302, 330, 335
Schissler, D. O., 43, 54, 71, 72, 77, 113, 184, 213, 221, 222, 274, 276, 292, 335, 336
Schlier, C., 5, 345, 358, 379
Schmeltekopf, A. L., 5, 180, 197, 301, 308, 315, 379
Schmidt, T. W., 180
Schulz, G. J., 302, 358, 379
Schumm, R. H., 358
Schweinler, H. C., 379, 385
Scott, J. T., 180
Searcy, A., 346
Sholette, W. P., 304
Siegel, M. W., 369, 379
Simons, J., 160
Simpson, J. A., 379
Slater, J. C., 158
Smith, A. C. H., 305
Smith, D., 180, 301, 308
Smith, D. L., 330
Smith, F. T., 180
Smith, P. T., 56, 61
Smith, R., 155
Smith, R. A., 158
Smith, S. J., 346
Smyth, H. D., 4, 9, 23, 24, 43, 74
Soa, E. A., 379
Sommer, L. H., 203
Sommerfeld, A., 28
Sood, S. P., 317
Sorenson, T. S., 203
Spangenberg, K. R., 379
Spence, D., 379
Spialter, L., 203

Spohr, R., 358
Staley, R. H., 386
Stammreich, H., 379
Stanton, H. E., 342
Starr, W. L., 307
Stebbings, R. F., 230, 256, 305, 308
Stephens, R. M., 330
Stevenson, D. P., 43, 54, 61, 71, 72, 74, 77, 113, 121, 141, 143, 180, 184, 197, 203, 213, 221, 222, 274, 276, 285, 292, 330, 335, 336, 337
Stewart, D. T., 335
Stewart, G. W., 203
Stockdale, J. A. D., 379, 385
Strunck, H. J., 345, 358, 379
Stueckelberg, E. C. G., 155
Su, T., 141
Sun, C. E., 119
Sutton, M. M., 330
Szwarc, M., 42

Tal'rose, V. L., 5, 43, 71, 77, 197, 222, 316, 349
Tanaka, 31
Tanaka, Y., 291
Tate, J. T., 38, 42, 61
Taubert, R., 275, 282
Taylor, H. S., 50, 113, 274, 335
Teller, E., 315
Theard, L. P., 75, 121, 141, 330
Thomas, D. W., 221
Thrush, B. A., 180
Tide, D. R., Jr., 141
Tiernan, T. O., 5, 203, 261, 379, 385
Toschek, P., 345, 358, 379
Trodden, W. G., 385
Truhlar, D. G., 86, 260, 358, 379
Tsao, C., 230, 240, 260, 270
Tulley, J. C., 260
Turner, B. R., 203, 230, 256, 308, 332, 379
Turner, D. W., 141, 300, 315
Tutte, W. T., 74
Tyndall, A. M., 161

Urena, A. G., 203
U. S. Energy and Development Administration, 203

Van Raalte, D., 321
Vance, D. W., 222, 239
Vdorin, V. M., 203
Vedeneyev, V. I., 320
Vestal, M. L., 330
Vogt, D., 358, 379
Volpi, G. G., 316, 330
Von Keussler, 33
Vorachek, J. H., 330

392

SUBJECT INDEX

About The Editor

JOE L. FRANKLIN is Emeritus Professor of Chemistry at Rice University. From 1963 until his retirement in 1976, he was Robert A. Welch Professor of Chemistry at Rice; from 1972 to 1976, he was chairman of the department. Prior to joining the Rice faculty, Professor Franklin served for nearly 30 years as a member of the research staff of Humble Oil and Refining Company (now Exxon) where he served in several capacities. In the course of his career in industry, he became interested in ions in gases; since the early 1950s, he has made this interest the principal part of his research program. His early studies of appearance potentials of ions and the thermochemical values that can be deduced from them led directly to the exploration of the effect of pressure and so to ion-molecule reactions. Professor Franklin and his early collaborators, F. H. Field and F. W. Lampe, were among the pioneers in the reawakening of interest in ion-molecule reactions, and he has continued an active program in this field ever since.

Professor Franklin received all his university education at the University of Texas, where he was awarded the B.S. and M.S. in chemical engineering in 1929 and 1930, and the Ph.D. in physical chemistry in 1934. He received the Southwest Regional Award of the American Chemical Society in 1962.

Where There's Smoke

Also by Sandra Brown
in Thorndike Large Print ®

The Silken Web
Fanta C
Whole New Light
22 Indigo Place
Texas! Lucky
Texas! Chase
Texas! Sage

This Large Print Book carries the
Seal of Approval of N.A.V.H.

WHERE THERE'S SMOKE

Sandra Brown

Thorndike Press • Thorndike, Maine

Thorndike Large Print® Basic Series edition published in 1993 by arrangement with Warner Books, Inc.

The tree indicium is a trademark of Thorndike Press.

Set in 16 pt. News Plantin by Heidi E. Saucier.

This book is printed on acid-free, high opacity paper. ∞

Library of Congress Cataloging in Publication Data

Brown, Sandra, 1948–
   Where there's smoke / Sandra Brown.
     p.   cm.
   ISBN 1-56054-781-2 (alk. paper : lg. print)
   ISBN 1-56054-782-0 (alk. paper : lg. print)
   1. Large type books. I. Title.
  [PS3552.R718W46 1993b]
  813'.54—dc20   93-23766
                                       CIP

Acknowledgments

During the writing of this book, I relied heavily on the assistance of experts in widely varied fields of endeavor, all of whom were cooperative in spirit and generous with their time. I wish to extend many thanks to:

Mr. Bob McNeece, who knows more about the oil business than I could ever comprehend;

Mr. Larry Collier, a "pumper" whom I stumbled upon by accident, but who proved to be such a valuable source of information;

Dr. Ernest Stroupe, M.D., an emergency room physician, who, along with the friendly and helpful people of Mother Frances Hospital, Tyler, Texas, talked me through the medical sequences of the story;

And a pilot who shared not only his knowledge of aviation, but a candid account of events he'd rather be forgotten.

SANDRA BROWN
October 26, 1992

Chapter One

He'd never particularly liked cats.

His problem, however, was that the woman lying beside him purred like one. Deep satisfaction vibrated through her from her throat to her belly. She had narrow, tilted eyes and moved with sinuous, fluid motion. She didn't walk, she stalked. Her foreplay had been a choreographed program of stretching and rubbing herself against him like a tabby in heat, and when she climaxed, she had screamed and clawed his shoulders.

Cats seemed sneaky and sly and, to his way of thinking, untrustworthy. He'd always been slightly uncomfortable turning his back to one.

"How was I?" Her voice was as sultry as the night beyond the pleated window shades.

"You don't hear me complaining, do you?"

Key Tackett also had an aversion to postcoital evaluation. If it was good, chatter was superfluous. If it wasn't, well, the less said the better.

She mistook his droll response as a compliment and slithered off the wide bed. Naked, she crossed the room to her cluttered dressing table and lit a cigarette with a jeweled lighter. "Want one?"

"No, thanks."

"Drink?"

"If it's handy. A quick one." Bored now, he gazed at the crystal chandelier in the center of the ceiling. The fixture was gaudy and distinctly ugly. It was too large for the bedroom even with the light bulbs behind the glass teardrops dimmed to a mere glimmer.

The shocking pink carpet was equally garish, and the portable brass bar was filled with ornate crystal decanters. She poured him a shot of bourbon. "You don't have to rush off," she told him with a smile. "My husband's out of town, and my daughter's spending the night at a friend's house."

"Male or female?"

"Female. For chrissake, she's only sixteen."

It would be unchivalrous of him to mention that she had acquired her reputation for being an easy lay long before reaching the age of sixteen. He remained silent, mostly from indifference.

"My point is, we've got till morning." Handing Key the drink, she sat down beside him, nudging his hip with hers.

He raised his head from the silk-encased pillow and sipped the straight bourbon. "I gotta get home. Here I've been back in town for . . ." he checked his wristwatch, "three and a half hours, and have yet to darken the door of the family homestead."

"You said they weren't expecting you tonight."

"No, but I promised to get home as soon as

8

I could manage it."

She twined a strand of his dark hair around her finger. "But you didn't count on running into me at The Palm the minute you hit town, did you?"

He drained his drink and thrust the empty tumbler at her. "Wonder why they call it The Palm. There isn't a palm tree within three hundred miles of here. You go there often?"

"Often enough."

Key returned her wicked grin. "Whenever your old man's out of town?"

"And whenever the boredom of this wide place in the road gets unbearable, which, God knows, is practically every day. I can usually find some interesting company at The Palm."

He glanced at her abundant breasts. "Yeah, I bet you can. Bet you enjoy getting every man in the place all worked up and sporting a hard-on."

"You know me so well." Laughing huskily, she bent down to brush her damp lips across his.

He turned his head away. "I don't know you at all."

"Why that's not true, Key Tackett." She sat back, looking affronted. "We went through school together."

"I went through school with a lot of kids. Doesn't mean I knew all of them beyond saying hello."

"But you kissed me."

9

"Liar." Chivalry aside, he added, "I didn't like standing in line, so I never even asked you out."

Her feline eyes squinted with malice that vanished in an instant. As quickly as she extended her claws, they were retracted. "We never actually went on a date, no," she purred. "But one Friday night after a victory against Gladewater, you and the rest of the football team came strutting off the field. My friends and me — with just about everybody else in Eden Pass — lined up along the sideline to cheer as you went past on your way to the field house.

"You," she emphasized, digging her fingernail into his bare chest, "were the outstanding stud among all the studs. You were the sweatiest, and your jersey was the dirtiest, and of course all the girls thought you were the handsomest. You thought so too, I think."

She paused for him to comment, but Key regarded her impassively. He was remembering dozens of Friday nights like the one she had just described. Pregame jitters and postwin exhilaration. The glare of the stadium lights. The cadence of the marching band. The smell of fresh popcorn. The pep squad. The cheering crowds.

And Jody, cheering louder than anybody. Cheering for him. That had been a long time ago.

"When you went past me," she continued, "you grabbed me around the waist, lifted me clean off the ground, hauled me up against you,

10

and kissed me smack on the mouth. Hard. Kinda barbariclike."

"Hmm. You sure?"

"Sure I'm sure. I creamed my panties." She leaned over him, pressing her nipples against his chest. "I waited a long time to have you finish what you started then."

"Well, I'm glad to have been of service." He swatted her fanny and sat up. "Scoot." Reaching around her, he retrieved his jeans.

"You really are leaving?" she asked, surprised.

"Yep."

Frowning, she ground out her cigarette in a nightstand ashtray. "Son of a bitch," she muttered. Then, taking a different tack, she came off the bed and swept aside his jeans before he could step into them. She bumped against his middle seductively.

"It's late, Key. Everybody out at your mama's house will be sound asleep. You'd just as well stay with me tonight." She reached between his strong thighs and fondled him, with audacity and know-how, boldly looking into his face as her fingers coaxed a response. "You haven't lived until you've partaken of one of my breakfast specialties."

Key's lips twitched with amusement. "Served in bed?"

"Damn straight. With all the trimmings. I even —" She broke off suddenly, her hands reflexively clenching hard enough to cause him to grimace.

"Hey, watch out. Them's the family jewels."

11

"Shh!" Releasing him, she ran on tiptoe toward the open bedroom door. As she reached it, a male voice called out. "Sugar pie, I'm home."

"*Shit!*" No longer languid and seductive, she turned toward Key. "You've got to get out of here," she hissed. "Now!"

Key had already stepped into his jeans and was bending down to search for his boots. "How do you suggest I do that?" he whispered.

"Sugar? You upstairs?" Key heard footsteps on the marble tiles of the entry below, then on the carpet of the stairs. "I got away early and decided to come on home tonight instead of waiting for morning."

She frantically motioned Key toward the French doors on the far side of the room. Scooping up his boots and shirt, he pulled open the doors and slipped through them. He was outside on the balcony before he remembered that the master bedroom was on the second floor of the house. Peering over the wrought-iron railing, he saw no easy way down.

Swearing beneath his breath, he quickly reviewed his options. What the hell? He'd faced worse situations. Typhoons, bullets, an earthquake or two, acts of God, and man-made mayhem. A husband coming home unexpectedly wasn't a new experience for him, either. He'd just have to bluff his way through and hope for the best.

He stepped back into the bedroom but pulled up short on the threshold of the French doors.

12

The nightstand drawer was open. His lover was now reclining in bed clutching the satin sheet to her chin with one hand. With the other, she was aiming a pistol straight at him.

"What the hell are you doing?"

Her piercing scream stunned him. A second later, a blast from her pistol shattered his eardrums. It was a few pounding heartbeats later before he realized that he'd been hit. He gazed down at the searing wound in his left side, then raised his incredulous eyes back to her.

The running footsteps had now reached the hallway. "Sugar pie!"

Again she screamed, a bloodcurdling sound. Again she aimed the gun.

Galvanized, Key spun around just as she fired. He thought she missed but couldn't afford the time to check. He tossed his boots and shirt over the railing, threw his left leg over, then his right, and balanced on an inch of support before leaping through the darkness to the ground below.

He landed hard on his right ankle. Pain shimmied up through his shin, thigh, and groin before slamming into his gut. Blinking hard, he gasped for breath, prayed he wouldn't vomit, and strove to remain conscious as he swept up his boots and shirt and ran like hell.

Lara jumped at the sound of hard knocking on her back door.

She'd been absorbed in a syrupy Bette Davis classic. Muting the television with the remote control, she listened. The knocking came again, harder and more urgent. Throwing off the afghan covering her legs, she left the comfort of her living room sofa and hurried down the hallway, switching on lights as she went.

When she reached the back room of the clinic, she saw the silhouette of a man against the partially open miniblinds on the door. Cautiously she crept forward and peered through a crack in the blinds.

Beneath the harsh glare of the porch light his face looked waxy and set. The lower half of it was shadowed by a day-old beard. Sweat had plastered several strands of unruly dark hair to his forehead. Beneath dense, dark eyebrows, he squinted through the blinds.

"Doc?" He raised his fist and pounded on the door again. "Hey, Doc, open up! I'm making a hell of a mess on your back steps." He wiped his forehead with the back of his hand, and Lara saw blood.

Putting aside her caution, she disengaged the alarm system and unlocked the door. As soon as the latch gave way, he shouldered his way through and stumbled, barefoot, into the room.

"You took long enough," he mumbled. "But all's forgiven if you still keep a bottle of Jack Daniel's stashed in here." He moved straight to a white enamel cabinet and bent down to open the bottom drawer.

"There's no Jack Daniel's in there."

At the sound of her voice, he spun around.

14

He gaped at her for several seconds. Lara gaped back. He had an animalistic quality that both attracted and repelled her, and although she was inured to the smell of fresh blood, she could smell his.

Instinctively she wanted to recoil, but not from fear. Her impulse was a feminine one of self-defense. She held her ground, however, subjecting herself to his disbelieving and disapproving stare.

"Who the hell are you? Where's Doc?" He was scowling darkly and holding the bloodied tail of his unbuttoned shirt against his side.

"You'd better sit down. You're hurt."

"No shit, lady. Where's Doc?"

"Probably asleep in his bed at his fishing cabin on the lake. He retired and moved out there several months ago."

He glared at her. Finally, in disgust, he said, "Great. That's just fuckin' great." He muttered curses as he shoved his fingers through his hair. Then he took a few lurching steps toward the door and careened into the examination table.

Reflexively Lara reached for him. He staved her off but remained leaning against the padded table. Breathing heavily and wincing in pain, he said, "Can I have some whiskey?"

"What happened to you?"

"What's it to you?"

"I didn't just move into Dr. Patton's house. I took over his medical practice."

His sapphire eyes snapped up to meet hers.

15

"You're a doctor?"

She nodded and spread her arms to indicate the examination room.

"Well I'll be damned." His eyes moved over her. "You must be a real hit at the hospital wearing that getup," he said, lifting his chin to indicate her attire. "Is that the latest thing in lady doctor outfits?"

She had on a long white shirt over a pair of leggings that ended at her knees. Despite her bare feet and legs, she assumed an authoritarian tone. "I don't generally wear my lady doctor outfits past midnight. It's after hours, but I'm still licensed to practice medicine, so why don't you forget how I'm dressed and let me look at your wound. What happened?"

"A little accident."

As she slipped his shirt from his shoulders, she noticed that his belt was unbuckled and only half the buttons of his fly were fastened. She prized his bloody hand away from the wound on his left side, about waist level.

"That's a gunshot!"

"Naw. Like I told you, I had a little accident."

Clearly, he was lying, something he seemed accustomed to doing frequently and without repentance. "What kind of 'accident'?"

"I fell on a pitchfork." He motioned down at the wound. "Just clean it out, put a Band-Aid on it, and tomorrow I'll be fine."

She straightened up and unsmilingly met his grinning face. "Cut the crap, all right? I know

16

a bullet wound when I see one," she said. "I can't take care of this here. You belong in the county hospital."

Turning her back on him, she moved to the phone and began punching out numbers. "I'll make you as comfortable as I can until the ambulance arrives. Please lie down. As soon as I've completed the call, I'll do what I can to stop the bleeding. Yes, hello," she said into the receiver when her call was answered. "This is Dr. Mallory in Eden Pass. I have an emer—"

His hand came from behind her and broke the connection. Alarmed, she looked at him over her shoulder.

"I'm not going to any damn hospital," he said succinctly. "No ambulance. This is nothing. Nothing, understand? Just stop the bleeding and slap a bandage on it. Easy as pie. Have you got any whiskey?" he asked for the third time.

Stubbornly, Lara began redialing. Before she completed the sequence of numbers, he plucked the receiver from her hand and angrily yanked it out of the phone, leaving the cord dangling from his fist.

She turned and confronted him, but, for the first time since opening the door, she was afraid. Even in this small East Texas town, drug abuse was a problem. Shortly after moving into the clinic, she had installed a burglar alarm system to prevent thefts of prescription drugs and narcotic painkillers.

He must have sensed her apprehension. With

17

a clatter, he dropped the telephone receiver onto a cabinet and smiled grimly. "Look, Doc, if I'd come here to hurt you, I'd have already done it and gotten the hell out. I just don't want to involve a bunch of people in this. No hospital, okay? Take care of me here, and I'll be on my merry way." Even as he spoke, his lips became taut and colorless. He drew an audible breath through clenched teeth.

"Are you about to pass out?"

"Not if I can help it."

"You're in a lot of pain."

"Yeah," he conceded, slowly nodding his head. "It hurts like a son of a bitch. Are you going to let me bleed to death while we argue about it?"

She studied his resolute face for a moment longer and reached the conclusion that she either had to do it his way or he'd leave. The former was preferable to the latter, in which case she would be risking the patient's health and possibly his life. She ordered him to lie down and lower his jeans.

"I've used that same line a dozen times myself," he drawled as he eased himself onto the table.

"That doesn't surprise me." Unimpressed by his boast, she moved to a basin and washed her hands with disinfectant soap. "If you know Doc Patton well enough to know where he stashed his Jack Daniel's, you must live here."

"Born and raised."

"Then why didn't you know he'd retired?"

"I've been away for a while."

"Were you a regular patient of his?"

"All my life. He got me through chicken pox, tonsillitis, two broken ribs, a broken collarbone, a broken arm, and an altercation with a rusty tin can that was serving as second base. Still got the scar on my thigh where I landed when I slid in."

"Were you called out?"

"Hell no," he replied, as though that were beyond the realm of possibility. "More than once I've come through that back door in the middle of the night, needing Doc to patch me up for one reason or another. He wasn't as stingy with the medicinal whiskey as you are. What's that you're fixing there?"

"A sedative." She calmly depressed the plunger of a syringe and sent a spurt of medication into the air.

She then set it down and swabbed his upper arm with a cotton ball soaked in alcohol. Before she knew what he was about to do, he picked up the syringe, pushed the plunger with his thumb and squirted the fluid onto the floor.

"Do you think I'm stupid, or what?"

"Mr. —"

"If you want me anesthetized, get me a glass of whiskey. You're not pumping anything into my bloodstream that'll knock me out and give you an opportunity to call the hospital."

"And the sheriff. I'm required by law to report

19

a gunshot wound to the authorities."

He struggled to sit up and when he did, the open wound gushed bright red blood. He groaned. Lara hastily slipped on a pair of surgical gloves and began stanching the flow with gauze pads so that she could determine how serious the wound was.

"Afraid I'll give you AIDS?" he asked, nodding at her gloved hands.

"Professional precaution."

"No worry," he said with a slow grin. "I've been real careful all my life."

"You weren't so careful tonight. Were you caught cheating at poker? Flirting with the wrong woman? Or were you cleaning your pistol when it accidentally went off?"

"I told you, it was a —"

"Yes. A pitchfork. Which would have punctured instead of tearing off a chunk of tissue." She worked quickly and effectively. "Look, I've got to trim off the rough edges of the wound and put in some deep sutures. It's going to be painful. I must anesthetize you."

"Forget it." He hitched his hip over the side of the table as though to leave.

Lara stopped him by placing the heels of her hands on his shoulders. The fingers of her gloves were bloody. "Lidocaine? Local anesthetic," she explained. She took a vial from her cabinet and let him read the label. "Okay?"

He nodded tersely and watched as she prepared another syringe. She injected him near the wound. When the surrounding tissue was deadened, she clipped the debris from around the wound, irrigated it with a saline solution, sutured the interior, and put in a drain.

"What the hell is that?" he asked. He was pale and sweating profusely, but he had watched every swift and economic movement of her hands.

"It's called a penrose drain. It drains off blood and fluid and helps prevent infection. I'll remove it in a few days." She closed the wound with sutures and placed a sterile bandage over it.

After dropping the bloody gloves into a marked metal trash can that designated contaminated materials, Lara returned to the sink to wash her hands. She then asked him to sit up while she wrapped an Ace bandage around his trunk to keep the dressing in place.

She stepped away from him and looked critically at her handiwork. "You're lucky he wasn't a better marksman. A few inches to the right and the bullet could have penetrated several vital organs."

"Or a few inches lower, and I couldn't have penetrated anything ever again."

Lara gave him a retiring look. "How lucky for you."

She had remained professionally detached, although each time her arms had encircled him while bandaging his wound, her cheek had come close to his wide chest. He had a lean, sunbaked, hair-spattered torso. The Ace bandage bisected

21

his hard, flat belly. She'd worked the emergency rooms of major city hospitals; she'd stitched up shady characters before — but none quite this glib, amusing, and handsome.

"Believe it, Doc. I've got the luck of the devil."

"Oh, I believe it. You appear to be a man who lives on the edge and survives by his wits. When did you last have a tetanus shot?"

"Last year." She looked at him skeptically. He raised his right hand as though taking an oath. "Swear to God."

He eased himself over the side of the examination table and stood with his hip propped against it while he rebuttoned his jeans. He left his belt unbuckled. "What do I owe you?"

"Fifty dollars for the after-hours office call, fifty for the sutures and dressing, twelve each for the injections, including the one you wasted, and forty for the medication."

"Medication?"

She removed two plastic bottles from a locked cabinet and handed them to him. "An antibiotic and a pain pill. Once the lidocaine wears off, it'll hurt."

He withdrew a money clip from the front pocket of his snug jeans. "Let's see, fifty plus fifty, plus twenty-four, plus forty comes to —"

"One sixty-four."

He cocked an eyebrow, seeming amused by her prompt tabulation. "Right. One hundred and sixty-four." He extracted the necessary bills and laid them on the examination table. "Keep the

22

change," he said when he put down a five-dollar bill instead of four ones.

Lara was surprised that he had that much cash on him. Even after paying her, he still had a wad of currency in high denominations. "Thank you. Take two of the antibiotic capsules tonight, then four a day until you've taken all of them."

He read the labels, opened the bottle of pain pills and shook out one. He tossed it back and swallowed it dry. "It'd go down better with a shot of whiskey." His voice rose on a hopeful, inquiring note.

She shook her head. "Take one every four hours. Two if absolutely necessary. Take them with water," she emphasized, seriously doubting that he'd stick to those instructions. "Tomorrow afternoon around four-thirty, come in and I'll change your dressing."

"For another fifty bucks, I guess."

"No, that's included."

"Much obliged."

"Don't be. As soon as you leave, I'm calling Sheriff Baxter."

Crossing his arms over his bare chest, he regarded her indulgently. "And get him out of bed at this time of night?" He shook his head remorsefully. "I've known poor old Elmo Baxter all my life. He and my daddy were buddies. They were youngsters during the oil boom, see? It was kinda like going through a war together, they said.

"They used to hang out around the drilling

23

sites, came to be like mascots to the roughnecks and wildcatters. Ran errands for them to buy hamburgers, cigarettes, moonshine, whatever they wanted. He and my daddy probably procured some things that old Elmo would rather not recall," he said with a wink.

"Anyway, go ahead and call him. But once he gets here, he'll be nothing but glad to see me. He'll slap me on the back and say something like, 'Long time no see,' and ask what the hell I've been up to lately." He paused to gauge Lara's reaction. Her stony stare didn't faze him.

"Elmo's overworked and underpaid. Calling him out this late over this piddling accident of mine will get him all out of sorts, and he's already cantankerous by nature. If you ever have a real emergency, like some crazy dopehead breaking in here looking for something to stop the little green monsters from crawling out of his eye sockets, the sheriff'll think twice before rushing to your rescue.

"Besides," he added, lowering his voice, "folks won't take kindly to you when they hear that you can't be trusted with their secrets. People in a small town like Eden Pass put a lot of stock in privileged information."

"I doubt that many even know the definition of privileged information," Lara refuted dryly. "And contrary to what you say, in the time I've been here, I've learned just how far-reaching and accurate the grapevine is. A secret has a short life span in this town.

24

"But your message to me about Sheriff Baxter came through loud and clear. What you're telling me is that he enforces a good ol' boy form of justice and that even if I reported your bullet wound, that would be the end of it."

"More'n likely," he replied honestly. "Around here, if the sheriff investigated every shooting, he'd be plumb worn out in a month."

Realizing that he probably was right, Lara sighed. "Were you shot while committing a crime?"

"A few sins, maybe," he said, giving her a slow, lazy smile. His blue eyes squinted mischievously. "But I don't think they're illegal."

She finally relinquished her professional posture and laughed. He didn't appear to be a criminal, although he was almost certainly a sinner. She doubted that he was dangerous, except perhaps to a susceptible woman.

"Hey, the lady doctor's not so stuffy after all. She can smile. Got a real nice smile, too." Narrowing his eyes, he asked softly, "What else have you got that's real nice?"

Now it was her turn to fold her arms across her chest. "Do these come-on lines usually work for you?"

"I've always thought that where boys and girls are concerned, talk is practically unnecessary."

"Really?"

"Saves time and energy. Energy better spent on doing other things."

"I don't dare ask 'Like what?'"

"Go ahead, ask. I don't embarrass easily. Do you?"

It had been a long time since a man had flirted with her. Even longer since she had flirted back. It felt good. But only for a few seconds. Then she remembered why she couldn't afford to flirt, no matter how harmlessly. Her smile faltered, then faded. She drew herself up and resumed her professional demeanor. "Don't forget your shirt," she said curtly.

"You can throw it away." He took a step away from the table, but fell back against it, his face contorted in pain. "Shitfire!"

"What?"

"My goddamn ankle. I twisted it when I . . . Hell of a sprain, I think."

She knelt down and as gently as possible worked up the right leg of his jeans. "Good Lord! Why didn't you show me this sooner?" The ankle was swollen and discolored.

"Because I was bleeding like a stuck hog. First things first. It'll be all right." He bent over, pushed aside her probing hands, and pulled down his pants leg.

"You should have it X-rayed. It could be broken."

"It isn't."

"You're not qualified to give a medical opinion."

"No, but I've had enough broken bones to know when one's broken, and this one isn't."

"I can't take responsibility if —"

"Relax, will you? I'm not going to hold you responsible for anything." Shirtless, shoeless, he hopped toward the door through which he'd entered.

"Would you like to wash your hands before you go?" she offered.

He looked down at the bloodstains and shook his head. "They've been dirtier."

Lara felt derelict in her duties as a physician treating him this way. But he was an adult, accountable for his own actions. She'd done as much as he had permitted.

"Don't forget to take your antibiotics," she cautioned as she slipped under his right arm and fit her left shoulder into his armpit. She placed her left arm around him for additional support as he hopped through the door, his right arm across her shoulders. A pickup truck was parked a few yards from the back steps. Its front tires had narrowly missed her bed of struggling petunias.

"Do you have some crutches?"

"I'll find some if I need them."

"You'll need them. Don't put any weight on your ankle for several days. When you get home, put an ice pack on it and keep it elevated whenever possible. And remember to come in at —"

"Four-thirty tomorrow. I wouldn't miss it."

She looked up at him. He tilted his head down to look at her. Their gazes came together and held. Lara felt the heat emanating from his body.

He was muscular and fit, and she was certain that his vital body would heal quickly. He was a physical specimen, which she had tried, not entirely successfully, to regard through purely professional eyes.

His eyes scanned her, looking intently at her face, her hair, her mouth. In a low, rough voice he said, "You sure as hell don't look like any doctor I've ever seen." His hand slid from her shoulder to her hip. "You don't feel like one either."

"What is a doctor supposed to feel like?"

"Not like this," he rasped, gently squeezing her.

He kissed her then. Abruptly and impertinently, he stamped her lips with his.

Gasping in surprise, Lara disengaged herself. Her heart was knocking and she felt hot all over. A thousand options on how to react flashed through her mind, but she considered that the best one was to pretend the kiss hadn't happened. Taking issue with it would only give it importance. She would be forced to acknowledge it, discuss it with him, and that, she hastily reasoned, should be avoided.

So she assumed a cool, haughty tone as she asked, "Would you like me to drive you somewhere?"

He was grinning from ear to ear, as though he saw straight through her attempt to conceal her discomposure. "No, thanks," he replied cockily. "This truck's got automatic transmission. I'll manage with my left foot."

She nodded brusquely. "If I hear of any crimes that occurred tonight, I'll have to report this incident to Sheriff Baxter."

Laughing even as he grimaced in pain, he climbed into the cab of the pickup. "Don't worry. You're not obstructing justice." He drew an imaginary X over his left breast. "Cross my heart and hope to die, stick a cross-tie in my eye." The engine sputtered to life. He dropped the gear shift into reverse. "Bye-bye, Doc."

"Be careful, Mr. —"

"Tackett," he told her through the open window. "But call me Key."

Everything inside Lara went very still. It seemed her heart, which had been racing only moments earlier, ceased to beat at all. Blood drained from her head, making her dizzy. She must have gone drastically pale, but it was too dark for him to notice as he backed the pickup to the end of the driveway. He tapped his horn twice and saluted her with the tips of his fingers as the truck rumbled away into the darkness.

Lara plopped down onto the cool concrete steps, which were speckled with drying drops of blood. She covered her face with damp, trembling hands. The night was seasonably warm and balmy, but she shivered inside her loose white shirt. Goose bumps broke out along her legs. Her mouth had gone dry.

Key Tackett. Clark's younger brother. He'd finally come home. This was the day she'd been

29

anticipating. He was essential to the daring plan she'd spent the past year developing and cultivating. Now, he was here. Somehow, some way, she must enlist his help. But how?

Dr. Lara Mallory was the last person Key Tackett wanted to know.

Chapter Two

As she did every morning of her life, Janellen Tackett left her solitary bed the instant the alarm went off. The bathtub faucets squeaked, and the hot-water pipes knocked loudly within the walls of the house, but these sounds were so commonplace she didn't even notice them.

Janellen had spent all of her thirty-three years in this house and couldn't imagine living anyplace else, or even wanting to. Her daddy had built it for his bride over forty years ago, and although it had been redecorated and modernized with the passing decades, the indelible marks Janellen and her brothers had left on its walls and the scarred hardwood floors remained. These flaws added to its character, like laugh lines in a woman's face.

Clark and Key had regarded the house as merely a dwelling. But Janellen considered it an integral member of the family, as essential to her heritage as were her parents. With a lover's attention to detail, she had explored it so many times she intimately knew it from attic to cellar. It was as familiar to her as her own body. Maybe even more so. She never focused her thoughts on her body, never contemplated her own being, never stopped to consider her life and wonder whether she was happy. She simply accepted things as they were.

Following her shower, she dressed for work in a khaki skirt and a simple cotton blouse. Her hosiery had no tint; her brown leather shoes had been designed for comfort, not fashion. She pulled her dark hair into a practical ponytail. Her only article of jewelry was a plain wristwatch. She applied very little makeup. One quick whisk of powder blusher across her cheeks, a little mascara on the tips of her eyelashes, a dab of pink lip gloss, and she was ready to greet the day.

The sun was rising as she made her way down the dim staircase, through the first-story hallway, and into the kitchen, where she switched on the overhead light fixtures, filling every nook and cranny with the blue-white light of an operating room. Janellen despised the invasive cold glare because it kept the otherwise traditional kitchen from being cozy.

But Jody liked it that way.

Mechanically, she started the coffee. She had religiously kept to this morning routine since the last live-in housekeeper had been dismissed. When Janellen was fifteen, she had declared that she no longer needed a baby-sitter, that she was capable of getting herself off to school and of cooking her mother's breakfast in the process.

Maydale, their current housekeeper, worked only five hours a day. She did the heavy cleaning and the laundry and got dinner started. But for all practical purposes, along with her responsibilities at Tackett Oil and Gas Company, Janellen managed the household.

She checked the refrigerator to make sure there was a pitcher of orange juice ready and poured half-and-half into the cream pitcher. Jody wasn't supposed to be drinking half-and-half in her coffee because of the fat content, but she insisted on it anyway. Jody always got her way.

While the coffeemaker gurgled and hissed, Janellen filled a watering can with distilled water and went out onto the screened back porch to sprinkle her ferns and begonias.

That's when she saw the pickup truck. She didn't recognize it, but it was parked as though it belonged in that particular spot near the back door. It was parked right where Key had always —

She did an about-face, almost spilling the contents of the watering can before returning it to the counter. She raced from the kitchen and down the hallway, grabbed the newel post and executed a childlike spiral around it, then charged up the stairs. Reaching the second floor, she dashed to the last bedroom on the right and, without pausing to knock, barged in.

"Key!"

"*What?*"

Running his fingers through his dark, tousled hair, he lifted his head off the pillow. He blinked her into focus. Then he moaned, clutched his side, and flopped back down. "Jesus! Don't sneak up on me like that. Had a bedouin do that to me once, and I almost gutted him before realizing he was one of the few friendly to us."

Heedless of his reprimand, Janellen quickly threw herself across her brother's chest. "Key! You're home. When did you get here? Why'd you sneak in without waking us? Oh, you're home. Thank you, thank you, thank you for coming." She hugged his neck hard and pecked several kisses on his forehead and cheeks.

"Okay, okay, I get it. You're glad to see me." He grumbled and staved off her kisses, but as he struggled to a sitting position, he was smiling. "Hiya, sis." Through bloodshot eyes, he looked her over. "Let's see. No gray hairs. You've still got most of your teeth. Haven't put on more'n five or six pounds. Overall, I'd say you look no worse for wear."

"I haven't put on a single ounce, I'll have you know. And I look just like I always have. Unfortunately." Without coyness, she added, "You and Clark were the pretty ones of the family, remember? I'm the plain Jane. Or in this case, Janellen."

"Now why would you want to piss me off first thing?" he asked. "Why go and say something like that?"

"Because it's true." She gave a slight shrug as though it was of little or no consequence. "Let's don't waste breath talking about me. I want to know about you. Where'd you come from and when did you get in?"

"Your message was channeled to me through that London phone number I gave you," he told her around a huge yawn. "It caught up with me in Saudi. Been traveling for three, maybe four days. Hard to keep track when you're crossing that many time zones. Came through Houston yesterday and dropped off the company plane. Got into Eden Pass last night sometime."

"Why didn't you wake us up? Who's truck is that? How long can you stay?"

He raked back his hair and winced as though each follicle were bruised. "One question at a time, please. I didn't wake you up because it was late and there was no point. I borrowed the truck from a buddy in Houston who has to deliver a plane to Longview in a couple of days. He'll pick it up then and drive it back. And . . . what was the last one?"

"How long can you stay?" She folded her hands beneath her chin, looking like a little girl about to say her bedtime prayers. "Don't say 'just a few days.' Don't say 'a week.' Say you're staying for a long time."

He reached for her folded hands and clasped them. "The contract I had with that oil outfit in Saudi was almost up anyway. Right now I haven't got anything cooking. I'll leave my departure date open. We'll wait and see how it goes, okay?"

"Okay. Thank you, Key." Tears glistened in her fine blue eyes. When it came to that family trait, she hadn't been passed over. "I hated to bother you with the situation here, but —"

"It was no bother."

35

"Well it *felt* like a bother. I wouldn't have contacted you if I didn't think that having you here might somehow make things . . . better."

"What's going on, Janellen?"

"It's Mama. She's sick, Key."

"Is her blood pressure kicking up again?"

"It's worse than that." Janellen twisted her hands. "She's started having memory blackouts. They don't last long. At first I didn't even notice them. Then Maydale mentioned several instances when Mama lost things and accused her of moving them. She introduces topics into conversations that we've already talked about."

"She's getting up there in years, Janellen. These are probably nothing more than early signs of senility."

"Maybe, but I don't believe so. I'm afraid it's more serious than just aging because there are days when I can tell she doesn't feel well, much as she tries to cover up."

"What does the doctor say?"

"She won't see one," she exclaimed with frustration. "Dr. Patton prescribed medication to control her blood pressure, but that was over a year ago. She browbeats the pharmacist into refilling the prescription and says that's sufficient. She won't listen when I urge her to see another doctor for a checkup."

He smiled wryly. "That sounds like Jody all right. Knows better than anybody about everything."

"Please, Key, don't be critical of her. Help her. Help me."

He cuffed her chin gently and said, "You've carried the responsibility alone for too long. It's time I gave you some relief." His lips narrowed. "If I can."

"You can. This time it'll be different between you and Mama."

Grunting with skepticism, he threw off the sheet and swung his feet to the floor. "Hand me my jeans, please."

Janellen was about to turn and reach for the jeans bunched up on the seat of the easy chair when she noticed the bandage around his middle. "What happened to you?" she exclaimed. "And look at your ankle!"

He nonchalantly examined his swollen ankle. "It was kind of a rowdy homecoming."

"How'd you get hurt? Is it serious?"

"No. The jeans, please."

Still sitting on the edge of the bed, he extended his hand. Janellen recognized the stubborn set of her brother's scruffy jaw and handed him his pants, then knelt to help guide his bare feet through the legs.

"Your ankle's swollen twice its size," she muttered with concern. "Can you stand on it?"

"My doctor advised me not to," he answered dryly. "Give me a hand."

She helped support him as he put all his weight on his left foot and eased the jeans up his legs and over his hips. As he buttoned his fly, he

37

gave her the naughty smile that had wreaked havoc on a legion of virtuous reputations.

Janellen couldn't began to guess how many women her brothers had worked their magic on, especially Key. She'd always entertained a fantasy of spoiling a mixed blend of nieces and nephews, but it remained an unfulfilled dream. Key liked women, a wide assortment of them. She saw no indication that he'd soon settle down into marriage.

"You're pretty good at helping a man into his pants," he remarked teasingly. "Been helping one out of his lately? I hope," he added.

"Hush!"

"Well?"

"No!" She could feel herself blushing. Key had always been able to make her blush.

"Why not?"

"I'm not interested, that's why," she replied loftily. "Besides, no one's been swept off his feet by my dazzling face and form."

"There's nothing wrong with either," he said staunchly.

"But they're hardly dazzling."

"No, because you've got it into your stubborn head that you're plain Jane, so you dress the part. You're so . . ." disdainfully, he gestured at her prim blouse, "buttoned up."

"Buttoned up?"

"Yeah. What you need to do is unbutton. Unhook. Unstrap. Get loose, sis."

She pretended to be aghast. "As an old maid,

38

I take exception to such trashy talk."

"Old maid! Who the hell . . . ? You listen to me, Janellen." He pointed his index finger at the tip of her nose. "You're *not* old."

"I'm not exactly an ingenue either."

"You're two years younger than me. That makes you thirty-four."

"Not quite."

"Okay, thirty-three. Far from over the hill. Hell, broads these days wait until they're forty to start having kids."

"Those who do wouldn't appreciate your referring to them as 'broads.' "

"You get my drift," he insisted. "You haven't even reached your sexual peak yet."

"Key, please."

"And the only reason you're still a 'maid,' *if* you are —"

"I am."

"More's the pity . . . is because you clam up and shy away from any guy who even thinks about getting into your pants."

Janellen, stricken by his crudeness, stared at him speechlessly. She worked around men eight hours a day, five days a week, and occasionally on weekends. As a rule, their language was colorful and to the point, but they monitored it when Miss Janellen was within hearing. When her employees addressed her, they cleaned up their act.

Of course Jody would shoot on sight any man using vulgarities in either her or her daughter's presence. Paradoxically, Jody herself had an ex-

39

tensive vocabulary of obscenities and blasphemies, an irony that seemed to escape her.

The fact that Janellen emanated an invisible repellent against casual and unguarded behavior didn't please her. In fact, she considered this characteristic a liability. It set her apart and proved that she didn't attract men in any way, on any level including friendship. She couldn't even be one of the boys, although she'd grown up having to contend with two older brothers.

She wasn't so much affronted by Key's salty language as she was stunned. In a way she took it as a compliment. Key, however, couldn't guess that.

"Oh, hell," he muttered remorsefully and stroked her cheek. "I'm sorry. I didn't mean to say that. It's just that you're too hard on yourself. Lighten up, for chrissake. Have some fun. Take off a year and go to Europe. Raise hell. Create a ruckus. Scare up a scandal. Broaden your scope. Life's too short to be taken so seriously. It's passing you by."

She smiled, clasped his hand, and kissed the back of it. "Apology accepted. I know you didn't mean to hurt my feelings or insult me. But you're wrong, Key. Life isn't passing me by. My life is here, and I'm content with it. I'm so busy, I don't know how I'd fit in another interest, romantic or otherwise.

"Granted, my life isn't as exciting as yours, but I don't want it to be. You're the globe-trotter. I'm a homebody, not at all suited to hell-

40

raising and ruckuses and scandals."

She laid her hand on his forearm. "I don't want to argue with you on your first day home since Clark's . . ." She couldn't bring herself to complete the sentence. She dropped her hand from his arm. "Let's go downstairs. The coffee should be ready by now."

"Good. I could use a cup or two before facing the old lady. What time does she usually get up?"

"The old lady is up."

In the doorway stood their mother, Jody Tackett.

Bowie Cato came awake when he was nudged hard in the ribs with the toe of a boot. "Hey, you, get up."

Bowie opened his eyes and rolled onto his back. It took him several seconds to remember he was sleeping in the storeroom of The Palm, the loudest, raunchiest, and seediest tavern in a row of loud, raunchy, and seedy taverns lining both sides of the two-lane highway on the outskirts of Eden Pass.

As the recently hired janitor, Bowie did most of his work after 2:00 A.M., when the tavern closed, and that was on a slow night. In addition to the piddling salary he earned, the owner had granted him permission to sleep on the storeroom floor in a sleeping bag.

"What's goin' on?" he asked groggily. It seemed he hadn't slept for more than a few hours.

"Get up." He got the boot in the ribs again,

41

more like a bona fide kick this time. His first impulse was to grab the offending foot and sling it aside, throwing its owner off balance and landing him flat on his ass.

But Bowie had spent the last three years in the state pen for giving vent to a violent impulse and he wasn't keen on the idea of serving another three.

Without comment or argument, he sat up and shook his muzzy head. Squinting through the sunlight coming from the window, he saw the silhouettes of two men standing over him.

"I'm sorry, Bowie." Speaking now was Hap Hollister, owner of The Palm. "I told Gus that you'd been here all night, didn't leave the premises once since seven o'clock last evenin', but he said he had to check you out anyway on account of you being an ex-con. He and the sheriff asked around last night and, best as they can tell, at the present, you're the only suspicious character in town."

"I seriously doubt that," Bowie mumbled as he slowly came to his feet. "It's all right, Hap." He gave his new employer a grim smile, then faced a bald, bloated, burly sheriff's deputy. "What's up?"

"What's up," the deputy repeated nastily, "is that Ms. Darcy Winston nearly got herself raped and murdered in her own bed last night. That's what's up." He gave them the details of the attempted break-in.

"I'm awful sorry to hear that." Bowie divided

his gaze between the uniformed deputy and Hap, but they continued to stare back at him wordlessly. He raised and lowered his shoulders in a quick, quizzical motion. "Who's Ms. Darcy Winston?"

"Like you don't know," the deputy sneered.

"I *don't* know."

"You, uh, were talking to her last night, Bowie," Hap said regretfully. "She was here while you were on duty. Redheaded, big tits, had on those purple, skinny-legged britches. Lots of jewelry."

"Oh." He didn't recall the jewelry, but those tits were memorable all right, and he figured that Ms. Darcy Winston knew it better than anybody. She'd been guzzling margaritas like they were lime-flavored soda pop and giving encouragement to every man in the place, including him, the lowly sweep-up boy.

"I talked to her," he told the deputy, "but we didn't get around to swapping names."

"She was talking to everybody, Gus," Hap interjected.

"But only this 'un has a prison record. Only this 'un is out on parole."

Bowie shifted his weight and ordered his tensing muscles to relax. Dammit, he knew instinctively that trouble was just around the corner, barreling full steam ahead, ready to knock him down. He hoped to hell he could get out of its path, but the odds didn't look good.

This two-hundred-fifty pound sheriff's deputy

43

was a bully. Bowie had tangled with too many in his lifetime not to recognize one on sight. He'd seen them large and muscular; he'd seen them small and wiry. A man's size and strength had nothing to do with it. The common denominator was a meanness-for-meanness' sake that shone in their eyes.

Bowie had first encountered it in his stepfather soon after his desperate, widowed mother had married the drunken son of a bitch who got off by slapping him around. Later, he'd recognized it in the junior high school boys' gym teacher who daily, deliberately, humiliated the kids who weren't natural athletes.

Standing up to his abusive stepfather and defending those pitiful kids against the gym teacher had been the start of the troubles that had eventually landed Bowie in county jail as a juvenile offender. Slow to learn, years later he'd graduated to state prison.

But this wasn't his fight. He didn't know Darcy Winston and couldn't care less about the attack on her. He told himself that if he just stayed cool it would be all right. "I was here at The Palm all night, just like Hap told you."

The deputy surveyed him up one side and down the other. "Take off your clothes."

"Excuse me?"

"What, are you deaf? Take off your clothes. Strip."

"Gus," Hap said apprehensively. "You sure that's necessary? The boy here —"

44

"Back off, Hap," the deputy snapped. "Let me do my job, will ya? Ms. Winston shot at the intruder. We know she hit him 'cause there was blood on her balcony railing and on the pool deck. He left a trail of it as he ran off through the bushes." He hitched up his gun holster, which fit in the deep crevice beneath his overlapping beer belly. "Let's see if you've got a bullet wound anywhere. Take off your clothes, jailbird."

Bowie's temper snapped. "Go fuck yourself."

The deputy's face turned as red as a billiard ball. His piggish eyes were almost buried in narrowing folds of florid fat.

Now there'd be hell to pay.

Making an animalistic grunt, the officer lunged for Bowie. Bowie dodged him. The deputy took a wild swing, which Bowie also deflected. Hap Hollister shouldered his way between them. "Hey, you two! I don't want any trouble here. I'm sure y'all don't either."

"I'm gonna break every bone in that little cocksucker's body."

"No, you ain't, Gus." Gus struggled against Hollister's restraining arms, but Hap had tussled with angry drunks many times and was no small man himself. He could handle the deputy. "Sheriff Baxter would have your ass if you harassed a suspect."

"I'm not a suspect!" Bowie shouted.

Still restraining Gus, Hap glared at Bowie over the deputy's meaty shoulder. "Don't go shooting

45

off your mouth like that, kid. It's stupid. Now, apologize."

"Like hell!"

"Apologize!" Hap roared. "Don't make me sorry I stood up for you."

While the deputy seethed, Hap and Bowie exchanged challenging stares. Bowie reconsidered. If he didn't keep a job, his parole officer would be after him. It was a lousy, goin'-nowhere job, but it was gainful employment that demonstrated his desire to reintegrate into society.

He for sure as hell wouldn't go back to Huntsville. Even if he had to kiss the ass of every thick-necked meathead with a badge pinned to his shirt, he wouldn't go back to prison.

"I take it back." For good measure, he unbuttoned his shirt and showed his chest and back to the deputy. "No bullet holes. I was here all night."

"And there's probably three dozen or so witnesses who can testify to that, Gus," Hap said. "Somebody else tried to break into Ms. Darcy Winston's bedroom last night. It wasn't Bowie."

Gus wasn't ready to concede, although it was obvious that he had the wrong man. "Funny that as soon as this parolee hits town, we get the first report of a serious crime in as long as I can remember."

"Coincidence," Hap said.

"I reckon," the deputy grumbled, although he continued to glare suspiciously at Bowie.

Hap diverted him with a piece of local gossip.

46

"By the way, guess who else blew into town last night. Key Tackett."

"No shit?"

Hap's maneuver worked. The deputy relaxed his official stance and propped his elbow on a shelf, for the time being forgetting Bowie and the purpose of his visit to the honky-tonk. Bowie just wanted to return to the sleeping bag and get some rest. He yawned.

The deputy asked, "What'd old Key look like? Gone to fat yet?" Laughing, he slapped his belly affectionately.

"Hell, no. Hasn't changed a smidgen since his senior year when he led the varsity team all the way to the state playoffs. Tall, dark, and handsome as the devil hisself. Those blue eyes of his still spear into everything they land on. Still the smartass he always was, too. First time he's been back to town since they buried his brother."

Bowie's ears perked up. He remembered the man they were talking about. Tackett was the kind of man who made a distinct impression on folks — male and female alike. Men wanted to be like him. Women wanted to be with him. He'd no more than sat down on a barstool when Ms. What's-her-name with the red hair and big tits had grafted herself to him. They'd been real friendly, too, for more than half an hour. Tackett had left within minutes of her slinking exit.

Interesting coincidence? Mentally Bowie

scoffed. He didn't believe in coincidence. But they could cut out his tongue and feed it to a coyote before he'd tell the deputy what he'd seen.

"Clark's passing — that was a tough time for ol' Jody," Gus was saying.

"Yeah."

"She ain't been the same since that boy died."

"And on top of that, that woman doctor moved into town and got the gossips all stirred up again."

The deputy stared into near space for a moment, sorrowfully shaking his head. "What possessed her to come to Eden Pass after what happened between her and Clark Tackett? I tell you, Hap, folks nowadays ain't worth shit. Don't care nothin' about nobody's feelings but their own."

"You're right, Gus." Hap sighed and slapped the deputy on the shoulder. "Say, when you get off duty, come have a beer on the house." Bowie was impressed by Hap's diplomacy as he steered the deputy out of the storeroom and through the empty bar, expounding as he went on the sad state of the world.

Bowie lay back down on the sleeping bag, stacked his hands behind his head, and stared up at the ceiling. Cobwebs formed an intricate canopy across the bare beams. As Bowie watched, an industrious spider added to it.

Momentarily Hap returned. Taking a seat on a case of Beefeater's, he lit a cigarette, then offered one to Bowie, who accepted and tipped his head forward as Hap lit it for him. They smoked in companionable silence. Finally Hap said, "Might ought to think about looking for another job."

Bowie propped himself up on one elbow. He wasn't surprised, but he wasn't going to take the news lying down literally. "You firing me, Hap?"

"Not outright, no."

"I had nothing to do with that bitch."

"I know."

"Then why am I catching the flap? Who is she anyway? You'd think by the way y'all talked about her that she's the Queen of Sheba."

Hap chuckled. "To her husband she is. Fergus Winston is superintendent of our school system. Owns a motel on the other end of town and does pretty good with it. He's 'bout twenty years older than Darcy. Ugly as a mud fence and not too bright. Folks figure she married him for his money. Who knows?" He shrugged philosophically.

"All I know is, anytime Darcy can shake Fergus, she's out here looking for action. Hot little piece," he added without rancor. "Had her myself a time or two. Years back when we were just kids." He pointed the lighted end of his cigarette toward Bowie. "If a thief did break into her bedroom last night, she might have shot him for *not* raping her."

Bowie shared a laugh with him, but the humor

49

was short-lived. "Why are you letting me go, Hap?"

"For your own good."

"As long as I don't personally serve liquor, my parole officer said —"

"It's not that. You do the work I hired you for." He regarded Bowie through world-weary eyes. "I run a fairly clean place, but lots of low-lifes come through the door every night. Anything can happen and sometimes does. Take my advice and find a place to work where you ain't so likely to run into trouble. Understand?"

Bowie understood. It was the story of his life. He just seemed to attract trouble no matter what he did or didn't do; and an honest, hardworking sort like Hap Hollister didn't need a natural-born troublemaker working in his bar. Resignedly he said, "Employers ain't exactly lining up to offer jobs to ex-cons. Can you give me a few days?"

Hap nodded. "Until you find something else you can bunk here. Use my pickup to get around if you need to." Hap anchored his cigarette in the corner of his lips as he stood. "Well, I got a stack of bills to pay. Don't be in a hurry to get up. You had a short night."

Left alone, Bowie lay down again but knew he wouldn't go back to sleep. From the start he'd known that there was little future in working at The Palm, but the job had also provided lodging. He had thought — hoped — that it would be a temporary respite, like a halfway house between prison and life on the outside. But no. Thanks to a broad he didn't even know, and to some son of a bitch committing a B and E, he was back to square zero.

Where he'd been stuck all his life.

Chapter Three

Jody Tackett and her son gazed at each other across the distance that separated them. It was a gulf that hadn't been spanned in thirty-six years, and Key doubted it ever would be.

He forced a smile. "Hi, Jody." He'd stopped using any derivative of Mother years ago.

"Key." She turned a baleful gaze on Janellen. "I guess this is your doing."

Key placed his arm across his sister's shoulders. "Don't blame Janellen. Surprising y'all was my idea."

Jody Tackett harrumphed, her way of letting Key know that she knew he was lying. "Did I hear you say the coffee was ready?"

"Yes, Mama," Janellen replied eagerly. "I'll cook you and Key a big breakfast to celebrate his homecoming."

"I'm not so sure his homecoming is cause for celebration." Having said that, Jody turned and walked away.

Key let out a deep sigh. He hadn't expected a warm embrace, not even an obligatory hug. He and his mother had never shared that kind of affection. For as far back as he could remember, Jody had been unapproachable and inaccessible to him, and he'd taken his cues from her.

For years they had coexisted under an undeclared truce. When they were together, he was polite and expected the same courtesy to be extended to him. Sometimes it was, sometimes it wasn't. This morning she had been flagrantly hostile, even though he was her only living son.

Maybe that was why.

"Be patient with her, Key," Janellen pleaded. "She doesn't feel well."

"I see what you mean," he remarked thoughtfully. "When did she start looking so old?"

"It's been over a year, but she still hasn't fully recovered from . . . you know."

"Yeah." He paused. "I'll try not to upset her while I'm here." He looked at his sister and smiled wryly. "Is there a pair of crutches in the house?"

"Right where you left them after your car wreck." She went to the closet and retrieved a pair of aluminum crutches from the rear corner.

"While you're at it, get me a shirt, too," he told her. "Mine didn't make it home last night."

He ignored her inquisitive glance and pointed at the shirts hanging in the closet. She brought him a plain white cotton one that smelled faintly of mothballs. He put it on but left it unbuttoned. Securing the padded braces of the crutches in his armpits, he indicated the door with a motion of his head. "Let's go."

"You look pale. Are you feeling up to this?"

"No. But I sure as hell don't want to hold up Jody's breakfast."

53

She was already seated at the kitchen table sipping coffee and smoking a cigarette when Key hobbled in. Janellen went unnoticed as she began preparing the meal. Key sat across from his mother and propped his crutches against the edge of the table. He was keenly aware of his bearded face and mussed hair.

As always, Jody was perfectly neat, although she wasn't an attractive woman. The Texas sun had left her complexion spotted and lined. Having no tolerance for vanity, her only concession to softening her appearance was a light dusting of dime store face powder. For all her adult life she had kept a standing weekly appointment at the beauty parlor to have her hair washed and set, but only because she couldn't be bothered to do it herself. It took twenty minutes for her short, gray hair to dry under the hood dryer. During that twenty minutes a manicurist clipped and buffed her short, square nails. She never had them polished.

She wore dresses only for church on Sundays and when a social occasion absolutely demanded it. This morning she was wearing a plaid cotton shirt and a pair of slacks, both crisply starched and ironed.

As she ground out her cigarette, she addressed Key in a tone as intimidating as her stare. "What'd you do this time?"

Her words were accusatory, clearly implying that Key was responsible for his misfortune. He was, but it wouldn't have mattered if he had been a victim of whimsical fate. Accidents had *always* been his fault.

When he'd fallen from the branches of the pecan tree that he and Clark had been climbing together, Jody had said that a broken collarbone was no better than he deserved for doing such a damn fool thing. When a Little League batter hit him in the temple with a bat, giving him a concussion, he'd been lectured for not keeping his mind on the game. When a gelding stepped on his foot, Jody had accused him of spooking the horse. When a firecracker exploded in his hand and busted open his thumb on the Fourth of July, he'd been punished. Clark had gotten off scot-free, although he'd been shooting off firecrackers alongside his brother.

But there was one time when Jody's wrath had been justified. If Key hadn't been so drunk, if he hadn't been driving ninety-five on that dark country road, he might have made that curve, missed that tree, and gone on to fulfill his mother's ambitions for him to be the starting quarterback on an NFL team. She would never forgive him for messing up her plans for his life.

Based on past experience, Key knew better than to expect maternal sympathy. But her judgmental tone of voice set his teeth on edge.

His reply was succinct. "I twisted my ankle."

"What about that?" she asked, raising her coffee cup toward the wide Ace bandage swathing his middle.

"Shark bite." He threw his sister a wink and a grin.

"Don't smart-mouth me!" Jody's voice cracked like a whip.

Here we go, Key thought dismally. Hell, he didn't want this. "It's nothing, Jody. Nothing." Janellen sat a cup of steaming coffee in front of him. "Thanks, sis. This is all for me."

"Don't you want anything to eat?"

"No, thanks. I'm not hungry."

She masked her disappointment behind a tentative smile that wrenched his heart. Poor Janellen. She had to put up with the old lady's crap every day. Jody had an uncanny talent for making every inquiry an inquisition, every observation a criticism, every glance a condemnation. How did Janellen endure her intolerance day in, day out? Why did she? Why didn't she find herself a respectable fellow and get married? So what if she wasn't madly in love with him? Nobody could be as difficult to live with as Jody.

Then again, Jody wasn't as critical of Janellen as she was of him. She hadn't been that way with Clark either. He seemed to be cursed with a talent for inciting his mother's anger. He figured it was because he was the spit and image of his father, and God knows Clark Junior had provoked Jody till the day he died. She hadn't shed a single tear at his funeral.

Key had. He had never cried before, or since, but he'd bawled like a baby at Clark Junior's grave, and not because his daddy had always been an attentive parent. Most of Key's recollections of him centered around farewells that had always left him feeling bereft. But whatever rare, happy memories of childhood Key had revolved around his daddy, who was boisterous and fun, who laughed and told jokes, who always drew a crowd of admirers with his glib charm.

Key was only nine years old when his father was killed, but with the inexplicable wisdom of a child, he'd realized that his best chance to be loved was being buried in that grave.

As though reading his mind, Jody suddenly asked, "Did you come home to watch me die?" Key looked at her sharply. "Because if you did," she added, "you're in for a big disappointment. I'm not going to die anytime soon."

Her expression was combative, but Key chose to treat the riling question as a joke. "Glad to hear that, Jody, 'cause my dark suit is at the cleaners. Actually, I came home to see how y'all are getting along."

"You've never given a damn how we were getting along before. Why now?"

The last thing Key felt like doing was tangling with his mother. He wasn't exactly in top physical form this morning, and Jody always disturbed his mental state. She was lethal to a sense of humor and an optimistic outlook. He'd wanted to make this reunion easy, if for no other reason than to please his long-suffering sister. Jody, however, seemed determined to make it difficult.

"I was born here," Key said evenly. "This is

my home. Or it used to be. Aren't I welcome here anymore?"

"Of course you're welcome, Key," Janellen said urgently. "Mama, do you want bacon or sausage?"

"Whatever." Jody gestured irritably, as though brushing off a housefly. As she lit another cigarette, she asked Key, "Where've you been all this time?"

"Most recently Saudi Arabia." He sipped his coffee, recounting for Jody what he'd told Janellen earlier, omitting that it had been Janellen's request that had brought him home.

"I was flying wild well-control crews to and from a burning well. Hauled supplies every now and then, had a few medical emergencies. But they were finishing up there, and I didn't have another contract pending, so I thought I'd hang around here for a spell. You might find this hard to believe, but I started missing Eden Pass. I haven't been home in more than a year, not since Clark's funeral."

He sipped his coffee again. Several seconds passed before he realized that Janellen was staring at him like a nocturnal animal caught in a pair of headlights and that Jody was scowling.

Slowly he returned his coffee cup to the saucer. "What's the matter?"

"Nothing," Janellen said hastily. "Do you need a refill on coffee?"

"Yeah, but I'll get it. I think the bacon's burning." Smoke was rising from the frying pan.

58

Key hopped to the counter and poured himself a coffee refill. He needed another pain pill, but he'd left them upstairs in his bathroom. In spite of the doctor's orders, he'd washed down two of the tablets with a tumbler of whiskey before going to bed. That had gotten him through the night.

Now, the pain was back. He wished he had the gumption to take the bottle of brandy that Janellen used for baking from the pantry and lace his coffee with it. But that would only give Jody another reason to harp on him. For the time being, he'd have to live with the throbbing pain in his side and the heavy discomfort in his right ankle.

As cavalier as he'd been about his injuries, he winced involuntarily as he hopped back to his seat. "Are you going to tell us how you got so banged up?" Jody asked.

"No."

"I don't like being kept in the dark."

"Believe me, you don't want to know."

"I have little doubt of that," she remarked sourly. "It's just that I don't want to hear the sordid facts from somebody else."

"Don't worry about it. It's not your concern."

"It'll be my concern once it gets around town that on your first night home you wound up in the hospital."

"I didn't go to the hospital. I went to Doc Patton's place and found a lady doctor there who's pretty as a picture," he said with a wide

59

grin. "She treated me."

Janellen dropped a metal spatula, which clattered onto the top of the cooking range. At first Key thought that hot bacon grease had popped out of the skillet and burned her hand. Then he noticed the hard, implacable expression on Jody's face and recognized it as fury. He'd seen it often enough to know it well.

"What's going on? How come y'all are looking at me like I just pissed on a grave?"

"You have." There was a low, wrathful hum behind Jody's words. "You've just pissed on your brother's grave."

"What the hell are you talking about?"

"Key —"

"The doctor," Jody said, angrily interrupting Janellen and banging her fist on the table. "Didn't you notice her name?"

Key thought back. He hadn't been so badly hurt that certain attributes had gone unnoticed — things like her expressive hazel eyes, her attractively disheveled hair, and her long, shapely legs. He'd even made a mental note of the color of her toenail polish and the fragrance she wore. He recalled these intimate details, but he didn't know her name. What could it matter to Jody and Janellen? Unless they were prejudiced against all women in the medical profession because of one.

As he considered that thought, he began to experience a sick gnawing in his gut. Jesus, it couldn't be. "What's her name?"

Jody only glared at him. He looked to Janellen for an answer. She was nervously wringing out a dry dish towel, misery etched on each feature of her face. "Lara Mallory is the name she goes by professionally," she whispered. "Her married name is —"

"Lara Porter," Key finished in a low, lifeless voice.

Janellen nodded.

"Christ." He raised his fists to his eyes and mentally pictured the woman he'd met the night before. She didn't match the bimbo featured in all the tabloid photographs. None of her deft mannerisms or candid expressions corresponded with the mental images he'd painted of Lara Porter, the woman who'd been his brother's downfall, the woman who some political analysts hypothesized had changed the course of American history.

Finally Key lowered his hands and gave a helpless, apologetic shrug. "I had no way of knowing. She never gave me her name, and I didn't ask. I didn't recognize her from the pictures I'd seen. That all happened . . . what? — five, six years ago?"

He hated himself for babbling excuses, knowing full well that the damage had been done and that Jody wasn't going to forgive him no matter what he said now. So he took another tack and asked, "What the hell is Lara Porter doing in Eden Pass?"

"Does it matter?" Jody asked brusquely.

"She's here. And you're to have nothing to do with her, understand? By the time I get finished with her, she'll tuck tail and slink out of town the same way she slunk in.

"Until that time, the Tacketts and anybody who wants to stay on speaking terms with us are to treat her with nothing except the contempt she deserves. That includes you. That *especially* includes you."

She jabbed her cigarette toward him to make her point. "Have all the sluts you want, Key, as I'm sure you will. But stay away from her."

Key immediately went on the defensive and raised his voice to match his mother's. "What are you yelling at me for? I wasn't caught humping her, Clark was."

Jody rose slowly to her feet and leaned on the table, bearing down on her younger son over bottles of catsup and Tabasco sauce. "How dare you speak that way about him. Don't you have an ounce of decency, a smidgen of respect for your brother?"

"*Clark*," Key shouted, rising and squaring off against Jody across the table. "His name was Clark, and what kind of respect do you pay him by not even speaking his name out loud?"

"It hurts to talk about him, Key."

"Why?" He rounded on Janellen, who'd timidly made the comment.

"Well, because . . . because his death was so untimely. So tragic."

"Yes, it was. But it shouldn't cancel out his life." He turned back to Jody. "Before he died, Daddy saw to it that Clark and I shared some good times. He wanted us to be close in spite of you, and we were. God knows Clark and I were poles apart in everything, but he was my brother. I loved him. I mourned him when he died. But I refuse to pretend that he didn't exist just to spare your feelings."

"You aren't fit to speak your brother's name."

It hurt. Even now it cut him to the quick when she said things like that. She left him no recourse except to lash back. "If he was so bloody perfect, we wouldn't be having this conversation, Jody. There would never have been a Lara Porter in our lives. No bad press. No scandal. No shame. Clark would have remained the Golden Boy of Capitol Hill."

"Shut up!"

"Gladly." He shoved the crutches under his arms and headed for the back door.

"Key, where are you going?" Janellen asked in a panicked voice.

"I've got a doctor's appointment."

Defiantly he glared at Jody, then let the door slam behind him.

Lara had spent a restless night. Even under the best of circumstances she wasn't a sound sleeper. Frequently her sleep was interrupted by bad dreams and long intervals of wakefulness. She listened for cries that she would never hear again. Sorrow was the basis of her habitual insomnia.

Last night, meeting Key Tackett had made sleep particularly elusive. She had awakened with a dull headache. Encircling her eyes were dark rings, which cosmetics had helped to camouflage but hadn't eliminated. Two cups of strong black coffee had relieved the headache, but she couldn't cast off the disturbing thoughts about her late-night caller.

She hadn't believed it was possible for any other man to be as attractive as Clark Tackett, but Key was. The brothers were different types, certainly. Clark had had the spit-and-polish veneer of a Marine recruit. There was never a strand of his blond hair out of place. His impeccably tailored clothes were always well pressed; his shoes shone like mirrors. He had epitomized the clean-cut guy next door, the all-American boy whom any mother would love her daughter to bring home.

Key was the type from whom mothers hid their daughters. Although just as handsome as Clark, he was as dissimilar to his brother as a street thug to an Eagle Scout.

Key was a professional pilot. According to Clark, he flew a plane by instinct and put more faith in his own judgment and motor skills than he did in aeronautical instruments. He relied on technology only when given no other choice. Clark had boasted that there wasn't an aircraft made that his brother couldn't fly, but Key had opted to freelance rather than work for a commercial airline.

"Too many rules and regulations for him," Clark had said, smiling with indulgent affection for his younger brother. "Key likes answering to no one but himself."

Having met him and experienced firsthand the compelling allure of his mischievous smile, Lara couldn't imagine Key Tackett dressed in a spiffy captain's uniform, speaking to his passengers in a melodious voice about the weather conditions in their destination city.

Sitting in cockpits a great deal of the time had left him with attractive squint lines radiating from his eyes — eyes as blue as Clark's. But Clark had been blond and fair. Key's eyes were surrounded with thick, blunt, black eyelashes. He was definitely the black sheep of the family, even in physical terms. His hair was thick and dark and as undisciplined as he. Clark had never sported a five o'clock shadow. Key hadn't shaved for days. Oddly, the stubble had contributed to, not detracted from, his appeal.

The brothers were fine specimens of the human animal. Clark had been a domesticated pet. Key was still untamed. When angered — or aroused — Lara imagined he would growl.

"Good morning."

She jumped as though she'd been caught doing something she should feel guilty for. "Oh, good morning, Nancy. I didn't hear you come in."

"I'll say. You were a million miles away." The

nurse/receptionist placed her handbag in the file case and put on a pastel lab coat. "What happened to the telephone in the examination room?" She had come in through the back door before joining Lara in a small alcove where they kept supplies, beverages, and snacks. The kitchen of the attached house remained for Lara's personal use.

"It was flimsy, so I decided to replace it."

Because she hadn't yet sorted out her feelings about Key Tackett's visit to the clinic, she wasn't ready to discuss it with Nancy. "Coffee?" She held up the carafe.

"Absolutely." The nurse added two teaspoons of sugar to the steaming mug Lara handed her. "Are there any doughnuts left?"

"In the cabinet. I thought you were dieting."

Nancy Baker found the doughnuts and demolished half of one with a single bite, then licked the sugar glaze off her fingers. "I gave up dieting," she said unapologetically. "I'm too busy to count calories. And if I dieted from now till doomsday, I'd never be a centerfold. Besides, Clem likes me this way. Says there's more to love."

Smiling, Lara asked, "How was your day off?"

"Well," Nancy replied, smacking her lips, "all things considered, it was okay. The dog's in heat, and Little Clem found a pair of his sister's tap shoes, put them on, and wore them all day long on the wrong feet. When we tried to take them off, he screamed bloody murder, so it was easier just to let him wear them and look goofy. Tapping feet I can live with, screaming I can't."

Nancy's stories about her chaotic household never failed to be entertaining. She complained good-naturedly about her hectic routine, which revolved around three active children, all of whom were going through a "stage," but Lara knew her nurse loved her husband and her children and wouldn't have traded places with anyone.

Nancy had responded to an ad Lara had placed in the local newspaper, and Lara had hired her after their first interview, partially because Nancy was her sole applicant. Nancy was well qualified, although she'd taken time off from nursing to have Little Clem two years ago.

"Now that it's time to potty train him," she'd told Lara, "I'd rather go back to work and let Granny Baker do the honors."

Lara had liked her instantly and was even a little jealous of her. She'd had chaos in her life, too, but it hadn't been the crazy, happy kind that Nancy experienced daily. It had been the life-altering kind, the kind that wounded and left deep scars. Her calamities had been irrevocable.

"If it weren't for Clem," Nancy was saying as she finished her second doughnut, "I'd have killed the dog, possibly the kids, too, and then pulled my hair out. But when he came home from work, he insisted we drop the kids at his mother's house and go to dinner by ourselves. We pigged out on Beltbusters and onion rings at the Dairy Queen. It was great.

"After Little Clem went to sleep, I hid the tap shoes in the top of the closet so he wouldn't be reminded of them today. This morning Big Clem dropped the dog off at the vet, where she'll either get laid or spayed. By the way, if they've got a willing sire available, do you want dibs on a puppy?"

"No, thanks," Lara said, laughing.

"Don't blame you. I'll probably be stuck with the whole damn litter." She washed her hands in the sink. "I'd better go check the appointment book to see who's coming in today."

Both knew that the appointment book wasn't filled. There were far more empty time slots than confirmed appointments. She had been in Eden Pass for six months but was still struggling to increase her practice. If she hadn't had a savings account to fall back on, she would have had to close the clinic long before now.

Greater than the financial considerations were the professional ones. She was a good doctor. She wanted to practice medicine . . . although she wouldn't necessarily have chosen to do so in Eden Pass.

Eden Pass had been chosen for her.

This practice had been a gift handed to her when she least expected it, though it facilitated a plan she'd been formulating for some time. She had needed a viable excuse to approach Key Tackett. When the opportunity to place herself in his path had presented itself, she had seized it. But not without acknowledging that being the only GP in a small town would be a difficult transition for her.

She had also known it would be an even greater adjustment for the townsfolk who were accustomed to Doc Patton and his small cluttered office in the clinic. She had earned the diplomas now adorning the walls. The medical books on the shelves belonged to her. But the office still bore the former occupant's masculine imprint. As soon as it was economically feasible, she intended to paint the dark paneling and replace the leather maroon furniture with something brighter and more contemporary.

These planned changes would be only cosmetic. Changing the minds of people would take much more time and effort. Before his retirement, Dr. Stewart Patton had been a general practitioner in Eden Pass for more than forty years and in that time he had never made a single enemy. Since taking over his practice, Lara was frequently asked, "Where's Doc?" with the same suspicious inflection as Key Tackett had used when he posed the question to her last night, as though she had displaced the elderly doctor for self-gain.

Dr. Lara Mallory had a long way to go before earning the same level of confidence as Doc Patton had held with the people of Eden Pass. She knew she could never cultivate the affection of her patients that Doc Patton had enjoyed, because she was, after all, the scarlet woman who'd

69

been involved with Clark Tackett. Everyone in his hometown knew her as such. That's why her arrival had taken them by surprise. Lara had wishfully reasoned that once they recovered from the initial shock and realized that she was a qualified physician, they would forget the scandal.

Unfortunately, she had underestimated Jody Tackett's staggering influence over the community. Although they'd never met face to face, Clark's mother was crippling her attempts to succeed.

One afternoon when she was feeling particularly despondent, she'd brought it up with Nancy. "I guess it's no mystery why people in Eden Pass are willing to drive twenty miles to the next town to see a doctor."

"Course not," Nancy said. "Jody Tackett has put out the word that anybody who comes near this office, no matter how sick, will be on her shit list."

"Because of Clark?"

"Hmm. Everyone in town knows the scintillating details of y'all's affair. It had almost been laid to rest when Clark died. Then you showed up a few months afterward. Jody got pissed and set her mind to making you an outcast."

"Then why are you willing to work for me?"

Nancy took a deep breath. "My daddy was a pumper for Tackett Oil and Gas for twenty-five years. This was years ago, when Clark Senior was still head honcho." She paused. "You know that Clark — your Clark — was a third-generation

70

Clark Tackett, don't you? His granddaddy was Clark Senior and his daddy Clark Junior."

"Yes. He told me."

"Okay, so anyway," Nancy resumed, "there was an accident at one of the wells and my daddy was killed."

"Did the Tacketts admit culpability?"

"They did what they had to do to cover themselves legally. Mama got all the insurance money she was entitled to. But none of them came to the funeral. Nobody called. They had the flower shop deliver a big spray of chrysanthemums to the church, but none of them saw fit to visit my mama.

"I was just a kid at the time, but I thought then, and still think, that it was rotten of them to be so standoffish. True, Daddy's death didn't make a ripple in one barrel of their filthy oil, but he was a loyal, hardworking employee. Since then I've had a low opinion of all the Tacketts, but particularly of Jody."

"Why particularly of Jody?"

"Because she only married Clark Junior to get her greedy hands on Tackett Oil." Nancy inched forward in her chair. "See, Clark Senior was a wildcatter at the height of the boom. He struck oil the first time he drilled and made a shitload of money virtually overnight, then kept right on making it. Clark Junior came along. His main ambition in life was to have a good time and spend as much of his daddy's money as he could, mostly on gambling, whiskey, and women."

She sighed reminiscently. "He was the best-looking man I ever laid eyes on. Women from all over mourned his passing. But Jody sure as hell didn't. When he died she got what she'd wanted all along."

"Tackett Oil?"

"Total control. The old man was already dead. When Clark Junior slid off that icy mountain — in the Himalayas, I think it was — and broke his neck, Jody rolled up her shirtsleeves and went to work."

Nancy needed no encouragement to talk.

"She's tough as boot leather. Came from a poor farming family. Their house got blown down by a tornado. They all got killed except her. A widow lady took her in and finished raising her.

"Jody was as smart as they come and got a scholarship to Texas Tech. Straight out of college she went to work for Clark Senior. She was a land man and acquired some of his best leases even after everybody thought all the oil in East Texas was spoken for. The old man liked her. Jody was everything that Clark Junior wasn't — responsible, ambitious, driven. I think Clark Senior was the one behind the marriage."

"What do you mean?"

"The story is that Clark Junior had knocked up a debutante from Fort Worth. Her daddy had mob connections and, for all his money and social standing, was nothing but a glorified pimp. Clark Senior wanted no part of that, so he rushed Clark Junior into marriage with Jody.

"I don't know if that's true, but it's possible. Clark Junior loved to party. He could have had his pick from hundreds of women. Why would he agree to saddle himself with Jody if not to get out of a scrape with a mobster?

"Anyhow, they got married. Clark the Third didn't come along for years. The nastier gossips said it took Clark Junior that long to work up a hard-on for Jody, who never was a beauty. In fact, she goes out of her way to be plain. I guess she thinks that brains and beauty cancel out each other."

"Didn't she mind Clark Junior's womanizing?"

Nancy shrugged. "If she did, she didn't let on. She ignored his philandering and concentrated on running the business. I guess she didn't care about him nearly as much as she did the price of crude. Left to him, he probably would have bankrupted Tackett Oil. Not Jody. She's prospered when others have fallen by the wayside. She's a ruthless businesswoman."

"I'm getting a taste of her ruthlessness," Lara said quietly.

"Well, you have to understand where she's coming from about that." Nancy leaned forward and lowered her voice, although there was no one around to overhear them. "The only thing Jody loved better than Tackett Oil was her boy, Clark. She thought the sun rose and set in him. I guess he never crossed her. Anyway, she had his future all mapped out, including a stint in the White House. She blames you for destroying that dream."

"She and everyone else."

After a reflective moment, Nancy said, "Be careful, Dr. Mallory. Jody has money and power and an ax to grind. That makes her dangerous." She patted Lara's hand. "Personally, I'm rooting for anyone outside her favor."

Nancy was in the minority. In the months since that conversation, there'd been no discernible increase in the number of patients who came to the clinic. Only a few people in Eden Pass had risked Jody's disfavor by seeking Lara's professional services. Ironically, one of them was Jody's own son.

Surely by now Key Tackett had discovered his blunder. Her name had probably ricocheted off the walls of the Tacketts' house with the ferocity of a racquetball.

Let them curse her. She had come to Eden Pass with a specific goal in mind, and it wasn't to win the Tacketts' regard. She wanted something, but it wasn't approval.

When it came time for her to demand of them what they owed her, she didn't care if they liked her or not.

Relatively speaking, this was a busy morning. She was scheduled to see five patients before noon. Her first was an elderly woman who rattled off a litany of complaints. Upon examining her, Lara discovered she was as healthy as a horse,

74

if a bit lonely. She prescribed some pills — which were really multivitamin tablets — and told the woman about the fun exercise classes at the Methodist Church.

Nancy ushered in the next patient, a cantankerous three-year-old boy with an earache and a fever of one hundred two. Lara was getting the specifics of his illness from his frazzled mother when she heard a commotion coming from the reception area at the front of the building. Returning the squalling three-year-old to the arms of his mother, she excused herself and stepped into the hallway.

"Nancy, what's going on?" she called out.

It wasn't her receptionist who came crashing through the connecting door, but Key Tackett. His crutches didn't slow him down as he stormed toward her. Clearly he was furious.

Even though he came to within inches of her before stopping, Lara held her ground. "Your appointment isn't until this afternoon, Mr. Tackett."

The mother had followed Lara into the hallway and was standing behind her. The child's wailing had risen to a deafening level. Nancy had come up behind Key, looking ready to do battle in Lara's defense. She and Key were between them, but only Lara felt trapped.

"Why didn't you tell me last night who you were?"

Ignoring his question, she said, "As you can see, I'm very busy this morning. I have patients waiting. If there's something you wish to discuss with me, please make an appointment with my receptionist."

"I've got something to discuss with you, all right." A bead of sweat rolled down his temple. Most of the color had been leeched from his face. Both were manifestations of pain.

"I think you should sit down, Mr. Tackett. You're in a weakened state, certainly in no condition to —"

"Cut the medical bullshit," he shouted. "Why didn't you tell me last night that you're the whore who ruined my brother's life?"

Chapter Four

The ugly words struck like blows. Feeling light-headed, Lara took a deep breath and held it. The floor and walls of the corridor seemed to tilt precariously. She reached out and braced herself against the wainscoting.

Nancy elbowed her way past Key. "Now see here, Key Tackett, you can't barge into a doctor's office and create a ruckus like this."

"I'd love to chat with you, Nancy, and reminisce about old times, but I'm here to see the doctor." He spoke the last word like an epithet.

By now Lara had regained some composure. She motioned Nancy toward the mother and crying child. "Please see to Mrs. Adams and Stevie. I'll be with them as soon as possible."

Nancy was reluctant to comply, but after giving Key a threatening look, she shooed the woman and child back into the examination room and soundly closed the door behind them.

Lara stepped around Key and addressed the curious patients who were huddled in the doorway, peering down the hall. "Please take your seats," Lara said in as calm a voice as she could muster. "We've had a slight disruption in our office procedure. As you can see, Mr. Tackett is hurt and needs immediate medical attention, but he'll be taken care of and on his way shortly."

"Don't bet on it."

The waiting patients heard him and regarded her uncertainly. "I'll be with you as soon as I can," she reassured them. Then, confronting Key, she said, "I'll see you in my office."

The moment she closed her office door behind them, she vented her anger. "How dare you speak to me like that in front of an office full of patients! I ought to have you arrested."

"That scene could have been avoided," he said, motioning toward the hallway with his head, "if you'd told me who you were last night."

"You didn't ask for my name, and I didn't learn yours until seconds before you left."

"Well, you know it now."

"Yes, I know it now, and I'm not at all surprised to discover that you're a Tackett. Arrogance is a family trait."

"This isn't about the Tacketts. This is about you. What the hell are you doing in our town?"

"*Your* town? That's a curious choice of words for someone who spends very little time residing here. Clark told me that you're rarely in Eden Pass. To what do we owe the honor of this visit?"

He came a menacing step closer. "I told you before to cut the bullshit. I didn't come here to play word games with you, Doc, so don't try to divert me from the point."

"Which is?"

"What the hell you're doing here!" he shouted. Suddenly the door swung open and Nancy poked her head around. "Dr. Mallory? Want me to . . . do something?"

He didn't move a muscle, didn't indicate in any way that he had even heard her or noticed the interruption.

Subconsciously Lara had been preparing herself for this clash, so she wasn't that surprised at his angry appearance. Since it seemed inevitable that they have a showdown, she decided just to get it over with.

She glanced at the nurse. "No, thank you, Nancy. Try to keep the patients pacified until I can get to them." Then, looking up into Key's enraged face, she added, "I'll try to keep Mr. Tackett's unreasonable temper under control."

Nancy obviously had misgivings about Lara's decision, but she left them alone. Lara gestured toward a chair. "Please sit down, Mr. Tackett. You're ashen."

"I'm fine."

"Hardly. You're swaying."

"I said I'm fine," he repeated testily, raising his voice again.

"All right, have it your way. But I don't think either of us wants repeated what we say to each other. Will you kindly keep your voice lowered?"

Leaning on his crutches, he bent forward until his face was within inches of hers. "You don't want what we say repeated because you're afraid that the few people who don't already know will find out that your husband caught you butt naked in the sack with my brother."

79

She had heard the accusation many times before, and there seemed to be no antidote for its vicious sting. Time hadn't diminished its effect.

Turning her back to him, she moved to the window, which offered a view of the gravel parking lot. One of the patients who'd been waiting in the reception area was getting into her car. She couldn't have looked more sheepish if she were leaving an adult bookstore with a brown paper bag full of dirty magazines. Her retreating car raised a cloud of dust.

Watching her had given Lara time to form a response. "I'm trying very hard to put the incident with your brother behind me and get on with my life."

She turned to face him again and felt much more comfortable with space between them, although, even from a distance, his presence was potent. He still hadn't shaved and he looked more disreputable than he had the night before. Most disquieting was the raw sexuality he emanated. She sensed it. Keenly. Doing so seemed to give credence to his low opinion of her, and that bothered her tremendously.

Lowering her gaze, she said, "Don't I deserve a second chance, Mr. Tackett? It happened a long time ago."

"I know how long it's been. Five years. Everybody in the nation knows exactly when it happened, because the morning you were caught in bed with my brother marked the beginning of the end for him. His life was never the same."

"Neither was mine!"

"I guess not," he snorted sarcastically. "Not after you became the nation's most celebrated femme fatale."

"I didn't want to be."

"You should have thought of that before you sneaked into Clark's bedroom. Jesus," he said, shaking his head in bafflement. "Didn't you have any better sense than to commit adultery while your husband was sleeping in the room down the hall?"

Learning to conceal her emotions had become a matter of survival. At the height of the scandal, she had developed a means of stiffly setting the features of her face so they would reveal nothing of what she was thinking or feeling. She resorted to the technique now. To keep her voice from betraying her, she said nothing.

"Some of the details are a little hazy," he said. "Clear them up for me."

"I don't choose to discuss it with you. Besides, I've got patients."

"*I'm* a patient, remember?" He propped his crutches against the edge of her desk and, using it for support, hopped on his left foot toward her. "Give me your full treatment."

The innuendo wasn't accidental. His wicked grin reinforced it. Lara didn't respond, at least not visibly.

"Come on, Doc. Fill in the blanks. Clark had hosted a dinner party the night before, right?"

Lara remained stubbornly silent.

"I've got all day," he warned softly. "Not a damn thing to do but stay off my ankle. I can do that someplace else, or I can do it right here. Makes no difference to me."

Calling the sheriff and having him physically removed was a possibility, but he'd already told her that Sheriff Baxter was an old family friend. Involving him would only create more of an incident than this already was. What was the point of prolonging the situation except to save face? That had been sacrificed years ago. Since, she'd become a pro at swallowing pride.

"Clark had invited a group of people out from Washington to spend an evening in the country," she told him. "Randall and I were among those guests."

"It wasn't the first time you'd been to Clark's cottage in Virginia, was it?"

"No."

"You were familiar with the house."

"Yes."

"In fact, because Clark was a bachelor, you'd served as his official hostess lots of times."

"I had helped him organize several dinner parties."

"And that sort of put you two together."

"Naturally, we had to plan menus —"

"Oh, naturally."

"Clark was a public official. Even casual gatherings involved planning and preparation."

"Have I disputed anything?"

His condescension was as infuriating as his

angry accusations. Lara suddenly realized that her hands had clenched into tight fists. She willed them to relax.

"Arranging all these dinner parties," he continued, "planning and preparing and such, must have taken up a lot of your time."

"I enjoyed it. It was a welcome break from my duties at the hospital."

"Uh-huh. So while you two — you and Clark — had your heads together making all these plans, you became very, uh, close."

"Yes," she answered softly. "Your brother was a charismatic man. He had a magnetic personality. I don't believe I've ever met anyone who could match his energy, his verve. He appeared to be in motion even when standing still. He got excited about things and had such high ideals, such ambitious goals not only for himself but for the nation. It was no mystery to me why the voters of Texas elected him to Congress."

"Fresh out of law school," he told her, although she already knew that. "He served only one term in the House of Representatives before deciding to try for the Senate. Beat the incumbent by a landslide."

"Your brother was a man of vision. I could listen to him talk for hours on any subject. His enthusiasm and conviction were contagious."

"Sounds like love."

"I've admitted that we were very close."

"But you were married."

"Actually, Clark and Randall were friends before I ever met him. Randall introduced us."

"Ahh." He held up his index finger. "Enter the husband. The poor cuckold. What a cliché. Always the last to know that his wife is screwing around. And with his best friend to boot. Didn't ol' Randall become suspicious when you insisted on spending that night in Virginia instead of returning to Washington with the other guests?"

"It was Clark's idea. He and Randall were scheduled to play golf the following day. It would have been ludicrous to drive back to D.C., then return early the next morning. Randall saw the logic."

"That must have been real convenient for you, Doc. I mean, to have your husband accommodate you like that. Did you also fuck him that night just to throw him off track?"

She slapped him, hard. The slap startled her as much as it did Key. In her entire life she'd never struck anyone. She wouldn't have thought she was capable of it.

Learning to control herself had been a critical part of her upbringing. Giving over to one's emotions had been unthinkable in her parents' house. Crying jags, uproarious laughter, any form of unbridled emotional expression was considered unacceptable behavior. That ability to detach herself had served her well in Washington.

She didn't know how Key had managed to breach her conditioned restraint, but he had. If the palm of her hand hadn't been smarting so badly, she wouldn't have believed she'd really slapped him.

Faster than her thoughts could register this, he encircled her wrist, drew her against him, and pushed her arm up behind her back. "Don't ever do that again." The words were precisely enunciated through straight, thin lips that barely moved. His eyes were as direct and brilliant as laser beams.

"You can't talk to me like that."

"Oh yeah? Why not?"

"You haven't got the right to judge me."

"The hell I don't. In some parts of the world they still stone women for being unfaithful to their husbands."

"Would it have evened the score for you if I'd been stoned? Believe me, being brutalized by the media is just as deadly." The hand within his grip was becoming numb. She flexed her fingers. "You're hurting me."

Slowly he released her and took a step back. "Reflexes."

That was as close as she was going to get to an apology. Strange under the circumstances, but she thought he sincerely regretted hurting her.

He winced and pressed his hand against his side.

"Are you in pain?"

"It's nothing."

"Do you want something?"

"No."

As a physician, her instinct was to reach out

85

and lay her hands on him, render assistance. But she didn't. For one thing, he would shun her concern. But primarily she was apprehensive about touching him for any reason. Only now that the contact had been broken did she realize how closely he'd held her against him.

As she massaged circulation back into her hand, she tried to make a joke of it, as much to reassure herself as him. "I don't ordinarily slap my patients."

The attempted levity didn't work. He didn't even smile. Indeed, he was single-mindedly scrutinizing her face. "I didn't recognize you last night from the pictures I'd seen," he said. "You look different now."

"I've aged five years."

He shook his head. "It's more than that. Your hair's different."

She touched her hair self-consciously. "I don't lighten it anymore. Randall liked my hair lighter."

"Back to the husband. Poor Randall. Guess he felt like the rug had been yanked out from under him, huh? Wonder why he stayed with you?" His voice had regained the underpinnings of sarcasm.

"I mean there you were, Randall Porter's lawfully wedded wife, featured on the cover of the *National Enquirer,* being exposed as Senator Clark Tackett's married lover. The photos showed Randall hustling you away from the cottage, wrapped up in your nightie."

"You don't need to reacquaint me with the reports. I remember them well."

"And what does Randall do?" he asked as though she hadn't spoken. "He's with the State Department, right? A diplomat. He's supposed to have a way with words, a glib answer for everything. But does he deny the allegations? No. Does he step forward and defend your honor? No. Does he renounce you as a cheating slut? No. Does he proclaim that you've realized the error of your ways and become a born-again Christian? No."

He planted his hands on his knees and leaned forward. "Randall makes like a goddamn clam. Says nothing for the record before hightailing it off to that banana republic and hauling you with him. 'No comment' was all the media ever prized out of him."

He shrugged ruefully. "But then I guess there's not much you can say when your wife is caught screwing your best friend right under your nose and their affair becomes a political incident of national importance."

"I guess not." She was determined not to lose control again, no matter how provocative he became.

"Even though Randall died a martyr's death in service to his country, if you ask me, he was a coward."

"Well, I didn't ask you, Mr. Tackett. Furthermore, I refuse to discuss my late husband and our personal life with you. But while we're on the subject of cowardice, what about your brother's? He didn't go on the record with a denial or defend my honor, either." Like her husband, Clark had failed to make a statement of apology or explanation. He'd forsaken her to confront the disgrace alone. Their combined silence was as good as an indictment and had been the most humiliating indignity she'd had to bear, both publicly and privately.

"The jig was up. What could he do?"

"Oh, he did plenty. Do you really believe that Randall was assigned to Montesangre on a whim?"

"I never thought about it."

"Well think about it now. That country is a hellhole," she said emphatically. "A cesspool. An ugly, dirty, corrupt little republic. Politically speaking, it was a powder keg of violence ready to explode.

"Randall didn't choose to go there, Mr. Tackett. He didn't ask for the assignment. Your brother saw to it that we were sent," she said disdainfully. "His way of dealing with the scandal wasn't to confront it but to sweep it under the rug."

"How'd he manage that? Thanks to you, no one wanted to know him. His friends turned out to be the fair weather variety."

"But several people over at State owed Clark favors. He called them in, and — presto! — Randall was assigned to the most potentially dangerous area in the world at that time.

"Do you know the Bible story of David and Bathsheba?" Giving him no time to answer, she explained. "King David sent Bathsheba's husband to the front lines of battle, virtually guaranteeing that he would be killed. And he was."

"But that's where your parallel ends," he said, sliding off the edge of the desk and moving to stand directly in front of her again. "King David kept Bathsheba with him. Doesn't speak very well of you, does it?" he asked with a sneer. "Clark didn't value you enough to keep you around. You must have been a lousy mistress."

Spots of fiery indignation appeared on her cheeks. "Following the scandal, Clark and I had no future together."

"He had no future, period. You cost him his career in politics. He didn't even embarrass his political party by running again. He knew that Americans had had their fill of statesmen getting caught in compromising positions with bimbos."

"I am not a bimbo."

"Exception noted. You can probably type," he said caustically. "The point is that until you came along, my brother was Washington's golden boy. After that morning in Virginia, he became a pariah on Capitol Hill."

"Don't cry 'Poor Clark' to me! Your brother knew the potential consequences of his actions."

"And was willing to take the risks, is that it?"

"Precisely."

"Just what is it you do in bed that's so damn

89

great it can separate a man from his better judgment?"

"I won't even honor that with a response," she shot back angrily. "Do you think Clark was the only one to suffer consequences?" She splayed her hand over her chest. "I suffered losses too. My career, for instance, which was as important to me as Clark's was to him."

"You left the country."

"What did it matter? Even if I hadn't gone to Montesangre with Randall, I never would have had an opportunity to practice medicine in and around Washington. I'd still be struggling to practice anywhere if Clark's guilt hadn't compelled him to buy me this place."

"What?" His head snapped back.

Lara sucked in a sharp little breath. Her lips parted in amazement. She could tell that his stunned expression was authentic. "You didn't know?"

His eyebrows came together in a steep frown above the bridge of his nose.

"I can't believe it," she murmured. Carefully gauging his reaction, she said, "Clark bought this place from Dr. Patton when he retired, then deeded it over to me."

He stared at her for several ponderous moments, his gaze so intense it was difficult for her to meet it, but she did so unflinchingly. Confusion and suspicion waned within his eyes. "You're lying."

"You don't have to take my word for it. It's

90

a matter of public record."

"I was there when Clark's will was read. There was no mention of you. I would have remembered."

"He arranged it that way. Ask your sister. Ask your mother. She's repeatedly threatened to contest the legality of my ownership, but Clark saw to it that it's ironclad." She drew herself up straight and tall. Key's ignorance of this one fact had given her a distinct edge.

"I didn't learn about it myself until after his death. His attorney notified me. I was dumbfounded and thought there had to be some mistake because Clark and I had had no contact whatsoever since the scandal."

"You expect me to believe that?"

"I don't give a damn whether you believe it or not," she snapped.

"So, out of the blue, my brother buys a piece of property worth . . . what? A couple hundred grand? And gives it to you." He made a scoffing sound. "Bullshit. You must have put him up to it."

"I tell you, I hadn't seen or spoken to him in years," she insisted. "I didn't want to. Why would I want to see the man who had let me take the fall for a public scandal, who'd exiled me to that godforsaken place, who'd been indirectly responsible for the death of my —" She broke off.

"Your husband?" Key smiled slyly. "Ah, how soon they forget."

91

"No, Mr. Tackett, my daughter." She turned away only long enough to lift a picture frame from her desk. Holding it at arm's length, she thrust it at him so that he was nose to nose with the face in the photograph.

"Meet Ashley. My baby. My beautiful baby girl. She was also killed in Montesangre. Or, as you so eloquently put it, she died a martyr's death in service to her country." Tears filled Lara's eyes, blurring her image of Key. Then her arms sprang back with the impetus of pistons, and she clutched the picture frame to her chest.

Key muttered an expletive. After a long moment he said, "I'm sorry about your kid. I was in France at the time and read about it in an English newspaper. I also remember reading that Clark attended the memorial service for Porter and your daughter."

"Yes, Clark attended, but I wasn't there. I was still in the hospital in Miami, recovering from my injuries." Wearily she brushed back a loose strand of hair and returned the frame to her desk. "Your brother made no effort to contact me, and I was relieved. For his part in banishing us to Montesangre, I think I could have killed him if I'd seen him then."

"You didn't resent him to the point of rejecting his bequeathal."

"No, I didn't. Because of my notoriety, I was turned down for job after job. In all the years since my recovery, I wasn't able to hold a position for very long — only until the hospital bigwigs

92

linked Dr. Lara Mallory to Lara Porter. It didn't matter how capably I fulfilled my duties, I was invited to leave.

"Clark must have known that and obviously felt that he owed me something for all that I'd lost. He tried to secure my professional future. Otherwise, why would he buy this facility for me, completely furnished, ready to occupy if I chose to?"

Speculatively she tilted her head to one side. "It's curious that he drowned only days after adding that codicil to his will."

His reaction was fiercely defensive. She could see that even before he spoke. "What the hell are you suggesting with that remark?"

"Surely the rumors regarding Clark's drowning death reached you. There's speculation that it wasn't an accident at all, but a suicide."

"You're full of shit," he said, his lip curling. "And so is anybody who gave that rumor a second's thought. Clark took the boat into the lake to fish. Knowing him, he was too damned hardheaded to keep his life vest on. I wouldn't have been wearing one of the damned things either."

"Clark was a strong swimmer. He could have saved himself."

"Ordinarily," he said curtly. "Something must have happened."

"Like what? There was no storm that day, no evidence of trouble with the outboard motor. The boat didn't capsize. What do you think happened?"

He worked his inner cheek between his teeth but didn't come up with an answer. "All I know is that my brother wouldn't have taken his own life. And whatever reasons he had for giving you this place, he took to the grave."

"His reasons don't really matter, do they? I'm here."

"Which brings me back to my original question. Why would you want to come here? Clark was Eden Pass's favorite son. You're considered nothing but a whore who destroyed his political future. My mother will see to it that no one forgets that."

Considering the angry mood of the moment, this wasn't the time to divulge her real reason for coming to Eden Pass. That could wait until their mutual hostility eased — if that was possible. It was safer now to address his last statement.

"I'm sure she'll try."

"Is this place," he said, indicating the office with a sweep of his hand, "worth the grief? And believe me, Jody can dish it out."

"I want to practice medicine, Mr. Tackett. I'm a good doctor. All I ask is to be allowed to run my medical practice without interference."

"Well, it isn't going to be easy," he said slowly. "In fact, I think your life here in Eden Pass will make Hell pale by comparison."

"Should I take that as a threat?"

"Just stating the facts, Doc. Nobody in Eden Pass will dare offend Jody by becoming your patient. You can count on that. Too many families depend on Tackett Oil for their livelihoods. They'll drive forty miles for an aspirin before they'll darken your door."

He grinned. "It's going to be amusing to sit back and watch how long it takes you to fold up and go back where you came from. Before it's all over, there'll be fireworks. Guess you should be thanked for relieving the boredom around here." He slipped his crutches under his arms and limped toward the door.

Turning back, he gave her a slow, insulting once-over. "Clark was a damn fool to throw everything away for a woman. All I can figure is that you must be really hot in the sack. But is a roll with you worth losing all he lost? I seriously doubt it." His eyes moved down her body. "You're not even that good-looking."

He left the door open behind him, a clear indication of his contempt. Lara waited until she heard him leave through the front door, then sat down behind her desk. Her knees felt rubbery. Placing her elbows on the top of her desk, she rested her forehead on the heels of her hands. They were cold and clammy, yet her face and chest were emanating fiery heat.

Lowering her hands, she gazed at Ashley's photograph. Smiling sadly, she reached out to stroke her daughter's chubby cheek, but touched only cool, unyielding glass. From that drooling smile, those laughing eyes, Lara fed her resolve. Until she had from them what she wanted, she could and would withstand any hardship the Tacketts might impose.

Nancy came rushing in. "Dr. Mallory, are you all right?"

"I don't recommend a daily dose of him," Lara replied, forcing a smile. "But, yes, I'm fine."

The nurse disappeared and returned seconds later with a glass of ice water. "Drink this. Probably ought to be something stronger. Key has a knack for turning people inside out."

"Thank you." Lara drank greedily. "Just so you'll know, Nancy, he was here last night. He had sprained his ankle and came here expecting to find Dr. Patton." To protect Key's privacy and her own culpability, she didn't tell her nurse about the gunshot wound she had declined to report to the authorities.

Without being invited, Nancy plopped down in the chair facing Lara's desk. "Key Tackett always has been meaner than sin. I remember he once brought a live rattler to school in a toe-sack and terrorized all us girls with it. God knows how he kept from being bit himself. I guess that snake had better sense than to tangle with him.

"He's drop-dead gorgeous, but I'm sure he knows it. Those blue eyes and lazy smile have admitted him to many a set of parted thighs. I'm sure he's good at it, too. God knows he's had plenty of practice. Scores of women would line up to prove me right, but personally I've always thought he was a prize asshole."

Forcing a professional-looking smile to her

lips, Lara said, "Give me a few minutes, please. I need to collect my thoughts and freshen up, then I'll resume seeing the patients."

"Dr. Mallory," Nancy said kindly, "one by one your patients suddenly remembered 'important' things they had to do." Dropping the formality, she sympathetically added, "Honey, there's not a soul waiting out there to see you."

Chapter Five

Janellen was seated behind her desk in the business office of Tackett Oil and Gas Company. The square brick building had been designed by men, built by men, and furnished for men back in Clark Senior's heyday. Jody hadn't given a flip about decor. Most of the men who worked for Tackett Oil had been employees for years and they were accustomed to the office, comfortable with it. So even though Janellen spent more time there than anyone else, it never occurred to her to renovate or otherwise enhance the appearance of the office merely to please herself.

The only personal touch she had added was an ivy plant that she'd potted in a clay container shaped like a bunny. It was crouched on the corner of her desk, partially hidden by correspondence, invoices, and other paperwork.

Managing the office with unstinting efficiency was a matter of pride for Janellen. She opened it every weekday morning at nine sharp, checked the answering machine for messages and the FAX machine for overnight transmissions, then consulted the large calendar on which she jotted down notes to herself ranging from "call church re: altar flowers" — to commemorate her late father's birthday — to "dentist appointment in Longview."

This morning, however, she was preoccupied with her mother's health and the pervasive antagonism between Jody and Key. They hadn't raised their voices to each other since the morning following Key's unexpected homecoming, but the atmosphere crackled with hostile static whenever they were in the same room.

Janellen did her best to act as a buffer but was largely unsuccessful. Through Eden Pass's active grapevine, Jody had heard about Key's return visit to Dr. Lara Mallory's office. She accused him of flagrantly disobeying her; he reminded her that he was no longer a kid who needed to be told what to do and what not to do. She said he'd made an ass of himself; he said he'd learned to do that by example.

And so it went.

Mealtimes were torturous. The burden of carrying on a conversation fell to Janellen, and it was an exhausting challenge. Jody never had been an avid conversationalist at the dining table and was even less so now.

To his credit, Key made an effort. He regaled them with anecdotes of his adventures. Jody didn't think his stories were funny. She shot down all his attempts at humor and consistently turned the topic back to Dr. Mallory, which never failed to inflame Key's short temper. As soon as he finished eating, he invented an excuse to leave the house. Janellen knew he went out drinking because he rarely returned until the wee

99

hours of the morning, and his tread on the stairs was usually unsteady.

He probably was womanizing, too, but the town grapevine was stumped when it came to who might be receiving his favors.

He'd been home a week, but his return had fallen far short of Janellen's expectations. Instead of brightening Jody's outlook, Key's presence in the house had only made her more short-tempered. Which was puzzling. When he was away, Jody fretted over not hearing from him and worried for his safety. She was never demonstrative, but Janellen had seen the relief that registered in her face whenever they received a card from him letting them know that he was all right.

Now that he was home, nothing he did pleased her. If he was taciturn, she rebuked him. If he attempted conciliation, she rebuffed him. She took issue with the slightest provocation, and, Janellen conceded, her brother could be provoking. Like oil and water, his moods never seemed to mix with Jody's.

Things had really turned nasty the evening he'd confronted her about the codicil to Clark's will. "Why wasn't I informed that he'd bought and deeded that property to Lara Mallory?"

"Because it was none of your business," Jody retorted. What Clark had done was incomprehensible, especially to his mother. Janellen knew she had agonized over it. She wished Key had never learned of it. Barring that, she wished he'd never raised the subject with Jody.

'None of my business?" he repeated incredulously. "Don't you think such a stupid decision on his part should have been brought to my attention? It affects all of us."

"I don't know Clark's reasons for doing what he did," Jody shouted. "But I won't have you, of all people, calling your brother stupid."

"I didn't. I said his decision was stupid."

"Same difference."

Their heated argument had lasted for half an hour and only left Key furious and Jody's blood pressure skyrocketing. No one would ever know what had prompted Clark to do what he'd done. Janellen thought it futile to surmise his motivations. What she knew for certain was that her older brother would have been greatly distressed by the friction he'd unwittingly caused. Their home was a gloomy, antagonistic environment that Janellen wished desperately and vainly to change.

"Ma'am?"

Janellen had been so lost in thought that she jumped at the unexpected sound of a man's voice. He was standing just inside the doorway, backlighted by the sun, his face in shadow.

Embarrassed at being caught daydreaming, she surged to her feet and ran a self-conscious hand down the placket of her blouse. "I'm sorry. Can I help you?"

"Maybe. I hope."

He removed a straw cowboy hat and ambled

closer to her desk. His legs were slightly bowed. He was much shorter than Key, not even six feet would be her guess. He wasn't muscle-bound but seemed tough, strong, and wiry. His clothes were clean and appeared new.

"I'm looking for work, ma'am. Wondered if y'all had any openings."

"I'm sorry, we don't at present, Mr. . . . ?"

"Cato, ma'am. Bowie Cato."

"Pleased to meet you, Mr. Cato. I'm Janellen Tackett. What kind of job are you looking for? If you're new to Eden Pass, I might be able to refer you to another oil company."

"Thank you kindly for offering, but it wouldn't do any good. I've already asked around. Saved the best till last, you might say," he added with a fleeting grin. "Seems nobody's hiring."

She smiled sympathetically. "I'm afraid that's all too true, Mr. Cato. The economy in East Texas is tight, especially in the oil industry. Practically nobody's drilling. Of course, a lot of existing wells are still producing."

His woebegone brown eyes lit up. "Yes, ma'am, well that's mostly what I did before — that is, I was a pumper. Maintained several wells for another outfit."

"So you have experience? You know the business?"

"Oh yes, ma'am. Out in West Texas. Grew up in a pissant, uh, I mean a *small* town close to Odessa. Worked in the Permian Basin fields since I was twelve." He paused, as though giving

102

her an opportunity to change her mind after hearing his qualifications. When she said nothing, he bobbed his head in conclusion. "Well, much obliged to you anyway, ma'am."

"Wait!" As soon as Janellen realized that she had reflexively extended a hand to him, she snatched it back and, flustered, clasped it with her other and held them against her waist.

He regarded her curiously. "Yes, ma'am?"

"As long as you're here, you could fill out an application. If we have an opening anytime soon . . . I'm not expecting one, you understand, but it wouldn't hurt to leave an application in our files."

He thought it over for a moment. "No, I reckon it wouldn't hurt."

Janellen sat down behind her desk and motioned him into the chair facing it. In her bottom drawer, along with other business forms, she kept a few standard employment applications. She passed one to him. "Do you need a pen?"

"Please."

"Would you like some coffee?"

"No, thanks."

Picking up the pen she had given him, he lowered his head and proceeded to print his name on the top line of the application.

Janellen judged him to be about Key's age, although his face was marked with more character lines, and there was a sprinkling of gray in his sideburns. His hair was brown. It bore the imprint of his hatband in a ring around his head.

Suddenly he looked up and caught her staring at him. Before thinking, she blurted, "W-would you care for a cup of coffee?" Then she remembered that she'd offered him one less than thirty seconds ago. "I'm sorry. I already asked you, didn't I?"

"Yes, ma'am. I still don't care for any. Thanks, though." He bent back to the application.

Janellen fidgeted with a paper clip, wishing she had left on the radio after listening to the morning news, wishing there were some form of noise to fill the yawning silence, wishing she weren't so miserably ill-equipped when it came to making small talk.

At last he completed the form and passed it and the ballpoint pen back to her. She scanned the top few lines and was astounded to find that he was much younger than Key, actually two years younger than herself. It had been a rough thirty-one years for him.

Her eyes moved down the form. "You're currently employed at The Palm? The honky-tonk?"

"That's right, ma'am." He cleared his throat and rolled his shoulders self-consciously. "I grant you, it's not much of a job. Only temporary."

"I didn't mean to put it down," she said hastily. "Somebody has to work in those places." That came out sounding insulting, too. Her teeth closed over her lower lip. "My brother goes there all the time."

"Yeah, he's been pointed out to me. I don't recall ever seeing you there."

She got the distinct impression that he was trying to suppress a smile. In a nervous gesture, she moved her hand to the placket of her blouse and began fiddling with the buttons. "No, I've . . . I've never been there."

"Yes, ma'am."

Janellen wet her lips. "Let's see," she said, referring again to the application form. "Before The Palm you were working at the state —"

She faltered over the next plainly printed word. Too appalled by her blunder even to look at him, she stared at his application until the lines and words ran together.

"That's right, ma'am," he said quietly. "I did time in Huntsville State Prison. I'm on parole. That's why I need a job real bad."

Mustering all her courage, she lifted her eyes to meet his. "I'm sorry that I don't have anything for you, Mr. Cato." To her consternation, she realized she meant it.

"Well," he said, rising, "it was a long shot anyway."

"Why do you say that?"

He shrugged. "Being I'm an ex-con and all."

She wouldn't lie and tell him that his prison record would have no bearing on his chance for employment at Tackett Oil. Jody wouldn't hear of hiring him. However, Janellen was reluctant to let him leave without some word of encouragement. "Do you have other possibilities in mind?"

"Not so's you'd notice." He replaced his hat and pulled it low over his brows. "Thank you for your time, Miss Tackett."

"Goodbye, Mr. Cato."

He backed out the office door, closing it behind him, then sauntered across the concrete porch, jogged down the steps, and climbed into a pickup truck.

Janellen shot from her chair and quickly moved to the door. Through the venetian blinds, she watched him drive away. At the highway, he turned the pickup in the direction of The Palm.

More depressed than before, she returned to her desk. The paperwork was waiting for her, but she was disinclined to approach it with her usual self-discipline. Instead she picked up the application form that Bowie Cato had filled out and carefully reread each vital statistic.

He had put an X beside "single" to designate his marital status. The space for filling in next of kin had been left blank. Suddenly, Janellen realized that she was being a snoop. It wasn't as though she were actually considering him for a job. She didn't have one to offer him, and even if she did, Jody would have a fit if she hired an ex-con.

Impatient with herself for lollygagging away half the morning, she shoved Bowie Cato's application into the bottom drawer of her desk and got down to business.

★ ★ ★

"Not that tie, Fergus. For God's sake." Darcy Winston cursed with exasperation. "Can't you see that it clashes with your shirt?"

"You know me, sugar pie," he said with an affable shrug. "I'm color-blind."

"Well, I'm not. Switch it with this one." She pulled another necktie from the rack in his closet and thrust it at him. "And hurry up. We're the main attraction tonight, and you're going to make us late."

"I've already apologized once for running late. A busload of retirees from Fayetteville made an unscheduled stop at The Green Pine. Thirty-seven of them. I had to help check them in. Nice bunch of people. They'd been down in Harlingen for two weeks, building a Baptist mission for the Mex'cans. Holding Bible schools and such. Said those Mex'can kids took to snowcones like —"

"For chrissake, Fergus, I don't care," she interrupted impatiently. "Just finish dressing, please. I'm going to hurry Heather along."

Darcy stalked along the upstairs hallway of their spacious home toward their only child's bedroom. "Heather, are you ready?"

She knocked out of habit but entered without waiting for permission. "Heather, hang up that damn phone and get dressed!"

The sixteen-year-old cupped the mouthpiece. "I'm ready, Mother. I'm just talking to Tanner until it's time to leave."

"It's time." Darcy snatched the receiver away

from her, sweetly said, "Goodbye, Tanner," then dropped it back into its cradle.

"Mother!" Heather exclaimed. "How rude! I could just die! You're so mean to him! Why'd you do that?"

"Because we're expected at the schoolhouse right now."

"It's not even six-thirty yet. We're not scheduled to be there until seven."

Darcy wandered over to her daughter's dressing table and rummaged among the perfume bottles until she found a fragrance she liked, then sprayed herself with it.

Piqued, Heather asked, "What's wrong with your perfume? You have dozens to choose from. Why do you use mine?"

"You spend too much time on the phone with Tanner," Darcy said, ignoring Heather's complaint.

"I do not."

"Boys don't like girls who are too available."

"Mother, please don't meddle in my jewelry box. You leave it in a mess every time you open it." Reaching around Darcy, Heather flipped down the lid.

Darcy pushed her aside and defiantly reopened the lavender velvet box. "What have you got stashed away in here that you don't want me to see?"

"Nothing!"

"If you're smoking joints . . ."

"I'm not!"

Darcy riffled through the contents of the jewelry box but found only an assortment of earrings, bracelets, rings, pendants, and a strand of pearls that Fergus had bought for Heather the day she was born.

"See? I told you."

"Don't sass me, young lady." She slammed down the lid and scrutinized Heather with a critical eye. "And before we leave, wipe off about half of that eye shadow. You look like a tramp."

"I do not."

Darcy popped a Kleenex from the box and shoved it into Heather's hand. "You're probably behaving like one, too, every time you're out with Tanner Hoskins."

"Tanner respects me."

"And pigs fly. He wants to get in your pants, and so will every other man you ever meet."

Dismissing Heather's protests to the contrary, Darcy left the room and went downstairs. She felt pleased with herself. She believed parents should never let their kids get the upper hand and so she stayed on Heather like fleas on a hound. Every minute of Heather's day was reported to Darcy, who insisted on knowing where her daughter was, whom she was with, and for how long she was with them. According to Darcy Winston, only an informed parent could exercise the control necessary to raise teenagers.

By and large, Heather was obedient. Her active school schedule didn't allow much time during which she could get into trouble, but in the summer, when free time was easier to come by, opportunities for mischief-making were plentiful.

Darcy's vigilance wasn't based so much on maternal instinct as it was on memories of her own adolescence. She knew all the tricks a youngster could pull on gullible parents because she had pulled each one herself. Hell, she'd invented them.

If her mother had been more strict, more observant of her comings and goings, Darcy's youth might not have been so short-lived. She might not have been married at eighteen.

Her father had deserted her mother when Darcy was nine, and although she was at first sympathetic with her mother's dilemma, Darcy soon became contemptuous of her. Over the years, her contempt grew into open rebellion. By the time she was Heather's age she was running with a wild crowd that got drunk every night and frequently traded sex partners.

She graduated high school by the skin of her teeth — actually by giving a blow-job to a biology teacher with thick glasses and damp hands. During the summer following commencement, she got pregnant by a drummer in a country-western band. She tracked him to De Ridder, Louisiana, where he denied he'd ever met her. In a way, Darcy was glad he claimed no responsibility. He was a no-talent loser, a dopehead who spent his piddling portion of the band's earnings on substances he could smoke, snort, or shoot into his veins.

When she returned to Eden Pass, her future looked dim. Fortuitously, she stopped for breakfast at The Green Pine Motel. Flashing his horsy, toothy grin, Fergus Winston, who was settled into middle-aged bachelorhood, greeted her at the door of the busy coffeeshop.

Instead of perusing the menu, Darcy watched Fergus ring up the cash register receipts. Halfway through her first cup of coffee, she reached a life-altering decision. Within two hours she had a job. Two weeks beyond that, she had netted a husband.

On their wedding night, Fergus believed with all his heart that he'd married a virgin, and several weeks later, when Darcy announced that she was pregnant, it never occurred to him that her child had been sired by anyone except himself.

In all the years since, it still hadn't occurred to him, although Heather had been almost eight weeks "premature" and had still weighed in at a healthy seven and a half pounds.

Fergus didn't have time to dwell on these inconsistencies because Darcy kept his mind on the motel. Over the years she had convinced him that a clever businessman spent money in order to make it. He had revamped the food service, updated the motel's decor, and leased billboards on the interstate.

On one point Fergus stood firm. Only he had access to The Green Pine Motel's ledgers. No matter how persuasively Darcy cajoled, he alone did the bookkeeping. She surmised that he wasn't

111

reporting all his profits to the IRS, which was all right with her. What annoyed her was that, given access to the books, she probably would have been able to find loopholes that he'd overlooked. But in sixteen years of marriage he hadn't budged from his original position. It was one of the few arguments between them that Darcy lost.

Having remained a bachelor for so long, he was totally smitten with his young, pretty, redheaded wife and their daughter and considered himself the luckiest man alive. He was a generous husband. He'd built Darcy the finest house in Eden Pass. She'd had carte blanche to furnish it out of design studios in Dallas and Houston. She drove a new car every year. He was an adoring parent to Heather, who had twined him around her little finger as easily as her mother had.

He was unflappable and unsuspecting, even when Darcy took her first extramarital lover three months after Heather was born. He was a guest at the motel, a saddle salesman from El Paso on his way to Memphis. They used room 203. It had been easy to tell Fergus that she was going to visit her mother for a few hours.

In spite of her frequent infidelities, Darcy was sincerely fond of Fergus, mainly because his position in the community had considerably elevated hers and because he gave her every material thing her heart desired. She smiled at him now as he came downstairs arm in arm with Heather. "You two make a handsome pair," she said. "Everybody in town is going to be at that meeting tonight, and all eyes are going to be on the Winston family."

Fergus placed his arm across her shoulders and kissed her forehead. "I'll be pleased and proud to stand at the podium with the two prettiest ladies in Eden Pass."

Heather rolled her eyes.

Fergus was too earnest to notice the gesture. "I'm just sick about the reason for this town meeting, though." He sighed as he gazed into his beloved wife's face. "I shudder when I think that a burglar could have harmed you."

"It gives me goosebumps, too." Darcy patted his cheek, then impatiently squirmed free of his embrace. "We'd better go or we'll be late. On the other hand," she added with a smug laugh, "they can't start without us, can they?"

Chapter Six

Lara had specific reasons for wanting to attend the town meeting.

If Eden Pass was experiencing a crime wave, she needed to be aware of it. She lived alone and needed to take precautions to protect herself and her property.

It was also important to her future in Eden Pass that she become actively involved in all facets of community life. She'd already bought a season ticket to the home football games and had contributed to the fund to buy a new traffic light for the only busy intersection downtown. If she was seen frequently in everyday settings, like the Sak'n'Save grocery store and the filling station, maybe the townsfolk would stop perceiving her as an outsider. Maybe they would even accept her, in spite of Jody Tackett.

Her third reason for wanting to attend the meeting was far more personal. She found it curious that the outbreak of crime coincided with Key Tackett's coming to her back doorstep with a bleeding bullet wound. It was highly unlikely that he'd been breaking into the Fergus Winston home with burglary in mind, but it was a jarring coincidence that, for her peace of mind, she wanted laid to rest.

The high school auditorium, the pride of the consolidated school's campus, was frequently used as a community center. Lara arrived early, but the parking lot was already jammed with cars, minivans, and pickup trucks. The meeting had been deemed "vitally important" by the local newspaper. In it Sheriff Elmo Baxter had been quoted as saying, "Everybody ought to be at this meeting. It's up to the citizens of Eden Pass to stop this rash of crime before it gets out of hand. Nip it in the bud, so to speak. We have a clean, decent little town here, and as long as I'm sheriff, that's how it's going to stay."

His urging had yielded a good turnout. Lara was just one of a crowd who flocked toward the well-lighted building. As she entered the auditorium, however, she was singled out. In her wake she left whispered conversations. They were absorbed by the din created by the crowd, but she was aware of them nevertheless.

Trying to ignore the turned heads and gawking stares, she smiled pleasantly, greeting those she recognized — Mr. Hoskins from the supermarket, the lady who clerked in the post office, and a few who'd been brave enough to cross Jody Tackett's implied picket line to seek her professional services.

Rather than taking one of the available seats in the rear of the auditorium, which would have been convenient but cowardly, Lara moved down the congested center aisle. She spotted Nancy and Clem Baker and their brood. Nancy motioned for her to join them, but she shook her head and found a chair in the third row.

Her courage in the face of so much adverse attention was a pose. It was discomforting to know that tongues were wagging and that dozens of pairs of eyes were aimed at the back of her head, most of them critically. She knew that personal aspects of her life were being reviewed in hushed voices so that the children wouldn't hear about the brazen hussy in their midst.

Lara could not control what people thought or said, but it still hurt to know that her character was being bludgeoned and there wasn't a damn thing she could do to prevent it. Her only means of self-preservation would be to remain at home, but to her that was not a viable option. She had every right to attend a community function. Why should she be cowed by gossips and people so spineless they allowed themselves to be influenced by an aging old bitch, as she had come to think of Jody Tackett.

Obviously Mrs. Tackett had a much higher opinion of herself. When she made her fashionably late entrance, she strode down the center aisle looking neither right nor left. She felt that friendliness was either a waste of time or beneath her dignity. In any case, she didn't stop to chat even with those who spoke to her.

Her bearing was militant, but she wasn't as physically imposing as Lara had expected. Clark had described his mother in such elaborate terms that Lara recognized her, but she had formed a mental picture of Jody that fit midway between Joan Crawford and Joan of Arc.

Instead, Jody was a short, stocky, gray-haired woman who was average in appearance and attired in clothing that was high in quality but low in fashion flair. Her hands were blunt and unadorned. Her features were harsh to the point of appearing masculine, and she embodied the iron will for which she was known.

A hush fell over the crowd as she moved down the aisle. Her arrival was as good as an announcement that the meeting could begin. Indubitably she was Eden Pass's number-one citizen, deferred to by all.

Lara was perhaps the only one in the auditorium who realized that Jody Tackett was seriously ill.

She had the telltale wrinkles of a heavy smoker around her mouth and eyes. Beyond that, her skin was friable. Bruises and splotches dotted her arms. As she extended her hand to the mayor, Lara noticed that her cuticles were thick. Such clubbing was symptomatic of pending arterial problems.

Following on Jody's heels was a woman who appeared to be about Lara's age. Her smiles were genuine but uncertain. She seemed uncomfortable with sharing her mother's limelight. Janellen perfectly matched Clark's description. He had once referred to his sister as "mousy," but he hadn't meant it unkindly.

"Daddy doted on her. Maybe if he hadn't died

when she was so young, she would have eventually blossomed. Mother didn't have much time to cultivate her. She was too busy keeping the business together. I guess growing up around Key and Mother and me, all Type A's, made sis shy and softspoken. She rarely got a word in edgewise."

Janellen had a delicate face and a fair complexion. Her mouth was too small, and her nose was a trifle long, but, like her brothers, she had spectacular blue eyes that more than compensated for her unremarkable features.

Since Jody had obviously influenced her, her lack of style was no surprise. But even Jody's clothing made more of a fashion statement than Janellen's. She was downright dowdy. Her severe hairstyle was sorely unflattering. It was as though she worked at making herself unattractive so that she would go unnoticed and remain in the large shadow that Jody cast.

Key brought up the rear. Unlike his mother, he didn't march down the aisle undeterred. He stopped frequently along the way to swap greetings and anecdotes with people he obviously hadn't seen in a while. Lara picked up snatches of these friendly exchanges.

"As I live and breathe! It's Key Tackett!"

"Hey, Possum! You ugly son of a bitch, how's life treating you?"

While someone named Possum was expounding upon his successful feed and fertilizer business, Key happened to glimpse Lara. When he did a double take, her stomach muscles tightened. They held each other's stare until Possum, so nicknamed no doubt because he bore an unfortunate resemblance to the marsupial, asked him a direct question.

"Sorry, what?" Key pulled his stare away from Lara, but not before Possum and others sitting nearby noticed who had momentarily captured his attention.

"Uh, I said . . ." Possum was so busy shifting his beady eyes between Lara and Key that he couldn't restate his question.

Thankfully, the high school principal chose that moment to approach the lectern on the stage. He spoke into the microphone. It was dead. After fiddling with the controls, he blasted everyone's eardrums with, "Thank y'all for coming out tonight." He finally adjusted the volume and repeated his welcome.

Key promised to meet Possum the next day for a beer, then joined Jody and Janellen in the front row where the mayor had saved seats for them.

The meeting got under way, the school principal presiding. He introduced the Fergus Winston family, who emerged as a unit from behind the gold velvet curtains. Lara observed them with interest. The teenage girl, who was introduced as Heather, seemed mortified to be seen with her parents in such a public arena. Mrs. Winston didn't appear to be on the verge of collapse as the school principal's solemn tone suggested. A picture of health, she was fairly bursting with vitality. The stage lights made her red hair look like flames. She demurely slid her hand into the crook of her husband's elbow.

Lara instantly distrusted her.

Fergus was a tall man with a perpetual stoop. Thinning gray hair inadequately covered his pointed, balding head. There were deep laugh lines around his wide mouth, but he wasn't smiling as he took the high school principal's place behind the lectern and gave his account of their harrowing experience.

By angling slightly to her left, Lara could see Key Tackett in the chair next to his sister's. His elbows were propped on the armrests, and he was tapping his steepled fingers against his lips. His ankle — the one he'd sprained — was propped on the opposite knee. He was slouching in his seat, and his eyes moved restlessly about as though he was finding the proceedings exceedingly dull, as eager for them to conclude as a young boy in church.

Lara looked again toward the stage and saw that she wasn't the only one watching Key. Mrs. Winston had him locked in her sights, too. Her expression was sly, almost smug.

"Well, that's all I've got to tell y'all," Mr. Winston concluded, "except to say to be on the lookout for any suspicious characters, any strangers around town, and to report any unusual happenings to the sheriff." To applause, he relinquished the microphone to the sheriff.

Elmo Baxter was a slovenly man who moved with the speed of a slug and had the world-weary expression of a basset hound. "I 'preciate Fergus and Darcy sharing their experience." He shifted his weight. "But don't y'all get the fool notion of sleeping with a loaded gun under your pillow. If you see signs of a break-in or notice a stranger hanging around your neighborhood, report it to my office. Me or Gus'll check it out using proper police procedure.

"Don't go taking the law into your own hands, y'all hear? Now, me and the city council decided we need a Crime Watch committee like they have in big cities. This committee would organize folks in the different neighborhoods to keep a lookout on goings-on and help everybody stay informed. Naturally it'll need a chairman. I'll take nominations now."

"I volunteer myself," Darcy Winston announced in a clear, loud voice.

She received a burst of applause. Fergus squeezed her hand and looked down at her with naked adoration.

"And I'd like for Key Tackett to serve as co-chairman," Darcy added.

Key jerked to attention. His boot landed hard on the floor, and Lara saw him wince. "What the hell did she say?" Everybody laughed at his stunned reaction. "I don't even live here anymore. Besides, what do I know about committees?"

The amused sheriff tugged on his elongated

earlobe. "I reckon knowing about committees isn't a requirement, but if a man ever knew about taking care of hisself, it's you. Right, Jody?"

She looked across Janellen at her son. "I think you ought to do it. Since when have you performed a community service?"

"Since he led the fighting Devils to the state championship!" Possum leaped into the center aisle and began waving his hands high over his head. "Let's hear it for the fearsome number 'leven, Key Tackett!"

Others stood and joined the cheering. Antsy children used the interruption as an opportunity to escape their parents. Rowdy teens gave one another high fives as they raced for the exits. Regaining control was out of the question, so Sheriff Baxter placed his lips close to the mike and said, "All in favor say 'aye.' Motion carries. Y'all are dismissed. Be careful driving home."

Lara was swept along into the aisle. Standing on tiptoe, she was able to see Darcy Winston imperiously motioning for Key to join her on the stage. She looked like a woman fully capable of shooting a fleeing lover in order to prevent getting caught with him. There was calculation in her perpetually pursed lips and tilted eyes.

"Excuse me."

Lara responded to the polite request coming from behind her and stepped aside, then turned to apologize for dawdling. She came eye to eye with Janellen.

Janellen was caught in a hesitant smile that

quickly turned into a small, round O of dismay. Unabashedly she gaped at Lara.

"Hello, Miss Tackett," Lara said politely. "Excuse me for blocking the aisle."

"You're . . . you're . . ."

"I'm Lara Mallory."

"Yes, I . . ."

Even if Janellen could have formed an appropriate response, Jody gave her no chance to speak. "What's the holdup, Janellen?" When she too noticed Lara, her expression hardened with malice.

"At last we meet, Mrs. Tackett," Lara said, extending her right hand.

Jody acknowledged neither her outstretched hand nor the greeting. She only nudged her daughter forward. "Move along, Janellen. I suddenly feel the need for some fresh air."

For several moments, Lara was immobilized by Jody's angry stare. But the chance meeting hadn't gone unnoticed, and soon she became aware of the studious avoidance of the crowd. Self-consciously she retracted her right hand. As she moved up the aisle, she was given a wide berth. She might as well have had leprosy. No one even looked at her.

At the exit, she paused and glanced back at the stage. Key had joined Mrs. Winston there. Scornfully, Lara turned away. They deserved each other.

Since Darcy was about as subtle as a carnival

barker, Key was given no choice but to join her on stage. After making such a production of flagging him up there, it would have aroused curiosity if he hadn't heeded her request.

As he had moved toward the stage, he had tried to locate Lara Mallory in the crowd, and was shocked to see her talking to his mother.

He watched as Jody spurned her handshake and brusquely herded Janellen up the aisle. To her credit, Dr. Mallory didn't quail or lose her composure. She didn't burst into tears or shout epithets at their retreating backs. Instead she held her head high as she moved gracefully toward the exit.

Key was tempted to charge after her and — do what?

Ask her why she had picked on his brother when there were thousands of randy young bucks in Washington, D.C., just itching to get laid?

See if she could clarify for him the haunting circumstances surrounding his brother's death?

Demand that she leave town by dawn, or else?

He would look like a damn fool and he didn't want to give her that satisfaction. Besides, he had a matter to settle with Darcy. Best to get that out of the way before tackling another crisis.

He climbed the steps to the stage. "Just what the hell are you up to, Darcy?"

"Hi, Key!" She was all smiles, and, despite his angry scowl, she manipulated him into an introduction. "Have you met my daughter? Heather, this is Mr. Key Tackett."

"Hello, Mr. Tackett." The girl spoke politely, but she obviously had other things on her mind. "Tanner's waiting for me," she told her mother. "Can I go now?"

"Come straight home."

"But everybody's going out to the lake."

"At this time of night? No."

"Mo-*ther!* Everybody's going. Please."

The stare Darcy fixed on Heather conveyed unspoken warnings. "Be home by eleven-thirty. Not a second later."

Heather protested sulkily. "Nobody else has to come in that early."

"Take it or leave it, young lady."

She took it. After bidding Key an obligatory goodbye, she joined a handsome young man waiting for her in front of the stage.

While Darcy had been arguing with Heather over the girl's curfew, Key had been watching Lara Mallory's solitary progress up the aisle. There was something very noble about her carriage. Before she went through the exit, she turned and looked toward the stage.

"Key?"

"What?" Only after the doctor disappeared did he turn his attention back to Darcy. Having followed the direction of his gaze, she too was focused on the exit doors at the rear of the auditorium.

"So, our scandalous new doctor put in an appearance tonight," she remarked cattily. "Have you had the honor of making her acquaintance?"

"Fact is, I have. She patched me up after you shot me." Key got a kick out of wiping off Darcy's complacent smile.

"You went to *her?*" she exclaimed. "Have you lost your freaking mind? I thought you'd have the good sense to go to the hospital, where you'd be known, but at least it's out of town."

"I was looking for Doc Patton. Nobody told me that he'd retired."

"Or that your brother set up his ex-mistress in business here?"

"No. Nobody told me that either."

He tried to keep his voice free of telltale inflection, but Darcy wouldn't have noticed anyway. He could tell the wheels of her scheming brain were in full gear.

"She could report the gunshot wound to the sheriff," she said worriedly.

"She could, but I doubt she will." He glanced toward the exit. "She's got enough to worry about. Besides, she couldn't prove anything. No bullet. It tore off a chunk of flesh on its way through." He leaned down and spoke softly so they wouldn't be overheard by those loitering about. "I ought to skin you alive for shooting at me. You could have killed me, you dumb bitch."

"Don't talk to me like that," she hissed, which was hard to do while keeping her deceptively friendly smile in place. "If I hadn't acted quickly, Fergus would have caught us mother-nekkid and screwing like rabbits. He could have killed us, and no jury in this state would have convicted him."

"Sugarplum?"

She spun around at the sound of her husband's voice. Key hitched his chin at him. "Hey, Fergus. It's been a long time."

"How're you doin', Key?"

"Can't complain."

Years ago there had been a rift between Fergus and Jody. It had something to do with the Tackett oil lease adjacent to Fergus's motel property. The details were murky, and Key had never wanted to know them badly enough to ferret them out. He figured that Jody, in her lust for oil and the power and money that went with it, had somehow cheated Fergus.

Their dispute was none of his business, except that Fergus had always looked at him like he was lower than buzzard shit, but that might have had more to do with how he had conducted himself during his youth. More than once he and Possum and their crowd had nursed their hangovers in the coffeeshop of Fergus's motel. He vaguely remembered puking up pints of sour mash in the rosebushes in front of The Green Pine after a particularly wild bacchanal.

Anyway, Fergus Winston didn't like him, but Key had never lost sleep over it.

"I'm not real excited about this committee job your wife just roped me into. By the way," he said to Darcy, "I'm resigning. Effective immediately."

"You can't resign. You haven't even started."

"All the more reason. I didn't ask to be part of any Crime Watch committee. I don't want to be. Find yourself another co-chairman."

She flashed him her most dazzling smile. "Obviously he wants to be begged, Fergus. Why don't you bring the car around to the front door? I'll meet you there. In the meantime, I'll do my best to change Key's ornery mind."

Key watched Fergus amble into the wings of the stage, calling good night to the custodian who was patiently waiting for everybody to leave so he could secure the building.

Darcy waited until her husband was out of earshot before turning back to Key. Keeping her voice low, she said, "Can't you see an opportunity when it all but bites you in the ass?"

"What do you mean, sugarplum?" he asked with mock innocence.

"I mean," she stressed, "that if we're on the same committee, people won't think anything about our being seen together." His stare remained opaque. Exasperated, she spelled it out. "We could get together anytime we wanted and wouldn't have to sneak around in order to do it."

He waited about three beats before bursting into laughter. "You think I'd sleep with you again?" As suddenly as it had started, his laughter ceased, and his face became taut with anger. "I'm royally pissed at you, Mrs. Winston. You could have killed me with that damn handgun of yours.

As it is, I can barely climb into a cockpit with this bum ankle."

She gazed at him through eyes gone smoky. "Small price to pay for the fun we had, wouldn't you say?"

"Not even close, sugarplum. You act like that's the golden fleece," he said, glancing pointedly at her crotch, "but I've had better. Lots better. Anyway, if you think I'd touch it again after this stunt you've pulled, then you're as crazy as you are easy."

The smoke in her eyes cleared. He saw fire. "I wouldn't fuck *you* again, either!"

"Then from what I hear, I'm in a minority of one."

Darcy was livid. "You're a son of a bitch and always have been, Key Tackett."

"You're right on the money there," he said with a terse nod. "In the most literal sense of the words."

"Go to hell."

Since there were still people milling about and visiting in the aisles of the auditorium, there was nothing more she could do except conceal her wrath, turn on her heel, and flounce away. She gave clipped replies to those who bade her good night as she stormed up the aisle.

Key followed at a more leisurely pace, feeling amused, pleased, and vaguely dissatisfied all at the same time. Darcy deserved his digs, but he hadn't derived as much pleasure from insulting her as he had anticipated.

Like a dutiful servant, Fergus was waiting for her beside their El Dorado, holding the passenger door open. As Darcy slid into the seat, Key overheard her say, "Hurry up and get me home, Fergus. I've got a splitting headache."

Key felt sorry for Fergus, but not because he'd slept with his wife; hell, just about everybody in pants had at one time or another. But even though his motel made money, he would never be an entrepreneur. That required a certain attitude that was clearly lacking in his long, thin face, his bad posture, and in his conservative approach to business. There were the Jody Tacketts of the world, and there were the Fergus Winstons. The aggressors and the vanquished. Some steamrollered their way through life while others either moved aside for them or got rolled over. In life and in love, Fergus fell into the latter category.

Such passivity was beyond Key's understanding. Why would Fergus ignore Darcy's unfaithfulness? Why was he willing to be an object of scorn? Why did he accept and forgive her infidelity?

Love?

Like hell, Key scoffed. Love was a word that poets and songwriters used. They vested the emotion with tremendous powers over the human heart and mind, but they were wrong. It didn't transform lives like the saccharine lyrics claimed it could. Key had never seen any evidence of its magic, unless it was black magic.

Love had caused his young heart to break when his father was killed, leaving him without an ally in a hostile environment. Love had kept his sister emotionally and psychologically chained to their mother. Love had cost Clark his promising career as a statesman. Had love also compelled Randall Porter to stay with his whoring wife?

Not for me, Key averred as he crossed the parking lot, his stride as long as his injured ankle would allow. Love, forgiveness, and turning the other cheek were concepts that belonged in Sunday school lessons. They didn't apply in real life. Not in his life, anyway. If, during a mental lapse, he ever got married, and if he ever found his wife in the arms of another man, he'd kill them both.

Reaching his car, he jammed the key into the lock.

"Good evening, Mr. Tackett."

He turned, stunned to find Lara Mallory standing beside him. A breeze was gently tugging at her clothing and hair. Her face was partially in shadow, the remainder bathed in moonlight. Although she was the last person he wanted to see at the moment, she looked damned gorgeous and for a moment he felt as though he'd been poleaxed.

His reaction was irritated, as was evident in his voice. "Did you follow me out here?"

"Actually I've been waiting for you."

"I'm touched. How'd you know where to find me?"

"I've seen you driving around town in this car. It's distinctive, to say the very least."

"It was my daddy's."

The Lincoln was a mile-long gas-guzzler almost two decades old, but Key had left instructions at Bo's Garage and Body Shop that it always was to be kept in showroom condition. He drove it whenever he was home and by doing so felt connected to the father he had lost.

The car had mirrored Clark Junior's flamboyant personality. Yellow inside and out, it sported gaudy gold accents on the grille and hubcaps. Key affectionately referred to it as the "pimpmobile." Jody frowned on the car's nickname, possibly because she knew it to be fairly accurate.

"You're still limping," Lara said. "You should be using your crutches."

"Screw that. They're a pain in the ass."

"You could do your ankle irreparable damage."

"I'll take my chances."

"How's your side? You didn't come back to the clinic."

"No shit."

"That drain should be removed."

"I pulled it out myself."

"Oh, I see. A tough guy. Well, at least you've shaved . . . with a butter knife, I suppose."

He said nothing because he had the uncomfortable impression that she was mocking him.

"Are you changing the dressing regularly? If

not, it could still become infected. Is the wound healing properly?"

"It's fine. Look," he said, propping his elbow on the roof of the car, "should I consider this a house call? Are you going to bill me for a consultation?"

"Not this time."

"Gee, Doc, thanks. Good night."

"Actually," she said, taking a step toward him, "I have something else to speak to you about and thought you would rather I do it here where we can't be overheard."

"Guess again. Whatever you want to talk about, I'm in no mood to hear. In fact, my mood tonight is what you might call fractious. Do yourself a favor and make yourself scarce."

He was about to duck into the driver's seat when she surprised him further by grabbing his arm. "You've got gall, Mr. Tackett. I give you credit for that. Or was it Mrs. Winston's idea to fake a break-in rather than get caught in adultery?"

Key was taken aback, but only momentarily. She was gazing at him solemnly, so solemnly that he smiled. "Well I'll be damned. The Whiz Kid thinks she's got it all figured out."

"Mr. Winston interrupted you while you were in bed with his wife, didn't he?"

"Why ask me? You've got all the answers."

"While escaping you sprained your ankle. To cover your tracks, Mrs. Winston shot at you. It's a scene straight out of a bad movie. Did you know she was going to shoot at you?"

"What the hell do you care?"

"That means you didn't."

"Don't put words in my mouth," he said crossly. "My question stands. What do you care? Or do you just have an unnatural curiosity about the love lives of other people?"

"The only reason I care," she said heatedly, "is because you barged into my clinic and called me a whore for doing the same thing you did."

"It's not quite the same thing, is it?"

"Oh really? How is it different?"

"Because Darcy and I weren't hurting anybody."

"Not hurting anybody!" she cried. "She's married. You claimed that was my most grievous sin."

"No, your most grievous sin was getting caught."

"So as long as her husband remains in the dark, it's okay for you to have an affair with her?"

"Not okay, maybe. But not catastrophic. The only ones suffering any consequences are the sinners."

"Hardly, Mr. Tackett. You've whipped an entire town into a panic over a 'crime wave' that doesn't even exist."

"That wasn't any of my doing. Fergus freaked out when he heard Darcy screaming and firing that pistol. He got a little carried away."

"Or maybe he used the mythical intruder to conveniently allay his own suspicions."

That possibility also had occurred to Key, but he wasn't going to admit it. "I'm not responsible for what went on inside his head."

"Doesn't it bother you that you've instilled fear into a whole town?"

"Fear?" he scoffed. "Hell. Folks are loving the scare. Eating it up. They have something to keep their minds off the heat during these last dull weeks before Labor Day. Sheriff Baxter told me that attempted break-ins and window-peepers have been reported all over town."

He chuckled. "Take Miss Winnie Fern Lewis for example. She lives in a spooky old three-story house over on Cannon Street. We used to tear down her clothesline every Halloween because she was mean and stingy and handed out only penny candy.

"Anyway, just yesterday Elmo told me that Miss Winnie Fern's reported a man standing outside her bedroom window watching her undress for six nights straight. She claims she can't describe or identify him because he always hides behind her rose o' sharons where he 'manipulates himself to sexual climax,' is the way she put it to Elmo. If he kept a straight face it's better than I could do.

"There's no window-peeper jacking off behind Miss Winnie Fern's rose o' sharons any more than there's a man in the moon, but she hasn't had a thrill like that in years, so what's the harm?"

"In other words, you feel that you've provided

135

a community service?"

"Could be. People in a small town like Eden Pass need something to generate excitement." He moved closer, close enough to catch the scent of her perfume. "What about you, Doc?" he asked in a low pitch. "What are you doing to generate some excitement, seeing as how Eden Pass doesn't have any legislators to seduce?"

She shuddered with indignation, and immediately Key realized he had lied when he told her he didn't see what had attracted his brother. Anger flattered Lara Mallory. With her head thrown back in that haughty angle, she could have been the proud bust on the prow of a sailing ship.

Except that she was softer. Much softer. He thought of softness each time the south breeze flattened her clothes against her body or lifted strands of hair away from her cheeks. She also had a very soft-looking mouth.

Not liking his train of thought, he asked, "Picked out your next victim yet?"

"Clark wasn't my victim!"

"You're the only married woman he ever got mixed up with."

"Which indicates that he was more discriminating than you."

"Or less."

Furious, she turned on her heel and would have stalked away if his hand hadn't shot out and brought her back around. "Since you started this, you're damned well going to hear me out."

She shook back her hair. "Well?"

"You said that my accusations were unfair."

"That's right. They're grossly unfair. You don't know anything about my relationship with Clark, only what you've read in the tabloids or deduced in your own dirty mind."

He grinned. She had just placed her slender foot into the snare. "Well, you don't know doodle-dee-squat about my relationship with Darcy, or with anyone else for that matter. Yet you ambush me out here and start preaching sin like a fire-breathing Bible thumper. If it was wrong for me to jump to conclusions about you, shouldn't it be just as wrong for you to hang me without a trial?"

Before she had time to reply, he released her, slid into the front seat of the yellow Lincoln, and started the motor. Through the open window he added, "You're not only a whoring wife, you're a goddamn hypocrite."

Chapter Seven

Lara drove aimlessly. The night was clear and warm. The breeze served only as a conveyer of the heat that emanated from the earth of this vast, hard place.

Texas.

"Texas isn't just a place," she had heard Clark say many times. "It's a state of mind. Xanadu with cowboy boots."

Lara had never set foot on Texas soil until six months ago, when she claimed the gift he had bequeathed her. She had brought with her preconceptions influenced by Hollywood — the barren, windswept landscapes interrupted only by rolling tumbleweeds like in *Giant*, and *Hud*, and *The Last Picture Show*. Those movies had accurately depicted Texas, but only the western portion of it.

East Texas was green. The verdant forests were comprised of some hardwoods but mostly pines, their trunks dark and straight and aligned so perfectly that Nature could have used a ruler to space them. In the springtime these forests were dappled with patches of pastel color from blooming dogwood and wild fruit trees. Herds of beef and dairy cattle grazed in lush pastures. Lakes brimming with fish were fed by rivers and creeks that had a history of overflowing their banks.

And everywhere there was space, large tracts of land that Texans took for granted if they had never traveled to the crowded Northeast, which most of them scorned as a breeding ground for perverts, pinkos, and pansies.

They had no use whatsoever for Yankees.

Their children pledged allegiance to the flag of the United States of America, but the native-born considered themselves Texans first, Americans second. The blood of the heroes of the Alamo flowed in their veins. Their heritage was rich with larger-than-life characters, and although their state carved a prominent notch in the Bible Belt, they were conversely boastful of bandits and ne'er-do-wells who had become folk heroes. The more notorious the character, the more popular the legends.

If Lara was having a difficult time understanding the people, she had instantly admired their land. County roads radiated from Eden Pass like the spokes of a wheel. Upon leaving the high school, she had selected one at random and had been driving without a destination for about an hour. She was well outside the city limits, and although she couldn't pinpoint exactly where she was, she didn't feel lost.

Steering her car onto the gravel shoulder, she cut the engine. As the motor noise died, she was engulfed by the sound of a discordant choir of cicadas, crickets, and bullfrogs. The wind rustled the leaves of the cottonwood saplings growing on the banks of the shallow ditches that lined the road.

She folded her hands over the steering wheel and rested her forehead on them, berating herself for letting Key Tackett get the best of her.

She had done exactly as he'd said: She'd cast stones without knowing all the facts. There were a thousand extenuating circumstances that could put a different complexion on what appeared a shabby affair. She realized that circumstances were not always what they seemed. Unknown factors often made the difference between right and wrong, good and evil, innocence and guilt. Shouldn't she know that better than anyone?

Her thoughts made her claustrophobic, so she left the car. An open meadow extended as far as she could see on either side of the road. In the near distance, beneath a sprawling pecan tree, a small herd of cattle was settled for the night. Several oil wells, pumping rhythmically, were eerily silhouetted as dark, moving shadows against the night. Rhythmically, they dipped their horse-shaped heads toward the earth, paying it homage like faithful disciples at prayer.

She supposed they were Tackett wells.

It hadn't rained in over a week, so the ditch was dry. She crossed it easily and moved to the wire fence that surrounded the pasture. Being careful of the sharp barbs, she leaned against a rough cedar post and, tilting her head back, gazed at the panoply of stars and a bright half-moon.

"What are you doing here, Lara?"

It was a question she frequently asked herself. Even before Clark's death, she had grappled with the idea of coming here and confronting him with her terms for settling their account. She'd planned to present him with a bill for repayment for all that she'd lost.

He died before she had implemented her plan. Although, tragic as his death was, it had little bearing on her achieving her goal. Clark wasn't essential to her plan. Key was.

Key. He despised her. Because of that, her task wasn't going to be easy. However, the difficulty didn't dampen her determination. Medical training had taught her that in order for things to get better, they often had to get worse. Before wounds could heal, they had to be lanced and the poison excised. She was willing to endure anything, no matter how painful, in order to lay to rest the ghosts that haunted her.

Only then would she finally have the peace that had escaped her since her daughter's death. Only then would she be able to put the tragedies of the past behind her and get on with the remainder of her life, either in Eden Pass or somewhere else.

The years following her return from Montesangre after the deaths of Randall and Ashley had been a wasteland of time. She hadn't lived; she'd existed. Full of despair and heartache and loneliness, she had moved through the days without connecting with anything around her. Work might have salved her heartache, but she'd been denied the opportunity. She was a pariah, an object of curiosity and ridicule, Clark Tackett's whore.

That's what Key had called her. A whore. Jody thought of her that way, too. Lara had seen the unmitigated contempt in her eyes. She'd expected nothing else, really.

Even her own parents had condemned her. They never had shared a warm relationship with their only child, but it had been especially strained since the scandal. They certainly couldn't understand why she would want to set up her medical practice in an out-of-the-way place like Eden Pass, Texas, particularly since that was Tackett territory.

"They need a doctor there," Lara had told them when they voiced their incredulity over her decision.

"Doctors are needed everywhere," her father had argued. "Why go there?"

"Because she always places herself in the worst possible situation, dear." Her mother spoke softly but coldly. "It's a habit she's acquired strictly to annoy us."

Her father added, "Taking the path of least resistance isn't a crime, Lara. After all that's happened, I would think you'd have learned that."

They would have been aghast if she'd told them the real purpose behind her move to Texas, so she didn't confide it. Making a futile attempt at self-defense, she'd said, "I know it won't be

142

easy to establish a practice there, but it's the best opportunity I've been offered."

"And you have only yourself to blame for that, and for all your other misfortunes. If you had listened to your mother and me in the first place, your life wouldn't be in shambles now."

She could have reminded them that they had encouraged her to marry Randall Porter. Even before meeting him, they'd been impressed by his credentials. He was charming and urbane and cosmopolitan. He was fluent in three languages and held a promising position in the State Department, an attribute they liked to throw up to their society friends.

They still regarded Randall as a saint for remaining married to her after the spectacle she'd made of herself with Senator Tackett. Would it make any difference to them, she wondered, if they knew how unhappy she'd been with Randall long before he introduced her to Clark?

Uncomfortable with her memories, Lara retraced her steps to her car and was about to get in when she became aware of a sound coming from overhead. Looking up, she spotted an airplane. It was nothing but a blinking dot of light on the horizon, but it came closer, flying low. In fact, it was cruising at a dangerously low altitude, barely clearing the treetops of the forest bordering the pasture. The aircraft was small — a single-engine plane, she guessed, with her limited knowledge of aviation.

It swooped in low over the pasture and crossed

the road about a hundred yards from Lara's parked car. She sucked in her breath as the plane approached the far woods. Only seconds before it reached the tree line, the plane's nose reared back at a drastic angle as it went into a steep climb, then banked to the left and gradually ascended to a safer altitude. Lara watched it until she could no longer see the lights.

Would someone be crop dusting at this time of night? Would chemicals be dusted over pastures where cattle were grazing? No, this had to be a stunt flyer.

"Fool," she muttered as she got into her car and turned on the ignition.

Of course, most considered her a fool for coming to Eden Pass and effectively waving a red flag at the Tacketts. But when one has absolutely nothing to lose, one isn't so shy of taking tremendous risks. What could the Tacketts say or do to her that hadn't already been said and done?

Once they had met her demand, she would gladly leave them to their town. In the meantime, she didn't care what they thought of her. She must, however, get them past their aversion even to talk to her. But how?

Jody was unapproachable.

Key was snide and abusive, and she didn't welcome subjecting herself to more of him until absolutely necessary.

Janellen? She had sensed in Clark's sister a spark of curiosity before Jody interceded. Could that curiosity be a chink she could use to pierce the Tackett armor?

It was worth a try.

Janellen was vexed with herself. She'd designated today to pay bills and had organized her desk accordingly. But when she reached for the folder in which she filed their accounts payable, she remembered having taken it to the shop the day before, wanting to compare the invoices with the equipment they had received to make certain that everything was in order. It wasn't like her to be so absentminded, and she chastised herself for it as she drove the mile from the office to the shop, as the workers called it.

The shop was actually uglier than the headquarters. As the company grew, the original building had been added onto several times to accommodate an ever-increasing inventory of equipment, supplies, and vehicles. Since it was Saturday, the building was deserted. Janellen pulled her car around back and parked near a rear door that opened directly into a tiny cubicle of an office. Here the men had access to a telephone, refrigerator, microwave, coffeemaker, bulletin board, and individual pigeonholes labeled with their names into which Janellen placed their paychecks twice a month.

Using her key, she let herself in and, ignoring the pin-up calendars and the odor of stale tobacco smoke, she moved behind the metal desk where she remembered last having the folder. When she found it, she tucked it under her arm, and

145

was about to leave when she heard movement beyond the door that connected the office with the garage. She opened the door and was about to call out when the unusual situation stopped her from speaking.

The oversized garage door was closed and the building, having few windows, was dim. A pickup had been squeezed between two Tackett company trucks. Into the pickup one of her men was loading small machinery, pipe, and other supplies that were the tools of their trade. He was checking the items against a list that he carried in the breast pocket of his shirt. Consulting it one last time, he climbed into the cab of the pickup.

Janellen scrambled from her hiding place and rushed forward to block his exit, placing herself between his bug-splattered grille and escape.

"Miss Janellen!" he exclaimed. "I . . . I didn't know you were here."

"What are you doing here on a Saturday morning, Muley?"

His face turned red beneath his tan, and he tugged on the bill of his cap with the blue Tackett Oil logo on it. "You know as well as I do, Miss Janellen, that I ran my route this morning."

"After which you're officially off."

"Just thought I'd get a head start on Monday morning. Came by to pick up some stuff."

"With the garage door shut and all the lights out?" She pointed at the back of the truck. "And you aren't loading that equipment into a company truck, but your own pickup, Muley. You're stealing from us, aren't you?"

"That's old equipment, Miss Janellen. Nobody's using it."

"So you decided to help yourself."

"Like I said, nobody's using it. It's going to waste."

"But it was bought and paid for by Tackett Oil. It's not yours to dispose of." Janellen drew herself up and took a deep breath. "Take the things out of the truck, please."

When he was finished, he hooked his thumbs into his belt and faced her belligerently. "You gonna dock my pay or what?"

"No, I'm not going to dock your pay. I'm firing you."

He underwent an instantaneous attitude change. His thumbs were removed from his belt loops. His hands clenched into fists at his sides. He took two hulking steps toward her. "The hell you say. Jody hired me and only she can fire me."

"Which she'd do in a heartbeat when she found out you were stealing from her. After she cut off your hand."

"You don't know what she'd do. Besides, you can't prove a goddamn thing. For all you know, I was going to offer to buy this stuff from you."

She shook her head somewhat sadly, feeling betrayed. "But you didn't, Muley. You made no such offer. You sneaked in here on a Saturday when you didn't think anyone would be around

147

and loaded the stuff into your pickup truck. I'm sorry. My decision is final. You can pick up your last check on the fifteenth."

"You rich bitch," he said with a sneer. "I'll go, but only because I think this company is in deep shit. Everybody knows Jody is on her last leg. You think you can run this company as good as her?" He snorted. "Nobody ever takes you seriously. We laugh at you, did you know that? Yeah, us guys come in here after our shifts and talk about you. It's amusing how you're trying to take over for your mama 'cause you ain't got nothing better to do with your time. Like fuck, for instance. We've got a running bet, you know, on whether or not you've still got your cherry. I say it's in there as solid as cement. Even if you are heir to all that Tackett money, who'd want to fuck a woman so brittle she'd break when you mounted her?"

Janellen reeled from the ugly insults. Her ears rang loudly and her skin prickled as though stung by a thousand fire ants. Miraculously, she held her ground. "If you're not out of here in ten seconds, I'll call Sheriff Baxter and have you arrested."

He flicked his middle finger at her and got back into his truck. He turned on the motor, gunned it, and shot from the garage like a rocket.

Janellen stumbled to the switch on the wall and quickly lowered and locked the garage door, then ran into the office and locked that door, too.

148

She crumpled into the chair behind the desk and, bending slightly from the waist, hugged her elbows. She'd stood up to a two-hundred-thirty-pound brute, but now that it was over, she was shaking uncontrollably and her teeth were chattering.

In hindsight, confronting Muley had been foolish. He could have harmed her, even killed her, and never come under suspicion. It would have been believed that a vagrant thief had killed her — perhaps the one who had broken into the Winstons' home.

She rocked back and forth on the cracked vinyl cushion. What had possessed her to challenge him? She must have a bravery gene she didn't know about. It had produced that spark of temerity when she'd needed it.

It took her a half-hour to calm down. By then she had begun to realize the ramifications of her impulsiveness. Her spontaneous decision to fire Muley had been correct. Now, however, she must inform Jody. She had little doubt that Jody would back her decision, but she dreaded telling her. Perhaps she wouldn't tell her until she had found a replacement. But how would she go about doing that? It wouldn't be easy to find a man as qualified. Muley was a good pumper —

Bowie Cato.

His name sprang into her mind and caused her heart to flutter. She'd thought about him a lot, more than just in passing, more than was decent, more than she liked to admit. Frequently she'd found herself daydreaming about his bowlegged gait and recalling the way his brown eyes viewed the world with a sad cynicism.

Dare she call him and ask if he was still interested in a job?

He'd probably left town.

Besides, what kind of fool would hire an ex-con after firing an employee for stealing?

Jody would have a tizzy. Her blood pressure would soar, and it would be Janellen's fault if she became seriously ill.

She enumerated a dozen solid objections but reached for the phone book and looked up the number of The Palm. Her call was answered on the first ring.

"Is . . . Yes, I'm calling for . . . Who is this please?" Her brave gene had returned to hibernation.

"Who did you want?"

"Well, this is Janellen Tackett. I'm looking for —"

"He ain't here."

"I beg your pardon?"

"Your brother's not here. He came in last night after that town meeting. Stayed 'bout half an hour. Knocked back three doubles in record time. Then he left. Said he was going flying." The man chuckled. "I sure as hell wouldn't have got into an airplane with him. Not with all that scotch sloshing behind his belt and considering the mood he was in."

150

"Oh dear," Janellen murmured. The pimp-mobile hadn't been in its usual place this morning. She had hoped it signified that Key was up and out early, not that he hadn't come home at all.

"This is Hap Hollister, Miss Janellen. I own The Palm. If Key comes in, can I give him a message for you? Want him to call home?"

"Yes, please. I'd like to know that he's all right."

"Aw hell, you know Key. He can take care of himself."

"Yes, but please have him call anyway."

"Will do. Bye-bye."

"Actually, Mr. Hollister," she cut in hastily, "I was calling for another reason."

"Well?" he said when she hesitated.

Janellen dried her sweating palm on her skirt. "Do you still have a young man working for you named Bowie Cato?"

Lara was weeding her petunia bed when a blue station wagon careened around the nearest corner, hopped the curb, sped up her driveway, and screeched to a halt in the loose gravel. The driver's door burst open and a young man dressed in swimming trunks clambered out, his eyes wild with fright.

"Doctor! My little girl . . . she . . . her arm . . . Jesus, God, help us!"

Lara dropped her trowel and came out of the flower bed like a sprinter off the starting blocks.

She stripped off her gardening gloves as she ran to the passenger side of the car and opened the door. The woman inside was even more hysterical than the man. She was holding a child of about three in her lap. There was a lot of blood.

"What happened?" Lara leaned into the car and gently prized the woman's arms away from the girl. The blood was bright red — arterial bleeding.

"We were on our way to the lake," the man sobbed. "Letty was in the backseat, riding with her arm out the window. I didn't think I was that close to the corner when I turned. The telephone pole . . . oh, God, oh, Jesus."

The child's arm had been almost severed. The shoulder ball joint was grotesquely exposed. Blood was spurting from the severed artery. Her skin was virtually blue, her breathing shallow and rapid. She was unresponsive.

"Hand me a towel."

The man yanked one from a folded stack of beach towels on the backseat and shoved it toward Lara. She pressed it firmly against the wound. "Hold it in place until I get back." The mother nodded though she continued to sob. "Apply as much pressure as you can." To the father she said, "Clear out the back of the car."

She raced for the door of her clinic. Even as she gathered up the paraphernalia for a glucose IV, she called the Flight for Life number at Mother Frances Hospital in Tyler.

"This is Dr. Mallory in Eden Pass. I need a

helicopter. The patient is a child. She's in shock, cyanotic, unresponsive, significant loss of blood. Her right arm is almost severed. No sign of head, back, or neck injury. She can be moved."

"Can you get her to the Dabbert County landing strip?"

"Yes."

"Both choppers are currently out. We'll dispatch to you asap."

Lara hung up the phone, grabbed her emergency bag, and ran back outside. In what must have been a frenzy, the panicked father had emptied the back of his station wagon. The driveway was now littered with deflated air mattresses and inner tubes, a picnic basket, six-packs of soft drinks, two Thermoses, an ice chest, and an old quilt.

"Help me get her into the back."

Together Lara and the child's father lifted her from her mother's lap and carried her to the rear of the car. Lara climbed in and guided the child's body down as her father laid her on the carpet. The mother scrambled in and hunkered down on the other side of her daughter.

"Get me the quilt." The man brought it to her, and Lara used it to cover the child to retain her body heat. "Drive us to the county landing strip. I hope you know where it is."

He nodded.

"A helicopter will soon be there to take her to Tyler." He slammed the tailgate and ran to the driver's side. Within two minutes of their

153

arrival, they were under way.

Working quickly, Lara removed the blood-soaked towel from the girl's shoulder and replaced it with small 4 × 4 sterile gauze pads. She pressed them into the wound, then tightly bound the child's shoulder with an Ace bandage. The bleeding could be fatal if it wasn't stanched.

Next she began searching the back of the child's hand for a vein. The patient began to retch. Her mother cried out in distress. Calmly, Lara said, "Turn her head to one side so she won't choke on her vomit." The mother did as she was told. The child's air passage was clear, but her breathing was thready, as was her pulse.

The father drove like a madman, honking wildly at every other car on the road, racing through intersections, and cursing through his tears. The mother cried noisily and wetly.

Lara's heart went out to them. She knew how it felt to watch uselessly while your child died a bloody death.

Dissatisfied with the small vein she'd located in the back of the girl's hand, she made a swift decision to do a cut-down. She pulled the child's foot from beneath the quilt and, as the mother watched in horror, used a scalpel to make a small incision in her ankle. She located the vein, made a small nick in it and inserted a thin catheter, through which she connected the IV apparatus. Her fingers moving hastily but skillfully, she closed the tiny incision with a suture to secure the catheter in place.

She was dripping with perspiration and used her sleeve to mop her forehead. "Thank God," she murmured when she saw that they had arrived at the landing strip.

"Where's the helicopter?" the father screamed. "Honk the horn."

A rheumy-eyed man in greasy overalls came hurrying out of the corrugated tin hangar and went straight to the driver.

"You Doc Mallory?" he asked.

The father pointed toward the rear of the station wagon. The mechanic bent down and gaped at the gory scene. "Doc?"

Lara opened the tailgate and got out. "Have you heard from Mother Frances Hospital?"

"They had one chopper picking up a man having a heart attack out at Lake Palestine and the other at a wreck on Interstate 20."

"Did they notify Medical Center?"

"Their chopper's at the same wreck. Hell of a pileup, I guess. Said they could dispatch one from somewhere else. They're putting out the call now."

"She's got no time!"

"Oh, God, my baby!" the mother wailed. "She's going to die, isn't she? Oh, God!"

Lara looked at the tiny body and saw the life ebbing from it. "God help me." She covered her face with her gloved hands, which smelled of fresh blood. This was her recurring nightmare. Watching a child die. Bleeding to death. Incapable of doing anything to prevent it.

"Doctor!"

The child's father grabbed her arm and shook her. "What now? You gotta do something! Our baby's dying!"

She knew that all too clearly. She also knew she alone couldn't handle an emergency of this magnitude. She could control the shock temporarily, but the girl would most certainly lose her limb if not her life if she didn't get emergency treatment immediately. The small county hospital wasn't equipped to handle trauma of this magnitude. A nasty cut, a broken radius, yes, but not this. Taking her there would be a waste of valuable time.

She rounded on the awestruck mechanic. "Can you fly us there? This is a life-or-death situation."

"I just tinker on 'em. Never learned to fly 'em. But there's a pilot here who might fly you where you need to go."

"Where is he?"

"In yonder." He hitched his thumb in the direction of the hangar. "But he's feeling right poorly hisself."

"Is there a plane available? Better yet, a helicopter?"

"That pro golfer that retired here a while back? He keeps a chopper here. Fancy one. Flies it back and forth to Dallas once or twice a week to play golf. He's a regular Joe. Don't reckon he'd mind none you using it, considering it's an emergency and all."

"Hurry, hurry!" the mother pleaded.

"Can this pilot fly a helicopter?" Lara asked the mechanic.

"Yeah, but like I said he ain't —"

"Keep the IV bottle elevated," she said to the mother. "Monitor her breathing," she told the father. She was taking a chance by leaving her patient but didn't trust the loquacious mechanic to convey to the pilot the urgency of the situation.

She rushed past him and entered the building at a run. Several disemboweled aircraft were parked inside. She didn't see anyone. "Hello? *Hello?*"

She went through a door on her left, entering a small, stuffy room. In the corner was a cot. A man was lying on his back, snoring sonorously. It was Key Tackett.

Chapter Eight

He smelled like a brewery. Lara bent over him and shook him roughly by the shoulder. "Wake up. I need you to fly me to Tyler. Now!" He mumbled something unintelligible, shoved her away, and rolled onto his side.

Inside a rusty, wheezing refrigerator Lara found several cans of beer, some foul-smelling cheese, a shriveled orange, and a plastic container of water, which was what she had hoped for. Gripping the handle, she removed the lid and tossed the entire contents into Key's face.

He came up with a roar, hands balled into fists, eyes murderous. "What the *fuck!*" When he saw Lara holding the dripping pitcher, he gaped at her with speechless incredulity.

"I need you to fly a young girl to Mother Frances Hospital. Her right arm is hanging on by a thread and so is her life. There's no time to argue about it or explain further. Can you get us there without crashing?"

"I can fly anywhere, anytime." He swung his legs to the floor and picked up his boots.

Lara spun around and left the building. The father rushed up to meet her. "Did you find him?"

"He's coming." She didn't elaborate. He was

better off not knowing that their pilot had been sleeping off a drinking binge. The mechanic was standing beside a helicopter, giving them the thumbs-up signal. "What's your name?" she asked the young father as they hurried across the tarmac.

"Jack. Jack and Marion Leonard. Our daughter's Letty."

"Help me get Letty to the helicopter."

Together they lifted her out of the station wagon and rushed her toward the helicopter. Marion trotted along beside them, holding up the bag of glucose. By the time they reached the chopper, Key was in the pilot's seat.

He'd already started the engine; the rotors were turning. The Leonards were too worried about their daughter to notice that his shirt was unbuttoned and that he desperately needed a shave. At least his bloodshot eyes had been concealed with a pair of aviator sunglasses with mirrored lenses.

Once they were aboard, he swiveled his head around and looked in Lara's direction. "All set?"

She nodded grimly. They lifted off.

It was too noisy to carry on a conversation, but there was nothing to say anyway. The Leonards clung to each other while Lara monitored the girl's blood pressure and pulse. She trusted that Key knew how to reach the heliport at Mother Frances Hospital. He had slipped on a headset; she saw his lips moving against the mouthpiece.

He turned and shouted back at her, "I found their frequency and am talking to the trauma team. They want to know her vital signs."

"Blood pressure fifty over thirty and falling. Pulse one forty and thready. Tell them to alert a vascular surgeon and an orthopedic specialist. She'll eventually need both. I've started an IV."

"Did you give her an anticoagulant?"

She'd debated that but had decided against it. "She's too young. The bleeding is temporarily under control."

Key transmitted the information. Lara continued to check Letty's blood pressure, breathing, and pulse. She strove for objectivity but it was difficult when the patient was this young, this helpless, and this seriously injured.

Occasionally Marion would reach over and touch her unconscious daughter's hair or stroke her cheek. Once she ran her thumb across Letty's plump toes. That distinct maternal gesture wrenched Lara's heart.

As the outskirts of the city slid beneath them, Key spoke again. "The trauma unit is standing by. They've given us permission for a hot landing."

Letty's shallow breathing stopped suddenly. Lara dug her fingers deep into the child's neck but couldn't feel a pulse.

Jack Leonard cried out in alarm. "What is it? Doctor? *Doctor!*"

"She's arrested."

"My baby! Oh, God, my baby!" Marion

screamed hysterically.

Lara bent over the girl and placed the heels of her hands just beneath her sternum. She pushed hard several times, trying to stimulate the heart with chest compressions. "No, Letty, no. Fight. Please. How much farther, Key?"

"I can see the hospital."

She sealed her mouth over Letty's nostrils and mouth and blew air into them. "Don't die. Don't die, Letty," she whispered fervently.

"Oh, Christ!" Jack cried hoarsely. "She's gone."

"Letty!" Marion screamed. "Ah, God, please. No!"

Lara didn't even hear their hysterical cries. Her attention was focused on the small body as she pushed rhythmically on the narrow chest and alternately rendered mouth-to-mouth resuscitation.

When she felt a blip of a pulse, she gave a shout of relief. The child's chest rose and fell as her breathing resumed. Lara continued to render CPR. The pulse was feeble but her heart was beating again.

"We've got her back!"

Key set the chopper down.

The trauma team approached, ducking the rotor blades. Lara relinquished her patient and helped hold Marion back as they hustled the child onto a gurney and into the emergency room. They followed, but a nurse intercepted them and directed them into a waiting area.

"I want to be with my baby." Marion strained toward the disappearing gurney.

"I'm sorry, ma'am, you have to wait out here. She's getting the best medical attention possible."

Lara nodded understanding to the nurse. "I'll see to her. Thank you."

Together, she and Jack got Marion into the waiting area. He spoke to her soothingly. "I've got to go call our folks, Marion."

"Go ahead. I'll stay with her."

"No," Marion said, firmly shaking her head. "I want to be with Jack." She couldn't be dissuaded. Supporting each other, the couple shuffled off to locate the public telephones.

"Is the kid going to make it?"

At the sound of Key's voice close behind her, Lara turned. He was watching the Leonards as they moved down the corridor.

"It'll be touch and go."

"You almost lost her, didn't you?" His gaze shifted to her. "And you fought like hell to get her back."

"That's my job."

After a moment he asked, "What about her arm?"

"I don't know. She may lose it."

"Shit." He slipped his sunglasses into the breast pocket of his shirt, which he'd taken time to button before following them into the hospital. "I need some coffee. Want some?"

"No, thank you."

162

"Whenever you're ready to go back to Eden Pass —"

Lara was shaking her head. "I'll wait here with them. At least until she's out of surgery. Feel free to leave whenever you like. I'll find a way back."

He gave her a hard look, then said curtly, "I'm going for coffee."

Lara watched him as he moved down the sterile corridor, his gait straight and steady except for a slight limp that favored his right ankle. In spite of his dishevelment, one would never guess she had roused him from a drunken stupor a short while ago.

He'd set the chopper down between a multi-level parking garage and the hospital building. It was tricky piloting. His boast of being able to fly anywhere at any time wasn't an empty one.

The Leonards returned from making their telephone calls and began their long vigil. When Key returned, he brought with him several cups of coffee and vending machine snacks. Lara introduced him to the anxious couple.

"We can never thank you enough," Marion told him tearfully. "No matter how it turns out, if you hadn't gotten us here, Letty . . . she . . ."

He squeezed her shoulder reassuringly, rather than diminish the gravity of the situation with empty platitudes. "I'll be back in a while." With no further explanation, he left.

Reports from the operating room were agonizingly slow in coming. Each time the OR nurse approached the waiting area, the three of them tensed. But her message on these brief and periodic visits was that the surgeons were doing all they could to stabilize Letty and save her arm from amputation.

It was busy in the ER that morning. Several people had sustained serious injuries in the wreck on the interstate. It had involved three vehicles, including a van filled with senior citizens on a field trip. The staff was harried, but from what Lara could see they were competent.

Key returned about an hour later, bringing with him a large shopping bag from Walmart. He extended it to Lara and Marion. "I thought y'all'd be more comfortable if you got out of those clothes."

Inside the sack they found slacks and T-shirts. Their clothes had grown stiff with Letty's blood. They used the nearest restroom to wash up and change. When Jack tried to reimburse Key, he wouldn't hear of it.

"You're Barney Leonard's son, aren't you? You run the laundry and dry cleaners for your daddy now, don't you?"

"That's right, Mr. Tackett. I didn't figure you knew me."

"You're doing a hell of a job on my shirts. Just the right amount of starch," Key told him. "That's repayment enough."

Jack solemnly shook his hand.

Their kinfolk arrived about an hour later, along with the Leonards' pastor. The subdued group huddled together and prayed for Letty's life. During her medical career Lara had witnessed many such scenes and no longer felt uncomfortable in the face of personal tragedy.

But Key obviously felt out of place. He paced the hallway and frequently disappeared. Each time he left, Lara figured he had flown the borrowed helicopter back to Eden Pass, but he always returned and asked if there had been any news on Letty's condition. During one of these unspecified absences, he had shaved and tucked in his shirttail. The improvements made him look marginally respectable.

Almost seven hours after Letty was wheeled into surgery, a paunchy, middle-aged man in blue scrubs entered the waiting room and called their name. The Leonards stood and grasped each other's hands, bracing themselves for what they were about to hear.

"I'm Dr. Rupert." He introduced himself as the vascular surgeon. "Your little girl is going to be fine. Unless there are unexpected complications, she should pull through."

Marion would have collapsed if her husband hadn't been there to support her. She began weeping in hard, dry sobs. "Thank you. Thank you."

"What about her arm?" Jack asked.

"We managed to save it, but at this point I can't tell you how much use it will be to her.

Full circulation has been restored, but there might have been nerve and muscle damage that won't show up until later. Dr. Callahan, the orthopedic surgeon, will be out shortly to speak with you. He'll talk to you about physical therapy. The important thing now is that she's alive and her vital signs are good."

"When can I see her?" Marion asked.

"She'll be kept in an ICU for several days, but you can see her at intervals. The nursing staff will let you know. Dr. Callahan'll be right out."

When their relatives swarmed forward to embrace Jack and Marion, the surgeon turned to Key. "Dr. Mallory?"

"Not me."

"I'm Dr. Mallory." Lara extended her hand. "I'm a GP in Eden Pass."

"You did some fine work considering what you were dealing with. Got her here in the nick of time."

"I'm glad," she said with a weary smile. Lowering her voice, she asked, "Any professional guesses on how much use she'll have of her arm?"

"If I were a betting man, I'd say better than fifty percent recovery. She's young enough to learn to compensate for any disability. If use is fully restored, she won't remember when this happened." He smiled wanly, the strain of the grueling surgery showing in his face. "But I bet she won't be poking her arm through any more open car windows."

They shook hands again. After exchanging a few final words with the Leonards, he retreated down the hallway. The Leonards hugged Lara, then left to phone other relatives and friends with the good news that the crisis had passed.

Awkwardly Lara looked over at Key. "I guess I'm finished here."

"Ready when you are, Doc."

Once they were airborne, Lara's stress evolved into profound fatigue. The day's events had taken their toll. Her body ached of muscle strain. She rolled her head, trying to work out the knots in her neck.

Viewed from above, the deepening twilight was beautiful, but she couldn't enjoy it for thinking about how close she had come to losing Letty Leonard.

Life's fragility was fully realized when a child died. Any death affected her, but a child's death made a shattering impact because she always equated it with the tragic way in which Ashley had been snatched from her. One moment her sweet daughter had been cooing and gurgling happy baby sounds, the next she lay bloody and limp.

Tears filled Lara's eyes. Her throat felt achy and tight. Had it not been for Key Tackett sitting beside her in the close confines of the cockpit, she would have wept bitterly.

Instead she forced herself to retain control. She remained stoic until he set the helicopter down at the Dabbert County landing field. The mechanic greeted them.

"How's the little girl?" he asked as Lara alighted.

"She's alive, and they saved her arm."

"Praise be. I'd've thought she was a goner. Hey, Key. It's a beauty of a chopper, ain't it?"

"First class, Balky," he conceded, passing the mechanic the keys.

Lara pointed at the Leonards' station wagon. "Would you please see that their car is cleaned up before they come to retrieve it?"

"Already did," the mechanic told her. "Bo done sent a boy from his garage to wash out the blood."

"That was very kind of you, uh . . . Balky, is it?"

He nodded. "Balky Willis. Pleasure, ma'am." He extended his hand to Lara.

She shook it. "Dr. Lara Mallory."

"Yes, ma'am, I figured you was her."

"I'm sure the Leonards will appreciate your thoughtfulness about their car."

"Weren't my idea. Key called from Tyler and suggested it."

Surprised, Lara looked at him. He shrugged indifferently. "Either way it went, I figured they didn't need any unpleasant reminders. Ready to go?"

"Go?" Only then did she realize she was without transportation. "Oh, would it be an imposition —"

168

He indicated the yellow Lincoln parked on the far side of the hangar.

Lara asked Balky to thank the golfer who had lent them the helicopter. "Tell him to send me a bill for any expense that was incurred."

"Sure thing." He saluted her and bade goodbye to Key.

"I'll expect a bill from you, too, Mr. Tackett," she said as they approached the Lincoln. "How much do you charge?"

He pulled open the wide passenger door and held it for her. "Depends on what service I've rendered."

Unsmiling, she slid into the car and sat staring straight ahead through the windshield.

Once they were on the highway headed toward town, Key remarked, "You know, your sense of humor ain't for shit. Don't you ever laugh?"

"When I hear something funny."

"Oh, I get it. I don't amuse you."

"Sexual innuendoes have lost their charm for me. I've been the subject of too many to find any humor in them."

He stretched his long body, adjusting his bottom more comfortably in the seat. The leather squeaked agreeably. "I guess that's the price one pays when she's caught up in a sex scandal."

"That's only one price she pays."

He gave her a frankly appraising stare, then returned his attention to the road. They drove in silence, the car gliding along the two-lane stretch of highway through the deepening dusk.

"Are you hungry?"

She hadn't thought about it, but now that he'd asked, she realized she was famished. All she'd had that morning before going out to weed her flower bed was some yogurt and two cups of black coffee.

"Yes," she admitted.

"Do you like ribs?"

"Why?"

"I know where you can find the best in the world. Thought we'd stop for some."

She glanced down at the clothes he'd brought to the hospital. "Much as I appreciate the change of clothing, I'm not really dressed for going out."

He barked a laugh. "You're almost overdressed for Barbecue Bobby's."

"He's aptly named."

"He didn't get his name from barbecuing, but for being barbecued." She looked at him quizzically. "See, one night Bobby Sims got on the wrong side of a bull rider named Little Pete Pauley. They were at a postrodeo dance and got in a fight over a woman. Bobby came out on top and humiliated Little Pete — who always was real touchy about being only five feet four standing in his boots.

"Later that night, Little Pete got revenge by setting fire to Bobby's house. Bobby made it out okay, except that most of his hair got singed off. Went around for six months as hairless as a lizard and smelling faintly of wood smoke. Everybody started calling him Barbecue. From there on, his life's work just naturally evolved."

Lara suspected he was spinning a yarn, but before she could express her doubts, he pulled into the parking lot of a tavern. "Hmm. Crowded tonight."

"This is a beer joint," she protested. A single strand of yellow lights, many of them burned out, had been strung along the roofline. They were the building's only decoration. "I'm not going in there."

"How come?" He turned to her. "Are you too prissy for us?"

He had backed her into a corner. If she refused to go in with him, he would once again accuse her of being a hypocrite, a holier-than-thou snob who couldn't rightfully throw stones when she herself had been caught transgressing.

On the other hand, she didn't want the rumor mill to grind out that she was being squired around town by Key Tackett. How tongues would wag! The lady doctor had corrupted Senator Clark Tackett, people would say, and now she had her hooks in his younger brother.

But facing down the gossip was a future possibility. Key's scorn was a sure thing in the here and now. She opened her door and got out. He was wearing an insufferably smug grin when he joined her at the entrance and pulled open the door.

The interior of the honky-tonk was no sightlier than the exterior. A pall of tobacco smoke clung to the ceiling, making the dim lighting dimmer.

171

The smell of beer was almost as strong as the bass being pumped from the gaudy jukebox in the corner. Several couples were two-stepping around a tiny dance floor. A long bar comprised one entire wall, and tables were scattered around the murky fringes of the room.

Every head turned toward the door when they walked in. The women inspected Key; Lara was a target for the men. Self-consciously she let him lead her to a table.

"Do you drink beer?"

She rose to meet the challenge in his voice. This was another test. "With barbecue? of course."

He placed two fingers in his mouth and whistled shrilly. "Hey, Bobby, two beers."

"Well, I'll be a cross-eyed billy goat!" the bartender boomed. "Two beers coming up for the long-lost Key Tackett."

Key sat down across from Lara and pushed aside the condiments in the center of the table. "Saving a kid's life and drinking beer with me all in the same day. You really enjoy living on the edge, don't you, Doc?"

He didn't expect an answer, and she didn't have time to offer one before a rotund man wearing a white apron stained with meat juices and barbecue sauce sauntered over carrying two longneck beer bottles in one hand. With the other, he whacked Key between the shoulder blades.

"Long time no see." He set the beer bottles

on the table. Lara quickly reached out to catch hers before it toppled over. Bobby didn't notice. He was still greeting Key.

"Heard you just got back from one of them A-rab countries. Heard if you look sideways at their women, they cut off your dick. That true? Wondered how a horny bastard like you could survive over there. Wondered when you were going to get out here to see me, you asshole."

"The place looks great, Bobby. Still doing a land office business."

"Hell, yes. As long as folks eat, drink, and screw, they know the best place to come to find all three. One-stop shopping. That's my business philosophy! Who's this?" He jabbed a finger in Lara's direction.

Key introduced her. The tavern owner didn't even attempt to hide his surprise. "So you're the shady lady I've heard so much about. Son of a bitch." He looked her over with a candor she appreciated after being eyed covertly by so many others.

"You hung your shingle out in town. Old Doc Patton's place, 's that right?"

"That's right." Lara smiled, noticing the burn scars above his eyebrows and along his hairline.

"Will wonders never cease." He shifted his gaze between the two of them. "Didn't reckon y'all would be on speaking terms."

"We're not," Key replied. "But we were hungry at the same time, so here we are. You going to serve us or jaw all night?"

Barbecue Bobby grinned. "Hell, yeah, I'm going to serve you. Can't wait to get my hands on your money. What'll ya have?"

"Two rib platters. No sauce on mine."

"I'll bring the sauce on the side and y'all can suit yourselves. A couple more beers?"

"When you bring the dinners."

"Sure hope I get sick real soon," Bobby said, winking at Lara. Then, shaking his head over the vagaries of life, he lumbered back to his bar.

Key took several long swallows of his beer. Lara sipped hers. "Did you go flying last night?"

He stopped drinking, but held the spout of the bottle against his lips and idly rubbed it across them. "Why?"

Lara looked away from his mouth and the beer bottle. "Just wondering."

"Yeah, I flew last night. Took out a Piper Cub. Know what that is?" She shook her head, although she now had a vague idea of what one looked like. "Nice little kite if you're going out for a spin. Why'd you ask?"

She wouldn't admit that in order to clear her head after their altercation in the school parking lot she had taken a drive in the country, or that she had watched a foolhardy, but highly skilled, pilot flirt with death and destruction.

"I was thinking about your ankle," she said. "Since you're still favoring it when you walk, I wasn't sure you could fly."

"It still gets sore. But I couldn't remain grounded any longer or I'd have gone crazy."

"Then this hiatus is unusual for you?"

"Flying's my business. I fly for hire. For whoever has a job that sounds interesting."

"That's your criteria? Whether it's interesting?"

"And well paying," he said with a grin. "I don't fly for chicken feed."

"You can pick and choose your clients?"

"Pretty much. Some outfits are top notch. Their planes are slick and expensive. They even enforce a few rules and regulations about how many hours a pilot can fly without sleep and how long it's been since his last beer. They expect you to fill out all the paperwork required by the FAA.

"But there are just as many outfits whose planes aren't as well maintained. Sometimes the landing strips at the destination aren't ideal. And about their only restriction on a pilot is that he's able to open one eye."

"You've flown under those conditions?"

" 'Under those conditions' I've earned some of my best money."

Having listened to him talk about it, she decided that money was the least of his motivators. "You love it, don't you?"

"Second only to sex. Sometimes it's even better than sex because there's no foreplay and airplanes can't talk."

She didn't take the bait.

He went on. "Up there, everything's so clean. There's no bullshit to cloud your thinking." He squinted as though searching for the appropriate description. "In the sky, things are uncomplicated."

"It looks extremely complicated."

"Flying's a motor skill," he said with a brusque shake of his head. "You're either born a flyer or you aren't. It comes from your gut, not your head. You're either good or bad. Decisions are either right or wrong. You fuck up, you die. It's that simple. There're no gray areas, no time for analysis. Only quick judgment calls that you hope to God are right."

"It wasn't that simple today," she reminded him.

"For me it was. I wasn't involved in the emergency. My job was to pilot the craft. That's what I did."

Lara didn't believe he was as nonchalant as all that. He had been more emotionally involved with saving Letty Leonard's life than he wanted to admit and would have been terribly upset if she had died en route to the hospital.

Barbecue Bobby served their beers and rib platters. On each was a side of succulent baby back ribs, french fries cooked in their jackets, creamy coleslaw, a slice of red onion, two slices of white bread, and a jalapeño pepper the size of a small banana. Key bit into his as though it were a piece of fruit. Just the scent of it brought tears to Lara's eyes, so she avoided it. The ribs, however, tasted as good as Key had promised. The pork, smoked for hours over mesquite wood, virtually melted off the bone.

"Did you always want to be a pilot?" Lara asked between bites.

"Did you always want to be a doctor?"

"I can't remember wanting to be anything else."

He shot her a wicked grin. "When you were a kid and played doctor, you played it for real, huh?"

"Actually yes," she returned with a smile. "Although not as you mean. My friends would eventually tire of the game and wanted to move on to playing 'teacher' or 'movie star' or 'model.' I never wanted to stop bandaging them until they looked like mummies. I took their temperatures with Popsicle sticks and gave them shots with meat basters."

"Ouch."

"It was a preoccupation my parents desperately hoped I would outgrow. I never did."

"They didn't cotton to you going into medicine?"

"Not at all. They wanted me to be a lady of leisure who does lunch with friends, holds office in service clubs, and organizes charity functions. Not that there's anything wrong with doing those things. For a lot of women that represents challenge and fulfillment. But it wasn't the life for me."

"Mama and Daddy couldn't understand that?"

"No, Mother and Father couldn't." He acknowledged the distinction with raised eyebrows.

177

Lara explained. "I came late in their marriage. In fact I was an unexpected and unpleasant surprise.

"But, since they were stuck with me, my parents decided to make the best of the situation and plotted the course of my life. Because I didn't want to follow the path they had carefully chosen, they've never let me forget what a burden I've been to them. And sometimes I was," she added with a reflective laugh.

"I once kept a friend in 'intensive care' for hours until her concerned parents came looking for her. They found her in my bedroom breathing through drinking straws that I'd poked up her nostrils. It's a wonder she didn't suffocate. I prepped another friend for brain surgery by giving her a very short haircut."

Chuckling, Key blotted his mouth with a napkin.

"Then there was Molly."

"What'd you do to her?"

"I cut her open."

He choked on his swig of beer. "You *what?*"

"Molly was our next door neighbor's golden retriever. She was a beautiful dog that I'd played with since I could toddle in the yard between our houses. Molly got sick and —"

"You operated?"

"No, she died. Our neighbor was disconsolate and couldn't bear to bury her the same day she expired. So they wrapped her in plastic and left her in the carriage house overnight."

"Good God. You performed an autopsy?"

"A crude one, yes. I coerced a friend of mine, who claimed to want to be a nurse, to sneak into the carriage house with me. We took along our housekeeper's kitchen utensils."

He laughed, running his hand down his face. "Most girls I knew played with Barbie dolls."

Defensively, Lara said, "As long as Molly was feeling no pain, I didn't see the harm in cutting her open and taking a look inside. I wanted to learn something about anatomy, although at the time I didn't even know the word."

"What happened?"

"As I began to remove Molly's organs, my so-called friend started screaming. Hearing the screams, Molly's owner called the police. They arrived roughly at the same time my parents missed my friend and me. They stormed the carriage house, saw the carnage, and all hell broke loose.

"Naturally, my parents were horrified and began accusing each other of having undisclosed 'bad seeds' in their family trees. The neighbor declared she would never speak to any of us again. My friend's parents told mine that there was obviously something dreadfully wrong with me and that I should have psychiatric care before I became a real danger to myself and others.

"My parents agreed. After weeks of expensive and extensive psychiatric sessions, the doctor's analysis was that I was a perfectly normal eleven-year-old. My only unusual trait was an obsessive interest in human anatomy from a strictly medical viewpoint."

"Bet your folks were relieved to know they hadn't raised a ghoul."

"Not really. They continued to believe that my desire to become a doctor was strange. To some extent, they still do." With her finger she absently traced a bead of condensation that trickled down her beer bottle.

"My parents are very social. Appearances are important to them, and they resent cogs in their well-oiled lives. I've provided many, beginning with my birth and ending —" She raised her eyes to meet his. "Ending with the scene at Clark's cottage. Like you, Mr. Tackett, they didn't chasten me for having an affair. Only for making it public knowledge."

At that moment, a body landed in the middle of their table.

Chapter Nine

Dirty dinner dishes clattered to the floor while rib bones scattered across the grimy planks like clumsy Pick-Up-Sticks. Four bottles of beer toppled. One broke, the others rolled away.

The man's weight had tipped the table to a forty-five-degree angle. He was bleeding from his nose. Grunting curses, he struggled to his feet and charged the man who had punched him.

"Time to go." Key calmly stood up and encircled Lara's upper arm. "Your first time at Barbecue Bobby's ought not to be spoiled with a fight."

She was spellbound by the sudden outbreak of violence. As the two young men continued to slug it out, a ring of onlookers formed an arena for them, shouting encouragement. She watched and listened in honor as blood splattered and cartilage crunched.

"They're hurting each other!" As Key ushered her toward the door, she tried to dig in her heels. He ignored her attempts and moved inexorably toward the exit, pausing only long enough to hand Bobby a twenty-dollar bill. "Still up to standard. Thanks."

"Sure thing. Y'all come back."

Bobby didn't take his eyes off the fight, which

had intensified. The fighters were throwing vicious punches and shockingly obscene insults at each other.

"I should stay," Lara protested. "They'll need medical attention. I could help."

Key gave the fighters an indifferent backward glance as he pushed her through the door. "They wouldn't welcome your help, believe me. Especially those two. They don't appreciate others poking their noses into family affairs."

"They're related?" Lara asked, aghast.

"Brothers-in-law." By now they were in the car, pulling out of the parking lot onto the highway. "Lem and Scoony have always been best friends. A few years back, Scoony's little sister started looking real good to Lem. They began dating. That didn't set too well with Scoony, having seen Lem in action with other girls. Scoony warned him that if he knocked up his sister he'd beat the shit out of him."

He concentrated on passing a loaded logging truck.

Impatiently Lara asked, "Well, what happened?"

"Lem knocked her up, and Scoony beat the shit out of him."

"And they've been enemies ever since?"

"No, they're still best friends. Missy, that's Scoony's sister, heard that Scoony was out to throttle Lem. She tracked them down — at The Palm, I believe it was — and joined the fracas. Kicked them both where they're most vulnerable.

"By the time the sheriff got there, both boys were in tears, cradling their privates, and blubbering like babies. Missy told Lem he could either marry her or she'd permanently emasculate him and told Scoony that if he didn't like it he could . . . Well, Missy never has been known for her ladylike language. Anyway, Lem and Missy got married, had a little boy, and everybody was happy."

"Happy?" Lara exclaimed. "What about tonight?"

"Oh, hell, that was nothing. They were just blowing off steam. By now they're probably buying each other a drink."

Lara shook her head in dismay. "This place. These people. I always thought tales of Texas were exaggerations to perpetuate the state's mystique. Like Barbecue Bobby. What you told me is really the way it happened, isn't it? A bull rider named Little Pete Pauley set fire to his house, his hair got singed, and that's how he got his nickname."

He looked surprised. "Did you think I was lying?"

"I don't know what to think."

She gazed through the windshield as if viewing the landscape of an alien planet. Although she would never admit it to him, she felt bewildered and overwhelmed. Would she ever fit in? Had she been deluding herself that she could? Eden Pass was as peculiar and at times as intimidating as a foreign country.

"It's so different here," she said lamely.

"True enough. Different for you, anyway." He pointed through the windshield at the approaching lights of town. "For every person living in Eden Pass, there's a story. I could spend all night with you and still not get around to all of them."

She reacted, turning her head quickly. His choice of words had been calculated. She could tell that by the way he was looking at her.

In a sexy voice he added, "But I don't guess we'll be spending any nights together, will we, Doc?"

"No, we won't."

"Because you and I don't have a damn thing in common, do we?"

"Only one thing. Clark. We have Clark in common."

At the mention of his brother's name, his sultry gaze instantly turned cold. His expression changed completely.

"Well, he and I didn't have much in common except our two parents and a home address. We loved each other, even liked each other. But Clark obeyed all the rules. I broke them. I grudgingly respected him for being good all the time, and I think he harbored a secret envy for my ability not to give a damn. We were as different as brothers could be and still be kin." His eyes moved over her. "Where we really differed was our taste in women."

"I doubt the two of you would appeal to the

same woman," she said stiffly.

"Right. It would either be one of us or the other. For instance, if Clark had taken you to dinner tonight, you wouldn't have had the pleasure of Barbecue Bobby's. You'd have dressed up fit to kill and gone to the country club. You'd have rubbed elbows with the upper crust, the social climbers, pillars of the community.

"They're still drinkers, liars, cheaters, and fornicators, but they're less honest about their failings than the folks out at Bobby's." He angled his head to one side. "Come to think of it, you'd've fit in much better out there at the country club with all those other hypocrites."

Lara took the insult with equanimity. "What is it about me that really bugs you, Mr. Tackett?" Once today she had slipped and called him by his first name. That had been at the height of the crisis with Letty Leonard. Last names seemed more appropriate now. It reestablished the breach.

He brought the Lincoln to a halt in her driveway, barely missing the Leonards' picnic paraphernalia still scattered about.

Laying his arm on the back of the seat, he turned to face her. "What really bugs me is that the whole world knows you're a whore. Your own husband caught you whoring. But you don't own up to what you are. You pretend to be another kind of woman entirely."

"What do you suggest I do, brand a letter A on my chest?"

185

"I'm sure many would pay for the pleasure. Me, for one."

"How dare you judge me? You don't know the first thing about me, and you know even less about my relationship with your brother." She shoved open the car door. "What do I owe you for today?"

"Forget it."

"I don't want to be obligated to you."

"You already are," he said. "You cost Clark everything that was important to him. He's no longer around to call in the marker, but I am. And when I do, it's going to be expensive."

"You've got it backward, Mr. Tackett. I'm the one who's holding the IOU, and you are the one who's going to pay."

"How do you figure?"

She gave him a level look. "You're going to fly me to Montesangre."

His arrogant grin collapsed, and for a moment he stared blankly at her. Then he cupped his hand around his ear. "Come again?"

"You heard me."

"Yeah, I heard you, but I can't believe it."

"Believe it."

He was incredulous. "Does the expression 'not in this lifetime' mean anything to you?"

"You'll take me there, Mr. Tackett," she said confidently as she alighted. "I'll see to it."

"Yeah, right, Doc." He was laughing as he backed the Lincoln out of the driveway. It fishtailed as he sped away.

"I love you."

"I love you, too."

Heather Winston and her boyfriend, Tanner Hoskins, were entangled on the quilt they'd spread out in the tall grass. Nearby the lazy waters of the lake slapped against the rocky beach. The moon had risen and was reflected on the water.

Even on the hottest evenings there was always a cool breeze around the lake, which made it more comfortable for the young lovers who parked there. In Eden Pass the lake was the most popular make-out spot. If you went to the lake with someone, everyone assumed the relationship was serious.

Heather and Tanner had a serious relationship, now four months old. Previously she'd gone out with Tanner's best friend, who, she came to find out, was fooling around with another girl. Following the much-publicized breakup scene outside the chemistry lab, Tanner went to her house to console her.

He'd been very sweet, calling his friend a stupid jerk-off and taking Heather's side on all points. Heather took a closer look at Tanner and decided that he was much more handsome than the creep who'd cheated on her.

After polling her best friends and discovering that they too thought Tanner was a good catch, she changed the tenor of the time they spent together. Soon it was known around school that she was "with" Tanner. She couldn't have been happier with the way things had turned out.

Since Heather Winston was the most sought-after girl in the junior class, Tanner was also walking on air. The first time he kissed her they'd frenched, and it nearly took the top of his head off. All the guys agreed that she had a body — taking after her mama, who was indisputably the hottest-looking bitch in Eden Pass. There was a lot of good-natured speculation in the locker room as to just how much of Heather's delights ol' Hoskins had sampled.

Tanner's responses to these teasing jeers were deliberately vague. Most of the guys chose to think he was getting all he wanted but was protecting Heather's reputation with gallant silence. Those more cynical figured he hadn't seen or touched anything that a swimsuit would cover.

The truth lay somewhere in between.

Tonight, he had unbuttoned her blouse and gotten his hand inside her brassiere. Heather permitted him to fondle her anywhere above the waist. Below it was where she customarily drew the line.

They were on the brink of a breakthrough, however. The gentle feathering of his tongue across her nipples had pushed Heather to a sexual height she'd never achieved before. Yearningly, she brushed her hand across the fly of his shorts.

He made a strangled, groaning sound. "Please, Heather."

Tentatively she pressed her palm against the

bulge in his crotch. Her friends had warned her that "it" got huge and hard. Even so, she was timid of his erection. Yet curious. And desirous. And her friends were going to start believing she was weird if she didn't move things farther along.

"Tanner, do you want me to?"

"Oh God," he moaned and began frantically grappling with his zipper.

He shoved her hand beneath the waistband of his underwear, and before she was quite prepared for it, her hand was filled with pulsing, adolescent lust.

Tanner muttered incoherently as she timorously explored his shape. She knew how this monstrous organ was supposed to couple with her body, although she didn't understand how it possibly could. Still, it was exciting to imagine. Her mind drifted through an array of erotic images, intensified by recollections of some of Hollywood's recent renditions of sex, movies that her mother had forbidden her to see.

Then he ruined it.

"Oh, God!" she cried. "What . . . ? *Tanner!* Oh, puke!"

"I'm sorry, I'm sorry," he panted. "I couldn't help it. Heather, I —"

She leaped up and headed for the lake at a run, refastening her bra and buttoning her blouse as she went. When she reached the pebbled beach, she knelt and swished her hand in the water. She was repulsed, not so much by the substance on her hand but by necking in general.

It was so juvenile, so common, so unromantic. Nothing like the misty love scenes in the movies.

She moved along the beach until she reached the fishing pier, then walked out to the end of it, sat down, and stared out over the water. Tanner caught up with her there a few moments later. He lowered himself beside her.

For a moment he said nothing. When he did speak, his voice was ragged with emotion. "I'm sorry. Christ, I didn't mean to. Are you going to tell?"

Heather saw that he was humiliated, and she regretted her adverse reaction to what she knew wasn't entirely his fault. She stroked his hair. "It's all right, Tanner. I didn't expect it and over-reacted."

"No, you didn't. You had every right to be disgusted."

"I wasn't. Truly. Anyway, it's okay. Of course I won't tell anybody. How could you think I would? Just forget about it."

"I can't, Heather. I can't because . . ." He hesitated as though to gather courage, then blurted, "Because if we'd been doing it right in the first place, it wouldn't have happened."

Heather returned her gaze to the moonlit water. He'd never come right out and said he wanted to go all the way. He wanted to — she knew that. But knowing it and hearing him say it were two different things. Hearing it was much scarier because it forced her to make a decision.

"Don't get mad," he said, "but hear me out.

Please. I love you, Heather. You're the prettiest, sweetest, smartest girl I've ever met. I want to, you know, know everything about you. Get inside you," he added softly.

His words shocked her in a pleasant way. They made her body tingle in secret places. "That's sexy talk, Tanner."

"I'm not just feeding you a line. I mean it."

"I know you do."

"Look around." He gestured back toward the parked cars. "Everybody else does it."

"I know that, too."

"Well, do you think . . . I mean, don't you want to?"

She gazed into his fervent eyes. Did she want to? Maybe. Not because she was passionately in love with him. She didn't see herself spending her life with Tanner Hoskins, the grocer's son, having children and grandchildren with him, growing old together. But he was sweet, and he clearly adored her.

She gave him a qualified yes.

Encouraged, he scooted closer to her across the rough boards. "It's not like you could get AIDS or anything because we're not strangers. And I'd make damn sure you wouldn't get pregnant."

Amused by his earnestness, she took his hand and squeezed it between her own. "I'm not worried about any of that. I'd trust you to take precautions."

"Then what's stopping us? Your folks?"

Her smile faded. "Daddy would probably shoot you if he knew we were even having this conversation. Mother . . ." She sighed. "She thinks we've already done it."

That was the crux of Heather's hesitation. Her mother. She didn't want to validate Darcy's low opinion of her.

Her relationship with her father was uncomplicated. He thought the sun rose and set on her. She was his pride and joy, his precious little girl. He would gladly die for her. She was confident of his unconditional love.

Her relationship with her mother wasn't as clearly definable. Darcy had a volatile and unpredictable personality. She wasn't as easy to love as her unflappable father. If Fergus was as constant as sunrise and sunset, Darcy was as changeable as the weather.

Some of Heather's earliest memories were of Darcy dressing her up and taking her downtown. She would parade her up and down the sidewalk of Texas Street, in and out of shops, making sure that everyone saw them and stopped to speak. Darcy had always liked to show her off.

But once they returned home, her mother's indulgent affection ceased. She withdrew the love she showered on Heather in public and began preparations for their next outing.

"Practice your piano, Heather. You won't win any blue ribbons in the competition if you don't practice."

"Stand up straight, Heather. People will think

192

you have no pride if you slouch."

"Stop biting your nails, Heather. Your hands look horrible, and besides, it's a terrible habit."

"Wash your face again, Heather. I can still see blackheads around your nose."

"Your jumps need work, Heather. You won't get reelected cheerleader next year if you start shirking."

Although Darcy professed to push her because she wanted her to be and to have the very best, Heather suspected that her accomplishments were more for her mother's sake than for her own. She also suspected that underlying Darcy's maternal love was a deep resentment that bordered on outright jealousy. It puzzled Heather. Mothers weren't supposed to be jealous of their children. What had she done or failed to do to provoke this unnatural emotion?

As Heather matured, their riffs had become more frequent and virulent. Darcy imagined that Heather was sexually misbehaving. She persistently made veiled accusations and sly innuendoes.

What a laugh, Heather thought scornfully.

Her mother was the one guilty of sexual misconduct. Everybody knew her reputation, even the kids at school, although no one had ever confronted Heather with it because they didn't dare. She was too popular.

But the whispered rumors reached her. It was a struggle to ignore them, especially at home when her mother was being particularly nasty. Countless times she could have used the latest gossip about Darcy to shut her up. But she hadn't and she wouldn't because of Fergus. She wouldn't do or say anything that might indirectly hurt her father or cause him embarrassment.

So when Darcy railed at her about her relationship with Tanner, and hounded her with questions about the depth of it, she withstood the inquisition in sullen silence.

Beyond petting, she hadn't done anything shameful. The fundamental reason for her abstention was that she didn't want to become like her mother. Obviously she had inherited Darcy's robust sexuality, but she didn't have to act on it. The last thing she wanted was a reputation for screwing around — like mother, like daughter. Nor would she betray her father's love the way her mother did.

Tanner had been sitting quietly at her side, patiently giving her time to sort through her misgivings. "I feel everything you do, Tanner. Honestly," she said. "Maybe not as urgently," she added with a gentle smile. "But I love you enough to want to have sex with you."

"When?" he asked thickly.

"When we feel the time and mood are right. Okay? Please don't pressure me about it."

His disappointment was plain, but he smiled and leaned forward to give her a tender kiss. "I'd better take you home before it gets any later. Your mother will have a shit fit if you're thirty seconds late."

They arrived punctually. Nevertheless, Darcy

194

was waiting for them at the front door with a glare for Tanner and a lecture for Heather on how a girl couldn't be too protective of her good name.

"Good morning."
"Good morning."
Bowie Cato and Janellen Tackett faced each other across the desk in the cramped office at the shop. He was surprised to notice that her eyes were on a level with his. He hadn't realized when they met the first time that she was almost as tall as he. She had looked so dainty, frail even, sitting behind that large desk, looking as nervous as a whore in church.

Now why would an analogy like that pop into his head when he was in the presence of a lady like her? As though he'd spoken his thoughts out loud, he hastened to make amends.

"I'm sorry I wasn't around when you called The Palm. Hap — Mr. Hollister — gave me your message to come by when it was convenient. Is now convenient?"

"Yes, and it was kind of Mr. Hollister to remember."

"He's been real decent to me."

"Well, thank you for coming. Have a seat, please."

She indicated the metal chair behind him. He lowered himself into it as she resumed her seat behind the desk. She carefully smoothed the back of her skirt and sat down in one fluid motion.

195

Some motions like that she carried off gracefully, without thinking about them. At other times, particularly when she was looking directly at him, her movements were as jerky and uncoordinated as a newborn colt's. She had the jitters worse than anyone he'd ever met. If he said "boo!" she'd probably faint dead away.

He couldn't imagine why Miss Janellen Tackett was nervous over this interview. She was the one holding all the aces. He needed her; his future hung in the balance, not the other way around.

"I . . ." She got a false start and began again after clearing her throat. "We've had a job become available."

"Yes, ma'am."

Her large blue eyes opened even wider. "You knew about it?"

When would he learn to keep his fat trap shut? "I, uh, heard you fired a man after accusing him of stealing."

"He *was* stealing!" Her loud exclamation startled them both. She appeared mortified by her outburst. Bowie decided to make it easier on her and in the process chalk up a few points for himself.

"I don't doubt it for a minute, Miss Tackett. You don't appear the kind of person who would make accusations until you were sure you were right."

Bowie had overheard the man everybody called Muley virtually bragging about being fired by "that skinny Tackett bitch." The harsh names the redneck had called Janellen and the unflattering way he'd talked about her hadn't jived with Bowie's memory of the softspoken, selfconscious lady he'd met.

He'd asked around, subtly, and found that the Tacketts had a reputation for fairness. They expected an honest day's work from their employees, but paid well. Miss Tackett was known to be especially reasonable and to cut her people a lot of slack. Muley Bill was obviously a liar as well as a thief.

"That Muley character is a loudmouthed bully, Miss Tackett," Bowie said. "So I didn't put too much stock in what he spouted off. I'm only wondering why we're wasting your valuable time talking about him."

"He was a pumper."

"Yes, ma'am."

"I'm offering you his job."

His heart lurched, but he kept his expression unreadable. He'd hoped her summons meant a job offer, but he was suspicious of being handed good fortune, fully expecting the other hand to slap him. "That sounds real fine. When do I start?"

She fingered the buttons on her blouse. "What I have in mind," she said haltingly, "is a probationary position. To see how . . . how you get along here."

There it was. The slap. "Yes, ma'am."

"This is my family's business, Mr. Cato. I'm

197

the third generation and feel a responsibility to protect —"

"Are you scared of me, Miss Tackett?"

"Scared? No," she replied with a lying little laugh. "For heaven's sake, no. It's just that you might not like working for Tackett Oil. Steady employment might require some difficult adjustments since you were recently released from . . ."

She shifted in her seat. "If, after a time, both parties agree that it's working out, I'll offer you a permanent position. How does that sound?" She gave him a wavering smile.

Bowie also shifted in his chair and carefully regarded his hat as he threaded the brim through his fingers. If anybody else had offered him a temporary job until he proved himself worthy, he'd say "screw you" and stomp out. But he recognized the chip on his shoulder for what it was and curbed his temper.

"Do all your new employees go through this, uh, probationary thing?"

She wet her lips and fiddled some more with the buttons on her blouse. "No, Mr. Cato. But frankly you're the first person I've ever considered hiring who is on parole from prison. I'm responsible for the daily operation of the business. I don't want to make a mistake."

"You won't."

"I'm certain of that. If I weren't, I wouldn't have called you for an interview."

"You can check my record with the Department of Corrections. I got a lot of time off for good behavior."

"I've already spoken with your parole officer." His eyes snapped up to hers and she blushed. "I felt I had to. I wanted to know what you . . . what you had done."

"Did he tell you?"

"Assault and battery, he said."

He looked away and pulled his lower lip through his teeth several times. Again, he was tempted to walk out. He didn't owe her a goddamn thing, and surely not an explanation. He didn't feel he had to justify himself to anyone.

But, oddly, he wanted Janellen Tackett to understand why he'd committed the crime. He couldn't pin down exactly why he wanted her understanding. Maybe it was because she looked at him like he was an actual person and not just an ex-con.

"The bastard had it coming," he said.

"Why?"

He sat up straighter, preparing to lay out the facts and let her read them as she pleased. "He was my landlord. He and his wife lived in the apartment below mine. It was a dump, but the best I could afford at the time. She — his wife — was as kind a woman as I ever knew. Ugly as sin but a good heart, you know?"

Janellen nodded.

"She'd do favors for me. Sew on shirt buttons, stuff like that. Sometimes she'd bring me leftover stew or a slice of pie because she said bachelors never ate right and a body couldn't survive only on Wolf Brand Chili."

He bounced his hat on his knee. "One day I met her on the stairs. She had a black eye. She tried to hide it from me, but the whole left side of her face was swollen. She made up an excuse, but I knew right off that her old man had worked her over. I'd heard him yelling at her plenty of times. I didn't know he'd started using her as a punching bag.

"I cornered him and told him if he wanted a fistfight I could give him a hell of a good one. He told me to mind my own business. Then he beat her again a couple of weeks later. That time we had more than words. I slugged him a few times, but she intervened and begged me not to hurt him."

He shook his head. "Go figure. Anyway, I warned him then that the next time he hit her, I'd kill him. A few months went by, and I thought he'd gotten the message. Then one night the racket downstairs woke me up. She was screaming, crying, begging for her life.

"I ran down to their apartment and kicked the door in. He had thrown her against the wall hard enough to put a hole in the sheetrock and to break her arm. She was cowering against the wall, and he was whipping her with a leather belt.

"I remember sailing through the air and landing square in the middle of his back. I beat the holy hell out of him. Almost killed him. Luckily one of the other tenants called the police. If they hadn't gotten there when they did, I'd've been sent up for manslaughter." He stopped, thinking back. "I'd had to deal with bullies like him all my life. I'd had enough of it, I guess, and just sorta snapped."

He was silent for a moment and stared at his hands. "At my trial, he broke down and cried, made his apologies to God and man and swore he'd never raise a hand to his wife again. My lawyer advised me to tell the jury that I didn't remember the attack, that I'd gone temporarily wacko, that I was too enraged to realize what I was doing.

"But, seeing as how I'd sworn on the Bible to tell the truth, I told them in all honesty that I wished I'd killed the son of a bitch. Any man who beats a defenseless woman like that needs killing, I said, and I meant it." He shrugged resignedly. "So he walked, and I went to the pen."

After another silence, Janellen's chair creaked slightly as she got up and moved to a tall metal filing cabinet. From it she withdrew several forms. "I'll need you to fill these out, please."

He remained seated and looked up at her. "You mean I'm hired?"

"Yes, you're hired." She quoted him a starting salary that flabbergasted him.

"And after hearing your story," she said, "I'm willing to waive the probationary period. It was a silly idea anyway."

"Not so silly, Miss Tackett. You can't be too careful these days."

201

His smile seemed to fluster her. She hesitated a moment, then leaned down to lay the forms on the desk in front of him. "These are tax and insurance forms. A nuisance, I'm afraid, but necessary."

"I don't mind the paperwork if it means a job."

As she talked him through the forms, Bowie tried to concentrate on them, but it was tough to do with her standing so close. She smelled good. Not overwhelmingly perfumed like the whores he'd gone to following his release.

She smelled clean, like soap and bedsheets that had dried in the sunshine. Her hands were slender and delicate and pale. They entranced him as she sorted through the documents and pointed out the dotted lines on which he signed his name.

From the corner of his eye, he could see her in profile. She wasn't beautiful, but she wasn't downright ugly, either. Her skin was smooth and fair, practically translucent. There was no wiliness in her expression, not like some women who you could tell were calculating their next move on you. Instead she seemed to be straightforward and honest and kind, qualities he'd rarely run across. He liked listening to her voice, too. It was as soft and soothing as he imagined a mother's lullaby would be.

And her eyes . . . Hell, those eyes could have dropped a man at fifty paces if she'd chosen to use them that way.

He didn't know why Muley, or any other man, would refer to her as a "stick of a woman." Of course, even in profile, it was obvious that she wasn't fleshed out and curvy. She was slender-hipped, narrow-waisted, and small-breasted. Just the same, he took several surreptitious glances at those buttons she had a habit of fiddling with and discovered, to his chagrin, that he wouldn't mind fiddling with them himself. He knew from experience that small-breasted women sometimes had the most sensitive nipples.

Mentally, he yanked himself away from his erotic thoughts. What the hell was the matter with him, thinking about Miss Janellen's nipples? She was a prim and proper lady. If she could read his mind, she'd probably call the law on him.

"Thank you, Miss Tackett, I think I can handle it from here," he said gruffly and hunched over the desk, blocking his view of her.

When he had completed all the forms, he pushed them across the desk and stood up. "There you go. When do you want me to start?"

"Tomorrow if you can."

"Tomorrow's fine. Who'll I report to?"

She gave him the name of his supervisor. "He's been with us a long time and knows how we like things done."

"Does he know I served time?"

"I thought it fair to tell him, but he's not the kind to hold it against you. You'll like him. He'll meet you here in the morning and drive you to all the wells you're responsible for. He'll probably run your route with you for several days.

You'll have use of a company truck, of course. I assume you have a driver's license?"

"Just got it renewed."

"How can we get in touch with you?"

"That could be a problem. I haven't got a permanent address yet. Hap's been letting me sleep in his back room, but I can't do that indefinitely."

She opened her desk drawer and withdrew a large business checkbook. "Find a place to live and have a telephone installed so that we can reach you at any time. We never know when an emergency will arise. If the phone company requires a deposit, have them call me." She wrote out the check, tore it from the book, and handed it to him.

Three hundred dollars, made out to him, just like that! He didn't know whether to be elated or affronted. "I don't take charity."

"Not charity, Mr. Cato. An advance. I'll take fifty dollars out of your first six paychecks. Will that be satisfactory?"

He wasn't accustomed to kindness and trust and didn't know how to respond. With Hap it was easy. Generally men didn't have to express themselves to other men. They seemed to understand one another's feelings without having to vocalize them. But with a woman it was different, especially when she was looking at you with crystal blue eyes the size of fifty-cent pieces.

"That's fine," he said, hoping he didn't sound

as awkward as he felt.

"Good." Coming to her feet, she smiled and extended her hand. Bowie stared at it for a moment and had an insane impulse to wipe his hand on his pants leg before touching hers. He gave it a swift shake and immediately released it. She quickly reclaimed it. There was a second or two of uncomfortable silence, then they both began to speak at once.

"Unless you —"

"Until —"

"You go ahead," she said.

"No. Ladies first."

"I was just going to say that unless you have any questions, we'll look forward to your reporting to work tomorrow."

"And I was going to say 'until tomorrow.' " He pulled on his hat and moved toward the door. "It'll feel good to be doing real work again. I sure appreciate the job. Thank you, Miss Tackett."

"You're welcome, Mr. Cato."

Halfway through the door, he halted and turned back. "Do you call all the men who work for you by their last names?"

The question seemed to catch her off guard. Rather than speak, she shook her head rapidly.

"Then call me Bowie, okay?"

She swallowed visibly. "Okay."

"And it's Boo-ie, like Jim Bowie and Bowie knife. Not Bowie like David, the rock star."

"Of course."

Feeling dumb for bringing it up — what the hell difference would it make to her how he pronounced his name? — he touched the brim of his hat and made tracks.

Chapter Ten

"Is the roast too dry, Key?"

Janellen's question roused him from his deep brooding. He sat up straighter, looked across the dinner table at her, and smiled. "Delicious as always. I'm just not very hungry tonight."

"That's what happens when you fill up on whiskey," Jody interjected.

"I had one drink before dinner. And so did you."

"But I'll stop with one. You'll go out and get drunk tonight, like you do every night."

"How do you know what I'll be doing tonight? Or any other night? Furthermore, what do you care?"

"Please," Janellen exclaimed, covering her ears. "Stop shouting at each other. Can't we have one meal together without an argument?"

Knowing his sister's anxiety was deeply felt, Key said, "I'm sorry, Janellen. You've served a great meal. I didn't mean to spoil it."

"I don't care about the meal. I care about the two of you. Mama, your face is as red as a beet. Did you take your medication today?"

"Yes I did, thank you kindly. I'm not a child, you know."

"Sometimes you act like one when it comes

to taking medicine," Janellen gently chastised. "And shouting across the dinner table is something you never allowed us kids to do."

Jody pushed aside her plate and lit a cigarette. "Your father didn't allow arguments at the dinner table. He said it spoiled his digestion."

Janellen brightened at the mention of their father. She had only foggy memories of him. "Do you remember that, Key?"

"He laid down the law about such things," he replied, smiling for his sister. "Sometimes you remind me of him, you know."

"You're kidding?" A blush of pleasure crept up her slender throat and over her face. She was pathetically easy to please. "Really?"

"Really. You've got his eyes. Doesn't she, Jody?"

"I suppose."

She wouldn't even agree with him on an obvious and insignificant point, but he refused to let it bother him. "All three of us kids inherited the Tackett blues. I used to hate it when people said to Clark and me, 'You boys have the prettiest eyes. Just like your daddy's.'"

"Why did you hate it?" Janellen asked.

"I don't know. Made me feel like a sissy, I guess. Being told that anything attached to him is 'pretty' isn't what a little boy wants to hear."

"Your father didn't mind hearing it," Jody said crisply. "He loved having people fawn over him. Especially women."

Ever guileless and naïve, Janellen said, "You

must have been very proud to have such a handsome husband, Mama."

Jody rolled the smoldering tip of her cigarette against the rim of the ashtray. "Your father could be very charming." Her face softened. "The day Clark the Third was born, he brought me six dozen yellow roses. I fussed at him for being so extravagant, but he said it wasn't every day that a man had a son."

"What about when Key was born?"

Jody's misty vision cleared. "I didn't get any flowers that day."

After a tense silence, Key said very quietly, "Maybe Daddy knew you wouldn't like them. That you'd only throw them out."

Janellen reacted quickly. "Mama explained why she threw out your flowers, Key. They made her sneeze. She must have been allergic to them."

"Yeah, she must have been."

He didn't believe it for a minute. Earlier in the week, vainly looking for a way to make peace with Jody, he'd brought her a bouquet. Janellen had arranged the flowers for him in a vase and placed it on the dresser in Jody's bedroom while she was out with Maydale.

The next morning, he'd found the flowers in the garbage can outside the back door. It wasn't so much that she'd thrown them out that had rankled him, but that she hadn't even acknowledged them until he presented her with the wilted evidence and asked for an explanation.

Calmly, coldly, she'd told him the bouquet had

given her hay fever. She hadn't said that they were pretty and that it was a pity she couldn't enjoy them. She hadn't thanked him for the gesture.

Not that he wanted or needed her thanks. He would survive without it. It just made him damn mad that she thought him stupid enough to accept her lame excuse for rebuffing a gift from him. Rather than give her the satisfaction of seeing him hurt and angry, he acted as nonchalant now as he had that morning he'd tossed the bouquet back into the trash can.

Jody broke another lengthy silence. "How's the new man doing?"

Janellen practically dropped her coffee cup. It clattered noisily against the saucer. "He . . . he's doing fine. I think he's going to work out well."

"I still haven't seen his references."

"I'm sorry. I keep forgetting to bring them home. But his supervisor reports that he's doing the job well. He's never late and is very conscientious. He gets along with the other men. Doesn't make trouble. I've had no complaints."

"I still can't figure why Muley up and quit without giving notice."

Janellen had told Key the circumstances of Muley's severance but had asked him not to tell Jody. Her reaction to a trusted employee turning thief was likely to be volatile and a threat to her high blood pressure. Key had agreed.

He also knew that Bowie Cato was an ex-con who'd barely had time to lose his prison pallor.

Even before Janellen introduced them, Key had seen him at The Palm. Hap had given him the scoop on Cato.

Key nursed no prejudice against former inmates. He'd spent a few days in an Italian jail himself a few years back. Cato was friendly but not ingratiating. He kept to himself, did his job, and avoided trouble. That could not be said of very many men who didn't have prison records.

Jody's viewpoint on social reform wasn't exactly liberal. She had a low tolerance for mistakes. She wouldn't welcome having an ex-con on the payroll, so the less she knew about Cato's background, the better for everybody. Muley was gone; Janellen had found a qualified replacement. That was the bare-bones story they'd given her. But apparently Jody smelled a rat. This wasn't the first time she'd broached the subject.

Key kept his expression impassive and hoped Janellen would do the same. But lying didn't come easily to her. Under her mother's incisive stare, she fidgeted with her silverware.

"Cato isn't from around here?"

"No, Mama. He grew up in West Texas."

"You don't know who his people are?"

"I think they're deceased."

"Is he married?"

"Single."

Jody continued staring at her daughter as she puffed on her cigarette. After what seemed an endless silence, Janellen glanced nervously at

Key. "Key's met him. He thought he was all right."

Damn! He didn't want to get caught in the cross fire. But he went to his sister's rescue. "He's a nice guy."

"So's Santy Claus. That doesn't mean he knows an oil well from his asshole."

Janellen flinched at her mother's crude phraseology. "Bowie knows a lot about oil, Mama. He's worked in the business since he was a boy."

As long as he'd already been drawn into it, Key furthered his sister's cause. "Cato is doing his job. Janellen likes him and so do the other men. What more could you want?" He knew, of course, what his mother wanted: Jody wanted to be young, healthy, and strong; she wanted to be at the controls of Tackett Oil and Gas and resented Janellen's hiring an employee without consulting her. If she'd hired a reincarnation of H. L. Hunt, Jody wouldn't have liked him.

"He's been on the payroll for . . . what, Janellen, two weeks?"

"That's right."

"And he hasn't caused a single mishap," he continued. "So it looks to me like Janellen made a sound business decision."

Jody turned to him, her contempt at full throttle. "Like your opinion counts for something where Tackett Oil is concerned."

"I wasn't speaking as an expert on the oil business," Key returned evenly. "Just as a guy who shook hands with another guy. Cato looked me straight in the eye, like he didn't have anything to hide. I met him at the end of the day. He was sweaty and his clothes were dirty, which indicated to me that he'd been working his ass off outdoors in the heat."

Jody sent a plume of cigarette smoke toward the ceiling. "Sounds as though you could learn a lesson or two about the work ethic from this Cato fellow. It wouldn't hurt you to sweat a little, get dirty, do some work around here."

"Key's been working, Mama. He fixed the latch on the gate."

"That's tinkering. I'm talking about sweat-of-the-brow, damned hard work."

"On your oil wells, you mean." Despite his best intentions to hold his temper, Key's voice was rising.

"It wouldn't kill you, would it?"

"No. It wouldn't kill me, but it isn't my gig. It's yours."

"Ah, that's why you never wanted to be part of the business. Because I was there first? You didn't want to play second banana to a woman."

Key, shaking his head, laughed ruefully. "No, Jody. I never wanted to be a part of the business because I'm not interested in it."

"Why not?"

Jody never accepted a simple answer at face value. He didn't remember a time when he hadn't been required to justify, explain, and account for his opinions, especially if they differed from hers. It was no wonder to him that his daddy

213

had turned to other women. With Jody, everything was a contest to see who could best whom. It wouldn't take long for a man to grow tired of that.

Forcing himself to remain calm, he said, "Maybe if we were still drilling for oil, if there was a challenge involved, I'd consider going into the business."

"You crave excitement, is that it?"

"Routine holds no appeal for me."

"Then you should have lived during the boom. It attracted your kind of people. East Texas was crawling with gamblers and con artists and crooks and whores. All living on a wing and a prayer. Taking high-stakes risks. Saying to hell with tomorrow, let the devil take it.

"That's the life for you, isn't it? You're not happy unless you're walking a tightrope with crocodiles on both sides ready to eat you if you fall. Just like your father, you thrive on adventure."

Key was clenching his teeth so tightly that his jaw ached. "Think whatever you want, Jody." Then, leaning forward, he stabbed the table with his index finger to emphasize each word. "But I never did and never will want to baby-sit a bunch of stinking oil wells."

"Key," Janellen groaned miserably.

She could barely be heard over Jody's chair scraping back. Her face was florid. "Those stinking oil wells allowed you to live high on the hog all your life! They provided food for your belly, clothes for your back, bought you new cars, and paid your way through college!"

Key rose, too. "For which I'm grateful. But am I supposed to become an oilman just to pay you back for upholding your responsibilities as a parent? If you and Daddy had been plumbers, would I be obligated to shovel shit the rest of my life? It was never expected of Clark to go into the oil business, so why me?"

"Clark had other plans for his life."

"How do you know? Did you ever ask him his ambitions? Or did he only follow *your* plans for his life?"

Jody drew herself up. "He had his career mapped out and would have followed it, had it not been for that whore of a doctor that you've been jockeying around the countryside."

"That was an emergency situation, Mama," Janellen interjected. "That little girl would have died if it hadn't been for Key."

Letty Leonard's accident had been a headline story in the local newspaper.

"Thank you, Janellen," Key said, "but I don't need you to defend what I did. I would have done it for a dog, let alone a little girl."

Jody was fixed on only one aspect of the drama. "I told you to stay away from Lara Porter."

"I didn't hightail it to the emergency room for her, for chrissake. I did it for the kid."

"Were you thinking of the kid when you bought the doctor's supper?"

Rather than appear surprised or guilty that she

215

also knew about his and Lara Mallory's barbecue dinner, he shrugged. "I hadn't eaten all day. I was hungry. She happened to be along when I stopped."

Jody's stare was hot with wrath. "I'm telling you one last time. Stay away from her. Do your drinking and whoring with somebody else."

"Thanks for reminding me. I'm getting a late start tonight." He strode to the sideboard, poured himself a shot of whiskey, and tossed it back defiantly.

Making a sound of disgust, Jody turned and left the dining room with a militant tread, climbing the stairs to the second story.

"Why can't you two get along?"

Key rounded on his sister, prepared to make a defensive comeback. Her remorseful expression stopped him.

"Jody starts it, not me."

"I know she's difficult."

He laughed sardonically at her understatement.

"Thank you for keeping my secret about Mr. Cato. Mama wouldn't want an ex-con on the payroll, even if he has turned out to be an exemplary employee."

Key cocked an eyebrow. "Exemplary employee? Isn't it too soon to tell?"

"Mr. Cato isn't the subject here," she said primly before switching subjects. "Did you really take her to dinner?"

"Who? Lara Mallory? Jesus, what's the big deal? I popped into Barbecue Bobby's for some ribs. She happened to be along because I was giving her a ride home from the airstrip. That's all there was to it. Is that a hanging offense?"

"She called me."

His anger evaporated. "She what?"

"She called me last week. Out of the blue. I answered the company phone, and she identified herself. She was very gracious. She invited me to lunch."

He laughed. "She invited you to lunch?" The notion was ludicrous.

"I was so taken aback, I didn't know what to say."

"What did you say?"

"I said no, of course."

"Why?"

"Key! This is the woman who ruined Clark's political future."

"She didn't rape him at gunpoint, Janellen," he said wryly. "I doubt if she tied him to the bedpost, either. Unless it was for recreational purposes."

"I don't see how you can joke about it," she said crossly. "Whose side are you on?"

"I'm on our side. You know that." He stared into near space for a moment, bouncing the empty shot glass in his hand. "It might have been interesting if you'd accepted her invitation, though. I'd like to know what she's up to."

"Do you think she's up to something?"

He thought about it for a moment. Admittedly his estimation of Lara Mallory had risen when he witnessed the determination with which she'd struggled to save the Leonard child's life. He'd seen military medics less committed to saving a patient.

However, despite the courage and skill she'd demonstrated in that crisis, she was still the key player in the scandal that had irreparably compromised Clark. She wouldn't have come to Eden Pass without strong motivation. She wanted something. She'd said as much when she told him she was holding an IOU she intended to collect.

You're going to fly me to Montesangre.

He hadn't believed for one second that she was serious. She'd made clear her low opinion of that country. Wild horses couldn't drag her back there.

So why had she said that? To get a rise out of him? To throw him off track and keep him guessing about her true motives?

"She wouldn't have called you unless she wanted something from you," he told Janellen irritably.

"Like what?"

"Who the hell knows? Possibly something as Mickey Mouse as a keepsake of Clark's childhood. Or something as abstract as public approval. You're a well-respected member of the community. Maybe she thinks that being seen with you would give her the acceptance she needs to make a go of her practice. Next time she calls —"

"If she does."

"I think she will. She's a gutsy broad. When she calls, reconsider. Lunch with her might be interesting."

"Mama would have a fit."

"She doesn't have to know."

"She'd find out."

"So what? You're a grown-up. You're allowed to make your own decisions even if they don't set well with Jody."

She placed her hand on his arm and spoke earnestly. "Please, Key, for both your sakes, make peace with her."

"I'm trying, Janellen. She doesn't want to make peace with me."

"That's not true. She just doesn't know how to give in graciously. She's old and crotchety. She's lonely. She doesn't feel well, and I think she's afraid of her mortality."

He agreed on all points, but that didn't solve the problem. "What do you want me to do that I haven't already tried? I've bent over backward to be polite and pleasant. I even brought her flowers. You see how much good that did," he said bitterly. "I'll be damned if I'm going to bend at the waist and kiss her pinky every time I see her."

"I'm not asking you to pamper her. She'd see straight through any insincerity and only resent you for it. But you could be less prickly. When she began talking about work, you could have told her about some of your recent jobs."

"I shouldn't have to display my achievements

like merit badges. I'm not out to impress her. Besides, she's not interested in what I do. She thinks flying is a hobby. If I was the pilot of *Air Force One*, it wouldn't be good enough for her."

He returned the shot glass to the tray, his motions slow and heavy with discouragement. "Jody doesn't want me here. The sooner I leave, the better she'll like it."

"Please don't feel like that. And don't go away with this thing festering between you. She's still devastated over Clark's death, and because she can't tolerate that weakness in herself, she overcompensates by lashing out at you."

"I've always been a convenient whipping boy. She hasn't liked me since the day I was born and Daddy failed to send her six dozen yellow roses."

"He hurt her, Key. She loved him, and he hurt her."

"Loved him?" he repeated with a bitter laugh.

Janellen looked serious and a bit puzzled. "She loved him very much. Didn't you realize that?"

Before he was able to refute her, the doorbell rang.

"It's going to get better between you. You'll see." She pressed his arm before releasing it. "I'll get the door."

Rejecting his sister's optimism, he decided to have another whiskey. He swallowed the shot whole. It stung his throat, seared his esophagus, and in all probability would upset his stomach.

He didn't enjoy drinking as much as he once had.

He didn't enjoy most things as he once had. When had taking a woman to bed become more trouble than it was worth? He was soured on life in general and didn't know why.

He had blamed his recent disenchantment on his sprained ankle and the bullet wound in his side. But his ankle only bothered him occasionally now, and his wound had healed, leaving only a little tenderness and a pink scar to remind him of it.

So what was wrong with him?

Boredom.

He had too much idle time in which to think. His thoughts invariably turned to Clark's accidental drowning and all the loose ends of the theories dangling like the ragged hem of a shroud. Key wanted the facts, yet was cautious not to root them out, afraid he'd learn something he didn't want to know. Every rock he'd overturned lately had ugly worms beneath it. He decided that some things were best left undisturbed.

Thank God he was actively flying again. He hadn't flown Letty Leonard to Tyler for the publicity it would generate, but since then his phone hadn't stopped ringing. He'd already flown some good contracts and had scheduled even more. He didn't particularly need the money, although it was always welcome. What he desperately needed was the activity and the sense of freedom that only flying afforded him.

For his peace of mind, he was in the wrong state, the wrong town, and the wrong house. He wanted to find a place that was completely different from anything he'd experienced, where the language was foreign and the food was strange. Some exotic place where the people had never even heard of the Tacketts.

He'd traveled all over the world searching for a place where nobody knew that he was Clark Tackett's brother. It was an ongoing quest. Eventually strangers would put two and two together. "Tackett? Any kin to the former senator from Texas? His kid brother? Well, I'll be damned."

Clark had been the measuring stick by which Key had been judged all his life.

"Key is almost as tall as Clark now."

"Key can run almost as fast as Clark."

"Key isn't as well behaved as Clark."

"Key didn't make the honor roll, but Clark always does."

He'd eventually exceeded his brother in height. During adolescence, he'd surpassed him as an athlete. But unfavorable comparisons had followed him into adulthood. Incomprehensible as it seemed, he'd never been jealous of Clark. He'd never wanted to be like his brother, but everyone else thought Clark was the example to which he should aspire. Jody thought so more than anyone.

As a kid, it had hurt him that she so obviously favored Clark. She'd bandaged his skinned knees but never kissed them. Rather, she'd rebuked his recklessness. His small gifts, the pictures he'd colored at school, were glanced at and set aside, never cherished, never taped to her vanity mirror.

When he was a teenager, he resented her coldness toward him. Blatant disobedience and rebellion had been his way of dealing with her favoritism for Clark. She only approved of him when he was throwing touchdown passes for the Eden Pass Devils, but that was self aggrandizement and had little to do with him personally.

Off the gridiron, he went out of his way to show her just how little he cared one way or another, although deep down he cared a great deal and couldn't understand why he was so unlovable.

But with maturity came the acceptance that his mother simply didn't love him. She didn't even like him. Never had. Never would. He'd given up trying to analyze why, and, frankly, he didn't much care anymore. That's just the way it was. Clark had been caught in a bedroom scandal involving a married woman, but Key was the one accused of "whoring."

Several years ago, having finally reached the conclusion that winning his mother's tolerance, if not her love, was a lost cause, he'd decided that it would be to everyone's advantage if he made himself scarce, a decision that also satisfied his innate wanderlust.

Now, even that was being stymied.

He was restless and bored, and the questions surrounding his brother's death were tethering him to their home. He needed to go looking for anonymity again, but whenever he was tempted to pack up and truck it, a vision of his sister's imploring face saddled him with guilt.

Her concerns were valid and justified. Aging and the loss of control that accompanied it were frightening to a woman as strong-willed as Jody. In good conscience, Key couldn't leave Janellen to handle her alone. He'd come to agree with Janellen's fear that Jody's forgetfulness and confusion were harbingers of something much more serious than senility. If a medical crisis did occur, he'd never forgive himself if he were thousands of miles away and unreachable. No matter that he wasn't her ideal of a son, Jody was still his mother. For the time being, he belonged in Eden Pass.

"Key?"

Lost in thought, he turned at the sound of his sister's hesitant voice.

"There's someone at the door to see you." She was looking at him in a peculiar, quizzical manner.

"Who is it? What does he want?"

"It's a woman."

Chapter Eleven

Lara arched her back, stretching the stiff muscles and holding the position for several moments. Gradually she relaxed and rubbed her eyes before repositioning her reading glasses on the bridge of her nose.

After eating an early dinner while watching the evening news, she had forgone watching prime time TV because it offered nothing enticing. Any enjoyment derived from reading fiction had been sadly reduced since that morning in Virginia. No novelist could conjure up a plot with as many twists, pitfalls, and calamities as those in her life the last five years. It was difficult to sympathize with a protagonist whose dilemma was mild when compared to her own.

With nothing to do for entertainment, she had decided to read through her patient files. The intricacies of medicine never failed to engross her.

While other students in her class had complained all through medical school, for Lara it had been like a vacation. She relished the required hours of study. Having unlimited access to textbooks and perplexing case histories was a luxury. She gorged on them like a gourmand with an endless supply of delicacies.

Unlike her parents, none of her instructors or classmates berated her for her unquenchable thirst for knowledge, or repeatedly told her that the study of medicine was unsuitable for a well-bred young woman and that there were much more acceptable avenues of interest to pursue.

She'd graduated third in her class at Johns Hopkins, excelled as an intern, and had been offered her pick of hospitals in which to serve her residency. Naturally, she'd enjoyed the grudging admiration of her colleagues, but the real reward lay in healing. A grateful patient's simple "thank you" surpassed the accolades of her associates.

Heartbreakingly, those rewards came few and far between now. That's why Lara enjoyed perusing her files, charting a patient's progress from diagnosis to cure.

She was roused by an approaching car. Expecting it to drive past, she watched with puzzled interest when it entered her driveway and wound around to the rear of the clinic. She laid aside her reading material and quickly left her office. As she made her way through the clinic, she experienced a twinge of déjà vu. This was disturbingly similar to the night Key Tackett had appeared on her threshold, his side bleeding from a gunshot wound.

It was so similar that she barely registered surprise when she opened the door to find him standing on her back steps. Only this time he wasn't alone.

Lara gave the girl a curious glance, then looked

226

at him. "I keep regular office hours, Mr. Tackett. That's something you seem to forget. Or ignore. Or is this a social call?"

"Can we come in?"

He wasn't in a mood to spar with her. A frown was pulling his eyebrows together, and his lips were compressed into a stern, narrow line. If he had come alone, Lara probably would have slammed the door in his face. She was on the verge of doing so anyway when she gave the girl a closer look and saw that she'd been crying. Her eyes and nose were moist and red, and her face was mottled. She was clutching a damp tissue so tightly her knuckles had turned white.

Beyond these visible signs of distress, she appeared to be a perfectly healthy girl in her late teens. She was stoutly built, with a deep bosom and full hips. Her face was pretty, or would have been if she'd been smiling. Her shoulder-length hair was straight and dark. Because of the bleak expression in her brown eyes, coupled with her obvious misery, Lara couldn't shut her out.

She stepped aside and motioned them in. "What can I do for you?"

The girl remained silent. Key said, "This is Helen Berry, Dr. Mallory. She needs a doctor."

"You're ill?" Lara asked the girl.

Helen glanced furtively at Key before saying, "Not exactly."

"I can't help you unless you tell me what the problem is. If it's a general checkup you need, you can be the first patient I see tomorrow morning."

"No!" the girl protested. "I mean . . . I don't want anybody to know . . . I can't . . ."

"Helen needs you to examine her."

Lara turned to Key, who'd spoken for the girl. "Examine her for what? If she's not ill —"

"She needs a gynecological examination."

Lara gave him a wide, inquiring stare that demanded further explanation. He remained silent, his expression immutable. The girl was anxiously gnawing her lower lip.

"Helen," Lara asked gently, "were you raped?"

"No." She gave her head a hard shake. "Nothing like that."

Lara believed her and was greatly relieved.

"I'll wait out here." Key executed an abrupt about-face and stalked down the hallway to the dark waiting room.

His exit created a soundless vacuum. It was several seconds before Lara let out her held breath. She gave Helen Berry a reassuring smile and said, "This way, please." The girl followed her into an examination room, where Lara pointed her onto the table.

"Don't you want me to undress first?"

"No," she replied. "I'm not going to do a pelvic examination until I have more information. Besides, my nurse isn't here to assist me. I never conduct an examination like that without an assistant."

That was for her protection as well as the patient's. In a sue-crazy society, doctors were paranoid about malpractice suits. Because of the scandal that haunted her, she was more vulnerable than most.

Her patient's eyes filled with fresh tears. "But you gotta examine me. I gotta know. I gotta know right now so I can decide what to do."

Obviously distraught, she was shredding the soggy tissue. Lara clasped her hands to keep them still. "Helen." She spoke gently but with authority. Her primary objective was to calm the patient. "Before we can proceed, I must get some information from you."

She reached for a chart and a pen and asked Helen for her full name. The paperwork could have been postponed, but doing it now forced the girl to compose herself. Working her way down the standard form, Lara learned that Helen was a local girl who lived in a rural area. She was eighteen years old and had graduated from high school the previous May. Her father worked for the telephone company. Her mother was a homemaker. She had two younger sisters, one brother. There was no history of serious illness in the family.

"Now," Lara said, setting the chart aside, "why did Mr. Tackett bring you to see me?"

"I asked him to. I had to." Her face crumpled and her lower lip fell victim to more brutalizing. Tears streamed down her plump cheeks.

Lara, believing she knew the cause of Helen's

distress, cut to the heart of the problem. "Do you think you're pregnant?"

"Oh, jeez. I'm so stupid!" With that, Helen flung herself onto the examination table, drew her knees to her chest, and began sobbing uncontrollably.

Lara moved swiftly to her side and took her hand again. "Helen, calm down. We don't know anything for certain yet. You might be crying over nothing. A false alarm."

She kept her voice calm and soothing, but she wanted to grind her teeth. She wished she had a double-barreled shotgun, loaded and aimed at Key Tackett's testicles. Bedding wayward housewives like Darcy Winston was one thing; seducing high school girls was quite another.

Lara smoothed back strands of Helen's hair. "When was your last period?"

"Six weeks ago."

"So you've only missed one? That doesn't necessarily mean you're pregnant."

Helen bobbed her head emphatically. "Yes it does. I'm never late."

Perhaps, Lara thought, but there were myriad reasons for delayed menses, only one of which was pregnancy. Still, she had learned that patients were often the best authorities on their own bodies. She couldn't blithely dismiss Helen's conclusion. "Have you had sexual intercourse?"

"Yes."

"Without using any contraception?"

Helen's head wobbled up and down in answer.

230

Lara was dismayed that high school students were still negligent in their use of condoms, which were the simplest and least expensive, yet reliable, protection against unwanted pregnancy and sexually transmitted diseases. In a community like Eden Pass, open discussion about these safeguards was certain to generate opposition from conservative parents and religious groups. Nevertheless, it was vital — indeed a life or death matter — to acquaint teenagers with the risks they were taking if they were sexually active and didn't take precautions.

"Any breast tenderness?"

"Some. No more than usual. But anyway, I did one of those home pregnancy tests."

"It was positive?"

"No question."

"They're fairly reliable, but there's always a margin for error in any test." Lara gave her a hand up. "Go into the bathroom and get a urine specimen. I can do a preliminary test tonight."

"Okay. But I know I'm pregnant."

"Have you ever been pregnant before?"

"No. But I know. If I am, he'll kill me."

She retreated into the adjoining toilet. Thinking of Key Tackett sitting complacently in her waiting room made Lara want to confront him immediately and convey her disgust. But her patient came first.

"I left it on the lid of the tank," Helen said when she emerged.

"Fine. Lie down on the table and try to relax."

In a few minutes, Helen's worst fear was once again confirmed. "I knew it," she wailed when Lara told her that the indications were positive. She began to cry again. Lara placed her arms around her and held her until the sobs became dry, racking hiccups.

"Until your pregnancy is confirmed beyond any doubt, I'd rather not give you a sedative. Would you like something to drink?"

"A Coke? Please."

Lara left her alone only long enough to fetch the soft drink. When she returned, Helen was weeping quietly but was more composed. She took several greedy sips of the cola.

"Helen, is marrying the child's father out of the question?"

"Yes," she mumbled. "A baby is the last thing he wants or needs."

Angry heat spread throughout Lara's body. "I see. What about your parents? How supportive will they be?"

"They love me," she said as more tears filled her eyes. "They won't kick me out. But Daddy's a deacon in our church. Mom's . . . Oh, God, they'll just die of shame."

"Do you intend to have the baby?"

"I don't know."

"You could always make it available for adoption."

She shook her head morosely. "I don't think he'd let me. Besides, if I had it, I could never give it away."

"Have you considered abortion?"

"That's probably what I'll have to do." She sobbed and blotted her nose. "Except . . . except I love him, you know? I don't want to kill his baby."

"You don't have to make that decision tonight," Lara said softly as she stroked the girl's hand.

"If that's what I decided, would you do it so nobody would know?"

"I'm sorry, Helen, no. I don't perform D and Cs to terminate pregnancy."

"How come?"

Having watched her own child die, aborting living tissue was something Lara simply couldn't do unless the mother's life was at risk. "That's just my policy," she told the girl. "However, if you are pregnant and that's the alternative you choose, I'll make the arrangements for you."

Helen nodded, but Lara doubted that she was retaining much of this conversation. Dismay had numbed her. Lara patted her hand and told her she would come for her in a few minutes. "Lie quietly and finish your drink."

Stepping into the hallway, she bolstered herself for the coming encounter. As she entered the waiting room, she flipped on the light switch and flooded the area with a cold, unforgiving, fluorescent glare. Key was slouched on one of the short sofas. Blinking to adjust his eyes to the sudden brightness, he slowly came to his feet.

"Why did you bring her to me?" Lara demanded angrily.

"I figured you needed the business."

"I appreciate your thoughtfulness," she said caustically, "but I would rather not have been drawn into another of your intrigues."

He folded his arms across his chest. "From your tone of voice, I gather Helen was right. She's pregnant?"

"It appears so."

His head dropped forward, and he swore elaborately.

"I take it you don't welcome this news."

His head snapped up. "Damn right, Doc. It sucks."

"You should have thought of that before sleeping with an unsophisticated girl like Helen. And why didn't you take precautions? Surely a man of the world like you keeps a handy cache of condoms. Or does using one hamper your macho image?"

"Now just a frigging minute. You —"

"Clark told me all about your satyric reputation. I thought he was exaggerating, but apparently he wasn't. 'Key Tackett's women.' Around here, it's like a club, isn't it? The only requirement for membership is to have gone to bed with you." She looked at him contemptuously.

"Maybe they should change the name to Key Tackett's *girls*," she said with a sneer. "What's the matter with you? Are you losing your boyish charm? Has aging bruised your ego? Are you so insecure over your fading youth that you've resorted to bedding high school girls?"

"What difference does it make to you?" With his eyes half-closed, he added softly, "Jealous?"

Lara drew herself up, angry for having stooped to his level. By doing so, she'd left herself open to counterattack. In a cool, professional voice she said, "Helen is seriously considering abortion. Until she reaches a firm decision, I'll be happy to give her prenatal care, provided she comes here alone, without you."

"She won't be coming here at all. All we wanted from you tonight was a yea or nay." Angrily, he reached into the hip pocket of his tight, worn jeans and fingered out his money clip. "How much do I owe you?"

"This one's on me, but I want something in exchange."

"Like what? No, let me guess. Let's see . . . a free flight to Timbuktu?"

She had wondered if he would make reference to their last conversation and wasn't surprised that his remark was sarcastic. She didn't take the bait. "What I want is your promise —"

"I don't make promises to women. While Clark was filling your ear about my sex life, did he fail to mention that?"

She strove to keep her voice even. "I don't want you dumping any more of your garbage at my back door. This is the second time I've had to clean up one of your messes. Leave me out of them, please. I want no part of your juvenile, romantic escapades."

"Is that right?"

"That's right."

Menacingly, he came nearer, until he was standing so close that their clothing was touching. She could feel his body heat, feel his breath on her uplifted face. His rage, too, was palpable. Only sheer determination kept her from backing down.

"That's funny, Doc," he whispered huskily. "I'd've thought this kind of romantic escapade was right up your alley."

She held her ground and his blue stare for as long as she could stand them, then backed up a few steps and turned away. "I've tried to make Helen reasonably calm, but she's still upset," she said over her shoulder. "If you have a smidgen of decency you'll be gentle with her tonight. No blame. No recriminations. Until she decides how to resolve this crisis, she's going to need patience and understanding."

"Well that's just fine, because I fairly ooze the milk of human kindness."

Lara shot him a fulminating look, then gave him her back and walked down the hall. She tapped on the examination room door before going inside. Helen was lying on her back on the padded table, staring at the acoustical tiles in the ceiling. Lara was relieved to see that she was no longer crying.

She plastered on a smile she hoped didn't look

236

too false. "How are you feeling?"

"Okay, I guess."

"Good. Key's waiting for you."

She assisted Helen off the table and they moved into the hallway. He was waiting at the back door, as though ready to make a quick getaway. To say he had the morals of an alley cat would be doing alley cats a disservice. It was a pity that his character didn't match his good looks.

The open collar of his shirt provided only a glimpse of what Lara knew was a broad chest. His jeans fit his sex, narrow hips, and long thighs like a second skin. Clark had rarely worn casual clothes, never Levi's. She'd never seen either him or Randall in cowboy boots. Key's were well-worn victims of the elements.

Key Tackett's women, she thought scornfully.

Being so physically attractive, his success with women wasn't surprising. Within weeks, he had slept with Darcy Winston and this eighteen-year-old. How many others were there? His affair with Darcy wasn't as shocking as his dalliance with this girl so much younger and more innocent than he. For some vague and disturbing reason, she was disappointed in him.

To his credit, he opened his arms to Helen, who rushed into his embrace. He held her tightly against him for several moments, his head bent low over hers, whispering so softly into her ear that Lara couldn't distinguish the words. Between sobs, Helen nodded her head against his chest.

237

Then, setting her away, he said, "Wait for me in the car, sweetheart. I'll be right out."

On her way through the door, she gave Lara a hasty thank-you. Key said nothing until Helen was out of earshot. "I'll see that she gets proper prenatal care, but it won't be from you."

Lara lamented losing a patient, but reasoned that was the price she would pay for giving him a lecture on philandering. In lieu of saying anything she might later regret, she gave him a curt nod. At this point she was willing to leave well enough alone.

Not Key. He got in another parting shot. "On my way over here, I heard something on the radio that might interest you. Late this afternoon, Letty Leonard died."

Key wasn't the only one who had heard of the child's death. Jody had.

Eden Pass was situated midway between the Dallas/Fort Worth metropolitan area and Shreveport, Louisiana. Its location provided it with a large selection of television stations. All three networks had affiliates in those cities, which were carried by the local cable company, along with CNN and other major cable stations.

When it came to regional news, however, Jody relied on the station that broadcast from Tyler. She personally knew the owners and was familiar with the on-air talent. Watching their newscasts was like having a member of her family visit every night to deliver the news.

She was inordinately tired this evening. Her angry exchange with Key had sapped her energy. That, coupled with their conversation about Clark Junior, had taxed her mentally, emotionally, and physically. Even though he'd been dead more than two decades, thinking about her late husband always left her feeling resentful and depressed.

Immediately following her huffy departure from the dining room, she'd retired to her room to watch television and had barely managed to remain awake for the ten o'clock news. In fact she was in bed, propped against the pillows, dozing, when the story about Letty Leonard awakened her.

Instantly alert, she used the remote control to increase the volume on the set. It wasn't a lengthy story. The only visual was a snapshot of the child and a floppy-eared dog sitting in front of a Christmas tree surrounded by heaps of unwrapped presents.

The anchorman solemnly reminded his viewing audience of the tragic accident that had recently occurred in Eden Pass and of the highly specialized surgery that had temporarily saved Letty's life. Her sudden death had been caused by an embolism that had dislodged and moved to her lung. It had come as a shock to the attending physicians, as well as to her family, who had believed she was on her way to a full recovery. The story consumed no more than twenty seconds of air time.

Jody muted the sound, threw off the covers, and got out of bed. Then she lit a cigarette, and as she drew the smoke deeply into her lungs and exhaled slowly she began to pace.

The news story hadn't mentioned Dr. Lara Mallory or Key. As far as the general public was concerned, their joint involvement was inconsequential. But it was like a pebble in Jody's shoe, an aggravation she was unable to live with.

Dammit, she'd told Key to keep his distance from that woman. Not only had he disobeyed, he'd helped the doctor rescue a dying child. Jody couldn't sit by and let Lara Porter become a local heroine.

But would she be considered a heroine now that the child had died? Exactly what was an embolism? What might have caused it? What could have prevented it? She didn't know, but she would damn sure find out if Lara Mallory Porter was in any way responsible for the girl's death.

She was still mulling over her strategy when Janellen came in to say good night. She didn't return Janellen's embrace. She'd never been comfortable with outward displays of affection, even token ones, and considered sentiment a waste of time.

It was foolish to cling to memories like the six dozen yellow roses Clark Junior had brought her the day Clark the Third was born. Her memory of them should have withered and died just as the petals had. Why didn't she forget them?

What good had they done her?

"Good night, Mama. Try to get some rest. Don't get up again and don't smoke any more tonight. It's not good for you."

As soon as she was alone, Jody lit another cigarette. Having one in her hand enabled her to think better. She often lay awake for hours, smoking in the darkness. What Janellen didn't know, she couldn't hound her about.

Janellen. What was going on with her daughter? she wondered. She seemed to be distracted these days, often staring into space for long stretches of time, a goosey expression on her face. At other times, she became upset over the least little thing. Small hazards that wouldn't have ruffled her before now sent her into conniptions. She wasn't acting like herself at all. It was probably something hormonal.

But Jody couldn't waste worries on her daughter when fretting over Key was her full-time preoccupation. He was impossible and had been since birth, even before birth if you counted the twenty-six hours of difficult labor he'd put her through. Twenty-six long, agonizing hours that she'd endured alone because Clark Junior couldn't be located.

Key was born the moment his father, reeking of another woman's perfume, arrived at the hospital. That's when her difficulties with Key had begun. She was mad at him before he had drawn his first breath, and even as a newborn he had sensed it. Their dislike for each other had intensified during his childhood when it seemed that he was incapable of staying out of mischief.

She had wanted him to be a replica of Clark the Third, but two boys couldn't have been more dissimilar. Everything Clark did was motivated by an anxious desire to please her. Her approval was essential to his peace of mind. He was disconsolate if he thought he'd fallen out of favor.

Just as fervently as his brother tried to please, Key tried to provoke. Whatever Jody wanted or expected of him, he was bound and determined to do the opposite. He delighted in her disfavor; he nurtured it. She'd wondered many times if he had driven his car into that tree out of spite, just so he couldn't fulfill her dream of having him play professional football. He was hardheaded enough to risk his life rather than bow to her wishes.

She was secretly proud of his success, but acknowledging it would be tantamount to conceding that he'd made a better life for himself than she could have made for him.

One of the reasons he loved his work so much was because it kept him away from home. Although they'd denied it, she knew Janellen had called him home to watch her die. She resented that. If he didn't give a damn, he didn't give a damn. Never had, never would. It was that simple. Why pretend their relationship was something it wasn't? He and Janellen thought her death was imminent. She could see it in their eyes. They had another think coming!

She chuckled in the darkness, coughing on cigarette smoke. Wouldn't her immortality come as a nasty shock to them? She'd made a career of taking people by surprise. It didn't pay to be caught napping around Jody Tackett. They could ask Fergus Winston if they didn't believe it.

Again Jody laughed, and again she coughed, harder, reminding herself that where her mortality was concerned, she might not have a choice.

Frowning, she viciously cursed fate. She wasn't ready to die. She had things left to do, the main one being to drum that Porter bitch out of Eden Pass. Clark must have been out of his head or under the influence of some mind-altering substance to have purchased Doc Patton's clinic and then deed it to her. What had he been thinking, for chrissake?

More than Janellen and Key guessed, as long as Lara Mallory Porter remained in Eden Pass, she posed a serious threat to them and to all they held sacred.

Jody hadn't yet figured out the doctor's reason for moving here. However, she knew with the same certainty that the sun rose in the east that it was for more than to accept Clark's legacy. Unless she wanted something more, she would have turned that clinic for a quick profit and never set foot in Tackett territory. She was here for a reason. Jody dreaded learning what it was, but must before either she or one of her children walked into a trap laid by Lara Porter.

She, Jody Tackett, had come from poverty and married the richest man around. She hadn't remained at the helm of an independent oil company for years, hadn't become a woman to be feared and revered, by sitting on her ass trying to figure out other people's motives. She acted first, before they were given a chance. A rattler struck before he was stepped on.

Jody remained awake for a long time, smoking and plotting. By the time she'd smoked her last cigarette down to the filter, she had formulated her next move.

Darcy lowered her car windows. The wind punished her hairdo, but it would blow away the odor of tobacco smoke that she'd absorbed in the bar. That might make Fergus suspicious. Smoking wasn't allowed in the nursing home where her mother resided. Visits to the expensive facility provided her excellent excuses to go out at night. She'd been going out more frequently than usual because her ego needed boosting. Thanks to Mr. Key Tackett, her self-esteem was shaky.

Knowing that she'd been dumped gnawed at Darcy, eating away at her self-confidence like a vicious rat. That's why she wasn't having any fun lately. She couldn't concentrate on any other man and wouldn't until she'd repaid Key for slinging this shit on her.

She hadn't even had the satisfaction of showing him how little she cared. Oddly, he hadn't been hanging out at the popular watering holes. The word around town was that he was flying a lot, chartering flights for clients from Dallas to Little Rock and as far south as Corpus Christi. But he couldn't be flying all night every night. Where was he going in between jobs? How was he spending his free time?

With another woman? She hadn't heard any scuttlebutt, and surely she would have. His name hadn't been linked to any local woman except for . . .

Darcy reacted as though she'd been slapped. "But that's impossible," she protested out loud.

Key Tackett and Dr. Mallory? Their names had been linked when they'd flown that kid to Tyler, but that sure as hell hadn't been a lark.

On the other hand, the doctor was a renowned man-eater. She'd been carrying on with her lover right under her husband's nose. Even Darcy had more morals — and better sense — than to do that.

Some men, however, liked a woman with the spirit of adventure. It added spiciness and suspense. James Bond didn't fuck shrinking violets, did he?

She gripped the steering wheel tighter. If Key was having a secret affair with his brother's mistress, Darcy would make certain that everybody in East Texas heard about it. By the time she got through spreading tales, he'd be a laughingstock. Taking Clark's leftovers? Ha! That would serve the bastard right.

But the rumors should contain at least a grain of truth or the laugh would be on her. How could she make certain that he was sleeping with Lara Mallory? She'd never even met the doctor. Lara Mallory would see right through any friendly overtures. She was no fool.

How could she get close to Lara Mallory without putting her on guard? It warranted some thought, but she was confident that she'd think of a way.

Arriving home, she let herself into the house, tiptoeing and moving around in the dark to keep from waking Fergus and Heather, who were asleep upstairs. She didn't want to account for the lateness of the hour unless absolutely necessary. She hated lying to her husband and avoided doing so whenever possible.

Moving past the door to the family room, she noticed that the television set had been left on. She went in to turn it off. As she rounded the leather sofa, two startled people leaped up. There were exclamations of surprise as they grappled for loose articles of clothing.

Darcy switched on the lamp, took in the situation at a glance, and angrily demanded to know — although she already did — *"Just what the hell is going on here?"*

Chapter Twelve

The pastor of the First Baptist Church commended Letty's soul to the Lord and said a final amen over the small white casket. Marion Leonard's keening cry echoed across the windswept cemetery, raising goose bumps on all who heard it. Jack Leonard was silent, but tears rolled down his gaunt, pale cheeks as he pulled his grieving wife away from their daughter's coffin. It was a heartrending scene that deserved privacy. Mourners began to disperse.

Lara had kept to the fringes of the crowd, trying to be as unobtrusive as possible. As she turned to leave, the white-hot flash of a high-tech camera exploded near her face. Instinctively she threw up her arm for protection. The first blinding flash was followed by another, then a third.

"Mrs. Porter, will you comment on the Leonards' malpractice suit against you?"

"What?" A microphone was thrust against her mouth. She shoved it aside. "I don't know what you're talking about. And my name is Dr. Mallory."

As the violet spots receded, she saw a horde of reporters blocking her path. She switched directions. The band flocked after her. Some were obviously affiliated with TV stations — their video cameramen trotted along beside them, connected by cables. Others were from newspapers; with them were the still photographers and their despised flashes. Five years ago, she'd become well acquainted with the accoutrements of mass communication.

What was the media doing here? What did they want with her? She felt as if her nightmare was being reenacted.

"Please, let me by."

Glancing back, she saw that others attending Letty Leonard's funeral had gathered in clusters and were speaking in hushed but excited voices, gaping at the sideshow. She hadn't created the spectacle but was nevertheless its unwilling star.

"Mrs. Porter —"

"My name is *Mallory*," she insisted. "Dr. Mallory."

"But you were married to the late U.S. Ambassador Randall Porter?"

She hurried across the neatly clipped grass toward the gravel lane where her car was parked in a line of others behind the white hearse and the limousine.

"You're the same Lara Porter who was Senator Tackett's mistress, isn't that right?"

"Please move aside." Reaching her car at last, she fumbled in her handbag for her keys. "Leave me alone."

"What brought you to Eden Pass, Mrs. Porter?"

"It is true that Senator Tackett brought you

248

here before his death?"

"Were you still lovers?"

"What do you know about his accidental drowning, Mrs. Porter? Was it actually a suicide?"

"Did your negligence cause the Leonard girl's death?"

She had been asked the other questions a thousand times before and had become inured to them. They bounced off the armor of repetition. But the last question brought her around. "What?" Looking directly at the young female reporter who had posed the question, she repeated, "What did you say?"

"Did your negligence cause the embolism that killed Letty Leonard?"

"No!"

"You were the first doctor to attend her."

"That's correct. And I did everything possible to save her arm and her life."

"Apparently the Leonards don't think so or they wouldn't be suing you for medical malpractice."

Had Lara not had experience in masking her reaction to personal and probing questions and verbal salvos, she might have reeled under the impact of this one. Instead she gazed back at the reporter without revealing her inner turmoil. The muscles in her face felt wooden, but she managed to move her lips sufficiently to get out the words.

"I took drastic measures to save Letty Leonard's life. Her parents are well aware of that. I haven't been notified of a pending malpractice suit. That's all I have to say."

Naturally the news hounds didn't accept that as her final word. As she drove away, they were still aiming lenses and microphones at her, hurling questions like stones. She gripped the steering wheel with sweating hands, keeping her eyes forward, ignoring the curious onlookers as she drove past them.

It was a warm, humid morning, but she hadn't been uncomfortable with the heat until the reporters had resurrected the ugly past. Now her clothes were sticking to her damp skin, her head was pounding, and her heart was beating at an alarming rate. She felt nauseated.

What had initiated all this media attention? Her move to Eden Pass had gone unnoticed; she'd lived in relative anonymity for more than a year. There had been newer scandals to exploit, more sensational stories to expose, sinners more sinful than she caught sinning. The story of Lara Porter and Senator Tackett had been buried in the graveyard of dead stories ages ago.

Until this morning. Letty Leonard's death had exhumed her. Once again she was a notorious public figure.

Yet, the story of Letty's accident, tragic as it was, hadn't warranted statewide or national media coverage; only the local press had reported it. Naturally, her name would have been in Letty's medical file, but unless a reporter was very astute, he wouldn't have connected Dr. Lara Mallory of Eden Pass with Lara Porter, Senator Clark Tackett's mistress.

In subsequent stories about Letty's surgery and recovery, she hadn't been mentioned at all, for which she'd been glad. The less publicity she generated, the better. She wouldn't have cared if her name never again appeared in newsprint. But it was going to appear now, with the stigmatizing word *malpractice* shadowing it.

Through the entire incident with Clark, through the disaster in Montesangre, her proficiency as a physician had never come under fire. Her reputation as an accomplished doctor had withstood the bombardments to her character. She had clung to that last vestige of pride.

Now, if the Leonards even suggested they might pursue a medical malpractice suit, her work would be placed under a microscope. It would be laid bare and dissected just as her private life had been. Nothing incriminating would be found, but that didn't matter. The examination itself would create headlines. In the public's mind, being suspect was equivalent to being guilty.

Once again she would become fodder for the news mill. Her floundering practice — the only important thing left her — would suffer until it was extinct.

Someone must have tipped the media that the Dr. Mallory who had first attended Letty Leonard was none other than the infamous slut Lara Porter.

As she had feared, parked outside her clinic were cars and vans designated with call letters. When she pulled her car into the rear driveway, reporters swarmed her. She shoved her way through them and entered the clinic via the back door, which Nancy was holding open for her.

"What in hell is going on?" the nurse demanded as she slammed the door behind Lara.

"The rumor is out that the Leonards are suing me for malpractice."

"Have they lost their minds?"

"I'm sure they have. To grief."

"These people," Nancy said, indicating the reporters just beyond the closed door, "and I use the term loosely, showed up about an hour ago and started pounding on the door. I didn't know what to think. The phone hasn't stopped ringing." Sure enough, the phone rang.

"Don't answer it."

"What do you want me to do, Dr. Mallory?"

"Call Sheriff Baxter and ask him to remove these reporters from the premises."

"Can he do that?"

"He can keep them off my property. They can still park in the street, which I'm sure they'll do. For the next several days, we'll be under siege. Maybe you ought to take this week off."

"Not on a bet. I wouldn't desert you to fight off these jackals alone." As Lara slipped out of her suit jacket, Nancy took it from her and noticed the damp lining. "I've never seen you secrete a drop of sweat. I doubted you even had sweat glands."

"That's nervous perspiration. They ambushed me at the funeral."

"Those buzzards."

"Make up your mind. Buzzards or jackals." It was comforting to know she had retained her sense of humor.

"Doesn't matter. They're both scavengers. I ought to get Clem over here with his shotgun. That would scatter them."

"I appreciate the gesture, but no thanks. I don't need the bad publicity," Lara said grimly. "Before I even got a foothold on being Dr. Mallory, a small-town doctor, I'm once again Lara Porter, Clark Tackett's married lover."

Nancy's face reflected her regret. "It's such a damn shame. I'm sorry."

"Thanks. I'll need all the friends I can get." She sighed with consternation. "I wasn't actually in hiding, but I didn't want my whereabouts publicized for fear that something like this would happen. Someone deliberately stirred up this hornets' nest. I don't believe for a moment that it evolved on its own."

"Tackett's the name. Treachery's the game."

Lara looked sharply at her nurse. "Key?"

Nancy shook her head. "Isn't his style. My guess is the old lady. You're making headway here. Not in leaps and bounds, but in baby steps. She can't tolerate that. Jody heard about that little

253

girl dying, knew that you'd been the first attending doctor, and saw a chance to create a ruckus."

"She could have done that when I moved to town."

"But it would have come out that Clark set you up here. That would have implied that he was still emotionally attached to you. Jody didn't want to flatter you that much. This time, Clark's out of the picture."

What Nancy said made sense. Lara headed for her office. "I doubt any patients will even attempt to get in today, but I'll be in my office if I'm needed."

She pulled down the window shades so she wouldn't have to witness the destruction of her lawn under the trampling feet of eager reporters. Once seated at her desk, she consulted the telephone directory. Her personality had undergone some drastic changes since that morning in Virginia. She was older now, tougher, and she wasn't going to take persecution lying down. Reaching for the phone, she dialed a number.

"Miss Janellen?"

"Bowie! What are you doing here?"

She was seated at the kitchen table, staring at the telephone she'd just hung up. He had poked his head around the door. She signaled him in.

"Seems like I'm always sneaking up on you, pulling you out of deep thought. I don't mean to." He moved into the room, looking uneasy.

254

"The, uh, maid told me to come on back. If this is a bad time for you . . ."

"No, it's all right. I'm just surprised to see you here."

"I tried the office first, then the shop. They told me there that you'd knocked off early today."

"My mother wasn't feeling well this morning when I left for work, and I was worried about her." As usual, when in Bowie's presence, she felt tongue-tied. She indicated one of the chairs across the kitchen table from her. "Sit down. I was about to have some tea. Would you like some?"

"Tea?" Dubiously he glanced at the steaming kettle on the stove. "Hot tea? It's a hundred degrees outside."

"I know, but, well, I like tea," she said with an apologetic shrug. "It's soothing."

"I'll take your word for it."

"Something else then? Lemonade? A soft drink? A beer? Key keeps beer in the fridge."

"No, thanks. Besides, I can't sit down. I'm dirty."

He looked wonderful to her. Until he called her attention to it, she hadn't noticed the dirt smeared on his jeans and shirt. Hunks of it clung to the soles of his boots. It was embedded in the grain of his leather work gloves, which he'd stuck into his belt, and his hat, too, was dusty.

"Don't be silly," Janellen said. "Mama made my brothers work during their summer vacations. They used to come in all sweaty and stinky — not that you're stinky," she said hastily. "I just meant that this kitchen was built for working men to . . . you know, to enjoy and relax in."

Realizing that she was blabbering, she forced herself to stop. "You obviously came here to discuss something with me, so sit down, please."

After a moment's hesitation, he lowered himself onto a kitchen chair, balancing his buttocks on the edge of the seat.

"Wouldn't you like something to drink?" she repeated.

"Lemonade, I guess." He cleared his throat.

"You were a million miles away when I came in," he remarked after taking a long swallow of his drink.

"I'd just had a very disturbing telephone call." She debated whether she should discuss the call with him. He was looking at her expectantly, and it would be a relief to talk about it with someone who was uninvolved and therefore impartial.

"Have you been following the story of the little girl from Eden Pass who almost lost her arm?"

"I heard she died."

"Yes. Her funeral was today. Such a tragedy." She paused. "The doctor who treated her for shock and took her to Tyler —"

"Dr. Mallory."

"Yes. Well, she . . . she called just now. See, she was once . . . my older brother was . . ."

"I know."

She gave him a grateful smile. "Then you can imagine how embarrassing and uncomfortable it's been for us to have her here in Eden Pass."

"How come?"

The question was totally unexpected, and for a moment she was taken aback. "Because she brings back such bad memories for us."

"Oh."

He didn't seem convinced, so she felt compelled to explain. "Lara Porter ruined Clark's political career."

Bowie cocked his head to one side and lightly scratched his neck as though ruminating on her point. "She's not a husky old gal by any stretch. I don't figure she could wrestle him down, strip him naked, and force him into bed with her, do you?"

This wasn't the first time Janellen had considered that, but only privately. If she had verbalized her thoughts, Jody would have gone through the roof.

Prudently Janellen avoided further discussion in that direction. "Somehow the media found out that Lara Porter is in Eden Pass, passing herself off as Dr. Mallory. Apparently she was accosted by reporters at Letty Leonard's funeral this morning and had to call Sheriff Baxter to disperse those who've besieged her clinic."

Bowie smacked his lips with disgust. "Imagine them disrupting that little girl's funeral like that."

257

"I know. It was ghastly of them." For a moment she reflected on the continuing turbulence caused by her brother's affair with Lara Mallory Porter. "It's believed that the Leonards are going to file a medical malpractice suit against her," she told Bowie, then paused to take a deep breath. "She thinks my mother is responsible."

"Is she?"

"No."

"You don't sound too sure."

Her fingertips brushed her lips once before moving to her blouse. It didn't have buttons, so she fiddled with the fabric, then nervously laid her hand on the table near her untouched cup of tea.

"I don't know if she is or not," she admitted at last. "Dr. Mallory called to speak to her. Maydale told her that Mama was resting. She wouldn't take no for an answer and demanded to speak to whoever was available." She fidgeted with the salt and pepper shakers. "I wish Key had been here. He's a pro when it comes to confrontation. He would have known what to say to her."

"What did you say?"

"That I'm sure our family didn't cause her recent hardships."

"Think she bought that?" Bowie asked skeptically.

"She said she doubts that I would be that spiteful, but that she wouldn't put anything past my mother or my brother." In a small voice she

258

added, "I'd hate to think they could be that cruel."

She stared into space for a moment, then returned her attention to her guest. "I'm sorry, Bowie. I didn't mean to take up your time with my family's problems. What did you need to see me about?"

He rolled his shoulders. "It's probably nothing. In fact, I tried for several days to talk myself out of bothering you with it." He had set his hat on the table. Now he scooted it aside and leaned forward. "You ever notice anything peculiar about well number seven?"

"No, should I?"

"Probably not, but I figured I had to get it off my chest. See, it's not yielding as much natural gas as it should. At least, that's my opinion. Its production doesn't jive with comparable wells."

"All wells are different."

"Yes, ma'am, I know that. They have personalities and they're constantly changing. Kinda like women. Each well has its quirks and you've got to get to know it real good. Stroke it every now and then."

Janellen ducked her head so quickly that she didn't see that Bowie ducked his, too. Her cheeks turned warm, but since this concerned business, she felt it was imperative to keep the conversation going.

"What's the daily MCF?" Gas was measured in hundreds of thousands of cubic feet.

"Two fifty per day. I figure that well's output ought to be higher."

"We allow for a four to five percent loss, Bowie. Even up to ten. There's probably a small leak somewhere in the line and the gas is being absorbed into the atmosphere."

He gnawed his cheek for a moment, then shook his head stubbornly. "I think the loss is higher than the allowance. After recording that well for the last several weeks, I think it should be a high gas producer, especially considering the oil we get out of it. Instead, it's one of our lowest."

"You've spent a lot of time studying it."

"On my own time."

Her heart swelled with pride. He was a conscientious employee who did more than was required. Her decision to hire him had been justified.

Even though she appreciated his concern, she felt it was misplaced. "I don't know what to tell you, Bowie. Well number seven produces as we've come to expect from it."

"Well, I reported it to the foreman, but he just shrugged it off and said its rate of flow has always been low, long as he can remember. Damned if I can figure out why, though. Just one of those worries that grabs hold and won't let go, you know?"

"Yes, I know." She stared into her cup of tea. After a long moment of silence, she raised her head. "There I go again. I can't keep my mind on business. I keep dwelling on that little girl's family. Her daddy does all our dry cleaning. He's a nice, friendly man. I know how devastated he and his wife are, because we felt the same way when Clark drowned. I thought we'd have to bury my mother with him."

"I never had a kid, but if I did, I can't imagine having to put him in the ground."

Janellen looked at him searchingly. He'd never had a child, but she wondered if he'd ever been married. There were a thousand personal questions she wished to ask him, but couldn't bring herself to. Among those questions was where he had acquired his insight into people. He had an uncanny knack for seeing beyond affectations and straight into the heart and mind of an individual.

Trusting his instincts, she asked, "Bowie, do you think Dr. Mallory did something that caused that little girl to die?"

"All I know about medicine is that there's no real cure for either a cold or a hangover."

She smiled. "I've only seen Lara Mallory in person once, but she looked so . . . so . . . put together."

Everything that I'm not, she thought dismally. Having seen Lara Mallory, she was no longer surprised that Clark had risked everything to be with her. She wasn't merely beautiful. Her eyes reflected compassion and intelligence, and she exuded self-confidence and competency.

Janellen wanted to despise her. She knew that she wouldn't be feeling this ambiguity if Dr. Mallory had come across as an empty-headed sexpot, all fluff and no substance. Instead, it was quite the opposite.

"I don't believe the woman I met could be negligent." Her conviction surprised even herself and made her feel disloyal. "I know I'm supposed to hate her, but . . ."

"Who says?"

"My mother."

"Do you always do what your mother says? Don't you ever think different from her?"

"Rarely." The admission made her sound like a wimp. She was probably sacrificing any respect Bowie had for her as an individual and as an employer.

But Lara Mallory's call had upset her terribly. She was past the point of trying to hide her feelings. Propping her elbow on the table, she rested her forehead on her hand. "Oh, God, I wish her affair with Clark had never taken place. He would have enjoyed a successful political career like Mama wanted for him. He even might still be alive. Mama would be happy. And I —"

She caught herself before saying that if events had been different, she wouldn't feel so responsible for holding things together now. Seeing to everyone's happiness and well-being was exhausting. It was also impossible.

Ever since the night that girl had come to the door asking for Key, he'd been even more irascible than before. He and Jody hadn't quarreled any more, but that was because each went out of his way to avoid the other. Key answered direct questions in gruff monosyllables. He was preoccupied with only God knew what, and Janellen didn't dare guess. He stamped through the house with his shoulders angrily hunched, his expression belligerent. He was so unhappy at home that he often left as abruptly as he had appeared.

Now, Lara Mallory had just burdened her with a new source of worry. Before she realized that she was crying, a tear rolled down her cheek.

"Hey, what's this?"

She sensed the movement of Bowie's arm, but she didn't expect him to touch her. When she felt his callused fingertips against her cheek, she raised her head and looked at him, her lips parting in stunned bewilderment.

She was rarely touched by anyone, and, because she was starved for the touch of another, she reflexively raised her hand and folded it around his.

He went incredibly still. Nothing moved except his eyes. They went from hers, to her hand covering his, then back to her eyes. Janellen sat just as still as he, but inside she was all aflutter. Her lower body felt feverish, full, heavy. Her breasts tingled and tightened, making her want to press her palms over them to contain the rush of excitement.

How long they remained staring at each other she never knew. She was held in thrall by Bowie's sad, sweet eyes and the pressure of his fingertips, which were damp with her tears. If he hadn't heard Key's car approaching, she might still have been frozen in that tableau when her brother slammed in.

As it was, she hastily shot to her feet and whirled around to greet him. "Key! Hi!" Her voice was unnaturally high and thin. "What are you doing here?"

"When I left this morning I still lived here." He divided an inquisitive look between her and Bowie, who she hoped could conceal guilt better than she. Her face was fiery hot. She knew she must be flushed from her throat, where her pulse was pounding, up to her hairline.

Key took a beer from the refrigerator. "Hi, Bowie. Want a beer?"

"No, thanks."

Janellen said, "I already offered him one, but he wanted lemonade instead."

"I just stopped by to tell Miss Janellen that —"

"He thinks the MCF on well number seven is low and —"

"It's probably nothing, but —"

"He thought we ought to know in case —"

"So I brought it up with Miss Janellen and —"

"And that's what we've been doing. Talking about that," she finished lamely.

"Uh-huh." Looking amused, Key popped open the beer and tilted it toward his mouth. "Well, don't let me interrupt this high-level business conference."

"No, it's all right." Bowie snatched up his hat as though it were a piece of incriminating evidence. "I was just on my way out."

"Yes, he was about to leave when you came in. I'll . . . I'll just walk him to the door now." Flustered and unable to look at either her brother or Bowie, she fled the kitchen and was waiting for Bowie in the entry, holding the front door open for him. She kept her eyes averted as he joined her there. "Thank you for the information, Bowie."

He pulled on his hat. "Just figured it ought to be brought to your attention. It's your money."

"I'll check into it."

"I don't think that's such a good idea."

At the sound of her brother's voice, she swung around. His shoulder was propped against the arched opening of the dining room as he nonchalantly sipped his beer.

"What's not such a good idea?" she asked.

"You checking into a malfunctioning well."

"Why not?"

"As of today, the Tacketts are in the news again."

"So?"

"So reporters are going to be crawling over Eden Pass like ants on a picnic ham. Until a hotter story comes along, that is. When they don't get anything out of me — and they won't — they're likely to come sniffing after you for a statement. Bowie," he said, looking at the pumper, "keep an eye out for her, okay? If she inspects any oil

265

wells, you go with her."

Bowie glanced uneasily at Janellen. "Meaning no disrespect, Mr. Tackett, but she's the boss."

"Boss or not, do it as a favor to me. I'm asking as her brother."

Again Bowie's eyes darted toward Janellen. She was fuming and didn't trust herself to speak. With uncertainty, Bowie said, "Okay, Mr. Tackett."

"Call me Key."

"Yes, sir. Well, 'bye, y'all."

He wasted no time in getting to the company truck and driving away. In fact, he looked grateful to be escaping with his hide intact.

Janellen rounded on her brother. "I don't need a keeper!"

"Well, I do," he replied, unfazed by her anger. "If a reporter pesters you, I'll go after him wanting to kick ass. That'll create more news and make a bad situation worse."

She resented his taking charge of her employee, of making it appear that she was incapable of taking care of herself. But his explanation was well founded. If a reporter did ambush her demanding a statement, and Key found out about it, there was no telling what he'd do. Once, when she was in high school, she'd come home from a date in tears. Key had almost throttled her terrified escort before she could explain that they'd just seen a sad movie.

Knowing that he was looking out for her best interests, she let her anger subside. "The situation is already worse than you know," she told him. "Lara Mallory called here a while ago wanting to talk to Mama. Dr. Mallory thinks she tipped the media about her being here in Eden Pass."

Key ran a hand around the back of his neck. "Well I be damned."

"Does that surprise you?"

"No. What surprises me is that the doctor and I are beginning to think alike. I also figured Jody was at the bottom of this. I know plenty of smart reporters, but no more than a handful of them knew Lara was involved in the Leonard girl's case; it would have been a bizarre coincidence if one of them had added two and two and come up with four." He looked toward the second story of the house. "Shrewd old bitch."

"Don't talk like that about our mother."

"I meant it as a compliment. You've got to give her credit for creative thinking."

"Was it so creative?"

"Meaning?"

Worriedly, she said, "You were there, Key. You saw everything. Was Dr. Mallory negligent? Do the Leonards have grounds for a malpractice suit?"

"I was concentrating on flying the chopper, but from what I saw, Lara fought like hell to save the kid's life. According to the autopsy report, that embolism was a freak of nature. Could have happened anytime. And another thing — the Leonards didn't seem the kind of people who'd be vengeful. They're faithful Christians."

"So it surprises you that they're looking for a scapegoat?"

"Right. I wouldn't put it past Jody to circulate a rumor of a malpractice suit, whether or not there's any truth to it. Lara's an easy target." Janellen looked at him quizzically. "What?" he asked.

"Several times you've referred to her as Lara. It sounds odd."

He hesitated, then said querulously, "That's her name, isn't it?"

Janellen had too many other pressing matters on her mind to pursue something so trivial. "She sounded awfully mad, Key. She said to tell Mama and you that she wouldn't be driven out of town like she was before. What did she mean?"

"She's referring to when she and Randall Porter went to Montesangre." He frowned. "She's got it into her head that Clark engineered the appointment for Porter by flexing some muscle in the State Department. His appointment looked and sounded good, but it was practically legalized banishment."

Janellen was stunned. "Do you believe her? Could Clark have been that devious?"

"Devious is a strong word, but our big brother was fairly adept at weaseling his way out of trouble."

"He never really got out of this trouble, though, did he?"

"No, he didn't," Key said slowly. "And as long

as Lara's around to remind everybody of it, he never will."

"So you agree with what Mama did. *If* she did."

"No. I want Lara Mallory out of Eden Pass, but I want her to hang herself. Left alone, I think she eventually will." Once again he glanced upstairs. "But you know Jody. She's never been one to let things follow their natural course. If things aren't moving along according to her plan and her timetable, she plays God."

"Please don't be critical, Key. She's sick. Can't you try and talk her into seeing a doctor?"

He barked a laugh. "That'd be a surefire way to guarantee that she wouldn't. But I agree. She should have a complete checkup, have some tests run." He placed his hand on her shoulder. "But I'm afraid that persuading her to do it is up to you, sis. Stay after her." He squeezed her shoulder, then headed for the stairs, taking his beer with him.

"Are you going out tonight, Key?"

"As soon as I shower."

"Are you going out with Helen Berry?"

He stopped dead in his tracks and turned. "Why do you ask that?"

Gauging by his expression, Janellen knew she'd struck a nerve. She also realized why people were sometimes afraid of him. "Helen's been going steady with Jimmy Bradley since they were freshmen. The gossip is that . . ." she paused to wet her lips, "that Helen recently broke up with him, very sudden."

"So?"

"Oh, Key." Taking hold of her courage by both hands, she asked, "Why? Why, when there are so many other women for you to choose from, would you pick her? Helen's half your age."

"Careful, Janellen. If you start digging into my personal affairs, I'll have to start digging into yours." He moved down two steps and lowered his voice to a stage whisper. "For instance, I might ask what's going on between you and Bowie Cato."

Her stomach dropped. "Nothing's going on!"

"No? Then why all the rushed, breathless explanations when I came into the kitchen? I haven't heard such fast talking since Drenda Larson's daddy caught us in his hay barn when we were thirteen."

"Bowie's an employee. We were talking business."

"Okay, I'll believe that," Key said, his cocky grin back in place. "If you'll believe that all Drenda Larson and I were doing in that haystack was looking for a needle."

Lara's prediction proved correct.

A week following Letty Leonard's funeral, the media moved to greener pastures to graze on other personal disasters and dilemmas. During that week, however, Lara had been hounded each time she stepped across her threshold. Sheriff Baxter had done his official duty, albeit grudgingly, and seen to it that the reporters and cameramen stayed off her property. But their presence on the public street made her a virtual prisoner at the clinic.

The TV network affiliates from Dallas and Shreveport had filed stories that were aired on national newscasts, but Lara Porter and the key role she'd played in the downfall of Senator Clark Tackett five years earlier only rated fifteen seconds of air time in the last few minutes of the newscasts. She'd lost her rank as a lead story.

The Leonards too had been thrust into the spotlight but had hired an attorney to do their talking. He was a wet-behind-the-ears graduate of Baylor Law School who had only recently passed the bar. He rose to the occasion, however, and wasn't intimidated by being in the limelight. Stubbornly and repeatedly, he told reporters that his clients had no statements to make and were trying to deal with their bereavement before addressing the question of liability for their daughter's death.

Lara had done some intensive soul-searching. It had been a judgment call as to whether to use an anticoagulant on Letty. After hours of review and research, she stood by her original decision. However, in order to ease her mind, she conferred with the emergency room doctor who had next tended to the young patient. He backed her decision and assured her he would testify to such if it ever became a matter of litigation.

As days passed and Lara didn't hear from the Leonards' lawyer, she hoped that the rumor of the malpractice suit against her was just that — a rumor. No doubt it had been spawned by one of the Tacketts. Her repeated calls to them had rendered nothing and only increased her frustration. Jody Tackett was either indeed too ill to take a telephone call, or she had good liars protecting her.

Lara had spoken to the housekeeper and to Janellen, but she hadn't seen or spoken to Key since the night he'd brought Helen Berry to her. He probably thought she'd been joking when she mentioned his taking her to Central America. Another opportunity to broach the subject hadn't presented itself, but her determination hadn't wavered one iota. It was just that so many other events had temporarily distracted her.

When she had awakened this morning, the last of the TV vans was gone, but because of the negative publicity, the patients with appointments had called to cancel. It was difficult to remain optimistic about cultivating a practice when she couldn't get people inside her door. She and Nancy went through the motions of working, but they had more idle time on their hands than either wanted to acknowledge.

By midafternoon she left her private office with the intention of dismissing Nancy early. Nancy, surprisingly, was speaking to someone in the waiting room.

"We'd like to see the doctor right away. I

272

know we don't have an appointment, but then you're not exactly overflowing with patients, are you?"

The strident, condescending voice belonged to Darcy Winston.

Chapter Thirteen

"May I help you?"

When Lara spoke from the doorway, Darcy turned. She wasn't as flawless close up as she'd appeared on the school auditorium stage. There were faint crow's feet around her eyes and frown lines across her forehead. She had artfully applied cosmetics, but her face bore unmistakable traces of hard living and deep-seated bitterness.

Lara had formed an unflattering opinion of Darcy Winston's character, but knew from experience that such bias was unfair. Trying to keep an open mind, she smiled and extended her hand. "Hello, Mrs. Winston, I'm Lara Mallory."

Darcy raised one carefully penciled eyebrow inquisitively. Lara explained how she recognized her. "I heard you speak at the town meeting. You were extremely convincing."

Again Darcy communicated by using her eyebrow. She gave Lara an arch look, obviously trying to guess how much she knew about her "intruder."

Lara turned to the girl standing beside her mother. "And your name is Heather, isn't it?"

"Yes, ma'am."

"I'm pleased to meet you, Heather."

"Thanks."

"Heather's the reason we're here," Darcy said.

"Oh? What's the problem?"

"I want you to put her on birth control pills."

"Mo-*ther!*"

The girl was mortified, and Lara didn't blame her. Unfortunately, Darcy was living up to Lara's expectations. She was a first-class bitch. Wanting to spare Heather further embarrassment, Lara asked, "Nancy, which examination room is ready?"

Nancy was eyeing Darcy, her expression sour. "Three."

"Thank you. Heather?" Smiling, Lara pulled open the connecting door and held it for the girl. Darcy fell into line behind her.

"Mrs. Winston, you may wait out here where you'll be more comfortable. Nancy will need some information from you in order to start a patient file on Heather. If you like, she'll get you something to drink while you wait."

"She's my daughter." Her tone made it obvious that she was accustomed to intimidating people and getting her way.

"And this is my office," Lara said with matching imperiousness. "Heather is my patient. I respect and protect the privacy of my patients."

Without another word, she closed the door on Darcy's tight, angry frown and showed the girl into the examination room. She left her there with Nancy, who would see that she was undressed, draped, and weighed before taking her

275

blood pressure and collecting specimens of urine and blood.

Nancy summoned Lara from her office with a light tap on the door. As she moved back toward the waiting area, Nancy whispered, "How am I supposed to pacify Bat Lady?"

"Throw her a small rodent."

Nancy gave her a thumbs-up sign. She went into the examination room where Heather was warily perched on the end of the table. "Everything okay?"

"Fine, I guess. I don't like having my finger pricked."

"Neither do I."

"It's better than having blood drawn from your arm, though. I hate needles."

"They aren't my favorite things either."

"But you're a doctor."

"I'm a person, too."

The girl smiled, more at ease now.

"When do you start cheerleading practice?"

"How'd you know I was a cheerleader?"

"The booster club sent me an application for membership." Lara examined her eardrums with an otoscope. "I saw your picture."

"We start practicing next week."

"So soon? Say 'ah.' " Using a tongue depressor, she looked at Heather's throat. "School doesn't start for another month yet."

"Ahhhh. Yeah, but we want to be good. Last year we won several trophies."

"Swallow for me. Any tenderness in your

glands here?" Lara asked as she felt Heather's neck.

"No, ma'am."

"Good. Take care of your throat. If you notice any soreness, let me know. Sore throats and hoarseness are inherent to yell leaders."

"Okay. Sure."

Lara lifted the drape and placed the stethoscope beneath Heather's left breast. The girl gasped. "I know it's cold," Lara apologized with a smile. After listening to her heart, she moved to her back to listen to her lungs. "Take several deep breaths through your mouth, please. That's good." After a moment she moved once again to stand in front of the girl. "Are your periods regular?"

"Yes, ma'am."

"Heavy?"

"Usually just the first and second day. Not after that."

"Do you have cramps?"

"Yeah. Really bitchin'."

"Do you take something?"

"Midol, aspirin. Stuff like that."

"Does it help?"

"I'll live," she replied with a grin.

Making Heather as comfortable as possible on the table, Lara summoned Nancy from the waiting room to assist her with a breast check and a pelvic examination.

"This is gross," Heather said as Lara guided her feet into the stirrups.

"Yes, I know. Try to relax as much as possible."

"Right," Heather said sarcastically when Lara opened the speculum.

When she was finished, she left Heather to dress and returned to her office. Heather joined her there a few minutes later. Lara indicated the sofa and sat down beside her, creating a mood that was more friendly than clinical.

"Why do you want to go on birth control pills, Heather?"

"She wants me to."

"Your mother?"

"She's afraid I'll get pregnant."

"Is that a possibility?"

Heather hesitated. "Well, I guess. I mean, I have a boyfriend . . . and we, you know."

"I'm not asking to be nosy," Lara told her gently. "I make no moral judgments. I'm a doctor who needs to decide what's best for my patient. The only way I can do that is to have as much information as possible." She let that sink in, then asked, "Are you having sexual intercourse?"

Heather looked down at her tightly clasped hands. "Not yet." Then she furtively glanced at the closed door. "She thinks we already have. I've told her we haven't, but she doesn't believe me."

Once she began, the words poured from her, crowding one another in their rush to get out. "She caught Tanner and me making out in the living room. We weren't doing anything. I mean,

I had taken off my blouse and bra, and Tanner had taken off his shirt, but by her reaction you'd have thought we were totally naked, that she'd caught us actually *doing* it."

Suddenly her eyes swung up to Lara. "Oh, jeez, I'm sorry. I didn't mean to say it that way. I didn't link it to you and Senator Tackett."

"No offense taken," Lara said quietly. "This is about you, not me. When your mother found you and Tanner, she jumped to the wrong conclusion, is that right?"

"To put it mildly, she went totally apeshit," Heather said, rolling her eyes. "She screeched so loud she woke up my daddy. He ran downstairs, bringing his pistol, thinking the house had been broken into again." She shoved back a handful of glossy auburn hair. "It was awful. Tanner kept telling them that he wouldn't do anything to hurt me, but Mother threw him out of the house and hasn't let me see him since. I've been grounded. She took away my car keys and my phone."

Tears filled her eyes. "I might as well be in Siberia. It's awful! *And I didn't do anything!* She looks at me like I'm a, you know, a whore. Daddy's tried to make peace, but she doesn't easily forgive and forget. I've told her a million times that I'm still a virgin. Technically, that is. Tanner's, you know, used his finger, but not his . . ."

Lara indicated her understanding with a nod.

"But Mother doesn't believe that. This morning she told me we were coming to you, and I was going to start taking birth control pills whether I liked it or not. She said if I was going to screw around, at least she wasn't going to get stuck with a grandkid to raise."

Lara empathized with the girl because Darcy's sentiments echoed those of her own parents. The message had been: Do whatever you want, just don't get caught and thereby inconvenience us. Heather sniffed wetly. Lara passed her a box of tissues. "I miss Tanner so much. He loves me. He really does. And I love him."

"I'm sure you do."

"He's so sweet to me. Not like her. Nothing I do pleases her."

Lara waited while Heather noisily blew her nose, then said, "I see no problem in prescribing the pills for you. You appear to be in good health."

"They'll make me fat, won't they?"

Lara smiled. "Weight gain can be a side effect, but I doubt that will be a problem for a young woman as active and energetic as you." She looked intently into the girl's face. "Aside from the physical aspects, I want you to be psychologically prepared for this step. Are you certain that this is what *you* want, Heather?"

Again, her eyes darted toward the door. "Yeah, it is. I mean, Tanner's promised that he'll use something, but if I was taking pills, too, no way could I get pregnant."

"Just remember that the pills don't protect you

from sexually transmitted diseases. If you're going to be sexually active, I suggest using a condom every time, even with a steady boyfriend. Encourage your friends to do the same."

She wrote out the prescription form, then together they moved toward the waiting room. Darcy was impatiently thumbing through a magazine. She tossed it aside as soon as they entered.

"Well?"

"I've given Heather a prescription for oral contraceptives and asked her to come back in six months just to see that everything is okay. Of course, she's to call me if she has any negative side effects or discomfort."

"You were in there an awfully long time."

Lara refused to be defensive. "Your daughter is a delightful young woman. I enjoyed talking with her. Which reminds me, I'm interested in implementing some health education programs at the high school. As president of the school board, would Mr. Winston be open to hearing my ideas?"

"You'll have to ask him."

"Then I will," Lara replied graciously, in spite of Darcy's curtness. "I'll contact him as soon as the semester begins."

"How should I handle the bill?"

"Nancy will take care of it now." Lara turned to Heather. "Good luck with cheerleading. I'll be watching you from the grandstands."

"Thanks, Dr. Mallory. I'll wave at you." She

grinned, then added, "It still feels weird calling a lady 'doctor.' "

They were several blocks from the clinic before Darcy broke the antagonistic silence with her daughter. "Well, you two certainly seemed chummy when we left."

"She's nice."

Darcy snorted. "Clark Tackett thought so too, and look where that landed him. She's nothing but trash. And trouble."

Heather turned away to gaze out the window.

Most of Darcy's criticism arose from jealousy. She hadn't expected or wanted Lara Mallory to be so charming. She was cool and classy. Every subconscious gesture bespoke good breeding and social training. She was so damned tidy that she'd made Darcy feel like she needed a bath. She was slender as a reed, and probably had not an ounce of cellulite clinging to her thighs. Her hair was thick and healthy. Her seemingly poreless skin was taut. From a woman's standpoint, there was a lot there to envy.

But what would a man, specifically Key Tackett, see in her? Her figure wasn't voluptuous. She had a candid gaze, like a man's. Or did her eyes assume a sultry mystery when she was with a lover?

After making up her mind to visit Lara Mallory, Darcy had been forced to wait a week before doing so. Heather and Tanner had provided her the perfect excuse, but then the Leonard kid had died and the town had been in upheaval. Everyone was watching Lara Mallory. Darcy decided it would be smart to wait until the dust had settled. She wanted an up-close and personal look at Dr. Mallory, but without the whole town knowing she was curious.

Was Lara Mallory Key's new squeeze? Dammit, she'd come away as mystified as before. The doctor seemed too cool to appeal to Key's lusty nature, but looks could be deceiving. And there was no accounting for taste, particularly a taste for women, which she knew was unique to every man.

So all Darcy had to show for her meeting with Lara was Heather's dewy-eyed admiration for the woman who might have snatched Key away from her. Not that she'd actually been in a position to claim ownership of him. He had picked her up in a bar and slept with her only once, but she believed that they had a future as lovers. Without the interference of another woman, it could happen. Lara Mallory might jinx it.

"Did y'all talk about me?" Darcy asked Heather peevishly. "I'll bet you made me out to be a bitch."

"No, I didn't."

"What did you say about me?"

"Nothing. Except general stuff."

"Then what did you talk about that took so damn long?"

Heather sighed with adolescent resignation. "We talked about cheerleading and my periods and Tanner and becoming sexually active and stuff."

"What did she say about you becoming sexually active?"

"That she didn't make moral judgments."

"At least she's not a hypocrite. That'd be the pot calling the kettle black, wouldn't it?"

"I guess."

"I thought you'd probably hear a sermon against going on birth control pills at your age."

"No," Heather said wearily. "She only lectured about condoms."

"Condoms?"

"Uh-huh. Mom, can I please have my phone back now?"

"What did the doctor say about condoms?" Heather shot her a mutinous glare, then recited hurriedly, "That they're still the best protection from disease and that if me and my friends are going to sleep with our boyfriends, we should always use them."

"She told you to have a condom handy just in a case a date turned into sex?"

"Something like that," Heather said, shrugging with unconcern. "Can I please have my phone back, Mommy? Please? And my car keys?"

A glimmer of an idea winked on inside Darcy's head. She regarded it from all angles and decided it was worth saving and nurturing. Smiling and feeling more like her old self, she reached across the console and patted Heather's knee.

"Sure you can, sweetie. You can have them

back as soon as we get home. But first let's stop and have a piece of pie with Daddy. I've been a perfect grouch all week and want to make it up to y'all, starting now."

Bowie Cato turned off the highway onto the state road that ran along the north end of The Green Pine Motel, where Darcy was alighting from her late model Cadillac. " 'S that Mrs. Winston?"

"Yes." Janellen had turned to wave. "Do you know her?"

"I've seen her. Who's that with her?"

"Her daughter, Heather. She's about the most popular girl at the high school these days."

"Pretty," Bowie commented, glancing back at the two women as they entered the motel lobby.

"Very. She works part-time at the motel for her daddy. I see her whenever we go to the Sunday buffet after church. She's friendly and sweet and well liked."

Bowie wondered if the daughter was as "well liked" as the mother. He'd seen Darcy Winston in action plenty of times at The Palm, beginning that night Key Tackett had returned to town and as recently as last night when she'd been playing a rowdy game of billiards with three Shriners who were having a night out on the town without their wives.

Darcy was a tramp, and everybody knew it. Just like everybody knew that Janellen Tackett was a lady. That's why folks looked askance at

285

them whenever she was with him. They were wondering what Miss Janellen was doing with a no-account ex-con like Bowie Cato.

He'd been wondering that himself. He both thanked and cursed Key for asking him to keep an eye on her. He thanked him because being near Janellen was about as close to a class act as he was ever going to get. He cursed Key because he was beginning to like being near her too well.

He enjoyed seeing her every day and having a good excuse for it. But it was temporary bliss. Sure as God made little green apples, something would happen to put an end to it. Waiting for the inevitable and wondering what disastrous form it would take was driving him nuts. Right now he was living a fairy tale. Trouble was, he didn't believe in fairy tales. They were for kids and fools. He sure as hell wasn't a kid, but he was beginning to think he was a fool.

He was letting himself in for a fall. No two ways about it.

Damned if he could stop himself, though. Every chance he got to be with her, he took. Like today. When word reached him that she was going out to take a look at the number seven well, he'd jumped into the truck and driven like a bat out of hell to get to the office before she left.

He caught her just as she was leaving and reminded her that Key didn't want her to be alone. He also said that the truck was more suited to

286

the well site than her compact car. She'd conceded and climbed into the cab of the truck with him.

But she wasn't happy about it.

She was as jittery as a chihuahua passing peach pits and wouldn't look him in the eye. She was probably ashamed to be seen riding around with a convicted felon. Hell, who could blame her?

"It gets pretty rough from here," he warned.

"I know," she said acidly. "I've driven it myself plenty of times."

He ignored that and took the turnoff. The dirt track, carved into the earth by tire treads, ran parallel to the highway several hundred yards away. In between was The Green Pine Motel. He'd heard talk of how Jody Tackett, years ago, had swindled Fergus Winston out of his oil lease.

Fergus had come to Eden Pass as a young man, bringing with him a small legacy and big dreams. He bought a patch of land that didn't look like much on the surface but had highway frontage and rumors of oil underneath.

He met Jody, who at the time was working for Clark Tackett Senior and was already reputed as being a knowledgeable land man. Jody befriended him and offered to let a Tackett Oil geologist check out his lease and give him an expert opinion. After weeks of assessment, she sorrowfully told Fergus that it was doubtful his land had any significant deposit of oil.

Fergus, somewhat in love with her by then, believed her, but he decided he needed a second, bipartisan opinion. He retained the services of another geologist who sadly informed him that horny toads were about the only thing his patch of ground was likely to harvest.

Fergus was disappointed but had come to believe that his future lay not in the competitive oil industry but in providing temporary lodging for the folks who wheeled and dealed in it. Jody, still passing herself off as a concerned friend, told him she hated to see him getting stuck with land that wasn't good for anything. She offered to buy his lease for Tackett Oil, which could use it as a tax writeoff. Fergus would then have enough capital to begin building his motel.

Relieved to be unloading a white elephant and recovering some of his investment, he sold the land and all the mineral rights for next to nothing, keeping only the strip of property that fronted the highway, on which he planned to build his motel.

But Fergus's white elephant was sitting on top of a black lake of rich crude. Jody knew that, and so did the Tackett Oil geologist, and so did the one Jody bribed to back up the lie of the first. The ink wasn't dry on the deed before Tackett Oil erected a drilling rig. When the well came in, Fergus was fit to be tied. He accused Jody and the Tacketts of being thieves and liars. When she married Clark Junior, he cursed her even louder. But he never legally pursued his allegations of dirty dealing, so folks discounted his grievances as sour grapes and jealousy because

Jody had jilted him in favor of Clark Junior.

Fergus built his motel, and it was profitable almost from its opening day. But even if it had been as fancy as a Ritz-Carlton, he'd never be as rich as Jody Tackett. To this day, he carried a grudge.

Bowie parked the truck outside the chain-link fence that formed a neat square around the pumping well. He alighted and went around to offer assistance to Janellen, but she had already hopped down by the time he reached her. He used his key to unlock the gate.

The motor driving the horse head pump was chugging away. He'd been out hours earlier to check on it, which he did every day except for his days off, when the relief pumper ran the route. He and Janellen weren't interested in the pump or the storage tanks, but in the meter box where red, green, and blue pens recorded the line pressure, temperature of the gas, and rate of flow onto circular charts that were changed biweekly. Fortunately the meter box for well number seven was located only yards from the well itself. It could have been miles away.

Fifteen minutes later, he was feeling like a damn fool. There seemed to be nothing wrong with well number seven. The meter box was functioning properly. There were no discernible leaks between the well and the meter box. Everything appeared to be in perfect working order.

"I guess you think I'm crazy," he mumbled.

"I don't think you're crazy, Bowie. In fact, if it would relieve your mind, I'll authorize you to put a test meter between the well and the recorder."

He got the impression that he was being humored. "Okay, I will," he said, calling her bluff. "Do you know if there was ever a flare line off this well?"

"If there was, it was capped off when they became illegal. We don't waste gas that way anymore."

They retraced their steps back to the gate. Bowie locked it behind them. "Did you tell your mama about this?"

"No."

"You didn't think it was important enough?"

By now she had reached the passenger door of the truck and turned to face him, shading her eyes against the sun. "I'll thank you not to put words in my mouth, Bowie. It's just that these days I don't worry Mama with anything that I don't have to."

"You sure look pretty, Miss Janellen."

"What?" she exclaimed. Her hand remained where it was, with her index finger following the curve of her eyebrows and her palm sheltering her eyes.

Oh, hell. He'd gone and done it now. He reached beneath his hat to scratch the back of his head. He hadn't meant to say what he was thinking. The words just popped out. And now an explanation was called for.

"It just, uh, struck me all of a sudden how pretty you look standing there. With the sun shining in your eyes and the wind whipping your hair around."

The hot, arid wind had also plastered her clothes to her body, so, for the first time since meeting her, her shape was clearly defined for him. In his estimation it was a very nice shape, but he didn't indulge his curiosity for long because her face was crumbling and her eyes were filling up with tears that had nothing to do with the sun's glare.

"Oh!" she sobbed. "Oh, Lord! I could just *die!*"

Her reaction alarmed him. All a parolee needed was to have a hysterical woman on his hands, bawling and carrying on and saying she could just die. He anxiously rubbed his damp palms against his thighs.

"Hey, Miss Janellen, don't get yourself all worked up now." Nervously he glanced around, hoping no one was witnessing her distress. "When I said . . . well, I didn't mean anything disrespectful. You're safe with me and that's a fact. What I mean is, I wouldn't —"

"Just because he told you to keep an eye on me doesn't mean you have to shower me with compliments you don't mean."

Bowie squinted his eyes and cocked his head, unsure he'd heard her right. "Come again?"

"I don't need him watching over me, or you either."

" 'Him'? Are you referring to your brother? Key?"

"Of course Key," she said with annoyance. "Ever since he asked you to keep an eye on me, I can't turn around without bumping into you."

"Well, I apologize for any inconvenience it's caused you, but I promised Key I'd look out for you, and I keep my promises. I plan to keep on looking out for you until he tells me to stop."

"*I'm* telling you to stop. As of this minute. All the reporters have left Eden Pass. I'm in no danger of being ambushed by them, so you don't need to trouble yourself any longer."

"It wasn't any trouble to drive you around, Miss Janellen."

"I can drive myself! I have since I was sixteen."

"Yes, ma'am, I know that, but —"

"And I can read a meter box the same as any man. Alone, too."

"I'm sure you can."

"While you feel duty-bound to trail me everywhere, I certainly don't need you throwing out empty compliments that —"

"It wasn't empty."

"— that you can laugh over later."

"Laugh?"

"I know what the men think of me. They think I'm a dried-up old maid. Muley told me that they laugh at me behind my back. You're trying to suck up to my brother —"

"Now hold on just a goddamn minute," Bowie interrupted angrily. "I don't suck up to anybody.

Got that? And leave your brother out of this, because he doesn't have a friggin' thing to do with why I said what I said. And I don't give a rat's ass about what any of the other men think. I make up my own mind about things, and if somebody disagrees with my opinion, well then screw 'em. When I told you you looked pretty, it's because I really thought so.

"God a'mighty! Most women would have said, 'Why, thank you, Bowie. What a nice thing to say,' and let it go at that. But not you. No. You gotta read something into it 'cause you're prickly and prissy and have a burr up your butt the size of Dallas."

His words reverberated in the air between them before the wind snatched them away.

But not soon enough, Bowie thought dismally. His self-control had snapped, something he'd thought would never happen with her. He'd lost his temper and shot off his mouth. He'd fucked up major big this time. Now she'd fire him, and the fault was all his.

She faced him, wide-eyed, tremulous, and speechless. Tears had made pools of her blue eyes, pools deep enough for a grown man to drown in. A small shudder rippled through her. She drew in a quick little breath that brought her lower lip in fleeting contact with her teeth.

It was too damn much.

Figuring that at this point he'd just as well be hanged for a sinner as a saint, he bent his head and kissed her. It was a hard and swift kiss.

It had to be. Any minute now she might start screaming. Besides, he didn't trust himself to linger and taste. He might do something really stupid that would land his sorry ass right back in jail.

The instant he pulled back, he turned her about and shoved her up into the truck. He climbed in on the other side, turned on the noisy motor, engaged the grinding gears, and guided the truck over the deeply rutted track.

They rode in silence all the way back to the ugly company office, where he'd picked her up. After he killed the engine, the silence was as engulfing as the heat that rose from the ground in shimmering waves.

She was probably still too distressed to speak, so it was up to him to say something. For several moments he stared through the dirty windshield, then said, "I'll take the truck back to the shop and turn in the keys. You can mail me my final check."

He heard her swallow, but he didn't look at her. He couldn't bear to see her disgust.

Finally, in a feeble voice, she asked, "Are you leaving Tackett Oil?"

He looked at her then, turning his head so quickly that his neck popped. "Aren't I?"

"Do you want to?"

"Don't you want me to?"

She shook her head and, in a barely audible voice, said, "No."

He didn't dare move for fear of shattering the

fragile mood. "Those things I said, Miss Janellen . . . I never should have used that kind of language in front of you."

"I grew up with two brothers. I know all the words, Bowie. And what most of them mean."

She flashed a gamine smile, but he didn't return it. "That, uh, that other — kissing you — well, that's grounds for firing me for sure. But I want you to know that I only did it because I lost my head."

"Oh." After a moment, while the silence and tension and heat thickened, she added, "Then it was purely an impulsive gesture?"

Something in her eyes compelled him to answer truthfully. "No, I can't truly say that it was, Miss Janellen. I'd thought about doing it before."

"I'd thought about it, too."

He couldn't believe what she'd just said, yet he was looking straight at her. He'd watched her lips form the words, and because his loins had filled with liquid fire, he knew he wasn't dreaming.

But it only got better.

He shifted slightly. She tilted her head inquisitively. Then they met somewhere in the middle of the bench seat. Within seconds of her soft declaration, he was holding her against him, her arms were twined around his neck, and they were kissing madly.

Her lips were responsive but shy, which was okay because Bowie wasn't an experienced kisser anyway. He'd never had a woman of his own, and easy women and whores usually skipped the kissing part. So he and Janellen tutored each other, and when his tongue slipped between her lips and connected with hers, they both murmured in delightful discovery.

Was her mouth actually sweeter than any other woman's he'd kissed, or was it that she was the first he'd french kissed with caring and not only as a hasty prelude to getting laid?

He lowered his hand to her waist and pressed it. Another tiny shudder went through her. God, it was exciting. He wanted to chart that shudder from her breasts, up her throat, and across her mouth. But of course he didn't.

Eventually she angled her head back and gazed up at him with rapidly blinking eyes. She was embarrassed. Her cheeks were flushed. Her breathing was rapid and shallow. She rolled her lips inward, then released a breathy little laugh.

"I'd better go now. If I'm late for supper, Key's likely to come looking for me."

He scooted back behind the steering wheel. "Sure enough."

"I'll see you tomorrow."

There was the slightest inflection of inquiry attached. "Bright and early." He smiled, although it was a strain because his cock was throbbing like a son of a bitch.

She opened the door and was on the verge of getting out when she turned back and said in one gust of breath, "I love you, Bowie."

She slammed the truck's door, ran to her car,

scrambled into the driver's seat, and drove away. Bowie watched the cloud of dust she raised until it had dissipated. Even then he sat behind the steering wheel of the truck, staring through crusty insect carcasses and oilfield grime, unable to move, shell-shocked by her parting words.

Well, that explained the kissing spree, he thought. Janellen Tackett wasn't right in the head. In fact, she was plumb crazy.

Nobody had ever loved Bowie Cato.

Chapter Fourteen

"Are you awake?"

"I am now." Lara's nightstand clock registered 2:03 A.M. "Who is this?"

"Key Tackett."

She groaned, burrowing her head deeper into her pillow and almost letting the telephone receiver slip from her hand. "Is this another of your emergencies?"

"Yes."

Sensing the strain in his voice, Lara came fully awake. This wasn't a prank. She sat up and switched on the nightstand lamp. "What is it?"

"Are you familiar with the state highway everybody calls the Old Ballard Road?"

"I know where it is."

"Go south on it two miles beyond the Dairy Queen. On your right will be a cutoff. There's an old windmill there, so you can't miss it. A few hundred feet beyond that, on your left, there's a farmhouse. My Lincoln is parked out front. Bring your stuff."

"What stuff?"

"Doctor stuff. Hurry."

"But —"

The line went dead. She flung back the covers;

her feet hit the floor running. It was second nature to respond to an emergency call. She didn't pause to consider the advisability of responding to this one until she was speeding down the dark, deserted highway. If the Tacketts really wanted to get rid of her permanently, how better than to trick her into going out in the middle of the night on an emergency call from which she would never return?

She had pulled on the first clothes her hands had touched and shoved her feet into a pair of sneakers. In the clinic, she'd filled her medical bag with supplies that would handle most, but certainly not all, emergencies.

She might very well be walking into a trap, but she could not have said no to the summons. And, strange as it was, she believed the urgency in Key's voice had been genuine.

She sped past the windmill before seeing it. If his directions hadn't included it, she never would have spotted the narrow, unmarked road. She backed up and took the turn sharply. Moments later her headlights swept across a frame farmhouse. As promised, Key's yellow Lincoln was parked in front. She pulled in beside it, grabbed her bag, and alighted.

The dogs went berserk.

Key had been watching for her from the living room window. As soon as she wheeled in, he pulled open the front door. Unfortunately he didn't reach it in time to call off the hunting hounds who charged out from their various lairs to surround Lara with snarling maws. They raised a horrendous racket.

She jumped onto the hood of her car and thrashed her legs, trying to kick away the howling attackers. Key emitted a shrill whistle that brought a sudden halt to the barking. A few of the hounds whimpered as they slunk back to their hideouts.

"Good Lord! I could have been chewed to pieces."

"All's clear now. Hurry." He pushed open the screen door. Tentatively Lara placed one foot on the ground. Out of the darkness came a menacing growl, but when Key ordered, "Hush!" the dog fell silent.

She picked her way up to the porch. "Whose house is this? Why am I here?"

"Helen lost the baby."

She stopped dead in her tracks and looked at him meaningfully. He motioned her inside with a brusque movement of his head. By the light of the Berrys' homey living room, he noticed that Lara's face was free of makeup. She hadn't taken time to brush her hair. It was still pillow-tousled, reminding him of the first time he'd seen her. That night, she hadn't known his name. She'd smiled at him a couple of times, even when threatening to notify the sheriff of his gunshot wound. She wasn't smiling tonight. Her expression said she wouldn't waste spit on him if he was on fire.

"Where is she?"

"Back here."

"When did the spotting start?"

"Spotting?" he repeated. "She was goddamn near bleeding to death when I got here."

He led her through a long, narrow hallway. The walls were decorated with framed photographs that chronicled the growth of a family. Time had yellowed some of them. The most recent one was of Helen in her graduation cap and gown.

Key stood aside and let Lara precede him into the bedroom where Helen lay in a single bed, clutching a teddy bear to her chest and quietly weeping.

"Helen? The doctor's here." He moved to the side of the bed and took her hand. It was flaccid and cold. He pressed it between his own, trying to restore animation and warmth.

He didn't know which was worse, her abject despondency now or her previous hysteria. She had called him at The Palm. "It's a woman," Hap had said as he passed him the telephone receiver. "Says your sister told her to try and catch you here. She sounds stressed out."

That had been an understatement. He'd hardly been able to hear her above the din inside the bar, but her alarm came through loud and clear. When he reached her house and rushed into the bedroom, he saw a copious amount of dark, clotted blood on her sheets. He'd immediately called Lara Mallory.

"Hello, Helen," she said now, bending down and laying a gentle hand on Helen's brow. "Everything's going to be all right. I'll take care of it, okay?"

Her bedside manner was flawless, but Helen didn't buy it. "I lost my baby."

"You're sure?"

Helen nodded and glanced across the room. Lara followed her gaze to the soiled sheets which Key had stripped from the bed and piled up in the corner. Lara looked at him. "Will you excuse us, please?"

He gave Helen's hand a hard squeeze. "Hang in there, sweetheart. I'll be in the living room if you need me."

"Thanks, Key."

He backed out of the room. Lara was placing a blood pressure cuff around Helen's arm as he closed the door. In the living room he posted himself at the wide picture window and stared out into the night. Away from the lights of town, the stars were brilliant. It never failed to astonish him how many there were. That was one of the reasons he loved night flying. Only then could he fully appreciate the vastness of the sky and know peace.

He wished like hell he were up there now.

A hound dog loped up onto the porch, slurped water from a bowl, yawned broadly, then dropped its head onto its front paws and went back to sleep. A night bird called plaintively. Occasionally the old lumber inside the walls would shift with a groan and a creak. Other than that, the house was quiet.

He wondered what was going on in the bedroom. How long would it take for Dr. Mallory to do whatever she was doing? Time crawled. When the bedroom door finally opened, he turned away from the window and rushed to meet her halfway down the hall. She was wearing surgical gloves and carrying the bloody sheets.

"Seeing these is upsetting her. They need to soak."

He led her to a screened-in back porch that ran the width of the house. It was equipped with a deep utility sink, into which she put the sheets, and then turned on the cold water. "You know your way around the house very well."

"Her daddy's about the best hunter in East Texas. I've gone with him lots of times, ever since I was a kid."

"That's why you know how to call off the dogs."

"Yes. This is where we cleaned up after dressing our kills." He nodded down to the sink now filling with pink water.

The sight of blood had never bothered him. He'd seen ghastly war injuries, men whose flesh was melting off their skeletons following oil well fires, even the severed head of a Moslem woman caught in adultery. He'd thought he had a cast-iron stomach where violence was concerned, that nothing could make him queasy.

He was wrong. This blood bothered him tremendously. He ran his hand down his face and looked away from the sink.

"I examined the expulsion," Lara said as though reading his mind. "She miscarried the embryo."

He nodded.

"Where are her parents?"

"They took the younger kids to Astroworld today," he answered mechanically as he watched Lara peel off her surgical gloves. "Helen wasn't feeling well and begged off. It's a good thing, too. She hadn't told them about the baby yet. Imagine if this hadn't happened at home, in bed. Jesus," he added grimly, "it doesn't bear thinking about."

"Besides, the fewer people who know about this, the better, right? Especially for you. Look at it this way, you're off the hook now."

Although it took all the willpower he possessed, he let the insult pass.

When the sink was full, she turned off the faucet. "I've given Helen an injection to retard the bleeding and a sedative to help her sleep. In the morning she can come to the clinic and I'll do a D and C."

"Good. Her folks aren't supposed to be back until late tomorrow night."

"By then she'll be home, although I recommend a few days of bed rest. She can tell them she's got a severe case of cramps, which, unfortunately, is true." After a significant pause, she added, "I also highly recommend that sexual intercourse be suspended for several weeks. You'll have to take your fun with someone else."

His eyes homed in on hers. Matching her scorn measure for measure, he said, "Any suggestions?"

They didn't break eye contact until the dogs set up another howl. A car door slammed. There were running footsteps on the porch.

"Helen?"

Key moved around Lara and went through to the living room. Jimmy Bradley was standing there, frantically glancing around.

"Key?" he exclaimed. "What are you doing here? Me and some of the guys went to Longview to knock around tonight. When I got home my brother said you'd called. Said for me to haul ass over here. What's happened? Where is everybody? Where's Helen?"

"She's in her bedroom."

Jimmy noticed Lara, who had just entered the room, gave her a puzzled glance, then cut his eyes back to Key. "What's going on?"

"This is Dr. Mallory."

"A doctor? For Helen?" he asked with mounting alarm.

Key laid a hand on the young man's broad shoulder. "She had a miscarriage tonight, Jimmy."

"A mis— ?" He gulped hard, darted another look at Lara, then at Key. "Jesus." He broke away from Key, ran down the hall, and burst into the bedroom. "Helen?"

"Jimmy? Oh, Jimmy! I'm sorry!"

Key looked at Lara. She was staring at him, whey-faced, her lips parted in surprise. "I hate to disappoint you," he said dryly, "but the baby wasn't mine. Helen came to me for help because she knew she could trust me."

He allowed himself only a moment of self-righteous indignation before turning abruptly and following Jimmy to the bedroom. Jimmy was seated on the edge of the bed, clutching Helen to him, running his hands over her back and shoulders. Both were crying.

"Why didn't you tell me, Helen? Why?"

"Because I was afraid you'd give up your scholarship. I didn't want you to be stuck with me and a baby."

"Honey, as long as I can carry a damn football, I can go to school. That college doesn't care if I've got three wives and six kids. You should have told me. You went through hell all by yourself."

"Key helped." She sniffed. "I knew how much you respected him, so, when I didn't know where else to turn or what to do, I asked him for advice. He begged me to tell you, but he also promised to keep my secret."

"I didn't think I should keep the secret any longer, Helen," Key told her from the open doorway. "I felt Jimmy had a right to know so I called him tonight."

"I'm glad you did," Jimmy said fervently.

"So am I. Now," Helen added softly as she

nuzzled his chest. "I've missed you so much."

"Me, too. When you broke up with me, I got mad for a few days. Then the hurt set in. I couldn't figure why you'd stopped lovin' me all of a sudden like that."

"I didn't stop loving you. I never will. It's because I love you so much that I didn't want to be a burden to you, to hold you back or keep you from taking this opportunity."

"As if you could ever be a burden. You're my second half, Helen. Don't you know that?" Jimmy bent his head and kissed her softly on the lips, then pulled back and whispered, "I'm sorry about our baby."

When Helen began crying again, Key knew it was time to leave the young lovers to work through their reconciliation and regret alone. He stepped into the bedroom only long enough to retrieve Lara's black bag.

"Sometime before her folks come home, see to things on the back porch," he told Jimmy. "Take her to Dr. Mallory's office in the morning. No one else will ever know."

The younger man nodded. "Thanks, Key. You're the best." Key kissed the tip of his finger and pressed it to Helen's temple, then left the room.

He found Lara in the living room, seated on the sofa, hugging her elbows. She looked at him with cold reproach. "You could have told me."

"And spoiled your fun? Think of the hours

of pleasure you've had despising me."

"I'm sorry."

Suddenly he was very tired and didn't feel like dragging this out. Every time they were together, they were at each other's throats. The emotional events of tonight had left him feeling drained; the fight had gone out of him. "Forget it."

She stood and reached for her bag. He handed it over to her. It weighed down her arm like an anchor. "You okay?" he asked. "You don't look so hot." She too appeared tired, bone-weary, and dispirited. "You're pale."

"No wonder. You woke me out of a deep sleep, and I didn't take time to use my blusher." She moved to the front door. "Can I get out of here without being mauled by coon dogs?"

Key secured the front door and they left the house together. The dogs were roused, but Key gruffly ordered them to stay where they were. Once Lara was in the driver's seat of her car, she rested her forehead on the steering wheel.

"Are you sure you're all right?"

"Just tired." She raised her head and reached for the door. He moved aside and let her close it, then watched as she drove away. He kept her in sight as he climbed into the pimp-mobile. She drove slowly, as if it were a newly acquired skill.

At the crossroads, he debated over whether to return to The Palm. It was late. Only the drunkest of the drunks would still be there. He didn't feel like carousing. But he wasn't ready to go home, where he always felt claustrophobic.

In the opposite direction, the taillights of Lara's car disappeared behind a rise in the road. "What the hell," he muttered as he turned the Lincoln around.

In spite of her protests, she hadn't looked too chipper. He was responsible for getting her out at this time of night. The least he could do was follow her to see that she got home safely.

Lara didn't notice his headlights in her rearview mirror, so it came as an unpleasant surprise when the Lincoln pulled into her driveway as she was unlocking the clinic's back door.

"I'm closed!" she called. Undeterred, Key joined her on the back steps. "What do you want now? Why can't you leave me in peace?"

Her voice was beginning to fray. If she noticed the weakness, he was certain to hear it too. The tears she had managed to hold back during the drive home filled her eyes, making his image watery.

Turning her back to him, she inserted the key into the lock. At least she attempted to, but her vision was blurry and her hands were unsteady.

Key reached around her. "Let me."

"Go away!"

He took the key from her, pushed it easily into the lock, and opened the door. The alarm began its delay buzzing. He went in ahead of her and moved to the panel.

"What's the code?"

She wanted to tell him to go to hell, wanted

to forcibly remove him, but didn't have the strength for either. "Four-o-four-five." He punched in the code and the buzzing ceased. "It won't do you any good to know the code," she told him peevishly. "I'll change it tomorrow."

"Where's your coffeepot?"

"In the kitchen. Why?"

"Because you look like shit, like you could keel over any second now. A cup of strong black coffee would probably be good for whatever's ailing you."

"You're what's ailing me. Leave me alone, and I'll be fine. Can't you do that? Please? It's so simple! *Just go!*"

She didn't want to fall apart in front of him, but the choice was no longer left to her. Her voice cracked on the last two words. She raised her hand to indicate the back door, but it moved to her mouth instead and covered a sob as her knees buckled. She sank into the nearest chair. Tears overflowed her eyes. Her shoulders began to shake. Despite her best intentions, she couldn't contain the racking sobs.

Propping her arm on the back of the chair, she laid her head on the crook of her elbow and surrendered to the emotional outburst. Pride deserted her. Grief, bitterness, and pain had clawed their way to the surface and, having been tamped down for so long, would not be restrained.

To his credit, Key didn't interfere by asking questions or offering banalities. The light remained off; the concealment of darkness lent some comfort. She cried until her head ached. Then, for several minutes, she kept her face buried in her sleeve and suffered the aftershocks of the violent catharsis. The tremors came in waves, significant but not sufficient to produce another tidal wave of emotion.

Eventually she raised her head, expecting to see him standing there gloating. She was alone but noticed that a dim light from the kitchen spilled out into the hallway. Weakly coming to her feet, she smoothed back her hair and went to the kitchen.

He was leaning back against the range. Only the night light above the cook surface had been turned on. It cast dark shadows onto his face as he sipped from a steaming cup of coffee. He'd found her bottle of brandy. It was standing open on the counter. She could smell its pungent bouquet, enticingly mingled with fresh coffee.

As soon as he noticed her, he nodded toward the coffeemaker. "Want me to pour?"

"No, thanks. I can." Her voice sounded rusty from so many tears. It disturbed her that he was on her turf, making himself at home in her kitchen in the hours just before dawn. Key Tackett, her self-proclaimed adversary, had been rummaging through her pantry, handling her things, and was now offering to pour her coffee in her own kitchen.

"Feel better?"

She listened for sarcasm behind his seemingly innocent question but heard none. Nodding, she

311

carried her cup to the kitchen table and sat down. She took a sip. The coffee was scalding and potent, the way a man would brew it. "You can go now. You don't have to stay. I'm not self-destructive."

Ignoring what she'd said, he pushed himself away from the stove and, bringing the bottle of brandy with him, sat down across from her. He added a dollop of the liquor to her cup.

His eyes were steady and disconcertingly watchful. His fingertips moved up and down the glassy surface of the coffee mug cupped between his strong, tanned hands. She feared that if she watched them too long, they would have a hypnotic effect on her.

"What was that all about?"

Self-consciously, she hooked her hair behind her ear. "That's really none of your business, is it?"

His head dropped forward, and he cursed as he exhaled.

His hair grew in a swirling pattern around the crown of his head. Even in the dim light she could see the cowlicks. The most gifted barber would be challenged by them. Perhaps that's why he wore his hair long and loose and in no particular style.

When he raised his head, his eyes were angry. "You refuse to let me be a nice guy, don't you?"

"You're not a nice guy."

"Maybe I'm trying to change." She gave him a retiring look, which only heightened his anger.

"Bury the hatchet for once, okay? And bury it someplace besides my skull. Can't you forget my last name? Even temporarily? I'll try to forget yours. Deal?" He held her stare until she lowered her gaze.

Taking that as concession, he said, "Thanks for what you did tonight. I was out of my element and knew it the minute I saw the condition Helen was in, physically and emotionally. It was a scene out of hell, and you handled it like a real pro. You . . . were terrific."

Again Lara listened for sarcasm, but there was none. Those words, she knew, were difficult for him to say. It would be churlish of her not to accept the compliment. "Thank you." Then, with a self-deprecating laugh, she added, "Actually I'm great during emergencies. I never crack under pressure. Only afterward. Then I collapse."

It seemed a long time before he spoke again. When he did, it was in a hushed voice that invited confidence. "What was the crying binge about, Lara?"

She felt herself respond not only to his tone but to his speaking her name. Still she hesitated, unwilling to bare her soul to him. Although what did it matter now? He'd already witnessed her loss of self-control.

Her throat ached from so much crying. She cleared it before speaking. "My daughter. It was about my daughter."

"I guessed as much. Go on."

She threw back her head, then rolled it around her shoulders. "Sometimes when a case involves a child, it conjures up the nightmare. Ashley dies all over again." She sniffed and blotted her nose with a paper napkin from the dispenser on the table.

"There've been two in the last few days. First Letty Leonard. Now Helen's fetus. Knowing that a small, helpless, innocent life was needlessly lost . . ." She shrugged eloquently. "It still affects me. Deeply." She sipped from her coffee mug, which felt very heavy in her trembling hand. The brandy had been a good idea. It warmed and soothed all the way down.

"Tell me about her."

"Who, Ashley?"

"Pretty name."

"She was pretty." She laughed softly, with embarrassment. "Every mother thinks that about her child, I know, but Ashley *was* pretty. From the day she was born. Blond and blue-eyed, cherubic-looking. She had a perfectly round face and rosy cheeks. Truly a beautiful child. And she was a good baby. Content. She didn't cry much, even during the early months. She had an unusually happy disposition. Her smile was like sunshine. Even strangers commented on it. She . . . beamed. Yes, beamed," she said reflectively.

"She seemed destined to make everyone around her smile, to light up a room when she walked in. She certainly lit up my life." Her coffee was growing cold. She folded her hands

314

around the mug in a vain attempt to retain the warmth.

"Until she was born, I was desperately unhappy. Randall's job required all his time and concentration. Montesangre is a hideous place. I loathe it. All of it. The climate, the land, the people. Living there in banishment was the bleakest period of my life. Or so I thought at the time. I didn't learn what real despair is until I lost my child."

She paused for a moment to stave off another smothering attack of bereavement. She swallowed with difficulty and briefly mashed her fist against her lips. When she felt it was safe to speak, she cleared her throat again and continued.

"Ashley made even that horrid place bearable. When I nursed her, it was as nurturing to me as it was to her. For weeks after I weaned her, my breasts ached." She covered her breasts with her hands, feeling once again the pain of disuse and remorse. Then, remembering herself, she lowered her hands and glanced at Key. He sat unmoving, watching and listening. "And then she died."

"She didn't die. She was killed."

She sipped her coffee, but it was cold now so she pushed the mug aside. "That's right. There is a distinction, isn't there?"

"Definitely."

She waited for him to say more, but he didn't. "What do you need, a play-by-play account?"

"No," he answered quietly. "I think that's what *you* need."

It was on the tip of her tongue to tell him to go to hell, but the words died unspoken. She didn't have enough energy for defiance. Moreover, perhaps he was right. Perhaps she did need to talk about it.

"We were on our way to a party," she began. "A wealthy local businessman was throwing a birthday bash for one of his seven children. I didn't particularly want to go. I knew it would be an ostentatious affair. The way in which the wealthy Montesangrens flaunted their wealth made you almost sympathize with the rebels. Anyway, Randall insisted that we attend the party because the host was an influential man.

"I dressed Ashley in a new dress. Yellow. Her color. I put a yellow bow in her hair, on the top of her head where her curls were the thickest." She touched her own hair to demonstrate.

"Randall had arranged for someone on the embassy staff to drive us, thinking it would be more impressive if we arrived with a chauffeur. He was sitting in the front seat with the driver. Ashley and I were in the back. We were playing patty-cake. The car approached a busy intersection. Ashley was laughing, squealing. She was happy."

Lara couldn't go on. Resting her head in her palm, she pinched her burning eyes shut. After a moment, she forced herself to continue.

"The driver stopped for the traffic light. Sud-

316

denly, the car was surrounded by armed, masked guerrillas. I didn't realize this at the time. It all happened too fast. I didn't know anything was wrong until the driver fell forward against the steering wheel. He'd been shot through the head at close range. The second bullet shattered the front windshield. It struck Randall.

"The third bullet was intended for him too, but he had slumped to one side. Ashley was hit instead. Here." She touched the side of her neck. "Her blood splattered over my face and chest. I screamed and fell across her to protect her. That's when I was shot, in the back of my shoulder. I didn't even feel it."

She paused and sat staring into space. It was an effort to continue, but she knew that healing processes were customarily painful.

"Bystanders started screaming. People left their cars idling and scattered in every direction, seeking cover. They were safe. It was us the rebels were after. Three of them opened the passenger door and grabbed Randall. He shouted in pain and outrage. I believe one of the gunmen struck him in the temple with the butt of his pistol. Randall lost consciousness before they carried him to their waiting truck. I read all this later in the newspaper, after they had executed him. I knew nothing at the time of the kidnapping. All I knew was that my baby was dying.

"I knew it, but I couldn't accept it," she continued hoarsely. "I was screaming. I couldn't stop the bleeding. I pushed my finger into the bullet hole in her neck to try to stop it. The authorities arrived within minutes of the attack, but I was hysterical. They had to prize Ashley away from me. They dragged me to an ambulance. I don't remember anything after that. I lost consciousness. When I woke up, I was in a hospital in Miami."

She didn't realize that tears were rolling down her face until one ran into the corner of her lips. She licked it away. "The ambush on our car marked the official beginning of the revolution. The rebels attacked the birthday party, too. It was a bloodbath. Only a few survivors lived to tell about it. No doubt we would have been killed there. I don't know why they chose to ambush us en route.

"Because of what happened to Randall, the United States closed the embassy in Montesangre — what was left of it after it was ransacked — and abruptly discontinued diplomatic relations with their new government.

"Following his execution, the revolutionaries returned Randall's body to the States. It was more a gesture of contempt than largess, because they also sent gory photographs of the firing squad to the secretary of state. They didn't send back Ashley's remains, nor any pictures of her body or coffin. No death certificate. Nothing. They ignored all Washington's demands for either more information or the release of her body. After a while, Washington lost interest and stopped demanding. I've continued to badger

318

them, but as far as our government is concerned, the matter is closed.

"Oh, God." She covered her face with her hands. "My baby is still down there. I never got to touch her. Never got to see her face one last time. Never got to kiss her goodbye. She's somewhere down there in that wretched place. That —"

"Don't, Lara." He was there in an instant, standing beside her chair, smoothing back her hair. "You're right. It's a goddamn nightmare, but for Ashley it was over in a heartbeat. She didn't suffer any fear or pain."

"Yes, the pain has been all mine. I thank God for that. But at times it's so crushing that I don't think I can stand it anymore. There's no relief from it." She pressed her fist against her chest. "It hurts so bad. *I want my baby back!*"

"Shh. Don't do this to yourself. Don't." He pulled her to her feet. His arms went around her.

Instinctively, her fingers curled into the fabric of his shirt, and she pressed her face against his chest. "I'll never forget it. But there are parts of it that I can't remember. Like frames of a motion picture film, segments have been clipped out, and I'm afraid they're important. I want to remember the missing pieces, but my mind blocks them out. Sometimes I can almost grasp a lost memory, then it eludes me. It's as if I'm afraid to grasp it. I fear those things I can't remember."

"Shh-shh. It's all right. It's over and you're safe."

The assurances were whispered into her hair before his lips moved to her brow. Lara became aware of how good it felt to be held by someone physically stronger than herself. There had been no one with whom she could share this grief. Not her parents, who implicitly blamed her for everything that had happened, including Ashley's death. All her friends had deserted her when she made banner headlines for being Clark's mistress. For years she'd carried this burden alone. It was an unexpected luxury to lean on someone else and, for a few moments, relinquish a portion of the cumbersome weight.

Placing his fingertips beneath her chin, Key tilted her head up and grazed her lips with his. "Don't cry anymore, Lara." The raspy words were lightly ground against her mouth. "It's all right." Again, his lips rubbed hers. "Don't cry."

Then he kissed her, a deep, hot, wet, questing kiss.

Lara's eyes slowly closed. She swirled in a maelstrom of fluid heat. Her will was voluntarily surrendered, and her mind went on a sensuous ride where nothing mattered except the connection — mouth to mouth, tongue to tongue, man to woman. It fulfilled a primal need she wasn't even aware she possessed.

Her response was instinctual. Her hands clutched him yearningly. She tipped her middle up, a gesture purely feminine, a silent solicitation for intimacy.

As though from a distance she heard his soft curse, then felt his hands moving across her shoulders, down her back, over her hips, drawing her against him, pressing her close. Closer.

It was that sudden and shocking familiarity with his body, or perhaps a self-preserving resurgence of sound judgment, that jolted her out of the sensual mist and into cold reality.

She pushed herself away and turned her back to him. Seeking support, she leaned forward against the counter. She took several deep breaths and vainly tried to disregard the desire rioting through her.

"Take me there."

He said nothing.

She let go of the counter and faced him. "Take me there. I've got to know what happened to my child. I've got to see her death certificate, touch the soil in which she's buried. Grasp . . . something."

His face remained impassive.

"That closure, that final goodbye, is essential to one's survivors. That's why we have funerals and eulogies and wakes." Still he said nothing. "Damn you! Say something."

"You weren't bullshitting. You really intend to go back."

"Yes. And you're going to fly me."

He folded his arms across his chest. "Now why would I do something that dumb?"

"Because you're smart enough to realize that

321

I'm right. Clark was instrumental in getting Randall assigned to Montesangre. My baby died as a consequence of your brother's cowardly, political machinations."

"A debatable point at best," he said. "So, in order to make your argument more convincing, you decided to throw in some tongue-twisting kisses, right?"

Heat rushed to her face. "One has nothing to do with the other," she said gruffly.

He made a snide, scoffing sound. "You know, Doc, you've just lived up to all my expectations. In fact, you surpassed them." He whistled long and softly, wagging his hand as though he'd touched something hot. "One little kiss and you're ready, baby."

He snickered insultingly as he looked her over, then started toward the door. "Find yourself another sucker. I'll pass on taking a vacation to a war zone. I'm sure as hell not interested in fucking my dead brother's leftovers."

He was so angry, it was a life-threatening risk to drive, yet he pointed the Lincoln toward home and pushed it through the night like a Sherman tank. He was angry with her, but that was nothing new or surprising.

The surprise was that he was angry with himself. He, who never analyzed his actions or apologized for anything he did, was riddled with guilt because he wanted his late brother's mistress. If circumstances had been different, if she had given him the go-ahead, he'd be tugging off his boots right about now.

Jesus. Didn't he have any more character than to be craving a piece of the woman who'd caused his brother's downfall? Jody was right about him after all. Who better to know a child's character than his mother? He was rotten to the core, just like his old man. Where women were concerned he had no discretion and no conscience. If he did, his cock wouldn't be hard enough to drive nails, and the taste of Lara Mallory's mouth wouldn't still linger on his tongue.

When they were growing up, he and Clark had shared things, sometimes voluntarily, sometimes under parental duress. They swapped sweaters, shaving lotion, skateboards. But they'd never shared women. Not the easy girls at school. Not even whores.

This tacit agreement had evolved out of their adolescence, possibly because romance was one arena in which they didn't want to compete. As brothers, they were constant subjects of comparison, but they drew the line when it came to sexual aptitude. Key had never wanted a girl that Clark had dated before him, and, although he couldn't put thoughts into Clark's head, he figured his brother had felt the same way. That's why his desire for Lara Mallory was so puzzling and infuriating. It violated one of his own commandments.

He knew he had just as well get over this itch for her because he could never scratch it. To want the woman who had tainted his brother's name and destroyed his future was sinful. And while sin had never been a deterrent to his doing anything he wanted to do, stupidity certainly was.

That was the crux of his anger. He felt like a stupid fool for listening like a trusted old fogy while she poured out her tearful story. He'd brewed coffee, for chrissake! Then he'd gone one step farther and held her. Kissed her.

"Shit." He hit the steering wheel with his fist.

She was probably still laughing, knowing that she'd built a fire in his gut that he doubted ten other women could extinguish. A woman didn't let you make love to her mouth like that without knowing damn good and well what it was doing to you. No wonder she'd chosen that moment to make her pitch about a trip to Central America. She figured she had him so wound up he'd agree to take her to Mars if she asked.

Guess again, Doc, he thought with a smirk. He'd been hot for a lot of women, but even in the throes of passion he'd never taken a total departure from his reason.

On second thought, she hadn't looked particularly complacent when he left. She had seemed as confused and humiliated as he felt now. True enough, the story of her daughter's death had been heartbreaking. He still didn't trust her, but when it came to Ashley's murder, who could doubt that her suffering was genuine? The kid's death had shattered her, and she wasn't over it yet.

When I nursed her, it was as nurturing to me as it was to her.

She seemed destined to make the people around her happy.

She had adored that kid and had taken her death harder than Randall Porter's brutal execution. Of course, following the nasty scandal involving Clark, their marriage couldn't have been on solid ground. By her own admission, she'd been miserably unhappy in Montesangre. Only the birth of her daughter had made life there livable. To her, Ashley must have been like a consolation prize, a sign of God's forgiveness. Having lost Clark, she'd transferred all her love and attention to her baby.

Suddenly Key withdrew his foot from the accelerator. The Lincoln began to slow down. He stared sightlessly into the darkness that was gradually lifting on the eastern horizon. But the imminent sunrise didn't register on him. Nor did he realize that the Lincoln was straddling the center stripe as it rolled to a stop.

Other things Lara had said echoed in his head.

Blond and blue-eyed.

Her smile was like sunshine.

She beamed.

Key knew of only one other person who'd been described in such radiant, solar terms. Clark Tackett the Third.

"Son of a bitch," he whispered as his hands

heedlessly slipped from the steering wheel and landed in his lap.

Lara Mallory's beloved Ashley had been his brother's child.

Chapter Fifteen

Ollie Hoskins went to work with his feather duster on the cans of pork 'n' beans, chili, tamales, and tuna in aisle 6. As manager of the Sak'n'Save supermarket, he could have delegated dusting the shelves to one of the stockboys, but he enjoyed doing the menial tasks — pricing, stocking, sacking — because the work was clearly defined and easily dispatched. It was mindless labor that he could do while thinking about something else.

He'd served in the United States Navy for fifteen years before mustering out, and while he didn't miss the months at sea, he looked back fondly on the freedom from responsibility he'd enjoyed as a sailor. He'd never desired to be an officer and was still better at taking orders than issuing them.

One spring while on shore leave in Galveston, he'd met a young woman on the beach, fallen in love, and married her within a month. When it came time for him to reenlist, she urged him not to and relocated them to her hometown of Eden Pass so that she could be close to her mother.

They probably would have been better off staying in the service, Ollie thought now as he moved to aisle 5, where the shelves were neatly

327

stocked with flour, sugar, spices, and shortening. His wife's family had never welcomed him into the fold. Ollie hailed from "up north somewhar," and, in their estimation, the only thing worse than being a Yankee would be to have an ethnic heritage. That he was Anglo made him tolerable — barely.

After twenty years, he still wasn't crazy about his in-laws, and vice versa. The bloom of love had long since faded from his marriage. Now, about the only thing he and his wife had in common was their boy, Tanner.

In their individual ways, they doted on him. His mother frequently embarrassed him with her overt demonstrations of affection. She'd been unable to conceive after Tanner — a condition that she implied was Ollie's shortcoming, not hers — so she fussed over him like a mama bear with her cub. It tickled her pink that he was Heather Winston's steady. Having her son dating the most popular girl at the high school somehow elevated her social standing among her friends.

Ollie had nothing against Heather. She was as cute as a button, friendly, full of pep. He only hoped that Tanner didn't let the romance get out of hand. He'd hate to see his son's future compromised by healthy lust.

Frequently Ollie looked at Tanner and marveled over the genetic quirk that had produced from his seed, and his wife's lackluster bloodline, such a smart, good-looking boy. Thank God he was athletic. If he'd wanted to play an instrument in the marching band, or had aspired to be a chemist or a rocket scientist, his relatives would have shunned him as a weirdo. But Tanner could kick and throw and carry a football, so he was affectionately walloped and jabbed and hugged by his rowdy cousins and uncles. They claimed him as theirs and conveniently forgot that Ollie was physically responsible for his origin.

Ollie didn't mind. Tanner was his, and he nearly busted his buttons every Friday night when number twenty-two charged onto the football field wearing the crimson and black of the Fighting Devils. The approaching season promised to be Tanner's best one yet.

Ollie finished straightening the cans of Crisco, rounded the sale display of Nabisco cookies at the end of the aisle, and entered aisle 4 — coffee, tea, and canned beverages. Two women were moving along the aisle. The younger was pushing the cart while the older consulted a shopping list.

"Good morning, Miss Janellen, Mrs. Tackett," Ollie said pleasantly. "How are you this morning?" He'd never quite gotten the knack of saying "y'all." This deficiency in his vocabulary still branded him a Yankee outsider.

"Good morning, Mr. Hoskins," Janellen replied.

"Ollie, have the butcher cut us three T-bone steaks, one inch thick. And I don't mean seven-eighths. Last time they were cut much too thin and were so tough we couldn't chew them."

"I apologize, Mrs. Tackett. I'll make certain it's done to your liking this time." Just as Miss Janellen could always be counted on for a smile, he could depend on Jody Tackett to be a bitch. Lying, he said, "It's good to see you up and about."

"Why wouldn't I be?"

He was only trying make friendly conversation. By the way she snapped at him, you'd think he'd insulted her. "Why, no reason," he said, feeling his bow tie growing tighter around his neck. "I'd just heard you weren't feeling well these days. But you know how gossip travels."

"I'm feeling great. As you can see."

"Mama and I haven't been shopping together in a long time." Sweet Janellen was trying to smooth over the awkward moment. "We thought we'd treat ourselves."

"Well, it's certainly good to see you both. I'll go tell the butcher about those steaks and have them waiting for you at the checkout counter." He poked the handle of the feather duster into his rear pants pocket, turned, rounded the end of the aisle, and bumped into a grocery cart pushed by another woman.

"Dr. Mallory!" he exclaimed.

"Hello, Mr. Hoskins. How are you today?"

"Uh, fine." *Lord have mercy*, Ollie thought; Jody Tackett and Dr. Lara Mallory were on a collision course. He didn't want his store to be the scene of any trouble. "Did you see those watermelons in the produce section, Dr. Mallory?

330

They came in from South Texas early this morning."

"A whole watermelon is wasted on one person, I'm afraid."

"I'll slice one and sell you a portion."

"No, thanks. I'll stick to cantaloupe."

When she smiled, his heart sped up a little. Regardless of the reputation that stereotyped sailors, he'd never been a dedicated skirt chaser. But he'd have to be blind not to notice that Dr. Mallory was a real looker. Her face and figure turned heads. In Eden Pass her name was synonymous with *temptress*.

Frankly, he'd never seen that side of her. She was friendly but never flirtatious. Maybe he just wasn't her type, although a natural flirt usually flirted with everybody of the opposite sex. Like Heather's mother. Now that woman was a tart if he'd ever seen one. He hoped to goodness that Heather didn't take after Darcy in that respect. Tanner was a good boy, but it wouldn't take much encouragement from a pretty girl like Heather for him to do something he ought not.

"Let me know if there's anything you need, Dr. Mallory."

"Thanks, Mr. Hoskins. I will."

Regrettably, he saw no way to avoid disaster. He moved aside and let her pass, thinking that maybe he should warn her that Jody Tackett was in the next aisle. He hoped the doctor didn't need any coffee or tea. Fatalistically he watched as she wheeled her cart into aisle 4. He loitered at the end of it, pretending to rearrange packages of Oreos and Fig Newtons. He prayed that he wouldn't be called upon to referee a cat fight.

The squeaky front wheel on Dr. Mallory's cart rolled to a stop. For several moments there was silence, then he heard her say, "Good morning."

Janellen replied in her shy little voice, "Good morning, Dr. Mallory."

"I'm glad to see you're feeling better, Mrs. Tackett." Dr. Mallory gave Jody ample opportunity to respond. When she didn't, the doctor added, "I've called your house several times, hoping to speak with you."

"We have nothing to say to each other." Only Jody Tackett could have put that much venom into a few simple words. "Let's go, Janellen."

"Excuse me, Mrs. Tackett, but we have an awful lot to say to each other. I'd like very much to talk to you about Clark."

"I'll see you in hell first."

"Mama!"

"Hush, Janellen! Come along."

"Please, Mrs. Tackett. Mrs. Tackett? Mrs. Tackett!"

At first there was an underlying plea in the doctor's voice. Then inquiry. Then alarm.

"Mama!"

Ollie Hoskins knocked over several packages of Nutter Butters in his haste to get to aisle 4 to see what had happened. He arrived in time to see Jody Tackett reel sideways against her cart. She extended her arms at her sides, palms down, as though trying to regain her balance. The cart rolled forward; she lost her support and fell against the shelves stacked with boxes of Lipton's Tea. Several glass jars of instant decaf crashed to the floor, breaking on impact and spilling their fragrant powders. Jody fell backward against the shelf, then slid to the floor. She lay prone upon shattered jars and instant tea.

Janellen dropped to her knees. "Mama! Mama!" Lara Mallory didn't waste a second. She was beside Jody before Ollie could blink. "Call 911," she shouted back at him. "We need an ambulance."

He, in true military fashion, passed the command to one of his subordinates, a checker who happened to be restocking cigarettes in front of her register. She turned and ran toward the office phone. The aisle was now filling up with other shoppers who'd been alerted by Janellen's frantic screams. Deserting their carts, they converged on aisle 4 from every corner of the supermarket. Ollie ordered them to stand back so the doctor would have room to see to Mrs. Tackett.

"Hold her arms. She could break a bone."

Janellen tried to catch Jody's flailing arms so she wouldn't bang them against the shelves. Even if no bones were broken, she was going to be badly bruised.

Dr. Mallory dug into her handbag and produced a clear, acrylic key ring in the shape of a large key. She thrust it into Jody's mouth and used it to depress her tongue.

"It's okay, it's okay," she told Janellen. "Her breathing passage is clear now. I'm holding down her tongue. She can breathe."

"But she's turning blue!"

"She's getting oxygen now. Keep holding her arms. Mr. Hoskins, did you call for the ambulance?"

"Yes, ma'am," Ollie briskly replied. He turned to the checker, who nodded her head to confirm. "Anything else I can do?"

"Find my brother," Janellen said. "Get him here."

Jody was drooling from the corners of her mouth. Her legs were still thrashing. It took all Janellen's strength to confine her arms. Dr. Mallory kept her tongue depressed with the key ring, but her breathing sounded like a combine. Ollie didn't have a soft spot for Jody Tackett, but he figured the lady deserved some privacy.

"All you people, clear this aisle."

Of course no one moved. He shoved his way through the growing crowd and ran to his open, elevated office at the front of the store.

Knowing that Key Tackett was a pilot, Ollie called the county airstrip first. Key wasn't there, but old Balky Willis gave him Key's portable phone number. "He left here 'bout fifteen minutes ago. He had that hand-held gadget with him."

Twenty seconds later, Key answered his portable phone with a cheerful, "Pimp-mobile."

"Mr. Tackett?" Ollie said nervously. He'd

334

never had a run-in with Key, but he'd heard about the unfortunates who had. Even his brothers-in-law, all of them wild as March hares and ready to draw blood at the drop of a hat, spoke Key Tackett's name with reverence and respect. "This is Ollie Hoskins down at the Sak'n'Save and —"

"Hey, Ollie. I watched that Crimson-Black scrimmage the other night. Tanner's going to give 'em hell this season."

"Yes, sir, thanks. Mr. Tackett, your mother just collapsed here in —"

"Collapsed?"

"Yes, sir. Your sister and —"

"Is she all right?"

"No, sir. We've called for an ambulance."

"I'm on my way."

Ollie dropped the phone and rushed back to aisle 4. Clusters of shoppers blocked it at both ends. "Excuse me. Let me through." It pleased him to discover that he'd regained his military bearing sufficiently that he could make people heed him. "Please, everybody, stand back," he ordered with newfound confidence. He moved to stand directly behind Dr. Mallory.

"Is she having a stroke?" Janellen asked the doctor fearfully.

"Possibly a mild one. Tests will tell. Has she done this before?"

"No."

Dr. Mallory leaned down nearer the fallen woman. "Mrs. Tackett, an ambulance is on the

335

way. Don't be frightened."

Jody had ceased to struggle for breath. Her limbs had relaxed and now were limp. She rolled her eyes from side to side as though trying to orient herself. Lara gradually withdrew the large plastic key from her mouth. It had teeth marks deeply imbedded in it, which explained why Dr. Mallory hadn't used her fingers to clear Jody's breathing passages. She wiped saliva from Jody's chin with a Kleenex from her own purse.

"You had a seizure, but it's over."

"Mama? Are you all right?" Janellen clasped her hand.

"She'll be groggy for several minutes," Dr. Mallory said. "That will pass."

"Let me through. What are y'all gawking at? Don't you have anything better to do? Get the hell away from here."

Key plowed through the crowd of spectators. They parted for him. Ollie stepped forward. "You must have been close to have gotten here so fast."

"Thanks for calling me, Ollie. Clear these people out, will you?"

"Yes, sir!" Ollie barely stopped himself from saluting. Key Tackett had that effect on people. "Okay, everybody. You heard Mr. Tackett. Clear this area."

"Key! Thank heaven!" Janellen cried. "Mama had a seizure."

"Jody?"

"Don' le' 'er touch me."

336

He knelt beside his mother, but his piercing eyes were on the doctor. "What's the matter with her?"

"Just as your sister said, she had a seizure. Serious and scary, but not fatal."

Key bent over his mother. "They've called an ambulance for you, Jody," he said in a low, reassuring voice. "It'll be here soon. Hang in there."

"Ge' 'er away from me. Don' want 'er to touch me."

Her speech was slurred, but her message was clear.

"Dr. Mallory saved your life, Mama," Janellen said gently.

Jody tried to sit up but couldn't. She fixed a murderous stare on Dr. Mallory. Although she couldn't articulate her animosity, it was effectively conveyed.

Key made a swift motion with his head. "Take off, Doc. She doesn't want you near her. You're only making matters worse."

Janellen said, "Key, if she hadn't —"

"But —" the doctor interrupted.

"You heard me," he barked. "Get out of her sight."

They glared at each other for what seemed to Ollie a long time, as if there was a lot more there than the eye could see. Eventually Dr. Mallory came to her feet. She was visibly shaken and her voice was unsteady. "Your mother is gravely ill and needs immediate medical attention."

"Not from you."

Even though the words weren't directed at him, Ollie quailed at Tackett's fierce expression and bone-chilling tone.

"Thank you, Dr. Mallory," Janellen said quietly. "We'll see that Mama gets the medical care she needs."

Her services having been flatly rejected, she turned her back on the Tacketts and moved down the aisle toward the onlookers. They parted for her as they had for Key. She didn't return to her cart of groceries but headed straight for the exit.

Ollie watched her leave, his respect for her increasing. She had a lot of class. She hadn't slunk past the bystanders but had walked tall and proud. Neither the Tacketts nor the gawkers had daunted her. He resolved to personally deliver her groceries to her once this crisis was over.

The wail of a siren was heard outside and moments later paramedics rushed into the store. Mrs. Tackett was transported by gurney to the waiting ambulance, which sped away. Key and Janellen roared after it in his yellow Lincoln.

Long after all the instant tea in aisle 4 had been swept up and the shelves straightened, store customers lingered to discuss what they'd seen and heard, and the drama was re-created for new arrivals who had missed it. The seriousness of Jody Tackett's condition was speculated upon. Some said she was too mean to die and would live to be one hundred. Others surmised that she was only a breath away from death. Some wondered out loud about the future of Tackett Oil. Would Jody's death, whenever it occurred, also mean the end of the oil company, or would Key stop his globe-hopping and stay in Eden Pass to manage it, or was Miss Janellen strong and savvy enough to seize control? Opinions varied widely.

However, the juiciest gossip that day centered around Dr. Lara Mallory and how, even as she faced death, Jody Tackett had refused her assistance. The doctor's notorious affair with Senator Tackett was rehashed for those whose memories had faded.

Ollie was resentful of the clacking tongues. Not that his opinion mattered, but he didn't think Dr. Mallory was getting a fair shake. Hadn't she saved stingy, nasty Jody Tackett's miserable hide, when she'd probably just as soon have watched the old woman swallow her own tongue?

She was almost tearfully grateful when he delivered her groceries that afternoon. She thanked him profusely and offered him a cold drink for his effort. She might have been a fallen woman once, but a nicer lady you'd never find, was his way of thinking.

"Can you believe it? Old Jody was lying there on the floor of the Sak'n'Save, foaming at the mouth, they said, jerking and twitching something awful. But the old girl had enough fight left in her to refuse medical attention from Lara Mallory."

The Winstons' housekeeper had prepared a cheesy chicken casserole for dinner. Darcy was doing more talking than eating. Fergus was transferring food from his plate to his mouth with single-minded purpose. To Heather, the casserole looked like something that had already been regurgitated. She pushed the chunks of food around her plate, pretending to eat. Now that she was taking birth control pills, she counted every calorie and wasn't about to waste several hundred on this junk.

Besides, her mother's enjoyment of the gossip that had circulated through town about Mrs. Tackett's seizure had ruined Heather's appetite. Darcy had learned all the gory details at the beauty shop and recounted them with disgusting enthusiasm.

"She peed her pants. Jody Tackett peed her pants. Can you believe it?" Darcy chortled. "Incontinental, they call it."

"It's 'incontinent,' Darcy," Fergus corrected. "And it's hardly something I want to talk about over supper."

Heather reached for her glass of iced tea. "Tanner's daddy said Dr. Mallory saved Mrs. Tackett's life. If I were her, I'd've let the old fart die."

Darcy's fork clattered to her plate. "That's fine language for a proper young lady! And this juvenile crush you have on Lara Mallory has become annoying, Heather."

"I don't have a 'crush' on her. I just think it was stupid of Mrs. Tackett not to let the doctor help her. I mean, if you're dying, isn't any doctor, even one you personally dislike, better than none at all?"

"Not if you're Jody Tackett," Fergus remarked as he paused to blot his mouth. "That woman's heart is the hardest substance on earth. I agree with you, Heather. I'd have let her choke."

"As usual, you two are taking sides against me." Darcy angrily pushed her plate aside.

"Sides?" Fergus asked, bewildered. "I didn't know we were choosing up sides over this. What's it got to do with us?"

"Not a damn thing," Darcy snapped. "I just fail to see what makes Lara Mallory such a bloody heroine in Heather's eyes."

"May I be excused?" Heather asked in a bored voice.

"You may not! You haven't eaten a bite."

"I'm not hungry. Besides, this casserole is gross. It reeks with fat."

"I should have been so lucky to have a maid cook dinner for me when I was your age!"

"Oh, please." *Here we go,* Heather thought — *another sob story about Mother's deprived childhood.*

"She shouldn't have to eat it if she isn't hungry," Fergus said.

"Naturally, you let her have her way."

341

"Thanks, Daddy. Tanner and I will get something later."

"You're going out with Tanner again tonight?" Fergus asked.

"Of course." Heather looked at her mother and smiled smugly. "We're officially together now."

"Together?"

"Going steady," Darcy clarified impatiently, never taking her eyes off Heather. "I can't say I'm thrilled about it."

Heather, holding her mother's stare, took another sip of tea. Putting her on birth control pills had been Darcy's doing, but Heather was getting back. She seized every opportunity to remind her mother that whenever she and Tanner went out on a date, they could have sex without suffering any consequences.

Darcy couldn't say anything to her, especially in front of Fergus. He still didn't know about the contraceptives and would have raised hell with Darcy for encouraging them. He clung to the quaint notion that morality was a deterrent to premarital sex.

Heather took pleasure in keeping her mother perpetually miffed. Her sidelong glances and innuendoes conveyed that she was now sexually active. But she hadn't let Tanner go all the way yet. It wasn't because she didn't love him, or that she feared an unwanted pregnancy, and there certainly was nothing to fear in the way of parental reprisal.

Her reason for holding out was the same as

it always had been. She didn't want to become a replica of her mother.

Tanner was being very sweet about her abstinence. Since that night at the lake when he had disgraced himself, he was loving and patient, gratefully taking whatever crumbs of eroticism she chose to toss him and asking for nothing more.

Heather was still Fergus's little angel, and when she was with him she strove to maintain his image of her. Her relationship with her mother, however, had deteriorated. They were undeclared adversaries, two women in a silent face-off. The battle lines that had been suggested before were now clearly drawn.

"I didn't realize that you'd made Dr. Mallory an idol, Heather," Fergus observed as he stirred sugar into his coffee. "I didn't even know you'd met her."

"Mother took me to see her. Didn't she tell you?"

"For a checkup," Darcy said hastily. "She needed a physical exam for cheerleading, and it was going to be a month before she could see an out-of-town doctor. I decided it was silly to shun Dr. Mallory just because she was involved with Clark Tackett at one time. Who cares? It's ancient history. Besides, an enemy of Jody Tackett's is a friend of yours, right?"

"I must say Dr. Mallory showed a lot of gumption by moving to Eden Pass in the first place. She shoots straight from the hip, too. I like that."

"When have you talked to her?" Darcy wanted to know.

"Yesterday. She called me and asked for an audience with the school board. She wants to speak to the high school kids about sexual responsibility. I think the idea is a little bit radical for Eden Pass, but I told her we'd hear her ideas at the meeting next week."

For several moments Darcy regarded him without comment. "You're right, Fergus. She's got her nerve. She was caught in adultery. How sexually irresponsible can you get?"

"She emphasized that she wasn't concerned with the moral aspects. She only wants to alert the kids to the health risks involved."

"I doubt that'll go down well with the local preachers. And don't be so sure that morality doesn't figure in there somewhere. Lax morality, that is. She told Heather to make sure she always had a condom handy."

"That's not what she said!" Heather exclaimed.

"Same as," Darcy said curtly. "Before we know it, the kids at the high school will start packing rubbers in their lunch boxes and having quickies between classes."

"Darcy, please!" Fergus harrumphed. "Heather shouldn't be listening to this."

"Wake up and smell the coffee, Fergus. Kids nowadays know all about everything. Once Lara Mallory gives them the green light, they'll be screwing like rabbits."

344

Fergus flinched. "She's not going to encourage them to have sex. She wants to warn them of the possible consequences."

"Oh, brother! She really snowed you, didn't she? What I think she wants is an outbreak of teenage pregnancies in order to drum up some much-needed business."

"That's ridiculous, Mother."

"Shut up, Heather! I'm talking to your father."

"But you're twisting Dr. Mallory's words around. It's not fair."

"This is an adult conversation, and no one invited you to join in."

At that moment Heather hated her mother and wanted badly to expose her hypocrisy. But her love for her father guaranteed her silence. Darcy knew that and used it. She was the one now wearing the smug smile. Heather scraped back her chair and flounced from the dining room.

On her way out she heard her mother say, "Go ahead and grant Dr. Mallory an audience with the school board, Fergus. It'll be fun to sit back and watch the fur fly."

"I thought I'd . . . I probably shouldn't have come." Now that she was standing on the front porch of Lara Mallory's clinic, spotlighted by the overhead light fixture, Janellen felt like a fool. It wouldn't surprise her if the doctor slammed the door in her face. She wouldn't blame her, either.

"I'm glad you came, Miss Tackett. Come in."

345

Janellen stepped into the dim waiting room and glanced around. "It's late. I shouldn't have disturbed you."

"Quite all right. How is your mother?"

"Not too well. That's what I came to talk to you about."

Lara indicated the hallway that led to the rear of the building. With Janellen behind her, she moved out of the clinic and into her private living quarters.

"I was having a glass of wine. Will you join me?"

They entered a cozy den where magazines were scattered over tabletops and scented candles flickered in votives. The TV was tuned to a cable station that broadcast classic movies. The one currently being shown was in black and white.

"I'm a fan of old movies," Lara said with a self-deprecating smile. "Maybe because they usually have happy endings." She used the remote control to turn off the set. "Chablis is all I have. Is that all right?"

"I'd rather have a soft drink."

"Diet Coke?"

"Fine."

While Lara was getting her drink from the kitchen, Janellen stood as though rooted to the floor in the center of the room. She had invaded the enemy camp, but it was certainly a comfortable place. Two walls of the room were lined with bookshelves. Most of the reading material was related to medicine, but there was also a collection of hardcover and paperback fiction. Over the fireplace, where once had hung the stuffed head of a ten-point buck, there was now an Andrew Wyeth print. On the sofa table stood a silver-framed photograph of a baby girl.

"My daughter."

Janellen jumped at the sound of Lara's voice as she reentered the room carrying an icy glass of soda. "Her name was Ashley. She was killed in Montesangre."

"Yes, I know. I'm sorry. She was a beautiful child."

Lara nodded. "I have only two photographs of her. That one and another in my office. I have those because I reclaimed them from my parents. None of our personal effects were ever recovered from Montesangre. I wish I had something of Ashley's. Her teething ring. Her teddy bear. Her christening gown. Something." She shook her head slightly. "Please, sit down, Miss Tackett."

Janellen gingerly lowered herself onto the sofa. Lara sat in the easy chair she'd obviously been occupying when her doorbell rang. There was a crocheted afghan bunched up on the hassock in front of the chair and a glass of white wine stood on the end table.

"Is your mother in the hospital?"

Janellen shook her head.

"No?" That was obviously not the answer she had expected. "I thought for certain her condition would require at least one night in the hospital."

"She should be hospitalized." Janellen felt herself on the verge of tears. She picked at the cocktail napkin wrapped around the glass of soda. "I came because . . . because I wanted to hear what you had to say. You were there during my mother's seizure. I'd like to know your professional opinion of it."

"Your mother certainly didn't."

"I'm sorry about the way she behaved toward you, Dr. Mallory," Janellen said earnestly. "And if you ask me to leave, I'll understand."

"Why would I do that? I don't hold you responsible for what your mother said and did."

"Then please give me your opinion of her illness."

"It's unethical for me to second-guess another doctor's diagnosis when I haven't even examined the patient."

"Please. I need to talk to somebody about this, and there's no one."

"What about your brother?"

"He's upset."

"So are you."

"Yes, but when Key gets upset or worried, he . . ." She lowered her eyes to the glass in her hand. "Let's just say he's currently unavailable. Please, Dr. Mallory, give me your opinion."

"Based strictly on what I saw?"

Janellen nodded.

"With the full understanding that I could be incorrect?"

Again Janellen nodded.

348

Lara took a sip of wine. Looking toward the portrait of her daughter, she pulled in a deep breath, then released it slowly. Her eyes moved back to Janellen. "What treatment did your mother receive at the county hospital?"

"They examined her in the emergency room, but she refused to be admitted."

"That was foolish of her. Were you given a diagnosis?"

"The doctor said she'd had a mild stroke."

"I concur. Did they do a complete blood work?"

"Yes. She was prescribed medication that's supposed to thin her blood. Is that what you would recommend?"

"Along with extensive tests and observation. Did they do an EKG?"

"The heart thing?" Lara nodded. "No. They recommended it, but she wouldn't stay that long."

"Was a brain scan done?"

"Yes, but only after Key threatened to tie her down if she didn't consent. The doctor said he didn't find any significant cerebral infarction." She tried to quote him precisely. "I'm not certain what that means."

"It means that your mother has no significant amount of dead brain tissue due to a loss of blood supply. Which is good. However, that doesn't mean that the blood to her brain isn't being interrupted or completely blocked. Did he suggest doing sound wave tests on the carotid artery?

349

They're called Dopler studies."

"I'm not sure." Janellen rubbed her temple. "He was talking so fast, and Mama was complaining so loudly, and —"

"These tests would determine if there's an obstruction in the artery. If there is, and the blockage isn't eliminated, there's a very good possibility for infarction, resulting in permanent disability or even death."

"That's what they said, too," Janellen said hoarsely. "Something like that."

"No angiogram to see where the blockage might be?"

"Mama refused that. She ranted and raved and said she'd had a dizzy spell and that's all there was to it. Said she only needed to go home and rest."

"Did the impairment to her speech and muscle control last very long?"

"By the time we got her home, you couldn't tell anything had happened."

"That quick recovery fools patients into believing they've suffered only a dizzy spell." Lara leaned forward. "Does your mother frequently forget things? Does she sometimes have blurred vision?"

She told the doctor what she had shared with Key a few weeks earlier. "She never admits to any of this, but the spells have gotten noticeable. I tried persuading her to see a doctor, but she refused. I think she's afraid of what she'll hear."

"I can't be certain without examining her,"

Lara said, "but I think she's experiencing what we call TIAs, which stands for transient ischemic attacks. 'Ischemia' refers to insufficient blood circulation."

"I'm following you so far."

"When one of these occurs, it interrupts the blood supply to the brain. It's like an electrical blackout. The part of the brain that's affected is turned off. The dementia you described, blurred vision, slurred speech, and the dizziness are all symptoms, warning signals. If they're not heeded, the patient can suffer a major stroke. Today was probably the strongest warning yet. Has she complained of numbness in her extremities?"

"Not to me, but she wouldn't."

"Does she have high blood pressure?"

"Very. She takes medication to control it."

"Does she smoke?"

"Three packs a day."

"She should stop immediately."

Janellen smiled wanly. "Never in a million years."

"Urge her to eat properly and monitor her cholesterol intake. She should do moderate exercise. See that she takes her medication. Those precautions will help prevent a life-threatening stroke, but there are no guarantees."

"There's no complete cure?"

"For selected patients the arterial blockage can be removed surgically. It's a fairly routine procedure. Unfortunately, without the proper tests and your mother's full cooperation, that's not an option." Sensing Janellen's despair, she leaned forward and pressed her hand. "I'm sorry. And remember, I could be wrong."

"I doubt you are, Dr. Mallory. You've said essentially what the emergency room doctor told us. Thank you for discussing it with me. And for the soda." She set the untouched drink on the coffee table and stood to go.

"Under the circumstances, I doubt we can be friends, but I'd like us to be cordial. Please call me Lara."

Janellen smiled but remained noncommittal. When they reached the front door, both were surprised to see that it was raining. It was much easier to talk about something as banal as the weather. Finally, Janellen shook the doctor's hand.

"You had every right to be rude to me. Thank you for inviting me in."

"Thank you for giving my opinion credibility. The next time you visit, let's hope the reason for it won't be so serious."

"Next time? Are you asking me to come back?"

"Of course. Feel free to drop in anytime."

"You're very nice, Dr. . . . Lara. I can understand why my brother was so attracted to you."

Lara shook back her hair and, looking up at the rainy skies, laughed mirthlessly. "You're wrong. Key isn't the least attracted to me."

Janellen was stunned. "Key?" she repeated with puzzlement. "I was referring to Clark."

Chapter Sixteen

Bowie flipped up the collar of his denim jacket and huddled closer to the exterior wall of the house. The eaves provided scant protection from the blowing rain. He was getting soaked.

He really couldn't say why he was at the Tacketts' place at this time of night, standing outside in the rain. He should be stretched out in front of his secondhand TV set. His rented trailer had few amenities, but at least it was dry.

Whatever the weather, he had no business being here. Jody Tackett's health was a private family matter. They'd hardly want an outsider butting in. None of that had affected his decision to come; he had felt compelled. When he arrived, he noticed that Key's Lincoln was gone and so was Janellen's car. He parked the company truck out of sight behind the detached garage. The only car in the driveway belonged to the housekeeper.

He saw no need to announce his presence to her. What would he say? He supposed he could tell her the truth — that he was worried about Miss Janellen; how she was reacting to her mama's collapse in the Sak'n'Save. Then the housekeeper would probably want to know what business it was of his, and he'd have to say no business of his at all, and she'd shoo him off the porch and probably call the law.

So he lurked in the shadows, standing ankle deep in rainwater. He couldn't adequately justify his reason for being there. He just knew he had to be. Furthermore, he intended to stay right where he was, come hell or high water, until he saw for himself that Miss Janellen was holding together.

He hadn't laid eyes on her since that afternoon of their kiss, followed by her startling declaration that she loved him. He hadn't taken it seriously, of course. Something had caused her to blurt it out — PMS, or too much sun, or maybe an allergy pill that had made her a little goofy. In hindsight, she probably felt like cutting out her tongue.

Because he empathized with anyone who shot off his mouth without thinking, he'd been avoiding Janellen, sparing her the embarrassment of having to face him and offer an excuse for her bizarre behavior. Sure enough, she'd gone out of her way to avoid him, too.

They couldn't keep dodging each other forever, though. Sooner or later they'd meet, so it might as well be tonight when she had something even more terrible to fret over. He couldn't do anything about her mama's failing health, but he could relieve her of one concern. He could assure her that he didn't intend to take advantage of something she'd said during a mental lapse of unknown origin.

Headlights appeared at the end of the private drive. Bowie's gut clenched reflexively as he watched the car turn off the county road and onto Tackett property. He shrank back closer to the wall, not wanting to be seen until he was certain it was Janellen. Reputedly, Key kept a loaded Beretta beneath the driver's seat of his car. It could be gossip, but Bowie would just as soon not have it confirmed the hard way. If Key saw a prowler, he might shoot first and ask questions later.

The headlights, diffused by the rain, approached slowly. Bowie recognized Janellen's car. She parked in the driveway, got out, and dashed through the rain toward the back door. The screen door squeaked when she pulled it open. She had her key in the latch when he softly called her name.

Startled, she spun around. Rain fell on her pale face as she peered through the gloom. "Bowie! What in the world are you doing out here?"

"Are you okay?"

"I'm okay, but you're soaked. How long have you been out here? Come inside."

"No, I'll go on along home now." He knew he must be a sorry sight, what with the brim of his hat dripping rainwater and his pants wet from the knees down. "I just wanted to make sure you were all right, considering what happened this morning. Word around the shop is that Mrs. Tackett is feeling poorly."

"Unfortunately, that's true." She unlocked the door and insisted he follow her inside. Reluctantly he stepped into the kitchen, but stayed just inside the door.

"Take off your jacket," she said. "And your boots. They're sopping wet."

"I don't want you to fuss."

"No fuss. Let me check on Mama and send Maydale home, then I'll make some coffee." She moved through the dark kitchen, but turned when she reached the doorway. "Don't go away."

Bowie's heart swelled so large he could barely draw breath. She hadn't screamed or shuddered or puked when she saw him. That was a good sign. Now she was asking him, almost pleading with him, to stick around. "No, ma'am. I surely won't."

While she was gone, he removed his hat and his damp jacket and hung them on a wall peg near the back door. Balancing on one leg at a time, he tugged off his boots and placed them beside a pair that obviously belonged to Key. The toes of his socks were damp, but he was relieved to see that they didn't have holes.

He tiptoed across the vinyl tile floor. Leaving the lights off, he gazed through the window over the sink, watching the rain drip from the eaves. After several minutes he heard a muffled conversation at the front door, then watched through the window as Maydale picked her way around puddles to her car while trying to protect her beehive hairdo with a silly plastic bonnet.

At the sound of Janellen's approach, he turned. "How's your mama doing?"

"Sleeping."

"She's all right, then?"

"Not really. She won't follow doctor's orders. She's too hardheaded to heed the warnings, like the one she got this morning. She doesn't believe her condition is serious."

"From what I've heard, she's a stubborn old gal."

"To say the very least."

"Maybe her condition isn't as bad as the doctors say."

"Maybe."

"Sometimes they exaggerate to make their point and justify their bill."

Her wan smile indicated she didn't believe that and knew that he didn't either. "Well," she said, pulling herself up straighter, "I promised you some coffee."

"You don't have to bother."

"No. I want to. I'd like some, too. I won't be sleeping much tonight, so I might just as well."

She moved toward the pantry, but her footsteps were sluggish and her voice unsteady. She didn't turn on the lights, probably because she didn't want him to see the tears in her eyes. He saw them anyway.

The coffee canister almost slipped from her hands before she set it down on the counter. Peeling a single paper filter from the compressed stack proved to be a challenge. Once that was done, she spilled coffee grounds as she scooped them from the canister.

"Oh, dear. I'm making a mess." She began

twisting her hands and brutalizing her lower lip by pulling it through her teeth.

He felt about as useless as a teat on a boar hog. "Why don't you sit yourself down, Miss Janellen, and let me make the coffee?"

"What I'd really like you to do . . ." She struggled to get the next words out. "What I'd really like . . ."

"Yes, ma'am?"

She turned and looked at him imploringly. "If it's not too much to ask, Bowie."

"Name it."

She uttered a little squeaking sound, tilted her head to one side, then swayed forward. He caught her, encircled her with his arms, drew her against his chest, and hugged her close. She was so slight, he was afraid he might be holding her too tightly, but trustingly she laid her cheek on his shoulder.

"Bowie, what will I do if Mama dies? *What?*"

"You'll go right on living, that's what."

"But what kind of life will I have?"

"That depends on what you make of it."

She sniffed wetly. "You don't understand. Key and Mama are all that's left of my family. I don't want to lose them. If Mama dies, Key will go on about his business, and I'll be left here alone."

"You'll make out just fine by yourself, Miss Janellen."

"No, I won't."

"Now why would you say that?"

"Because I've never had an identity of my own. People only see me in relation to my family. I'm

Clark Junior's daughter. Clark and Key's little sister. Jody's girl. Even though I've been doing most of the work at Tackett Oil the last couple of years, everybody thinks I'm just Mama's puppet. I guess they're not too far wrong. She's always told me what to do, and I've obeyed her, partially because she's usually right, but mostly, I suppose, because I lack the self-confidence to stand up to her and offer a different opinion. I've never really minded answering to her, but when she's gone, what then? Who will I be? *Who am I?*"

He pushed her away and gave her a little shake. "You're Janellen Tackett, that's who. And that's enough. You're stronger than you know. When the time comes for you to stand up on your own, you'll do it."

"I'm afraid, Bowie."

"Of what?"

"Failing, I guess. Not living up to expectations." She laughed, but it was a sad sound. "Or, more to the point, I'm afraid that I will live up to everyone's expectations and land flat on my backside when Mama's not here to call the shots."

"It won't be that way," he said with a stubborn shake of his head. "You've got years of experience. The men are used to taking orders from you. You're smart as a whip. I always thought of myself as fairly clever. I've got some street smarts, but when I'm with you — and this is the God's truth — I feel dumber than dirt."

"You're not dumb, Bowie. You're very smart. Nobody else noticed the discrepancy in well number seven."

"Which turned out to be nothing."

"We didn't know that until you installed the test meter."

He'd put the test meter midway between the well and the recorder. The data registered had been the same. A leak could be anywhere along the line. In order to locate it, he'd have to move the test meter until a section of line was isolated. That could go on indefinitely. He'd checked the records and, sure enough, that well had had a flare line, but it had been capped off years ago. He felt like a fool for making such a big deal over something his bosses considered insignificant.

Janellen's hands were still riding on his waist, and that's all he could think about now. Finally he said, "I'm sorry about your mama, Miss Janellen, because I know how much you care about her. I hope she lives to a ripe old age so you'll be spared the grief of her passing. But with or without her, you're your own person. You don't have to be anybody's daughter or sister or . . . or wife. You're good enough all by yourself. You've got plenty on the ball and don't let anybody make you think different."

"You're good for me, Bowie," she whispered.

"Aw, hell, I'm not good for much of anything."

"That's not true! You are! You're very good

360

for me. You make me focus on my strong points instead of my weaknesses. Don't get me wrong. I know my limitations. I've lived with them all my life. I know I'm intelligent, but not exceptionally so. I'm not self-assertive, I'm timid, and I lack confidence. I'm not pretty. Not like my brothers."

"Not pretty?" Bowie was baffled, so baffled he didn't stop to wonder when he'd begun thinking of her as beautiful. "Why, you're the prettiest thing I've ever seen, Miss Janellen."

Flustered and confused, she ducked her head. "You don't have to tell me that. Just because of what I said the other day."

He cleared his throat uncomfortably. "I want you to know right now that I'm not holding you to that."

"You're not?"

"No, ma'am."

"Oh." The features of her face worked emotionally. Then she lifted her gaze back to his. "How come?"

He shifted his weight from one foot to the other. "Well, 'cause I know you didn't really mean it, that's how come."

She wet her lips and took a quick breath. "In fact I did, Bowie."

"You did?"

"I meant it from the bottom of my heart. And if you, well, you know, if you ever wanted to kiss me again, it would be all right."

The buzzing inside Bowie's head almost

drowned out the pounding rain on the roof. His heart was beating so hard and fast that it hurt. His throat was tight, but he managed to strangle out, "I do want to kiss you again, Miss Janellen. I surely do."

He slipped his hands beneath her hair and cupped her jaw, then drew her mouth toward his. Her parted lips responded warmly. This time they needed no warm-up, no rehearsal. They skipped getting reacquainted and picked up right where they had left off, engaging in a kiss that left them breathless when they at last pulled apart.

He pressed his mouth against her throat while her hands clutched at his back. "I never knew it could feel like this, Bowie."

"Neither did I. And I've been doing it for some time now."

They kissed again and again, each kiss piercingly sweet and increasingly intimate. They kissed until their lips were swollen, their passions brimming.

He longed to nestle his erection in the cleft of her long thighs, but he curbed the impulse. However, with an eagerness that was instinctual and almost childlike in its innocence, she arched her body against his, in effect accomplishing what he wouldn't do for himself.

The contact was erotically shattering. It would have evoked the animalistic urges of a saint, and that was something Bowie Cato had never claimed to be.

He fumbled beneath her skirt and grabbed a handful of her bottom, kneading the silk-covered flesh once, twice, while mashing his distended fly against her mound. It wasn't premeditated. He didn't weigh the benefits against the consequences. If he'd thought about it at all, he'd never have done it. It was an unthinkable thing to do.

Janellen's soft exclamation brought reality crashing down on his head, and along with it shame and self-disgust.

He released her immediately. Without a word, he crossed the kitchen in three strides, grabbed his boots, his hat, and his jacket, and stomped out the kitchen door and into the downpour.

The moment he reached the truck he'd left parked behind the garage, a jagged fork of lightning rent the darkness, connecting the firmament and the earth with a hot-white brilliance that crackled with wrath and seared the air with ozone.

Bowie figured it was God, meaning to strike him dead. His aim was just a little off.

Thunder rattled the liquor bottles and glassware behind the bar. "Brewing up a real storm out there," Hap Hollister observed as he poured Key another drink.

"Grounded me. I was supposed to be flying to Midland tonight, taking an oilman and his wife home."

"I'm right proud of you, Key. You've got better sense than to fly in this weather."

"Wasn't me who chickened out. It was the wife. Said she didn't want to die in a plane crash."

Hap, shaking his head over the younger man's derring-do, moved away to serve the other customers who had braved the storm to come to The Palm. Some were playing billiards, leaning on their cues and drinking longnecks as they awaited their turns. Others were watching a late-season baseball game on the large-screen TV mounted beneath the ceiling in one corner of the bar. Drinkers were grouped in twos and threes.

Only Key drank alone at one end of the bar. His dark expression and hunched shoulders signaled his mood. News of the incident at the Sak'n'Save had reached every ear in town, and so his silent request to be left alone was sympathetically honored by everyone in the tavern.

Jody was the subject on Key's mind as he sipped his fresh drink, but his thoughts weren't running toward the sympathetic. He'd like to give his mother a good swift kick in the butt. At the hospital and later, when he and Janellen had taken her home against the doctor's recommendation and their own better judgment, Jody had griped and complained and staved off all their attempts to make her comfortable.

"I'm hiring a live-in nurse for you, Jody," he'd told her as Janellen urged her to get into bed. "Janellen keeps office hours. I'm away a lot. Maydale's a good housekeeper, but we can't count on her to handle a medical emergency like the one that occurred this morning. You should have someone with you constantly."

"That's a wonderful idea, Key!" Janellen exclaimed. "Isn't it, Mama?"

Disregarding Janellen, Jody blew smoke at him from her fresh cigarette. "You took it upon yourself to hire me a nurse?"

"She'll be here around the clock to fetch and carry for you."

"I can fetch and carry for myself, thank you very much. I don't want a busybody fussing over me, bossing me, meddling in my things, and stealing me blind when I'm not looking."

"I went through a top-notch agency in Dallas," he patiently explained. "They won't send us a thief. I specified our requirements. I made it clear that you're not an invalid, that you're independent and value your privacy. They're checking their files to see who's available, but promised a nurse would be here no later than noon tomorrow."

Jody's eyes narrowed to slits. "Call them back. Cancel. Who the hell gave you the authority to make my decisions for me?"

"Mama, Key's only doing what he thinks is best for you."

"I'll tell him what's best for me. I want him to butt out of my life. And you too," she said, snatching her jacket away from Janellen, who had assisted her in taking it off. "Get out of my room. Both of you." At the risk of bringing on another attack, they had left her.

He was worried sick about her. When he'd seen her lying on the floor of the Sak'n'Save, spittle on her chin, her dignity gone, he'd almost passed out himself. But he could hardly remain compassionate when his every attempt at kindness was met with a scornful tongue-lashing.

Hell, he could take Jody's crap. He'd been taking it all his life. When weighed against her precarious health, their verbal skirmishes seemed petty. At issue now was that his mother refused to accept the seriousness of her illness. She could die if she didn't undergo the treatment prescribed for her. Only a fool would flaunt mortality like that.

Then, smiling wryly, Key reminded himself that he'd been willing to fly into a stormy cold front and would have done so if the passengers who'd chartered the plane hadn't nixed it.

But that was gambling, a game of chance with risks involved, the outcome uncertain. It wasn't like being told by medical experts that you were a time bomb with the clock ticking and that if you didn't take care of the problem you could die or, what to Key's mind would be worse, live in a vegetative state for the rest of your life.

The doctor at the county hospital had bluntly laid out the sobering facts of Jody's diagnosis to Janellen and him. He would have liked a second opinion. He would have liked having Lara Mallory's opinion.

"Shit." He signaled Hap for another hit.

The last thing he wanted to think about was

Lara Mallory. But, like the intoxicating whiskey, she had a way of infusing his head, permeating it, saturating it. Silent and invisible, she was always there, fucking with his mind.

Had his brother sired her child? Had her husband known? Had *Clark* known? Had knowing that his child died violently precipitated Clark's suicide?

If so, didn't he owe it to Clark — and to Lara — to go to Montesangre and find out the details of the child's death?

Hell, no. It was none of his business. Nobody had appointed him Clark's custodian. It was her problem. Let her deal with it. It had nothing to do with him.

But the more he thought about it, the more convinced he became that Ashley was his niece. He'd tried not to think about it at all, but that was impossible. Just as impossible was forgetting how devastated Lara had been when she recounted her daughter's violent assassination. God, how did anyone retain his sanity after experiencing something like that?

A few weeks ago, he would have bet his last nickel that he would never waste a charitable thought on Lara Mallory. After hearing her story, he would have to be a real bastard not to feel charitable. So he had held her. Comforted her. Kissed her.

Angrily, he drained his drink, then stared into the glass as he twirled it around and around over the polished surface of the bar.

He'd kissed her all right. Not a little, meaningless, charitable peck, either. He'd kissed his brother's married lover and the scourge of his family like it counted. She had accused him of taking advantage of her emotional breakdown, but she was wrong. Oh, he'd pretended that she had his motives pegged perfectly, but, honest to God, when he was kissing her, the last thought in his head was that she was a lying, cheating adulteress who had beguiled Clark. In his arms, with her mouth moving pliantly beneath his, she became only a woman he desperately wanted to touch. He'd abided by the ground rules he himself had stipulated — he'd forgotten her name.

"Haven't you got anything better to do than watch ice cubes melt? Like, for instance, buy a lady a drink?"

Frowning over the unwelcome interruption, Key lifted his gaze to find Darcy Winston seated on the barstool beside his. "Where'd you come from?"

"Just stopped to get in out of the rain. Do I get that drink or not?"

Hap approached. Key nodded tersely, and the bartender took Darcy's order for a vodka and tonic. Key declined when asked if he wanted another.

"Making me drink alone? How rude!" Darcy's carefully painted lips formed a pout.

"That was the idea. To drink alone. You didn't take the hint."

She sipped the drink Hap slid toward her.

368

"Worried about your mama?"

"For starters."

"I'm really sorry, Key."

He doubted that Darcy gave a damn about anybody's well-being except her own, but he nodded his thanks.

"What else is on your mind?"

"Not much."

"Liar. You're sulking. Does it have anything to do with Helen Berry going back to Jimmy Bradley? I hear they're more in love now than they were before you broke them up."

He lowered his head until his chin almost touched his chest. The breakdown of communication was so absurd that he chuckled.

"What's so funny?"

"This town. The other side of the world could blow up, stars could collide and cause another Big Bang, and folks here would still be scurrying around to find out who was screwing who."

"Who are you screwing?"

"That's my business."

"Bastard."

She glared at him with such ferocity that he laughed again. "You sure are dressed up for a Tuesday night, Darcy," he observed, taking in her conservative dress and plain high-heeled pumps. Of course nothing looked conservative or plain on Darcy. The dress was made of flaming pink silk, which she wore well despite her red hair. Her chest filled out the bodice and then some. She'd left the top three buttons undone to provide an enticing peek at cleavage. The high heels added length and shape to her already long and shapely legs. She looked hot — there was no doubt about that.

"I was on my way home from the Library Society meeting," she told him.

"Eden Pass has a Library Society? I didn't even know we had a library."

"Of course we do. And the society has forty-two members."

"No shit? How many of them can read?"

"Very funny." She finished her drink and slammed the glass onto the bar. "Thanks for the drink. Call me if you ever get your sense of humor back. You're a real drag these days."

"What'd you say to piss her off?" Hap asked after she had stalked out. He reached for her glass and dunked it in a basin of soapy water.

"Does it matter?" Key asked testily.

It was still raining, but Key didn't even duck his head as he walked to his car. His mind was on so many other things, the inclement weather was inconsequential.

He got into the Lincoln on the driver's side and had inserted the key into the ignition before he noticed her. She slid across the yellow leather seat and placed her hand high on the inside of his thigh.

"I know what's wrong with you."

"You don't have the foggiest notion, Darcy."

"I'm an expert at these things, you know. I was born with a sixth sense. I can tell what a

370

man wants and needs just by looking at him."

" 'S a fact?"

"That's a fact. When a man wants it, he gives off an odor just like a woman does."

"If that's true, there ought to be a pack of dogs after you."

Taking that as a compliment, she moved her hand to his crotch. "You want me, Key. I know you do. You're just too stubborn to take back the ugly things you said that night at the town meeting." She stroked him, and he had to admit that her technique was excellent.

"This is silly. Neither of us wants to make the first move to reconcile. There's no point in both of us being miserable over a little pride, is there?"

She began unbuttoning his jeans. Key, assuming the role of an impartial observer, let her. He was curious to gauge his response. She lifted him out of his jeans and massaged him between her hands. His cock began to grow hard.

"Oh, baby," she said with a sigh. "I knew that all you needed was Darcy's magic touch."

She smiled at him seductively, then lowered her head to his lap. Her tongue was alternately quick and light, then languorous and lazy. She licked him delicately and sucked him hard. Her teeth threatened pain before her lips kissed soothingly. She knew what she was doing.

Key rested his head against the seat and squeezed his eyes closed. He didn't desire Darcy and was therefore amazed that his body was functioning as it should. On the other hand, why should that surprise him? he wondered. He'd bedded women without ever learning their names. He'd forgotten more women than he remembered. They'd only done for him something he could have done just as well for himself. His body could do it without involving his mind.

He was glad that Darcy hadn't kissed him. That would have made it personal. He would have had to share a part of himself with a woman who meant nothing to him. He didn't even like her.

If Darcy had kissed him, her avaricious tongue might have swept away the taste of another kiss, which he wasn't ready to forget. He kept the memory of it under lock and key like an old man hoarding victory ribbons. On occasion, Key let himself think about that kiss, recall its sweet sexiness, just as that old man would take out his ribbons and finger them sadly while remembering past glories. Then, annoyed with himself and feeling like a fool, Key would shut out the memory, as the old man, ashamed of his sentimentality, would slam the drawer in which he kept his treasured ribbons.

It was pathetic, Key thought, the way individuals longed for something that could never be.

Now he let his mind go blank, disassociating himself from the act but granting his body permission to respond. He didn't touch Darcy, not even when he came. Instead, he clenched his

372

hands around the steering wheel until his fingers turned white. As soon as it was over he calmly rebuttoned his jeans.

Darcy sat up and opened her purse to get a tissue, then daintily blotted her lips. "You know how we know God is a man?" Key said nothing; he'd heard the joke. "Because if God were a woman, come would taste like chocolate."

"Charming."

Either she failed to catch the disgust underlying his comment or she chose to ignore it. Laughing, she rubbed her breasts against his arm. "Where do you want to go? Or should we use that lovely backseat?" she suggested, glancing behind her. "Pity they're not making big cars like they used to. Some of the best fucking I ever did —"

"Good night, Darcy. I'm going home."

"The hell you are! We're not finished."

"I'm finished."

"You mean to tell me that I —"

"You did exactly as you pleased. I didn't ask you to," he reminded her softly. "Now will you kindly haul your carcass out of my car so I can go home?"

She spat in his face.

As quick as a striking cobra, he grabbed a handful of her hair and yanked her head back. "I didn't kill you for shooting me, but I just might for doing that."

Chapter Seventeen

Darcy believed him. She was well aware of the murderous temper for which Key was famous. However, it went against her nature to back down once the die was cast.

"Let go of me, you son of a bitch."

He relaxed his fist, releasing her hair. "Get out," he said succinctly.

"I'm going. But not before I tell you exactly what I think of you. You're sick. Not just mean, *sick*."

"Fine. Now that we've established what's wrong with me, get out of my car."

"You're fucked in the head, and it's not that fat Berry girl who's doing it to you. It's Lara Mallory." His right eye twitched, but the rest of him went dangerously still. Knowing she'd struck a chord, she plucked it again. "Don't you feel just a teensy-weensy ridiculous, falling for your big brother's ex-bimbo?" She laughed derisively.

"Shut up, Darcy."

"The notorious lady doctor has big bad Key Tackett by his short-and-curlies. He didn't learn a thing from his brother's experience with her, did he?"

She knew she should stop while she was ahead,

374

but she couldn't resist making him squirm. Since adolescence, she'd been able to manipulate every man she'd met. Except Key. That had wounded her ego severely, but she knew it wouldn't prove fatal.

"Have you fucked her yet, Key?" she taunted, pushing her face close to his. "When she came, did she cry out your name or dearly departed Clark's? Who's the better lover, I wonder, Senator Clark Tackett or his baby brother? Is that what attracts you to her? Do you want to prove that you're every bit as good in the sack as Clark was?"

Key moved so suddenly, she flinched. He shoved open the driver's door and got out. Then, reaching in, he grabbed the front of her dress and pulled her out. The pink silk soaked up the rain. Her heels sank into the muck.

He ignored her screaming curses as he got back into his car and started the engine. When he reached for the door, Darcy grabbed the handle and wouldn't let go. "Where are you going, Key? To visit your brother's mistress? You're going to be a laughingstock when word of this gets around. And you can bet both balls on it getting around. I'll see that it does. As if it's not funny enough that she's a whore, she's your late brother's whore."

"At least whores put a price on it, Darcy. You can't give yours away." He jerked the car door closed, pushed the gear stick into reverse, and peeled away. The wheels slung wet gravel and mud onto Darcy's shoes and designer stockings.

She shouted dirty names after him. Then, standing there in the drenching rain, she resolved to teach the bastard a well-deserved lesson. She would find Key's greatest weakness and devise a way to pierce it. Only not tonight. She would wait until her anger cooled down and she could approach the problem analytically.

As she slogged toward her car, she was adamant on one point — nobody treated Mrs. Fergus Winston the way Key had and got away with it.

"Thank you, gentlemen," Lara said in conclusion to her address to the seven members of the Eden Pass school board. "I hope you'll give my proposal for some informal sex education seminars careful consideration. If you want any further information to facilitate your decision-making, don't hesitate to call me."

"You've made some very convincing arguments and raised some interesting points," Fergus Winston said. "It's a touchy topic. We've got a lot to mull over. It might take a week or two before we reach a decision."

"I understand. Thank you for allowing me —"

She broke off when the door behind her opened. All eyes swung to it, astonishment registering on every face. Lara swiveled around. Darcy Winston had entered the conference room. Accompanying her was Jody Tackett.

Lara almost recoiled from the malice in Darcy's

eyes when she looked at her. She also had an air of complacency, although she wasn't actually smiling. Jody didn't even deign to glance in Lara's direction.

Hastily the seven board members came to their feet. Only Fergus spoke. He addressed his wife by name, but his eyes were fixed on Jody Tackett.

"What are you doing here, Darcy? This is a closed session."

"Not anymore." Jody still looked unwell, but her voice was strong enough to penetrate matter.

"She insisted on coming," Darcy explained. Fergus finally tore his baleful stare from Jody and looked at his wife. "I'm sorry, Fergus. I know you asked me not to discuss the items on the school board's agenda until they were ready to be made public, but I felt so strongly about this particular issue that I had to do something."

Lara rose from her seat. "I presently have the floor, Mrs. Winston. If you want to address the school board, I suggest you go through the proper channels and petition for an audience the way I did. Or aren't the rules the same for everyone?" She turned and looked pointedly at Fergus.

He had been glaring at Jody Tackett as though she were poison. He looked ready to strangle his wife for bringing her into a chamber where he was in charge.

"Dr. Mallory's right," he said. "If you and Jody have something to call to this board's attention, do it in the proper manner. You can't just bust in like this and interrupt a meeting."

377

"Ordinarily we wouldn't," Darcy agreed. "But —"

"I'll speak for myself." Impatiently Jody approached the conference table. When she was certain she had the undivided attention of each board member, she asked bluntly, "Have y'all lost your senses?"

Eyes were averted. No one spoke. Finally Fergus stiffly invited her to take a chair.

"I'd rather stand."

"Suit yourself."

"I always have."

The animosity between them was palpable. The others seemed embarrassed by it and looked away, but Lara didn't let the awkwardness prevent her from speaking. "Mr. Winston, I insist that the board extend me the courtesy of concluding our meeting."

She was patently ignored.

Jody turned to Reverend Massey, pastor of a local church. "I can't understand you, preacher. Every Sunday you preach against fornication. Yet you're thinking of letting an adulteress talk to our young people about sex?" She sniffed with incredulity and disdain. "Makes me wonder why I'm giving my tithe to your church."

He smiled sickly. "We haven't reached a decision, Jody. We've merely listened to Dr. Mallory's proposal. Rest assured that she's not advocating sin."

"Is that right?" Jody looked toward Darcy. "Tell him what you told me."

She stepped forward, making certain to stand directly beneath the overhead light like an old pro of the boards locating center stage. In a rushed, breathless voice she said, "I took Heather in for a checkup a few weeks ago. Afterward, she told me that Dr. Mallory urged her to start having condoms handy whenever she went on a date."

"That's not what I said!" Lara cried. "I warned Heather about being sexually active without using condoms. Obviously what I told her was misconstrued. Either she didn't fully grasp my meaning, or Mrs. Winston is rearranging the words to suit her purpose here."

"I'm doing no such thing," Darcy shot back. Then, to the board, "Not only that, she told Heather to tell all her friends the same thing. Now if that's not goading teenagers to fool around, I don't know what is. All they need is the power of suggestion and they run with it. You know how kids are. Telling them to take rubbers on their dates is like handing them a license to . . . you know." Chastely she lowered her eyes.

Lara wanted to retaliate, to tell them that Darcy had brought Heather to her specifically to get a prescription for birth control pills. But she couldn't do so without violating patient confidentiality. The secret smile Darcy flashed her indicated that she was well aware of that.

"I cautioned Heather about promiscuity and a multiplicity of partners," she admitted. "I suggested she share the information with her friends. I in no way advocated sexual misconduct."

"Even though you're an expert on the subject?"

"Darcy, please," Fergus said with a soft groan. "Let's keep personalities out of this. Our focus here should be on the young people of our community."

"Amen," the reverend intoned. "Frankly, I have misgivings about holding such open discussions on human sexuality. Our youth have enough temptations to withstand as it is. Their minds are fertile. We should plant seeds that would yield strong spiritual fibers, not doubts and confusion over the devil's handiwork."

"Save the sermons for Sunday, preacher," Jody said. "But I'm glad to know I can count on your vote against this idea."

Her gaze moved down the long the table, pausing on each member of the board. She looked straight through Lara as though she weren't there.

"Once you've had time to think about it, I'm sure all of you will come to the same conclusion. If you don't, I'll have to reconsider my own plans."

"What plans?" one of the board members asked.

"My son Clark loved every day he spent in the Eden Pass school system and often credited it for preparing him for his political career. He would have liked having his name on a school facility. Something like the Clark Tackett the Third Gymnasium. It's getting to where I'm scared to go to the basketball games anymore, afraid I'll break my neck climbing into those rickety bleachers. Those computerized scoreboards are nice, too, aren't they? Wouldn't it be something if Eden Pass were the first school in the area to have one? We'd put the bigger schools to shame, wouldn't we?"

Lara lowered her head. In her mind she could hear the tap-tap of a hammer nailing the coffin shut on her proposal.

Jody let their greedy minds devour the bait before continuing. "I was born in Eden Pass. Lived here all my life. Went through twelve grades of public school here, and so did my three children. I've always boasted that our school system is one of the best in the state."

She leaned on the table and thumped it with the knuckles of her blunt, freckled hand. "I'll change my opinion in a New York minute if you let this woman speak one word under the schoolhouse roof. Why in God's name would you even consider it, knowing what everybody in the country knows about her? Do you want a woman like her having any influence over your kids?" Her face had turned red. She was laboring to breathe.

"I would rather die than let her lay a hand on me. And I'm not just throwing words around. Ask anybody who was in the Sak'n'Save last Tuesday morning."

"You've made your point, Mrs. Tackett." Lara was afraid that Jody was building up to another stroke. She didn't want to be blamed for bringing on the fatal one. "I'm sure everyone here knows that you resented my efforts to save your life. I'm not going to fight you on this because engaging in a contest like that is beneath my dignity. Secondly, I know I can't win. I don't have the resources to bribe the school board with new gymnasiums and state-of-the-art scoreboards."

"Now see here," the minister blustered, "I resent that implication."

Lara ignored him. "Primarily, I'm backing down because I'm afraid the fight might kill you."

Jody focused on her for the first time since entering the room. "Well you're wrong. I won't die until I see you on your way out of town. My town. Clark's town. I won't rest until you're gone and the air is fit to breathe again."

Lara calmly stacked the typed pages of her presentation and zipped them into a black leather portfolio, tucking it and her handbag under her arm. "Thank you, gentlemen, for giving me your attention this morning. Unless I hear from you otherwise, I'll assume that my proposal was rejected."

None of them had the guts to look her in the eye. She derived some satisfaction from that as she turned and walked from the room.

Darcy followed her out. Lara didn't stop until she had reached the main entrance of the building. There, she turned to confront Darcy. "I know why Jody Tackett hates me," she said. "But why do you? What have I ever done to you?"

"Maybe I just think people ought to stay where they belong. You had no business coming to Eden Pass. You don't fit in. You never will."

"What do you care whether I fit in? How am I a threat to you, Mrs. Winston?"

Darcy made a scoffing sound.

"That's it, I'm sure," Lara said. "For some unfathomable reason, you regard me as a threat." Could Darcy's hatred for her relate to Key Tackett? It was an uncomfortable thought, which she kept at arm's length. "Believe me, Mrs. Winston, you've got nothing that I want."

Darcy licked her lips like a cat over a bowl of cream. "Not even a daughter?"

Lara reeled, unable to grasp the extent of the other woman's cruelty. "I didn't give you enough credit," Lara said. "You're not only selfish and spiteful, you're deadly."

"Fucking-A, Dr. Mallory. When it comes to getting what I want, I pull no punches. I have absolutely no scruples, and for that reason I'm dangerous. You can pack up that bit of information and take it with you when you leave town."

Lara shook her head. "I'm not leaving. In spite of what you or Jody Tackett or anybody else says about me, no matter how vicious your threats become, you can't drive me out."

383

Darcy's lips broke into a beautiful smile. "This is going to be fun."

Laughing, she turned and retraced her steps to the administrative offices. Her laughter echoed eerily in the cavernous foyer.

Darcy blew her nose into a monogrammed handkerchief. "I can't stand having you mad at me, Fergus."

After seeing Jody Tackett home, she returned to her house to find Fergus lying in wait for her. She'd seen him this angry with other people, but never with her. It alarmed her. Fergus was her safety net. He was always there to fall back on if things went wrong.

"Please don't yell at me anymore," she begged tremulously.

"I'm sorry. I didn't mean to raise my voice."

Darcy sniffed, then blotted her running mascara. "What I did, I did for you."

"I fail to see that, Darcy."

"Dr. Mallory had placed you in an impossible situation. Because you're president of the school board, you had to be nice to her and honor her request for an audience. Right?"

"Right," he answered warily.

"But I knew you didn't want her conducting sex seminars and handing out rubbers to the high school kids, including our daughter. I was only trying to help you out of a tight spot."

"By dragging Jody Tackett into it? Jesus." He ran his hand over his pointed head. "Haven't you learned anything about me in the years we've been married? I want nothing to do with Jody. I sure as hell don't want her bailing me out of a jam. She's the last person on earth I want to be beholden to."

"I know. I know, Fergus." Her voice had taken on a wheedling tone. "But desperate times call for desperate measures."

"I'll never get desperate enough to send for Jody Tackett's help. The one time I trusted her, I was screwed, blued, and tattooed. For years afterward folks laughed over the way she'd duped me."

"They're not laughing at you anymore."

"That's because I've worked my ass off to make a success of my business. My name means something in this town in spite of Jody Tackett."

"So, relax. You've showed her up."

"It's not enough. It'll never be enough."

She exhaled with exasperation. "The feud is over, Fergus, and you've won. She's old."

"Only a few years older than me."

"Compared to you, she's in her dotage. Besides, she's incidental. Dr. Mallory is responsible for this mess."

"Most of what she said made good sense."

Darcy bit back a crude retort. In a measured tone, she said, "I'm sure it did. She's smart. She's got degrees and diplomas hanging on her office walls." She wiped her nose with the hankie. "I, on the other hand, am just an ignorant housewife. What do I know?"

"Oh, honey, I'm sorry."

Fergus lowered himself beside her on the edge of their bed and clasped her hand. Over the years she had led him to believe that she was more sensitive to her lack of higher education than she actually was. When the occasion called for it, she used it as leverage.

"I wasn't implying that Dr. Mallory was smarter than you."

One eloquent tear rolled down her cheek. "Well, she is. She's a manipulator, too. It probably comes from being around people in politics. She's maneuvered Heather into thinking that she hung the moon. Now you're taking her side over mine."

"No, sugar. That's not it at all. The point is that I hated your calling in Jody for reinforcement."

"It's not because I thought you needed it." She reached out and stroked his face. "God as my witness, that's not the reason I went to her."

"Then why?"

"Because I wanted to put Dr. Mallory in her place. And who better to do it than her archenemy? Don't you see, Fergus? Jody did the dirty work for you, but you, as president of the school board, will get the credit for warding off that Yankee doctor and her so-called progressive ideas."

Deep furrows appeared on his forehead as he reasoned it through. "I never thought of it like that."

Darcy glanced up at him from beneath her eyelashes. "Do you think Dr. Mallory's pretty?"

"Pretty? Well, yeah, I guess she is."

"Prettier than me?"

"No, sugar pie," he said, smoothing back her hair. "There's not a woman alive as pretty as you."

"And I belong to you, Fergus." Snuggling against him, she whispered, "You're the best husband in the whole world." Her hand curled around his neck. "Would you think I was terrible if I wanted to make love right now?"

"In the daytime?"

"It's naughty, I know, but, gosh, Fergus, I just love you so much right now, I want to show it."

"Heather might —"

"She'll be at cheerleading practice for another hour. Please, honey? When you show your strong side and shout at me a little, I get all weak inside. Seeing that macho side of you makes me so hot. I get . . . wet. Down there. You know."

His large Adam's apple slid up, then down. "I . . . I had no idea."

"Feel." She guided his hand beneath her skirt and pretended to swoon when he touched her between the thighs. "Oh, my, God!" she gasped.

Within minutes, Fergus had forgotten all about their quarrel and the reason for it. Darcy kissed and stroked and thrust and panted her way back into his good graces.

If Fergus knew he'd been had, he was content to ignore it.

It took a fortnight for Lara to admit that Darcy Winston and Jody Tackett's threats might have substance. After twenty-one days, she cried uncle. Following the Tuesday morning of Jody Tackett's collapse in the supermarket, Lara didn't see a single patient.

Nancy dutifully reported for work each day, creating busy work for herself to pass the sluggish hours until it was time to go home. Lara filled the days by reading current medical journals. She told herself that this time was valuable, that she was fortunate to have time to keep abreast of new developments and research. But she couldn't completely delude herself. Doctors with full patient loads rarely had time for reading.

She heard nothing from the young attorney retained by Jack and Marion Leonard. If they were pursuing a medical malpractice suit against her, she hadn't yet been notified. Should it come to that, she was confident that once the facts were known, she would be exonerated. However, the negative publicity generated by the litigation would be professionally devastating and emotionally demoralizing. She clung to the hope that they had reconsidered.

The school board never contacted her. Darcy had rallied friends and PTA members to petition the school board against allowing any offensive persons or projects to filter into the school system. Daily, the newspaper was filled with letters to the editor, written by parents and community leaders who were incensed by the proposal recently submitted to the school board by Dr. Lara Mallory. The consensus of the letters was that Eden Pass wasn't ready for such immoral programs to be incorporated into its school curriculum and never would be. The disapproval had been vocal and vehement.

Everywhere she went she was either ignored, sneered at, or leered at by rednecks who assumed she had loose morals because she'd openly discussed such a racy topic with the school board.

She was an outcast. Eden Pass's Hester Prynne. If she hadn't experienced it, she wouldn't have believed shunning this absolute was possible in contemporary America. She began to believe that Jody's prophecy might be fulfilled: she would live to see Lara Mallory leave town.

But not before she got what she came for.

The Tacketts had made her a pariah. They had sabotaged her medical practice. But she'd be damned before she let Key ignore her demand. He would take her to Montesangre. Now.

Chapter Eighteen

"Is he here?"

The yellow Lincoln was parked outside the hangar.

"No, Doc, he ain't," Balky said, earnestly trying to be helpful. "But he was s'posed to come back sometime this evenin'. 'Less he decided to stay in Texarkana. Can't never tell 'bout Key."

"Do you mind if I stick around for a while?"

"Not at all. Might be a waste of time, though."

"I'll wait."

He shook his head in a way that suggested people were mysteries to him. He had a much deeper understanding of engines and what made them tick. Muttering to himself, the mechanic ambled back to the gutted airplane he'd been working on when Lara arrived.

She preferred waiting outside the hangar where the air was slightly less stifling. It was half an hour before she saw the blinking lights of the approaching aircraft and heard the drone of its motor. The sky was clear, deep blue on the eastern horizon, lavender overhead, crimson fading to gold in the west. Key once had tried to explain the peacefulness he derived from flying. On nights like tonight, she could almost relate to his mystical bond with the sky.

He executed a faultless landing and taxied the twin-engine Beechcraft toward the hangar. She was standing on the tarmac when he climbed out of the cockpit. He saw her immediately, but his expression registered neither surprise, gladness, disappointment, nor anger, making it impossible for her to gauge his mood.

Flexing his knees and arching his back, he sauntered toward her. "In Hawaii when your arrival is greeted by a pretty girl, you get *leied.*" He smiled, his teeth showing white in the gathering dusk. "L-e-i-e-d, that is."

"I get it," Lara said dryly.

"Smart lady like you, I figured you would."

She fell into step with him as he moved toward the hangar's wide entrance. "What do you do now? I mean, now that you've landed and your job is finished."

"Hand the keys to Balky and walk away."

"That's it?"

"I'll pick up my money first."

"Who did you fly today?"

"A cattle rancher and his foreman from Arkansas came to look at a bull. I picked them up in Texarkana this morning. They spent most of the day negotiating a price with the owner of the bull, a man named Anderson who owns a large spread near here. It's his plane. He hired me to ferry them back and forth."

"It's a very nice plane," she said, glancing back at it.

"Worth about ninety-five grand. A Queen Aire."

"Sounds like a mattress."

"It does, doesn't it?" Grinning, he entered the building. "Hey, Balky." The mechanic turned and Key tossed him the keys to the airplane.

"Any problems?"

"Smooth sailing. Where's my money?"

Balky wiped his hands on a rag as he moved into the small room where Lara had found Key asleep the morning of Letty Leonard's accident. He went to the desk in the corner opposite the cot and switched on a gooseneck lamp. From a drawer he withdrew a standard white envelope and handed it to Key.

"Thanks."

"Sure 'nough."

Balky left them. Key opened the envelope and counted the bills inside, then stuck it in the breast pocket of his shirt.

"He paid you in cash?" Lara asked.

"Uh-huh."

"No invoice? No record of the transaction?"

"I struck a verbal agreement with my client. Why involve anybody else?"

"Like the IRS?"

"I pay taxes."

"Hmm. The FAA?"

"Mounds of paperwork for every little trip. Who needs it?"

"Don't you have to file a flight plan, stuff like that?"

"Up to twelve hundred feet is uncontrolled airspace. The 'see and avoid' rule applies."

"You always keep to the twelve-hundred-foot ceiling?"

He had tired of the patter. "Interested in flight instruction, Doc? I've got my instructor's license and could have you soloing in no time. I'm expensive, but I'm good."

"I'm not interested in flight instruction."

"You just happened by to shoot the breeze?"

"No, I wanted to talk to you."

"I'm listening." He took a beer from the refrigerator, propped one elbow on the top of the outdated appliance, tilted his head back, and took a long draft.

"It's about a job."

He lowered the can and looked at her with interest. "We've eliminated flying lessons, and I gather it's not another emergency flight to the hospital."

"No."

He regarded her for another long, silent moment before tilting the beer toward her and asking, "Want one?"

"No, thank you."

He took another swig. "Well? My curiosity's killing me."

"I want you to fly me to Montesangre."

He calmly finished his beer and tossed the empty can into the trash can with an accurate hook shot. He sat down in the swivel chair, leaned back, and propped his feet on the corner of the desk, pushing aside the gooseneck lamp with the heel of his boot.

Lara remained standing. There was no place for her to sit except on the cot. He didn't offer it to her, and even if he had, she would have declined.

"You've asked me that more than once, and I've said no. Is there something wrong with your hearing?"

"I'm not joking."

"Oh, you're not joking," he said, tongue-in-cheek. "Excuse me. Hmm. Well. Then are you figuring on parachuting out?"

She folded her arms beneath her breasts. "Of course not."

"Surely you aren't suggesting a landing on Montesangren soil. 'Cause to be suggesting that, you'd have to be plumb crazy."

"I'm serious."

"So am I, Doc. How's your Spanish? Maybe you need to brush up on it. Do you know how Montesangre translates?"

"Yes. 'Mountain of blood.' I know firsthand that it's a literal translation. I felt my daughter's blood running warm and wet over my hands."

He swung his feet to the floor and brought the chair upright. "Then why in hell do you want to go back?"

"You know why. I've been trying to go back for years, ever since I regained consciousness in that Miami hospital. I can't get into the country through proper channels. They're blocked."

"So you're looking at me as an improper channel."

"In a manner of speaking."

"In a manner of speaking, folks are getting blown away down there."

"I'm fully aware of that."

"And you still want to go?"

"I have to go."

"But I don't."

"No, you don't. I was thinking you might regard it as an adventure."

"Well, think again. I've been called many things, but never a fool. If you want to go down there and get your ass shot off, that's your business, but I'm kinda fond of my ass, so you can X me right out of your plans."

"Hear me out, Key."

"I'm not interested."

"You owe me this."

"As you've said. I don't buy it."

"You'll be gratified to know that I haven't had a single patient in the clinic since the morning of your mother's seizure. Jody fought off my attempts to help her. You brusquely denounced me in front of the crowd."

"I didn't have time to use tact. My mother was near death."

"Precisely. And when word got around that the Tacketts preferred death over my medical assistance, the few patients I had cultivated disappeared. Months of hard work was destroyed. The confidence that had been so hard won was invalidated with a few harshly spoken words. Since then I've twiddled my thumbs."

"You're breaking my heart."

She took a deep breath to curb her temper. "I wanted to conduct sex education seminars at the high school. They're vitally important, something that would have benefited the young people of the community."

"Yeah, I read all about it in the newspaper."

"What they didn't print is how Jody bribed the school board to disallow the program."

"You really know how to get folks fired up, don't you?"

"Compared to your mother, I'm an amateur. Once she got finished with me, what little credibility I had left, your lover Darcy ravaged."

"You know, I've heard about mental cases like yours. They're called persecution complexes."

She let that pass. "I've officially closed the doors of the clinic. I dismissed Nancy today. My career has been temporarily suspended. So, you got what you wanted. Your family has effectively demolished any chance I had of practicing medicine in Eden Pass. All things considered, I believe you owe me a concession."

"I owe you zilch."

"I've closed the clinic, but that doesn't mean I'm preparing to leave town." She was down to one final ace. She had to play it. "Your mother vowed she would live to see me leave Eden Pass in disgrace. I doubt she will. I can remain here without working until my savings run out, which, if I live frugally, could be several years."

"That's bullshit. You love medicine too much. You wouldn't give it up."

"I wouldn't want to, but I would."

"Just to spite us?"

"That's right. However, I'm willing to bargain. I'll spare your family any more discomfort and embarrassment, provided you fly me to Central America. As soon as we return, I'll leave. Believe me, I won't be that sorry to go. I'm tired of constant strife and petty gossip. I'm tired of examining myself every time I go out, hoping I'll pass muster.

"Let me tell you something," she said, leaning across the desk, "as far as I'm concerned, the people of Eden Pass have failed to pass muster. They're judgmental and narrow-minded hypocrites, cowards bending to the will of an embittered old woman.

"Take me to Montesangre, Key, and I'll leave this town to you, not because I'm not good enough for it, but because it's not good enough for me."

He said nothing for several moments, then spread his arms out from his sides. "Is that everything?"

She gave a terse nod.

"Good," he said, rolling off his spine and coming to his feet. "I gotta run. I'm hungry as a bear, and Janellen is expecting me for supper."

Lara caught his sleeve as he rounded the desk. "Don't patronize me, you son of a bitch. You've

397

trashed me and my practice, but I won't let you ignore me."

He flung off her hand. "Look, I don't give a damn about local politics and gossip. What my mother does with the school board or anybody else is her business. Unless it involves me directly, I stay out of the boiling pot.

"I guess you're a pretty good doctor, and your clinic has come in handy on occasion, but I couldn't care less if you do brain surgery there, or twiddle your thumbs, or shut it down entirely. Darcy Winston is *not* my lover. And if you've got a hankering to sneak into a country that's on our government's shit list, fine. But count me out."

"How conveniently you turn ethical," she said heatedly, indicating his shirt pocket. "You run illegal charters on a daily basis!"

"Turning you down has nothing to do with ethics. I'm not looking to get killed. Beyond that, I don't trust your motives any farther than I can throw you. So you wasted —"

"What if Ashley is still alive?"

He fell silent and regarded her with piercing intensity.

"Uh, 'xcuse me, Key?" Balky was standing in the doorway, his rheumy eyes darting between them with uncertainty. "I'm leaving for the night. Will you lock up?"

"Sure thing, Balky. Good night."

"Night. Night, Doc."

"Good night."

They listened to his departure. The interruption defused the tension, but only marginally. Keys turned his back on her and ran his fingers through his hair. "Is that a possibility?"

"Probably not. The point is that I don't *know*. I guess in the back of my mind I've clung to the faint hope that she somehow survived. "Her body was never shipped back like her father's." Wearily, she rubbed the back of her neck. "Of course, as a physician and considering the severity of her wound, I know that's highly improbable. She died and was buried. Somewhere alien and unknown to me. I can't live with that. If nothing else, I want to bring back her remains and bury them in American soil."

He turned to face her, but said nothing.

"I need you to do this," she pressed. "One way or another I want to take my daughter out of that place and bring her home. But I can't get into the country. Even ally nations have very few airlines that serve Montesangre because the government is in such constant upheaval. When and if I did get through, as an American citizen I'd be denied entrance into the country and shipped out on the next flight."

"I'd say that's a fairly accurate guess."

"More than a guess. I've been in contact with people in similar circumstances. Many Americans have loved ones in Montesangre whose fates are unknown. Their fact-finding missions have been futile. If they got as far as Ciudad Central, they were dealt with harshly. A few were imprisoned for hours, even days, before being returned to the airport to await the next outbound plane. Some claimed they barely escaped with their lives, and I believe them."

"That's why I don't want to fly over the place, much less land, get out, and walk around," Key said.

"If anyone can get an airplane in and out of there, it's you. Clark constantly bragged about your flying skills. He told me how you've flown into impossible situations to deliver supplies or make rescue attempts, and that you thrive on taking risks — the more dangerous the circumstances, the better." She paused for breath. "Supposing you agreed to do it, could you get an airplane?"

"That's a broad supposition."

"Go with it for the sake of discussion. Could you get a plane?"

He thought it over for a minute. "I know a guy who once asked me to crash a plane for him so he could collect the insurance. He was that badly in debt. He offered to give me thirty percent of his take. If I lived."

"Can you do that? Deliberately crash a plane and live?"

"If you do it right," he said with a fleeting grin. "His offer was tempting. Hell of a chunk of cash. But it wasn't worth the risk."

"Is he still in financial straits?"

"Last I heard."

"Does he still have the airplane?"

"Last I heard."

"So he might be agreeable to your flying it into a potentially dangerous situation. If it never came back, he could collect his insurance money and keep one hundred percent of it. If we did make it back, he'd have the money we paid him to use the plane. How much would he charge to lease it?"

"It's a sweet plane. Cessna 310. Not that old. Taking into consideration the distance . . . say twenty thousand."

"Twenty thousand," she repeated softly. "That much?"

"Ballpark. In addition to my fee."

"Your fee?"

"If my ass is going to be target practice for a guerrilla with an automatic rifle, you're damn right there's a fee."

By the expression on his face, she knew she wouldn't be able to afford him. "How much, Key?"

"One hundred grand." At her shocked expression, he added, "Payable the day before we leave."

"That would be almost every cent I've got."

He shrugged. "Tough luck. Guess we won't have to get shots after all. I'm glad. Hate needles."

Once again he tried to go past her. This time she blocked his path and placed her hands on his arms. "I really hate that. I think you know how much I hate it or you wouldn't do it."

"Do what?"

"Act cavalier. Talk down to me. Damn you! I won't let you joke about this. You know how important it is to me."

Using her restraining hands to his advantage, he moved forward until he'd backed her into an army-surplus file cabinet. "Just how important is it to you?"

"Extremely. Otherwise do you think I would have asked a Tackett — any Tackett — for a favor?"

The pressure of his body against hers was exciting. So were his smoldering eyes. But she wouldn't give him the satisfaction of knowing that. She kept her chin defiantly high, her gaze steady.

"You could even go so far as to say that I'm your last resort, couldn't you, Lara?"

"You're the reason I came to Eden Pass." The statement took him aback, as she had guessed it would. "Clark handed me a golden opportunity to reestablish a medical practice, but I would have turned it down if not for you. I wanted to meet his daredevil brother, the one who could 'fly anytime, anywhere,' to quote you.

"I knew you were away most of the time, but I also knew you'd return sooner or later. I resolved to get you to take me to Montesangre, one way or another. In a very real sense, yes, you're my last resort."

He had listened with rapt attention, obviously

402

stunned by her admission. He recovered quickly. A slow grin spread across his mouth. "So I can name my price, right?"

"You already have. One hundred thousand dollars."

He reached and idly stroked her cheek. "Which I'd be willing to waive in exchange for fucking you."

Her hand flew up to bat his away from her face, but instead she gripped his wrist, closing her fingers tightly around it as far as they would reach. "I should have known you would turn this into something ugly. I tried to appeal to your decency, but you have none. You feel no sense of responsibility to anyone except yourself."

"Now you're catching on, Doc," he whispered. "You can't imagine how liberating it is to be completely free from obligation."

"Free from obligation? Your brother is partially responsible for Ashley's death. Out of all us sinners, my daughter was the only blameless victim of the whole mess. I hold Clark accountable. Just as I hold myself responsible."

She dropped her hand from his wrist. "Where Ashley's concerned I have no pride. I won't ever see her turn a cartwheel, or hear her run scales on a piano, or kiss her skinned knees, or listen to her bedtime prayers. I want only what I can have, and that's to see her buried in American soil. If sleeping with you is the only way I can accomplish that, then it's a small price to pay."

The passionate glow in his eyes cooled to a

cynical frostiness. He backed away, but in slow degrees, so that it seemed to take forever before they were no longer touching.

"As you said, Doc, I have no sense of decency. I'd help an old lady across the street if a Mack truck were bearing down on her, but that's about as noble as I get. I'm not my brother in any way, shape, or form. I left all the good deeds to him. Curious as I am to know what made your snatch so irresistible to him, I'll pass."

As he moved through the door, he called over his shoulder, "Lock up on your way out, will ya?"

"You're late."

"I know."

"We didn't hold supper."

"I'm not hungry anyway."

Key and Jody exchanged words like gunfire. He went straight to the sideboard and poured himself a stiff drink.

"We're having black-eyed peas and ham, Key," Janellen said. "You love black-eyed peas. Please sit down and let me fill you a plate."

"I'll sit down, but I don't feel like eating."

He'd been in a rotten mood since Lara Mallory had asked him to help her retrieve the remains of a little girl, who was probably his own flesh and blood, from Montesangre. Could Clark's guilty conscience have driven him to take his own life? Key had previously denied the rumors of suicide. They no longer seemed so farfetched.

404

He brought the liquor decanter to the table with him. Defying Jody's critical glare, he poured himself another drink. "How was your day, Jody? Feeling better?"

"There's nothing wrong with me. Never was. I got short-winded and everybody made a big deal of it."

He declined to argue with her at the risk of raising her blood pressure. Since her stroke, he'd walked on eggshells around her, doing whatever was necessary to placate rather than provoke her.

He still thought having a live-in nurse was a good idea, but he hadn't broached the subject again. He'd dodged every verbal missile she'd fired at him, knowing that her rotten disposition stemmed largely from fear. Hell, if he'd had a seizure like the one she'd suffered, he'd be on edge, too.

"How about you, Janellen? Anything exciting happen to you today?"

"No. Business as usual. What did you do today?"

He told them about the rancher from Arkansas. "Anderson paid me well. It was easy work. Boring as hell, though."

"And to you that's the most important thing, isn't it?" Jody said. "God forbid you ever get bored."

Raising his glass of whiskey, Key saluted her accuracy.

"Just like your father." Jody sniffed contemptuously. "You're always looking for adventure."

"What's wrong with that?"

"We've got tapioca pudding for dessert, Key. Would you like some?"

"I'll tell you what's wrong with that." Jody ignored Janellen's desperate attempt to avoid a quarrel. "You're a big baby, living in a dream world. Isn't it time you grew up and committed yourself to something worthwhile?"

"He's flying for one of the timber companies, Mama. They're using him to spray the trees for pine beetles. Saving forests is worthwhile."

Jody didn't hear her daughter. She was focused on Key. "Life isn't made up of adventures. It's working at something day in and day out, rain or shine, good times or bad, whether you feel like it or not."

"That doesn't sound like 'life' to me," he said. "That's my definition of drudgery."

"Life isn't always fun."

"Exactly. That's why you have to look for it. Or make it."

"Like your father did?"

"Yes. Because he couldn't find it at home." By now his temper was at the breaking point. "He searched for it in other places, with other women, in other beds."

Jody came out of her chair like a shot. "I won't have you talking that filth at my dinner table."

Key stood, too, squaring off across from her. "And I won't have you bad-mouthing my father."

"Father?" she said scornfully. "He was no fa-

406

ther. He left you for months at a time."

It hurt, that reminder of the countless times he'd watched his father's car disappear around the bend in the road, knowing in his breaking young heart that it would be endless days before he would see him again.

He wanted to hurt her back. "He left to escape you, not us kids."

"Key!" Janellen cut in.

Again, she went unheeded. Now that the well of his resentment had been tapped, he couldn't control the gush of angry words. "You never offered me a kind word or a soft touch. Did you treat Daddy any differently? Did you ever talk to him without making it a goddamn lecture on his faults? Did you ever stop thinking about crude oil long enough to laugh with him, to tease and act silly just for the hell of it? When he was depressed, did you draw him to your breast and comfort him? Not that your bosom would have been comforting, or even yielding. It's as hard as a drill bit."

"Key!" Janellen cried. "Mama, sit down. You look —"

"Your father didn't need my love. He got it from whores all over the world. And he flaunted them in my face. He was with one the day you were born." She drew herself up and took several labored breaths. "The only good thing that came out of my marriage to Clark Tackett Junior was your brother."

"Saint Clark," Key said with a sneer. "Maybe

he wasn't as saintly as you think. Tonight I was talking about him with his former mistress. Seems Dr. Mallory blames Clark for packing her and her family off to Central America and getting them shot. She asked me to take her down there and help bring back her daughter's remains. Ain't that a bitch?"

"You aren't considering it, are you?" Janellen looked at him aghast.

"Why shouldn't I? Her money's green."

"There's still a revolution going on down there. People are being slaughtered every day."

Although he'd responded to Janellen, his eyes never left Jody. "Dr. Mallory thinks we Tacketts owe her this. In exchange for my services, she's agreed to leave Eden Pass and never come back."

"You are not to do it, do you understand me?" Jody's voice quivered with wrath.

"Even if it means ridding us of Lara Mallory?"

"You can't trust her to keep her word. Under no circumstances are you to even consider going to Central America with her."

He placed his hand over his heart. "Why, Mother, your concern for my safety is touching."

"I don't give a goddamn about your safety. My only concern is to protect the remaining shreds of Clark's reputation. If you go anywhere with that whore, you deserve no better than to get your damn-fool head blown off."

Janellen covered a gasp with her hand and sank back into her chair.

"Why don't you go ahead and say it, Jody?"

Key shouted. "If you can't have Clark, you'd just as soon see me dead, too."

Jody swept up her pack of cigarettes and lighter, turned, and marched from the dining room.

For the longest time his rigid arms braced him against the back of his chair. His knuckles turned white against the polished oak, as though at any second he might pick up the chair and heave it through the dining room window.

Until she spoke, he'd forgotten that Janellen was there. "What you said was so . . . so horrible, Mama was too angry to refute you."

He looked at her bleakly. The muscles in his arms relaxed, and his hands dropped to his sides. Turning on his heel, he started for the door. "You're wrong, Janellen. She didn't refute me because I spoke the truth."

The lamp on the nightstand came on. Lara woke up instantly and rolled toward the light, then sprang to a sitting position, her heart in her throat. "What are you doing in here? How'd you get in?"

"I picked the lock on the back door," Key replied. "You forgot to change the code on your alarm."

His eyes were drawn down to her bare breasts. Lara, still trying to orient herself, didn't scramble for cover. His gaze remained fixed on her for several moments. Then, swearing softly, he snatched up the robe lying across the foot of her bed and tossed it to her.

"Put that on. We need to talk."

Still dazed from awakening to find him in her bedroom, she followed his instructions without argument. She sat on the edge of the bed.

Key paced along the footboard, gnawing on his lower lip. Suddenly he stopped and looked at her. "We'd never get clearance to land. Have you thought of that?"

She was muzzy from the abrupt manner in which she'd been awakened. "No. I mean, yes." She drew a head-clearing breath and pushed her hair off her face. "No, we'd never get clearance to land, and yes, of course I've given it a lot of thought."

"Well?"

"I've got a map marking a private landing strip."

"A WAC?"

"A what?"

"A World Aeronautical Chart. A map specifically for pilots."

"I don't think so. It looks like an ordinary map."

"Better than nothing," he said. "Where'd you get it?"

"It was sent to me."

"By someone you trust?"

"A Catholic priest. Father Geraldo. He befriended us while we where there. Randall made him the official embassy chaplain."

"I thought the rebels had executed all the clergymen."

"They've murdered many of them. He's managed to survive."

Key ruminated on that as he sat down in an easy chair beside the bed, so close to her that their knees almost touched. "Sounds to me as though your priest might being playing both ends against the middle."

"Very possibly," Lara admitted with a weak smile. "He claims to be bipartisan."

"He goes with the flow."

"That's the only way he can continue to do the Lord's work."

"Or save his own skin."

"Yes," she admitted reluctantly. "But I have no reason to mistrust him. Anyway, he's all we've got."

Key blew out his breath. "Okay. Let's temporarily shelve that and move to point B. Do you know if they have radar?"

"I'm sure they do, but it couldn't be very sophisticated. Nothing there is. Technologically they're decades behind the rest of the world."

"How far from Ciudad Central is this landing strip?"

Mentally she converted the kilometers. "About forty miles."

He whistled. "That'd be close. How am I supposed to avoid their radar?"

"There must be ways. Drug smugglers do it all the time."

He looked at her sharply. "I've never smuggled dope."

"I didn't mean to imply —"

"Sure you did." He held her gaze, then shrugged impatiently. "Fuck it. Believe what you want to."

He left the chair and began to pace again. Lara had a thousand questions to ask but didn't dare. She mainly wanted to know why he'd changed his mind. Like a caged animal, he restlessly prowled her bedroom.

"*If* we can slip through their radar, *if* this landing strip is where it's supposed to be . . ."

"Yes?"

"How do we get around?"

"I can make arrangements for Father Geraldo to pick us up."

"Go on."

"There's an underground organization that manages to slip supplies, letters, and such into and out of Montesangre. That's how the map got to me. I waited a year for it, but I've had it for several months. Utilizing this underground, I can have Father Geraldo notified when to meet us."

"It'll take another year?"

"No. I put everyone on alert. They're standing by."

"You were that sure I'd agree?"

"I was that sure I'd do anything to see that you did."

They paused, watching each other.

Key was the first to shake himself free. "Does this priest speak English?"

"Actually his name is Gerald Mallone. He's an American."

He swore. "Which means he's doubly suspicious and is probably being tailed everywhere he goes."

"I doubt it. He's steeped in Montesangren culture, more Latin than Irish in temperament. Besides, he's fully aware of the dangers. He's been living with them for years and knows how to avoid them. The landing strip should be fairly safe. I've been told it's on the coast, at the foot of a heavily vegetated mountain range."

"Safe! Jesus. I'll have to fly in at night, over open sea, dodging radar, and set that puppy down in the middle of a goddamn jungle, hoping all the while that we won't run into a mountain or get blown out of the sky." He saw her about to speak and raised both hands. "I know, I know. Drug smugglers do it all the time. No doubt on this very strip."

He paced another few minutes. She didn't interrupt his thoughts.

"Okay, say we land without crashing and burning, say we manage to leave the plane without having an army of rebels or contras shooting us on sight, say this semitrustworthy priest is there, where does he take us?"

"Ciudad Central."

He dragged his hand down his face. "I was afraid you'd say that."

"That's probably where my daughter is buried."

His eyes moved to her tousled tawny hair. "You'll stick out down there like a polar bear in the Sahara. Aren't you afraid of attracting someone's attention when you take a shovel into the graveyard and start digging?"

She took a swift breath.

"I'm sorry. Strike that for insensitivity." He returned to the chair and continued in a kinder tone of voice. "I doubt very seriously they'll let you exhume the casket, Lara. Do you know which cemetery your daughter would be buried in?"

"No."

"How about Father what's-his-name?"

She shook her head. "The last word I had from him is that he's checking into it. Civil records have been haphazardly kept the last several years. By the time we get there, I hope he's uncovered a clue." She smiled apologetically. "That's the best I can do."

"What if he can't obtain any more information?"

"I'll do the detective work myself."

"Christ. That's impossible."

"It's not as hopeless as it sounds," she said with as much conviction as she could garner. "There's a Montesangren who worked in the embassy, a savvy young man who knew his way around. He was initially hired to do clerical work, but soon became invaluable to Randall by translating official documents. Randall had only a rudimentary understanding of Spanish.

Emilio is smart and intuitive. If I can find him, I know he'll help us."

"If you can find him?"

"He might not have escaped the attack on the embassy. His name didn't appear on the casualty lists, but I doubt the lists were complete. If he wasn't killed, he's probably in hiding. Anyone who'd worked in the American embassy would be regarded as a traitor by the rebels."

"Suppose he's dead or otherwise unavailable. What then?"

"Then I'm truly on my own."

"You're willing to take that risk?"

"I'll go to any lengths to bring Ashley back."

"Right," he said. "You're even willing to offer your sweet body to dirty old me." He was staring at her thighs, where the robe had parted a few inches above her knees.

Lara said nothing and sat very still.

Abruptly he stood. "Tap in to this underground network. Gather all the information you can. Don't discount anything. Don't trust your memory, either; take copious notes. I want to know everything. Time of sunrise, sunset, temperature, population, the speed limit, every frigging fact you can think of. Let me be the judge of what's significant and what isn't. In situations like this you never know what scrap of information might mean the difference between living and dying.

"We'll travel light. Take only one bag you can carry easily. Don't take anything you value, nothing you couldn't drop and run away from, literally. Keep in mind that if we're successful, we'll be carrying out a casket. That may be all we can handle. Questions?"

"What about the airplane?"

"I'll arrange for it and the weapons."

"Weapons?"

"You didn't think I'd go to a turkey shoot without a gun, did you? Can you shoot?"

"I can learn."

"We'll start lessons as soon as I've got the guns. I'll handle the transactions alone, but I expect to be reimbursed for all expenses."

"Of course."

"There's only one condition: Don't ask me any questions about the arms or the plane. If the feds get curious and start asking questions, you can honestly say you don't know."

"What will you say?"

"I'll lie. Convincingly. When do you want to go?"

"As soon as you can get an airplane."

"I'll be in touch."

Lara stood. "Thank you, Key. Thank you very much."

He came to stand directly in front of her, his movements and speech no longer brisk. "As to my fee, does your offer still stand?"

She gazed into his dark, brilliant eyes and tried to convince herself that the weakness in her knees was caused by relief over his agreeing to make the trip, that it wasn't a reaction to the sexual energy he radiated.

Lowering her head, she pulled apart the ends of her sash. The robe separated. She waited only a moment before peeling it from her shoulders and letting it fall onto the bed behind her.

She stood before him naked.

The silence was dense, the tension tangible. Although she wasn't looking at him, she felt his eyes moving over her. Her skin tingled, as though his gaze were actually touching her, leaving brush strokes of heat. Breasts, belly, sex, thighs, all were touched with his eyes.

She turned warm. She grew damp. The tips of her breasts tightened and strained. Her earlobes pulsated feverishly. And somewhere deep inside her she throbbed with carnal awareness.

"Look at me."

She raised her head.

"Say my name."

"Key." At first a whisper, she repeated it. "Key."

He slid his hand around the back of her neck and lowered his head. His kiss was rough and possessive. Behind each thrust of his tongue was a hint of anger . . . at first. Then it seemed to be searching for something it couldn't find. Perhaps a desire as thick as his own.

He found it. Only he never knew. Because as abruptly as it began, it ended.

"I'll take ten thousand now." His voice was amazingly calm, but there were lines of strain around his lips, which moved woodenly. "We'll negotiate the balance of what you owe me when and if we come back alive." He turned away.

She whipped the robe from the bed and held it against her. "Key?"

He stopped on his way through the door and, after a long hesitation, turned around.

"I know why I'm doing this, but why are you?" She shook her head with misapprehension. "What changed your mind? What have you got to gain?"

"Except for a measly ten grand, absolutely nothing. The point is, like you, I haven't got a goddamn thing to lose."

Chapter Nineteen

"Did you love my brother?"

The question came out of nowhere.

Lara had closed her eyes, but she wasn't dozing. She was too nervous to sleep, though her eyelids were gritty from lack of it. She hadn't slept well for the last several days before their departure.

It had been at least a half-hour since Key and she had exchanged a word. There'd been no sound in the cockpit except the drone of the two engines. They'd left Brownsville, Texas, late that afternoon. For hours thereafter, the rugged terrain of the interior of Mexico had stretched to the horizon. After crossing the Yucatán peninsula, Key had flown out over the Pacific Ocean and made a wide U-turn. No land was yet in sight as they approached Montesangre from the sea.

There was only a sliver of moon; Key had planned their trip around the lunar cycle. He'd eliminated the lights on the wingtips of the craft. The stygian darkness was relieved only by the muted illumination of the instrument panel.

She had sensed his mounting tension as he mentally prepared for the difficult landing and hadn't distracted him with meaningless conversation. They'd left Eden Pass at noon and flown to Brownsville, where they'd eaten. She'd had no appetite, but Key had insisted she clean her plate. "You don't know how long it'll be before your next meal," he'd said.

He'd refueled the airplane, which she assumed belonged to the man in serious debt since it was a Cessna 310. As agreed, she didn't ask. In preparation for the trip, Key had removed all but two of the five seats — in order to make room for the casket, she assumed. He'd also equipped the plane with a navigation aid radio.

"It's called 'loran,' " he explained. "I can set the latitude and longitude of the landing strip and this baby finds it for me. Can you get me the coordinates?"

Through the underground, she had obtained this vital information, but they had experienced some anxious days before it arrived. "We can't go during a damned full moon," Key ranted. "If your priest doesn't come through by the twenty-fifth, we'll have to wait another month."

They could have waited a month, but mentally they were geared up to go. Waiting longer would have increased their stress. They had talked the topic to death. Their nerves were raw. Fortunately, barely making it under the deadline, the priest came through with the coordinates Key needed.

Behind their seats he'd stowed the duffel bags in which they'd packed a few changes of clothes and toiletries. Her doctor's bag had been packed to capacity. Key had also brought along a camera bag carrying a 35mm camera and several lenses. If they were questioned by anyone in authority — and he assured her that wasn't likely — they would pretend to be a couple on their way to Chichén Itzá to photograph the pyramids.

There was a hidden compartment in one of the wing lockers. He'd placed a rifle there. He'd kept the two handguns in the cockpit. She had recoiled the first time she saw the weapons.

"This one's yours." He held a revolver.

"I can barely lift it."

"You'll be able to if you have to, believe me. Grip it with both hands when you fire."

"Randall wanted to teach me to fire a gun when we moved to Montesangre, but I didn't want to learn."

"You don't have to be a good marksman with this. It's a Magnum .357. Just point it in the general direction of your target and pull the trigger. Consider it a hand-held cannon. Whatever you shoot at, you'll destroy or severely damage."

She shuddered at the thought. Ignoring her aversion, he'd given her a crash course on how to fire and load the revolver.

They were as prepared as they would ever be. Now they were close to their destination. A million things could go wrong: some of them he'd shared with her, many he had probably kept to himself, she thought.

Was his unheralded question about her loving Clark his way of diverting his mind from the dangers they faced?

She turned and looked at him in profile. He hadn't shaved in a week. "Built-in camouflage," he'd said when she mentioned the darkening stubble. The beard only intensified his good looks, adding the dubious charm of disreputability.

"Did I love Clark?" she repeated. Facing forward again, she stared through the windshield into the unrelieved blackness. She tried not to think about this flying island of technology being all that was between her and the Pacific Ocean. To her mind, aerodynamics defied logic. The craft seemed awfully small and terribly vulnerable in this vacuum of black.

"Yes, I loved him." She felt the sudden movement of his head as he turned to look at her. She kept her gaze forward. "That's why his betrayal was so devastating. He threw me to the wolves and watched from the safety of his elected office while they ripped me to shreds. Not only did he fail to come to my rescue, but, by his silence, he denounced me. I wouldn't have thought that Clark was capable of such disloyalty and cowardice."

"He showed no lack of courage when he took his lover into his bed while her husband slept down the hall," he observed. "Or was that stupidity? Sometimes there's little distinction between bravery and ignorance. What made you do it when there was such a good chance of getting caught?"

"Love is a powerful motivator. It makes us its victims and causes us to do crazy things, things we wouldn't ordinarily do. During that weekend at the cottage, the atmosphere was . . . charged. Expectant."

She looked down at her hands, rubbed her palms together. "Desire that strong obscures conscience and better judgment. It overpowers the fear of discovery." She sighed and raised her head. "I should have read the warning signs. They were glaringly apparent. In hindsight, I realize that disaster was inevitable and imminent. I just wasn't paying attention."

"In other words, you were so eaten up with animal lust that common sense didn't stand a snowball's chance in hell."

"Don't sound so superior. Your 'animal lust' for a married woman got you shot! Besides, that's ancient history. Why bring it up now?"

"Because if I don't make it out of this godforsaken banana republic, I'd like to think I died for a noble cause. I'd like to believe that you were more than a roll in the sack for my horny brother, and that for you he wasn't just a convenient diversion from an unhappy marriage."

It was on the tip of her tongue to tell him to go to hell. But, in effect, she had placed her life in his hands. Without him, her chances of surviving this trip were nil. Like it or not, they were comrades with a common goal. Infighting should be kept to a minimum.

"Despite the way our relationship ended, I loved Clark," she said. "I believe with all my heart that he loved me. Does that make this mission noble enough for you?"

"Was he Ashley's father?"

She hadn't seen that curveball coming. For a moment she was dumbstruck. She had never hinted that Clark had fathered her child. Not even the news hounds with the sharpest teeth had sunk that particular fang into her. On second thought, she realized, she shouldn't be surprised that Key was the first to raise the question. It was characteristically shocking.

"I can't answer that."

"You mean you don't know? You were screwing them at the same time?"

"I'll rephrase," Lara said heatedly. "I *won't* answer. Not until we've done what we came down here to do."

"What difference does it make?"

"You're the one who asked about Ashley's parentage. You tell me if it makes a difference."

"Oh, I see. You think I might try harder to find her remains if she was a Tackett." He made a disagreeable sound. "Your opinion of me must be even lower than I thought. Exactly where do I rank on your scale of life forms? A notch above pond scum? Or a notch below?"

Anger was a supreme waste of energy considering the ordeal facing them. "Look, Key, we've certainly had our differences. We've both slung more than our share of mud. Some of it was warranted. Some of it was spiteful. But I trust you. If I didn't, I wouldn't have asked you to bring me down here."

"You had no other options."

"I could have hired a mercenary."

"You couldn't afford the going rate."

"Probably not, but shortage of funds wouldn't have stopped me. Eventually I would have gotten the money, even if I'd had to wait for my inheritance."

"But you felt that we Tacketts owed you this."

"That wasn't it entirely." She hesitated; he looked over at her. "True, I came to Eden Pass specifically to coerce you into bringing me down here. But I didn't expect to feel this confident about my choice."

Their eyes locked and held for several moments. Finally Lara turned away. "Once we're safely on our way back home, I promise to tell you anything you want to know. In the meantime, don't throw any more poison darts, okay? I won't throw any either."

He said nothing for several minutes. When he did, he spoke in a gruff voice on a topic unrelated to Ashley's origins. "One way or another, we'll be going down soon."

"One way or another?"

"We'll either reach the coast and find the landing strip, or we'll run out of fuel and ditch into the ocean. In the meantime, why don't you try to get some sleep."

"Is that supposed to be a joke?"

He grinned. "Yes."

"Not funny."

She searched the horizon but didn't see even a seam in the darkness. Key carefully monitored the instruments. She noticed the decrease in their altitude.

"You're going down?"

"Below five hundred feet, just in case their radar is more sophisticated than you think. You're sure the priest will be there?"

"I don't have an ironclad guarantee." He'd grilled her on this a thousand times. She was as sure as she could be under the circumstances. "He's been given our estimated time of arrival. When he hears the airplane approaching, he's to light torches on the landing strip."

"Torches," he said scoffingly. "Probably tomato soup cans filled with kerosene."

"He'll be there and so will the torches."

"The wind's picked up to twenty knots."

"Is that bad?"

"Less than ten would be ideal. Forty would be impossible. I'll settle for twenty. Crosswinds are always a factor along a seacoast. I wonder how close the jungle is to the shore?"

"Why?"

"This late at night it could produce ground fog, which could mean that we'd miss not only the torches but the mountain. Until we ran into it, of course."

Her palms began to sweat. "Can you think of

anything encouraging?"

"Yes."

"What?"

"If I die, Janellen will be doubly rich."

"I thought you were the fearless pilot," she said with exasperation. "The Sky King of the nineties. You told me you could fly anything, anywhere, anytime."

He wasn't listening. "There's the shore." He checked the loran. "We're here. Start watching for the lights. It's up to you."

"Why me?"

"Because I've got to keep us from crashing into those goddamn mountains while keeping below five hundred feet. It's dicey. At least there's no fog."

The rocky shore could vaguely be detected on the horizon. Eons ago, a chunk of mountain had broken away from the strip of Central America that is now Montesangre. That chunk had drifted into the Pacific ocean where it became an island three hundred and eighty miles offshore. In a geological time frame, this had been a recent event. The jagged tear in the mountain range hadn't had time to erode into sandy beaches. Thus, the mountains dominated Montesangre's coast and formed an inhospitable shore.

Consequently, the country had not enjoyed the healthy tourist trade of its more fortunate neighbors who depended on vacationers from North America and Europe to support their national economies. Such economic deprivation had

427

caused more than one armed conflict between Montesangre and surrounding Central American republics.

From the air, the mountain range resembled the letter C, which curved from the interior of the country, forming a northern border with the neighboring nation, then running parallel to the shore for miles before tapering off. In the hollow of that C nestled the capital city, Ciudad Central. Ninety-five percent of Montesangre's population was concentrated in the city proper or in scattered villages surrounding it.

Beyond those villages in all directions stretched miles of dense jungle, populated only by wildlife, vegetation, and several tribes of Indians who lived very much as they had for centuries, without the enlightening, or corrupting, elements of modern civilization.

Lara had flown into Montesangre only once before; after her arrival she hadn't left the country until the day she was transported out, injured and unconscious. As the shore became hastily more distinguishable, she was filled with a sense of dread. She recalled how miserably unhappy she had been when she arrived with Randall. On that day, she'd had only the knowledge of the life growing inside her womb to sustain her and buoy her ravaged spirit. Ashley was the only reason she ever would have returned.

"Also keep an eye out for other aircraft," Key said. "I can't do any sightseeing."

"No one knows we're coming."

"You hope. Just in case, I don't want an army helicopter flying up our ass, do you?"

Lara glanced at him. The cockpit's temperature was comfortable, but a trickle of sweat was running down his bearded cheek. Her skin too was damp with nervous perspiration.

"We've got nowhere else to go but down," he muttered as he read the gauges. "I couldn't even make it out of Montesangren airspace. We're shit-out of fuel. Where're the goddamn torches?"

Frantically Lara leaned forward and scanned the coastline. She saw nothing but a narrow stretch of beach that bled into the tree line. The mountains loomed darkly above it.

What if Father Geraldo wasn't there? What if he'd been tortured until he divulged information? What if it was known by the rebel commanders that the widow of the late U.S. ambassador was returning? Not only her life but Key's would be in peril. There would be no one to help them. They would be at the mercy of their captors and, as Lara knew, the Montesangrens were not a merciful people. Their best hope would be to crash and die instantly.

"Shit!"

"What?"

"I've got to pull her up. Hold on." He pushed forward on the throttle quadrant and the craft went into a hard climb. Lara looked below. They barely cleared the crest of the mountain. Key banked to the left and skimmed the steep, vegetated walls before swinging back out over the surf.

"Where's the padre, Lara?"

"I don't know." Anxiously she pulled her lower lip through her teeth. She'd been confident that their escort would be there.

"See anything?"

"No."

"Wait! I think I see —"

"Where?"

"Four o'clock."

He executed another drastic maneuver that sent her stomach plunging. She closed her eyes to regain her equilibrium. When she opened them, the horizon was back in place and three small dots of light were glimmering below and ahead of them. Then a fourth flickered on.

"That's him!" she cried. "He's here. I told you he would be."

"Hang on. We're going in."

He leveled the aircraft and decreased their altitude and air speed. Sooner than Lara anticipated, the spots of light were rushing toward them. They landed with a hard bump. The plane bounced along the uneven dirt strip. Key put all his strength into pushing the throttle forward. He practically stood on the foot pedals. The landing strip was built on an incline to assist slowing them down and facilitating a short landing. Still, it seemed to take forever to stop. They came breathtakingly close to the trees at the end of the crude runway.

He turned off the motor. They sighed with relief. Key placed his hand on her knee. "Okay?"

"Okay." Since she had to alight before he could, she reached for the door.

"Wait." He sat tense and still, his eyes sweeping the black curtain of darkness outside the airplane. "I want to see who our welcoming committee is."

They sat in silence. Behind them, the six torches, three on each side of the landing strip, were extinguished one by one.

Key kept his right hand on her knee. With his left, he reached for the handgun beneath his seat. He'd told her it was a Beretta 9mm. He slid back the top, automatically loading the first bullet into the chamber. It was now cocked and ready to fire.

"Key!"

"We're sitting ducks. I'm not going to be snuffed out without putting up at least token resistance."

"But —"

He held up his hand for silence. She heard it, too — an approaching vehicle. Looking back, she saw a jeep emerging from the darkness and slowly taking shape. It pulled up behind the aircraft and stopped. The driver stepped out and moved toward the plane.

Key aimed the Beretta at the shadow figure.

Lara released a gasp of relief. "It's Father Geraldo. He's alone."

"I hope to hell he is."

Lara opened her door and gingerly stepped out of the plane, climbing down using the footholds in the wing. "Father Geraldo," she said as she jumped to the ground. "Thank God you're here."

He extended his hands. "Indeed. It's good to see you again, Mrs. Porter."

She extended her hand, and he enfolded it in a warm, damp clasp. "You're looking well," she said.

"And you."

"Have you learned anything about where my daughter is buried?"

"I'm afraid not. I've made inquiries, but to no avail. I'm sorry."

The news was disappointing but not surprising. "I knew it wasn't going to be easy." Just then Key stepped off the wing. "This is Key Tackett."

"Father," he said in a clipped voice. "Thanks for sending those coordinates. Without them, we'd never have found you."

"I'm glad they were useful."

"Are you sure you weren't followed?"

"Reasonably sure."

Key frowned. "Well, let's get this baby out of sight before we attract company."

"I assure you," the priest said, "for the time being, we're safe."

"I don't like to take chances. Which way?"

"Because of the revolution, the drug traffic has slacked off considerably. The strip hasn't been used in a while. I brought along a machete, and while I was waiting for you I cleared out some brush." He indicated what appeared to be an impenetrable wall of jungle.

"Let's get to it."

After hacking away some of the densest brush, the three pushed the airplane off the landing strip. They retrieved the few items they'd brought with them, including the hidden rifle, then covered the plane with the brush.

"This is a remote spot," the priest said to Key, who was surveying the camouflaged aircraft from every angle. "Even in daylight I don't think it'll be detected. Allow me, Mrs. Porter."

He picked up Lara's duffel and the camera bag and headed for the jeep. Hoisting his own duffel and the rifle to his shoulder, Key spoke to Lara in an undertone.

"You failed to mention that the padre is a drunk."

"He's been conducting Mass. That's sacramental wine on his breath."

"Like hell. It's Jamaican rum. I've vomited up enough of it to know how it smells."

"Then you're in no position to judge."

"I don't care if he guzzles horse piss, so long as he's reliable."

Before she could defend the charge, they reached the jeep. Father Geraldo, who wore his forty years as though they were sixty, helped Key stow their gear in the back. "If you don't mind riding back here, it will be more comfortable for Mrs. Porter in front."

"I don't mind," Key said, easily swinging himself up into the backseat. "From here I can guard our rear."

"Well said." The priest smiled at him. "We live in turbulent times."

"Right. Over drinks some time I'd love to philosophize with you. Now, I think we'd better relocate. Pronto."

If the priest took umbrage at Key's reference to drinks, he didn't show it. After assisting Lara into the passenger seat, he climbed behind the steering wheel. "Best to leave the lights off until we approach the city. The roads are sometimes patrolled at night."

"By whom?" Key wanted to know.

"By whoever wants to patrol them. It changes on a daily basis."

"What's the political climate like now?" Lara asked.

"Volatile."

"Terrific," Key muttered.

"The old regime wants to regain control. President Escávez is still in hiding, but rumor is that he's trying to assemble an army and reclaim his office."

"The rebels won't allow it without a bloodbath," Lara said.

"No doubt," the priest agreed, "but Escávez isn't their primary concern. He believes the people still love him, but he's wrong. No one wants to return to the days of his despotism before the revolution. He's just an old man deluding himself, more a nuisance than a threat. The rebels have bigger problems to worry about."

"Such as?" Key asked. He'd worked up a sweat swinging the machete and moving the plane. He removed his shirt and used it to mop his face, neck, and throat. Lara envied him that freedom. She was sweltering. Her blouse clung to her skin.

"Lack of money is their primary problem," the priest replied to Key's question. "Lack of supplies. Lack of zeal. The men are disenchanted. After living in armed camps in the jungle for years, revolution isn't nearly as exciting as it seemed in the beginning.

"They're tired of fighting, but they fear their leaders too much to return home. They're hungry, diseased, and homesick. Some haven't seen their families since Escávez was overthrown. They hide in the jungle and come out only to wreak havoc on small villages and scavenge for food. Mostly they fight among themselves. Since Jorge Pérez Martínez was assassinated —"

"He was? We didn't hear about that in the States," Lara said, surprised. Pérez had been a general in Escávez's army who had staged the military coup to overthrow him. The rebels had regarded him as a savior of an oppressed people.

"He was killed by one of his own men more than a year ago," the priest told her. "For months the leadership was up for grabs. First one lieutenant, then another proclaimed himself Pérez's successor, but none could hold the rebels together. There were many factions with no cohesiveness. As a result, the counter-revolutionaries, among them Escávez, began to make inroads.

"Then, one of Pérez's protégés emerged and declared himself the new general of the rebel army. Over the last several months he's gained support, I think chiefly because his men fear him. He's supposedly ruthless and will stop at nothing to cement his position as leader. *El Corazón del Diablo*. The Devil's Heart. That's what they call him." He glanced sideways at Lara. "He passionately hates Americans."

Saying anything more would have been superfluous. She looked back at Key to find his eyes on her, piercing and intent. "It's no worse than we expected," she said defensively.

"No better, either."

"I brought some clothes," Father Geraldo said, gesturing at the soft bundle at Lara's feet. "Before we reach the outskirts of the city, you'd better put them on."

They'd been following a rutted dirt road that snaked through the jungle, seemingly without destination. Each time a night bird screeched, Lara's skin broke out in goose bumps, though the humidity was stifling. Her hair felt heavy on her neck, more so when she placed a scratchy scarf over her head as was customary of the matrons of Montesangre, except for the progressive generation of women who fought alongside their male comrades in arms.

In the bundle of clothing she also found a

shapeless cotton print dress. She gathered it into her hands and stepped into it, working it up her legs and over her hips before placing her arms through the sleeves. She tied it at her waist with a sash.

For Key the priest had brought the muslin tunic and pants of a farmer and a straw hat. As he placed it on his head, the jeep topped a hill. Ciudad Central was spread out below them, a blanket of twinkling lights.

At the sight of the city she despised, fear and loathing filled Lara's heart. If she'd had a choice at that moment, she might have given up her insane objective and returned to the airplane. But somewhere in that urban sprawl her daughter was buried.

As though sensing her trepidation, Father Geraldo pulled the jeep to a halt. "What you intend to do will be extremely dangerous, Mrs. Porter. Perhaps you should reconsider."

"I want my daughter."

Father Geraldo engaged the gears and switched on the headlights. They started down the curving road. The narrow shoulder dropped off into nothingness. Fearfully Lara wondered how much rum Father Geraldo had consumed that evening. Whenever the wheels sank into the soft shoulder, she gripped the edge of her seat.

As it turned out, the condition of the road and Father Geraldo's level of inebriation were inconsequential. As they came around a bend, they were impaled by blinding spotlights and deafened by a chorus of shouting voices. *"Alto!"*

A platoon of guerrillas surged forward to surround them, guns aimed and ready to fire.

Chapter Twenty

Jody knocked on Janellen's bedroom door. "Mama?"

Jody opened the door but remained standing on the threshold. She didn't remember the last time she'd been in Janellen's room, and some of the furnishings were unfamiliar to her. However, she recognized the cherrywood fourposter bed, chest of drawers, and dresser; they'd belonged to her daughter since she graduated from the crib.

The drapes and wallpaper were new, or at least it seemed they were. The pale gold and china-blue print combinations were too festive and feminine for her taste. She vaguely recalled granting Janellen permission to redecorate but couldn't remember when that had been. Five years ago? Yesterday?

Janellen was lounging in an easy chair upholstered in floral chintz, her feet resting on the matching ottoman, a paperback novel lying open in her lap. A small brass lamp at her elbow cast soft, flattering lighting over her. It came as an unpleasant shock that Janellen looked almost pretty.

During her childhood Jody realized that her daughter was not going to be a raving beauty.

Rather than finding this regrettable, she was glad and had done everything possible to guarantee Janellen's homeliness. She'd never dressed her in anything bright or sassy and she had styled her hair in the least becoming way.

She firmly believed that desexing her daughter was the best thing she could do for her. Wishing to attract a man was a weakness inherent to women. Jody aimed to see that Janellen never fell into that trap.

Compliantly, Janellen had conformed to the mold her mother designed for her. She'd become an intelligent, competent woman who could never be accused of frivolity or flirtation. She'd been too reasonable to fall in love. Her plainness had spared her the deviousness of playboys, fortune hunters, and men in general. In that respect, Jody considered her daughter most fortunate.

There was one major drawback. Janellen had the Tackett eyes. *His* eyes. He'd been dead for years, but that living legacy, which all her children had borne, never failed to disconcert her. It was as if Clark Junior were in the room with her, watching her from behind their daughter's face.

"Mama, what is it? Are you feeling all right? Is anything wrong?"

"Of course I'm feeling all right. Why wouldn't I be?"

Janellen's curiosity was understandable. Jody never sought her daughter's company and certainly not at this hour. It was almost midnight.

Janellen had tucked Jody in hours ago, but she'd been unable to sleep. Smoking heavily, she'd paced the floor of her bedroom. Her body was tired, but her mind wouldn't relax and allow her to rest.

She'd always been an insomniac, even as a girl when frustration over her family's poverty had affected her sleep patterns. Night after night she had lain awake between two snoring siblings, scheming ways to free herself of poverty's stranglehold.

The tornado that had destroyed her house and killed her family had been a godsend.

Once she began working for Tackett Oil, the challenge of the job kept her clever mind too energized for sleep. Later, she'd spent years pacing the floor of her solitary bedroom while conjuring up infuriating, devastating scenarios of Clark Junior with other women.

Pushing that embittering thought aside, Jody said, "Where is your brother?"

"Key?"

She shot Janellen a retiring look. "Of course Key."

"He's out of town."

The problem with Janellen was that she'd learned her lessons too well. She'd conformed, she'd done what was expected of her, she'd never been rebellious, never created unpleasantness of any kind, but she was a titmouse. Sometimes her eager-to-please expression was too much to stomach. This was one of those times. Jody wanted to shake her hard.

"He's gone to Central America, hasn't he? He took that bitch down there just to show me that he didn't give a damn how I felt about it."

"Yes, he went to Montesangre with Dr. Mallory, but not because —"

"When did he leave?"

"Today. They planned to arrive tonight. He said he would call if he had a chance, but he didn't think it was likely."

Jody's posture remained rigid. The folds of her housecoat hid her hand from Janellen. Otherwise her daughter would have seen how hard she was gripping the crystal doorknob.

"He's a goddamn fool. She crooked her finger at him and he went runing." Her lips curled contemptuously. "Just like your father, he can't resist a chance in a woman's bed, no matter who she is or what it costs him."

"Key went because Dr. Mallory wants to bring back her baby girl's remains."

The sentimental implications didn't soften her. "When are they due back?"

"He didn't know." Janellen's eyes filled with tears. "He left some papers with me. I'm supposed to open them if he doesn't . . . if they don't . . ."

If she hadn't been holding on to the door with such determination, Jody might have collapsed from the impact of her emotions. She had to get out of there before she made a fool of herself.

Without a word, she backed into the hallway

442

and pulled the door shut with a decisive click. Only then did she give vent to her inner turmoil. Her shoulders slumped forward. Bowing her head, she raised her fist to her lips and mashed them hard in order to keep from uttering an anguished sound.

After a time, she returned to her bedroom feeling alone and very frightened.

Reaching between the front seats of the jeep, Key thrust the Magnum against Lara's side. "Take it," he whispered. "Don't be skittish about using it if you have to."

She didn't argue. The guerrilla fighters had completely surrounded them. Their expressions were menacing. She clutched the revolver and placed it in her lap, hiding it in her voluminous skirt.

"Buenas noches, señores." Father Geraldo spoke pleasantly to the band of armed men. Key counted a dozen. Three times that many were probably keeping cover in the foliage. He didn't like the odds.

"¿Quién es?" One of the soldiers separated himself from the others. He was dressed in camouflage fatigues and armed to the teeth. His stance and tone were belligerent, his eyes hostile and suspicious.

The priest introduced himself. The soldier spat in the dirt. Unruffled, Father Geraldo said in fluent Spanish, "You know me, Ricardo Gonzáles Vela. I conducted your mother's funeral Mass."

443

"Years ago," the soldier growled, "when we still believed in such foolishness."

"You no longer believe in God?"

"Where was God when women and children begging for food were slaughtered by the swine under the command of Escávez?"

Father Geraldo was disinclined to engage in a theological or political debate, especially since the other soldiers cheered and raised their weapons to reinforce their comrade's opinion.

The angry young rebel glared at the priest, then his eyes shifted to Lara, who'd had the good sense to keep her head down to hide her Anglo features. "Who is this woman?" Ricardo jabbed the barrel of his rifle in her direction. "And him?"

"They live in a small village in the foothills. Her husband was killed defending the village from contra forces. She's pregnant. Her brother-in-law," he said, hitching a thumb toward Key, who'd remained slumped down and seemingly disinterested, "already has four sons. He cannot afford to feed two more mouths. I offered to bring her to the city and provide food and shelter in exchange for housekeeping duties at the rectory until she can find someone else to take care of her."

One of the soldiers made a crude comment about the kind of "housekeeping duties" she would be performing for the priest. Key had a basic understanding of Spanish. He didn't catch all the words, most of which were slang, but these

444

duties had something to do with her getting onto her knees.

Ricardo smiled hugely in appreciation of his comrade's ribald wit, then instantly sobered. He gave Key a contemptuous once-over. "You look strong and tall. Why aren't you fighting? El Corazón's army needs fighters."

Key's stomach tensed, but he pretended not to understand that the question had been directed to him. Thankfully Father Geraldo took his cue.

The priest motioned Ricardo closer. He approached warily, his military accoutrements making sinister jingling sounds in the darkness. Key heard several guns being cocked and wondered if he should do the same with the one hidden in the sleeve of his peasant shirt.

Lowering his voice to a confidential pitch and tapping his temple with his index finger, Father Geraldo whispered, "He's an idiot, good for milking goats and planting beans, but otherwise useless." He shrugged eloquently.

"But you said he has four sons," Ricardo argued.

"All of them nine months and ten minutes apart. The poor fool doesn't realize that rutting makes babies."

A roar of laughter went up from the guerrillas. Ricardo relaxed his vigilance. "When will he return to his village?"

"In a few days."

Ricardo leered. "Perhaps we should pay a visit

445

to his village while he's away. Maybe his wife will be lonely."

The others laughed, including Father Geraldo. "I am afraid you would find her unaccommodating, *amigo*. She was grateful for these few nights of rest."

Ricardo swept his arm toward the road ahead. "We will not detain you. You are no doubt eager to have the widow begin her housekeeping duties."

"*Gracias, señores,*" he said, addressing the laughing group. "God's blessings on you and on El Corazón del Diablo."

He put the jeep into first gear. Key's gut muscles began to unknot. The jeep had rolled forward only a few yards, however, before Ricardo commanded them to halt again.

"What is it, comrade?" Father Geraldo asked.

"An airplane was sighted tonight, flying low over the mountains from the coast. Did you see it?"

"No," the priest replied, "but I heard it. Unmistakably. About an hour ago. Back there." He pointed toward the mountains, but in a direction several degrees off the spot where they'd hidden the aircraft. "I thought it was delivering supplies to your army."

"And so it was." Ricardo lied as nonchalantly as the priest had. "The army of El Corazón del Diablo lacks nothing, especially courage. We'll fight with our bare hands if we must, to our deaths."

Father Geraldo saluted him and let off on the brake. They were allowed to proceed without further delay. None of them breathed easily until they were well away from the reconnoiters.

"Very well done, padre," Key whispered from the backseat. "I couldn't have lied more convincingly myself."

"Unfortunately this isn't the first time I've had to break a commandment in order to save lives."

"Lara, you okay?"

She nodded her covered head. "Do you think we'll be stopped again?" she asked the priest, her voice muffled by the scarf.

"Probably not, but if we are, we'll stick to the same story. Keep your head down and try to look like you're grieving."

"I am grieving," she said.

From the backseat Key told her to keep the pistol ready to fire if necessary. She nodded acknowledgment, but said nothing more.

At one time the population of Ciudad Central had exceeded one million. Key doubted that half that many lived there now. Even taking into account the lateness of the hour, the city appeared deserted. The streets were dark, as most city streets would be past midnight. But these streets were beyond dark and sleepy — they were dead.

Structures that had once been thriving businesses and gracious homes were now battle-scarred shells. Nearly every window in the city had been boarded up. No light shone through those few that hadn't been. Lawns that maraud-

447

ers hadn't completely trampled were in a sad state of neglect. Vines and undergrowth grew unchecked. The jungle was reclaiming territory that had belonged to it long before man had striven to tame it.

On walls and fences and every other conceivable surface had been scrawled graffiti advocating one junta or another. The only point on which all sides seemed to agree was their hatred for the United States. Cartoons depicted the president in all manner of disgusting and humiliating postures. The American flag had been desecrated in countless ways. Key had been in many countries hostile to the United States, but he'd never felt the antipathy as strongly as here, where it was as powerful as the stench of raw sewage.

"Oh, my God!"

Lara's gasp drew Key's attention forward. A woman's body was hanging by the neck from a traffic-light cable. Her mouth was a gaping, black, fly-infested hole.

"Some of El Corazón's handiwork," the priest explained to his horrified passengers as they passed beneath the swaying corpse. "Montesangren women are valued as soldiers. They're not spared military duty because of their gender. When they're found guilty of an offense, they're dealt with just as harshly as their male counterparts."

"What was her crime?" Lara's voice was husky with revulsion.

"She was exposed as a spy who carried secrets

448

to Escávez. They cut out her tongue. She drowned in her own blood. Then they hung her body in that busy intersection. It's a warning to everyone who sees it not to cross El Corazón del Diablo."

Considering the risks Father Geraldo was taking to help them, Key didn't blame him for his closet drinking.

"Here we are," he said as he pulled the jeep into a walled courtyard. "You'll find it changed since you were here, Mrs. Porter. The few Montesangrens who are still faithful to the church are afraid to have it known. I hold daily Masses, but more frequently than not, I'm the only one in attendance. That makes for empty offering plates."

Key alighted and looked around. The courtyard was enclosed on three sides by stone walls covered with bougainvillea vines. When Father Geraldo noticed Key's interest in the arched opening through which they'd entered, he said, "Until three years ago, there was a very beautiful and intricate wrought-iron gate. It was requisitioned by the rebels."

"Sounds like the Civil War when the Confederate army made cannonballs from iron fences. What'd the rebels use your gate for?"

"Pikes. They severed the heads of the generals of Escávez's army, impaled them on the pikes, and left them in the city square until they rotted. That was shortly after you left, Mrs. Porter."

She didn't quail or turn pale or faint. "I'd like

449

to go inside," she said in a level voice. "I'd forgotten how ferocious the mosquitoes here can be."

Key admired her fortitude. Maybe the danger they'd experienced tonight, coupled with seeing evidence of so many atrocities of war, had inured her. Then he reminded himself as they carried their gear toward the entrance of the rectory that she'd experienced an atrocity firsthand.

One of the encompassing walls of the courtyard doubled as the exterior wall of the church. It was taller by two-thirds than the other two walls. Typical of Spanish architecture, the sanctuary had a bell tower, although the bell was missing.

Another of the walls formed the exterior of the school, which Father Geraldo sadly explained was no longer used. "I wished to teach catechism, but all the various juntas wanted the children indoctrinated to violence and retaliation, which are incongruous with Christ's teachings. The nuns were faithful, but feared for their lives. Parents, under the threat of execution, were afraid to send their children to class. Eventually the enrollment dwindled to nothing. I closed the school and requested that the nuns be reassigned to the States. There had been so many clergymen executed that all elected to leave.

"For a while the vacant school was used to house orphans. There were dozens of them, victims of the war. Their parents had either been killed or had abandoned them to join the fighters. One day soldiers arrived in trucks and trans-

450

ported the children to another place. No one would ever tell me where they were taken.

"This," he said, unlocking a heavy wooden door, "is where I live and do what little work I'm still permitted to do."

To Key, the rectory was extremely claustrophobic, but he was accustomed to having the sky as his ceiling. The priest's quarters were a warren of small rooms with narrow windows and low, exposed-beam ceilings. Key had to duck his head to pass through the doorways. His shoulders barely cleared the walls of the dim corridors. More than once the toes of his boots caught on the seams of the uneven stone floor.

"I'm sorry," the priest said when Key tripped and bumped into a wall. "The rectory was built by and for European monks much smaller than you."

"No wonder they prayed all the time. They didn't have room to do anything else."

Father Geraldo indicated that they precede him through a connecting doorway. "I have refreshments in the kitchen. You'll be glad to know that it was modernized in the late fifties."

By contemporary American standards, the kitchen was woefully outdated, but it was centuries ahead of the other rooms of the rectory. They sat down at a round table while Father Geraldo served them fruit, cheese, bread, and slices of a canned ham one of his relatives in the States had smuggled to him. Out of deference to his meager hoard, they ate sparingly.

451

"The water is supposed to be sterilized, but I boil it anyway," he said as he removed a pitcher from the refrigerator. He placed lemon slices in their glasses. There was no ice. He also set a bottle of Jamaican rum on the table. Only after Key had helped himself to it did the priest pour a glass for himself.

"It helps me sleep," he said sheepishly.

Lara was polite enough to wait until they'd finished the meal before broaching the subject of her daughter's grave. "Where do we start our search, Father Geraldo?"

He looked at them uneasily. "I thought you might have a plan. All my inquiries have led to dead ends. This doesn't mean that no information exists. It simply means that no one is willing to impart it."

"The result is the same," Key said.

"Unfortunately, yes."

Lara, however, seemed undaunted. "I want to start by searching the American embassy."

"There's no one there, Mrs. Porter. It was looted and has remained vacant these past years."

"Do you remember my husband's aide and interpreter, Emilio Sánchez Perón?"

Key had traveled extensively in Central and South America and was familiar with the custom of tacking on the mother's maiden name to establish an individual's identity.

"Vaguely," the priest answered. He refilled his glass from the bottle of rum. According to Key's count, this was his third drink. "As I recall, he

452

was a quiet, intense young man. Slight in build. Wore glasses."

"That's Emilio. Have you seen or heard from him?"

"I assumed he was killed when the embassy was raided."

"His name didn't appear on the casualty list."

"That could have been an oversight."

"I realize that," Lara said, "but I'm clinging to the hope that he's still alive. The embassy library fascinated him. He spent most of his off-duty hours there. Do you know if the library was ransacked along with the rest of the building?"

Father Geraldo shrugged. "The rebels have very little time for recreational reading," he said with a wry smile. "But I wouldn't expect to find anything there intact, including the library. I haven't seen it, but from what I've heard, the building was destroyed."

The discouragement that settled on Lara's face was heartbreaking to see. "What about Ashley's death certificate?" Key asked. "Wouldn't a doctor have signed one before she was buried?"

"That's a possibility," the priest conceded. "If the certificate wasn't destroyed, if the doctor's name was recorded, and if we can locate him, he might know where her body is buried."

Lara sighed. "It seems hopeless, doesn't it?"

"Tonight it does." Key came to his feet and assisted her out of her chair. "You're exhausted. Where is she sleeping?"

"I need a bathroom first, please."

"Of course." Father Geraldo indicated a narrow passageway. "Through there."

While Lara was in the bathroom, which fortunately had plumbing, Key and the priest shared another drink. "If you're so limited in the work you can do here, why don't you return home?" Key asked. "Getting reassigned shouldn't be a problem considering the number of missionaries who've been slaughtered."

"I made a commitment to God," he replied. "I may not be very effective here, but I doubt I'd be much more effective elsewhere."

He raised his glass of rum and drank deeply. Father Geraldo knew that in the States he would be committed by the Church to an alcohol-addiction rehab facility. Staying in war-torn Montesangre was his self imposed penance for his weakness.

"You might die here if you stay."

"I'm well aware of the possibility, Mr. Tackett, but I'd rather die a martyr than a quitter."

"I'd rather not die at all," Key said somberly. "Not yet."

The priest looked at him with renewed interest. "Are you Catholic, Mr. Tackett?"

Key chuckled at the notion. There wasn't even a Catholic church in Eden Pass. The few Catholic families in town traveled twenty miles to worship. They were treated with only a little more tolerance than the Jewish families and were looked at askance by the Protestants of his hometown, where most folks erroneously assumed that if you were American-born you were automatically Christian.

"I was raised a Methodist, but don't hold that against them. They did their best. I was the scourge of every Sunday-school teacher unfortunate enough to have me in class. I eliminated any doubts they might have had as to the devil's existence. I'm living proof that Lucifer is alive and well. When it comes to righteousness, I'm a lost cause."

"I don't believe that." The priest raised his glass and looked through the rum as he spoke. "I'm not much of a priest, but I haven't forgotten all my training. I can still see into a man's heart and judge his character with a fair degree of accuracy. It took a man of courage and compassion to bring Mrs. Porter here, particularly when one considers her relationship with your brother."

Key let that pass without comment and leaned across the table so he could whisper. Water was running in the bathroom, but he didn't want to take a chance on Lara overhearing. "Since you claim to be a fairly good judge of character, would you say the soldier on the road was fooled by that crock of shit you fed him?"

The water in the bathroom stopped running.

The priest drained his glass. "No."

Father Geraldo and Key exchanged a stare rife with unspoken meaning. Lara rejoined them, fatigue weighing down her small frame.

"Bedtime," Key said, coming to his feet.

The priest led them through a maze of hallways. Entering a cloister, he smiled at Lara encouragingly and indicated the window. "It opens onto the courtyard. I thought you'd like that. But be sure to use the mosquito netting."

She didn't seem to notice that the cot beneath the crucifix was narrow, that the only lighting was a weak, bare bulb suspended from the ceiling, that the chamber was airless and hot, and that in lieu of a closet there were three wooden pegs extending from the wall.

"Thank you very much, Father Geraldo. You're placing yourself at tremendous risk in order to help me. I won't forget that."

"It's the least I can do, Mrs. Porter. More than once this church benefited from your generosity even though you aren't Catholic."

"I admired the work you were doing here. It superseded the arguable points of dogma."

He smiled poignantly. "I remember when your daughter was born. I happened to be visiting the hospital wards that day, heard you had just given birth, and stopped by your room to extend my congratulations."

"I remember. We had met socially on a few occasions, but you were wonderfully kind to visit me that day."

"That was the first time I'd ever seen you smile," he remarked. "And so you should have. Your Ashley was a beautiful baby."

"Thank you."

The priest took her hand. After giving it a

brief squeeze, he said good night and left the room. Having been reminded of her daughter's birthday, she looked forlorn and small, as though grief were shrinking her. Key wanted to alleviate her bereavement, to touch her with compassion and understanding as the priest had, but his hands remained at his sides.

"Do you still have the pistol?" he asked.

"I put it in the camera bag."

The bag was hanging by its strap from one of the wall pegs. Key removed the large revolver and handed it to her. "Sleep with it. Don't be without it."

"Did Father Geraldo tell you something I should know? Are we in danger?"

"I think we should be prepared for our situation to get worse before it gets better. If we have no trouble, it'll be a lucky break." He nodded toward the cot. "Try to get some rest. Tomorrow will be a long day. We'll start at the embassy."

She held him with a puissant stare that made him increasingly uncomfortable. "Tell me the truth, Key," she said softly. "Don't talk down to me as though I were a child. You think this is a wild goose chase, don't you?"

He did, but he didn't have the heart to tell her so. Father Geraldo had confirmed what he'd guessed — that the soldiers had let them into the city because they were curious to find out more about them and what they were doing there, not because they'd believed the priest's tale about a widow and her idiot brother-in-law.

Key believed they'd be lucky to escape Montesangre with their lives. He doubted very much that they'd fly away unscathed with the casket bearing Ashley Porter's remains.

But while he didn't have the heart to tell her the truth, he wouldn't insult her intelligence with a fatuous lie. Compromising by avoidance, he said, "Get some rest, Lara. I plan to."

Rather than go to bed, he returned to the kitchen, where he kept Father Geraldo company while the priest drank himself into a stupor. Leaving him slumped over the table soundly snoring, Key found a cot in the tiny room across the hall from Lara's. He stripped to his underwear, lay down between the scratchy muslin sheets, and dozed fitfully, his ears attuned to any noise.

He must have slept more deeply than he'd thought, because he awakened with a jolt when someone shook his shoulder. Reflexively he grabbed the Beretta, released the safety, and sat upright.

Lara stood beside the cot, washed and combed and dressed, her hand arrested in midair near his shoulder. The muzzle of the gun was only inches from her face.

"Jesus." Key exhaled shortly. "I could have killed you."

She was shaken and pale. "I'm sorry I startled you. I called your name several times. It . . . it wasn't until . . . I touched you . . ."

They stared at each other through the morning gloom. It became increasingly difficult to breathe the heavy, humid air. Her breasts rose and fell with the effort.

Sometime during the night, he'd kicked off the top sheet. Sweat trickled through his chest hair, rolled over his ribs, down his belly, and collected in his navel. An erection like a telephone pole had distended the front of his briefs.

"It's seven o'clock." She sounded as though she'd just run a mile uphill. "I've made coffee." She turned and fled.

Key dropped the gun and covered his face with both hands, dragging them down his haggard, bearded cheeks. Morning erections weren't uncommon, but this one was unusually hard.

As he pulled on his clothes, he stared at the open doorway through which Lara had hastily retreated.

"You were right. There's nothing here."

Lara kicked a chunk of ceiling plaster out of her path. What had been done to the American embassy library defied description. The crystal chandelier lay shattered on the quarry tile floor, which had been robbed of the Aubusson rugs that had once adorned it. The bookshelves had been stripped. Piles of ashes were mute testimony to the fate of the volumes.

The flag that had once stood in the corner was in tatters. Epithets to the United States had been spray-painted on the paneled walls. None of the tall windows remained intact. Apparently guns had been fired into the ceiling, because loose plaster and sections of molding were scattered over the floor. The furnishings had been confiscated. Rodents and birds now nested in the rubble.

"I'm sorry, Mrs. Porter."

"It's not your fault," she told Father Geraldo, who was hovering nearby. He was wan; his skin looked pasty, and his eyes were bloodshot. His hands were shaking so badly that he could barely drink the coffee she'd brewed before their departure from the rectory. She pretended not to notice when he laced his coffee with rum. "You tried to warn me that this was what I'd find."

"Is there anything else you'd like to see?"

"Randall's office, please."

"Make it quick," Key said.

He stood near a window, flattened against the wall. He could see out while remaining hidden. They had dressed in the clothing the priest had provided the night before and had parked the jeep off the main street before entering the building. Nevertheless, both he and Lara doubted that their disguises could fool anyone who looked closely at them.

Key was carrying the rifle. His handgun was tucked into the waistband of his pants. From the moment they'd entered the ravaged building, he'd been more interested in what was going on in the streets than in what she might discover inside.

He turned his head away from the window.

"The same jeep has driven past three times. There are two soldiers in it. They're flying El Corazón's flag. I don't trust their nonchalance."

"We'll be quick," she promised as she and the priest picked their way through the litter to the doorway of the library. Key followed but continued to glance over his shoulder as they made their way up the staircase to the room that had been the ambassador's office.

"Wait!" he cautioned as Lara reached for the closed door. She yanked back her hand, and he approached with the rifle. "Stand aside." She and the priest stood with their backs against the wall, out of the way of the door. Key pressed himself against Lara, then used the butt of the rifle to nudge open the door.

He hesitated a moment longer, then explained. "It was the only door in the building that was closed. It could have been booby-trapped."

Stepping around him, she moved into the office. At one time furnished to befit a United States ambassador, it had been ransacked as completely as the library. The desk was still there, but it had been bashed until it was barely standing. The top had been scarred by a knife, probably the same one used to slash the leather chair. White cotton stuffing sprouted from the gashes. The liquor cabinet had been raided; Waterford decanters and glassware had been shattered against the far wall.

Father Geraldo heaved a sad sigh. "It appears, that your husband's office suffered the same fate as the other rooms. He headed for the door, but Lara reached out and caught his sleeve.

"Wait. Maybe not." She moved to the far wall where there was a credenza that appeared not to have been disturbed. She opened one of the compartments and uttered a small exclamation.

"Look. Papers and files." She scanned one of the documents. "They're written in Spanish, but they look official."

Father Geraldo read them over her shoulder. "It's a trade agreement." He read further. "Basically, unrefined sugar in exchange for weapons. But it's dated several months before the coup was staged, so it can't be of much interest."

"It is to somebody." Reaching deeper into the credenza, she pulled out a pair of reading glasses and held them up for the priest to see.

"That looks like —"

"The kind that Emilio wore," she finished, her voice excited. "I knew it! I knew that if he was alive —"

Suddenly Key stepped forward and covered her mouth with his hand. He also motioned the priest to silence and angled his head toward the door, which they'd left open.

"Someone's out there," he mouthed.

He signaled Lara to crouch behind the credenza. She adamantly shook her head and headed for the door. He grabbed the back of the loose dress and brought her up short. Furious, she spun around and glared at him. But her glare fizzled beneath his, so she did as he instructed and crouched down at the end of the credenza. Father Geraldo knelt beside her.

By now she too heard the faint rustling of footsteps beyond the door. Key crept closer to it. He had propped the rifle against the desk, but was holding the handgun out in front of him as though he fully intended to use it.

What if they had caught Emilio off guard? What if he'd heard their approach and, fearing for his life, had hidden in another room? He was barely more than a boy, and he'd been loyal to Randall and her. *He* might know the location of Ashley's grave. Key, with his trigger-happy reflexes, could shoot him the instant he appeared in the doorway.

Lara held her breath and listened. Unmistakably the footsteps were coming nearer, although the one making them was trying to go undetected. His approach was halting, as if he, too, was pausing occasionally to listen. Finally the footsteps ceased. Unless her ears were playing tricks on her, the person had stopped just beyond the door, exactly as they had done before Key forced open the door with the butt of his rifle.

Lara watched in dread as he aimed the gun at the doorway.

There was movement in the opening.

Lara surged to her feet and rushed forward. "Emilio, look out!"

Chapter Twenty-One

Startled by her shout, Key spun around and back-handed her, knocking her to the floor. Then, hearing a sound in the doorway, he dropped, rolled, and fired three times.

The blast echoed in the empty building, causing Lara momentary deafness. She tasted blood. Woozy and stunned, she struggled to a sitting position and looked toward the doorway. On the threshold, one side of his body opened by gunshots, lay a goat.

"Fuck!" Key yanked Lara to her feet and shook her hard. "What the hell were you thinking?" He shoved her toward the door. "Let's get the hell out of here. Come on, padre. In a minute or less this place is going to be crawling with troops."

Stumbling from the room, she barely avoided stepping in the gore. Key splayed his hand on her back and pushed her ahead of him down the staircase and through the formal reception halls on the ground floor. Her lip was throbbing; she knew it was rapidly swelling.

When they reached the rear door through which they'd entered, Key jerked her to a halt. Cautiously he poked his head outside and surveyed the immediate area. Lara glanced at Father

Geraldo. Breathing heavily, he was supporting himself against the doorjamb. Sympathetically he passed her a handkerchief. She blotted her lip with it; it came away stained with blood.

Key said, "Let's go. But keep your head down and be ready to run for cover. There could be snipers on the roofs."

He gripped her hand and made a dash for the jeep. He hoisted her into the passenger's seat, then ran around to the driver's side, taking over Father Geraldo's position as driver. The priest didn't seem to mind. Without argument he scrambled into the backseat only seconds before the jeep lurched toward the nearest alley.

Key stayed off the main roads, driving at a breathtaking speed down one alley and up another, dodging heaps of garbage and warfare debris, unpredictably switching directions like a crazed animated character in a video game.

"Did I hurt you?" He gave Lara a swift glance.

"Of course you hurt me. You hit me."

"If you'd kept your butt where I'd told you to keep it, it wouldn't have happened." He swerved to avoid colliding with a youth on a bicycle. "Jumping up and hollering like that. Jesus Christ!" He banged his fist on the steering wheel. "You were a prime target for whoever was outside that door. I didn't have time to ask you nicely to duck. I knocked you down to save your life."

"From a *goat?*"

"I didn't know it was a goat and neither did you."

"I thought it was Emilio."

"And what if it had been? Were you hoping he'd kill me?"

"I was trying to keep you from killing him."

"I've got more self-control than that."

"Do you?"

He stopped the jeep so suddenly that she was pitched forward. "Yes, I do. And you, better than anybody, ought to know that." His eyes held hers for several telling seconds.

Finally she turned away.

Key whipped his head around. "Well, padre, what do you think of the day so far?"

Father Geraldo lowered a flask from his mouth and wiped it with the back of his hand. "It's a shame we had to leave the goat. It would have fed several families."

Key looked ready to throttle him, but the priest's droll comment struck Lara as funny, and she began to laugh. Father Geraldo laughed too. Eventually Key acknowledged the macabre humor of the moment with a taut smile.

"Ah, hell." He sighed, throwing back his head and gazing up at the patch of sky visible above the two buildings between which they were parked. "A goddamn goat."

Once their laughter subsided, he turned to Lara and touched her lower lip. He winced with regret when his fingertip picked up a bead of fresh blood. "It was reflex. I didn't mean to hurt you."

466

"It's nothing." She dabbed the cut with the tip of her tongue and tasted not only her blood but the slightly salty spot where his fingertip had been. "I don't want to stop the search now."

" 'Now'?"

"It's incredible to me that the credenza was spared. Either it's a miracle, or Emilio is alive and has recently been in that office setting things right. Those were his eyeglasses. I'd swear to it. He's been there recently."

"Well, he won't be back today. If he was lurking around somewhere, we surely scared the hell out of him."

He was probably right, Lara thought. Emilio was her best chance of gleaning information — if he was indeed still alive and if she could coax him out of hiding. She intended to return to the embassy later, with or without Key and Father Geraldo, and stay through the night if necessary in order to make contact with her husband's former aide. Key would have a litany of objections against that strategy, so she decided to postpone telling him her intentions for as long as possible.

There were, however, other avenues she could explore in the meantime. "Father Geraldo, wouldn't Ashley's death be a matter of public record?"

"Perhaps. Before the revolt, this nation made stabs at being civilized. If the records haven't been destroyed, they would be on file at city hall."

"What kind of red tape would you have to cut through to get to them?" Key asked.

"I won't know until I try."

"If it's known what you're looking for, we'd just as well raise a red flag."

The priest thought about the dilemma for a moment. "I'll tell them I'm looking for the records of someone named Portales. Portales, Porter. If the death certificates are filed alphabetically, Ashley's name should be in the same volume."

"Volume? Aren't they computerized?" Key asked.

"Not in Montesangre," Father Geraldo replied with a rum-induced smile.

It turned out to be remarkably simple. After the incident at the pillaged embassy, they almost didn't trust their good fortune.

Not quite half an hour after Father Geraldo had left them in the jeep, parked on a side street a couple of blocks from the courthouse, he returned, walking jauntily and wearing a happy grin. "God has blessed us," he told them as he climbed into the backseat.

Although he'd been gone only a short while, to Lara it had seemed like an eternity. She feared that no records would be found and that this errand would produce no new information. Key, pretending to take a siesta beneath his straw hat, had kept careful watch, fearing that they would attract attention.

Ciudad Central was a city in turmoil, but a fair amount of commerce was still being conducted. People moved from place to place in the lumbering city buses, in private cars, on bicycle, and on foot. For all the movement, however, one didn't get a sense of bustling activity.

The pervasive mood was one of wariness. People didn't collect in clusters to chat, lest their reason for gathering be misinterpreted by the soldiers in the military vehicles that imperiously sped along the thoroughfares. Children were kept near their nervous, cautious mothers. Shopkeepers transacted business without engaging their customers in lengthy conversations.

Lara and Key were relieved to see Father Geraldo return. "You found out where Ashley's buried?" Lara asked eagerly.

"No, but there was a death certificate. It was signed by Dr. Tomás Soto Quiñones."

"Let's go," Lara told Key, motioning for him to start the jeep.

"Hold on. This Soto," he said, turning to Father Geraldo, "who's side is he on?"

Lara was impatient to follow up on the clue. "It doesn't matter."

"The hell it doesn't."

"He's a doctor. So am I. That takes precedence over political affiliations. He'll extend me a professional courtesy."

"Will you grow up?" Key said with exasperation. "For all you know he's El Corazón's brother-in-law or a spy for Escávez. Either way,

469

if we go barging in there and say the wrong thing, we're screwed."

"Excuse me." Addressing Key, Father Geraldo played peacemaker. "In my work, I've crossed paths with Dr. Soto several times. I've never known him to profess allegiance to any particular faction. He treats the wounded of all sides, much as I do."

"See? Now can we go?"

Key ignored Lara. "Even if he's sympathetic, he'd be risking his neck to help us. The potential danger could make him reluctant to talk. He might outright refuse. Worst-case scenario is that he'll sic El Corazón's death squads on us."

"I'm willing to take the chance," Lara said adamantly.

"You're not the only one involved."

"If you won't go with me, I'll go alone."

Key tried to intimidate her with his stare. When she held her ground, he turned to Father Geraldo. "What's your gut instinct on *el doctor?*"

Indecision flitted in the priest's dark eyes. Finally he said, "Whether or not he consents to help us, I think we can trust him to secrecy."

Lara agreed.

"Okay, you two," Key said softly. "Have it your way, but we're going to go about it my way."

Lara and Key waited in the doctor's cramped hospital office while Father Geraldo once again acted as their mouthpiece. Even though Key had

470

closed the blinds against the afternoon sun, the room, without air-conditioning, was stifling. Lara's bodice clung to her damp skin. Perspiration had formed a dark wedge in the center of Key's shirt. He frequently used his sleeve to wipe his sweating forehead. They didn't waste either oxygen or energy on conversation.

Silence was also an added precaution. They didn't want their voices to attract anyone on the hospital staff to the doctor's private office. Explaining their presence there could prove tricky.

The waiting became interminable. Lara folded her arms beneath her head and laid it on the doctor's desk. They'd been there over two hours. What was taking so long? Her imagination began to run wild: They'd been discovered. Armed troops had been summoned and were taking up positions around the hospital. Key was probably right; Dr. Soto used his medical profession as a cover. He was actually a spy. He'd seen through Father Geraldo's ruse, tortured him into telling the truth and —

The instant she heard the approaching Spanish-speaking voices, she sat up. Key had heard them, too. He moved into position behind the door and signaled her to remain quiet and out of sight until the doctor was inside the room.

Her heart beat hard against her ribs. A trickle of sweat slid between her breasts. The doorknob turned and Dr. Tomás Soto Quiñones preceded the priest into his office. He reached for the light switch and flipped it on. "It was a routine birth, but these things can take —"

He spotted Lara and looked at her quizzically.

"Forgive me, Doctor," Father Geraldo said humbly as he ushered the doctor across the threshold. Still in Spanish, he explained, "I've been less than truthful. I do wish to discuss with you a soup kitchen for the starving. Perhaps at a later time?"

Key reached around them and closed the door, posting himself between it and the dumbfounded physician.

Father Geraldo apologized to Lara and Key for the delay. "He agreed to see me as soon as he delivered a baby. The labor stalled and took longer than he had estimated."

"You're Americans?" the doctor exclaimed in flawless English. "How did you get across the border? Please tell me what is going on." Uneasily he glanced at Key's stern visage and at the pistol tucked into his belt. He gaped at the priest, then at Lara, who was now standing at the edge of his desk. "Who are you?"

"My name is Dr. Lara Mallory." Although it hadn't bled for hours, her lip felt like it had an anvil attached. "Three years ago, I was living in Montesangre with my husband, Ambassador Randall Porter."

"Yes, of course," he said as recognition dawned. "Your picture was in the newspapers. Your husband was kidnapped and executed. Such a tragedy. Senseless violence."

"Yes."

"The medical community has continued to mourn the ambassador's death. Since diplomatic relations with the United States were suspended, it has been difficult to obtain pharmaceuticals and medical supplies."

"As a physician, I can appreciate your problem." She took several steps forward. "Dr. Soto, I'll personally see to it that you'll receive an abundance of supplies if you'll help me now."

The doctor glanced over his shoulder at Key, gave the priest another inquisitive look, then turned back to Lara. "Help you in what way?"

"Help me locate my daughter's grave."

Dr. Soto regarded her in stunned surprise, but he said nothing.

"When my husband was taken, she was killed in the gunfire. She was buried here. My government, and several Montesangren regimes, have ignored my repeated requests to have her remains exhumed and sent to the United States. I'm here to do it myself. But I don't know where she's buried."

Far down the corridor, rubber-soled shoes were squeaking on the vinyl floors. The clatter of metal servers and china announced that the dinner carts had arrived. But in this cubbyhole office next to the emergency exit door there was nothing but silence.

Finally the doctor cleared his throat. "You have my deepest sympathy. You're to be admired for undertaking such a dangerous mission. But I am at a complete loss. How would I know where

473

your daughter is buried?"

"You signed her death certificate." Lara moved closer to him. Key tensed and reached for his weapon, but her quick glance ordered him not to interfere. "Do you remember the incident?"

"Naturally."

"Her name was Ashley Ann Porter. She died on March fourth of that year, just hours before the revolution was officially declared."

"I remember distinctly when your daughter was killed and your husband taken captive. You too were injured."

"Then you must remember signing Ashley's death certificate and releasing her body for burial."

Sweat had popped out over his face. He was a stout man, solidly built, shorter than she. His face was square, with a broad, flat nose indicative of some Indian blood in his lineage. His hands looked too large and blunt to perform surgery, although Father Geraldo had said that he was well respected as a surgeon.

"Regrettably, I do not remember signing such a document."

She uttered a despairing cry. "You must!"

"Please understand," he said hastily, "those hours and days following the ambassador's abduction were the most turbulent in this country's history. There were hundreds of casualties. Our president and his family barely escaped with their lives. Anyone who had served his administration in any capacity was publicly executed. The streets ran with blood."

Lara had read the newspaper accounts from her hospital bed in Miami. She didn't doubt the accuracy of the doctor's description of the chaos.

Speaking for the first time since the doctor's arrival, Key was more skeptical. "You don't remember one little Anglo girl among all those other corpses?"

Soto shook his bald head. "I am sorry, señor. I know it comes as a disappointment."

Lara took several deep breaths to fortify herself, then extended her right hand to him. "Thank you, Dr. Soto. I apologize for the theatrical way in which we approached you."

"I understand the necessity for caution. Your husband was unpopular with the rebels who are now in power."

"My husband represented the United States, and they had taken a position that favored President Escávez. Randall was only doing his job."

"I understand," Soto said quietly. "Nevertheless, I can almost guarantee that the families and friends of men who were tortured and killed by Escávez's henchmen will not be so generous in their thinking."

"Can we trust you to keep your mouth shut about this?" Key asked abruptly.

"*Por supuesto*. I would not betray you."

"If you do, you'll regret it."

Father Geraldo stepped between them. "I think we'd better leave Dr. Soto to his duties."

"Yes," Lara agreed. "There's no point in involving you further."

Father Geraldo gave the doctor his blessing and asked forgiveness for tricking him. Dr. Soto assured the priest that he understood. As Lara moved toward the door, Soto laid a hand on her arm. "I am sorry, Señora Porter. I wish I could have been of more help. *Buena suerte.*"

"*Muchas gracias.*"

Replacing the scarf over her head, she followed Father Geraldo from the doctor's office. Key brought up the rear as the priest led them out the way they had come in, through a wing of the hospital that had been closed because the unstable government could no longer afford to keep it open. He knew the layout of the hospital very well, having spent years visiting sick parishioners there.

They emerged undetected. Lara was surprised to see that darkness had fallen while they'd been inside. Not that she cared whether it was daylight or dark. She could barely muster the energy to place one foot in front of the other and probably would have stopped dead in her tracks if Key hadn't herded her along.

After having her hopes raised by the discovery of Ashley's death certificate, the outcome of her meeting with Dr. Soto was a crushing disappointment. Fate had trampled her, and she lacked the initiative to continue.

She still planned to return to the embassy in the hopes of finding Emilio Sánchez Perón. First, however, she must rest. Rest would boost her morale. She knew that once she'd slept several hours, reviewed her options, and charted another course of action, she'd feel much more optimistic.

That was the pep talk she gave herself as she trudged toward the jeep.

She never made it that far. Key dragged her behind a dumpster at the rear of the hospital. "Pst! Padre!"

Father Geraldo turned. "What is it?"

"There's no reason for the cloak-and-dagger act," Lara complained. "No one spotted us."

Key motioned Father Geraldo closer. "What time will Soto be leaving the hospital?"

He shrugged. "I have no idea. Why?"

"Our doctor friend is lying."

"But I've known him —"

"Trust me on this, padre," Key interrupted. "You might be a good judge when it comes to saints, but I know sinners. He's lying."

"How?" Lara asked.

"I don't know, but I want to find out. He said he didn't remember your daughter. That's bullshit," Key declared. "That ambush made headlines all over the world. I was in Chad when it happened and it made the front pages there. It started a revolution, yes. Bodies passed through the city morgue like shit through a greased goose, yes. He might have been up to his armpits in corpses, but no way could he forget signing a death certificate for a U.S. ambassador's daughter killed in a bloody shootout. No way, José."

It was amazing how instinctively and completely Lara trusted Key. With the dark scruffy beard, he looked like the meanest of desperadoes, a man who attracted danger and thrived on it. His startling blue eyes moved like quicksilver as they surveyed the surrounding buildings. They didn't miss the smallest movement. His voice was quiet, urgent, compelling, and convincing.

"What are we going to do?" she asked.

Her unqualified trust must have silently communicated itself to him, because his alert eyes stopped their surveillance and fell on her.

"We wait."

At the sound of the fatal click, Dr. Soto came to a sudden standstill. Key thrust the barrel of the Beretta behind the doctor's ear and yanked his left arm behind his back, shoving his hand up between his shoulder blades.

"If you make a peep, you're history." His voice was a hiss in the darkness, so low it could have been mistaken for the rustle of leaves stirred by the faint breeze. "Walk."

The doctor didn't argue. He moved toward the jeep that rolled out from the deep shadows of the alleyway. Behind the wheel sat Father Geraldo, looking both excited and apprehensive. Lara was balanced on the edge of the backseat, gripping the seat in front of her, watching as Key approached with their hostage.

"Frisk him, Lara." She jumped to the ground

and ran her hands over the outside of the doctor's clothing.

"I am unarmed," he said with dignity.

"You're also a goddamn liar," Key said. With a nod, Lara confirmed that the doctor wasn't concealing a weapon, then returned to her place in the jeep. "Get in."

Soto did as Key ordered and climbed into the front seat. Key vaulted in to sit beside Lara, digging the muzzle of the gun into the hollow at the base of the doctor's skull. Father Geraldo put the jeep in gear and they took off.

"Where are you taking me? For God's sake, please . . . I don't know why you are doing this. What do you want from me?"

"The truth." Lara leaned forward so she could be heard. "You know more than you're telling about my daughter's death, don't you?"

Key nudged the back of Soto's head with the pistol. "No!" the doctor protested in a high, thin voice. "I swear I know nothing. As God is my witness,"

"Careful," Key warned. "There's a man of God present who tells Him everything."

"I cannot help you," he whimpered.

"Cannot or will not?" Lara asked.

"Cannot."

"That's not true. What do you know that you're holding back?"

"Mrs. Porter, I implore you —"

"Tell me," she insisted.

Father Geraldo drove down a dirt lane that

ended in a remote clearing above the river. The river began as a clear, rushing stream in the mountains, but by the time it had snaked its way down through the jungle and cut a swathe through Ciudad Central, where it swept up garbage and pollutants, it emptied sludge into the ocean. He brought the jeep to a stop but kept the motor idling.

"Were you on duty at the hospital that day our car was ambushed?" Lara asked.

He tried to nod but couldn't because of the revolver. "*Sí,*" he whispered in fear.

"Did you see my daughter?"

"*Sí.* She was critically wounded."

Lara swallowed, remembering the amount of blood gushing from the wound on Ashley's neck. The carotid artery had no doubt been severed. She closed her eyes in an attempt to stamp out that mental picture. Later she could grieve. Now, she didn't have the luxury of time. "What happened to my daughter's body?"

"Father," Soto pleaded, rolling his eyes toward the priest, "I beg you to intercede. I have a family to protect. God knows my heart goes out to Mrs. Porter, but I am afraid of reprisals."

"You damn sure should be." Key spoke in a near growl. "El Corazón isn't here, but I am. We haven't come a thousand miles to fuck around with you. Tell her what she wants to know, or you're no use to us. *¿Comprende?* In other words, you're dispensable."

Lara didn't approve of Key's fear tactics. They

480

had agreed that he would use them only when all else failed, or — and this was doubtful — when they became convinced that Soto was telling the truth and that he didn't know anything about Ashley's burial. She was reasonably sure he wouldn't make good on his implied threats, but hopefully Soto would fall for them before she had an opportunity to put it to test.

"Padre?" Soto begged, his voice cracking as he glanced fearfully at the murky, polluted waters below. *"¿Por favor?"*

Father Geraldo crossed himself, bowed his head, and began to pray softly. He couldn't have been more convincing.

"I'm tired of this shit." Key jumped over the side of the jeep and motioned with his head for the doctor to alight.

"Cementerio del Sagrado Corazón," he blurted.

"Sacred Heart. She's buried there?" Lara asked.

"Sí." The doctor expelled his breath and seemed to deflate like a balloon. "During those early days of fighting, they took most of the casualties there. Take me there, and I will show you."

Father Geraldo stopped praying and put the jeep in reverse. Key climbed back in. He had a warning for the doctor: "You'd better not be bullshitting us."

"No, señor. I swear it on the heads of my children."

The cemetery was located on the other side

of the city. It would have been a long drive under normal circumstances. The distance was increased by the circuitous route the priest took. He doubled back several times to make certain they weren't being followed. To avoid roadblocks and military convoys, he zigzagged through seemingly abandoned neighborhoods where streetlights remained dark and only alley cats were brave enough to show themselves.

Lara's nerves were jangling by the time they reached the cemetery gates. "It's locked!"

"But it's a low wall. Come on." Key was the first one out of the jeep. He motioned Soto down. "Keep both hands on your head. If you lower them, I'll shoot you."

"You cannot shoot me or you will not know where to look for the girl's grave."

The bluff didn't work on Key. He flashed a grin that showed up extraordinarily white against his black beard. "I didn't say I'd kill you. I just said I'd shoot you. For instance in the hand. You wouldn't be able to change a Band-Aid, much less do surgery." He stopped smiling. "Now move."

The four of them had no difficulty getting over the low stone wall. Soto indicated the direction in which they should go. They didn't risk using a flashlight. There was no moon, so they had to pick their way carefully around tombstones and over uneven ground.

The cemetery was situated on a hillside and offered a commanding view of the city with the mountains rising behind it. It had not escaped the effects of war. The grounds were no longer maintained. Very few graves appeared to have been tended since the revolution began. It broke Lara's heart to think of her daughter being buried in this desolate place that was overrun with weeds and inhabited by jungle reptiles that slithered unseen in the underbrush.

Ashley won't be here for long, she vowed silently.

Indeed, Dr. Soto had reached a shelf of land that rimmed a wide depression. There he stopped. Moving slowly so he wouldn't incite Key to make good his threat, he turned toward Lara. She was taken aback by the ghoulish appearance of his eyes until she realized that the wavering sheen in them was actually unshed tears.

"I would not have had you know this, but you insisted," he said. "It would have been much better if you had not forced me to bring you here. Better yet that you had forgotten what happened to you in Montesangre and stayed in America."

"What the hell are you jabbering about?" Key demanded.

Lara, more mystified than angry, moved closer to the edge and looked down into the depression. It was about twenty yards in diameter, roughly round in shape, and resembled a meteor crater, although vegetation had cropped up in spots.

Still perplexed, she turned to Father Geraldo. He was staring into the shallow bowl of earth.

His shoulders were hunched forward, and his arms hung loosely at his sides. He had a listless grip on his flask, but he wasn't drinking from it. Seeing the depression had stupefied him and supplanted his preoccupation with rum.

Key too was staring beyond the ledge as though demanding it to offer up an explanation. Then suddenly his whole body twitched as though a string coming out the top of his head had been jerked hard. He dropped the pistol in the dirt and grabbed the doctor by the lapels of his linen suit, lifting him until his toes dangled inches from the ground.

"Are you telling us —"

"*Sí, sí.*" Key had shaken the tears from the doctor's eyes. They coursed down his face. "*Doscientos. Trescientos. ¿Quién sabe?*"

"Two hundred or three hundred what?" Lara's voice rose in panic. "Two hundred or three hundred —"

When the answer struck her, she lost her ability to breathe. Her mouth remained open, but she couldn't exhale or inhale.

Key released the doctor and rushed toward her. "Lara!"

The most bloodchilling sound she had ever heard rose above the sepulchral silence of the cemetery. At first she didn't realize that the wail had been ripped from her own throat. Spreading her arms wide, she flung herself toward the rim of the depression and would have plunged to the bottom if Key's extended arm hadn't caught her at the waist. She bent double over it. He hauled her backward, but she fought him with the abnormal strength of the demented.

Finally managing to tear herself free, she crawled toward the edge, inexorably, clawing at the earth, uprooting clumps of grass, and all the while making that unnatural keening sound.

"No! God no! Please no! *Ashley!* Oh, Jesus, no."

Dr. Soto was blathering about the day the mass grave was ordered. It had been dug by bulldozers specifically to accommodate the enormous number of casualties. Morticians couldn't keep up with the demand, he said. When the morgue had filled to capacity, they'd begun placing cadavers wherever they could find space. Hundreds had died in the streets, where their bodies had been left to decompose. It became a health hazard to the living. There were outbreaks of typhoid and other contagious diseases. The rebel commanders dealt with the problem the most expeditious way they could devise.

"Lara, stop this!" Key's hands were on her shoulders, trying to pull her up, but she dug her fingers into the earth and wouldn't let go.

"I am sorry. So sorry," Dr. Soto repeated.

She understood now why he had been reluctant to tell her about this mass grave. He had feared reprisals, but not from El Corazón. From her.

"Leave me alone." As Key tried to pull her away from the brink of the macabre pit, her fingernails left bloody tracks down his forearm. He grunted in pain but only redoubled his efforts to bring her under control.

"Lara." Father Geraldo knelt beside her, speaking gently. "God in His infinite wisdom —"

"*NO!*" she screamed. "Don't talk to me about God!" Then in the next breath she entreated the deity for mercy.

"Who did this?" Key's hard hands were still bracketing her shoulders but he had fixed a murderous glare on Dr. Soto. "Who ordered that little babies be shoveled into a mass grave? Good God, are you people barbarians? I want a name. Who gave the order? I want that motherfucker's name."

"I am sorry, señor, but it is impossible to know who gave the order for a mass burial. Everything —" Dr. Soto's next utterance was a soft gasp. He dropped to his knees, clutching his chest, then collapsed onto his side.

Father Geraldo was into his third Hail Mary when he pitched forward and landed flat on his face in the damp soil near Lara's right hand.

In fascination and horror she watched a dark pool form beneath his head.

"Christ!"

Key reached for the Beretta he'd dropped earlier but wasn't fast enough. For his failed effort he got a boot in his ribs and went down with a grimace and a groan.

Crabbing backward, Lara tried frantically to move away from the gelatinous mess that had

486

once been Father Geraldo's head. She was yanked to her feet so swiftly that her teeth crashed together.

"*Buenas noches, señora.* We meet again."

It was the guerrilla leader from the roadblock outside Ciudad Central. Ricardo.

The military transport truck hit a chuckhole. Lara was thrown against the steel side of the "deuce and a half," which was the American slang for the tonnage of the truck. They'd been traveling for hours.

Almost before her brain had registered that they were surrounded by armed men, her hands had been roughly tied behind her. They were still bound, making it impossible to maintain her balance as the truck bounced along. She'd been thrown from side to side so many times, she would be covered with bruises. If she lived.

That was still open to speculation.

Father Geraldo was dead. Dr. Soto had died in midsentence. Key was very much alive. Thank God. He had kept up a litany of abusive curses as they were dragged from the cemetery and forced into the truck. Several soldiers had been riffling through their belongings left in the jeep. One had been fiddling with the camera and lenses in the camera bag. Key shouted at him. "Keep your goddamn hands off that!"

Like Lara, his hands were tied behind him, but he rushed forward and kicked the bag out of the soldier's hands. The hotheaded soldier cracked the butt of his pistol against Key's temple. Key staggered and dropped to his knees, but he wasn't cowed. He looked at the soldier and, with blood dripping from the wound on the side of his head, grinned and said, "Your mother got you by fucking a jackass."

Whether he understood English, the soldier interpreted the comment as an insult and lunged for Key. Before he could get retribution, Ricardo ordered the younger man to get them into the truck.

There was some discussion among them as to whether they should bring the jeep along or leave it at the cemetery gate. Ricardo decided to let one of the guerrillas follow them in it.

Lara and Key were hoisted into the back of the truck. Their belongings, including the camera bag and her doctor's bag, were tossed in after them. The soldiers climbed aboard, then lowered and latched the canvas canopy. They could see nothing, but their captors insisted that they be blindfolded. Naturally, Key didn't submit. It took three men holding him down before they could secure the dirty bandanna over his eyes. Lara knew that physical resistance would be futile, but her eyes conveyed the full extent of her contempt before she was likewise blindfolded.

The road was virtually impassable. The soldiers were unwashed. In the airless confines of the truck, the smell was overpowering. She was thirsty but knew that any request for water would go unheeded. Her butt was sore, as were her arms and legs. The bindings around her wrists were beginning to chafe.

She wanted to know where they were taking them and why. How much longer until they reached their destination? Did they even have a specific destination? When they reached it, what then?

She conserved the strength it would take to ask. No one would answer her. They had attempted to communicate only once. Key had been punished for it.

"Lara?" His throat had sounded as raspy and dry as hers. "You okay?"

"Key?"

"Thank God." He sighed. "Hang in there and —"

"*¡Silencio!*"

"Fuck you."

There was a scuffle, then a moan, and Key hadn't spoken to her since.

She tried self-hypnosis to remove her mind and body from the present situation. But each time she tired to conjure up mental pictures of a desert sunset, or a rolling tide, or drifting clouds, her focus returned to the mass grave in the cemetery where her daughter would be interred forever.

Accomplishing what she had set out to do was an impossibility. Why then didn't she try to escape, and let a soldier's bullet be her deliverance? Father Geraldo and Dr. Soto had felt no pain. Instant extinction. How lovely.

489

Why did she still have the will to survive?

No, it was stronger than will. It was a resolve to see the ones responsible for such an atrocity punished. Burying the daughter of a U.S. ambassador in such an unspeakable manner violated universally acknowledged human rights. If she lived, she would see to it that the world knew about the disgrace.

Lara had dealt with many terminally ill patients. Until tonight she had not understood their unwillingness to surrender life. How could one hang on, stubbornly clinging to life, knowing that the situation was hopeless? She'd often contemplated the human spirit's refusal to accept death. Now she understood that one could survive even the worst possible circumstances.

The survival instinct was stronger than she had believed. It preserved life, even when the mind had given up. If that were not so, she would have died upon seeing that mass grave and learning that her baby girl was buried there. That innate determination to live sustained her through the long night.

She must have dozed because she came awake when the truck ground to a halt and she heard sounds of activity outside the truck. She smelled wood smoke and cooking food.

"Here already?" Key quipped sarcastically.

She was brought to her feet and lifted out of the truck. Her limbs were stiff and sore. She stumbled when she was shoved forward, but the fresh air on her skin and in her lungs was welcome. She breathed deeply and tried to work circulation back into her legs.

Suddenly the blindfold was ripped off. Ricardo was standing close, smiling broadly. *"¡Bienvenido!"* She recoiled from his rancid breath. "El Corazón is anxious to welcome his special guests."

She was surprised at his command of English. "I have plenty to say to El Corazón, too."

He laughed. "A woman with a sense of humor. I like that."

"I wasn't being funny."

"Ah, but you were, señora. Very funny."

Just then a woman dressed in dirty fatigue pants and a sweat-stained tank top launched herself against him. After an embarrassingly passionate kiss during which he openly fondled her, she purred, "Come inside. I have food for you."

"Where is El Corazón?" he asked.

"Waiting inside."

Still groping each other, they ambled toward a crude shack and climbed the rickety steps to a shallow porch and a curtained doorway. The other soldiers were being similarly greeted by women in the camp and given bowls of food dished from a communal cooking pot suspended over the campfire. They drank fresh coffee from tin cups. Lara would have settled for a drink of water. Her lip was still tender and swollen.

Two men with semiautomatic weapons were standing guard over her and Key. When Lara first saw him, she gasped. He was sitting on the ground near her, but the guards stood between them. The wound on his temple had coagulated. It looked nasty and needed to be cleaned and disinfected, probably sutured. She wondered if she'd be given access to her doctor's bag, but thought not.

His eyes were ringed with shadows of fatigue, as she knew hers also must be. His clothes, like hers, were filthy and perspiration-stained. It was barely daylight, so the sun wasn't yet a factor, but the humidity was so high that a mist as dense as fog clung to the tops of the trees in the jungle that surrounded the clearing.

Key was looking at her with a stare that penetrated, but she didn't need this silent communiqué to realize how precarious their situation was. While he had her attention, he cut his eyes toward the camera bag. One of the soldiers had unloaded it and their other bags from the truck and dropped them near where she stood.

Lara cocked her head inquisitively, knowing he was trying to tell her something but unable to decipher what.

Then he mouthed, "Magnum." She glanced quickly at the camera bag. When she looked back at him, he nodded almost imperceptibly.

"Señora, señor." Ricardo swaggered from behind the curtained doorway and propped himself against one of the posts supporting the thatched roof. "You are very fortunate. El Corazón will see you now."

A respectful silence descended over the camp.

Those who were eating set aside their food. All eyes turned to the front of the shack. Even the children who'd been chasing one another and dodging toy machine-gun bullets ceased their play. The rebel soldiers stopped trying to impress the women with exaggerated tales of their exploits. Everyone's attention was focused on the porch of the shack.

Ceremoniously, the curtain was drawn aside, and a man emerged.

Lara sank to her knees. In a voice almost soundless, she exclaimed, "Emilio!"

Chapter Twenty-Two

"Excuse me, Miss Janellen?"

At the sound of Bowie's voice she almost jumped out of her skin, but she gave no sign of it. With the cool condescension of a Russian royal, she raised her head. "Hello, Mr. Cato. What can I do for you?"

He was standing in the doorway that connected the shop with the tiny office in its rear. The ugly, ill-formed building was quiet and, except for the two of them, deserted.

Bowie had brought in with him the scent of outdoors. The first hint of autumn was in the air, and she could smell it on his clothes. His hair had been mashed flat by his hat, the brim of which he was nervously threading through his fingers. His lips were chapped. She looked at him with concealed yearning.

"I was just wondering if you'd heard anything from your brother and Dr. Mallory?"

"No," she replied, feeling a pang of guilt. It was selfish of her to be so wrapped up in her heartbreak over Bowie when their lives could be in danger. Key had promised to call home if he was able, but there had been no communication from him since their departure three days ago. Janellen was sick with worry, and, although her mother hadn't admitted it, she was, too. She stayed in her bedroom except at mealtimes, when it seemed that even polite conversation was an effort.

"That's too bad," Bowie said. "I was hoping they'd be on their way back by now." He fiddled with a loose straw in the brim of his hat.

"Was there something else, Mr. Cato?"

"Uh, yes, ma'am. My paycheck. It wasn't in my box this morning. Any other week, I wouldn't bother you about it, but my rent's due tomorrow."

Knowing full well that he spoke the truth, she looked toward the empty pigeonhole labeled with his name. "My goodness. I apologize for the oversight, Mr. Cato. I must have left your paycheck in the safe."

The official company safe was a monstrosity that easily outweighed three pianos. It dominated one corner of the cramped room. The black steel facade was ornately trimmed with gold swirls and curlicues. It dated back to the days when her grandfather had paid his roughnecks in cash.

As she moved toward it, Janellen felt Bowie's eyes on her, and it was unnerving. Thankfully, the combination to the safe was second nature to her. She opened it and withdrew his check from the drawer where she'd intentionally left it that morning. Since he hadn't taken the initiative to approach her since the night they'd embraced in the kitchen, the night following Jody's seizure, she'd made it necessary for him to seek her out.

He'd fled during a thunderstorm, preferring the cold company of lightning and torrential rain to the warmth of her arms. Bowie might have been disappointed in her kisses, might have been disgusted by her eager response to his caresses, but she was not going to let him simply ignore her and pretend that they hadn't shared some degree of intimacy.

"There you are, Mr. Cato." As she handed him the check, she was careful not to let her fingers touch his. "I'm sorry I overlooked it."

She resumed her seat behind the desk and returned to the paperwork she'd been doing when he came in. Her heart was thudding so strongly and so loudly that she could count each beat against her eardrums. Whatever happened next was up to him. The next few moments were critical. If he turned and left without another word, it would break her heart. Her nonchalance was a pose she'd affected to hide her despair. If that tempestuous kiss at her kitchen sink was the extent of their love affair, she'd just as soon stop breathing.

Ten seconds ticked by. Twenty. Thirty.

Bowie shuffled his feet.

Janellen waited, making small notations in red ink on the invoice while her entire future and self-image dangled by a thread.

"How, uh, how come you've stopped calling me Bowie?"

Janellen looked up, feigning surprise to find that he was still there. She pretended to ponder her answer. "I didn't think we were on a first-name basis any longer."

"Why's that?"

"When two people address each other by first names, it implies friendship. Friends don't avoid each other. Friends call, drop by, pass the time of day together, make a point to see each other. Friends wave when they drive past; they don't turn their heads and pretend not to see." This last referred to the day before. He'd deliberately ignored her when they'd accidentally met on Texas Street.

"Now, Miss Janellen, I know you thought —"

"Even former friends don't pretend that the other person no longer exists." Her voice began to quaver and for that she hated herself. Whatever the outcome, she had vowed not to cry in front of him.

"Friends don't act like they've never been . . . friendly. Like they've never . . . Oh!" To her mortification, tears filled her eyes. She stood and turned her back to him, cramming a tissue beneath her nose.

"I'm no good at this," she said mournfully, blotting her eyes. "I can't play games like other women. That trick with your paycheck was stupid and juvenile. I know you saw right through it. I just didn't know any other way to force you to see me alone."

She turned to face him, knowing that she

497

looked her worst. She didn't cry prettily like the actresses in movies. When she cried, the whites of her eyes turned pink, her nose turned red, and her complexion got blotchy.

"I'm sorry, Bowie. I know this must be terribly embarrassing for you. Feel free to go. You don't have to stay. I'm fine. Honest."

But he didn't move. In fact, if there was anything redeemable in the last couple of minutes, it was that he appeared as miserable as she. "Truth is, Miss Janellen, I'm the one who's sorry that I put you through a scene like this."

She reasoned that since she had already made a fool of herself and had nothing more to lose, she might as well get to the bottom of it. "Why have you been avoiding me?"

" 'Cause I didn't think you'd want to see me after . . . Shit." Mumbling the expletive, he turned his head away. But when his gaze landed on a voluptuous calendar nude, he hastily looked back at Janellen. "I didn't think you'd want to see me after what I did to you. I didn't show you any respect, and I do respect you a hell of a lot."

Her cheeks grew warm as she recalled his hand moving beneath her skirt, clutching her bottom with what she'd thought was uncontrollable lust. It had been shocking, yes, but thrilling.

"Well, I wasn't behaving very respectfully myself, was I?" she asked a bit breathlessly. "But I assumed that our respect for each other had been well established. I thought that our

498

friendship had moved to another level. I thought you might want to, uh, maybe, you know, to fuck."

His hat landed on the top of the desk. He dropped into the chair facing it and planted his elbows among the invoices, holding his head between his hands. His cheeks puffed out, then his lips pursed as he blew out a gust of breath.

"I know that's the right word," Janellen said timidly. "Key says it all the time to mean . . . that."

"Yes, ma'am, it surely is the right word. It gets the message across, all right."

"Well then? Was I wrong?"

Bowie massaged the back of his neck. After what seemed to Janellen an eternity, he raised his head. "Fact of the matter is, it isn't the right word. If that's what I wanted, we could have done it on your kitchen linoleum. But I think too much of you to toss up your skirts and go at you like you're no better than a ten-dollar whore. See, Miss Janellen, you're quality and I'm trash, and nothing's ever going to change that."

"You're not trash!"

"Compared to you I am. Besides which, I'm an ex-con."

"You served time for doing something that needed to be done. In my opinion, the beast you assaulted deserved prison, not you."

He smiled indulgently at her vehemence. "Unfortunately, the state of Texas didn't agree." Turning serious again, he said, "Neither would the people of Eden Pass. If you were to take up with me, how do you think folks would react?"

"I don't care." She rounded the desk and knelt in front of the chair in which he sat, trustingly placing her hands on his thighs. "Bowie, all my life I've lived according to what other people wanted for me. I've done everything that was expected of me and nothing that would be looked upon with disfavor. But not too long ago, Key reminded me that life is passing me by." She inched closer. "I didn't realize how right he was until you kissed me. Then, for the first time in my life, I experienced a sense of bursting free. I don't want to grow old and then discover that I missed the best things life has to offer because I was afraid of offending someone else. For thirty-three years I've been the prim and proper Miss Janellen, and frankly I'm bored with her. The only fun and excitement she's ever had was with you.

"So what if the hometown folks raise their eyebrows over us? They've been asking for years over my spinsterhood, pitying me because I didn't have any beaux. Between pity and disapproval, I choose disapproval." Taking a deep breath, she added, "If you like me — even a little — don't back off because you're afraid of damaging my reputation."

"If I like you even a little," he repeated, smiling his sad smile. He pulled her up and settled her on his lap. "I like you so much my heart goes to aching every time I think about you, which is all the time."

He took her hand and stroked the back of it, his touch light, as though he feared breaking the fragile bones. "Folks aren't going to cotton to us being a pair, Janellen. You stand to lose so much. Me, I got nothing to lose. No money, no name, no family or friends or a position in the community. But you could be hurt bad."

She laid her fingers against his lips. "I won't be hurt, Bowie."

"Yes, you would. I'd hurt you, and I can't hardly bear to think about it."

Their faces were very close. His eyes were dark and intense, and she knew that he was no longer referring to the effect that their being together would have on her social standing. He was talking about the physical pain their coupling would cause her.

She whispered, "I wish for that hurt. I wish for it right now."

She fell against him softly. A low moan escaped her as his arms enfolded her. She tilted her head back against his biceps and welcomed his urgent kiss. They kissed hotly and hungrily, their mouths melding.

He stroked the side of her face, trailed a finger along the line of her jaw, touched her neck. Between fervent kisses Janellen whispered encouragement. When his hand moved to her breast and gingerly covered it, she lovingly murmured his name.

"I can't go on with that or I'll do what I swore I wouldn't do."

She opened her eyes and sat up straight. "What are you saying, Bowie?"

"That I'm not going to fuck you." She uttered a small sound of protest and dismay. He hastened to add, "What I want to do is make love to you. I want to do it proper, on a bed with sweet-smelling sheets, in a place that's clean and worthy of you."

She relaxed and laughed softly. "That doesn't matter to me, Bowie."

"It matters to me. I still think I'm the worst thing that's ever happened to you, but you're the best thing that's ever happened in my life, and that's for damn sure. I won't treat you like any old gal I could get off with."

While Janellen was disappointed that he'd stopped the foreplay, her heart filled with tenderness.

"I'm fairly certain you're a virgin." He glanced up and she nodded. "I can't imagine why that is, but I'm damned glad no other man beat me to you. It's an honor I don't take lightly, so when it happens, I want it to be like the first time for me, too. And in a way it will be. I've never done it with a woman I'd share my toothbrush with."

She giggled and nuzzled his shoulder. "Will you share your toothbrush with me?"

In reply he kissed her, pressing his tongue deep into her mouth. "I'll be looking for a place where we can go," he said hoarsely when the kiss ended.

"Your trailer," she suggested enthusiastically. "I'll come tonight after supper."

"That trailer is fine for me, but it's not fit for you to set foot in."

"Bowie!"

He stubbornly shook his head. "It's got to be someplace special. When I find it, you'll be the first to know."

"But when?"

"I don't know yet." His eyes burned with desire. "As soon as possible."

"Until then, you can come to the house every night after Mama goes to bed."

"I'd never sleep with you — sneaky like that — under your mama's roof."

"I didn't mean we'd sleep together, just be together. I can't leave Mama alone. Maydale would get suspicious if I asked her to stay late every night. I'd soon run out of excuses to be away from the house. If we're going to see each other, you have to come there."

He frowned. "That's tempting fate, Janellen. If we take chances like that, something terrible could happen."

"That's silly. Nothing's going to happen."

"Your mama could catch us together. Then the shit would hit the fan."

He was right on that score. But even her mother's disapproval wouldn't keep Janellen from seeing him. "We'll make doubly sure we don't get caught until we're ready to make our

503

'friendship' public." She grinned happily. "I'm ready to tell the whole world."

"I'd postpone that if I were you." He was as grim as she was ebullient. "Sooner or later, something's bound to go wrong. I'm unlucky like that."

"Everything's changing for both of us."

"Janellen." He held her face between his hands again and peered closely into her eyes. "Are you sure about this? Are you absolutely sure? 'Cause being with me isn't going to be a picnic for you. In fact, it's likely to be hell."

She covered his hands with hers. "Being without you would be hell. I'd sooner die. I love you."

"I love you too. And you can believe it or not, but I've never said that to another living soul."

They kissed again, and she wore him down until he gave her his promise to come to her back door at midnight that night.

Heather Winston had absolutely no interest in the search for the Northwest Passage. Irritably, she set aside her American history textbook and gave her mind over to the much more important matter of keeping Tanner Hoskins in line.

She was on duty at the check-in desk of The Green Pine Motel, as she was every weekday night from seven till ten. It wasn't hard work. It allowed her time to do homework and study for exams. But it also kept her from spending time with Tanner. Between cheerleading practice, the football team, and all their other extracurricular activities, they had very little time to be together except on weekends.

She didn't like it any better than he did, but he complained the most. "Lately your mother's got you on such a short leash, it's hardly worth going out at all."

Heather was afraid he would soon tire of their situation and seek the companionship of a girl with a lighter schedule and a more lenient curfew. Just that morning she'd caught him flirting with Mimsy Parker at her locker between second and third periods. Everybody had seen them together. By the time school was dismissed for the day, it was all over campus that Heather was on the verge of being dicked.

She wouldn't have it.

Recently Tanner had been elected student body president. He'd scored two touchdowns last Friday night. He was the most popular boy at school this year. She wasn't about to let Mimsy Parker have him.

As she was devising various methods of keeping him faithful, a bowlegged man entered through the automatic doors, removed his hat, and surveyed the lobby.

"Hi. Can I help you?"

"Evenin', Miss Winston."

"You know me?"

"I've seen you with your folks. My name's Bowie Cato."

She recognized his name. He was the ex-

505

convict now working for the Tacketts. Heather experienced a thrill of fear. Was he about to rob her? His darting eyes were wary and nervous. She was the only one on duty in the lobby. A waitress and a short-order cook were keeping the restaurant open, but they wouldn't be of any help to her if Bowie Cato had armed robbery and murder in mind.

"You might think this is a peculiar request," he said, after self-consciously clearing his throat. "But, well, I got, uh, kinfolks coming to spend the weekend. My trailer isn't big enough to sleep them and me, too, and anyway, these kin are kinda persnickety. So what I'm looking for is a place for them to stay. One night, maybe two."

"I'll be happy to make a reservation for you, Mr. Cato. Will they be here this weekend?"

"No, no, I don't need a reservation. What I mean is, I'm not sure what day they'll be getting here. They're sorta unpredictable."

"Oh." Heather was at a loss. He appeared harmless. She saw no sign of a weapon, although he could have a gun concealed inside his denim jacket, she supposed. He wasn't menacing, but she couldn't account for his jitters. "When you find out their date of arrival, you could call us and reserve a room. This time of year, we usually have vacancies."

"Yes, ma'am." Seeming reluctant to go, he looked through the brochures and state maps in the cardboard rack on the counter. "Uh, actually, I wondered if it would be possible for me to see a room. Like preview it, check it out. Your nicest room," he added quickly. "They like things fancy."

Heather laughed. "You want to see if our rooms are fancy enough for your relatives?"

"Meaning no offense, Miss Winston." He raised his hands and looked so disarming that Heather felt silly for being afraid of him. "These folks are like that. Uppity. Always wanting everything just so. I promised to check out the motel situation before they commit to a visit."

Heather moved to the drawer where keys were filed according to the room number. "The honeymoon suite is our nicest room."

"The honeymoon suite? I like the sound of that."

Heather put a sign on the counter that said BACK IN TEN MINUTES and hid her smile as she motioned him through a pair of wide glass doors. He didn't have relatives coming to visit any more than she had wings. He was planning a rendezvous with a lady friend. It was kind of sweet, Heather thought, the way he was making special plans for it.

"The suite is convenient to the swimming pool." She called his attention to it as they walked through a landscaped courtyard.

"Bit nippy for a swim."

"It's heated year-round."

"No foolin'?" He glanced dubiously at the water.

"No fooling. That pool is my daddy's pride

507

and joy. My mother talked him into installing it when they expanded and added this new wing. But it was Daddy's idea to heat it. The honeymoon suite was also my mother's idea. It's not as elaborate as ones you'd find in Dallas or Houston hotels, but it's pretty. Here we are."

She unlocked the door for him and stood aside. He hesitated on the threshold. "If you feel uneasy coming inside with me, Miss Winston, I can take a look-see by myself."

His eyes were so apologetic and earnest that Heather would have followed him into a dark alley wearing all Darcy's diamonds. "After you, Mr. Cato."

The "suite" was decorated in mint and peach, the quality of materials a notch above what was used in the other rooms. It had a sitting room and a bedroom with a king-size bed. The bathtub had a built-in whirlpool. Otherwise it was standard motel fare. Heather wouldn't want to spend *her* wedding night in it, but she supposed it would seem luxurious to the hicks in Eden Pass.

Bowie Cato nodded appreciatively to every amenity she pointed out, but remained noncommittal. "Where does that go?" he asked, indicating a door on the far side of the bedroom.

"The parking lot. If a guest wants to rent just the bedroom, we lock the door that connects to the parlor."

"Hmm. So you can come into the bedroom using the parking lot door without having to go through the lobby and around the pool?"

"That's right," she answered, suppressing another grin. Mr. Cato was having a secret affair. "The TV in the bedroom has a VCR, so you can bring your own movies to watch."

"Oh, I doubt we'll be watching —"

He broke off when he realized that he'd given himself away. Embarrassed, his ears turned red and he swallowed hard. She smiled to let him know that his secret was safe with her. "Like doctors and lawyers, people in the hotel business are very discreet."

"Yes, ma'am. Well, I think I've seen all I need to see. Thank you kindly. Can I go out through this door?" He moved to the one that opened directly onto the parking lot.

"I'll lock it behind you. Should I make a reservation for you?"

"Not tonight, thanks. I'll be in touch when, uh, a date's been set. Is that okay?"

"Sure."

Still looking sheepish, he replaced his hat and waved goodbye. Heather locked the suite and returned to the lobby. As far as she could tell, no one had been there during her absence, nor had the search for the Northwest Passage grown more interesting. She couldn't concentrate for thinking about Tanner. He'd told her he would be at home studying tonight, but was he?

On impulse, she dialed his number, asked his father if she could speak with him, and was relieved when Ollie told her to hold on while he called Tanner to the phone.

"Hi, it's me. Whachadoin'?"

"Studying history."

"Me, too. It sucks." She twirled the phone cord. "I'm sorry I totally bitched you out after school today."

"It's okay."

Heather could tell by his tone of voice that it wasn't. "Everyone was saying —"

"Don't believe everything you hear."

That was a little too glib a response, she thought. Why wasn't he denouncing the rumors and denying any interest in Mimsy Parker? *I'm losing him,* she thought in panic. She knew she'd never live it down. "Tanner, why don't you come drive me home when I get off at ten? Please? I want to see you."

"Don't you have your car?"

Since when did he need an excuse to see her? "I can tell my folks that it wouldn't start, so I called you."

"I guess I could."

"Okay." She consulted the clock. "I'll see you in thirty minutes. Unless you want to come now and keep me company until the night clerk gets here."

"I'll be there at ten."

Peeved, Heather hung up. She used the remaining thirty minutes of her shift to primp. The reflection in her compact mirror was reassuring. Mimsy Parker might have boobs the size of cantaloupes, but Heather still had the best hair, the best clothes, the best smile, the best eyes. Nor were her boobs anything to scoff at. Any bigger and they'd sag like Mimsy's in a few years.

Anyway, possession was nine-tenths of the law. Tanner was still hers. She just needed to guarantee that she kept him.

The night clerk, a pimpled geek who had a mad crush on her, arrived a few minutes early. When Tanner pulled his car into the porte cochere, in order not to appear overanxious she pretended to be busy behind the desk with the geek. After letting him wait a full five minutes, she joined him in his car.

"He's so dumb!" she exclaimed in exasperation as she slid into the passenger seat. "Honestly! He's in the National Honor Society but hopeless when it comes to common sense. Hi." She leaned across the console and kissed his cheek.

"Hi."

Heather pretended that the spat had never taken place and that Mimsy Parker didn't exist. She chatted nonstop about school and teachers, inconsequential things. "I've got to get something to wear for the homecoming game. I think Mother and I are going to Tyler Saturday to shop. If we can't find anything there, we'll go to Dallas the next Saturday. You're so lucky you don't have to worry about what you'll wear for the coronation during half-time. You'll be in your football uniform."

That was a subtle reminder that she had been nominated for homecoming queen and that he was damned lucky to be her official escort. "Your

511

football jersey will be all muddy, and when you take off your helmet, your hair will be sweaty. You always look so sexy like that. It makes me hot just thinking about it."

When she dropped her hand into his lap, she made it appear a casual gesture. She felt his instantaneous response. *What a goose I've been,* she thought. *What an idiot!* Sex was power. Look at how much mileage her mother got out of it: all she had to do was whisper something to Fergus and look at him seductively, and she got whatever her heart desired.

From the time Heather had been old enough to recognize that kind of manipulation for what it was, she'd been scornful of it. Maybe it was time for a change of heart. Her sexuality was an unlimited and as yet untapped resource.

What was she saving it for? Why not use it? Now. When it was needed. Every other woman did. Her mother. That slut Mimsy Parker. If she wanted to keep Tanner . . .

"Stop here," she said suddenly. They were still a block from her house. "I want to talk to you for a minute."

Tanner pulled the car to the curb, killed the engine, and cut the headlights. "What about?"

She wanted to slap that surly smirk off his face. Instead, she smiled beguilingly and drew him close. "I don't really want to talk." She pressed her open mouth to his and reached for his tongue with her own.

He was taken off guard but quickly recovered.

After a few tongue-twining kisses and some carefully choreographed moves, his erection was well defined behind his fly. She ran her hand up and down it, massaging.

He reached beneath her sweater and seized her breast. "What got into you?" he panted as he unsnapped the front closure of her bra.

Mimsy Parker, she thought. "I just love you so much. Oh, yes." When he lightly pinched her nipple, she placed her hand on the back of his head and guided it down to her. "Tanner, I had the best idea tonight. Listen." She outlined her plan as she slid her hand inside his jeans. "Doesn't that sound wonderful?"

"Yes. Oh, Jesus, oh God. Wait. I have a rubber. Want me to —"

"No. I want to see it."

"Faster, babe. Yes. Yes."

"Touch me, Tanner." She opened her thighs and guided his palm to her center.

After several steamy minutes of dual masturbation, he dropped her at her front walk. His eyes were still lambent, his face flushed; he was pathetically grateful and newly besotted.

Her confidence restored, Heather skipped up the steps of her house. Mimsy Parker didn't stand a snowball's chance in hell of stealing her boyfriend.

As she went inside, ready with an elaborate lie as to why Tanner had brought her home, she silently thanked that ex-con for giving her the idea that had saved her romance.

Chapter Twenty-Three

El Corazón del Diablo gave his prisoners his most ingratiating smile. His eyes flickered to Key, but after one curious glance they returned to Lara. Key doubted that she realized she had sunk to her knees.

No sooner had the thought crossed Key's mind than she slowly came to her feet. "I can't believe it. Emilio, what —"

"I am no longer Emilio Sánchez Perón," he snapped, his glassy smile vanishing. "I have not been that naive, idealistic youth in a long while. Certainly not since the revolution and your return to the United States." He almost snarled the last two words. "A nation I hold in utmost contempt."

Key hated what the young man said, but he was impressed by the manner in which he said it. He spoke fluent English without a trace of a Spanish accent, although he didn't use contractions.

The squalid backdrop made his neatness even more pronounced. He was smooth shaven and immaculately clean, not an easy condition to maintain in the middle of a jungle. His black hair had been pulled back so tightly that his head was as sleek and shiny as a bowling ball. He had a short queue at the nape of his neck. The style accented his high cheekbones, the lean angularity of his face, the hard, angry slash of his mouth. His eyeglasses had thin gold-wire frames.

Key had tangled with tough customers from all parts of the world. He couldn't recall one who had looked more chilling than Emilio Sánchez. He was slightly built, but the cold, dead quality in his eyes was symptomatic of unmitigated cruelty. The eyes of a snake.

"If you hate the United States so much, why were you working for my husband at the embassy?" Lara asked.

"My position there allowed me to receive information which others found very useful."

"In other words you were spying."

He flashed another grin. "Between you and your husband, I always considered you the more intelligent."

"You were using the embassy as a source of information. For how long?"

"From the beginning."

"You bastard."

A murmur arose from those around them who understood English. El Corazón's smile slowly dissolved, as though it were melting in the heat. "Having narrowly escaped with your life once, you were a fool to return to Montesangre, Mrs. Porter."

"I came to retrieve my daughter's remains. I wished to return them to the United States."

"You came in vain."

"I know that now. I condemn the Montesangrens who buried her in a pit." Tears formed in her eyes, but her posture was now unbowed. "God damn you all."

"You'll find it difficult to attract God's attention from here, Mrs. Porter. He hasn't listened to the people of Montesangre for decades. We no longer believe He exists."

"Is that why you found it so easy to murder Father Geraldo?"

"The drunken priest?" he said scornfully. Ricardo slapped him on the shoulder as though he'd told a joke. "He had outlived his usefulness long ago. He was merely another mouth to feed in a country of starving people."

"What about Dr. Soto? Surely he was useful to your regime."

"And also to Escávez."

"You are unforgivably wasteful. Dr. Soto was a healer. When it came to saving lives, he didn't think politically."

"Which was his downfall," El Corazón replied blandly. "In Montesangre one cannot have divided loyalties. Speaking of which," he said, his eyes moving to Key, "I'm curious about your loyalties, or lack thereof, Mr. Tackett. My curiosity alone has kept you alive."

"My life's an open book."

The soldiers guarding Key had allowed him to stand. His ribs hurt like hell. A couple of them had probably been cracked when he was kicked during the attack at the cemetery. His head hurt worse. The wound on his temple had scabbed over, but his whole cranium throbbed. He itched from having had so much sweat dry on his skin, leaving a salty, gritty residue. On top of everything else, he was hungry.

Sánchez said, "You are assisting the whore who unraveled your brother's political career. I find that peculiar. What would compel you to risk your life for her?"

"Not her. Her daughter. I believe she might have been my brother's child."

"Indeed?" El Corazón removed a folded white handkerchief from the rear pocket of his trousers and blotted his forehead. Even despots were victims of the jungle heat.

Key enjoyed knowing that the other man wasn't immune to discomfort. It made his own aches and pains more bearable. "Now that I know what happened to Ashley's body, I agree with Lara in her opinion of your country."

"Which is?" Sánchez asked as he meticulously replaced the handkerchief in his pocket.

"Montesangre is a shithole and El Corazón del Diablo is the toilet paper."

With lightning speed, Ricardo whipped a pistol from the holster around his hips and aimed it at Key. Languorously Sánchez raised his hand. Ricardo lowered the pistol but glared at Key murderously.

"You are either very foolish or very brave," Sánchez said reflectively. "I prefer to believe you are brave. Only a brave man would have dared

517

fly an airplane into my country without permission." He smiled his chilling, reptilian grin. "In spite of your clever piloting and the ridiculous charade enacted by you and the priest when my men stopped you on the road, we knew exactly where you landed your aircraft. I haven't seen it for myself, but Ricardo tells me that it is an excellent airplane. Well equipped. It will be useful as we continue our fight. Thank you very much for contributing it to our cause."

Key looked at Lara. When their eyes met, the best he could do was shrug helplessly. He had no tricks up his sleeve. Even if he could get to the Magnum pistol in the camera bag, he'd be gunned down before he could use it. Then they would murder Lara, too, and her death might not be so mercifully quick.

"Untie their hands."

Considering the gravity of Key's thoughts, El Corazón's brusque order came as a surprise. Ricardo voiced his objections, but Sánchez cut them short. "We are not savages. Give them water and something to eat."

Ricardo delegated the unwelcome responsibility to his subordinates, who roughly shoved Lara and Key to the ground. With heart-stopping ferocity and quickness, they severed the cords binding their hands. Key's wrists had been chafed raw. Lara's, he saw, were worse. The skin had cracked opened and she was bleeding.

They were brought crude bowls of a stew comprised mostly of rice and beans. The chunks of meat were scarce and unidentifiable. Key figured he was better off not knowing what it was. A young boy with a body as slender and tough as a jungle vine and eyes as hostile and flat as El Corazón's brought him a crockery pitcher of water. He drank greedily.

When he lowered the pitcher, he became aware of the nearby scuffle. Lara had dumped her portion of food onto the ground and was being jeered for pouring out the water that had been offered.

"How very childish, Mrs. Porter," El Corazón remarked. Someone had brought a chair for him. As he sat in the shade of the porch, two girls, one on either side, fanned him. "It surprises me that you would be so demonstrative. I remember you as a woman who displayed very little emotion."

"I would never accept your charity after what you did to Father Geraldo and Dr. Soto."

"As you wish."

She looked at Key, her irritation with him plain. He shrugged, knowing the insolent gesture would only increase her annoyance with him for eating and drinking what their captor had offered. If they stood a ghost of a chance to escape, they would need physical strength.

He wasn't as principled as Lara, maybe, but he was a hell of a lot more practical. Only moments before he'd been sympathetic to her physical discomfort. Now he could easily have throttled her for squandering food and water, which she desperately needed.

At a signal from Sánchez, several guerrillas detached themselves and moved out of sight behind the hut. Key finished his food and drank the remainder of the water. As the empty utensils were being taken from him, the soldiers returned, leading a man and a woman. Both had their hands tied behind their backs.

They were filthy. The stench of body odor and excrement was overpowering, a threat to Key's full stomach. The man had been beaten about his head. His hair was matted with dried blood. His features were so distorted by swelling, bruises, and abrasions that Key doubted his immediate family would have recognized him.

The woman had probably suffered more. As she was shoved forward, several of the soldiers in the camp whistled and called out Spanish insults that Key had learned as a boy in Texas. It was easy to conclude how she had been brutalized. The trauma had rendered her insentient. Her eyes were vacuous. She didn't respond to anything going on around her.

Sánchez left his chair in the shade and moved to the edge of the porch, where he looked beyond the bedraggled pair and addressed Key and Lara. "This man and woman were having sex while they were on watch. As a result of their carelessness, troops loyal to Escávez raided one of our camps. All of them died in the ensuing fight, but not before they killed two of my finest soldiers."

"*Por favor,*" the man blubbered through

swollen, discolored lips. *"El Corazón, lo siento mucho. Lo siento."* He repeatedly muttered the apology. She was his betrothed, he said. They had loved each other since they were children. Having explained that, he acknowledged that they were wrong to have jeopardized the lives of their comrades.

"She's a whore," Sánchez calmly countered. "She lay with fifty men last night."

The man sobbed but didn't argue. He begged for mercy, swearing on the graves of his mother and father that he would never be so negligent in his duties again. He dropped to his knees and crawled forward until he was inches from the toes of Sánchez's polished hoots, appealing to his commander to grant them forgiveness and mercy.

"You admit that it was lust which cost the lives of your comrades? You are weak. A stupid lecher, a slave to your selfish passions. She is a whore, a bitch in heat who would offer herself to anyone."

"Sí, sí." The accused bobbed his head rapidly.

"The liberation of Montesangre is the only thing for which one should feel such unrestrainable ardor. We must all be willing to make personal sacrifices."

"Sí, El Corazón, sí."

"I could have you castrated."

The sly, softly spoken threat sent the man into a paroxysm of pleading and promising, spoken in such rapid Spanish that Key had difficulty following it.

"Very well, I will not emasculate you." The man began to cry and whimper with relief, croaking elaborate accolades to El Corazón's greatness. "But such carelessness cannot go unpunished."

As a surgeon would extend his hand for a scalpel, Sánchez thrust out his hand. Ricardo slapped a pistol into his palm. El Corazón leaned forward, pressed the barrel of the gun against the groveling man's forehead, and pulled the trigger.

The woman jumped reflexively at the sudden racket but seemed impervious to the splattering of her fiance's blood and brain matter. At a signal from El Corazón, Ricardo stepped off the porch and moved behind her. He lifted her head by her long hair and, with a deft motion of his arm, cut her throat with a wicked-looking knife. When he released her hair, she crumpled to the ground beside her slain lover.

Key cut his eyes to Lara. She sat unmoving and silent. He admired her stoicism. This sideshow was for their benefit, but, like him, she refused to give El Corazón the satisfaction of seeing her react with revulsion and fear.

I might be next, Key thought, *but the tightassed little bastard won't see me on my knees begging for my life.*

A hush of expectation fell over the camp. Activity was suspended. Key guessed that the anticipation had nothing to do with the two grisly corpses being dragged away, but rather with what would be his and Lara's fate. Executions of enemies and traitors like those they'd just witnessed were probably commonplace, daily occurrences to enforce discipline and discourage disobedience. The camp followers, even the children, were inured to them. But having two American citizens to punish was a unique diversion that had captured everyone's imagination.

It was Lara, however, who began the offensive.

"You were an intelligent young man, Emilio Sánchez Perón." Her voice was soft with fatigue, but it carried to every ear in the camp. "You could have become a great man, an excellent leader, the leader who could have boosted Montesangre out of its rut of poverty and antiquity and into the twenty-first century. Instead you have regressed to what you accused me of being — a child. A petulant, cowardly, self-serving brat.

"You talk about freedom from oppression," she continued. Scornfully her eyes swept the camp. "This community is the most oppressed I've seen in Montesangre. You aren't a leader, you're a bully. One of these days one of your followers is going to tire of your bullying and show you no mercy. You're not to be feared but pitied."

Those who understood English gasped at her temerity. Those who didn't could accurately interpret the expression on El Corazón's face. It became suffused with color. His eyes glinted malevolently.

"I am not a coward," he said stiffly. "I killed

General Pérez because his resolve was weakening."

"I'll be damned," Key whispered. Sánchez was the usurper to whom Father Geraldo had referred. He was the soldier who'd murdered his own commander in order to seize control of the rebel forces.

"Yes, Mrs. Porter," Sánchez was saying. "I see you are surprised. I want you to understand how determined I am to become the undisputed leader of my country. I will do whatever is necessary, although sometimes the tasks are unpleasant." He glanced down at the fresh blood drying in the sun.

"Like shooting your own man point-blank?"

"Yes." He broke a smile that was so confident, so smug, that it was actually more bone-chilling than the brutal act had been. "Like that. And like organizing the ambush on Ambassador Porter's car."

Lara's body jerked. She blanched. Even her lips turned white. "You?"

"Under General Pérez's orders I coordinated the operation because I was familiar with the ambassador's agenda. You were not scheduled to attend the birthday party. You and Ambassador Porter quarreled over it. He insisted that you go with him.

"You should have followed your instincts and refused. He was our target, not you. If you had stayed at the embassy I might possibly have sneaked you out before it was attacked. As it turned out, my hands were tied. It was too late to call off the ambush."

"Ashley."

Key didn't actually hear her speak the name, but he saw her lips form it.

"Ashley." As the implications sank in, her voice gained strength and she screamed, "You killed my daughter!"

"I did no such thing," he said. "She was an unfortunate casualty of war. Actually I was rather fond of the child."

His cavalier dismissal of her daughter's violent death sent Lara into a frenzy. Suddenly she spun into motion, transforming into a whirling, ducking, rolling blur of limbs. The violent conversion was so instantaneous that it caught even her guards unaware. When they regained their wits, they naturally expected her to rush forward, toward Sánchez. They weren't prepared for her to move backward.

By the time she stopped moving, the contents of the camera bag had been dumped into the dirt and she was aiming the Magnum revolver at Sánchez. At least two dozen rifles and pistols were cocked and aimed at her.

"*No!*"

Key leaped to his feet and threw a body tackle at Lara, knocking her to the ground. The searing pain in his ribs almost caused him to black out, but he held on to her, trying to restrain her thrashing arms and gain possession of the weapon. Cruel irony that it was, Sánchez was their only hope of survival. If Lara killed him, they would be as good as dead, too. As long as they remained alive, there was hope of their getting out of Montesangre.

With surprising strength, she fought like a hellcat. "Let me go! I'll kill him!"

Several of the soldiers had joined the melee. Key was pulled away from her. He didn't know why the guerrillas hadn't opened fire on the two of them and dispatched the threat to El Corazón. Not until he saw him calmly approaching did Key realize that he was probably protected by a bulletproof vest. And, it seemed, unless the camp was under direct attack, no one fired a single round without a direct order from him.

"Release her."

At the sound of his voice, the guerrillas released Lara and backed away from her. She surged to her feet and, holding the Magnum in remarkably steady hands, pointed it at Sánchez.

"Lara, no!" Key hissed. He struggled with his captors, but to no avail. "Don't do it. For God's sake, don't."

"She will not kill me, Mr. Tackett." Although he was speaking to Key, Sánchez's eyes were fastened to Lara's.

She pulled back the hammer of the pistol. "Don't belittle me, Emilio. At this moment I'm capable of anything. Because of you, my baby died that morning. I'm going to kill you. Then I don't care what your ragtag band of butchers does to me."

526

"You will not pull the trigger, Mrs. Porter, because that would make you what you accuse me of being — a cold-blooded killer. You are a healer, someone sworn to extend life, not end it. You cannot kill me. It goes against everything you are."

You smart son of a bitch, Key thought. Sánchez was grandstanding for his troops. This was the stuff legends were made of, and the little prick knew it. He was gambling that Lara would not pull the trigger, and the odds were strongly in his favor. He'd had years to study her while working at the embassy. He knew the kind of woman she was, knew of her dedication to healing. The ability to kill wasn't within her.

"You bastard." Tears left muddy trails in the grime on her face. The heavy pistol began to waver in her hands. "My baby's dead because of you."

"But you cannot kill me."

"They put her sweet little body in a mass grave and covered it with dirt. I hate you!"

"If you hate me so badly, pull the trigger," he taunted. "An eye for an eye. I should think that your killing me would be just retribution."

Key refused to let Lara be made a fool. It would cost them their lives if she pulled the trigger, but he figured them for dead anyway. He decided to take out Sánchez with them.

"Call his bluff, Lara!" he shouted. "Blow him away. Aim for his smug puss."

Her trembling had become uncontrollable.

527

Even if she had been able to pull the trigger, her aim would have been off. Sánchez moved closer. "Stay where you are!" she yelled. "I'll kill you."

"Never."

"I will!" Her voice cracked with hysteria.

"You never could."

Confidently, Sánchez reached out and closed his hand over the gun. Lara put up token resistance, but he easily yanked it from her clutches. She covered her face with her hands and began to sob. Sánchez, smiling complacently, placed the barrel of the Magnum against the crown of her bowed head.

Key's savage bellow was a torturous cry, the kind one would imagine coming straight from the bowels of hell.

Sánchez grinned. "Your sentiment is touching, Mr. Tackett. I'm afraid this disproportionate respect for human life, any human life, will eventually be the downfall of America. How typically, sadly American you are. You choose to save the life of your brother's whore."

"If you kill her, you're history." He spoke the warning through clenched teeth.

"You are in no position to issue threats, Mr. Tackett."

"If I don't get you in this lifetime, watch your back in hell."

He struggled against the soldiers restraining him. He kicked backward and caught one in the kneecap. It crunched satisfyingly. He elbowed the other in the gut. Like his comrade, he went down. Freed, Key charged forward, but watched in impotent outrage and horror as Sánchez squeezed the trigger of the Magnum.

The empty chamber clicked.

Key skidded to a halt. Inertia propelled him off balance as his knees turned to gelatin. He pitched forward, landing hard in the dirt.

Sánchez laughed at the spectacle. "I am not a fool, Mr. Tackett. The bullets were removed when the gun was discovered in the camera bag. Your attempts to hide it were woefully amateurish."

He tossed the revolver back into the bag, then used the pristine handkerchief once again to wipe off his hands. "I am indebted to you and Mrs. Porter for providing us with a morning of entertainment."

"You fucking son of a bitch." Key struggled to his feet and staggered toward Lara. No one stopped him, which in itself was an insult. He must have seemed too pathetic to pose any real threat.

Little did they know.

He had been destructively livid many times. He'd used his fists in brawls, bashing bodies and furniture. But he didn't recall a single instance when he'd felt as though he could actually take another's life.

Until this moment.

Given the chance, he could have literally torn Sánchez apart with his bare hands. He wanted to sink his teeth into his throat, taste his blood. It was an animalistic, primordial reaction that he would never have thought himself capable of experiencing, and it was frightening in its intensity.

"Why don't you just kill us and get it over with?"

"I have no intention of killing you, Mr. Tackett. Is that what you thought?"

"You're going to keep us here indefinitely? Why, so we can provide you with entertainment every morning?"

Sánchez smiled. "That is a tempting proposal, but I cannot be that self-indulgent. Actually I am releasing you. You will be returned to Ciudad Central and given accommodation in the finest hotel. Tomorrow at noon, you will be placed aboard a commercial jet bound for Bogotá. From there you will make your own travel arrangements."

Key eyed him skeptically. "What's the hitch?"

"When you reach the United States — I will make certain that the media and proper authorities are apprised of your illegal visit to Montesangre — you can make plain my message to your government."

"Message?" By now Lara had stopped crying and was listening. Key had placed his arm around her shoulders, and she was leaning against him.

"The message is that I will stop at nothing to gain control of this country. President Escávez has neither the military muscle, the personal endurance, nor the public support to defeat me. His power is a thing of the past. In a few months his diminishing army will be completely destroyed. By the end of the calendar year, I plan to establish my government in Ciudad Central."

"What makes you think the United States gives a shit about you and your pissant government?"

Sánchez bared his small, sharp teeth in a gross travesty of a smile. "My countrymen are in dire need of supplies, food, medicine. I would like to reestablish diplomatic relations with the United States."

"I bet you would. What's to make the offer attractive to us?"

"I could also make the same request of several South American countries who need an impartial corridor through which to transport drugs. Montesangre's policy has been to resist this lucrative method of revenue, but these are desperate times."

"How trite. You're not going to say desperate times call for desperate measures, are you?"

Again Sánchez smiled his obnoxious smile. "We must consider all our options. Montesangre would be a convenient stopover between South America and the United States, and the dealers are willing to pay well for the privilege."

Key thought about the landing strip designed specifically for drug runners. He'd told Lara the truth when he said he'd never flown drugs, but that didn't mean he hadn't been asked or hadn't been tempted. Percentages were strongly in favor of never getting caught, and the money couldn't be topped.

But the thought of profiting creeps who turned adolescent girls and boys into prostitutes to support their habits went against his moral code. Contrary to what most people thought about him, he wasn't entirely without conscience.

"What makes you think that anyone will listen to Lara and me?"

"Your trip here will be well documented by the media. Even if the government slaps your hands, your courage will be lauded. The public will be sympathetic to your mission and its regrettable failure. You will be in the spotlight.

"Unfortunately, Mrs. Porter's reputation is dubious, therefore she does not inspire trust. But you are Senator Tackett's surviving brother. No doubt he still has some loyal colleagues in high places. They will listen to you."

"If I have an opportunity, I'll pass along your message," Key agreed tightly.

"You must do better than that, Mr. Tackett. You must give me your word."

He had no intention of getting involved in Montesangren politics even from a distance. Once Lara and he were safely out, the whole damn country could slide into the Pacific for all he cared. But until that time, he would promise Sánchez anything he wanted to hear. "You have my word."

Lara spoke for the first time. Some of her spirit had returned, though it was obvious she was functioning on adrenaline. "You'll burn in hell, Emilio."

"Still delusional," he said retiringly.

"Oh, hell is real, all right. I've been there. The day my husband was kidnapped and my daughter was killed, and again last night when I saw the place where she is buried."

"Such accidents occur during war."

"War?" She sneered. "You're the one nursing delusions. This isn't war, it's terrorism. And you're not a warrior, you're a hoodlum. You have no honor."

Honor was a sacred thing in the Montesangren culture. Key feared Lara might have gone too far, insulting Sánchez in the most offensive way before a crowd of disciples. He held his breath, thinking that El Corazón might rescind his offer to release them. But with a brusque motion of his hand he ordered that they be returned to Ciudad Central.

Key didn't give him time to change his mind. He climbed into the truck, then leaned down to assist Lara up. To his relief their hands were left unbound. The camera bag, their duffels, and Lara's medical bag were tossed in behind them. Two soldiers took up positions on either side of the rear opening.

Key sat down and leaned against the interior wall. He guided Lara down beside him. "Where are the others?" she asked in a whisper. "He's sending back only two to guard us?"

"Seems so."

The truck's noisy engine was coaxed to life. With a screech of gears, it moved forward. Through the opening in the back, they watched the camp roll past. When they last saw Emilio Sánchez Perón, the dreaded El Corazón del Diablo, he was seated on the porch of his ramshackle hut, consulting with his lieutenants while being fanned by adoring young girls.

"He's so damn smug," Lara angrily observed. "He thinks we no longer pose a threat to him."

Key cupped her chin and brought her head around. "Do we?"

She considered the question, then slowly shook her head as tears began to slide down her cheeks. "No. Even if I'd been able to kill him, his death wouldn't have brought back Father Geraldo, or Dr. Soto, or Randall, or Ashley."

He whisked a tear off her cheek. "No, it wouldn't."

"Then what would be the point? I'd be a killer, no better than he."

"I haven't had a chance to say anything about what we found last night. I'm sorry, Lara."

She nodded her thanks, but hadn't the strength to say more. Within moments, she succumbed to exhaustion. Her eyes closed, and her head fell back against the wall of the truck. Almost immediately she was breathing evenly, having found release in sleep.

One of their guards approached with blindfolds. "Bug off, Bozo," Key said to him. "We're going to sleep. Our eyes will be closed."

The guerrilla consulted his comrade. The other shrugged indifferently. The blindfolds were withdrawn and the soldier returned to sit near the tailgate with his counterpart. They lit cigarettes.

Despite his aching ribs, Key slipped his arm around Lara so her head wouldn't bump against the truck. He positioned her against his side. She turned and settled her head on his shoulder.

One of the soldiers made a crude comment about the instinctive way she nestled the cleft of her thighs against his hip. The two laughed, flashing Key lewd grins.

He gave them the finger before surrendering to his own exhaustion.

Chapter Twenty-Four

At sunset they arrived at the hotel. It once had been a showcase, but, like everything else in Ciudad Central, it had suffered the effects of war. Lara had attended diplomatic receptions and parties held in its ballrooms in bygone days. Now the staff was inadequate and unfriendly, acting more like surly soldiers obeying orders than like hosts.

After spending hours in the back of the bouncing truck, Lara was so relieved to have reached her destination that the hotel's notable lack of amenities didn't bother her. The formality of registering was waived. She and Key were promptly escorted under armed guard to the third floor.

The hallways were deserted. There was only silence behind the numbered doors. Lara guessed that this floor was reserved for "special guests," and that it could rightfully be called a detention center. Essentially, anyone given a room on the third floor was under house arrest.

"Señora Porter." The bellman handed Lara a room key. He gave Key another. "I trust your stay with us will be comfortable." Under the circumstances, his hospitality was a parody. Nevertheless, he bowed to them, then he and the two guards retreated to the elevator. Only the bellman got in. The guards posted themselves outside the sliding doors. There were also soldiers at the emergency exit doors at both ends of the corridor.

Lara unlocked the door to her room. Key followed her inside. The room was clean but tacky. Through an open door she saw the flamingo-pink tiles of the bathroom and a plastic shower curtain with lurid hibiscus blossoms. She dropped her doctor's bag and duffel at her side and stood in the center of the room, too dispirited to take another step.

Key was behind her. He touched her gently. Turning, she looked at him, and, for the first time since leaving El Corazón's camp, she really saw him. He looked battered and beleaguered. She reached up to touch the wound on his temple, then, realizing that the gesture wasn't professionally motivated, she lowered her hand.

Softly he said her name. As they stood facing each other, he asked, "Are you all right?"

"Yes." Her voice was hoarse from screaming at Sánchez, whose only reaction to her accusations had been a gloating smile. He'd demonstrated no remorse for Ashley's death. Remembering, tears came to her eyes. She inclined toward Key and began shaking her head mournfully. "No, no, I'm not all right. My baby is dead, forever lost to me."

His arms encircled her and held her protectively. "Shh. Don't cry. He can't hurt you anymore. We're safe."

537

Suddenly she wanted very badly to be convinced of that. Her fingers curled inward, digging hard into the muscles of his chest. She desperately needed to touch, to be touched, and apparently Key was just as eager to allay his own fears.

He tipped her head up as his descended. Simultaneously a violent hunger was unleashed, and they aggressively sought to satisfy it. He claimed her mouth with a frantic, needful thrust of his tongue.

Lara arched against him and locked her arms around his neck. He pulled her shirttail from the waistband of her pants and impatiently tore the buttons from their holes. Reaching behind her, he unfastened her bra, then slid his hands forward to cover her breasts. His strong fingers pressed into her flesh.

His name drifted across her lips — a question, a profession, a prayer.

Responding, he lowered his head and took her nipple between his lips. Her head fell back upon her shoulders, and she gave herself over entirely to the hot urgency of his caress. He pulled her deeply into his mouth, the flexing of his jaws strong and possessive. Then he kissed her mouth again, moving his head from side to side, changing angles, testing positions, tasting her completely.

At last he raised his head and looked at her, his eyes feverish and painfully blue. His eyebrows were pulled into a frown of determination above his straight, narrow nose. His lips were a thin, firm line of resolve set between his bearded cheeks.

Lara wanted him with the purest, most undiluted sexual desire she'd ever experienced. Yet she closed her eyes, shaking her head in denial. "I don't want to be one of Key Tackett's women."

"Yes, you do. Tonight you do."

He carried her to the bed and laid her down against the pillows. He must have known her mind better than she knew it herself, because she reached for him eagerly when he followed her down. His lips tasted salty with sweat and were slightly gritty, but she couldn't get enough of them.

He pushed aside her blouse and the cups of her brassiere and moved his hand across her breasts, lightly grinding her nipples beneath his palm until they were stiff and so sensitive that his merest touch caused her back to arch above the bed.

She did nothing to stop him from unfastening her pants and pushing them down, along with her panties, until they were gathered around her ankles. He undid his trousers, but it was Lara's hands that shoved them over his buttocks.

He entered her.

She received him.

He was incredibly firm. She was wet and snug. His head sprang up, and he looked down into her flushed face. She could feel the color in her cheeks, hear her own quick, soughing breath.

His eyes locked with hers as he pushed deeper. She clamped her lower lip between her teeth to keep from crying out.

When he was fully seated inside her, he grimaced with pleasure. Then, with a moan, he pressed his forehead against hers. "Oh, Christ. A fantasy fuck."

He began to move; she raised her hips to meet his smooth thrusts. Each one took her breath, but she couldn't deny herself the overwhelming sensations they evoked.

He waited for her. When she climaxed, he sank all ten fingers into her hair and held her head between his hands, kissing her mouth as thoroughly and intimately as their coupling. Her orgasm was long and strong and more than he could endure. Allowing himself to come, he buried his face in her neck and drew a patch of her skin against his teeth.

It was a long time before either of them moved.

They did move, eventually, from the bed and from her room into his. Their dirty clothes and muddy boots had made a mess of her bed. Defying the curiosity of their guards as they crossed the hall, Key led her into his room, a mirror image of hers except that the tiles in his bathroom were turquoise and the shower curtain was decorated with smiling seahorses.

They removed their clothing and stepped beneath a shower from which they coaxed only rusty, tepid water. Scanty bars of soap were wrapped in green cellophane. They used up three of them to wash the grime off each other.

The water cooled but they stayed beneath the spray, exploring. She examined the gash on his temple and told him that she could put a butterfly clamp on it.

He said, "Don't bother. I'll live."

She examined his bruised ribs and told him that several were probably cracked.

He admitted that they hurt but wouldn't consent to her binding them. "The night we met, you mummified me. Damn bandage nearly drove me crazy. I took it off the next day."

She called him hardheaded as she combed her fingers through his chest hair. She cupped his weighty sex in her palms and sipped water from the delta-shaped hollow at the base of his larynx.

He covered the scar on her shoulder with tender kisses and called it beautiful when she demurred and tried to hide it. "Besides, it's hardly a scratch compared to mine."

With her finger, she followed the raised, red surgical scar that ran up his left leg from knee to groin. "What happened?"

He told her about the car wreck that had ruined his leg and all hopes for a career in the NFL. "Were you terribly disappointed? Is that what you wanted?"

"It's what Jody wanted. We'd never been pals. But after the accident . . ." He shook his head. "I don't want to talk about Jody." He touched her everywhere, giving and taking pleasure in equal portions. He was indulgent and sensual, more so than she would ever have believed. She thought that surely she was dreaming, although she had never dreamed this erotically about her husband. And never about Clark.

They finally left the bathroom and were foraging through their duffel bags for clean clothes when someone knocked on the door. "What do you want?" Key asked brusquely.

"*Tengo la comida para ustedes.*"

Cautiously he eased open the door. A soldier held a room service tray perched on his shoulder. "*Gracias.*" Key took the tray of food from him and, without giving him time to argue, slammed the door in his face and slid the chain back into the track.

He set the tray on the table. "I hope it's better than the fare at Sánchez's camp."

"It could be poisoned." Lara approached the table, pulling her hairbrush through her wet hair.

"Could be, but I doubt it. If he wanted to kill us, he wouldn't be that subtle. He'd have done it when he had an audience."

On the tray were an assortment of fruits and cheeses, cold roasted chicken, and bottled water. Key got a drumstick from the platter and without much interest took a bite. "Wonder why he let us go."

She began to peel an orange. "Odd, isn't it?"

"Damned odd. I don't know what I expected, but not this." He used the drumstick to point out their surroundings. "Not exactly The Plaza, but better than a bamboo hut with a dirt floor."

He chewed thoughtfully. "Bottom line. Our lives in exchange for my taking his 'message' to the States? Nope. Doesn't jive. Too easy. If he wanted to convey a message to our government, he could have used someone more influential than us, the head of state of an ally nation, for instance." He tossed aside the chicken bone and opened a bottle of water. "Why didn't he kill us, Lara?"

She returned the half-peeled orange to the tray. "I don't know." Moving to the windows, she parted the drapes and gazed out over the city.

"That orange would do you good. You haven't eaten all day."

She glanced back at the table with revulsion. "I don't want to feel obligated to Emilio Sánchez for anything."

"Don't cut off your nose to spite your face. You should eat."

"I'm really not hungry, Key. My mind isn't on my stomach." There was an edge of impatience in her voice, most of it self-directed. "I've been trying to sort through things."

"What things?"

"I don't know. Things. Everything. About what happened here three years ago. Randall. Ashley. If I dwell on that . . . that mass grave she's buried in, I'll probably go mad." She clutched a handful of drapery. "So I can't. I must concentrate on my memories of when she was alive. I must remember how bright and happy

543

she was, how much joy she gave me during the short time I had her."

Her hoarse voice began to waver. She paused to compose herself. "My daughter is lost to me, but if I focus on her life rather than her death, it doesn't matter so much where her body is buried. Her spirit is still alive. In that respect, this isn't a failed mission after all."

"You had to return here in order to come to terms with it."

She nodded. "Yes. That episode of my life — all of it, beginning with the scandal — has been governing my life for far too long. I accused everyone else of identifying me with tabloid headlines, but I'm the most guilty. I can't continue regarding myself a victim. It's time I got on with the rest of my life."

"In Eden Pass?"

"I haven't had much success there," she remarked as she turned to face him.

"Not because you aren't a good doctor, but because of us Tacketts. We've given you a hell of a hard time."

Suddenly reluctant to look at him, she averted her head.

"Key, why did this happen between us?"

"The animosity? Or the other?"

"The other."

He took a deep breath and held it, saying nothing for several moments. Finally: "You're the doctor. Got any theories?"

She did, and indicated so with a slight motion

of her shoulders. "People who've survived a life-threatening ordeal," she began slowly, "frequently want sex directly afterward." He raised one eyebrow, either with inquisitiveness or skepticism. She wasn't sure. "It makes sense. Sex is the ultimate release of emotion, a means of unequivocally affirming life.

"I've had shamefaced patients confess to me that immediately following a funeral, they made love. With extraordinary passion. Human beings have an innate fear of death. Sex is instant confirmation of survival.

"After the harrowing experiences we've been through the past few days, it follows that we'd expend our pent-up fears and emotions with sex. Fierce, aggressive sex. We're a classic example of this psychological phenomenon."

Key had listened politely. Now he walked to her, coming so close that she had to tilt her head back in order to look into his face. "Bullshit. It happened because we wanted it to." He kissed her hard and quick, stamping an impression of his lips on hers. "Damned if it needs any more justification than that."

The clothes they had so recently put on were discarded as they made their way to the bed. When the backs of his knees touched it, he sat down and guided Lara to stand between his thighs. He lifted her breast to his mouth and flicked the nipple with his tongue.

Her eyes fluttered closed and choppy little breaths issued from her throat. She wound strands of his hair around her fingers but allowed his head to move freely over her breasts and down the center of her body. His beard rasped her belly, eliciting exciting and forbidden sensations. Between her thighs she began to ache, deliciously. The lips of her sex became swollen and warm.

Key splayed his hands over her bottom and tilted her middle up against his face. He nuzzled her. He kissed her navel. He kissed the soft skin beneath it. With little puffs of heat, his breath stirred her pubic hair.

Then he turned her, and she landed on her back on the bed, the juncture of her thighs forming a cradle for his lowering head. He kissed her with unapologetic carnality. His mouth gently drew on her while his nimble tongue taught her things about herself she didn't know. As though inside her head, taking directions from her thoughts, he knew exactly when to probe, when to stroke, when to sink his mouth into her, and when to withdraw and caress her with the very tip of his tongue.

By the time he rose above her, she was sated, replete, dewy with perspiration, and drunk with passion. Nevertheless, her slack lips awakened beneath his searching kiss. When he entered her, it was a beginning, not a benediction.

Tenderly he traced the scar on her shoulder with his fingertip. "It was bad, huh?"

"Very bad. For a while the doctors believed

that I'd be extremely lucky to regain only partial use of my arm."

"Knowing you, you were determined to prove them wrong."

"After the wound healed, I spent months in physical therapy."

For a moment he watched her reflectively. "I think you should stop punishing yourself for not dying with the rest of your family, Lara."

"Is that what you think I'm about?"

"To an extent, yes."

She came up on an elbow and surveyed his lean, naked body. In addition to the scar on his leg, there were many on his torso. "What about you? You're reckless. You take senseless chances. What are you punishing yourself for?"

"It's not the same thing," he answered crossly. "I'm a thrillseeker for the sake of the thrill, that's all."

She gave him a look that said she wasn't buying it. Her eyes wandered from one scar to the next. There was a particularly wicked one cutting a jagged line across his ribs beneath his right arm.

"Knife fight," he said when she looked at him with a question in her eyes.

"Obviously you lost."

"Actually I won."

As to the fate of the loser, she was afraid to ask. "And this?"

"Plane crash. I walked away, but tore open my arm on a piece of fuselage."

She marveled at his nonchalance. "Other than today, have you ever been in real danger of losing your life?"

"Once."

"Tell me about it."

"I got shot. Here," he said, touching his newest scar, the one she was familiar with. "Nearly bled to death."

Laughing, she tossed her hair over one shoulder. "It was more than a scratch, but certainly not a mortal wound."

"I know that. But I wasn't talking about the wound itself," he said. "See, I stumbled into Doc Patton's place, expecting him, but finding someone else. A woman."

Lara became transfixed by his eyes and the hypnotic quality of his voice. "How was that life-threatening?" she asked huskily.

"I turned around and looked at her and thought, 'Shit, Tackett, you're a dead man.' "

She swallowed with difficulty. "We're grown-ups, Key. Beyond the age of consent and too old to play games. I don't expect hearts and flowers from you. You don't have to profess —"

He laid his index finger vertically against her lips. "I'm not telling you this to get you into bed. You're already here and I've already had you. I'm telling you because it's the truth, and you know it as well as I do. We're here together, like this, because we've wanted it from the beginning. We've both known that it was only a matter of time."

He reached up to stroke her cheek. "Once we looked at each other, I didn't stand a chance and neither did you. I wanted to fuck you on the spot."

"Until you discovered who I was."

"I wanted to fuck you anyway." Reaching behind her head, he clutched a handful of her hair and drew her face close to his. "Damn me to hell, I still do."

Key reached for her as she scooted off the bed and began gathering her clothes. "Where are you going?" he mumbled sleepily.

"To my room."

"What for?"

"A bath."

"We have a tub in here."

"But we used all the soap. Besides, I need to organize my things so that when they come to take us to the airport I'll be ready." She dressed hastily.

"What time is it?"

"Nine."

"Nine! We slept that long?" He sat up and ran his fingers through his shaggy hair.

"You don't have to get up. We've got plenty of time before noon."

"No, I'm getting up. I don't want to give the bastards any reason to delay our departure. As soon as I shower, I'll see if they'll bring us some coffee."

"I'll have everything ready by then." She

549

smiled at him, checked to make certain she had her key, then unlocked the door and stepped into the hall.

Contrary to what he'd said, Key didn't get up immediately, but lay back down and stared sightlessly at the ceiling. Last night Lara had confessed to some confusion. Being less straightforward then she, he hadn't admitted to his own ambiguity.

To assuage her conscience, she had dredged up a psychological explanation for going to bed with him, although he doubted that she believed her own sales pitch. He didn't think lust needed analysis or rationalization. It was a call to action all by itself.

His confusion was centered not on why it had happened but on how he felt about it — about her — now that it had.

He'd never enjoyed a woman more. Physically, they were a good fit. She had matched him in passion and skill. Despite all the tabloid journalism written about her, he hadn't expected her to be so sexually liberated. Memories of their love play now sent heat surging through his loins. Even after their marathon of sex, he was far from satisfied. He wanted more of her.

That, too, was unexpected and disconcerting. Usually the chase was most of the fun. Once caught, a woman's charms rapidly diminished. It bothered him greatly to realize that Lara had become only more intriguing. She had layers and dimensions he was eager to explore. Customarily, women were as disposable as razor blades. When one got dull, he threw it away and replaced it with another. He wasn't eager to dispose of and replace Lara.

Not that she was his to do with as he pleased.

Ah! He'd finally acknowledged the crux of all these niggling misgivings. She didn't belong to him. Furthermore, if circumstances had been different, she might still belong to his brother.

Clark had had her first.

That alone had prevented last night from being the most satisfying night of sex he'd ever engaged in. Inadvertently he must have conveyed his uneasiness about it. Either that or Dr. Mallory was damned perceptive.

She brought it up, after they had nibbled on the remainder of the food and decided that they should try to sleep. She lay on her side, facing away from him, her folded hands supporting her cheek. He'd been absently rubbing a strand of her hair between his thumb and index finger, thinking that she'd been luckier at falling asleep than he. He was surprised to hear her drowsily say, "I know what you're thinking about."

He moved his knee against the back of her thigh. "Okay, smarty, what am I thinking about?"

"Clark."

His smile receded and the strand of hair sifted through his fingers. "What about him?"

"You're wondering if I'm comparing the two of you, and, if so, how you measure up."

"I didn't know you were a shrink, too."

She turned her head and gazed at him over her bare shoulder. "I'm right, aren't I? Isn't that what you were thinking?"

"Maybe."

Smiling sadly, she gave her head a small shake. "You and Clark . . . you're two different people, Key. Equally attractive, both charismatic, each of you a natural leader, but so very different. I loved your brother, and I believe he loved me." She reduced her voice to a whisper. "But it was never like tonight." She rolled away from him and returned her cheek to her hands. He had thought she was finished, but she repeated, "Never."

He'd lain there for a while, steeped in jealousy, wanting desperately to believe her. Soon, however, desire superseded envy. Or maybe it wasn't so much desire as jealous possessiveness.

Moving suddenly, he placed his arm around her and roughly pulled her closer until her bottom was firmly pressed against his belly. He entered her with one hard thrust. He took a love bite from the back of her neck and held it between his teeth, feeling the need to dominate and control.

There was no need for it. She was receptive and giving and so erotically charged that he had only to press his open palm against her mound and the inner walls of her body contracted around his cock like a magic fist, massaging him, milking him of semen and of doubts.

It took a while for their breathing to return to normal. Their bodies glistened with a fine sheen of sweat. When he finally withdrew from her, she turned to face him and nuzzled his chest with her open mouth.

She said, "Shameless."

"I've never claimed to be otherwise."

"Not you. Me."

He'd fallen asleep with her in his arms, secure in the knowledge that their lovemaking had gone beyond mutual satisfaction. It had been in another league.

But now it was day, and his doubts were encroaching like the tropical humidity that accompanied the rising sun. He thought back to all that she'd said, to all her sensual responses, to her bold caresses. Surely it couldn't have been any better for her with his brother.

Had she ever ridden Clark until she collapsed, exhausted, on his chest?

Key's fists clenched at his sides.

Had she blissfully tortured Clark to climax with her sliding, kneading hand?

He cursed obscenely.

Had she permitted Clark to kiss her between her thighs, to separate and taste . . .

A bloodcurdling scream brought him bolt upright.

By the time the second one shattered the morning stillness, he had put on his pants and was at the door, all but pulling it from its hinges in his haste to get it open.

553

"Buenos días," Lara said to the guards as she left Key's room. Undaunted by their leers, she crossed the hall and entered her room, carefully locking the door behind her.

Their boots had tracked mud onto the carpet, and, as Key had pointed out, they'd ravaged the bed. He'd joked, telling her that regardless of what she might have heard about Texans, that was the first time he'd ever made love with his boots on.

Made love? Had he used those exact words, or was her memory being kind?

She shrugged off the disturbing thought, having had enough self-analysis for one twenty-four-hour period. The conclusions she'd reached last night had been positive. The rest of her life had begun when she fell into Key's embrace. The experience had been cathartic. Why try attaching a name to it? Her mood and her body spoke for themselves. She felt wonderful. For once, why not let it go at that?

Taking her duffel with her, she went into the bathroom. When she saw her reflection in the mirror over the basin, she laughed with self-deprecation. She had on no makeup, and, though her hair was clean, it had been washed with bar soap and looked it.

He hadn't seemed to notice. Or care.

A blush spread up from her chest to her neck and face. Unbuttoning the first few buttons of her blouse, she glanced down at her breasts and, as expected, saw that they were whisker-burned. Before they slept together again, she'd insist that he shave.

If they slept together again.

To her chagrin, she found herself hoping desperately that they would. Soon.

Smiling with anticipation, she pulled back the shower curtain and reached for the water taps.

Her scream reverberated off the flamingo-pink tiles.

Lying in the bathtub, beaten and bleeding but very much alive, was Randall Porter.

Her husband.

Chapter Twenty-Five

"How charming you look." The former United States ambassador to Montesangre stood as his wife entered the parlor. "Although I liked your hair better when you lightened it. When did you stop?"

"While I was recuperating in Miami. Those were difficult months for me. Hair color wasn't a priority."

Lara glanced at Key. Declining to stand when she came in, he was slumped in an upholstered chair, one ankle balanced on the opposite knee, his foot rapidly jiggling up and down. His steepled fingers tapped his lips in time to the movement of his foot. The posture would have looked insouciant on anyone else, but Lara sensed that he was on the verge of exploding.

If Randall noticed Key's tenuously controlled rage, he gave no indication of it. "Would you like something to drink, darling? We have a few minutes before going downstairs."

"No, thank you. I don't want anything to drink. And I don't see why it's necessary for me to participate in this news conference."

"You're my wife. Your place is by my side." At the bar, Randall poured himself a club soda.

"Mr. Tackett? Anything?"

"No."

Randall returned to the sofa where he'd been sitting when Lara joined them from the bedroom of the Houston hotel suite. The well-appointed rooms were a considerable improvement over the accommodations in Montesangre.

Well-wishing floral arrangements crowded every available surface. Their mingled scents were sweet and cloying and had given Lara a dull headache. She thought these expressions of congratulations ludicrously hypocritical, having been sent by many of the same bureaucrats and political figures who, five years ago, had been relieved to see Randall and his cheating wife shuttled off to Montesangre, thereby sparing Washington the embarrassment of having them underfoot.

Technically, Randall was still a United States ambassador. When the media was notified by news services in Colombia of his shocking resurrection, the story took precedence over all others and earned the banner headline of virtually every newspaper in the world. His return to life sent the entire nation into a tailspin, the press into a frenzy.

In Bogotá he'd been treated for his wounds, which were more superficial than they'd first appeared. Key had relented and had his ribs X-rayed. Three were cracked, but he'd sustained no internal injuries.

Lara's injuries were as severe, but not as evident. For fatigue she was prescribed hot, healthy meals and two nights of drug-induced sleep. She'd eaten and slept but continued to look shell-shocked. Her movements were disjointed, her speech distracted. A husband she believed dead had suddenly returned to life. Her entire system had been thrown into shock.

Neiman Marcus had generously offered to outfit her for her first public appearance following her return to American soil. For the newsworthy occasion the store had donated a silk and wool blend two-piece suit, matching Jourdan pumps, and suitable accessories and costume jewelry. The hotel salon had sent the staff to her suite to do her hair, nails, and makeup. On the surface, she was well turned out and appeared ready to accompany her husband to the news conference that was scheduled to begin in half an hour in the hotel's largest ballroom.

She'd just as soon face a firing squad, she thought.

In a very real sense that was exactly what it would be. Too jittery to sit, she moved aimlessly about the room among the furniture cluttered with floral bouquets. "You know what they'll dredge up, Randall."

"Your affair with Clark," he replied without a qualm. They had informed him of Clark's death on the flight from Montesangre to Colombia, but he already knew about it. World news filtered in, although little was filtered out.

"I'm afraid that's unavoidable, Lara," he continued. "I'll try to distract them with my story of the last three years."

"You don't look all that worse for wear." Key ceased wagging his foot and tapping his lips. "You look tan, fit, and well fed."

Lara too had noticed Randall's superior physical condition. He looked even better than when she'd met him seven years ago, as if he'd enjoyed several months' vacation in Hawaii rather than three grueling years as a political prisoner.

He pinched up the creases of his new suit trousers, also a gift from Neiman's. "After the first few months of my captivity, I was treated very well.

"At first, the rebels beat me unmercifully," Randall told them. "For several weeks they ritualistically whipped me with pistols and chains. I thought this was preliminary to their killing me."

He finished his soda and checked the time. Seeing that he still had a few minutes, he continued. "One day they hauled me into General Pérez's quarters. I say 'hauled' because I couldn't walk. They carried me like a sack of potatoes.

"Pérez was pleased with himself. He showed me photographs of my 'death,' as they'd staged it. They'd executed a man, God knows who, shooting him in the head so many times it was little more than pulp."

Lara hugged her elbows. The room was frigid. After sweltering in the tropics for three years, Randall had said he wanted to keep the air conditioning as high as possible.

"You can imagine how devastating it was for me to see those photographs. They also showed me American newspapers reporting my death. They had photos of my funeral. I realized the hell you must be going through." He looked at Lara with commiseration. "I thanked God you were safe but knew you would be agonizing over the violent way in which I'd died. Knowing that no one would be sent to rescue me was the worst torture of all. As far as anyone knew, I was dead."

"Did they tell you about Ashley?"

"No. I didn't learn that she'd been killed in the ambush until I read the newspaper accounts of my funeral. The only comfort I could derive was knowing that you had miraculously survived. If it hadn't been for the priest —"

"Priest? Father Geraldo?"

"Of course. He got you on one of the last American-bound planes to leave Montesangre. I thought you knew."

"No. I didn't," she said in a subdued voice. "I should have thanked him."

"It was certainly an act of bravery," Randall said. "Emilio harbored a grudge against him for facilitating your escape. I suppose that's why he ordered Father Geraldo's murder."

Key cursed beneath his breath. "So good of you to tell her that."

"Lara's a realist, aren't you, darling? Nevertheless it's a pity about the priest. And about Dr. Soto."

"I can never atone for involving them," she said quietly. "I'll always feel partially responsible for their deaths."

"Don't do that to yourself," Key said insistently. "They'd been pegged for elimination, with or without us. Sánchez said as much."

She threw him a grateful look for the sentiment but knew she would carry the guilt of their murders to her own grave.

"You were incredibly brave to return to Montesangre, Lara," Randall said. "Thank God you did. If you hadn't, I'd still be a hostage."

Key surged to his feet. He'd shaved his dark beard, but his hair was still overly long and contributed to his look of a caged wild animal. Disdaining the role of national hero in which he now found himself, he'd declined Neiman's offer to provide him with new clothes. On his own, he'd bought new jeans, a sport coat, and cowboy boots.

"I don't get it," he said. "Lara and I arrive unannounced in Montesangre, and thirty-six hours later your captors up and decide to let you go?" He spread his arms away from his body. "Why? What does one have to do with the other?"

Randall smiled indulgently. "Obviously you have something to learn about the mind-set of these people, Mr. Tackett."

"Obviously I do. Because your story sounds like a big pile of *caca* to me."

Randall's eyes narrowed marginally. "You

saved my life and Lara's. Therefore I'll extend you the courtesy of overlooking your unnecessary vulgarity."

"Don't do me any favors."

Randall dismissed him and addressed his next words to Lara. "Emilio likes to play mind games. Remember the chess tournaments we hosted at the embassy?"

"This is more serious than chess, Randall."

"To you and me. I'm not so sure Emilio makes the distinction between a board game and the little dramas he plays out for his own amusement using human lives as the stakes. He thanked you for providing entertainment to his camp that morning, remember?"

"*I* remember," Key said. "And I'm glad you brought that up because something else has been bugging me. You said you were inside the shack while all that was going on, right?"

Randall nodded. "I was bound and gagged, unable to alert you to the fact that I was still alive. That was Emilio's inside joke."

"When did you first learn that I was in Montesangre?" Lara asked.

"The morning following your arrival. I knew something was afoot because my guards were brusque and wouldn't look me in the eye. We'd developed a grudging respect for one another over the years. Suddenly they were hostile and taciturn again.

"After Ricardo intercepted the jeep on the road, it was only a matter of hours before they deduced who the 'widow' was. There was some speculation about the idiot brother-in-law." He looked pointedly at Key. "But once Emilio learned your name, he put two and two together. He knew about Lara's . . . friendship with Clark.

"The more you snooped around, the more volatile the situation became. The night before you were brought to the camp, I was transported there. Emilio taunted me with the threat of killing you slowly and painfully while I watched. I was beaten, but not severely. He wanted me conscious for the next morning's theatrics.

"After you were taken away, I was beaten again, then driven to Ciudad Central. We were probably only an hour behind you, but my guards and I spend the night in the truck. The last thing I remember is being knocked unconscious shortly after dawn. Your scream when you found me in the bathtub roused me. I was as shocked as you to find myself still alive."

He stood and slipped on his suit coat. "Well, I think it's time to go."

"I still can't comprehend Emilio's strategy," Lara argued, making no move to join him at the door.

"We'll talk about it later."

"No, we'll talk about it now, Randall. If you insist that I face the press, I need to fully understand the situation. They'll ask me about my dealings with El Corazón del Diablo. I'll gladly tell them everything I know about the slender, bookish young man who worked as a translator

563

at the embassy, and about the cold-blooded murderer I met this week. But I can't expound on foreign policy without having a clearer picture of what was in Emilio's mind. Why did he let us go? Why did he keep you alive but imprisoned for three years and then suddenly release you?"

Randall gnawed the inside of his cheek, apparently annoyed by her confusion. He decided to humor her. "I've had three years to ruminate on why my death was staged. The savagery of it was to demonstrate how much Montesangre resented the United States' intervention into its internal affairs."

"Why didn't they kill you for real?" Key asked.

"I assume they wanted to keep me as a trump card. Had the U.S. decided to send troops into Montesangre, as they did into Panama, they could have used me as a hostage."

"So why were you released now?"

"That's simple, Lara. They're starving. Montesangre relies entirely on imports for virtually everything. Under the embargo enforced by the United States, and adhered to by the nations who are either allied with or fearful of us, their resources were quickly exhausted. Frankly, I'm amazed that they've held out this long. They probably wouldn't have if Pérez were still their leader. They would have relaxed their political position long before now without someone as ruthless as Emilio at the helm. He's made himself into a demigod."

"What are you, president of his fan club?" Key asked caustically.

"Certainly not," Randall coldly countered. "He was my jailer for three years. However, I've witnessed firsthand the suffering of the Montesangrens. I have tremendous sympathy for them and wish to help their plight. For all his ruthlessness, Sánchez is the best hope for pulling the country together, feeding the hungry, ending the chaos, and establishing some semblance of order. And, putting personal considerations aside, I must admire his tenacity.

"He's inordinately determined and patient. Using your venture to release me was a brilliant stroke of ingenuity. He knew the human-interest value of this story, knew it would gain the attention of the American people. It's his invitation to the United States to reopen diplomatic dialogue."

"That's the message he gave me to deliver. Why use his ace in the hole?"

Randall smiled as though amused by Key's naïveté. "He knew I would have more credibility in Washington than a cowboy."

"I'm not a cowboy."

"Of course you are." His eyes slid over Key's jeans and boots, making plain his low opinion of them. "The only difference is that you ride airplanes instead of horses. Otherwise, you're a range bum. Even your brother thought so."

Key lunged for him, but Lara stepped between them. Putting her back to Key, she angrily faced

565

Randall. "Clark thought no such thing! He loved Key very much."

Randall smiled and said softly, "I bow to your superior knowledge of whom and what Clark loved." He extended his hand. "We really must go, darling. Ready?"

Disregarding his proffered hand, she moved stiffly toward the door. Sensing that Key wasn't following, she turned to him. "Coming?"

"No."

She panicked. The only thing that would hold her together during this press conference was knowing that Key was beside her. He couldn't lend her physical support, of course, but she'd relied on his strong presence to bolster her.

Gauging by the resolve in his expression, she knew arguing would be futile, but still she had to try. "You're expected."

"They'll just have to be disappointed. The newspapers are hinting that I took you to Montesangre to rescue him." He hitched his head toward Randall. "That's not why I went, and I'm not going to pretend that it was."

"They'll think you're only being coy, Mr. Tackett."

Key glared at her husband. "I can't control what they think. The only thing I have any real control over is myself, and I'm not going to be carrion for a flock of vultures with cameras. If you want a quote, write that one down." Looking at Lara again, he said, "You don't have to go either. No one can force you."

She fought the magnetic pull that would have drawn her to him. There were so many things to say, so many explanations to make, but in order not to cause more damage than had already been done, she had to remain silent.

Naturally she was glad that Randall hadn't died a brutal death. She celebrated his release from a long and hellish captivity. From a very selfish viewpoint, however, his deliverance couldn't have come at a worse time. Randall had been liberated, but her imprisonment was just beginning.

Tears filled her eyes. One rolled down her cheek. Seeing it, Key started to say something, but obviously thought better of it. They gazed at each other in mute misery.

"Well, well," Randall said around a dry little cough. Not knowing that he was echoing Lara's thoughts, he said, "It appears that the husband's resurrection from the dead has come at an inopportune time."

She quickly turned away from Key. "As you said, Randall, we're going to be late. Let's go."

He held up his hand to forestall her. "They'll wait. This, on the other hand, demands immediate attention."

"There is no 'this.' "

"You always were a terrible liar, Lara." He chuckled. "Out of deference to the shock you've sustained, I haven't imposed my marital rights these past few nights. It's a good thing I didn't.

Undoubtedly I would have found your bedroom door locked."

She gave him a fulminating look but said nothing.

He laid his finger lengthwise against his lips and fixed an appraising gaze on Key. "He's such a contrast to Clark, I'm amazed you find him attractive. He's certainly not as polished as his older brother. Still, he does emanate a hot-blooded, animalistic quality that I suppose a woman like you would find appealing."

"I'm not deaf and dumb, you son of a bitch," Key said. "If you've got something to say, say it to me directly."

"All right," he said pleasantly. "Didn't you feel the least bit foolish fucking a woman known nationwide as your brother's whore?"

Even Lara couldn't have stopped Key then. He sidestepped her and encircled Randall's throat with his hands.

"Key, no!" She tried to pry his fingers off Randall's neck, but they were unyielding. He backed him into the door; Randall's head made connection with a solid *thunk*. Frantically, he clawed at Key's fingers, but they squeezed tighter.

"Please. Key!" she cried. "Don't make matters worse! Don't make me another tabloid headline!"

Her shouted plea registered. She saw him blink rapidly as though to dispel a fog of rage. When her words sank in, his fingers began to relax.

He released Randall with an abrupt gesture of contempt.

Randall recovered himself and, with a semblance of dignity, straightened his coat and necktie. "I'm glad cowboys no longer carry six-shooters. I could be dead."

Key was still breathing hard and looking dangerous. "You talk about Lara and me that way again, and I'll kill you."

"How chivalrous," Randall said scornfully. He turned to her. "Well, Lara. For the final time, shall we go?"

Key rounded on her and gripped her by the shoulders. "You don't have to do what he says." He gave her a little shake. "You *don't*."

"Yes, I do, Key." She spoke quietly but with steely conviction.

At first he was incredulous. Then his bafflement turned to anger. She watched his face grow taut with fury. She knew he wouldn't understand her decision, and she couldn't explain it. So she had no choice but to withstand his disgust.

He released her, turned on his heel, yanked the door open, and stalked out. Hopelessly, she watched him go.

"I thought it went very well, but after all that talking, I could stand a drink." Randall slipped out of his suit jacket and carefully laid it across the back of a chair as he moved to the bar. "Want something, darling?"

"No, thank you."

He mixed a scotch and soda and smacked his lips appreciatively after the first sip. "One of the many things I missed during my captivity." Sitting on the sofa, drink in hand, he unlaced his shoes. "You're subdued, Lara. What's wrong?"

"What's wrong? I'm fair game and this is the first day of hunting season." She rounded on him. "I hate being put on display, and I bitterly resent you for forcing me to reopen my life to public scrutiny."

"You should have thought of the consequences before you finagled Key Tackett into taking you to Montesangre."

"I tried every other resource I knew of before asking Key. He was my last hope. I've explained why I went. Why I had to go."

"And your noble motivation was duly noted by the press. You were quite effective when you described the mass grave. You'll probably be nominated for Mother of the Year." He took another sip of scotch. "I honestly don't know why you're so upset."

"Because to even recount the incident at the cemetery is an invasion of my privacy, Randall. And while my motives were pure, the reporters' weren't. They were only politely interested in the events of our trip, and the ruthless despot, El Corazón, and what effects your release might have on foreign policy.

"What they really wanted was dirt. 'Why did you team up with Senator Tackett's brother, Mrs. Porter?' 'Does Key Tackett resent the role you played in Senator Tackett's downfall?' 'Was his death a suicide?' 'How did you feel when you discovered your husband is still alive, Mrs. Porter?' What kind of questions are those?"

"Profound, I would say." With deceptive calm, he set his drink on the coffee table. "How *do* you feel about your husband's return from the dead, Mrs. Porter?"

She avoided his goading glance. "I prefer being addressed by my professional name, Randall. I've been Dr. Mallory for a long time. 'Mrs. Porter' has negative connotations for me."

"Yes, like the fact that you're married," he said with a snide laugh. "You aren't very lucky, are you, Lara? It was so damned untimely for you to fall in love. And with Clark's brother, no less." He threw back his head and laughed harder. "The irony of it is so rich."

She refused to give him the satisfaction of denying or confirming his assumption. Her relationship with Key, which was indefinable even to herself, was none of Randall's business, except insofar as she was still legally his wife. Emotionally, she hadn't felt conjugally linked to him since before that disastrous weekend in Virginia.

He finished his drink. "It's getting late. We'd better get some rest. We're booked on a ten o'clock flight to Washington tomorrow morning."

"I'm not going to Washington."

He had bent down to pick up his shoes. Slowly he straightened. "The hell you're not. It's all arranged."

571

"Then unarrange it. I'm not going."

"The President of the United States is scheduled to receive us in the Oval Office." His face had become flushed.

"Extend him my regrets. I won't be able to make it."

She headed for the bedroom. Randall stormed off the sofa, grabbed her arm, and brought her around. "You'll be there with me every step of the way through this, Lara."

"No, I won't, Randall," she averred, pulling her arm free. "Frankly, I'm surprised you want to share the limelight. When you left Washington, you were a cuckold, a laughingstock. You're returning a hero. You'll probably be invited to appear on all the TV talk shows, to write a book — there might even be a movie-of-the-week in your future. Your credibility has been fully restored and once again you've got the ear of the president. Why would you want me there, stealing a few rays of your spotlight and reminding everyone of that large, dark blot on your career?"

"To keep up appearances," he said with a cold smile. "You are still my wife. I'm willing to overlook your sleeping arrangements with Key Tackett. After all, you thought I was dead."

"Don't assume that moral posture with me, Randall. The martyred husband who continues to forgive his wayward wife." Her words were laden with contempt. "That's the pose you struck when photos of me being hustled from Clark's cottage hit the newsstands. Little did anyone guess that you'd been having affairs almost from the day we married."

"I've never confessed to that," he replied blandly. "You surmised it for your own benefit."

"I also surmise that you didn't live a celibate life in Montesangre. If you were chummy with your guards, I'm certain they made arrangements for you."

"A very astute guess, Lara. In fact I did enjoy a satisfying physical relationship while I was in captivity. She was a beautiful girl, petite and delicate with ebony eyes. She was pathetically willing to please me no matter what I asked of her. She was hardly suited to guerrilla warfare, although she was dedicated to the cause and to her second cousin, Emilio Sánchez Perón.

"When he found out she'd become my lover, he had her disemboweled. I believe he was jealous. During their youth they'd been very close. Or maybe he was afraid that her devotion to me would divide her loyalties. Either way, he brought an end to a very gratifying diversion."

Lara was sickened by the story and the cavalier manner in which Randall related it. She said, "I should have divorced you before we went to Montesangre."

"Possibly. But by then you were pregnant. That made things difficult for you."

"Yes, because you threatened to take the baby away from me unless I stayed with you."

"I could have, too. You were an adulterous wife, hardly a model parent. What court in the

573

land would have awarded custody of a newborn to Clark Tackett's whore?"

He'd posed the same question five years earlier. She'd known it wasn't an empty threat. Had she pursued a divorce and refused to go with him when he left the country, he would have exhausted every effort to win legal custody of the child. She would have fought him to the Supreme Court, except for one major consideration — Ashley. During the years most vital to her development, she would have been shuttled between them, more an object under dispute than a human being. That would have made it almost impossible to raise a contented, well-adjusted child. She hadn't wanted that for her baby.

"Your insults can't hurt me, Randall, because I don't love you. You don't love me. Why perpetuate this myth any longer?"

"Appearances are very important in my line of work," he said with exaggerated patience. "You are garnish, Lara. You always have been. Most wives are. The smarter and prettier they are, the better, but all are little more than what parsley is to prime rib."

Disgusted, she backed away from him.

"Your objections have been noted," he said in a condescending way that further infuriated her. Then he smiled. "Actually I find this new rebellious streak of yours rather exciting, but I'm tiring of it. Save it for a more convenient time, hmm? You'll follow me to Washington and stand meekly by my side just as you followed me to

Montesangre and fulfilled your duties as my official hostess."

"The hell I will." She confronted him defiantly and fearlessly. "Because of the terrible ordeal you'd been through, I gave you the benefit of the doubt. But your three years of confinement haven't changed you, Randall. You're as selfish and manipulative as you ever were. Maybe even more so because you now feel the world owes you for what you endured.

"I'm glad you're alive, but I want nothing to do with you. Don't think you can persuade me otherwise. It's over and has been for years.

"I went to Central America with you in exchange for Ashley. I agreed to stay for one year following her birth. We were only weeks away from the deadline when she was killed. I lost her anyway," she said with rancor. "Now that she's dead, your threats are worthless. You have no bargaining power because I've already lost everything that was valuable to me."

"What about Tackett brother number two?"

"You can't harm Key."

"No?" he asked silkily. "Reading between the lines, I'd say he held his brother in very high regard. Think about it, Lara."

The threat was very subtle, but very real. She schooled her features not to give away her alarm. "You wouldn't say anything to him."

He laughed. "Just as I guessed. He doesn't know. It's still our little secret."

She regarded him for a moment, then snickered. "This time, Randall, I'm calling your bluff." She moved toward the bedroom but at the door turned back. "I don't give a damn what you do so long as you stay away from me. Go to Washington. Make headlines. Rub elbows with the president. Become a celebrity. Have all the affairs you want. The divorce I threatened you with years ago is going to become a reality. I'm filing for it immediately. And from now on, if you want a response, address me as Dr. Mallory. I won't answer to your name."

She slipped into the bedroom and slammed the door.

Chapter Twenty-Six

Janellen shielded her eyes from the sun as she impatiently kept a lookout for the pimp-mobile. When she spotted it turning off the main road, she cried, "Mama, he's here!"

Key had called from the landing strip to notify them that he'd just flown in and would be home shortly. The evening before, he'd called from Houston. "The prodigal has returned. Kill the fatted calf."

Janellen hadn't gone to quite that extreme, but she'd told Maydale to prepare a special dinner. Key was alive and well! He was back!

She skipped down the steps and planted herself directly in the path of the approaching Lincoln, forcing him to stop. Flattening her hands on the hood, she smiled at him through the windshield, then ran to the driver's side and launched herself into his arms as he alighted.

"Whoa, there! Watch those cracked ribs." He regained his balance and gave her a hug, then held her at arm's length. "Damn my eyes! You look gorgeous!"

"I do not," she coyly protested.

"I know gorgeous when I see it. What's new? Something."

"I got a haircut and body wave, that's all. In

fact I was under the dryer at the beauty parlor when somebody thumped on it and pointed at the TV. They were doing a news bulletin about you, Dr. Mallory, and her husband leaving Montesangre and returning home via Colombia. When I saw y'all on that screen, my heart nearly stopped."

His smile faltered. "Yeah, it's been an eventful week." Then, tweaking her cheek, he said, "I like the new hairdo."

"Mama hates it. She said it's too frivolous for a woman my age. Do you think so?" she asked worriedly.

"I think it's sexy as hell."

"Why, thank you kindly, sir." She bobbed a curtsy.

"Hmm. You've learned to flirt, too." He placed his hands on his hips and tilted his head as he eyed her up and down. "Is there something going on that I should know about?"

"No." Her answer had been too quick and too emphatic. If her cheeks looked as hot as they felt, her brother would know instantly that she was lying.

"Cato's still sniffing, huh?"

She tried to keep from smiling but was helpless to contain the joy that infused her at the very mention of his name. It conjured up memories of the hours they'd spent necking in the parlor late at night, arguing in whispers over the right-ness and wrongness of their romance — she advocating the former, he the latter — planning on a future that she insisted they had and he insisted they didn't.

For all their quarrels about the nature and life span of their affair, it *was* an affair. Short of having it consummated and being with Bowie twenty-four hours a day, Janellen couldn't have been happier.

That happiness was transparent, especially to her intuitive brother. He broke a wide smile. "He'd better treat you right. If he doesn't and I hear of it, I'll chase him down, tear off his nuts, and feed them to a dog. You can tell him I said so."

"I wouldn't tell him any such thing!" she declared. "It'd be unladylike." Then she laughed at her private joke, remembering the shocking vocabulary she'd used with Bowie to assure that she got his attention. She didn't regret it. It had worked.

Linking arms with Key, she turned him toward the house. "You must be exhausted. I had Maydale put fresh linens on your bed. You can climb between them as soon as you've had dinner and a long, hot bath."

When he came to a sudden standstill, Janellen glanced up. Jody was watching them from the porch. She looked very well. Apparently the doctors had been alarmists after all, and, as usual, Jody had been right. She was getting better in spite of their dire prognosis.

In the last few days there'd been visible signs of improvement. She claimed to feel better and had more energy. She'd been alert and hadn't fussed when it came time to take her medication. She'd even cut back to two packs of cigarettes a day. Yesterday she'd resumed her standing appointment at the beauty shop.

Janellen doubted it was coincidental that Jody had begun perking up on the day they learned that Key had left Montesangre. Despite their frequent quarrels, her mother and brother cared deeply for each other.

"Hello, Jody."

His tone was reserved, cautious. He was remembering the hurtful, thoughtless things Jody had said to him before he left. Jody too must have been remembering her searing words. Her thin lips twitched once, as though she experienced an uncomfortable twinge.

"I see you made it back in one piece."

"More or less."

Janellen's eyes darted between them, wanting desperately to keep this unspoken truce in force. "Let's go inside and have a drink together before dinner."

Jody preceded them into the parlor. She declined a drink but lit a cigarette. "I read that the rebel army confiscated your airplane." She aimed a plume of smoke toward the ceiling.

"That's right. Thanks, sis." He took the scotch over rocks his sister had poured for him. "Doesn't matter. The guy who rented it to us was hoping we'd crash or that something catastrophic would happen so he could collect the

580

insurance. He needed the cash more than the airplane."

"I figured it was something like that. You deal with such unscrupulous characters."

"Speaking of unscrupulous characters," Janellen said, trying to avoid any nastiness, "Darcy Winston was at the Curl Up and Dye the day I got my perm. She was going on about her daughter Heather and how she and Tanner Hoskins can't keep their hands off each other. She said before it was over, she might have to turn the garden hose on them."

Key laughed. Janellen looked at him with perplexity. "Everyone else laughed when she said that. I don't get it."

"Oh, for heaven's sake, Janellen," Jody said impatiently.

"What?"

"Never mind," Key said. "Go on. What else did Mrs. Winston have to say?"

"When the news bulletin about you and Dr. Mallory came on, she elbowed everybody else out of the way and hogged the TV. When they announced that Mr. Porter wasn't dead after all, she made a spectacle of herself."

"In what way?" Key was no longer smiling.

"By laughing. No one else thought it was funny. She crowed. Honestly, that woman gives 'tacky' a bad name."

"She's a hot little tramp," Jody said as she flicked ashes into the ashtray. "Fergus thought that marrying a white-trash slut would automatically make her respectable. It didn't, of course. Underneath her fancy designer clothes, she's still trash. Fergus has always been a fool."

Maydale called them to supper and served Key his favorite foods: chicken-fried steaks and roast beef with all the trimmings. For dessert there were two pies — one peach, one pecan — and homemade vanilla ice cream.

Janellen expected him to wolf down the banquet she'd ordered for him, but he ate sparingly. He smiled when talking to her and answered all her questions, but with little elaboration. He was polite to Jody and said nothing to goad or provoke her. For a man who had narrowly escaped death at the hands of guerrilla rebels, he was abnormally subdued.

During lapses in conversation, he stared broodingly into space and had to be forcibly drawn back into the present when talk resumed.

Following the meal, Jody excused herself to go upstairs to watch TV in her room. Before she left the dining room, she looked at him and said, "I'm glad you're all right."

He stared after her thoughtfully.

"She means it, you know," Janellen said quietly. "I think she was more worried about you than I was, and I was crazy with it. She had a real turnaround the day we heard that you were alive and on your way home."

"She looks better than when I left."

"You noticed!" she exclaimed. "I think so, too. I think she's getting well."

He reached out and stroked her cheek, but his smile was sad.

"There's something else, Key. Something about Mama. Yesterday when I came home from work, I couldn't find her and went looking through the house. Guess where she was. In Clark's room, going through his things."

No longer distracted, he was suddenly alert and interested.

"To my knowledge she hasn't been in that bedroom since we picked out his burial suit. What possessed her to go in there now?"

"She was going through his things?"

She nodded. "Papers, certificates of merit, yearbooks, memorabilia, memos he'd written while he was a senator. And she was crying. She didn't even cry when he was buried."

"I know. I remember."

It struck her then that Key looked very much now as he had at their brother's grave site. Although his actions and verbal responses appeared normal, she got the sense that he was only going through the expected motions, just as he had following Clark's death. He wore a shattered and lost look, as though something incomprehensible had happened.

During the days following their brother's funeral, she'd been too engulfed in her own sorrow to deal with Key's, although even if she'd tried, he probably would have rebuffed her. Besides, she would have felt inadequate. She still did. Nevertheless, she laid her hand on his arm and pressed it compassionately.

"I read a book on bereavement to help me get through Clark's death. According to the author, who's a psychologist, grief can be a delayed reaction. Sometimes a person can deny it for years. Then one day it hits them, and they let it all out. Do you think that's what happened with Mama?"

Key remained thoughtful and didn't say anything.

"I think it's a breakthrough," Janellen said. "Maybe she's finally come to grips with losing him. Now that she's sorted out her feelings, maybe she won't be so angry anymore. You two got along well at dinner. Did you notice the difference in her attitude?"

Key smiled at her affectionately. "You're the eternal optimist, aren't you?"

"Don't make fun of me," she said, wounded.

"I'm not making fun of you, Janellen. It was an observation meant to compliment. If everyone were as guileless as you, the world wouldn't suck nearly as bad as it does."

He playfully tugged on one of her new curls, but his grin was superficial. "Who knows what compelled Jody to pick through Clark's things? It could mean anything or nothing. Don't expect too much from her. Things don't change that drastically, that quickly. Some things never change. You're in love. You're happy and want everybody else to be."

She laid her head on his chest and hugged him

tightly. "It's true, Key. I'm happier than I've been in my entire life. Happier than I believed possible."

"It shows, and I'm damned glad for you."

"But I feel guilty."

Roughly he pushed her away. "Don't," he said angrily. "Milk it for all it's worth. Squeeze every single drop of pleasure from it. You deserve it. You've put up with shit from her, from me, from everybody for years. For chrissake, Janellen, don't apologize for finding happiness. Promise me you won't."

Stunned by his vehemence, she bobbed her head. "All right. I promise."

He pressed a hard kiss on her forehead, then set her away from him again. "I gotta go."

"Go? Where? I thought you'd want to stay home tonight and get some rest."

"I'm rested." He fished in his jeans pocket for his car keys. "I've got a lot of catching up to do."

"Catching up on what?" He shot her a telling look and headed for the door. "Key, wait! You mean like drinking?"

"For starters."

"Women?"

"Okay."

She intercepted him at the front door and forced him to look her in the eye. "I haven't asked because I figured it was your private business."

"Asked me what?"

"About Lara Mallory."

"What about her?"

"Well, I thought, you know, that the two of you might . . ."

"You thought I might take Clark's place in her bed?"

"You make it sound so ugly."

"It was ugly."

"Key!"

"I gotta go. Don't wait up."

Before she opened the door, Lara peered through the blinds to see who had rung the bell, then hastily undid the locks. "Janellen! I'm so glad to see you. Come in." She stood aside and ushered her unexpected guest into the waiting room.

"I hope I'm not disturbing you. I always seem to drop in without calling first. I acted on impulse again."

"Even if you'd called, you wouldn't have been able to get through. I took my phone off the hook. Some reporters don't take 'no' for an answer."

"They've been calling Key, too."

Hearing his name was like getting an arrow through her heart. Trying to ignore the pain, she removed a box of books from the seat of a chair. "Sit down, please. Would you like something to drink? I'm not sure what's in the house —"

"I don't care for anything, thank you." Janellen glanced around at the disarray. "What's all this?"

"This is a mess," Lara said with a wry smile as she sat down on a crate. Wearily, she pushed back a loose strand of hair. Since her return, even involuntary motions seemed to require a tremendous amount of energy. "I'm packing."

"What for?"

"I'm leaving Eden Pass."

Janellen was possibly the only person in town who didn't welcome the news. Her expression was a mix of dismay and despair. "Why?"

"That should be obvious." There was a bitterness in Lara's voice that she couldn't mask. "Things didn't work out here as I had hoped. Clark was wrong to deed me this place. I was wrong to accept it."

She was touched to see tears in Janellen's eyes. "The people in this town can be so stupid! You're the best doctor we've ever had."

"Their opinion of me had nothing to do with my qualifications as a physician. They bowed to pressure." It was unnecessary to cite Jody Tackett as the party responsible for the shunning.

Janellen already knew, and felt guilty by association. "I'm sorry."

"I know you are. Thank you." The two women smiled at each other. If circumstances had been different, they could have become very good friends. "How is your mother doing? Has the medication been effective?"

Janellen told her about Jody's marked improvement. Lara didn't want to dampen her optimism, but felt it was her professional duty to interject some realism. "I'm glad to hear that she's feeling better, but stay vigilant. She must continue taking the medication until her doctor instructs otherwise. I recommend frequent, periodic checkups. And before you completely reject the idea of angioplasty to dilate the carotid, I recommend another round of extensive testing."

"I don't think Mama would agree to it, but if I notice signs of stress or — heaven forbid — another seizure, I'll insist."

They chatted for a few minutes more, then Janellen rose to leave. At the door she said, "I saw your husband on *The Today Show* this morning. They had videotape of him being greeted by the president."

"Yes, I saw it, too."

"The interviewer asked why you weren't with him. He said you were so overwrought from your experiences in Montesangre that you were unable to accompany him to Washington."

It rankled that Randall was serving as her mouthpiece and giving out false information. She had made her position unequivocally clear to him when they were in Houston and had remained locked in her bedroom of the suite until she was certain he had left the hotel for the airport to catch his Washington flight. They hadn't said goodbye.

His excuses for her absence in Washington were self-serving, but, other than confronting him about it, there was nothing she could do to stop him. The issue wasn't worth having another private encounter. Their next one would be in a divorce court, and then she would have an attorney speaking for her.

"It must have been . . ." Janellen hesitated, then plunged ahead. "Well I can't even imagine how you felt when you discovered that he had been alive all this time."

"No, I'm sure you can't imagine."

Introspectively, Lara again saw Randall lying in the bathtub. She heard her screams echoing off the gaudy tile walls, heard the crunch of breaking wood as Key kicked his way through the door, felt his arms closing around her. She had buried her face against his chest. At first they had thought Randall was dead.

But he'd come back to life.

Key hadn't touched her since, not even casually.

There were no words to describe the enormity of the shock caused by Randall's resurrection, so she simply said, "I was astounded to see him alive."

"I'm sure you were, but you don't appear overwrought. Why didn't you go to Washington with him?" On the heels of her blunt question, Janellen quickly withdrew it. "I'm sorry. That was unforgivably rude."

"No need to apologize. You asked a legitimate question. The answer is simply that I chose not to go. Politics is Randall's arena, not mine. What he does with his recent celebrity is up to him.

I want to ignore mine, and I wish that everyone else would."

"So does Key."

The arrow in her heart twisted. "He seemed extremely uncomfortable to find himself suddenly in the spotlight."

Janellen's sweet face puckered with anguish as she blurted out, "He's going away again. To Alaska. He told me this morning. He's been offered a job as a spotter along the pipeline. That's a pilot who checks for leaks."

Lara nodded vaguely.

"He says it's good money and that he needs a change of scenery. I reminded him that he'd just had a change of scenery, but he said the trip to Central America didn't count. I don't want him to go," she said, her anxiety plain. "But now that Mama's in better health, I guess there's nothing to keep him here."

"I guess not." Her voice had a hollow ring.

"I'm so worried about him," Janellen went on. "At first I thought he was just tired from the ordeal, but you've been back a week and he hasn't snapped out of it yet."

Lara was instantly alarmed. "Is he ill?"

"No, he's not sick. Not physically. He's withdrawn. His eyes don't sparkle anymore. He doesn't even yell when he gets mad. That's not like him."

"No, it isn't."

"It's like somebody pulled the plug on the electricity that kept him charged."

Lara didn't know how to respond.

"Well," Janellen concluded awkwardly. "I just thought I'd tell you."

She hesitated, as though there was more she wanted to say. Lara wondered if she knew that they'd slept together. Surely she couldn't know . . . but maybe she'd guessed.

"Well, uh . . . When are you leaving town?"

"I don't have a timetable, just whenever I get everything packed. I haven't yet made arrangements with a realtor to handle the sale of this building."

"Will you be moving to Washington?"

"No," she answered sharply. Ameliorating her tone, she added, "I haven't made any specific plans."

"You're going to pack up and leave, and you don't even know where you're going?"

"That's the gist of it," Lara replied with a weak smile.

Janellen was flabbergasted, but common courtesy kept her from prying further. "When you know your new address, would you please send it to me? I realize there's bad blood between you and us Tacketts, but I'd like to stay in touch."

"You had nothing to do with the 'bad blood,' " Lara said gently. "I'd love to hear from you."

Janellen seemed to debate whether it was the proper thing to do, but in the end she gave Lara a quick hug before rushing down the walk to her car.

Lara watched until she drove out of sight. Slowly she closed the door, symbolically ending a chapter of her life. This visit with Janellen was probably the last contact she'd have with the Tacketts.

Later, Janellen and Bowie were cuddled up on the parlor sofa. All the lights were out. Jody had retired to her room hours earlier. Key, as usual, was out.

Bowie was semireclined on the corner cushions with Janellen sprawled across his lap. She was using his shoulder as a pillow for her head while she mindlessly strummed his bare chest through his unbuttoned shirt.

"It was so sad," she whispered. "She was standing there surrounded by all those boxes, looking like she was at a complete loss about what to do next."

"Maybe you read her wrong."

"I don't think so, Bowie. She looked like she didn't have a friend in the world."

"Doesn't make sense. She just found out her dead husband is alive."

"It doesn't make sense to me, either. Why isn't she with him? If I had believed you were dead, and discovered you weren't, I never would let you out of my sight again. I love you so much that —" She raised her head. "Well, I'll be. That's it. Dr. Mallory doesn't love her husband anymore. Maybe she's fallen in love with somebody else."

"Calm down now. You're cooking up something in your mind that ain't necessarily so."

"Like what?"

"Like there's something brewing between the doctor and your brother."

"You think so too?" she asked excitedly.

"I don't think anything. I think that's what you think. Flying off to Central America alone together and getting captured by guerrilla fighters is pretty romantic stuff. Sounds like a movie. But don't go reading anything into it that's not there."

She looked chagrined and admitted that a romance between Key and Lara had crossed her mind. "Both of them seem so wretchedly unhappy since they got back. Key's itching to leave."

"He's always been a drifter. You told me so yourself."

"It's more than wanderlust this time. He's not rushing toward a new adventure, he's running away from something. And that describes Dr. Mallory, too. She didn't act like a woman who's beloved has suddenly returned from the dead." She made a face. "From what I saw of him on TV, I can't say I blame her. He sounded like a real jerk. Besides, he's not nearly as handsome as Key."

Bowie chuckled. "You've got a romantic streak a mile wide, you know that?"

"Key said that I'm in love and want everybody else to be as happy as I am. He was right."

"About you wanting everybody to be happy?"

"About my being in love." She gazed into his soulful eyes, her love exposed. Cupping his face, she asked earnestly, "When, Bowie?"

This subject often came up. Each time it did, it either fanned their passions or squelched them. Tonight it caused a physical breach. Frowning, he disengaged himself from her embrace, stood, and began rebuttoning his shirt.

"We have to talk, Janellen."

"I don't want to talk anymore. I want to be with you. I don't care where it has to be as long as we can be together."

He averted his eyes self-consciously. "I found a place I think might do."

"Bowie!" She had a hard time keeping her voice to an excited stage whisper. "Where is it? When can we go? Why didn't you tell me?"

Choosing to answer her last question first, he said, "Because it isn't right, Janellen."

"You don't like the room?"

"No, the room is all right. It's . . ." He paused and shook his head with exasperation. "I hate sneaking in here every night like a damn kid, fumbling around in the dark, copping feels, having to whisper like we're in the goddamn library, then leaving by the back door. It's no damn good."

"But if you've found a place where we can go —"

"It would only be worse. You're too fine a

woman to be snuck through the back doors of motels for a quick toss." He held up his hands to stave off her protests. "And another thing, you might think we could carry on without anyone finding out, but you're fooling yourself. We couldn't. I've lived in Eden Pass long enough to know how fast and accurate the grapevine is. It's too risky to take a chance.

"Sooner or later word would get back to your mama. She'd probably come after me with a shotgun or sic the law on me. Hell, I've been in scrapes before. If she didn't flat-out kill me, I'd survive. But not you. You haven't had a troubled day in your life. You wouldn't know how to handle it."

"I've had lots of trouble."

"Not the kind I'm talking about."

She'd learned from her brothers that men hated when women cried, so she tried her best to keep from bursting into tears. "Are you trying to get out of it, Bowie? Are you making up excuses when actually you just don't want me? Is it my age that's turned you off?"

"Come again?"

A small sob escaped. "That's it, isn't it? You're trying to worm out of it because I'm older than you."

He was equally vexed and incredulous. "You're older than me?"

"Three years."

"Who's counting?"

"Apparently you. That's why you're trying to

back out. You could have a woman much younger than I."

"Shit!" He paced in a small circle, swearing under his breath. Finally he came back around and looked down at her with annoyance. "How long did it take you to dream up that crap? For chrissake, I didn't even know how old you were, and even if I had known, it wouldn't have made any difference. Don't you know me better than that? Shit."

"Then why?"

His aggravation dissolved, and he knelt in front of her, clasping her hands. "Janellen, as far as I'm concerned, you're way up there above any other human who's ever drawn breath. I'd rather lose my right arm than hurt you. That's why I never should have let this get started. The first time I felt that yearning for you, I should have packed up and left town. I knew better, only I couldn't help myself."

He paused, searching her face with such intensity that he seemed to be memorizing it. He ran his thumb across her trembling lips. "I love you better'n I love my own self. That's why I won't sneak you in and out of rented bedrooms, hide you like you were a floozie, and have you gossiped about like you're white trash."

He came to his feet and reached for his hat. "I won't do that to you. No way in hell. No, ma'am." He placed his hat on his head and gave the brim a firm tug. "That's the end of it."

* * *

Lara weakly leaned her head against the doorjamb. "This isn't a good idea, Key."

"Since when has anything involving you been a good idea?"

He forced his way past her. She closed the back door behind him after checking to make sure no one was around to see his arrival. It was a futile precaution. Having the distinctive yellow Lincoln parked in her driveway was as good as announcing it on local radio.

When she turned back into the room, he was leaning against a supply cabinet. His shirttail was hanging loose outside his jeans. He was an untidy, disturbing, sexy reminder of the first time she'd seen him in this same room.

That night he'd asked her for whiskey. This time he'd brought his own. The liquor sloshed inside the bottle when he raised it to his mouth and took a drink. The gash on his temple had closed, but the skin around it was still bruised. So were his ribs. His expression was insolent, his complexion flushed.

"You're drunk."

"You're right."

She folded her arms across her middle. "Why'd you come here?"

"Can Ambassador Porter come out and play?" he asked mockingly.

"He's still in Washington."

"But he'll be here tomorrow. They printed a story about it in the evening edition. 'Hero

597

Statesman Visits Eden Pass.' Big fuckin' deal."

"If you knew he wasn't here, why'd you ask?"

He grinned. "Just to get a rise out of you. To see if your heart would go pitter-pat at the mention of his name."

"I think you'd better go." Coldly turning her back to him, she opened the door.

His hand shot forward from behind her and slammed it shut; then he kept his palm flattened against the wood, trapping her between himself and the door. In the small wedge of space, she turned to face him.

"You never did answer my question."

"What question?"

"About your daughter. Since we made it back alive, I want to know. Was she Clark's kid?"

What did he want to hear? she wondered. What did she want to tell him?

The unvarnished truth.

Oh, God, what a liberating luxury that would be. She could fully explain the situation, fill in all the unknown details, and, by doing so, perhaps make Key feel more charitable toward her.

The mitigating circumstances were the critical ones. Ironically, because they were so very critical, they must remain a secret.

Especially from Key. Especially now that she knew she loved him.

"Randall was Ashley's father."

Regret flickered in his eyes. "You sure?"

"Yes."

She could see that it made a difference to him,

but he tried not to show it. "So you suckered me into risking my life for nothing."

"I didn't persuade you to go to Montesangre, you persuaded yourself. I never even suggested that Clark was Ashley's father."

"You never denied it, either." He leaned in closer. His whiskey-scented breath felt hot on her face. "You're a real piece of work, aren't you? A clever manipulator. A tricky chick.

"At first I couldn't understand how my rational brother could have such a careless affair with his best friend's wife. You did a real seduction number on him, didn't you? Pussy-whipped him till he didn't know which end was up. Then dopey ol' Randall stayed with you. What a sap. He's a prick, probably a liar, but even he doesn't deserve your royal treatment."

His hands clasped her waist and with one swift motion yanked her against him. He nuzzled her neck beneath her ear. "You're good at getting what you want from a man, aren't you, Doc? You mind-fuck him real good before he even gets his cock out."

Lara squeezed her eyes shut. The accusations were ugly. They hurt, especially coming from Key. Key, who more than once had risked his life to save hers, who had been tender and passionate, ardent and loving, whose touch she still craved and whose voice haunted her dreams.

Based on the facts, *as he knew them,* he had cause to insult her. His scorn was founded on what he believed was truth. It was a miscalculation she couldn't rectify — far more for Key's sake than her own.

She wanted him desperately. But not this way. She'd conditioned herself to tolerate the world's contempt, but she refused to nurture his.

"I want you to leave."

"Like hell." He dropped the liquor bottle, slipped his hand beneath her skirt, and tugged on her panties. "You're all I can smell. All I can taste. All I think about." His mouth covered hers and ground an angry kiss into it. "Jesus, I gotta get you out of my system."

"No, Key!" She pressed her thighs together.

"How come? It's not like you haven't been unfaithful before."

She swatted away the hand groping at her breasts. "Stop this!"

"You owe me, remember? Either the ninety thousand balance of my hundred grand. Or this." He forced his hand between her thighs and fondled her intimately. "I choose this."

"No!"

"Don't worry, I'll leave before sun-up. Your husband won't catch you in the act this time. I'm smarter than my brother. I'm also better. *Aren't I?*"

"No, you're not," she cried. "Clark never had to resort to rape!"

That sobered him as instantly as the cold water she'd once thrown in his face. He released her and staggered backward, his breath coming harsh and loud.

Knowing the root of his aggression, Lara felt more sorrow than anger. She longed to touch his face, run her fingers through the damp strands of hair clinging to his forehead, placate him, tell him she regretted having to hurt him in the worst possible way — by unfavorably comparing him to Clark.

Instead, she had to let her statement stand and watch his lip curl with repugnance for his brother's cast-off, adulterous whore.

He looked her over and made a scornful sound. "No, I'm sure he didn't. Relax, Doc. You're safe from me."

He reached around her and pulled open the door. The liquor bottle almost tripped him. He kicked it out of his way. It crashed against the wall and shattered.

He stormed through the door, leaped over the steps, and climbed into the Lincoln. He gunned it; the tires spun in the gravel before gaining traction. He sped away.

Lara closed the door and, with her back to it, slid to the floor. Folding her arms across her lap, she bent at the waist and released a keening cry.

Chapter Twenty-Seven

"So this is it? This is what you're so reluctant to leave?"

Randall had strolled through the rooms of the clinic and wound up in Lara's private office, where she'd been packing books and files. He'd flown from National Airport to Dallas/Fort Worth and leased a car for the two-hour drive to Eden Pass.

For hours before his arrival, media vans had been cruising the street in front of the clinic on the lookout for him. When he arrived, reporters and cameramen flocked to him in impressive numbers.

His ordeal in Montesangre had atoned for the scandal involving his wife and Senator Tackett. Like a wayward child who'd taken his punishment and turned over a new leaf, he'd been warmly received by the president and the Department of State. Having experienced the Montesangren culture from the inside out, he was its reigning expert on Capitol Hill. He was newsworthy.

Lara remained indoors while Randall conducted an impromptu press conference. After fielding questions for several minutes, he begged to be excused.

"My wife and I have had very little time alone since our return. I'm sure you can understand."

After some good-natured snickering, they reloaded their Betacams and microphones into their vans and left. Many honked and waved as though bidding goodbye to a chum.

Now dusk was gathering outside, but Lara hadn't turned on the lamps in her office. The semidarkness was more in keeping with her mood. It also hid the dark circles beneath her eyes.

Knowing she would never see Key again, she had cried herself into a stupor following his angry departure the night before. He'd left hating her. Her sense of loss was wrenchingly painful and came close to how she'd felt when she regained consciousness in Miami and realized that the terrible nightmare she'd had was indeed real.

Finally, sometime around 2:00 A.M., she garnered the wherewithal to make her way to bed, where she'd lain awake until dawn. She'd spent the day packing her belongings, working feverishly between lapses of immobilizing depression in which her hands were rendered useless and she stared vacantly into space through dry, gritty eyes.

The gloaming made the office feel cozier, warmer, safer, a refuge for her abject despair. She had come to like Dr. Patton's paneled walls and masculine furniture and wished she could look forward to years of enjoying this office.

"It's so provincial," Randall observed as he

dropped onto the leather love seat.

"The equipment is modern."

"I'm talking about the whole setup. It's not like you at all."

He didn't have a clue as to what she was like. "Sick people aren't confined to cities, Randall. I could have had a good practice here." She folded down the flaps of a cardboard box and sealed it with duct tape. "That is, if I'd been given a decent chance to cultivate one."

"Tackett territory."

"Indisputably."

"I'm curious about something." He crossed his legs with the negligent elegance of Fred Astaire. "Why in God's name, when you had the whole continent to choose from, did you elect to practice here? In Texas of all places," he said with obvious distaste. "Why pick the town where you'd be most despised? Do you have a bent toward masochism?"

She had no intention of recounting for Randall the last three years of her life. In fact, she had no intention of letting him stay beneath her roof. Before sending him away, however, there was one thing she wanted him to know.

"It wasn't easy for me to pick up my career where it left off," she began. "Even though I had been badly injured and had lost my child and my husband to a bloody revolution, people were slow to forgive. I was still considered Clark's bimbo.

"I applied for staff positions at hospitals all

604

over the country. Some even hired me on my credentials alone before linking Dr. Lara Mallory with Mrs. Randall Porter, whereupon I was sanctimoniously asked to resign in the best interests of the institution. This happened a dozen times at least."

"So you finally decided to hang out your shingle. I suppose you used my life insurance money for financing. But that still doesn't explain why you chose to practice here."

"I didn't buy the practice, Randall. It was deeded to me free and clear. By Clark." She paused for emphasis. "It was one of the last official things he did before his death."

It took him a moment to assimilate the information. When he did, he sucked in a quick breath. "Well, I daresay. He was buying absolution for his sins. How touchingly moral."

"I can only guess at his motivations, but yes, I think he felt he owed me this."

"Now I suppose you're going to present me with a bill. What do I owe you for accompanying me to Montesangre?"

"A divorce."

"Denied."

"You can't deny me anything," she said vehemently. "Key and I saved you from imprisonment in that miserable place! Or have you already forgotten? Has your instant fame wiped your memory clean?"

Gradually a smile spread across his face. It was as patronizing as his tone of voice. "Lara, Lara.

605

So naïve. After all you've been through, you still fail to see beneath the surface, don't you? Hasn't experience taught you anything? Where there's smoke . . . and so on." His hand made a lazy circular gesture. "Haven't you learned to look beyond appearances and see things as they really are?"

"You've made your point, Randall. What the hell does it mean?"

"Do you honestly think that you and that hot-headed pilot precipitated my release?"

His voice had become soft, sibilant, and smug. It caused the hair on the back of her neck to rise. She had a premonition of dread. "What are you saying?"

"Put on your thinking cap, Lara. You passed medical school with flying colors. Surely you can figure this out."

"In Montesangre."

"Yes," he said encouragingly. "Go on."

"Emilio . . ."

"Very good. What else? Stretch your clever little mind."

The mental barriers were opaque, but once she broke through them, everything was crystal clear. "You weren't his prisoner at all."

He laughed. "Good girl! I hate to sound un-appreciative, but don't credit yourself with saving my life. My 'five-year plan,' as I like to think of it, was about to be realized in any event. Your comical misadventure with Key Tackett was merely a fortuitous development that Emilio and

606

I used as our catalyst. It made the denouement so much more convincing."

Lara stared at the man to whom she was legally married and knew she was looking into the eyes of a madman. He was perfectly composed, exceedingly articulate, and dangerously sly, the most frightening portrait of a villain.

"It was all a hoax?" she whispered.

Randall left the leather love seat and came to stand close to her. "Following that morning in Virginia, I was despised in Washington. Clark had powerful allies, including the president. He was no doubt embarrassed over Clark's conduct, but he stood by his protégé. To a point, anyway.

"At Clark's request, he appointed me ambassador and called in favors in the Senate to have my approval rushed. On the surface, I accepted graciously, humbly, like they had done me a bloody favor. Actually, I despised it as much as you, knowing that it was a legal form of banishment.

"No sooner had I arrived at my post than I began to devise ways of returning to Washington a hero. Emilio was a bright boy who had his own ambitions, which were fulfilled with Pérez's death."

"Murder."

"Whatever. Together, we contrived a plot that would give each of us what he wanted. My 'escape' had to be carefully timed and fully capitalized upon. Once I returned to the U.S., rather than harboring a grudge toward my captors, I

607

would insist on being reassigned to Montesangre, reopening the embassy, and reestablishing diplomatic relations with the new regime."

Imperceptibly, Lara was edging toward the telephone. "Emilio's regime."

"Precisely. Upon my advice to the president, Emilio's government would soon be acknowledged. With the endorsement of the United States, he'd have absolute control of his republic. I'd be credited with restoring peace to a hostile nation which could be strategic in fighting the drug wars. After a suitable time, my endeavors surely would be rewarded either with a plum appointment abroad or in Washington. A far cry from the cuckold, hey?"

"You're crazy."

"Like a fox, Lara. It's been well thought out, I assure you. After years, the realization is unfolding even better than anticipated. What I need now is a loving wife to round out my image as a exemplary diplomat.

"So, darling, you will remain faithfully and meekly by my side, smiling at the press, waving to the crowds, until I say otherwise. Don't even think of doing anything to jeopardize this."

She began to laugh. "You're a traitor with delusions of grandeur, Randall. Do you honestly think I'm going to participate in this traitorous 'five-year plan' of yours?"

"Yes, I think you will," he replied calmly. "What choice do you have?"

"I'll blow the whistle. I'll tell them about

Emilio's brutality. I'll call —"

"Who would believe *you?*" He shook his head sadly over her delusions. "Who would trust anything said by the woman caught in adultery with Senator Tackett? You have no more credibility now than you did that morning we left his cottage."

He indicated the telephone she'd been inching toward. "I can see you re itching to call for help. Go ahead. You'll only make a laughingstock of yourself. Who's going to believe that a U.S. ambassador started a revolution which was contrary to the interests of the country he served?"

" 'Started a revolution'? What do you mean? The revolution started when . . . when our car was . . . No, wait." She held up her hand as though to ward off a barrage of confusing thoughts. They were crowding her mind so quickly she couldn't arrange them.

"You're slipping, my dear," he said silkily. "The mental sluggishness must come from living on the frontier. Think, now. I said *five*-year plan. It took root when we reached Montesangre, not when I was kidnapped."

Her heart began to beat faster; she clutched her throat, which had suddenly gone dry. Something was just beyond her grasp. Something she should remember: Something —

The truth struck her with the impact of a bullet. The fog lifted from her memory and those forgotten instants immediately preceding the ambush were replayed in slow motion in her mind.

She was playing patty-cake with Ashley in the backseat. The car approached the intersection. As it slowed down, armed men rushed forward, surrounding it. The driver was shot and slumped forward over the steering wheel.

She cried out. Randall turned to look at her. "Goodbye, Lara." Unafraid, he smiled.

Her breath rushed out in a gust. "You knew!" she screamed. "You and Emilio arranged the ambush on our car! You had our daughter killed!"

"Shut up! Do you want the whole neighborhood to hear you?"

"I want the whole world to hear me."

He struck her across the mouth. Talking rapidly, quietly, he said, "You fool! I didn't intend for the child to be killed. The bullets weren't meant for her."

Lara didn't even stop to consider what that statement implied. She lunged for the camera bag. It was on her desk, where she had left it, undisturbed, since the day she returned from Montesangre.

Under the concealment of darkness, she plunged her hand into the bag. Her fingers closed around the butt of the revolver. She withdrew it and swung around, aiming the barrel at the center of Randall's chest.

"This is your last chance to change your mind."

Janellen smiled at Bowie. "I'm not going to change my mind. I'm absolutely, positively, one

610

hundred percent sure of my decision. Besides, you were the one with cold feet, the one dead set against it. I finally wore you down, so I'm not about to back out or let you, either." She linked her arm with his and nestled her head on his shoulder. "Just drive, Mr. Cato. I'm anxious to get there."

"If anybody sees me driving your car —"

"It's dark. Nobody's going to see us. If someone does, they'll probably think that Key asked you to protect me from reporters again."

"Yeah, I saw them all over town today."

"They were hoping to catch a glimpse of Mr. Porter." The reminder intruded on Janellen's happiness and caused her to frown. "Mama watched him on the news. Seeing him really upset her."

"Why should it?"

"Because it calls to mind the scandal, Clark, all that. She skipped supper and went upstairs to her room."

"You waited until Maydale got there before you left?"

As prearranged, he and Janellen had met at the Tackett Oil office. "Yes. She came to spend the night. I told her I was going to Longview to attend a self-improvement seminar."

"What about Key?"

"Key never gets home before noon, sometimes not even then. He claims he's playing poker till dawn with Balky out at the landing strip. It's easier to sleep out there than to drive home, he says. Anyway, he'll never know I'm gone."

Bowie glanced nervously at every car that passed. "This sneaking around doesn't feel right. Something terrible is bound to happen."

"Honestly, Bowie." She sighed with affectionate exasperation. "You're the most pessimistic, fatalistic person I've ever met. A few months ago you were the one with the record, but I was living in a kind of prison. Both our fortunes have changed."

"Yours will if you stick with me long enough," he said glumly. "You'll lose your fortune."

"I've told you a million times that I don't care if I do. My family had lots of money, but we weren't happy. There was no love between my parents. That antagonism affected my brothers and me. We felt it even before we were old enough to understand it.

"It made Clark an overachiever who couldn't forgive himself even the most insignificant mistake. Key went too far the other way and lives like he doesn't give a damn about anything, although I believe that's a defense mechanism. He doesn't want anyone to guess how deeply he was hurt by our father's death and Mama's rejection.

"And I became a shy, introverted dullard, afraid to voice an opposing opinion on anything. Believe me, Bowie — money doesn't buy happiness and love. I'd rather have your love than all the riches in the world."

"That's 'cause you've never had to do without the riches."

They'd been over this ground so many times they'd trampled it to death. She was determined not to let an argument cast a pall over the happiest night of her life.

"I know exactly what I'm doing, Bowie. I'm beyond the age of consent. I love you to distraction, and I think you love me the same."

He glanced at her and answered with deadpan seriousness. "You know I do."

"That gives us the strength to face anything. What can possibly happen to us that we can't combat?"

"Oh, damn," he groaned. "You've just tempted Fate to show us."

"Bowie," she said, laughing and nuzzling his neck, "you're a sight."

Darcy spotted Key the moment she entered The Palm. He sat alone at the end of the bar, hunched over his drink like a stingy dog with a bone.

She was in a buoyant mood. Fergus was at a school board meeting, which traditionally dragged on for hours. She loved school board meetings. They liberated her for an evening out.

Heather was on desk duty at the motel. Odds were highly in favor of her taking home the crown of homecoming queen this coming Friday night.

Darcy had spent over seven hundred dollars to outfit Heather for the occasion. Fergus would have a fit if he knew, but she considered the

613

expenditure a good investment. If Heather won homecoming queen, it would boost her chances of getting into the best sorority when she went to college. Fergus might not appreciate the subtle way these things worked, but Darcy did.

Although she drove a new car every other year, belonged to the country club, wore expensive clothes, and lived in the largest house in Eden Pass, she still was excluded from the inner social circles.

She was determined that Heather would reverse that. Heather would be her ticket into every tight clique even if she would have to enter through the back door.

Key's posture smacked of potential danger, but she decided to approach him anyway. So what if the last time she'd seen him she'd spat in his face and he'd threatened to murder her? Things weren't going so well for him these days. Having been brought to heel, he might be in a more receptive mood.

She slid onto the barstool next to his. "Hi, Hap. White wine, please. Put some ice cubes in it." The bartender turned to get her drink. She glanced at Key. "Still mad at me?"

"No."

"Oh? You've learned how to forgive and forget?"

"No. In order to be mad, you have to give a shit. I don't."

She quelled her anger, smiled at Hap as he served her wine, and took a sip. "I'm not sur-

prised that you're in such a bear of a mood."
As she turned toward him, she brushed his knee
with hers. "Must've been quite a shock to discover the dead husband was alive."

"I don't want to talk about it."

"I guess not. It's a touchy subject. Did you
at least get to screw her before Ambassador Porter got dumped in her bathtub?"

Key's muscles tensed, telling Darcy he had.
She was treading on thin ice, but the one thing
she couldn't tolerate from a man was indifference. She'd rather be verbally or physically
abused than ignored. Besides, she was curious.

"Was she as good as you expected? Not as
good? Better?"

Better, she would guess by the way he tossed
back the remainder of his drink and signaled for
Hap to pour him another. Gossip around town
was that you'd have to be real stupid to cross
Key Tackett these days. He was truculent. Testy.
Spoiling for a fight.

Just yesterday, at noon, right in the middle
of Texas Street, he'd threatened to shove a
journalist's camera up the guy's ass if he didn't
get it out of his face. Later, he'd gotten into a
fight at Barbecue Bobby's with a redneck from
out of town who'd parked his pickup too close
to the Lincoln to suit Key. Witnesses said it'd
be a while before the redneck ventured into Eden
Pass again.

Reputedly, he was on the brink of drunkenness
at any time of the day or night, and he spent

hours at the county airstrip with that dimwit Balky Willis. Someone said he was taking target practice at 4:00 A.M. on the lights at the football stadium, but that was unsubstantiated.

If Lara Mallory's performance in bed had disappointed him, he wouldn't care that her husband had turned up alive and well. On the contrary, the better he liked her, the angrier he'd be over the turn of events.

From what Darcy had heard and could now see for herself, Key was good and pissed.

Jealousy made her reckless. She dared to probe another tender spot. "Guess you know now why your brother was willing to risk his career for her." His jaw flexed. "Wonder how she compared the two of you and which one earned the most points. Did y'all discuss your merits?"

"Shut the fuck up, Darcy."

She laughed. "You did, then. Hmm. Interesting. Three people in one bed can get awfully crowded."

Key turned his head and fixed a heavy-lidded, bloodshot stare on her. "From what I hear, you've been one of a trio more than a few times."

Darcy's temper flared, then instantly subsided. Her laugh was low, seductive. She leaned closer, mashing her breast against his arm. "Damn straight. Had quite a time for myself, too. You ought to try it sometime. Or have you?"

"Not on this continent."

Again she laughed. "Sounds fascinating." She

trailed her finger up his arm. "I'm dying to hear all the slippery details."

He didn't dismiss the suggestion out of hand. Encouraged, Darcy reached for her handbag and took out a latchkey. She dangled it inches beyond his nose.

"There are distinct advantages to being a motel proprietor's wife. Like having a skeleton key that'll open the door to every room." She ran her tongue along her lower lip. "What do you say?"

She leaned back a fraction so he'd be certain to see that contact with his biceps had aggravated her nipples to stiff points. "Come on, Key. It was good between us, wasn't it? What else have you got going?"

He finished his drink in a single draft. After tossing enough money on the bar to cover his drinks and Darcy's wine, he pushed her toward the door.

He said nothing until they were outside. "Your car or mine?"

"Mine. You can spot that yellow submarine of yours a mile away. Besides, if my car's seen at the motel, nobody thinks twice about it."

As soon as they were seated in the El Dorado, she leaned across the console and brushed a light kiss across his lips. It was an appetizer, a teaser for good things yet to come. "You've missed me. I know you have."

He remained slumped in his seat, staring balefully through the windshield.

Darcy smiled with feline complacency. He was

sulking, but she'd have him revved up in no time. If it was the last thing she did, she'd prove that Lara Mallory was forgettable.

The Cadillac sped in the direction of The Green Pine Motel.

Jody knew Janellen well. The girl wasn't nearly as clever as she thought she was. Ordinarily, any alternation in her routine sent Janellen into a tailspin. She would cajole her to eat, beg her not to smoke, insist that she go to bed, implore her to get up. She hovered like a mother hen.

But tonight when she declined supper, Janellen's nagging had lacked its customary fretfulness. Even before tonight, Jody had detected remarkable changes in Janellen. She fussed over her appearance like never before. She'd begun wearing makeup and had had her hair screwed into that curly, bobbed hairdo. She dressed differently. Her skirts were shorter and the colors brighter.

She laughed more. In fact her disposition was cheerful to the point of giddiness. She went out of her way to be friendly to people she had shied away from before.

Her eyes twinkled with something akin to mischief, which disconcertingly reminded Jody of Key. And of her late husband. Janellen was keeping a secret from her mother for the first time in her life.

Jody guessed it was a man.

She'd overheard Janellen tell Maydale that

618

cock-and-bull story about a seminar in Longview, when it was obvious she was keeping a rendezvous with her fellow, probably at the same motel where her father had entertained some of his tarts. The sordidness of it left a bad taste in Jody's mouth. Hadn't the girl learned anything she'd tried to teach her? Before some fortune-hunting Casanova ruined Janellen's life, she'd have to attend to it.

All the important family issues were her responsibility and had been since she said "I do" to Clark Junior. Where would the Tacketts be today if she hadn't helped maneuver their destiny? Never content to let events evolve on their own capricious course, she handled all the crises herself.

Like the one she was scheduled to take care of tonight.

Of course, first she had to sneak past Maydale.

Fergus Winston's mind was pleasantly drifting.

The school board treasurer was a soprano soloist in the Baptist church choir. She so enjoyed the sound of her own voice that she detailed each entry on the budget report instead of distributing copies and letting the other board members read it.

As she itemized the entries in her wavering falsetto, Fergus hid a private smile, reflecting on his own healthy financial report. Thanks to a relatively temperate summer that had attracted fishermen and campers to the lakes and forests of

East Texas, the motel had enjoyed its best season yet.

He was seriously considering Darcy's suggestion of using some of the profits to build a recreation room with workout equipment and video games. Darcy hadn't steered him wrong yet, not since he'd hired her to coordinate his coffeeshop. She had a knack for money-making ideas.

She also had a knack for spending every cent he made. Like most folks, she didn't think he was too astute. Because he loved her, he let her live under the illusion that he didn't know about her extramarital affairs. It hurt that she sought the company of other men, but it wasn't as painful as living without her would be.

He'd heard a radio psychologist spouting off about deep-seated psychological reasons for aberrant human behavior that had roots in childhood. No doubt Darcy was such a case. It made him sad for her, made him love her even more. As long as she continued to come home to him, he would continue to turn a blind eye to her infidelities and a deaf ear to the ridicule of his friends and associates.

She thought he didn't know about the lavish amounts of money she spent on herself and Heather, but he did. His wife had a creative mind, but he was a bean counter. He knew down to the penny what the motel was worth. Over the years he had learned where to hide profits from the IRS, where to be extravagant, where to cut corners.

He smothered a chuckle behind a cough. Thanks to Jody Tackett, he saved thousands of dollars each year. He'd always hoped he would live to see his old enemy die. Before her health got any worse and she became insentient, he must decide whether to let her in on his little secret.

Timing would be critical. After all, he would be confessing a crime. He wanted her lucid enough to grasp the full impact of his admission, but incapable of doing anything about it.

Maybe he should put it in the form of a thank-you note. *Dear Jody, Before you take up residence in eternal Hell, I want to thank you. Remember hew you screwed me out of the oil lease? Well, I'm pleased to inform you that —*

"Fergus? What do you think?"

The soprano roused Fergus from his wool-gathering. "I think you've been comprehensive. If there are no corrections or questions, I suggest we move on."

As the vice president introduced the first item of business on that evening's agenda, Fergus returned to his satisfying fantasies of vengeance.

"Your treachery killed my daughter." Lara's voice remained as steady as her extended hands cupping the Magnum .357. "You bastard. You killed my baby. Now I'm going to kill you."

Having the gun leveled at him gave Randall pause, but only momentarily. He recovered admirably. "You tried this dramatic posturing in

Montesangre and it didn't play. Emilio saw through it just as I do. You're a healer, Lara, not a killer. You value human life too highly to ever take one.

"However, not everyone shares your elevated regard for his fellow man. Such lofty ideals prohibit you from seizing what you want. The final step is the only one that really counts, Lara. Whether or not you take it determines success or failure. One must be willing to take the final step or he might as well not put forth the effort. In this particular scenario, pulling the trigger is the final step, and you'll never do it."

"I'm going to kill you."

His composure slipped a fraction, but he continued with equanimity. "With what? An empty revolver? The bullets were removed, remember?"

"Yes, I remember. But they were replaced. Key had hidden extra ammunition in a secret pouch of the camera bag. The soldiers missed it during their search. He reloaded the gun before we left the hotel to catch the plane to Colombia." She pulled back the hammer. "I'm going to kill you."

"You're bluffing."

"That's the last judgment call you'll ever make, Randall. And it's wrong."

The racket was deafening. The darkness was splintered by a brilliant orange light as Lara was flung backward against the wall. The heavy revolver fell from her hand.

★ ★ ★

He inserted the latchkey into the lock. Unseen, they entered the honeymoon suite and closed the door behind them. He reached for the light switch, but when he flipped it up, nothing happened.

"Bulb must be burned out," he said.

"There's a lamp on the end table."

She crossed the sitting room, feeling her way in the darkness. His curiosity about mechanical things compelled him to try the light switch once again.

The light bulb wasn't at fault, but rather an electrical short in the switch. When he flipped it up again, it sparked.

The room exploded.

Chapter Twenty-Eight

Lara had the breath knocked out of her when she hit the wall. Collecting herself, she stumbled to the window. It seemed the whole north side of Eden Pass was ablaze.

Grabbing her medical bag, she raced from the house and ignored traffic laws in her haste to reach the roiling column of black smoke. She quickly determined that the site of the explosion was The Green Pine Motel.

She arrived within seconds of the fire truck and the sheriff's patrol car. One wing of the building was engulfed in flames. Periodic explosions within the conflagration sent plumes of fire into the night sky. Damage to the property would be extensive. The casualty rate would depend on the number of rooms occupied. Lara mentally prepared herself for the worst.

"Any signs of survivors?"

Sheriff Baxter had to strain to hear her over the roar of the flames. "Not yet. Jesus Christ. What a mess."

For all their valiant efforts, Lara knew that Eden Pass's fire department, which depended largely on community volunteers, didn't have a prayer of bringing this blaze under control. The fire chief was smart enough to realize that. He didn't send his willing but ill-equipped men into the fire, but gave them orders to try to keep it from spreading. He put in calls for assistance to the larger fire departments within driving distance.

"And call somebody at Tackett Oil," Sheriff Baxter shouted. "That well is too damn close for comfort." The deputy, Gus, got on his police radio.

"Sheriff, can I use the cellular phone in your car to call the county hospital?" The sheriff bobbed his head.

She slid into the driver's seat of the patrol car and placed her call. Luckily she was put through to an efficient emergency room nurse. She explained the situation.

"Dispatch your ambulances at once. Send extra emergency supplies, painkillers and syringes, bandages, portable oxygen canisters." They only had two ambulances, so she suggested that reinforcements be called from surrounding counties. "Also, alert Medical Center and Mother Frances Hospital in Tyler. We'll probably need their helicopters to take the most seriously injured to their trauma centers.

"Tell them to put their disaster teams on standby. Notify all regional blood banks that extra units of blood might be needed, and get an inventory of what types are immediately available. They'll also need extra staff. It's going to be a messy night."

"Over there!" Sheriff Baxter was wildly gesturing to the firemen when she rejoined him.

Shouts could be heard coming from the wing of the motel that hadn't been demolished by the original blast. Lara watched fearfully as a group of volunteer firemen entered the burning building. At any second, another explosion might take their lives.

After several tormenting moments, they began leading out survivors. Two of the firemen were carrying victims on their shoulders. Others were walking under their own power, but Lara could see that they were dazed, burned, and choking from smoke inhalation.

She instructed the firemen to line them up on the ground, then she moved among them, assessing their injuries, mentally noting the ones who were the most critically injured, dispensing the only medicine she had at the moment — encouragement.

The wail of sirens had never been so welcome. The first of the ambulances arrived and disgorged three paramedics. Working quickly with them, she started IVs, began giving oxygen, and indicated which of the injured should be taken immediately to the hospital. Paramedics unloaded several boxes of emergency supplies for her use, then sped away with their injured passengers.

The others looked at her through pain-glazed eyes. She hoped they understood how difficult it was to play God, to decide who would go and who would stay.

The firemen made other forays into the blaze.

The number of survivors increased, but that made it more difficult for Lara to deal with everyone. Two were in shock. Several were crying, one was screaming in agony. Some were unconscious. She did what she could to administer essential first aid.

She was kneeling beside a man, applying a tourniquet to a compound fracture of his ulna, when car tires screeched dangerously close. She turned her head, hoping to see another ambulance.

Darcy Winston stumbled from the driver's side of her El Dorado. *"Heather!"* she screamed. "Oh my God! Heather! Has anybody seen my daughter?"

She charged toward the building and would have rushed headlong into the inferno if one of the firemen hadn't caught her and pulled her back. She fought him. "My daughter's in there!"

"Oh, no," Lara groaned. "No." Had the girl with whom she'd developed an instant rapport been a casualty? She looked for Heather Winston among the rescued, but she wasn't there.

"Sweet Jesus."

At the sound of Key's voice, Lara turned and realized with lightning clarity that he had arrived with Darcy. Shoving personal considerations aside, she said, "Help me, Key. I can't handle this alone."

"I'll get a chopper. On the way I'll call my sister and get her over here to help you." He glanced in the distance. "Christ, that well —"

"They've already notified someone at Tackett Oil."

"That's number seven. It's on Bowie's route, I believe. He should be along shortly. Once he caps off the well, he'll pitch in and help, too."

He had remained in motion since alighting, rounding the hood of Darcy's car and moving toward the driver's side. "You okay?"

"I'm fine. Just please help me get these people to the hospital."

"Be right back." He jumped behind the wheel and sped away even before closing the car door. Moments following his departure, three more ambulances arrived.

The volunteer firemen carried five more victims from the building, replacing the ones Lara had dispatched to the hospital. An elderly woman succumbed to smoke inhalation a few minutes after her rescue. Her daughter held her lifeless hand and sobbed.

A toddler, who appeared unharmed, was crying for his mother. Lara didn't know to whom he belonged, or if his mother had even been rescued.

"I'll take care of him."

The offer came from Marion Leonard. Lara's lips parted in surprise, but she didn't waste time on questions. "That would be very helpful. Thank you." She passed the crying child to Marion, who carried him away, speaking soothingly.

Jack Leonard was there too. "Tell me what to do, Dr. Mallory."

"I'm sure the firemen could use some help dispensing oxygen." He nodded and went to do as she suggested.

Fergus Winston had arrived, Lara noticed. He was holding his wife in his arms. Darcy was gripping the lapels of his coat and crying copiously. "You're sure, Fergus? You swear to God?"

"I swear. Heather called to tell me that they were having an extra cheerleading practice tonight. I gave her permission to leave her shift early."

"Oh, Jesus, thank you. Thank you." Darcy collapsed against him.

He held her close, smoothing back her hair, stroking her tear-ravaged cheeks, assuring her that their daughter was safe. But his long, sad face and woebegone eyes reflected the light from the fire that was rapidly consuming his business.

When the clap and clatter of helicopter blades reached her ears, Lara looked skyward. A Flight for Life helicopter had arrived. Minutes later it lifted off with two patients aboard. Shortly after that, Key landed the private helicopter he'd borrowed before to transport Letty Leonard. Lara directed two women who had sustained severe cuts and bruises from a blown-out window to the chopper.

"Have you seen Janellen?" he shouted over the racket. Lara shook her head. "Our housekeeper said she went to Longview." He shrugged. "No one at Tackett Oil can locate Bowie either."

"If she shows up, I'll tell her you're looking for her."

He gave her a thumbs-up sign. "I'll be back when I can." The chopper lifted off.

Lara returned to her task, which she worked at unceasingly until time had no relevance. She measured it only by the number of survivors she could keep alive or make more comfortable until they could be transferred to a hospital. She tried not to think about those whom she could not save.

She wasn't without volunteer help. Jimmy Bradley and his wife of two weeks, Helen Berry, arrived and offered her their assistance. So did Ollie Hoskins. Her former nurse, Nancy Baker, was a most welcome sight. She was able, quick, and experienced enough to handle even the most gruesome injuries. Other townsfolk who had previously shunned her volunteered their services. She didn't refuse anyone's help.

That night the motel had been staffed by six employees. The total number of guests occupying rooms was eighty-nine — and two that no one knew about.

Bowie Cato carried his bride over the threshold of the honeymoon suite in the downtown Shreveport hotel.

"Oh, Bowie, it's beautiful." Janellen admired the skyline view as he set her down in the center of the room.

630

"I shopped around. When I heard about this place, I had to get written permission from my parole officer to come over here on account of it being in Louisiana."

"You went to a lot of trouble."

"It was worth it if you like it."

"I love it."

"For what it's costing, we might not eat for the first month of our married life."

She laughed and placed her arms around his waist. "If you ask your boss nicely, I bet you'll get a raise."

"There's not going to be any favoritism to me just 'cause I'm the boss lady's husband," he said sternly. "I'm no gold-digger. I made that plain the night I talked myself right out of an affair and into an elopement." He shook his head in bafflement. "Still can't quite figure how that happened."

"You refused to let me be gossiped about like I was trash. And I said the solution to that was for us to get married."

He worriedly gnawed the inside of his cheek. "Your mama might have it annulled."

"She can't. I'm a grown-up."

"Key might shoot me."

"I'll shoot him back."

"Don't joke about it. I hate like hell to come between you and your family."

"I love them, but nothing is as important to me as you are, Bowie. For better or worse, you're my husband now." She coyly ducked her head. "Or you will be as soon as you stop talking and take me to bed."

In high heels, she was as tall as he. Leaning forward, she kissed him lightly on the lips. He made a grunt of acquiescence and took her into his arms, drawing her close for a deep kiss. He became fully aroused almost immediately and stepped back self-consciously. "Want me to leave you alone for a while?"

"What for?"

Nervously, he rubbed his palms up and down his thighs. "So you can . . . Hell, I don't know. Do what brides do, I guess. I figured you wanted some privacy."

"Oh." She was crestfallen and it showed in her expression. "I thought you might want to undress me yourself."

"I do," he said in a rush. "I mean, if you want me to."

She seemed to think it over carefully before nodding.

He flexed his fingers like a safecracker about to attempt his personal best and reached for the buttons on her blouse — small pearl buttons very much like the ones that had engendered his first fantasies about her.

Their restraint diminished with each article that was removed. They undressed each other leisurely, allowing time to celebrate each discovery. Even though she'd grown up with two brothers in the house, she had a childlike curiosity about his body. Whispering in wonderment, she told him he was handsome, and he said he hadn't realized her eyesight was so bad. When he told her she was beautiful, she believed it, because his caresses were strongly convincing. He made her feel like a goddess of beauty and romance.

"I don't want to hurt you, Janellen," he whispered as he poised above her.

"You won't."

He didn't, even when he was deep inside her. She was awkward and perhaps too eager to please, so he told her to relax and let him do all the work. She did as he suggested, and to their mutual delight and surprise, her climax was as tumultuous as his.

Afterward, they drank the complimentary bottle of champagne that came with the room. She selected names for their first four children. He swore that by Valentine's Day he'd have enough money saved to buy her a wedding ring like a proper groom, but she argued that she didn't need anything tangible to symbolize his love. She felt it with every breath she drew.

Drowsy with love and champagne, he murmured, "Want to try out the whirlpool bath, or watch HBO, or something?"

"Or something." She flashed him a gamine smile that would have amazed the matrons of Eden Pass who had considered her a hopeless old maid, then slid her hand beneath the sheet and boldly fondled him.

"Good Lord have mercy on us all," he said,

633

gasping. "Miss Janellen's done turned into a regular sex fiend."

Had Bowie and Janellen turned on the television set in their honeymoon suite, they would have seen the news bulletins on the catastrophic fire in Eden Pass that had already claimed ten lives. All the victims had been identified and the authorities were notifying next of kin.

It was hours before the firefighters from six counties finally brought the flames under control. By dawn, the preliminary investigation into the cause of the explosion was under way. Inspectors began sifting through the smoldering ruin.

Early speculation was that Tackett Oil's well number seven might have been a contributing factor. Since Bowie couldn't be located, his supervisor had capped off the oil and gas lines.

Following that precaution, there had been no other explosions, indicating that the well had indeed been feeding the flame.

Key, the only Tackett readily available, was being questioned by federal agents from the Department of Tobacco, Alcohol, and Firearms.

"Y'all ever have any problem with that well leaking oil or gas, Mr. Tackett?"

"Not to my knowledge, but I'm not involved in my family's business."

"Who is?"

"My sister. She's out of town."

"I understood that your mother was the

634

ramrod of the outfit."

"Not for the last several years."

"I'd still like to talk to her."

"I'm sorry, but that's out of the question. She had a mild stroke a few weeks back and is virtually bedridden."

Lara, who was standing by listening, said nothing to contradict him. Neither did anyone else.

"All I can tell you," he said to the agents, "is that Tackett Oil has always been stringent about safety. Our record is unblemished."

The agents huddled together for another conference.

Scores of curious bystanders milled about, eager to survey the damage now that the threat of danger had passed. They consoled Darcy and Fergus Winston over their enormous loss.

Darcy, who still looked spectacular while everyone else was covered with grime, continually scanned the gathering crowd for sight of Heather. She'd asked Lara several times if she had seen her. She wept softly and daintily and kept repeating to those who offered words of encouragement, "I just can't believe that all our hard work went up in smoke. But of course we'll rebuild."

Fergus, however, seemed more nervous than disconsolate. Lara found his behavior puzzling. Perhaps he hadn't kept up his insurance premiums.

"She ought to be here," Lara overheard Darcy say to Fergus, her exasperation plain. Apparently she felt that Heather should be on the scene to round out the family image for the media.

Two shouts were uttered almost simultaneously.

Both came from the west side of the complex where the first explosion had occurred.

"Give me some help here!"

"Sir! Maybe you ought to look at this."

Lara and Key were among those who broke into a run. They and several others clustered around the man who'd shouted first. "There's a body underneath here."

Key helped him lift an iron support beam off the charred remains of a human being.

Before anyone had time to absorb that shock, one of the other agents said, "Christ. Here's another one." He'd made another grisly discovery several yards away.

"Sir!" The second agent who had shouted ran up to his superior. He was winded from his twenty-yard sprint. "I found something." He pointed toward an open field. "I think it's a gas line, but it isn't on the motel schematic. It's coming up vertically. My guess is that it's linked to an underground line that leads straight to that well."

Key shouldered his way up to the agent. "What are you saying?"

The senior agent frowned. "Mr. Tackett, it looks to me like somebody's been siphoning natural gas off your well."

Just then a scream rent the morning air. It

636

came from the crowd behind the sheriff's cordon. Darcy was clutching a teenage girl by the shoulders and shaking her until her head wobbled back and forth.

"What are you saying? You're a liar!" She slapped the girl hard. "Heather was at cheerleading practice. She told Fergus she was leaving early to go to cheerleading practice. I ought to kill you, you lying little shit!"

The girl blubbered, "I'm not lying, Mrs. Winston. Heather told me to cover for her if you called my house. We didn't have cheerleading practice. She said . . ." She hiccupped; the words came out choppily. "Heather said Tanner was going to meet her here and they were going to spend the night in one of the motel rooms." Misery contorted the girl's tear-bloated face. "She said it was going to be so romantic because they were going to sneak into the honeymoon suite."

Ollie Hoskins had worked tirelessly throughout the entire night doing whatever he could to help. He panicked upon hearing his son's name. "Tanner? Tanner? Tanner was here? No. It can't be. My boy, he . . . No!"

Darcy pushed aside Heather's sobbing friend and watched the grim firemen as they carried two stretchers from the smoking debris of what had been the honeymoon suite. On each stretcher lay a sealed black plastic bag.

"No. No. Heather? *NO!*"

Then Fergus stunned everyone by dropping to his knees and folding his arms over his head.

With an anguished cry, he fell face first onto the ground.

"I could use a cup of coffee." Key approached her as she moved toward her car. "Besides, I don't have a car here." He had arrived with Darcy, and that hadn't been coincidental. However, mentioning that now would have been petty, so neither did. "I'll call for a ride at your place if that's all right."

He was as grimy as she, his clothes sweat- and soot-stained. She'd lost count of how many times he'd taken off in the helicopter only to return as quickly as possible to transport another casualty.

When all the injured had been taken to area hospitals, he began helping the volunteer firemen. Lara too stayed at the site to administer first aid for their minor cuts and burns. Subconsciously she had found herself listening for Key's distinguishable voice. Even in the predawn gloom she could easily pick him out among the others.

She motioned with her head for him to get into her car. Once they were under way, she asked, "What do you think they'll do to Fergus?" He'd been taken away in handcuffs.

"He'll spend the rest of his life behind bars. Besides stealing from us, he's got twelve deaths to account for."

Lara shivered. "Including his own daughter."

"He'd better hope they never let him out.

Darcy threatened to kill him if she got the chance. She would, too." After a moment, he said, "I only slept with her that once. The night she shot me."

Apparently the look she gave him was inadvertently accusatory, because he added, "Last night, I'd just told her to take me back to my car, and we were arguing about it, when the explosion occurred."

"I did her a disservice," Lara admitted in a quiet voice. "I didn't credit her with loving anyone except herself. She loved her daughter very much. I know how it feels to lose a child. I can also relate to her wanting to kill Fergus for the role he played in Heather's death. It was accidental, but he was ultimately responsible."

She pulled into the rear driveway of the clinic, reluctant to go in and face what she'd left. "Randall is in there."

"One of my favorite people." He expelled a deep breath as he opened the car door. Together they went inside. "Unlocked," he remarked.

"I left in such a hurry, I didn't bother."

They moved through the silent, dim rooms. The ugly facts that had been revealed to her moments before the explosion came back now, enclosing her in rage.

"I don't think he's here," Key said.

"He wouldn't leave."

"Hey, Porter, where are you?" he called. He approached the doorway to Lara's private office.

The door was only halfway open. He gave it a slight push.

Apprehension crawled up her spine. "Key, before —"

"Porter?" He stepped into the room. "Holy shit!"

His expletive galvanized her. She bolted into the room but drew up short on the threshold. "Oh my God!"

Key knelt beside Randall's prone body. There was no question as to whether he was dead. A congealing pool of blood had formed beneath his head. His face was a frozen mask of surprise.

"I didn't do it!" Lara gasped. "I didn't. I didn't pull the trigger."

Key raised his head and looked at her. "What the hell are you talking about? Of course you didn't do it."

"I pulled a gun on him, but —"

"*What?*"

"The Magnum." He followed her pointing finger to the revolver lying where she'd dropped it. "But I never pulled the trigger." She covered her mouth with her hand, for once made sick at the sight of so much blood. "The concussion from the explosion knocked me against the wall. . . . But I didn't shoot him. Did I?" Near panic, she stretched forth her hand. "Key! Did I?"

He stood and nudged the Magnum with the toe of his boot. His expression was incredulous and bleak.

"I didn't," she said, vigorously shaking her

head. "I swear to God! I couldn't. I only wanted to frighten him. I wanted him to experience some of the fear he'd inflicted on me at Emilio's camp."

"Lara, you're not making sense."

"Randall was responsible for Ashley's death," she cried, desperate for him to understand.

"How?"

"He was allied with Emilio from the beginning." In disjointed sentences and broken phrases, she related to him what Randall had told her.

"I know it sounds inconceivable. But it's the truth! I swear it. Oh no," she cried, pressing the heels of her hands against her temples when she saw his skepticism. "Not again! I can't go through this again. I can't be blamed for something I didn't do!"

"I believe you. Calm down."

"Oh God, Key! I did not shoot him. I couldn't. I *didn't!*"

"No, I did."

The husky confession came from behind the wedge of space between the partially open door and the paneled wall. Key reached past Lara and closed the door in order to see who was hiding behind it.

Chapter Twenty-Nine

"Jody!"

Jody Tackett was sitting on the floor in the corner, her legs folded beneath her hip. A pistol, the obvious murder weapon, lay nearby. She was conscious, but had lost muscle control on the left side of her face. She had drooled on her blouse.

"She's had a stroke." Lara moved Key aside and knelt beside his mother. "Call 911."

"Don't bother. I'm dying. I want to. I can now." Jody's words were slurred, the consonants only partially formed, the sounds left open, like her lips. The vowels were guttural. But Jody was forcing herself to be understood. "Couldn't let him."

"Couldn't let him what, Jody?" Key knelt beside her. "Couldn't let him what?"

Lara called 911. For the second time in twelve hours she requested two ambulances — one for Jody, one for Randall. Then she returned to her place beside Jody and wrapped a blood pressure cuff around her upper arm. "She must have come in right behind me," she told Key. "He fell exactly where he was standing when I left."

"Couldn't let him tell about Clark." Jody struggled with the words.

"Don't talk, Mrs. Tackett," Lara said gently.

She released the cuff and firmly pressed her fingers into Jody's wrist to take her pulse. "Help is on the way."

"What about Clark?" Key supported the back of Jody's head in his palm. "What did Randall Porter know about Clark that you didn't want him to tell?"

"Key, this isn't the time. She's critically ill."

"She blew your husband's brains out!" he shouted at Lara. "Why, goddamnit? I want to know what drove my mother to murder. Do you know?"

"You're upsetting my patient," she replied tightly.

"Christ. You do know. What was it?"

She remained silent.

He looked down at Jody, realizing, as Lara did, that she was frantically trying to impart something before it was too late. "Jody, what was it? Did Porter know something about Clark's drowning? Was it a political assassination staged to look like an accident? Did Clark know that Porter was still alive?"

"No." Imploringly, Jody rolled her eyes toward Lara. "Tell him."

Lara shook her head slowly, then emphatically. "No. No."

"Lara, for God's sake. He was my brother." Key reached across Jody and took Lara's chin, forcibly turning her face toward him. "What do you know that I don't? What did Porter know that was such a threat to Clark, even dead? What-

643

ever it is, it's why Jody didn't want you in Eden Pass, right? She was afraid you'd leak a secret."

"Porter . . ." Jody wheezed. "Porter was . . ."

"No, Mrs. Tackett," Lara pleaded. "Don't tell him. It won't solve anything and will only hurt him." She looked at Key. "Don't ask her. It crushed her. She committed murder over it. Leave it alone. I beg you, Key, leave it alone."

Her pleas fell on deaf ears. He bent low over Jody, until his face was inches from hers. "Porter was what? Plotting something with Clark? Was Clark caught up in a political intrigue he couldn't get out of? An illegal arms deal? Drugs maybe?"

"No."

"Tell me, Jody," he urged her softly. "Try, please. Tell me. I've got to know."

"Randall Porter was —"

"Yes, Jody? What?"

"No, Key. Please. *Please.*"

"Shut up, Lara. Randall Porter was what, Jody?"

"Clark's lover."

For several seconds Key remained motionless. Then his head snapped erect and his eyes drilled into Lara's. "My brother and Porter . . . ?"

Lara sank against the wall, defeated. The secret she had wanted desperately to reveal for five years, she now wished could have died with Jody Tackett, so that she wouldn't have to watch the disillusionment spread over Key's face like a dark ink spill.

"They were lovers?" His voice was as brittle and dry as ancient parchment. It crackled on each word.

She nodded forlornly.

"That morning in Virginia, my brother was in bed with Porter, not you. *You* caught *them.*"

Tears ran down her cheeks. She rubbed them off with her fist. "Yes."

"Jesus," he swore, bearing his teeth. "Ah, Jesus." He propped his elbow on his raised knee and shoved his fingers through his hair, cupping his forehead in his palm. He held that anguished posture for ponderous moments.

Eventually he lowered his hand and looked down at his mother. "Clark confessed to you, didn't he?"

"When he gave . . ."

"When he bought this place for Lara," Key prompted. Jody nodded imperceptibly. Her eyes were swimming in tears. "You demanded to know why he'd do such a crazy thing for the woman who'd ruined his career. He broke down and told you the truth. You denounced him, probably disowned him. So he killed himself."

A terrible sound issued from Jody's chest.

"Key, don't do this to her," Lara whispered.

But it wasn't his intention to torment her. He slipped his arms beneath Jody and lifted her against his chest. She looked small and helpless in his brawny embrace, this woman who, using brains instead of beauty, had bagged the notorious playboy of Eden Pass, had driven Fergus

645

Winston to commit a criminal act to exact revenge, and had for decades instilled in her employees a fearful respect and in an entire town fierce loyalty.

Key wiped the saliva off her chin with his thumb, then rested his cheek on the top of her head. "It's all right, Mother. Clark died knowing you loved him. He knew."

"Key." She spoke his name, not reproachfully, but penitently. She managed to lift her hand and place it on his arm. "Key."

He squeezed his eyes so tightly shut, tears were wrung from them. When the ambulance arrived, he was still cradling her in his arms, cooing to her like a baby, rocking her gently.

But by then Jody Tackett was dead.

"Thank you, Mr. Hoskins." Ollie had personally carried her groceries out to her car and stowed them in the trunk.

"You're welcome, Dr. Mallory."

"How is Mrs. Hoskins?"

He pulled a handkerchief from his hip pocket and unabashedly dabbed at his eyes. "Not much good. She sits in Tanner's room a lot. Dusts it. Runs the vacuum over the rug so much, she's worn down the pile. Doesn't eat, doesn't sleep."

"Why don't you bring her to see me? I could prescribe a mild sedative."

"Thanks, Dr. Mallory, but her problem isn't physical."

"Grief can be physically debilitating. I know.

646

Encourage her to come see me."

He nodded, thanked her again, and returned to his duties inside the Sak'n'Save. This was one of the supermarket's busiest days of the year, the Wednesday before Thanksgiving. Texas Street was jammed.

A crew of volunteers was hanging Christmas decorations, stretching strings of multicolored lights across the street and mounting a Santa wearing a cowboy hat and boots on the roof of the bank building. Passersby offered unsolicited advice.

Despite the recent catastrophe, life went on in Eden Pass.

Lara was about to back her car out of the metered parking slot when Key's Lincoln loomed up directly behind her and blocked her exit. He got out and moved between her car and the pickup truck parked next to her.

Noisy honking and a shout drew his attention back to the street. "Hey, Tackett, you gonna move this piece of yellow shit, or what? It's blocking the whole damn street."

Key called back, "Go around it, you ugly son of a bitch." Wearing a good-natured smile, he flicked his middle finger at his friend, Possum. He was still laughing when he reached the driver's door of Lara's car. He knocked on the window and peeled off his aviator sunglasses. "Hey, Doc, how've you been?"

They hadn't been alone together since the day Jody died. If he could be cavalier, so could she,

647

although her heart was racing. "I thought you'd gone to Alaska."

"Next week. I promised Janellen I'd stick around till after Thanksgiving. She and Bowie will be celebrating their first one together. It's important to her that I be here to carve the turkey."

"She brought him to meet me."

"The turkey?"

She rolled her eyes, letting him know her estimation of his joke. "I like your brother-in-law very much."

"Yeah, so do I. I particularly like him because he's touchy about folks thinking he married Janellen for her money. He works like a Trojan to prove he didn't. He's inspecting every Tackett well for safety violations. He'd blame himself for the disaster caused by well number seven, only Janellen won't let him. He knew something was out of kilter. Time ran out before he located the problem, is all.

"Anyhow, they're gaga over each other. I feel like a fifth wheel. Once I'm gone, they'll have the house to themselves. I've deeded over my half of it to her."

"That was generous."

"That house didn't hold any good memories for me. Nary a one. Maybe they'll make it a happy place for their kids." Shaking his head, he chuckled. "Who'd've ever thought Janellen would elope?" In a quieter voice, he added, "Her timing was off a bit. She'll go to her grave blam-

648

ing herself for not being here when Jody had her stroke."

He was back to calling his mother Jody, but Lara remembered the tenderness with which he'd held her, calling her Mother as she died. "Did you tell Janellen about Clark?"

"No. What would be the point? It was hard enough on her to learn that Jody had murdered your husband."

There'd been an inquest. Key had cited Jody's dementia as the cause of her violent act. In her confusion, he told the judge, she'd linked Randall Porter's sudden reappearance with Clark's death. She killed him, thinking she was protecting her child. The court bought it. In any event, the killer was dead. Case dismissed. Sometimes the good ol' boy system was the fairest.

He turned his blue stare full force onto Lara. "You could have told the truth at the inquest."

"As you said, what would be the point? No one would have believed me five years ago. I couldn't prove anything then or now, and besides, it would only have dragged things out indefinitely. I was glad to finally see an end to it. The important thing to me was that Ashley's death was avenged."

She'd had Randall's body cremated. Since there had been a formal funeral for him years earlier, she didn't feel she owed the public another spectacle. She'd held a private memorial in Maryland for him. Only a handful of former colleagues had been invited to attend.

"What about the scheme Porter cooked up with Sánchez?" Key asked.

"When the president called to extend his condolences, I told him that I didn't agree with my late husband's assessment of the situation in Montesangre. I said that you and I had witnessed firsthand El Corazón's brutality to his own troops as well as his enemies. Speaking strictly as a citizen, I told him I wouldn't want my tax dollars to support his regime."

"He called me, too. I told him the same thing, in language a little more blunt."

"I can imagine."

He leaned against the rugged pickup parked beside her and raised one knee, flattening the sole of his boot against the dented door. He looked like he belonged there, comfortable in his Texas uniform — denim jeans and jacket. The brisk autumn wind tossed his dark hair around his head. His eyes were a few shades deeper than the sky.

She yearned for him.

"I thought you were leaving Eden Pass, Doc."

"I changed my mind and reopened the clinic. The people here have accepted me now. Business is so good, I've rehired Nancy. She's asking for an assistant."

"Congratulations."

"Thank you."

During a noticeable lapse in the conversation, neither knew quite where to look.

"Marion Leonard is pregnant," she told him.

"She wouldn't mind your knowing. They announced it immediately. She was among my first patients after I reopened."

"Ah, that's good." He nodded sagely. "Then there was never anything to that rumor of a malpractice suit?"

"I guess not."

They didn't go into the role Jody had played in starting the rumor.

"Did you read the TAF's report when they published it in the newspaper?" he asked.

After weeks of investigation, the federal agency had released their findings. The explosion at The Green Pine Motel had been caused by an illegal gas line running from Tackett Oil's well number seven to the motel. The gas was being used to heat and cool the motel. A leak in the line had filled the infrequently used honeymoon suite with odorless natural gas. It had compressed to a highly combustible level. The spark from the electrical short was enough to cause the blast.

Fergus Winston, against the advice of his attorney, pleaded guilty to all charges and was now weeks into his life sentence.

Darcy had closed their house and left town. Gossip was rampant. Some said she held vigil over Heather's grave by night and the prison by day, hoping for a chance to kill Fergus. Others said she had gone completely 'round the bend and had been committed to a psychiatric hospital. Still another tumor was that she'd latched on to a minor league baseball player and was shacked up with him somewhere in Oklahoma.

"As I understand it," Lara said, "Fergus tapped into the old flare line."

"Right. They were common. They burned off the gas from a well. Then Granddaddy decided to market the gas in addition to the oil. He tapped off that line. Anyway, flare lines became illegal. Fergus knew about the one on that well, reopened it, and extended it to his motel. He had free gas for years and probably laughed up his sleeve about it."

Again they ran out of conversation. When the silence became uncomfortable, Lara reached for her ignition key. "Well, I'd better run. I've got frozen things in the trunk."

"Before that morning, did you know that Clark and your husband were lovers?"

She didn't expect the question. Her hand fell away from the ignition.

He squatted down beside her car door so that their faces were on the same level. Loosely clasping his hands, he rested his wrists on the open window. "Did you?"

"I had no idea," she answered softly. "When I saw them, I went numb. But only for a moment. Then I went a little crazy. Became hysterical."

"Who called the press?"

She didn't even consider avoiding his questions or glazing her answers with euphemisms. "The phone on the nightstand beside my bed rang. I woke up and answered it. The caller identified himself only as one of Clark's close friends. He

652

called him a few ugly names." A spasm of pain flashed across Key's face, but Lara went on doggedly.

"He asked if I knew that Clark had dumped him in favor of my husband. Then he hung up. I took it for a crank call and turned to tell Randall about it. But he wasn't in the other twin bed. I got up and went looking for him."

She bowed her head and rubbed her forehead with her thumb and index finger. "I found them in Clark's bedroom. Later, I figured that same caller must also have notified the media and told them that an explosive news story was about to break at the cottage. Anyway, reporters arrived within minutes of my discovery. Clark became almost as hysterical as I. It was Randall's idea to make it look like . . ." She raised her shoulders and sighed. "You know the rest."

Key muttered epithets to Ambassador Porter. "Why didn't the guy on the phone come forward to contradict the tabloid stories about you?"

"I suppose he lost his courage," she replied. "Anyway, he accomplished what he wanted. He brought down Senator Tackett."

"You could have exposed them, Lara. Why didn't you?"

She laughed mirthlessly. "Who would have believed me? Randall had had affairs with women. Many of them. They would have sworn that he was wholly heterosexual, and he was."

His brows furrowed with perplexity.

"He knew about Clark's sexual preference, and

used it," she said. "One favor in exchange for another, I suppose. Randall wasn't above that sort of cruel manipulation. He used Clark. He used me. He'd do anything to get what he wanted."

"Like pretending to be dead for years."

"Yes. And it didn't bother him at all that our daughter was killed in a cross fire." She hesitated to broach the next subject because it was sensitive for several reasons. "Key . . ." She averted her eyes from his. "I didn't trust Randall to tell me the truth about his bisexuality. In fact, I suspect that he was also Emilio's lover. Anyway, I ran extensive blood tests on Randall and me while I was still in the first trimester of my pregnancy. I didn't want to transmit the AIDS virus to my child.

"Both of us tested negative, but I never took another chance. The night I conceived Ashley — which was only a few weeks before the incident — was the last time I slept with Randall." She met his direct gaze. "The very last."

"I didn't ask."

"But you have a right to know."

His unwavering gaze was disquieting. They were surrounded by noise and confusion, yet a ponderous silence stretched between them. She found comfort in the sound of her own voice.

"Back to my credibility — the concept of 'innocent until proven guilty' is a myth. Before I fully recovered from the shock of finding my husband in bed with another man, I was branded an adulteress who'd been caught in the act. If I'd come forward with the truth, it would have been regarded as nothing more than a vicious counterattack."

Sadly she shook her head. "Once I was photographed in my nightgown, being hustled from Clark's cottage by my husband, I was labeled."

"I thought my brother had more integrity than to let someone else take the rap for him."

"He got swept up into Randall's lie, just as I did. The consequences of it were so extreme that he really couldn't consider telling the truth.

"But, unlike Randall, it ate on his conscience. Giving me the medical practice here in Eden Pass was his way of making restitution, of telling me he was sorry." She smiled wanly. "Don't be too hard on him, Key. He'd lived as a closet homosexual for years. That must have been a terribly lonely and unhappy existence."

"I'm still wrestling with it, trying to reconcile the brother I knew with the man in bed with Randall Porter. I keep thinking about one summer when we went to camp together. Naturally, we did what adolescent boys do when they sneak off into the woods. We jacked off until we were sore. We had come-comparing contests, for chrissake. If we were that close, why couldn't he tell me?"

"Maybe he didn't know then."

"Maybe. But by the time he was elected senator, he did. On election night, after his opponent had conceded, and all the hoopla died down, we got stinking drunk to celebrate." He smiled at the fond memory. "The next morning, he had to meet the press with the worst hangover in history. He threatened to kill me for doing that to him. The last time I saw him alive, we still had a laugh over it."

Gradually his smile faded. He stared into near space. "I wish he'd had enough confidence in me to tell me."

"Would you have accepted it?"

"I'd like to think so." He pinched his eyes shut for a moment. "Jody's opinion of homosexuals was no secret," he said bitterly. "I think Hitler had more tolerance. It must have been quite a scene when Clark told her."

"I'm sure it was devastating to them both."

"Whatever she said to him pushed him over the edge." He stood up and slid his hands into the rear pockets of his jeans, palms out. He looked down at his feet, rolled back on the heels of his boots, then let them fall forward to slap the pavement.

"She was good at that, you know, pushing people to the edge. Good, hell." He scoffed at his understatement. "She wrote the book on it. She knew exactly which screw to turn, and when, and how tight to turn it. She just couldn't leave people in peace to be what they were. Not Clark, or Janellen, or me, or my daddy." He glanced up suddenly. "She left me a letter."

Lara cleared her throat. "Yes, Janellen mentioned it."

"Did she tell you what she wrote?"

"No. Only that each of you found a letter to be opened on the occasion of Jody's death."

"Yeah, well the date on mine indicated that she wrote it while we were in Montesangre." His mouth turned down at the corners, and he raised his shoulders in a half-shrug. "She said that everybody was under the impression that she hated Daddy for chasing other women and leaving her for extended periods of time. But the truth of it, according to her letter, was that she loved him. To distraction, she said. Beyond reason. Those are quotes."

He kept his head down, his eyes on his boots. "She loved him, and he hurt her. Badly. The letter said that every time he, uh, took another woman, it was like a knife in her heart because she knew she wasn't pretty and vivacious. Not the sort of woman who could hold his interest. She knew that the only reason he married her was to get out of a scrape. But he never knew, or if he knew he didn't care, that she truly loved him.

"To his way of thinking it was a marriage of convenience. Jody got to run Tackett Oil like she wanted; he used his marriage as a safety net if his philandering got him into a fix. Not a bad bargain except that Jody loved him, so his infidelities hurt her."

He removed his hands from his back pockets and tubbed them together, then turned up one palm and studied it as though trying to make sense of the crisscrossing lines. "And," he said around a deep breath, "her letter said that the reason she was always so hard on me was because I was exactly like my daddy. Looked like him, had his temperament, liked nothing better than to have a good time. Later I even raised hell and womanized like him.

"She . . . she, uh, said she had loved me all along, but that it hurt her to even look at me. The day I was born, he was with another woman. I was a living reminder of that, so it was impossible for her to show me any love. Mostly, in an odd sort of way, she was afraid I'd reject her love, just like my daddy did. So she didn't chance it."

He rolled his shoulders, a brave attempt to appear indifferent. "That's what she wrote me. Crap like that."

"I don't think it's crap and neither do you." He raised his head and looked at her. "Jody loved both her sons, Key. She fought to the bitter end of her life to protect Clark from scandal."

"Then why'd she struggle with her last few breaths to tell me about him?"

"Because she wanted you to know that Clark had disappointed her. He'd always been her fair-haired child and you knew it. She refused to die until she'd balanced things out. That was a tremendous personal sacrifice for her, which should prove to you how much she loved you."

He squinted, but she couldn't tell if it was from the sun's glare or because he'd been struck by an enlightening thought. "This personal sacrifice stuff is a big thing with you."

She tilted her head, looking at him with misapprehension. He launched into an explanation. "You didn't keep Clark's secret because you were afraid no one would believe you. You kept quiet because you loved Clark. You told me so yourself on the way to Montesangre.

"It was friendship, never a sexual thing. Even though Randall Porter was a roach on pig shit, you wouldn't have cheated on him while you were legally married. I learned that for myself. But you respected Clark as a statesman and loved him as a friend. That's why you didn't squeal on him even though he'd betrayed you.

"Then you banished yourself to Montesangre with Porter for the sake of your baby. Another personal sacrifice. You have a habit of making sacrifices for the people you love, Lara."

He leaned forward and placed his hands on the open window, bracing himself against it. "When Jody wanted to tell me that Porter, not you, was Clark's lover, you begged her not to. You were given a chance to prove wrong all the ugly things I'd said to and about you. But you didn't take it. Because you wanted to protect me from knowing the truth about my brother, you refused to say a word." His eyes went straight through her. "And ever since then, I've wondered why that was."

Lara's throat ached with emotion. "Have you reached any conclusions?"

"I think I'm close to a breakthrough." Suddenly he opened her car door. "Get out."

"Pardon?"

"Get out." He reached inside and pulled her out. Backing her against the car, he slid his hands under her hair and trapped her head in place for his solid, searching kiss.

"I don't want to go to Alaska," he announced abruptly when he pulled back. "It's colder than a witch's tit up there, and they won't know a chicken-fried steak from an armadillo. I have more charter business here than I can handle. And there's a pretty piece of property out near the lake that I've had a hankering to buy for years. Just seemed wasteful to build a house only for myself, without a wife and kids."

She pressed her face into the open wedge of his jacket and breathed in his warm scent as the fabric of his shirt absorbed her glad tears. Then, angling her head back, she asked, "Will you ever tell me that you love me?"

"I already did. You just weren't listening."

"I was listening," she said huskily.

He lowered his voice to an urgent whisper. "Then talk me out of leaving, Doc."

Her fingertips feathered over his eyebrows, his nose; they traced the shape of his beautiful mouth. "What could I say that would make you stay?"

"Say yes."

"To what?"

"To everything. We'll fill in the questions later."